ADVANCED COMPOSITE MATERIALS AND STRUCTURES

ADVANCED COMPOSITE MATERIALS AND STRUCTURES

Proceedings of an International Conference

Howard Plaza Hotel, Taipei, Taiwan, Republic of China
May 19–23, 1986.

Edited by

G. C. SIH
Institute of Fracture and Solid Mechanics,
Lehigh University, Bethlehem, Pennsylvania, USA

S. E. HSU
Chung Shan Institute of Science and Technology,
Lung-Tan, Taiwan, Republic of China

VNU SCIENCE PRESS
Utrecht, The Netherlands
1987

VNU Science Press BV
P.O. Box 2093
3500 GB Utrecht
The Netherlands

First published in 1987

ISBN 90–6764–083–2

Printed in Great Britain

CONTENTS

PREFACE ix
by G. C. Sih, S.-E. Hsu and T. Y. Wung

SPONSORS xi

CONFERENCE PHOTOGRAPHS xiii

OPENING ADDRESS xxi
by Minister K. T. Li

OPENING REMARKS xxiii
by G. C. Sih

MESSAGE FROM THE PRESIDENT OF THE HOST SOCIETY xxv
by S.-E. Hsu

PLENARY LECTURES
Overview on the mechanics of composite materials and structures 3
by G. C. Sih
Research on special composites in CSIST, ROC 25
by S.-E. Hsu
New experimental approaches to stress analysis of composite structures 45
by J. T. Pindera and M. J. Pindera
Composite materials and repair 69
by R. Jones
Recent developments in advanced composites research 87
by T.-W. Chou
Acoustic emission and microfracture of a solid 99
by Y.-H. Pao
Sliding ellipsoidal inhomogeneities under shear 113
by T. Mura, R. Furuhashi and T. Mori

SECTION I PROCESSING AND MANUFACTURING
Graphite fiber reinforced ceramics, TiB_2/TiC/MoC by SHS process 125
by S.-E. Hsu, C.-I. Chen, C.-R. Lee and S.-J. Chen
A study on the magnetic properties of two-component magnetic composites 135
by S.-E. Hsu, K.-L. Wang, T.-I. Hou and L.-C. Su
Multi-directional carbon/carbon composites prepared by HIP and CVD processes 147
by K. Yang, T.-M. Wu and C.-P. Chang
Characterization of PZT-polymer piezoelectric composites 163
by C.-S. Hao, C.-W. Hsieh, P.-C. Yao, S.-J. Yang and S.-E. Hsu
Gas-permeability in graphite fiber composite as a result of CVD processing 173
by Y.-H. Lee
Property-temperature relationship of TSTR polyimide/graphite fiber composites 187
by S.-W. Leu, W.-Y. Chen, S.-E. Hsu and L.-C. Hsu
Effect of the microstructures and processings on the strength of carbon fiber 197
by S. C. Yu, Y. H. Yen, W. F. Chen and H. Y. Lin

Rolling and forging characteristics of aluminum matrix filamentary composites 211
by W. Wang and T. Erturk
Thermal annealing study of cold powder metallurgy processed Cu-Ni composites 219
by Y. D. Yao
Effect of formaldehyde/phenol ratio, viscosity and catalyst type on the characteristics of casting phenolic resins 229
by S.-L. Jwo and W.-Y. Chen

SECTION II THERMAL AND MOISTURE EFFECTS
Mechanical behavior of SiC/2014 Al composite at high temperature 247
by S.-C. Chou and J. L. Green
Flammability of epoxy 259
by J.-F. Chang and W.-Y. Chen
Water and solvent effects on carbon fiber/epoxy composites 271
by K. J. Chang and R. Y. Wu
Thermal and mechanical analysis of acid anhydrides cured epoxy resin systems 279
by H.-T. Chiu and C.-B. Chen
Vibrations of hygrothermal elastic composite plates 293
by L.-W. Chen and Y. M. Chen
Effects of fiber length distribution on elastic moduli of short fiber reinforced thermoplastic 311
by W. K. Chin, H. T. Liu and Y. D. Lee
Characterization of Mode I delamination of composite laminates at elevated temperatures 323
by S.-C. Yen and M. El-Zein
The effect of moisture absorption and physical aging on the physical and mechanical properties of composites 335
by M.-F. Sheu, C.-S. Yeh, Y.-P. Bian and C.-L. Ong
Influence of moisture on the mechanical properties of graphite epoxy system 347
by A. Ankara, D. Weisgerber and J. Vilsmeir

SECTION III FAILURE AND FATIGUE ANALYSIS
Stability of a crack interacting with fibers 365
by C. K. Chao
Local and global instability in fracture mechanics 375
by A. Carpinteri
Failure trajectory initiating from an elliptical inclusion 397
by R. C. Chu
Energy density redistribution caused by broken fiber 413
by D. M. Chou
Influence of stacking sequence on the instability of layered composite 425
by D. Y. Tzou
Fracture and strengthening mechanisms in discontinuous SiC reinforced aluminum composites 437
by T. G. Nieh, J. Wadsworth and W. H. Kao
Acoustic emission analysis of composite failure mechanisms with special reference to the modelling of amplitude distributions using a personal computer 453
by Y. H. Wu, M. H. Yang, J. Y. Kuo and S. P. Chang

Prediction of stiffness degradation of composite laminates under fatigue loading 467
by J. N. Yang, A. Meskini and S. H. Yang
Inner materials degradation effects on stiffness and dynamic crack propagation in composites subjected to cyclic loading 479
by A. Kobayashi and N. Ohtani
The fatigue residual strength in centrally notched composite materials 489
by M.-H. R. Jen

SECTION IV COMPOSITE LAMINATES
High speed asperity excitation over a layered medium 505
by F. D. Ju and J.-C. Liu
Analysis of asymmetric laminates 517
by H. Chin, J. K. Chen and C. T. Sun
The failure characteristics of a GR/EP laminate 529
by D.-G. Hwang and H. T. Hahn
The improvement of notched strength prediction for laminated composites 559
by T.-H. G. Hu and D. G. Goetschel
Free vibrations of elliptical laminated plates 573
by C.-C. Lin and S.-C. Lin
The flexural analysis of laminated composites using a triangular plate bending element 587
by A.-D. Han, F.-Y. Shieh and C.-C. Lin
Laminated composites of amorphous alloys for EMI shielding 599
by D.-R. Huang, P.-C. Yao and S.-E. Hsu
Stress and strength analysis of finite composite laminates with unloaded fastener holes 611
by C.-C. Ko, C.-C. Lin and C.-L. Ong
Free vibration and buckling analysis of thick initial stress laminate 623
by T.-Y. Ye and C.-L. Wu

SECTION V STRESS/STRENGTH AND INSTABILITY ANALYSIS
Analysis of stress singularities of the crack tip in the two-dimensional orthotropic elastic body 639
by Y. Fujitani
Prediction of tensile behavior of strain softened composites by flexural test methods 647
by T.-J. Chuang and Y.-W. Mai
Dynamic visco-elastic properties of glass fibre reinforced polypropylene 659
by L. W. Chen and P. H. Lin
Generic buckling of specially orthotropic cylindrical shells 667
by I.-H. Yang and W.-S. Kuo
High performance epoxy matrix: the properties of modified epoxies for graphite composites 677
by G. Lee and W.-Y. Chen

SECTION VI JOINT AND STRUCTURE DESIGN
Free and forced nonlinear dynamics of composite shell structures 693
by Y. C. Wu, T. Y. Yang and S. Saigal
Response of orthotropic cylindrical shells subjected to impact loading 705
by C. C. Chao, T. P. Tung and S. C. Lee
Evaluation on AT-3 composite speedbrake by static test and analysis 717
by R. C. Chu, L. J. Hsu and Y. J. Chiou

A simple numerical method for analyzing elasto-plastic stress distribution and progressive failure in adhesive layer of bonded joints 727
by S. Amijima, T. Fujii and T. Tanaka
Stress and strength analysis of single-lap adhesively bonded joints 741
by C.-C. Ko, C.-L. Ong and J. Q. Tarn
Analytical modelling of composite sandwich panel 759
by C. H. Lee

LIST OF PARTICIPANTS 771

PREFACE

This International Conference on Advanced Composite Materials and Structures has been organized jointly by the Chinese Society for Materials Science, the Society of Theoretical and Applied Mechanics in the Republic of China and the Institute of Fracture and Solid Mechanics of Lehigh University in the USA. It was held at the Howard Plaza Hotel in Taipei, May 19-23, 1986 and attended by approximately 250 participants from more than 12 different nations. The support from the various organizations associated with the government, industry and university was most encouraging and made the international event a great success, both technically and socially.

During the week-long meeting, approximately 75 papers were presented in connection with the analytical modelling, numerical calculation, material testing, non-destructive evaluation, design methodology, processing and manufacturing aspects of composite materials and structures. Attention was focused on the fundamental understanding of processing-structure-property relationships of advanced composite systems and the assessment of the state-of-the-art such that an account of future research topics can also be made. The theme, timing and location of the Conference are chosen with cognizance of the development of composite technology in countries such as Japan, Republic of China, South Korea, Singapore and Hong Kong. This is clearly evidenced by the contributions from the Republic of China (ROC), particularly those from the Chung-Shan Institute of Science and Technology. As marked by Minister K. T. Li from the Science and Technology Advisory Group of The Executive Yuan in his opening address, advanced composite products can enhance productivity in other industries and high-quality research in universities. Application of composites will, no doubt, find new horizons as more knowledge is gained in the variables that govern their complex behavior. To this end, the panel discussion attended by a number of leading scientists and engineers shed light on design know-how, processing and manufacturing, material characterization and the need for the producer, user and designer to consolidate.

The Conference participants also had an opportunity to experience an earthquake initiated from the mid-land of Taiwan that led to the cancellation of the tour to visit the Taroko Gorge, a disappointment to many of the far-distance visitors. Enlightment, however, was provided by Mrs. S. E. Hsu who gave a superb performance in Chinese art. Her action was captured in a couple of the photographs in this Volume. The banquet and entertainment were also delightful. Most memorable were the traditional customs and dances of the Cho aboriginal tribes. A special note of acknowledgement goes to the members of the local organizing committee. Their patience and long-working hours are most appreciated. To be recognized are the following individuals:

Shu-En Hsu (Co-Chairman)
Patzen Wu (Co-Chairman)
Tung-Ying Wung (Co-Chairman)
Guan-Chung Chang
Ching-Liang Yen
Chi-Cheng Hsien
Chia-Chi Chu
Chi-Jen Wei
Wen-Yih Chen
Hsiao-Feng Kuo
Wen-Huu Dow
Wu-Hsiun Chan
Guang-Shyang Ger
Ru-Chu Chu

In addition to the efforts made by the editors, this Volume could not have been completed without the assistance of Mrs. Barbara DeLazaro.

May, 1986
Taipei, Taiwan

G. C. Sih
S.-E. Hsu
T. Y. Wung

SPONSORS

Science and Technology Advisory Group, The Executive Yuan
National Science Council, The Executive Yuan
Ministry of Education
Chung Shan Institute of Science and Technology
Taiwan Power Company
China Ship Building Corporation
China Steel Corporation
Chinese Petroleum Corporation
Materials Science Center, National Tsing Hua University
Union Chemical Laboratories, Industrial Technology Research Institute
Materials Research Laboratories, Industrial Technology Research Institute
Institute of Aeronautical and Astronautical Engineering, National Cheng Kung University
Kunnan Enterprise Limited
Fiberglass Reinforced Plastic Society of the Republic of China
Institute of Applied Mechanics, National Taiwan University
Institute of Materials Engineering, National Taiwan University
Institute of Chemical Engineering, National Taiwan University
Institute of Materials Science and Engineering, National Tsing Hua University
Institute of Chemical Engineering, National Tsing Hua University
Institute of Power Mechanical Engineering, National Tsing Hua University
Institute of Fracture and Solid Mechanics, Lehigh University, Bethlehem, Pennsylvania 18015 USA
Graduate School of Mining, Metallurgy and Materials Science, National Cheng Kung University
Graduate School of Mechanical Engineering, National Cheng Kung University
Fishing Boat and Marine Engineering Research Center, National Cheng Kung University
Department of Earth Science, National Cheng Kung University
National Sun Yat-Sen University
National Tsiao Tung University
Feng Chia University
National Taiwan Institute of Technology
Transactions of the Aeronautical and Astronautical Society of the Republic of China

His Excellency, Minister K. T. Li of the Science and Technology Advisory Group of The Executive Yuan in the Republic of China, delivered the inaugural speech at the International Conference on Advanced Composite Materials and Structures.

Professor G. C. Sih, Director of Institute of Fracture and Solid Mechanics, Lehigh University, USA, commented on the role of advanced composite material and structures in the development of modern technology.

At the opening session from left to right: Minister K. T. Li, G. C. Sih and T. Y. Wung, one of the Conference Co-Chairmen.

From left to right: award recipients F. D. Ju and Y. H. Pao, and Conference Co-Chairman S. E. Hsu.

Staff at the registration desk.

Minister K. T. Li at the entrance hall of conference site in the Howard Plaza Hotel.

Co-Chairman S. E. Hsu chats with guests at the welcoming cocktail.

Co-Chairman G. C. Sih welcomes the Pindera's from Canada.

A well-attended plenary lecture.

A session on composite technology.

Mrs. S. E. Hsu captures a most enthusiastic audience on Chinese art.

Mrs. G. C. Sih expresses delight on the artistic work of Mrs. S. E. Hsu.

A toast to the success of the conference on advanced composites.

Exchange of greetings between S. E. Hsu and G. C. Sih.

Panel discussion on the advancement of composite technology in the Republic of China from left to right: S. C. Chou, T. W. Chou, T. Y. Wang, S. E. Hsu, G. C. Sih, J. T. Pindera and Y. H. Pao.

A farewell portrait from left to right: D. W. Ding, D. G. Hwang, L. C. Hsu, S. C. Chou, R. Y. Wu, R. Jones, C. C. Chao, G. S. Ger, S. E. Hsu, W. Y. Chen, G. C. Sih, W. H. Chan, J. T. Pindera, Y. H. Pao, F. D. Ju, C. C. M. Ma and S. N. Tong.

OPENING ADDRESS

Distinguished Guests, Ladies and Gentlemen:

It is my privilege to be the honorable chairman of the International Conference on Advanced Composite Materials and Structures, 1986. On behalf of the Organizing Committee, I sincerely welcome all the distinguished guests, scientists, engineers, experts and specialists who spare their precious time, some even travelled thousands of miles, to attend this Conference.

The Conference was first proposed by Professor George Sih, Director of the Institute of Fracture and Solid Mechanics, Lehigh University, USA. Dr. Shu-En Hsu, Chairman of the Chinese Society for Materials Science, and Dr. Tung-Ying Wang, Chairman of the Society of Theoretical and Applied Mechanics, ROC, both concurred in jointly sponsoring the Conference with Lehigh University. The aim of the Conference is to provide an opportunity for scientists and engineers to present their research results, to meet and exchange opinions, and to discuss problems on procedures for characterizing the fundamental properties and behavior of advanced composites and for designing high-performance composite structures. Special emphasis is laid on the processing-structure-property relationships for advanced composites with constituents made of metallic and non-metallic materials.

As you all realize, the so-called composite materials are really nothing new. Clay bricks strengthened with straw are typical composites which have been used for centuries by our ancestors as good building material. However, composite materials have not been upgraded to a new field of application until the development of material science and technology which advanced composite research from macro- to micro-era. FRP are the most important composite among others. In the past decades, FRP industry boomed very rapidly and its products covered a broad spectrum that closely related to our everyday life.

Following the progress of civilization, the requirements of materials become more and more severe, and designers are very selective in materials. They asked new materials be invented to fulfill their design conception instead of using existing conventional materials. To date, aero and space industries, weapon systems, etc., all call for new materials to be stronger, lighter, cheaper, more enduring and more resistant to temperature and corrosion. To meet these requirements, advanced composite materials are no doubt the best answer. Thanks to the generous supports of governmental and industrial agencies and continual efforts of scientists and engineers, the development and application of new composites as well as the research and investigation of their mechanics and fracture behaviors advanced rapidly in the recent years. Their achievements have made the advanced composites one of the most important materials for aero, space, weapon and many other high-tech related industries. Ladies and gentlemen, you are also credited with this achievement and should be proud of yourselves. Please accept my admirations.

The FRP industry of the Republic of China also grew rapidly in the recent two decades. Yearly, FRP products consumption increased from 300 tons in 1956 to 25,000 tons last year. In the first ten years, FRP products covered merely boats, bath tubs, corrugated plates, cooling towers, furniture, etc., and were almost all made by hand lay-up method. In the second decade, however, the products were upgraded to a higher level and a broader scope thanks to the introduction of new techniques and equipment, such as filament winding process to fabricate pipes, preform hot press method to produce helmets and chairs, spray-up method to make containers, bath tubs and many SMC and BMC hot pressing products. Recently, more advanced equipment such as pultrusion machine, resin injection machine and matrix transformation machine were imported for more advanced composite products.

While the composite industries kept on booming, domestic research institutions such as the material science department in various universities, the Industrial Technology Research Institute and Chung Shan Institute of Science and Technology also sped up their pace in composite related R&D, and assisted the industries with their expertise through know-how transfer, technical consulting, personnel training, etc. The cooperation was mutually beneficial and helped boost the composite industries significantly.

In the future, demands of composites will no doubt keep growing. Innovation of the composite materials and advancement of the state-of-art call for your continual efforts. I sincerely hope, and believe, that in the next few days, all the participants will benefit through paper presentations, problem discussions and experience exchange.

Holding such a high level international academic conference with so many renowned scholars and specialists sharing their expertise with us is indeed very fruitful to the promotion of our technological level. We also enjoy the opportunity to get in touch with the composite experts from all over the world, to establish a channel for further cooperation and information exchange, and, most of all, to promote mutual understanding and build up everlasting friendship.

In addition to the above-mentioned three sponsors, many cooperative sponsors including governmental agencies, private enterprises, societies, universities and research institutions also provided supports of various kinds: academical, financial, consulting or advisory. This Conference would not have been possible without their enthusiastic assistance. We all owe them many thanks. The list is too long to read here, and so we have included them in relevant documents of the Conference to express our appreciation.

I wish you every success and good luck. To those from abroad, have a nice stay! Thank you.

May, 1986
Taipei, Taiwan
Republic of China

K. T. Li
Minister Without Portfolio

OPENING REMARKS

It gives me a great honor and pleasure to welcome you to this International Conference on Advanced Composite Materials and Structures which is unique in two respects. I believe my Co-Chairman, Dr. Hsu Shu-En, will agree with me that this is the first time that an international conference of a technical nature with *open registration* is held in the Republic of China. The overwhelming number of distinguished Chinese-American scholars and researchers supporting this event is also very heartwarming. In this respect, I assure you that I have no intention to undermine those equally distinguished engineers and scientists from the other countries including Australia, Canada, France, Federal Republic of Germany, Greece, Italy, Japan, Singapore, Turkey, United Kingdom and United States of America. The organizing committee is indeed honored by their presence.

Those of you who are frequent visitors of the Republic of China would agree with me that this country has made striking progress in economic growth and technological development. This success, in my opinion, lies in the high quality manpower and flexible response to new developments in technology and science. The large number of technical papers contributed by the Republic of China on composites is indicative of her commitment in this field. In the area of sporting goods, Taiwan has already captured a significant portion of the market while future growth lies in more technology extensive products. The increasing use of these materials in the automotive industry alone is eye-catching, not to mention the aircraft and aerospace applications. Outlook for RP/C is a 17% per year increase. Some 132,000 lbs of thermoset RP/C were used in passenger cars in 1983, increasing to 174,000 lbs in 1984. The amount of 336,000 lbs and more is expected by 1989. The major area of application is body panels. Others will include underbody structural parts (chassis, frame and suspension, springs, drive shafts, etc.), passenger compartments and electrical systems. Market competition will be keen. Composites should be manufactured more economically and used to the greatest advantage. The challenge will be to increase processing speed, reduce cost of production, control quality and improve reliability of the finished product. An understanding of the scientific principles that govern their behavior is not only necessary but essential.

Much has yet to be learned about the relationships among processing, structure and performance. Composites apparently made from identical raw materials but processed differently can have various different properties and react differently under load. Damage in the form of flaws, voids, etc., therefore, is of great concern in addition to their interactions with the size, space arrangement and chemical connectivity of the constituents.

It is with these objectives in mind that this Conference has been organized which will serve as a forum for identifying critical problem areas that are in need of further development and/or more in-depth research. Not to be overlooked is the friendship developed among the participants during this week. The exchange of new ideas and information can play a key role in technological advancement. I wish you a pleasant stay on this beautiful island known as "Formosa" and also take the liberty to offer the service of the local organizing committee to you.

May, 1986
Taipei, Taiwan
Republic of China

G. C. Sih
Lehigh University

MESSAGE FROM THE PRESIDENT OF THE HOST SOCIETY

The Society of Theoretical and Applied Mechanics in ROC and the Chinese Society for Materials Science have accepted the challenge of hosting the International Conference on Advanced Composite Materials and Structures, which was proposed by Professor George C. Sih of the Institute of Fracture and Solid Mechanics, Lehigh University. As the President of the Chinese Society for Materials Science, I take pleasure in welcoming you to this Conference which will be rewarding in many different ways.

Just as what Professor Sih has said, this is the first time that an international conference of a technical nature with open registration is held in the Republic of China. Ever since Professor Sih made the proposal in March last year, the ROC government, the industries and academic groups have echoed in support favorably. It was finally decided that Lehigh University, the Society of Theoretical and Applied Mechanics in ROC and the Chinese Society for Materials Science jointly sponsor this Conference. I would like to take this opportunity to express to Dr. Sih my deep gratitude and high esteem for his enthusiasm. Without his international academic reputation and his instruction, neither would this Conference be held successfully nor are there so many reputed scholars abroad present in this Conference, which will promote both ROC's academic position and international exchange in advanced knowledge related to composites.

It has been several centuries since composites were used. The invention of advanced composites represents the breakthrough of modern materials technology and the milestones from material application, material design to material exploration. Generally speaking, the so-called composite, by means of micro or macro, is the combination of two kinds of materials to make use of their merits and make up their defects. Because of the advanced composites, some physical restrictions of mono-materials were broken. Therefore, the engineering technology limitations, such as speed, density, temperature, environment, etc., have been overcome and now we are living in the space age.

Today, composites cover a broad range and a variety of categories. Because of different matrixes, we have PMC (Plastic Matrix Composite), MMC and CMC. Because of different strengthener, we have particulate composites and fiber reinforced composites including FRP (Fiber Reinforced Plastic), FRM and FRC. According to the application, it can be divided into two major parts, structural and non-structural composites. The theme of the Conference focuses on structural materials, especially on the basic concept of its structure, process and properties. However, some physical properties of non-structural materials like magnetism, acoustics, electro-magnetism, heat transfer, etc., are also involved. There are twelve sessions included in the Conference under the major topics of materials, mechanics, structural design, experiment, process, manufacturing and application, which, I believe, will have profound influence on ROC's industrial development both theoretically and practically.

The purpose of the Host Societies and the Organizing Committee is to make it a successful conference both from a professional and social standpoint. In this five-day conference, besides opening session, there are seven plenary lectures and a panel meeting on May 23. A Taipei city tour is arranged to see its beautiful surroundings. You are also invited to a banquet on Thursday night. The Proceedings will be printed and worldwide published by VNU Science Press BV of the Netherlands. Because of the inconsistent schedule, I am regretful that this Conference is unable to be held in conjunction with the 1986 International Composite Materials Exhibition in Taipei early in May; otherwise, you may see the composites applications in this country.

As a closing remark, I would like to thank Minister Li for being our honorable chairman, our sponsors, supporters and the Organizing Committee for their dedication and hard work in making this Conference possible. We are very pleased to see you here in Taipei and extend a hearty welcome to you all. Wish you a wonderful time.

May, 1986
Taipei, Taiwan
Republic of China

S.-E. Hsu
Chinese Society for Materials Science

PLENARY LECTURES

OVERVIEW ON THE MECHANICS OF COMPOSITE MATERIALS AND STRUCTURES

G. C. Sih

Institute of Fracture and Solid Mechanics, Lehigh University
Bethlehem, Pennsylvania 18015 USA

ABSTRACT

With the discovery of new classes of fibers, resins and the ways with which they can be mixed to yield outstanding properties, the development of composite technology continues to rise at a rapid rate, particularly in processing and manufacturing. The ability to predict behavior from the component properties and an understanding of the scientific principles that govern composite behavior, however, have fallen far behind. This is mainly because of the diversified nature of composite behavior. Lacking in particular are the assessment of the various damage mechanisms that may involve matrix cracking, fiber breaking, fiber/matrix debonding, delamination, etc. Even less understood are failures affected by the fabrication process. Composites made from the same fibers and matrix materials but processed differently can have very different properties. In this context, structure refers to both macroscopic defects such as flaws, voids, unbonded interfaces, etc., and microscopic features involving not only smaller defects but also the formation of polymer chain-like network. The temperature, chemistry and internal stress can all interact and strongly affect the characteristics of the large interfacial area developed between the polymer and the fibers. This creates inherent difficulties for repeating experimental measurements.

One of the major stumbling blocks is the process of damage accumulation in composites. Affected by the load-time history will be the sequence of failure modes such as delamination, initiation and growth of cracks in the fiber and matrix and at the interface. This will alter the resulting energy dissipation. As the conventionally measured quantities such as compressive strength, tensile strength, fracture toughness, etc., tend to depend on internal damage, it raises a series of questions whether uniaxial and cracked composite specimen data could be used to predict composite structure behavior in situations other than those tested.

Notwithstanding the complexities of analyzing and understanding composite behavior, this presentation will address the load transfer characteristics of fiber reinforced composites as damage takes place in the form of matrix cracking, fiber breaking and/or interface debonding. New free surfaces are created to relieve the high concentration of energy within a volume of material that may or may not be nonhomogeneous depending on the scale level of observation. For instance, interfacial energy per unit volume can be related to the interfacial energy per unit area through the interfacial area per unit volume, a knowledge of which is essential for determining composite behavior in addition to the interaction of mechanical and physiochemical effects that prevail at the interface. The fluctuation of volume energy density dW/dV and/or surface energy density dW/dA in composites would therefore be regarded as the quantities to be analyzed. They are related by the rate change of volume with surface dV/dA which serves as a scaling length parameter and determines the degree of *homogeneity* of the system. Here, homogeneity includes the combined influence of material structure, geometry and load type. The overall instability of the composite system can be determined from the relative distances among the local and global stationary values of the volume energy density. Brief discussions will be given to several examples involving the normal and shear impact of layered media, influence of fiber on matrix cracking and vice versa, interface shear, etc., the details of which will be covered by other presentations in this Volume.

The understanding of composite failure is of great concern because it affects performance qualities. The current high rates of growth cannot continue unless manufacturers can produce composites more economically and can design products that take full advantage of how to best combine the constituents such that damage can be controlled for the load-time history considered. Better ways for reliable processing and predicting performance must be made available if composites are to be used to greatest advantage.

INTRODUCTION

The role with which high-performance composites play in commercial and military applications has long been recognized. More and more of these lightweight multi-component materials are being put to use in large-scale and load-bearing structural members while new classes of fibers, resins and ways for combining them are being discovered. Predictive capability of composite behavior, however, lacks far behind as the conventional methodology* is not able to consistently account for the hierarchy of interacting failure mechanisms that involve different rates of energy dissipation as the service conditions and/or internal structure of the composites are altered. Chemical, thermal and mechanical effects may also interact and it is not always clear how their influences can be best combined or described in the analysis, particularly when damage occurs at both the microscopic and/or macroscopic scale level. Even though there is still some distance from a full understanding of these damage processes and from models that describe them in a way that a designer can use, the need to develop a general approach and to explain the various observed failure mechanisms cannot be overemphasized.

Fundamentally speaking, all materials are nonhomogeneous and anisotropic. A polycrystal metal structure, when viewed under a microscope, is no less complicated than a composite material. Grain boundary conditions can affect the overall behavior of a polycrystal in a way similar to that of the fiber/matrix adhesion or interphase between the constituents. Whether modelling of the microstructure needs to be included in a given analysis or not depends strictly on the rate at which energy is being applied and dissipated to cause damage. Engineers gained more experience working with metals. For the majority of the situations, metals are manufactured at a scale level such that the grain boundary effects can be sup-

*The classical theories of elasticity, plasticity, viscoplasticity, etc., cannot account for the change in local strain rates in accordance with load-time history even if the material were *isotropic* and *homogeneous*. Their success for explaining composite material behavior is even more remote.

pressed and adequately reflected through the macroscopic parameters, such as yield strength, ultimate strength, etc. As the loading or energy transfer rate is lowered to that occurring in creep, microstructure details will have to be considered when addressing failure. Since composites are designed to be load-specific, their internal structures must be constructed to reflect the nonhomogeneous and/or anisotropic nature of the response. A knowledge of this response being damage dependent is a prerequisite for setting design limits on composites.

Despite the exhaustive research efforts [1-5] made on studying composite behavior, the use of specimen data to design composite structural components must be seriously questioned*. Data on composite tensile and compressive strengths in the longitudinal and transverse direction of the fibers and shear strength in the plane of the plate specimen may or may not apply to other composites that consist of several laminae to form an angle-ply laminate subjected to service loading conditions. Because nonhomogeneity and anisotropy can significantly affect the load transfer characteristics, *each composite specimen, regardless of loading, must be regarded as a structure*. Composite behavior should, therefore, be predicted from the constituents and interphase conditions** involving chemical and/or mechanical properties.

More effort needs to be devoted to control the interphase between the components. Energy transmitted across the dividing lines of dissimilar media is intimately related to that stored and dissipated within the system because the volume of each element changes continuously with the surface area***. This exchange is a basic mechanism of energy transfer across an interface and takes place nonuniformly throughout the composite. It can be quantified from a knowledge of the stationary values of the surface and volume energy density denoted, respectively, by dW/dA and dW/dV. The valleys and peaks of dW/dV provide information not only on the locations of possible failure initiation by yielding and/or fracture [8-10] but also on the stability of internal structure. Stacking sequence of a laminate, for example, can be optimized to control the rate of energy dissipation and hence the stability of failure involving fiber breaking, delamination, etc. Composite damage must also involve the nondestructive evaluation of thermal, mechanical and chemical effects. To this end, advancements have been made on the Electromagnetic Discharge Image (EDI) technique [11,12] using high voltage electrical discharge. This method is particularly suited for analyzing failure in composites. Changes in the microstructure, chemical composition, and interface condition can be identified with the mechanical behavior of composite so that a more precise distinction can be made between the intrinsic and apparent material

---

* Such a procedure has been established only recently [6,7] for isotropic and homogeneous materials based on the exchange of surface and volume energy. Plasticity theory is inadequate as it assumes the coincidence of uniaxial data with the effective stress and effective strain curve. Ignored is the influence of dilatation that is dominant in regions of failure initiation.

** Continuity of stress and displacement normally invoked in continuum mechanics analyses is not adequate as it avoids addressing one of the most important features of interface behavior that is energy dissipation.

*** Such an effect expressed by the rate change of volume with surface, dV/dA, is assumed to vanish for all classical theories of continuum mechanics as the element size shrinks in the limit.

parameters at a given scale level. A primary function of the EDI method is the evaluation of energy dissipated in various different forms. Reference can also be made to [13] that presents another way of monitoring of energy dissipation in composites. In what follows, emphases will be given to a consistent treatment on the mechanics of composites from both the theoretical and experimental viewpoint.

COMPOSITE MECHANICS

Composite mechanics is concerned with the analytical modelling of multi-component materials and the prediction of their behavior. The process of combining materials such as fibers and resins with disparate properties to form a single structure is, by no means, straightforward. Because of the dissimilar nature of the components, inconsistency in bonding can arise and alter the load transfer characteristics in addition to uncertainties in the alignment of fibers and the introduction of mechanical damage in fabrication. Particular precautions should be taken in modelling the initial state of the composites as failure can be affected by the residual stresses that arise during fabrication when differential rates of shrinkage between fiber and matrix can occur.

Properties of Constituents. Although the choice of the individual components in a composite is made in terms of the mechanical, thermal, physical and chemical properties, one of the major considerations is the load-bearing capability. Composite manufacturers inadvertently advertise their products in terms of mechanical properties such as strength and/or toughness. The idea is to combine one or more components into a single unit such that the individual constituents will retain their own specific properties and the composite will perform together as a structure in response to the prescribed load.

Figure 1 shows the uniaxial data for fiber reinforced composites with high fiber volume (HFV) and low fiber volume (LFV). The fiber being high strength and more stiff are susceptible to flaws or defects while the matrix is more forgiving. Shown in Figures 2 and 3 are the degradation of tensile strength as a function of length for the E-glass and boron fiber [14]. Note that the longer fibers tend to be weaker as they are more proned to be damaged by defects. Such an effect is not usually accounted for in the analysis. Similarly, the macroscopic features, such as flaws and voids and molecular features of polymeric matrix materials can also vary widely. They generally consist of small molecules joined chemically in chains or networks of repeating units to form huge molecules with an infinite variety of possible structures. The size, internal arrangement, chemical connectivity, and spatial distribution of the molecules have a direct influence on the final properties of the matrix, the polymer/fiber interphase, and/or the composite. Illustrated in Figures 4(a) and 4(b) are, respectively, ordered and disordered molecular networks as they are processed by combining the deuterium-labeled epoxy and hardener. Correlation of these networks with measured performance properties is vital as the real material is probably somewhere between the two ex-

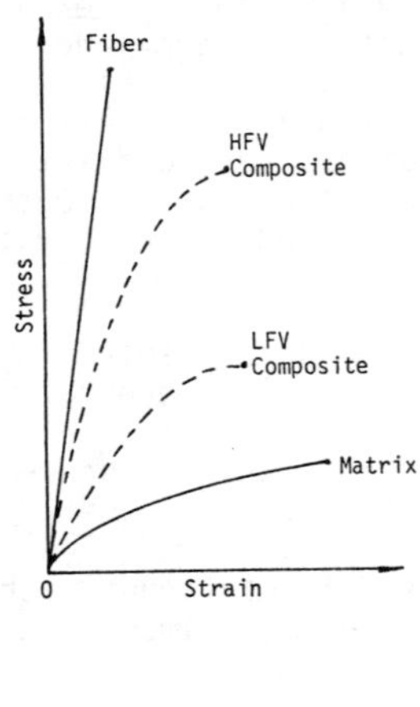

Fig. (1) - Uniaxial response of composite specimens

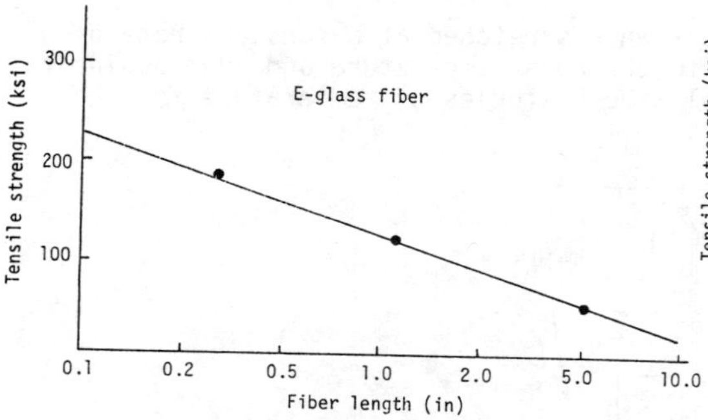

Fig. (2) - Degradation of strength with fiber length for E-glass [14] (p. 481)

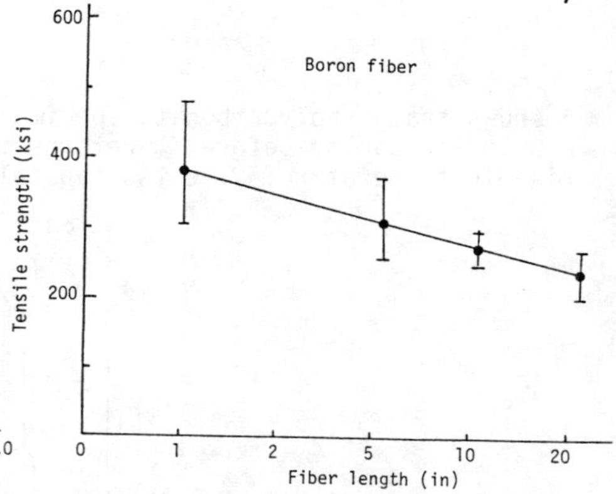

Fig. (3) - Degradation of strength with fiber length for boron [14] (p. 529)

(a) ordered network (b) disordered network

Fig. (4) - Schematics of order and disorder in the molecular networks of polymeric matrix

tremes shown in Figure 4. At present, there exists no mathematical models* for assessing the repeating units in the molecular chains that are arranged in space and linked chemically. This is mainly because of the lack of a generalized theory for treating the mechanical, thermal, physical and chemical behavior in an unique manner.

Recent works [15-17] have shown that the thermal/mechanical interactions are inherent of the energy dissipation process. It is not sufficient to characterize material behavior by its mechanical properties. The associated thermal fluctuations exhibit the true nature of irreversibility and the degree of disorder in the material. In other words, the thermal characteristics can differ widely for seemingly similar stress and strain curves of the same material. Contrary to the ordinary notion that a material gets heated when stretched monotonically, Figure

---

*Only in recent years has the method of neutron scattering emerged for determining the shape and structure of linear-chain macromolecules or lightly cross-linked rubbers in the solid state. The National Bureau of Standards has extended this technique to study the molecular networks of thermosetting materials.

5 shows that a polycarbonate specimen* when stretched at a constant rate of 2.0 in/min cools before it returns to the room temperature and then begins to rise in temperature. The two typical time histories of temperature for the

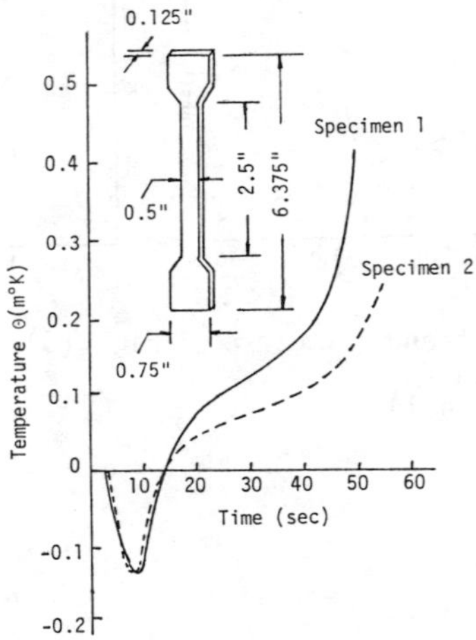

Fig. (5) - Cooling/heating of polycarbonate in uniaxial extension

polycarbonate specimens in Figure 5 are manifestation of the difference in local energy dissipation which is sensitive to changes in the molecular structure. They can serve as a measure of the degree of order in the network in terms of contributions from both an ideal, ordered network and a disordered network as those shown in Figures 4(a) and 4(b). The temperature fluctuation data also monitor the change in the chemical composition of the hardener molecules and the curing conditions such that network can be altered in predictable ways, both experimentally and theoretically. Such a capability can be used to improve composite matrix materials.

Classical continuum mechanics theories are not able to predict the phenomenon of heat transfer reversal** as the material is loaded monotonically. This is because they are unable to correctly account for the inherent thermal/mechanical coupling. Only the energy density theory [6,7] is thus far able to theoretically predict not only the feature of cooling/heating but also the time history of energy dissipation as a material is damaged gradually. Conventional approaches cannot simultaneously address the coupling*** between thermal, mechanical, physi-

* These specimens were made specially by the Dow Chemical Company for measuring the temperature change based on a technique developed at the Institute of Fracture and Solid Mechanics [18].

** This invokes a sign change in entropy which is not permitted in classical thermodynamics.

*** Moisture, temperature and deformation coupling in composites are discussed in [19]. Depending on the boundary conditions, strong coupling can occur and cannot be assumed to be negligible as an a priori.

cal and chemical effects. They deal primarily with mechanical properties which do not yield information on the way with which energy dissipation is affected by the microstructure.

Interface Integrity and Failure. Much has yet to be learned about characterizing the interface between the layers of fibers or along the fiber/polymer dividing line. The interaction energy across two dissimilar media can differ significantly depending on the bonding action that involves both electrochemical and mechanical interaction. The former arises by wetting the solid surface with a fluid. Bonding of the two phases is achieved through intermolecular forces. Theoretical strengths* of the order of 10^6 to 10^7 lb/in^2 can result within interatomic distances of a few Ångstroms as shown schematically in Figure 6. The Van der Waals bonds are slightly weaker with theoretical strengths of the order of 10^5 to 10^6 lb/in^2 and involve interatomic distances of 3 to 5 Ångstroms. Mechanical

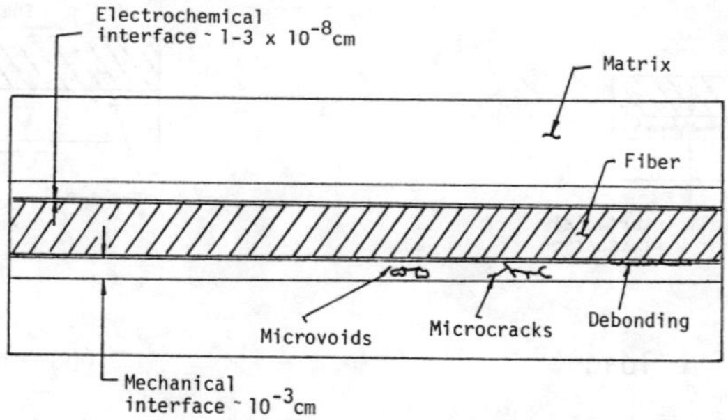

Fig. (6) - Schematic of electrochemical and mechanical interface

bonds are developed as a result of differential thermal expansion of the fiber and matrix. A compressive environment is created that results in high frictional adhesion. Compressive loading on an isolated filament in a glass fiber/epoxy composite can result in a frictional bond of 10^3 lb/in^2. This, however, is spread over a much larger distance that is four to five orders of magnitude larger than the chemical bond, Figure 6. Minute defects such as microvoids, microcracks, debonding, etc., can also prevail at the interface that further complicate the state of affairs at the fiber/matrix interface. Even though the initial interfacial energy states arising from electrochemical and mechanical bonding are essential, they alone are not sufficient for characterizing the interface integrity and/or failure. The true nature of the interphase depends simultaneously on the mechanics at a two-phase boundary and rates of energy transfer that are application-specific. No generalization is possible. Each case must be analyzed separately according to the specificity of the system.

The consideration of load transfer across interfaces becomes a major issue for composites in structural applications. Unless the relations between the overall bulk behavior of the composites and alleged interfacial response are known, no design guidelines can be laid down. Depending on the load type [20], interface properties will affect the composite behavior in different ways. This is illustrated in Figures 8(a) and 8(b) in which the interfacial energy oscillated much

---

*Actual strength of an interfacial bond is lower because the classical theoretical treatment is overly idealized and is not able to include the combined effects of material, geometry and load transfer.

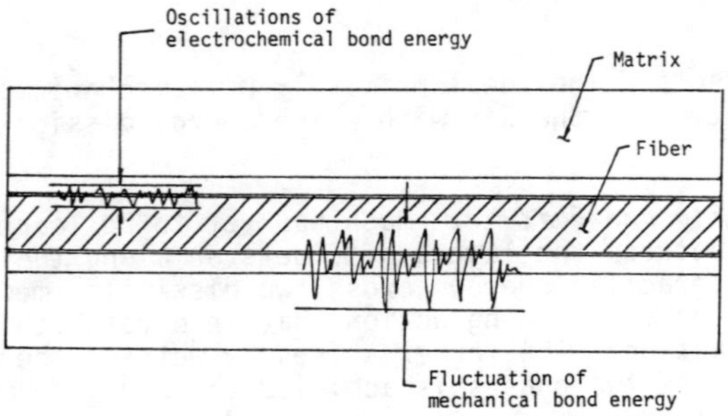

Fig. (7) - Initial interfacial energy states: electrochemical and mechanical

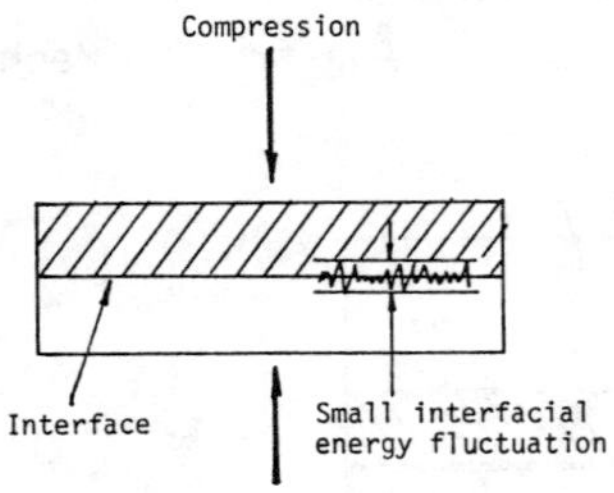

(a) compressive load

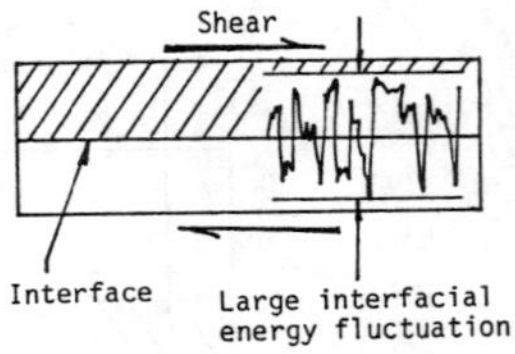

(b) shear load

Fig. (8) - Interfacial oscillation in energy due to loading: compression and shear

more widely in shear than that in compression. The local material elements respond at rates dictated by the properties of the two adjoining phases and the morphology in the region near the interface. Hence, the mutual synergistic effects must be regarded as unknowns and determined theoretically. The fundamental understanding on this subject has not been well developed. Apparent contradictions and/or inaccuracies can result if the interface properties are assumed to be known as a priori. This occurs even in the case of a single phase material where the same constitutive relation* is assumed to prevail at each point of the system. The stress and strain relations at elements near a boundary can differ widely from those inside a body [6,7]. They must be determined separately for each element and each time step of loading. Such an approach must be adopted for analyzing composite interface behavior.

Debonding of fiber/matrix interface should be distinguished from that of delamination between the laminae in a layered composite structure, Figure 9. The amplitude of energy oscillation in delamination is considerably larger than that associated with debonding given by Figure 7. Since both normal and shear stress

*Conventional mechanics theories assume not only that constitutive relations are known but they are the same at each point in the system. This invalidates their application to problems where the local strain rates undergo large changes.

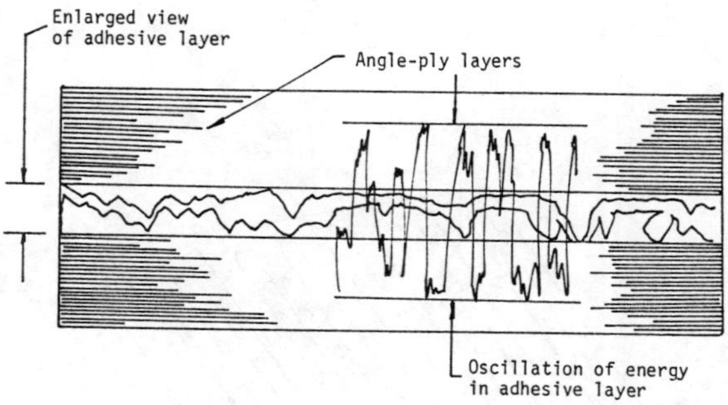

Fig. (9) - Delamination of a layered composite

may both be present, delamination can be best analyzed by application of the energy density function dW/dV [21-23]. The stationary values of dW/dV consider failure by both yielding and/or fracture as they account for dilatation and distortion.

TIME-DEPENDENT COMPOSITE DAMAGE

Much has been published on the resistance of composites to cracking and damage in general and yet, little has been done in coming forth with the scientific principle and framework to quantify the various observed failure modes. Tests, as well as analytical* results, are often reported with no theoretical justification nor physical understanding. Apparent contradictions are widely spread in the open literature and confusion continues to prevail. Frequent misuse of fracture mechanics concepts is not uncommon. Fracture toughness, for instance, is defined strictly for a homogeneous material where damage is dominated by the growth of a single crack. Failure in a composite is not so clear-cut as it may involve cracks in the fiber, matrix and along the fiber/matrix interface. Any empirically measured fracture toughness would depend on directional properties of the constituents with reference to the load. Such a quantity sheds little or no light on composite behavior as it is too sensitive to changes in the design parameters.

Composite failure can be best addressed in terms of damage over an area or volume. Above all, the properties of the constituents as measured separately must be correlated with the structural features of the composite. This requires a precise knowledge of how the time dependent energy is distributed and used to deform and damage the composite system.

Damage Characterization. Unlike the metals where failure may be dominated by the growth of macrocracks, fiber reinforced composites fail in a cumulative fashion that may involve a combination of different modes such as fiber breaking, matrix cracking and fiber/matrix interface debonding, Figure 10. Each of the modes may occur at a different time. Hence, the energy associated with creating

*The mere application of classical mechanics theories leaves much to be desired because many of the physical aspects of composite behavior have already been neglected in the original formulation, one of which is the interaction of surface and volume energy.

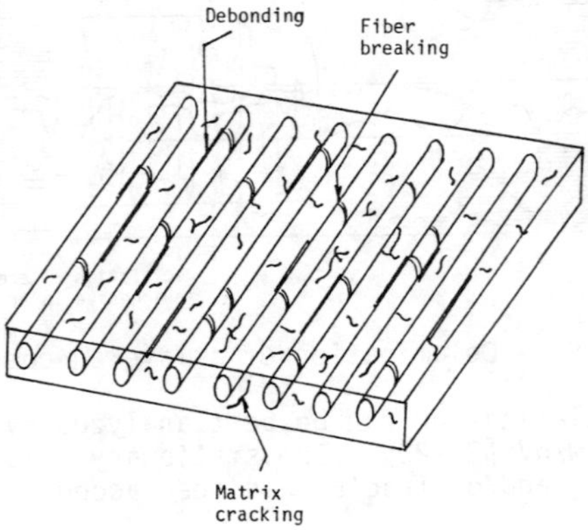

Fig. (10) - Failure modes in unidirectional fiber reinforced composite

new surfaces in a unit volume at a given *instance* may consist of*

$$\left(\frac{dW}{dA}\right)_p = \Sigma\left[\left(\frac{dW}{dA}\right)_f + \left(\frac{dW}{dA}\right)_{f/m} + \left(\frac{dW}{dA}\right)_m\right] \tag{1}$$

in which the subscripts f, m and f/m stand, respectively, for the fiber, matrix and fiber/matrix interface. Any one of the component surface energy density, say $(dW/dA)_{f/m}$, may be zero within certain time interval if debonding does not occur. In other words, the proportion of fiber breaking, matrix cracking and fiber/matrix debonding changes with time. What is available to create new damage surface at the next time increment is $(dW/dA)^*$. Hence, the total surface energy density is

$$\frac{dW}{dA} = \left(\frac{dW}{dA}\right)_p + \left(\frac{dW}{dA}\right)^* \tag{2}$$

The damage accumulated in a unit volume of composite element can be obtained from

$$\left(\frac{dW}{dA}\right)_i = \left(\frac{dV}{dA}\right)_i \frac{dW}{dV}, \quad i = \xi,\eta,\zeta \tag{3}$$

with dW/dV being the area under the true stress and true strain curve. The rates change of volume with surface $(dV/dA)_i$ are made proportional to the respective slopes of the stress and strain curfes. In this way, the energy state for

---

*Other forms of surface energy associated with damage may be involved such as $(dW/dA)_a$ that arises due to delamination or breaking of the adhesive in a layered composite.

each composite element can be uniquely determined and identified with data that are obtainable from uniaxial tests. This is accomplished by referring the components $(dW/dA)_\xi$, $(dW/dA)_\eta$ and $(dW/dA)_\zeta$ to the *damage planes* such that

$$A\left(\frac{dW}{dA}\right)_\xi = B\left(\frac{dW}{dA}\right)_\eta = C\left(\frac{dW}{dA}\right)_\zeta \tag{4}$$

The parameters A, B and C are determined by performing three different uniaxial tests along the axes of symmetry for the composite system. In this way, the constitutive relation for each composite element will be derived rather than assumed as it is now being done by the classical approach.

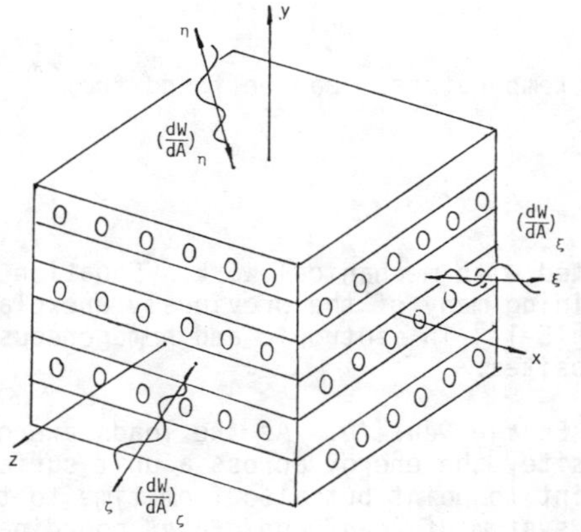

Fig. (11) - Schematic of a unit volume of composite element

Thresholds for characterizing the integrity of composite systems can thus be established by considering the rate at which the volume and surface energy density approach their respective critical values. Because of equation (4), it suffices to consider $(dW/dA)_\xi$. Failure of composite elements is assumed when

$$\left(\frac{dW}{dA}\right)_\xi > \left(\frac{dW}{dA}\right)_c; \qquad \frac{dW}{dV} \simeq \left(\frac{dW}{dV}\right)_c \text{ or } \frac{dW}{dV} < \left(\frac{dW}{dV}\right)_c \tag{5}$$

and/or

$$\left(\frac{dW}{dV}\right) > \left(\frac{dW}{dV}\right)_c; \qquad \left(\frac{dW}{dA}\right)_\xi \simeq \left(\frac{dW}{dA}\right)_c \text{ or } \left(\frac{dW}{dA}\right)_\xi < \left(\frac{dW}{dA}\right)_c \tag{6}$$

Depending on the rate of load transfer in the composite, the exchange of surface and volume energy density can vary over a wide range as indicated in equations (5) and (6). Numerical computation should start with a sufficiently small time interval such that either equation (5) or (6) is satisfied but not both at the same time. Otherwise, the predicted state of disturbance or damage would spread over a large portion of the system rendering inaccuracies in the results.

Not to be ignored is the energy dissipated to damage the composite. Let this quantity be designated by $\mathcal{D} = (dW/dV)_p$ which is to be distinguished from the available energy density $(dW/dV)^*$ as follows:

$$\frac{dW}{dV} = \left(\frac{dW}{dV}\right)_p + \left(\frac{dW}{dV}\right)^* \tag{7}$$

Energy dissipation in composite specimens has been measured experimentally* [13]. It, of course, must be positive definite and increase monotonically with time, i.e.,

$$\mathcal{D} \geq 0; \quad \frac{d\mathcal{D}}{dt} \geq 0 \tag{8}$$

Once $\mathcal{D}$ is known, the temperature Θ can be found from**

$$\frac{\Delta\Theta}{\Theta} = -\frac{\Delta M}{\Delta\mathcal{D}} \tag{9}$$

in which M is connected with mechanical work. Equation (9) has been used successfully for determining many of the previously unexplained thermal/mechanical interaction effects [15-17] in isotropic and homogeneous materials. It applies equally well to composites.

Local and Global Energy Density. As the loads are nonuniformly distributed throughout the composite, the energy across a unit surface and volume will differ not only from point to point but also from time to time. These variations can be referred to a system of *local* and *global* coordinates for predicting the failure path and stability of the composite system. The maxima and minima of dW/dA or dW/dV will be denoted, respectively, by $[(\ )_{max}]_L$ and $[(\ )_{min}]_L$ when referred to the local coordinates and $[(\ )_{max}]_G$ and $[(\ )_{min}]_G$ when referred to the global coordinates as shown in Figures 12(a) and 12(b). Locations of possible failure initiation due to excessive distortion and dilatation are assumed [8] to correspond to the maximum of $[(\ )_{max}]_L$ and $[(\ )_{max}]_G$ and maximum of $[(\ )_{min}]_L$ and $[(\ )_{min}]_G$. By restricting the subsequent discussion to failure by crack growth, only the pair $[(\ )^{max}_{min}]_L$ and $[(\ )^{max}_{min}]_G$ needs to be considered.

To begin with, consider the simplest example of a through crack loaded uniformly in an homogeneous material as shown in Figure 13(a). The locations of $[(dW/dV)^{max}_{min}]_L$ and $[(dW/dV)^{max}_{min}]_G$ are denoted by L and G, respectively, and the

---

* One of the experimental curves for $\mathcal{D}$ in [13] became negative which is a physical impossibility and obviously caused by error in measurement.

**Only when the energy is dissipated all in the form of heat Q will Θ reduce to the classical temperature T. In general, $\mathcal{D}$ is not equal to Q.

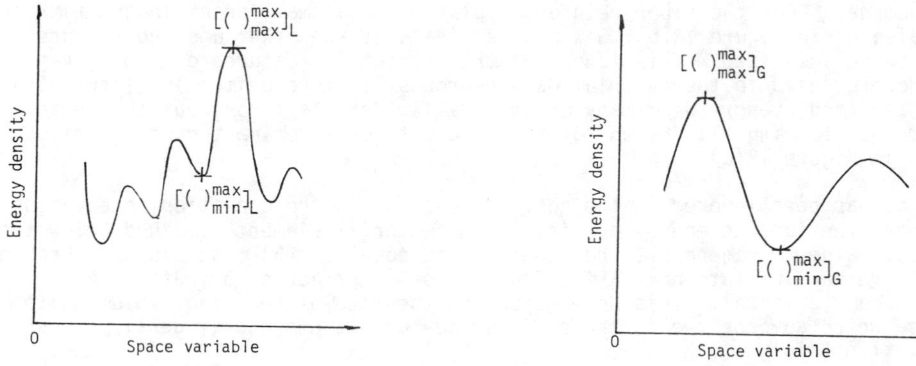

(a) local fluctuation (b) global fluctuation

Fig. (12) - Stationary values of local and global energy density

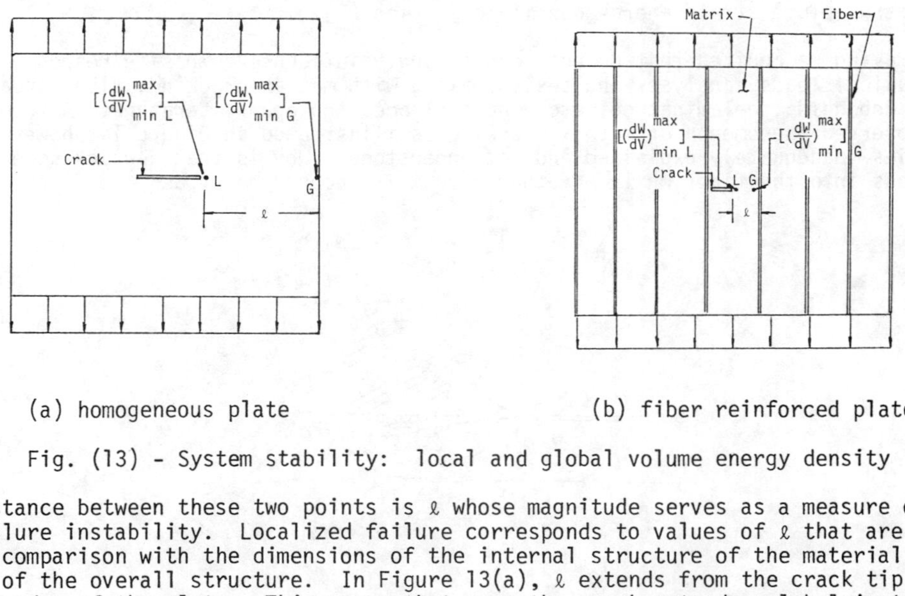

(a) homogeneous plate (b) fiber reinforced plate

Fig. (13) - System stability: local and global volume energy density

distance between these two points is ℓ whose magnitude serves as a measure of failure instability. Localized failure corresponds to values of ℓ that are small in comparison with the dimensions of the internal structure of the material and/or of the overall structure. In Figure 13(a), ℓ extends from the crack tip to the edge of the plate. This means that once the crack extends, global instability or fracturing of the entire plate is imminent. Local and global instability is said to coincide*. The plate dimensions and/or loading type for the homoge-

*This is known in fracture mechanics as unstable fracture. The idealized Griffith configuration consisting of a through crack in an uniformly loaded plate of infinite extent results in unstable fracture because ℓ is infinite.

neous plate in Figure 13(a) may be altered such that G will lie inside the plate boundary and making the system more stable insofar as failure by crack growth is concerned. For the fiber reinforced plate with a small crack in the matrix as indicated in Figure 13(b), G would terminate at the fiber and the failure tends to be localized. This is one of the attractive features of composites where energy used in the material damage process is more uniformly distributed and dissipated by spreading many of the smaller defects throughout the system rather than feeding all the energy to the growth of a dominant crack as illustrated in Figure 13(a).

What has been invoked in the above discussion is the condition in equation (6) where the surface energy density in the material element remained below the critical value. In general, the condition in equation (5) must also be checked as failure can initiate near a free surface or interface such that dW/dA will first become critical. This depends on how the rate of change of volume with surface which governs the exchange of surface and volume energy density as shown by equation (3).

Additional Illustrations. The ℓ-concept may also be used in conjunction with the theory of elasticity and plasticity. Failure must then be analyzed either by using dW/dA or dW/dV and not both simultaneously because they are regarded as independent quantities* in the classical continuum mechanics theories where dV/dA is taken to be zero in the formulation. In what follows, several of the frequently observed failure modes in composite systems will be predicted and explained from the volume strain energy density criterion [8].

Having recognized that defects can strongly influence composite response to mechanical loads, analyses and tests dealing with matrix cracking, fiber breaking, debonding, delamination, etc., have flooded the open literature. A frequently observed phenomenon of matrix cracking as illustrated in Figure 14, however, remains inadequately explained and not understood. Why is that a crack sometimes extends into the fiber while at other times, it chooses to bypass the fiber? An

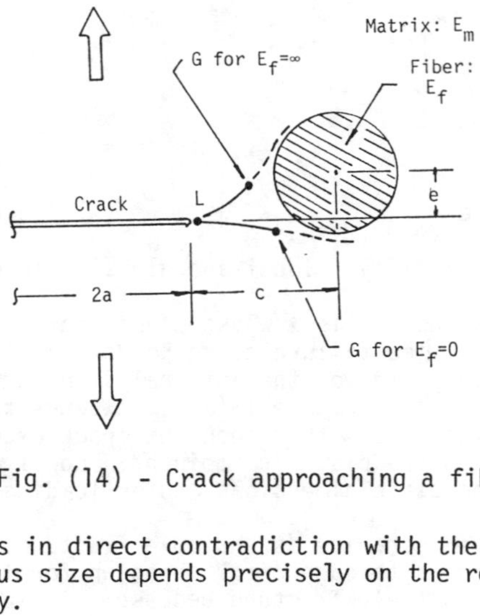

Fig. (14) - Crack approaching a fiber

*Such an assumption is in direct contradiction with the nucleation theory of Gibbs where the nucleus size depends precisely on the relative values of the surface and volume energy.

analysis of this problem has been made in [25] and showed that the direction of crack growth is very sensitive to changes in the relative moduli of the fiber and matrix, i.e., the ratio E_f/E_m with E_f and E_m being, respectively, the fiber and matrix modulus. Illustrated in Figure 14 is a fiber of radius R whose edge is at a distance c from the tip of a crack with length 2a. The center of the fiber is displaced slightly by the amount e from the crack plane. For typical values of c/a and R/a near unity, the location of G coinciding with $[(dW/dV)^{max}_{min}]_G$ is below the crack plane for $E_f = 0$ and above the crack plane for $E_f = \infty$. Hence, there exists a ratio of E_f/E_m for which G will lie on a failure path leading into the fiber. Depending on the geometric parameters a, c, e and R, variations of E_f along the fiber and E_m in the matrix, a crack in the matrix will, for the most part, propagate around the fiber except for situations where the morphology in the crack/fiber region and local material inhomogeneity governed by E_f/E_m are such that the path of $(dW/dV)^{max}_{min}$ connects the crack tip and fiber boundary.

Fiber breaking is another commonly observed failure mode that may or may not result in debonding of the fiber/matrix interface. This depends on the *rate* at which energy is released when a fiber snaps suddenly. Depicted in Figure 15 is

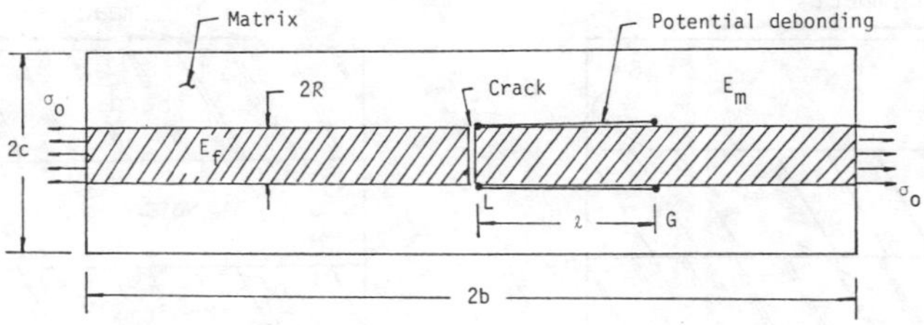

Fig. (15) - Fiber breaking and debonding

a circular fiber embedded in a cylindrical section of the matrix with radius c and length b. The system is loaded at a given rate up to a value σ_o until the fiber breaks. Such a situation has been analyzed in [26] by solving for the change in volume energy density fluctuation before and after the fiber breaks. A high elevation of dW/dV across the fiber/matrix interface was found for the case of a broken fiber. The results also suggest possible debonding along the interface from L to G as shown in Figure 15. The distance ℓ depends on the time rate history of σ_o and the properties of the fiber/matrix system.

Stacking sequence of a laminate has also been a subject of much past investigations. Displayed in Figure 16(a) to 16(d) inclusive are the normal and shear impact of a three-layered dissimilar material systems. Each layer has the same thickness. The stacking sequences of hard/soft/medium and soft/medium/hard analyzed in [27] yield, respectively, the smallest and largest value of ℓ for both normal and shear impact while other stacking sequences correspond to intermediate values of ℓ. Under normal impact, the hard layer should make initial contact with the load to absorb the shock such that failure can be confined to the surface layer where both points L and G lie within a layer distance, Figure 16(a). The same combination is also preferred for shear impact although ℓ is larger than that in

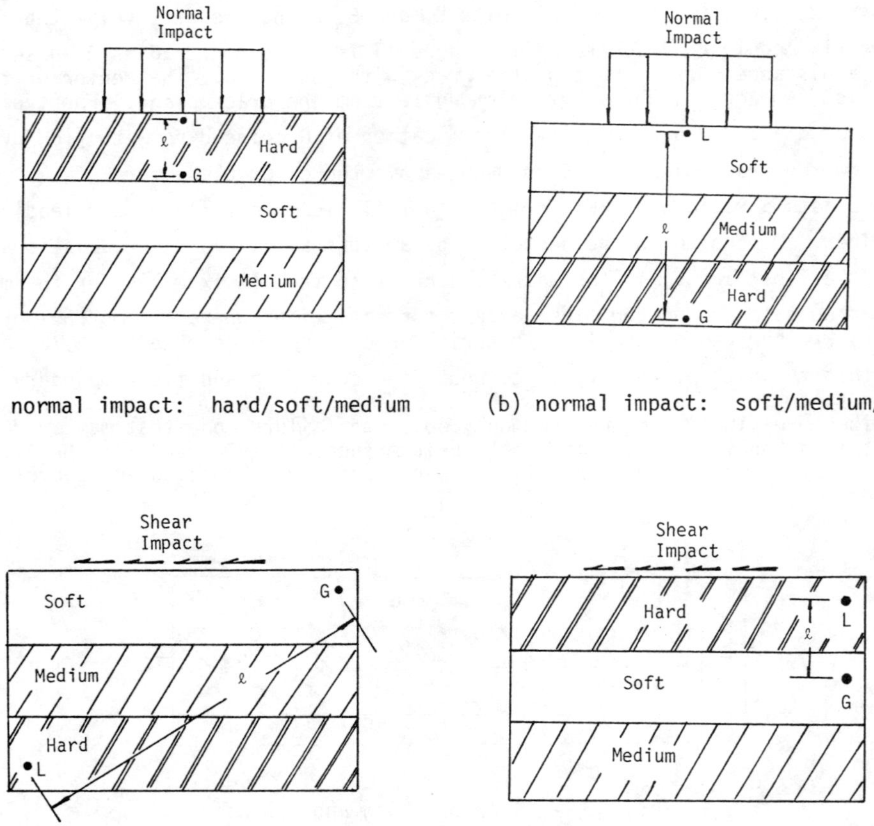

(a) normal impact: hard/soft/medium (b) normal impact: soft/medium/hard

(c) shear impact: hard/soft/medium (d) shear impact: soft/medium/hard

Fig. (16) - Schematics of a three-layered system subjected to normal and shear impact

normal impact because G extends into the second layer as shown in Figure 16(c). This confirms that composite laminate is weaker under shear, a well-recognized result. The worse stacking sequence for normal impact is given in Figure 16(b) where failure is considered to be globally unstable. This is because once cracking starts, all three layers will fracture. Figure 16(d) gives the most unstable combination of load type and stacking sequence as it yields the largest value of ℓ. The foregoing results are particularly relevant in the design of composite armors to provide maximum resistance against projectile penetration.

ELECTROMAGNETIC DISCHARGE IMAGING: NONDESTRUCTIVE EVALUATION

Because of the complex nature of composite failure behavior, a variety of analytical models have emerged in an attempt to explain the different failure modes. Since no unified treatment has come forth, damage inspection procedures have relied solely on the development of *hardwares* with the assumption that the data related to debonding, fiber and matrix degradation, etc., will eventually be understood and analyzed by theories. Among the techniques used are ultrasonics, radiography, acoustic emission, thermography, eddy current and a host of others. Such a procedure has been loosely referred to as Non-Destructive Testing (NDT). What is urgently needed is, of course, a consistent way of assessing composite damage so that the seemingly different failure modes can be weighed on a common basis. There is no other way for qualifying the integrity of composite materials and structures.

Damage being a time-rate dependent process is more widely spread throughout the composite simply because the load reacts more sensitively with material inhomogeneity. The idea of NDT is to correlate the difference between the emitted and received signal with the size and location of damage. Those energies that are already present within the system may also be released during the damage process to interrupt the NDT signal. The unique identification of energy dissipation rate with failure mode is a prerequisite for the quantitative assessment of damage, i.e., the quantity $d\mathcal{D}/dt$ or those in equation (1), namely $d(dW/dA)_f/dt$, $d(dW/dA)_m/dt$, etc. The Electromagnetic Discharge Image* (EDI) technique [11,12] commonly known as Kirlian photography, is particularly suited for incorporating the energy density theory [6]. Utilized in the process is high-voltage electrical discharge field. Electrons with the specimen are accelerated and multiply exponentially. Streamers are thus created that move at speeds of 10^7 to 10^8 cm/sec. In air, at high field strength, the streamers are blue while a reddish-purple glow appears in low electric fields. The intensity, color and pattern of the discharge image can be recorded photographically or digitally and identified with the mechanical, thermal, chemical and electrical disturbances in composites as they are damaged. Mechanical discontinuities in the range of 10^{-1} to 10^{-6} mm can be detected. Such a wide range of detection capability is unmatched by the other conventional methods. The flow chart in Figure 17 shows how the available (or dissipated) energy is monitored by EDI. A laboratory EDI model has been designed by the Institute of Fracture and Solid Mechanics at Lehigh University. The combined thermal, mechanical, chemical and electrical effects are recorded everywhere on a specimen as indicated in Figure 18(a). Spectral analysis of the pixel can then be made to obtain the optical intensity as a function of radiation wavelength, Figure 18(b). For each radiation wavelength λ_j, the optical intensity I_j can be isolated and identified with a given energy level U_j. This is accomplished by means of a plasma spectral analysis, the results of which are shown schematically in Figures 19(a) to 19(c). The results can be further processed and displayed by contour plots in two or three dimensions as illustrated in Figure 20 by which the individual thermal, mechanical, chemical or electrical effect

*This technique has been used extensively in the USSR in NDT and is rarely publicized. Preliminary efforts made by the U.S. Navy and Lawrence Livermore Laboratory lacked the theoretical support in NDT application.

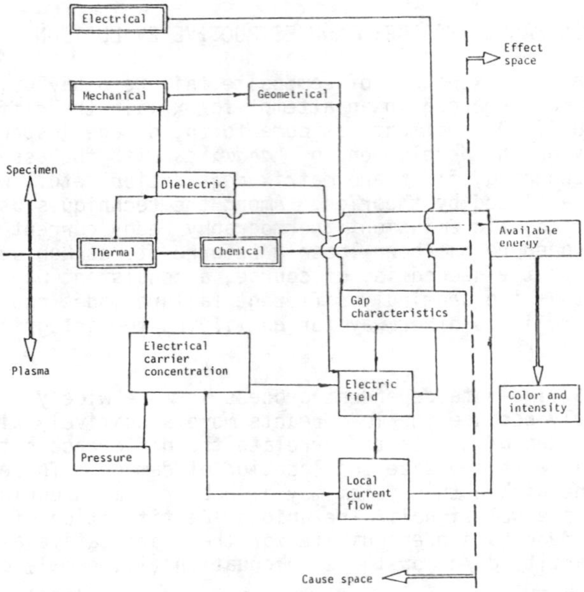

Fig. (17) - Cause and effect on intensity and color or emitted radiation

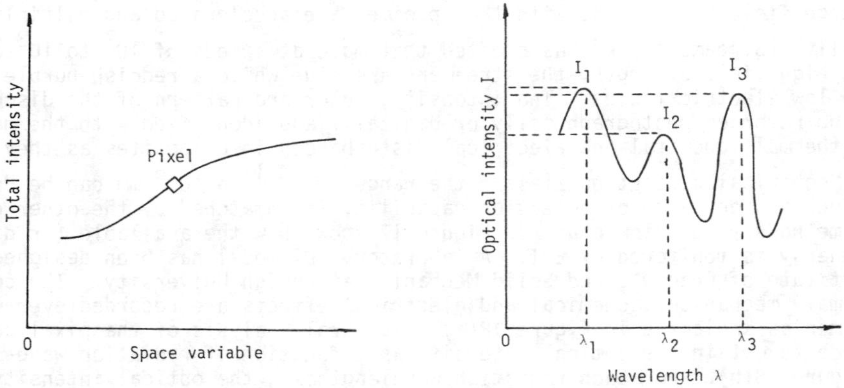

(a) combined intensity (b) individual wavelength

Fig. (18) - Intensity versus radiation wavelength

can be sorted out. The color, intensity and distribution of the contours provide information on the location and magnitude of the stationary values of the energy density whose relation with material damage has already been discussed earlier. These contours will change as a function of time, if load is present and hence the influence of time-rate dependency on damage can also be evaluated. To be more specific, oscillation of the electrochemical bond energy in Figure 7 may be displayed by contour plots in a two-dimensional space. Refer to Figure

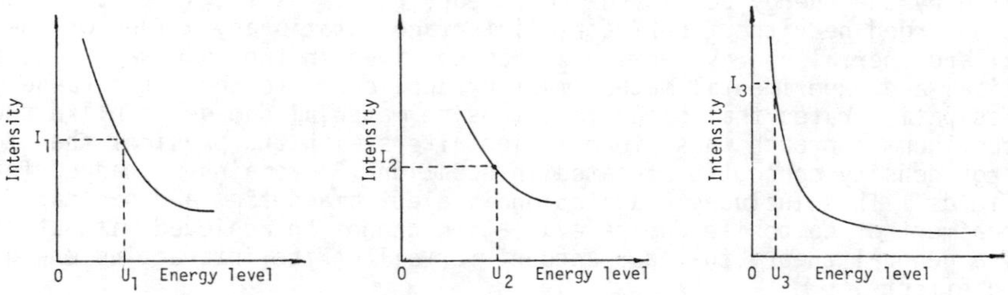

(a) energy U_1 (b) energy U_2 (c) energy U_3

Fig. (19) - Variations of intensity with energy level

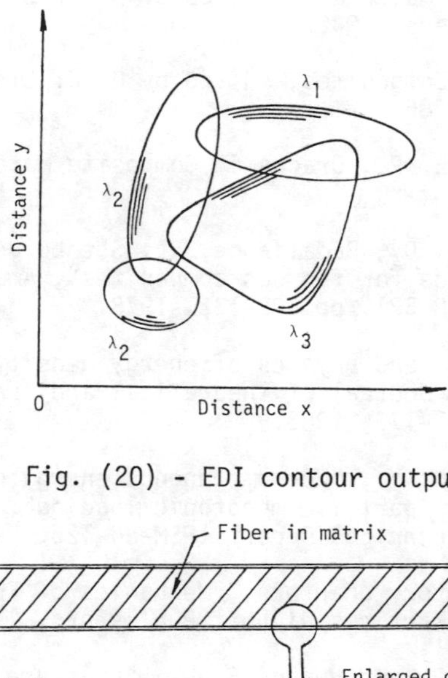

Fig. (20) - EDI contour output

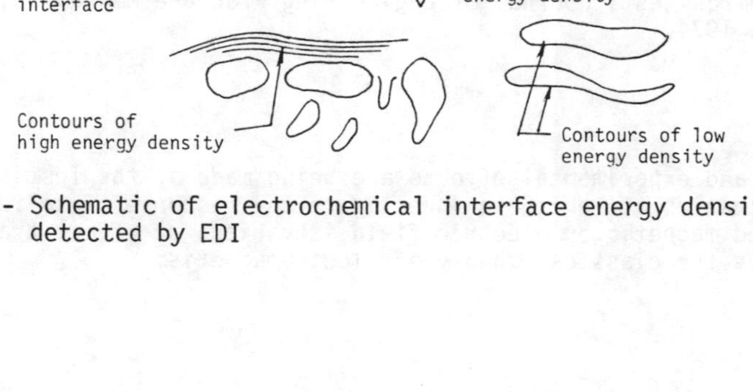

Fig. (21) - Schematic of electrochemical interface energy density contours detected by EDI

21 in which contours with high and low energy densities represent fluctuations in the electrochemical forces. They can be measured by the EDI technique and calculated by the energy density damage theory [28]. As the scale of investigation is enlarged near the fiber/matrix interface, stationary values of the mechanical and thermal energy density can be obtained in the same way. The *same* theoretical and experimental method must be applicable to cover the range of energy dissipation rates that occur in composite material damage. Unlike the classical continuum approach whose formulation alters with the physical phenomenon, the energy density concept postulated in equation (3) remains unchanged for all force fields. This includes gravitational, electromagnetic* and nuclear. Unified treatment of composite damage evaluation cannot be achieved without the support of a general theory that can account for all of the interacting energies and coupling effects.

REFERENCES

[1] Fracture Mechanics of Composite Materials, edited by G. C. Sih and V. P. Tamuzs, Martinus Nijhoff Publishers, 1979.

[2] Fracture of Composite Materials, edited by G. C. Sih and V. P. Tamuzs, Martinus Nijhoff Publishers, 1982.

[3] Failure Mechanics of Composites, edited by G. C. Sih and A. M. Skudra, North-Holland Publishers, 1985.

[4] Sih, G. C. and Chen, E. P., Cracks in Composite Materials, Martinus Nijhoff Publishers, 1982.

[5] Sih, G. C., Hilton, P. D., Badaliance, R., Shenberger, P. S. and Villarreal, G., "Fracture mechanics for fibrous composites", American Society for Testing and Materials, STP 521, pp. 98-132, 1973.

[6] Sih, G. C., "Mechanics and physics of energy density and rate of change of volume with surface", Journal of Theoretical and Applied Fracture Mechanics, Vol. 4, No. 3, pp. 157-173, 1985.

[7] Sih, G. C. and Tzou, D. Y., "Nonhomogeneous energy dissipation ahead of a slowly growing crack: part I: monotonic loading", Institute of Fracture and Solid Mechanics Technical Report IFSM-84-126, December 1984.

[8] Sih, G. C., "Introductory chapters", Mechanics of Fracture, Vol. I to VII, edited by G. C. Sih, Martinus Nijhoff Publishers, 1972-1982.

[9] Sih, G. C., Chen, E. P. and Huang, S. L., "Fracture mechanics of plastic-fiber composites", Journal of Engineering Fracture Mechanics, Vol. 6, pp. 343-359, 1974.

*Theoretical and experimental efforts are being made at the Institute of Fracture and Solid Mechanics at Lehigh to investigate the response of solids as the rate of the applied magnetic or electric field is varied, an effect that cannot be accounted for by the classical theory of electromagnetism.

[10] Sih, G. C., Chen, E. P., Huang, S. L. and McQuillen, E. J., "Material characterization on the fracture of filament-reinforced composites", Journal of Composite Materials, Vol. 9, pp. 167-186, 1975.

[11] Sih, G. C. and Michopoulos, J. G., "Nondestructive detection of damage in aluminum: electromagnetic discharge imaging", Journal of Theoretical and Applied Fracture Mechanics, Vol. 5, No. 1, pp. 23-30, 1986.

[12] Michopoulos, J. G. and Sih, G. C., "Electromagnetic discharge imaging technique as a means of nondestructive evaluation of material imperfections", Institute of Fracture and Solid Mechanics Technical Report IFSM-85-132, 1985.

[13] Mast, P. W., Beautien, L. A., Clifford, M., Mulville, D. R., Sutton, S. A., Thomas, R. W., Tirosh, J. and Wolock, J., "A semi-automated in-plane loader for material testing", Journal of Experimental Mechanics, Vol. 23, No. 2, pp. 236-241, 1983.

[14] Composite Materials: Testing and Design, American Society of Testing and Materials, STP 460, 1969.

[15] Sih, G. C. and Tzou, D. Y., "Heating preceded by cooling ahead of crack: macrodamage free zone", Journal of Theoretical and Applied Fracture Mechanics, Vol. 6, No. 2, pp. 103-111, 1986.

[16] Sih, G. C. and Tzou, D. Y., "Irreversibility and damage of SAFC-40R steel specimen in uniaxial tension", Journal of Theoretical and Applied Fracture Mechanics, Vol. 7, No. 1 (in press)

[17] Sih, G. C., Lieu, F. L. and Chao, C. K., "Thermal/mechanical damage of 6061-T6 aluminum tensile specimen", Journal of Theoretical and Applied Fracture Mechanics, Vol. 7, No. 2 (in press).

[18] Precision Temperature Measurement Laboratory, Institute of Fracture and Solid Mechanics, Lehigh University, November 1986.

[19] Sih, G. C., Michopoulos, J. G. and Chou, S. C., Hygrothermoelasticity, Martinus Nijhoff Publishers, The Netherlands, 1986.

[20] Sih, G. C., "Influence of interface modelling on composite failure", Continuum Models of Discrete Systems, University of Waterloo Press, Waterloo, Ontario, Canada, pp. 837-864, 1979.

[21] Sih, G. C., "Fracture mechanics of composite materials", Fracture of Composite Materials, edited by G. C. Sih and V. P. Tamuzs, Martinus Nijhoff Publishers, The Netherlands, pp. 111-130, 1979.

[22] Chou, S. C., "Delamination of T300/5208 graphite/epoxy laminates", Fracture of Composite Materials, edited by G. C. Sih and V. P. Tamuzs, Martinus Nijhoff Publishers, The Netherlands, pp. 247-263, 1982.

[23] Davis, M. J. and Jones, R., "Damage tolerance of fiber composite laminates", Fracture Mechanics Technology Applied to Material Evaluation and Structure Design, edited by G. C. Sih, N. E. Ryan and R. Jones, Martinus Nijhoff Publishers, The Netherlands, pp. 635-655, 1983.

[24] Sih, G. C., "The strain energy density concept and criterion", Journal of Aeronautical Society of India, Vol. 37, No. 1, pp. 43-60, 1985.

[25] Chao, C. K., "Stability of a crack interacting with fibers", Advanced Composite Materials and Structures, edited by G. C. Sih and S. E. Hsu, VNU Science Press BV, The Netherlands, pp. 365-374, 1987.

[26] Chou, D. M., "Energy density redistribution caused by broken fibers", Advanced Composite Materials and Structures, edited by G. C. Sih and S. E. Hsu, VNU Science Press BV, The Netherlands, pp. 413-423, 1987.

[27] Tzou, D. Y., "Influence of stacking sequence on the instability of layered composite", Advanced Composite Materials and Structures, edited by G. C. Sih and S. E. Hsu, VNU Science Press BV, The Netherlands, pp. 425-435, 1987.

[28] Sih, G. C., Energetics of Irreversibility, VNU Science Press BV, The Netherlands (in preparation).

RESEARCH ON SPECIAL COMPOSITES IN CSIST, ROC

S.-E. Hsu

Chung Shan Institute of Science and Technology
P.O. Box 1-26, Lungtan, Taiwan 32500, Republic of China

ABSTRACT

Most of the advanced composites are developed for structural use in order to improve its specific strength. Currently, a few special composites were developed in the Materials R&D Center, Chung Shan Institute of Science and Technology (CSIST) aiming at improving their physical behaviors. In this paper, six special composites are to be discussed to study their magnetic, electromagnetic, acoustic and thermal behavior in certain specific applications.

A STUDY ON THE MAGNETIC PROPERTIES OF TWO-PHASE MAGNETIC COMPOSITES

According to our previous published papers [1,2], the magnetic properties of two-phase particulate magnetic composites with hard ferro-magnetic components can be predicted if the individual properties of remanence, coercivity, the shape of hysteresis loop and volume fraction are known. Experiments showed good agreements between theoretical calculation and experimental results for isotropic low (-BH) volume magnetic materials. Recently, two rare earth hard magnets, $SmCo_5$ and Nd-Fe-B, were successfully developed in this laboratory. In this paper, we will present the results of experiments for two-phase particulate magnetic composites to testify the theoretical feasibility in the case of anisotropic composites which consist of at least one rare-earth hard magnetic component. It is concluded that the hysteresis loop can be designed for specific engineering applications based on the concept of magnetic composites.

Three kinds of rare-earth components, including $SmCo_5$, $Sm(Co,Fe,Cu,Zr)_{7.4}$ and $Nd_{15}Fe_{77}B_8$ are tried. Three systems of two-component magnetic composites, namely, $SmCo_5$-Ba-Ferrite, $SmCo_5$-$Sm(Co,Fe,Cu,Zr)_{7.4}$ and $Nd_{15}Fe_{77}B_8$-$SmCo_5$ are studied. Since each magnetic component has its specific sintering temperature, two-component magnetic composite cannot be prepared by conventional sintering process. In this experiment, two kinds of binding processes are used to replace the sintering procedure to settle the preform. The first method is by resin-infiltration and the second method is by explosive compaction. The magnetic properties of the composites are measured by using a hysteresis loop tracer. The experimental results of the demagnetization B-H curves of composites for $SmCo_5$-$BaO.6Fe_2O_3$, $SmCo_5$-$Sm(Co,Fe,Cu,Zr)_{7.4}$ and $SmCo_5$-$Nd_{15}Fe_{77}B_8$ systems which are all prepared by the first method are shown in Figures 1, 2 and 3, respectively. For the purpose of comparison, the theoretical result based on calculation with the known

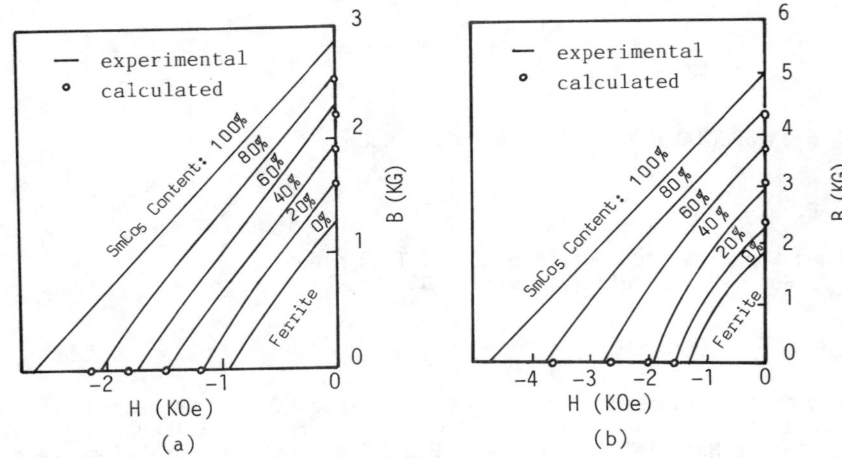

Fig. (1) - The demagnetization curves of composites and its constituents in $SmCo_5$-Ba-Ferrite system for (a) isotropic and (b) anisotropic cases

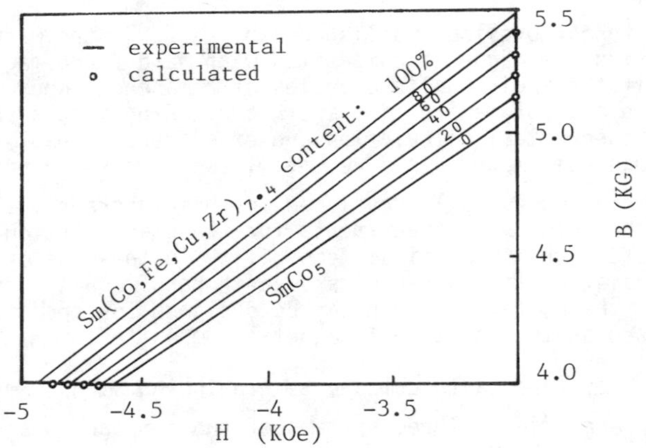

Fig. (2) - The demagnetization curves of composites and its constituents in $Sm(Co,Fe,Cu,Zr)_{7.4}$ - $SmCo_5$ system

remanence (Br) and coercivity (Hc) and volume fractions of each constituent are also shown in the same figures. In all cases, a good agreement has been obtained. These results indicate that the theoretical calculation can be equally applied to either isotropic or anisotropic cases. For those, the extensively used resin bonded magnets which consist of only one magnetic component are the simplest extreme case for two components magnetic composites.

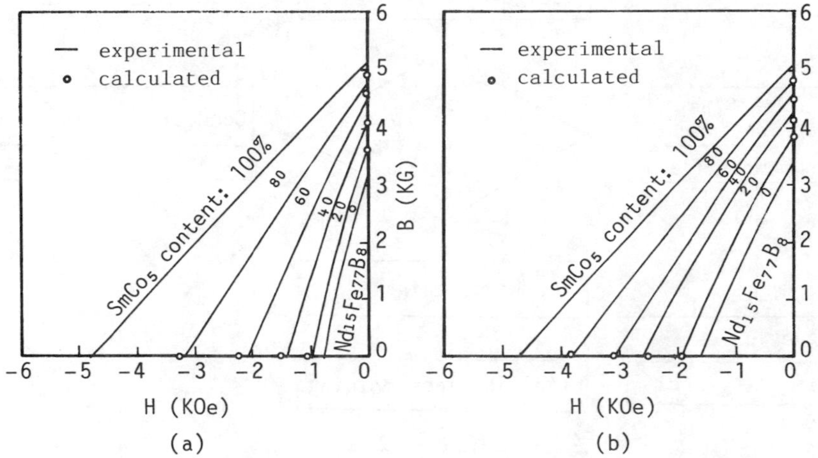

Fig. (3) - The demagnetization curves of composites and its constituents in $Nd_{15}Fe_{77}B_8$-$SmCo_5$ system. Average particle size of $Nd_{15}Fe_{77}B_8$ powder (a) 5 μm, (b) 630 μm

In Figure 3, the magnetic property of single component $Nd_{15}Fe_{77}B_8$ (i.e., 0% $SmCo_5$) is far below as expected even in resin-bonded condition. It is found the coercivity of Nd-Fe-B magnet is a strong function of particle size which is hardly compatible with the other component ($SmCo_5$) with the similar particle size.

LAMINATED COMPOSITES OF AMORPHOUS ALLOYS FOR EMI SHIELDING

Because of the excellent soft magnetic properties of amorphous alloys, many electro-magnetic interference (EMI) shields for protecting electronic devices, such as cathode ray tube (CRT), switching power supply, R-F filter as well as power line, etc., can be fabricated of amorphous alloys. A novel resin transfer molding (RTM) technique for making the funnel-shaped CRT shields was first adopted in this laboratory [3,4]. In contrast to the conventional Fe-Ni permalloy shields, all the magnetic, electric and mechanical properties of the laminated composite were significantly improved over a broad-band of 10 KHz-18 KHz.

The amorphous alloys (or metallic glass) were produced by multi-spinning technique under a quenching rate of 10^5°C/sec ∿ 10^6°C/sec. The funnel-shaped CRT shielding composites were laminated and molded by transferring epoxy resin and ferromagnetic powders into the laminated amorphous ribbons. The fabrication procedures are shown in Figure 4. The photograph of cross section for a lamination of six-layer ribbons stuck with epoxy resin is shown in Figure 5. It is found that the magnetic shielding effectiveness is a function of composition of the amorphous alloy, the thickness of the amorphous ribbons, magnetic annealing and weight fraction of the ferromagnetic powders.

Three compositions, $Fe_{40}Ni_{35}MoB_{18}$, $Fe_{40}Ni_{38}B_{17}Si$ and $(FeNi)_{78}SiMo_4B_{17}$ are melt-spinned as EMI shielding amorphous alloys. Magnetic shielding effectiveness

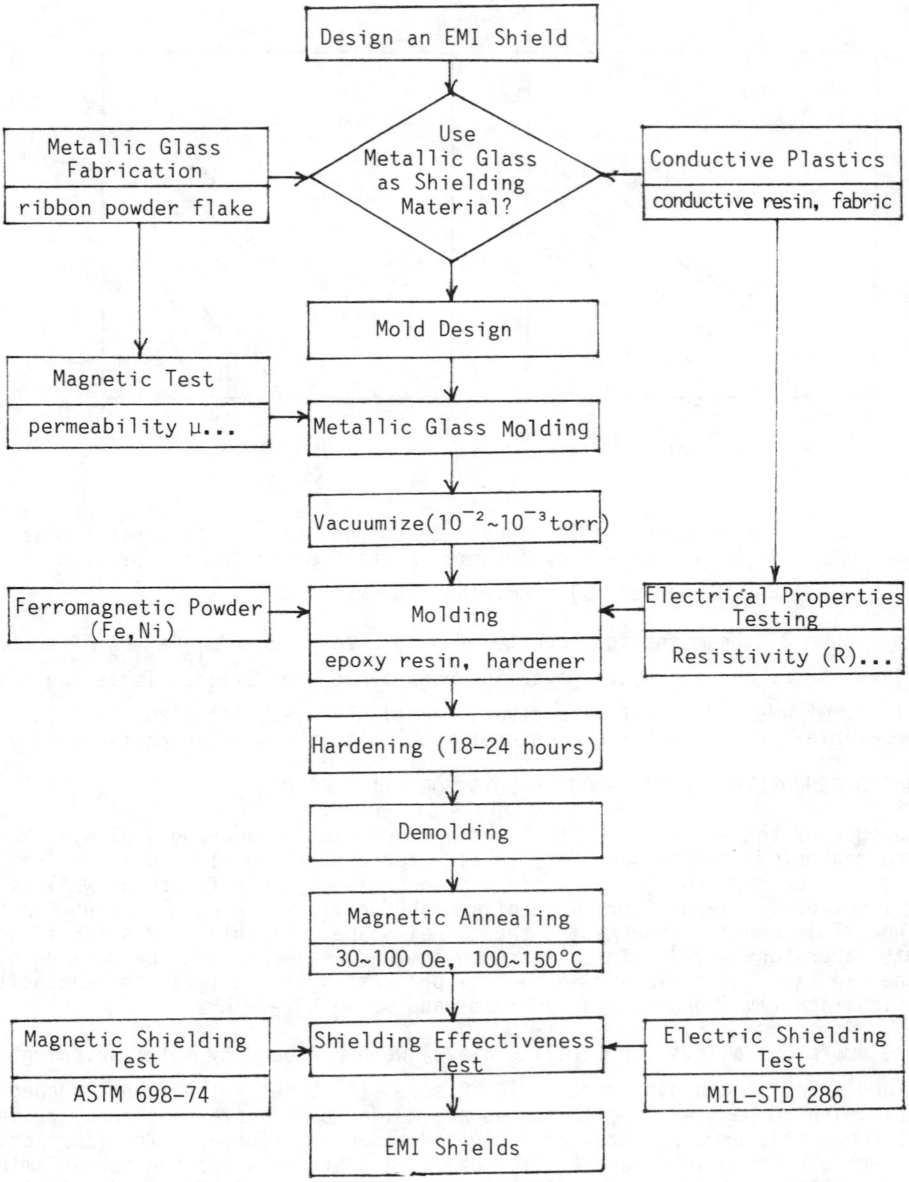

Fig. (4) - Fabrication procedure of laminated composite of amorphous alloys for EMI shields

of funnel-shaped CRT shields made of $(FeNi)_{78}Mo_4B_{17}Si$ ribbon composite is plotted in Figure 6 as a function of applied magnetic field. Another curve for shield

Fig. (5) - Photograph of cross section for a lamination of six-layer ribbons stuck with epoxy resin including glass fabric. (50X)

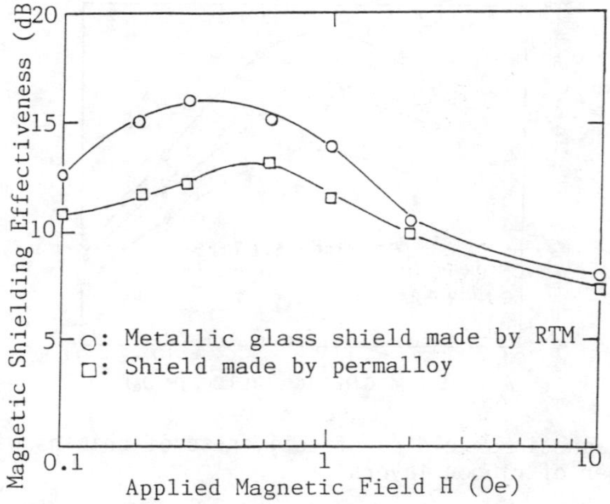

Fig. (6) - Magnetic shielding effectiveness of funnel-shaped EMI shields for 16" cathode ray tube

made of permalloy is also shown for comparison. Figure 7 shows the permeability of $Fe_{40}Ni_{38}Mo_4B_{18}$ as a function of applied magnetic field. It is noted that the permeability is not only as a function of applied field but also a function of magnetic annealing. It is advantageous that the permeability of ribbons is not obviously affected by bonding with epoxy resin. We confirm that amorphous ribbon molded epoxy composites are suitable for making a shield.

For a funnel shaped shield, the magnetic shielding effectiveness along the axis is shown in Figure 8. We find that the maximum shielding effectiveness is at the neck area of the shield near the minor diameter side. Also, the shielding

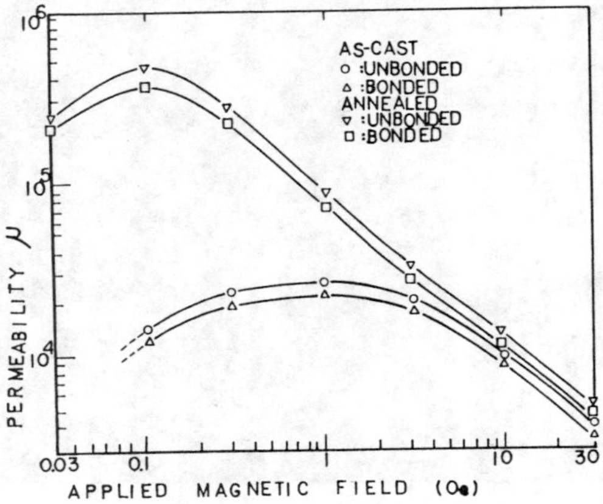

Fig. (7) - Permeabilities of $Fe_{40}Ni_{38}Mo_4B_{18}$ as a function of applied magnetic

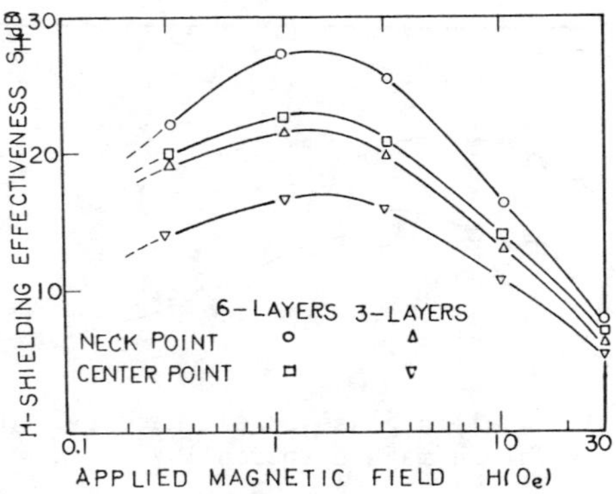

Fig. (8) - dc magnetic shielding effectiveness of shields with a different number of ribbon layers

effectiveness of six layers is larger than that of three layers. The theory of magnetic shielding, $S = \frac{1}{2}\,(t.\mu/R)$, where R is the diameter of a cylinder shield, μ is permeability and t is the thickness of the ribbons, is reasonably well satisfied.

Ferromagnetic powders, Fe or Ni, are added as a filler to the composite made of amorphous ribbons. It is found that the shielding of effectiveness increases and covers a broader range of magnetic field as shown in Figure 9.

Since the amorphous ribbons are good conductors, the shields show good electric shielding effectiveness over a broad spectrum. Figure 10 shows the plane-wave shielding efficiency in the microwave range (2-18 GHz). The attenuation losses are over 70 dB for three layers. It shows that the amorphous ribbon is efficient to shield against micro-wave interference.

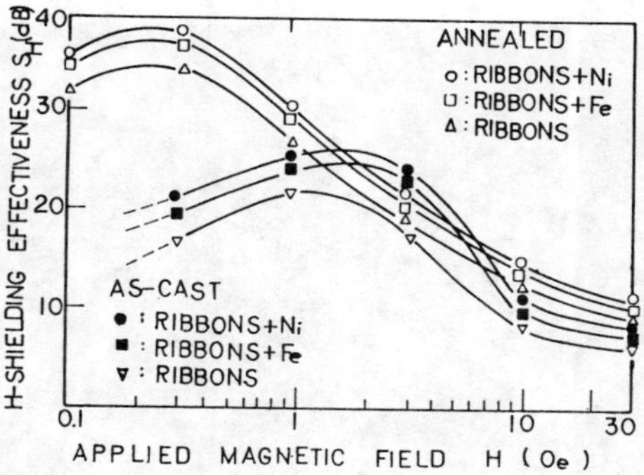

Fig. (9) - dc magnetic shielding effectiveness of shields made of amorphous ribbons and ferro-magnetic powders

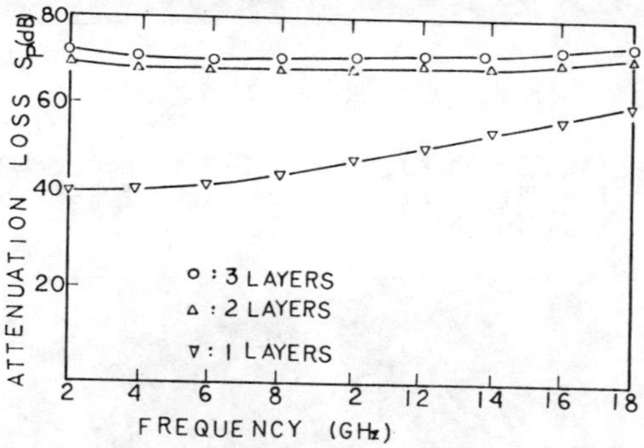

Fig. (10) - Attenuation losses of $Fe_{40}Ni_{38}Mo_4B_{18}$ amorphous ribbons in the microwave range

PIEZOELECTRIC COMPOSITES USED FOR UNDERWATER ACOUSTIC APPLICATIONS

PZT ceramics [$Pb(Zr,Ti)O_3$] are extensively used as piezoelectric transducer material in spite of their several disadvantages, namely, (1) high d_{33} but low d_h (= d_{33} + $2d_{31}$), where d_{33}, d_{31} are piezoelectric coefficient, and d_h is hydrostatic piezoelectric coefficient, (2) high permittivity, high density but low impedance matching with water and (3) brittle and low resistance to thermal shock. This paper is concerned with fabrication of PZT/polymer composite to reduce its permittivity (ε) and density (ρ), resulting in piezoelectric device, through which good impedance matching with water can be obtained [5,6].

In a composite, the electric flux pattern and electromechanical properties depend strongly on the manner in which the individual phase is interconnected. The connectivity of composite is defined as the number of dimensions in which each component phase is continuous. Each phase in a composite may be self-connected in zero, one, two, or three dimensions. What the connectivity we used includes 0-3 and 3-3 types of composites. The composite of 0-3 connectivity was made by uniform distribution of sintered PZT powder in a polymer matrix as shown in Figure 11. Whereas 3-3 connectivity was made by sintered porous PZT ceramic and back-filled with high polymer as shown in Figure 12. The hydrostatic properties for

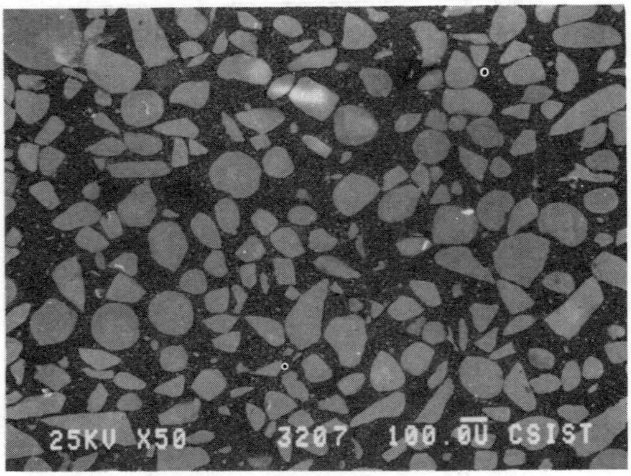

Particle Size: 74–140μm

Fig. (11) - The SEM photograph of 0-3 connectivity composites. The different particle size of PZT powder distribution in polymer matrix

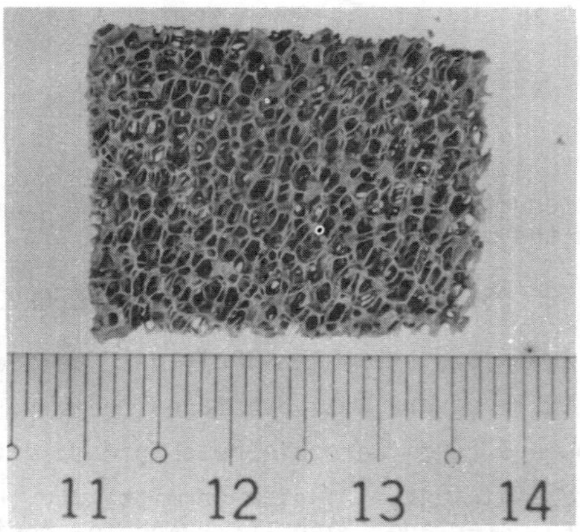

Fig. (12) - The photograph of 3-3 connectivity composite-porous PZT ceramic filled up with polymer

0-3 type composite is a function of particle size of PZT powder distribution, property of polymer and thickness of the material, whereas 3-3 type composite is a function of porosity in addition of other factors in 0-3 type.

For 0-3 connectivity composites, the density (ρ) will be remarkably reduced in comparison with the sole ceramics because the composite is fabricated with the different particles size of PZT powders and epoxy. Piezoelectric coefficient ($\overline{d}_{33}$) increases with particle size. The strength of electric field is a driving force which can re-orient the dipole moment at the period of poling, so the piezoelectric coefficient d_{33} increases also with field. For PZT/polymer composite however, the driving force of poling may easily focalize on the domain of PZT particles and reduce the dissipation in the polymeric phase. Therefore, the piezoelectric properties of PZT/polymer composite must depend both on electric property of polymer and of PZT ceramics. Due to the smaller resistance to electric field, the composite of 0-3 connectivity should be easily poled at the same field. The result is shown in Table 1 and Figure 13.

Table 1 - The piezoelectric properties of 0-3 type composites using polymer 1 in different particle size and poling field (thickness = 1.0 mm)

$\overline{K}_{33}$=dielectric constant, $\overline{d}_{33}$=piezoelectric coefficient (pC/N), $\overline{g}_{33}$=voltage coefficient (mV-m/N)

	25 Kv/cm			50 Kv/cm			75Kv/cm			100 Kv/cm			125 Kv/cm		
	$\overline{K}_{33}$	$\overline{d}_{33}$	$\overline{g}_{33}$	$\overline{K}_{33}$	$\overline{d}_{33}$	$\overline{g}_{33}$	$\overline{K}_{33}$	$\overline{d}_{33}$	$\overline{g}_{33}$	$\overline{K}_{33}$	$\overline{d}_{33}$	$\overline{g}_{33}$	$\overline{K}_{33}$	$\overline{d}_{33}$	$\overline{g}_{33}$
A	45.9	5	12.3	37.66	8	24.0	34.59	9	29.4	36.05	11	34.5	41.55	12	32.6
	-	-	-	-	-	-	-	-	-	38.03[a]	14[a]	41.6[a]	-	-	-
	-	-	-	-	-	-	-	-	-	49.5[b]	12[b]	27.4[b]	-	-	-
B	38.6	11	32.2	37.88	12	35.8	39.37	14	40.2	39.29	15	43.1	40.26	15	42.1
	-	-	-	-	-	-	-	-	-	37.87[a]	17[a]	50.7[a]	-	-	-
	-	-	-	-	-	-	-	-	-	56.79[b]	9[b]	17.9[b]	-	-	-
C	51.84	15	32.7	56.98	18	35.7	49.4	24	54.9	-	-	-	-	-	-

[a] the thickness of composite is 1.2mm.
[b] the thickness of composite is 1.5mm.

For 3-3 connectivity composites, the density (ρ) can be easily controlled by suitably selecting porosity of PU sponge. The composite having higher porosity or larger poles which have much more polymer phase should have lower density. To a higher porosity of composite, the piezoelectric coefficient, d_{33}, is lowering as the content of PZT decreases after being poled.

To the figure of merit of underwater applications, the polymeric phase plays an important role in PZT/polymer composites. The 3-3 connectivity composite which has skeleton of porous PZT ceramic is three dimensional continuity and with low density, the figure of merit of underwater applications is better than 0-3 connectivity composite. Figure 14 shows the hydrostatic pressure effect on the hydrostatic piezoelectric properties of various composites.

3-D C/C COMPOSITES PREPARED BY HIP AND CVD PROCESSES

3-D C/C composites are attractive candidate of structural materials for space and aeronautical engineering, especially for the components need to overcome high speed erosion at elevated temperatures [7,8]. 3-D C/C composites consist of a carboneous (or graphitized) matrix reinforced by a three dimensional structure woven by carbon (or graphite) fibers, Figure 15. High-temperature Isostatic Processing (HIP) and Chemical Vapor Deposition (CVD) process are utilized to prepare the carbon matrix.

In HIP process, the preform is impregnated with the pitch at 250°C in a pressurized vessel of 100 psi. The pitch-prepregnated preform is sealed in a stain-

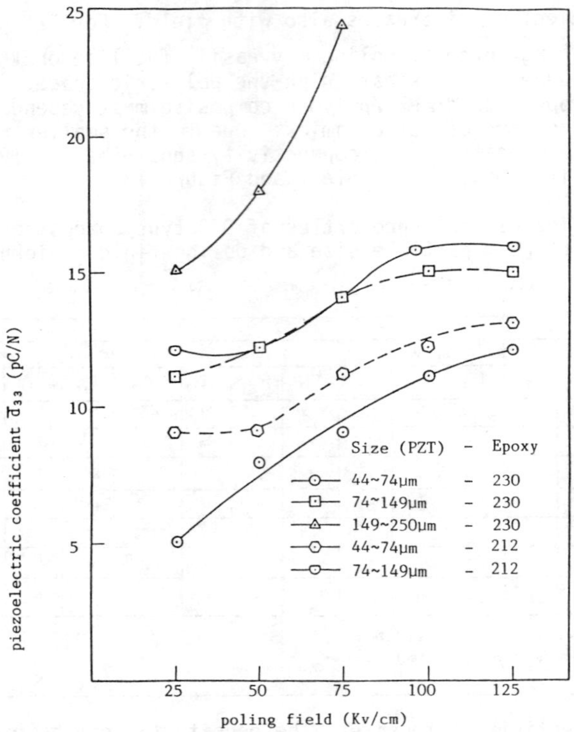

Fig. (13) - Relative $\overline{d}_{33}$ versus poling field, the 0-3 connectivity composite in various composition (particle size - polymer). Thickness: 1.0 mm

less steel container, and then hipped in the hipping autoclave at 650°C with 13,500 psi N_2 atmosphere. After carbonization at 1000°C following graphitization at 2200°C both in inert atmosphere, the C/C composite is rehipped in several cycles depending on required density. The density of the composites as a function of number of hipping cycle is plotted in Figure 16. Flexural strength of the 3D C/C composite versus density is shown in Figure 17. The final finished product prepared by HIP process is illustrated in Figure 18.

In CVD process, the graphite preform is impregnated in a high carbon content phenolic resin. The three-dimensional prepregnated preform is cured at 250°C and then carbonized at 1150°C. The process of carbon deposition is isothermally carried out in an acetylene cracking furnace. Carbon particles infiltrate into the open pores adsorbing onto the hot substrates and gradually depositing on the whole structure. Deposition rate is a function of various factors, such as temperature, pressure, gas flowing rate, carbon concentration, sample configuration and time of infiltration. The density change at different position as a function of infiltration time is plotted as shown in Figure 19. Figure 20 shows the microstructure of 3-D composite after 12 hours of CVD processing. Almost all voids and pores are closed or diminished.

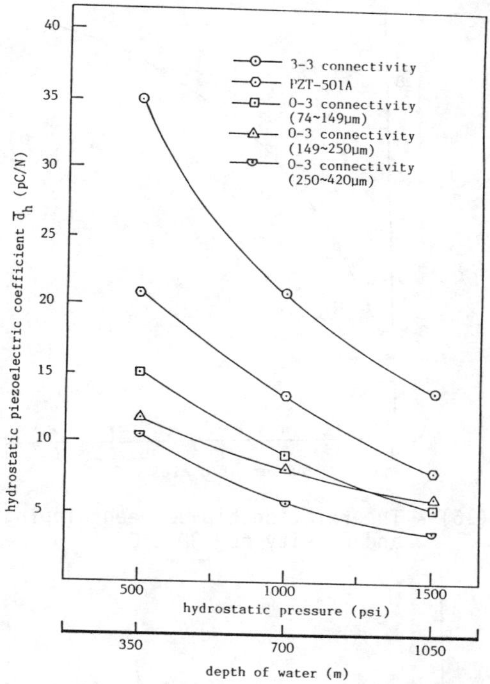

Fig. (14) - Effect of pressure on the hydrostatic piezoelectric properties of composite sample

Fig. (15) - 3D C/C preform

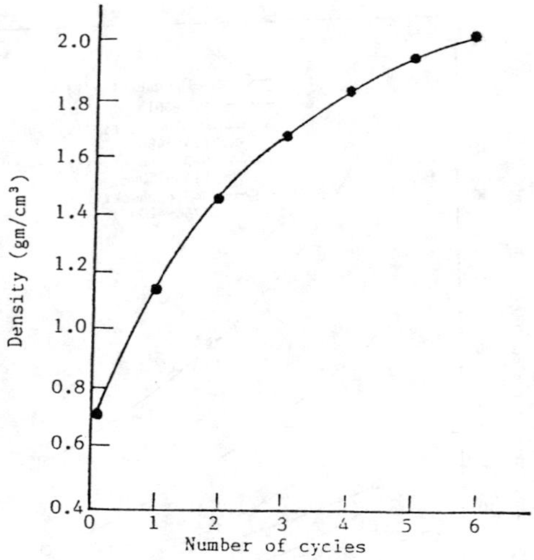

Fig. (16) - The relationship between hipping cycles and density for 3D C/C

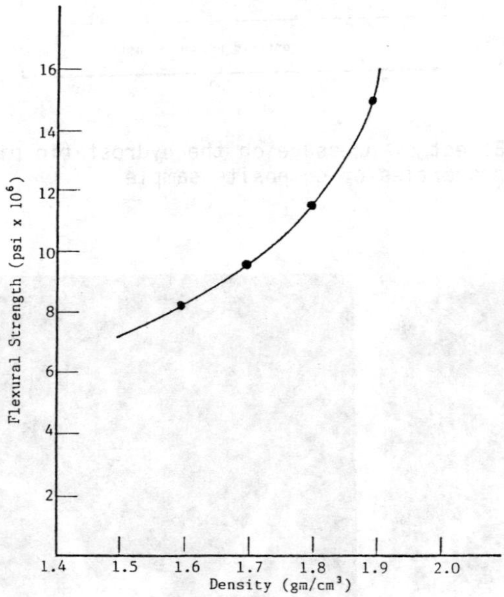

Fig. (17) - The relationship between density and MOR for 3D C/C (HIP)

The quality of 3D C/C composite can be further improved if the two processes are combined. Figure 21 illustrates the influence of process on flexural strength. It is noted that even a thin layer of carbon deposition (about 0.2 mm) by CVD process has significantly improved the mechanical strength of the hipped 3D composite.

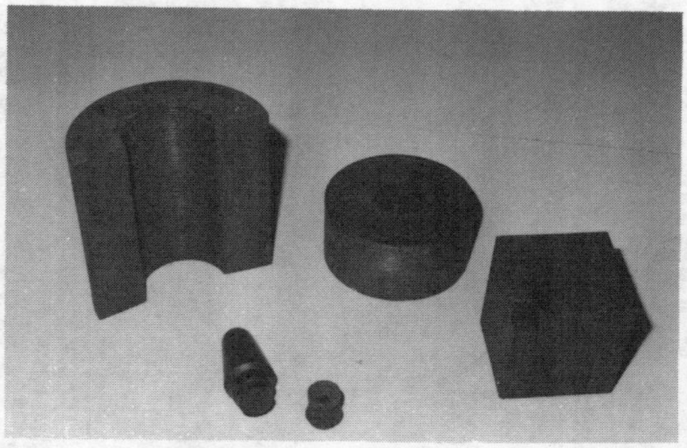

Fig. (18) - 3D C/C products

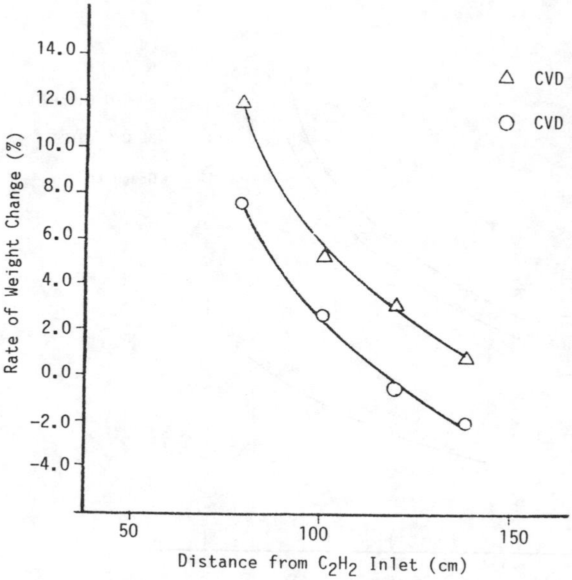

Fig. (19) - Relation between weight change and position

GRAPHITE FIBER REINFORCED CERAMIC COMPOSITES TiB_2/TiC/MoC BY SHS PROCESS

An economical processing sequence was developed involving self-propagation high temperature synthesis of TiB_2/TiC/MoC, containing up to 20% graphite fiber

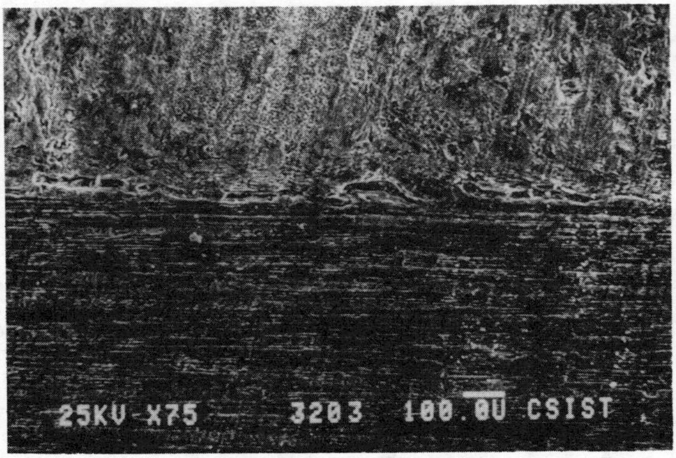

Fig. (20) - Microstructure of 3D C/C (HIP + CVD) with density 1.98 gm/cm^3

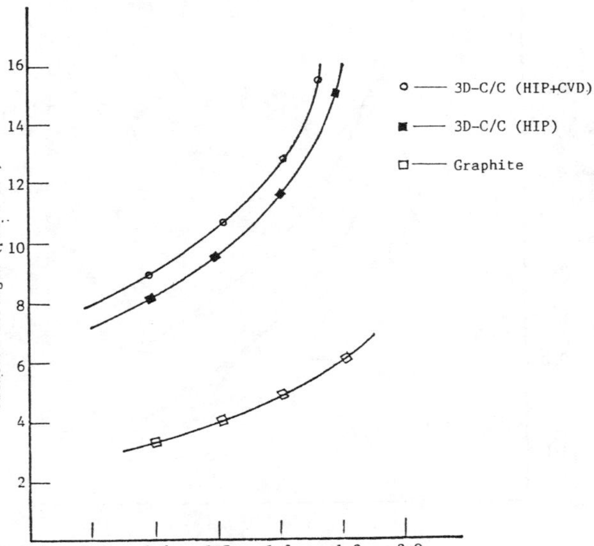

Fig. (21) - Comparison of flexural strength for 3D C/C (HIP), 3D C/C (HIP + CVD) and graphite versus density

by volume. Graphite fiber, having length of 100 ∿ 1000 μm and diameter of about 8 μm, was preformed in random orientation. Formation of porosity in the synthetic ceramic composite is crucial for SHS process. Efforts in this experiment were concentrated on the densification of the final product.

SHS is an unique method to synthesize refractory ceramics without sintering furnaces. The process is very simple, Figure 22. Many refractory powder are produced by this process [9,10]. Titanium diboride and titanium carbide have

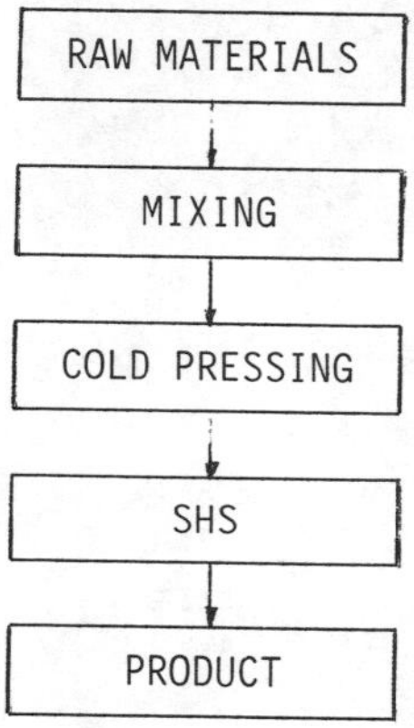

Fig. (22) - Specimen preparation flow chart

also been synthesized by this way for a long time [11,12]. However, none of them is successful in view of getting rid of crack and porosity. What we used in this experiment is Ti + B_4C system. The raw materials, titanium powder and B_4C powder are mixed stoichiometrically and pressed to form a green compact. After ignition, the SHS process proceeds violently and spontaneously. The reaction temperature is about 2800°C. Through SHS reaction, the green compact is converted into an irregular shaped ceramic as shown in Figure 23.

Fig. (23) - 3Ti + B_4C specimen after SHS

The reaction rate can be moderated by premixing some additives of heavy element such as nickel or molybdenum. The photograph of SHS synthetic specimen containing Ti, B_4C and Mo is shown in Figure 24. The X-ray diffraction shows that all the final products are converted to ceramic composite of TiB_2/TiC/MoB_2/MoC as shown in Figure 25. The hardness of the composite is of an inverse function

Fig. (24) - Mo added 5Ti + B_4C specimen after SHS

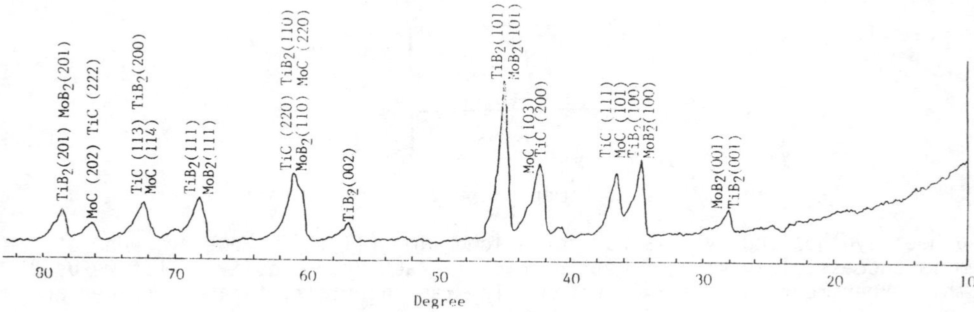

Fig. (25) - The X-ray diffraction pattern of SHS products.
Ti 60 wt% + B_4C 23 wt% + Mo 17 wt%

of addition of moderator, Figure 26. The hardness which is as high as 1900 Kg/mm^2 on the surface decreases as the moderator Mo (or Ni) increases.

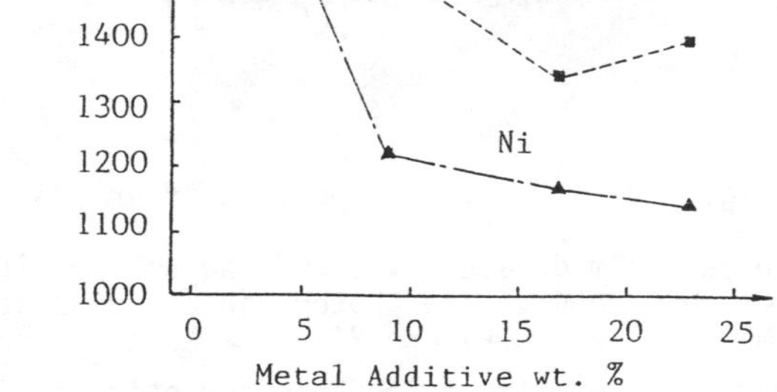

Fig. (26) - The influence of additive metal on the mechanical properties of SHS products

Attempt has been tried to reinforce the TiB_2/TiC/MoC composite by addition of random distributed graphite fibers. It is found that the graphite fibers almost vanished after SHS reaction if the fibers were not properly treated. In order to avoid direct contact between titanium and fibers, raw material of titanium and graphite fiber are coated with pitch. After SHS reaction, no shrinkage is observed although the number of pores and microcracks are apparently reduced. The microstructure of graphite reinforced TiB_2/TiC composite is shown in Figure 27. The SEM microstructure shows that the bonding strength between graphite fibers and matrix is rather weak. The strengthening effect of the fiber seems very subtle in comparison with the strength of the TiB_2/TiC/MoC composite without fiber.

Fig. (27) - Microstructure of SHS. Ti powder and graphite fiber coated with pitch. (a) Before SHS; (b) after SHS

HIGH-TEMPERATURE MECHANICAL BEHAVIOR OF TSTR POLYIMIDE/GRAPHITE FIBER COMPOSITES

Recent trend of composite development is to improve their specific mechanical strength at elevated temperatures for space industrial applications. Currently, Toho ST-3 graphite fiber reinforced Triaryl-S-Triazing Ring (TSTR) cross-linked composites were fabricated from aromatic nitrile terminated imide oligomers in this laboratory based on Dr. L. C. Hsu's patents [13].

The detailed procedure of fabrication of TSTR will be given subsequently in this book. Japan's Toho ST-3 graphite fibers were impregnated with an ethanol solution of TSTR prepolymer and wound on the mandrel of filament winding machine to form prepregs. The prepregs of graphite fiber reinforced TSTR composites were cured and post-cured following a conventional procedure.

Via an universal Instron 1125, the flexure properties of TSTR/ST-3 composites at various temperature up to 600°C are shown in Table 2, Figures 28 and 29. Those of PMR-15/ST-3 and Epoxy/ST-3 composites are also shown for comparison.

From Table 2, Figures 28 and 29, it is concluded that TSTR polyimide composites possess extremely high theroxidative stability - retain 67% of its room temperature strength at 400°C and 37% at 550°C, furthermore, retain 76% of its

Table 2 - Flexure strength versus temperature of TSTR/ST-3, PMR-15/ST-3, epoxy (MY720+DDS)/ST-3 composites

SPECIMEN $(0°)_8$	TEMPERATURE (°C)	RT	100	200	300	400	450	500	550	600
TSTR/ST-3	FLEXURE STRENGTH (k.s.i.)	209.5	176.1	157.0	151.8	141.3	132.3	110.3	76.7	61.8
	FLEXURE MODULUS (m.s.i.)	12.4	11.6	12.1	10.7	9.9	10.2	9.4	8.2	6.7
PMR-15/ST-3	FLEXURE STRENGTH (k.s.i.)	192.8	175.0	136.0	66.4	19.6	19.8			
	FLEXURE MODULUS (m.s.i.)	12.8	12.3	12.3	8.1	1.8	1.8			
EPOXY (MY720+DDS)ST-3	FLEXURE STRENGTH (k.s.i.)	184.0	121.5	27.2	9.76					
	FLEXURE MODULUS (m.s.i.)	15.9	14.9	5.3	1.9					

* unidirectional graphite fiber reinforced composite with resin content of 40% by weight .
* 5 to 6 samples were used for each test. C.O.V. is between 4-10%.
* Toho-ST-3 graphite fibers were used.

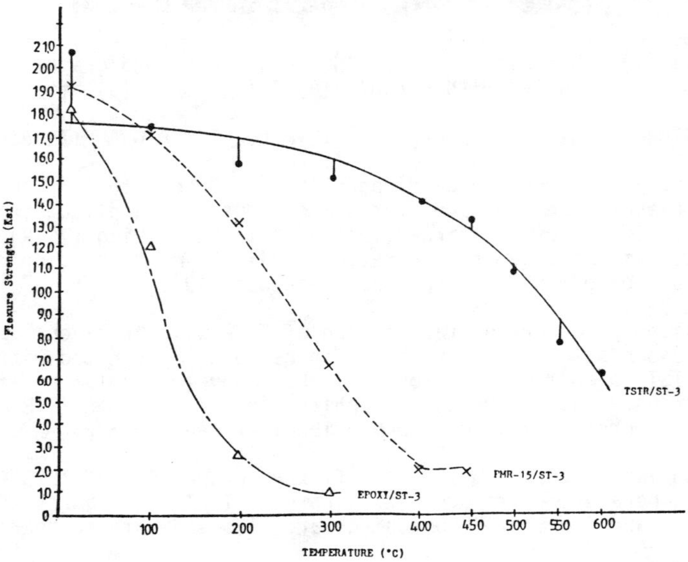

Fig. (28) - Flexure strength versus temperature of TSTR/ST-3, PMR-15/ST-3, epoxy (MY720+DDS)/ST-3, U.D. composites

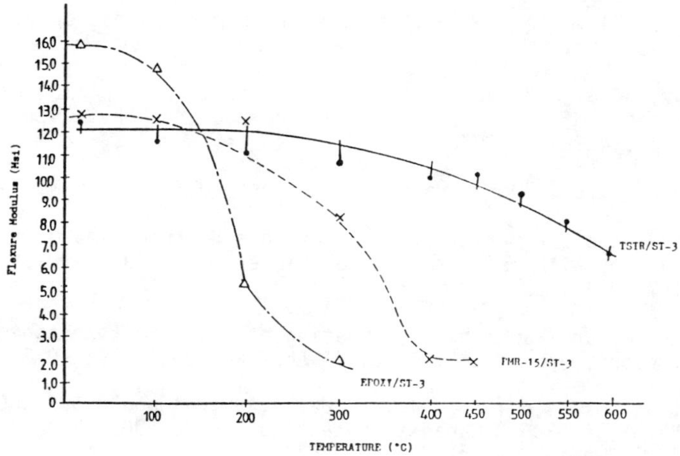

Fig. (29) - Flexure modulus versus temperature of TSTR/ST-3, PMR-15/ST-3, epoxy (MY720+DDS)/ST-3, U.D. composites

room temperature modulus at 500°C. These results indicate TSTR polyimide composites are the most promising structural materials known today for the development of aeronautical and space applications.

REFERENCES

[1] Wang, K.-L., Hsu, S.-E. and Chen, J., "A study on the magnetic properties of two-phase particulate magnetic composites", Journal of Magnetism and Magnetic Materials, Vol. 30, No. 1, pp. 37-49, 1982.

[2] Wang, K.-L., Hsu, S.-E. and Chen, J., "The hysteresis loop of two-phase particulate magnetic composites with non-linear hard magnetic components", Journal of Magnetism and Magnetic Materials, Vol. 37, No. 1, pp. 77-82, 1983.

[3] Huang, D. R., Dow, W. H., Yao, P. C. and Hsu, S.-E., "Electromagnetic shielding properties of amorphous alloy shields for cathode ray tubes", Journal Appl. Phys., 57(1), p. 3517, 1985.

[4] Huang, D. R., Dow, W. H., Yao, P. C. and Hsu, S.-E., "Composition dependence of magnetic shielding properties for amorphous ribbons $(FeNi)_{78}Mo_4B_{17}Si$", Proceedings of MRS, Boston, 1985.

[5] Rund, J. and Galgoci, E. C., Mat. Res. Bull. 19, 253, 1984.

[6] Yamada, T., Ueda, T. and Kitayama, T., Journal Appl. Phys., 53, 4328, 1982.

[7] Carnahan, K. R. and Kiger, R. W., "Carbon-carbon composites for space shuttle reentry thermal protection", ADv. Mater.: Composites and Carbon, Sym. of Am. Ceram. Soc., p. 171, 1971.

[8] Zweben, C., "Advanced composites for aerospace applications", Composites, pp. 235-240, 1981.

[9] Merzhanov, A. G. and Borovinskaya, I. P., "Self-propagating high temperature synthesis of refractory inorganic compound", Dokl. Chem. (Eng. Trans.), 204 [2], pp. 429-432, 1972.

[10] Zavitsanos, P. D., "Metal vapor generators", U.S. Patent No. 4092263, 1978.

[11] Takatsu, S. and Ishimatsu, E., "Sintering and properties of titanium diboride - alloy with refractory metal binder", Int. J. Refract. Hard Met. 1 (2), pp. 75-80, 1982.

[12] Person, H. O. and Randich, E., "Coating of metals with titanium diboride by chemical vapor deposition", Chemical Abstract 89:10319, 1978.

[13] U.S. Patent Nos. 4,061,856; 4,159.262; 4,458,041 and 4,555,565. British Patent No. 1,518.386. Canadian Patent No. 1,068,696. French Patent No. 2,287,448. Swedish Patent No. 7510867-0.

NEW EXPERIMENTAL APPROACHES TO STRESS ANALYSIS OF COMPOSITE STRUCTURES

J. T. Pindera

Institute for Experimental Mechanics and Department of Civil Engineering
University of Waterloo, Waterloo, Ontario, Canada, N2L 3G1

and

M. J. Pindera

Department of Engineering Science and Mechanics
Virginia Polytechnic Institute and State University
Blacksburg, Virginia 24061 USA

ABSTRACT

The rapidly increasing technological importance of composite materials and composite structures is leading to the development of new, more advanced models of their actual responses to mechanical and thermal loads. In this respect, the reliability and the predictive power of various methods and techniques of stress analysis become very important. Because of the inherently three-dimensional stress and strain states in composite materials and structures and the wide use of viscoelastic polymers as the matrix and some reinforcing fiber materials, a more rigorous modelling is needed of all the involved physical phenomena which influence the strain and stress states at the local and global levels. The influence of the above-mentioned factors on the applicability of analytical and experimental procedures is illustrated by examples of actual responses of homogeneous and composite structures.

This paper presents theories and techniques of three new methods of strain/stress analysis which have been developed on the basis of comprehensive physical models of involved phenomena: the isodyne, strain gradient and thermoelastic effect methods.

REVIEW OF COMMON METHODS

The traditional concepts of precision and accuracy of measurements applied together with the elementary statistical techniques no longer satisfy the requirements of modern technology. The problem of the resolution of measurement instrumentation, which in the past had been the major factor limiting the usefulness of measurements, practically ceases to exist if the related economic factors are not considered. For instance, the single mode lasers exhibit a coherence length which exceeds the most demanding requirements in engineering research; modern interferometric techniques allow routine measurements of displacements on the order of nanometers; high speed recordings can be performed in the picosecond range - the present limit in shaping laser pulses is in the region of a dozen of femtoseconds; electric signals in the nanovolt range are easily collected and processed; forces in the micronewton range can be measured and recorded, etc. These developments, together with application of modern methods and techniques of statistics, allow an engineering researcher to utilize signals which not so long ago were considered meaningless noise in the system - a nuisance at best [9].

The above-sketched progress in instrumentation and theories of measurement exposed a major, serious problem in experimental engineering research, namely, the problem of reliability of the methods and techniques used to evaluate the results of measurements; that is, the reliability of retrieving satisfactorily correct information from very accurate measurements. By extension, the same problem exists with regard to the analytical and numerical methods, solutions and procedures; that is, the question of reliability of analytical and numerical procedures with respect to the actual responses of engineering components and structures made of real materials. This fundamental problem of modern engineering mechanics, which can be conveniently - although rather roughly - reduced to the dichotomy between the physical-empirical and speculative-phenomenological understanding of reality, requires some comments. Otherwise, a rational assessment of the methods of experimental, analytical or numerical mechanics with regard to their usefulness in engineering practice would not be possible.

A. Modern Methodology

The concept of a model is the main concept in both physical-empirical and speculative-phenomenological approaches which, together with the concept of modelling criteria, represents the physical-theoretical basis of modern methodology in all methods of stress analysis: analytical, numerical, experimental. In accordance with both of these concepts, reality is represented by simplified models of it. Depending on the modelling criteria, different models of the same reality can be developed, where the process of model development can be presented by the following scheme [5,21]:

Reality → Physical Models → Mathematical Models → Various Analytical, Numerical and Experimental Procedures

The terms reality, physical models and mathematical models as used in the above scheme are defined in [5,21].

On the basis of accepted mathematical models, various analytical, numerical and experimental procedures are developed. Each of these procedures introduces simplifications or deviations from the chosen mathematical model in order to arrive at sufficiently simple relations between the input signals and response signals. For instance, in analytical solutions, it is often customary to limit analysis to very small values of the input/output signals, as is the case in the infinitesimal elasticity theory, and to apply the evaluated predictions to real systems loaded by forces of finite magnitudes. In experimental procedures which use iconic (or reduced) models made of matter and energy and having finite dimensions, it is customary to assume that such models respond in accordance with the models of Hooke or Prandtl and that the relations of the two-dimensional elasticity theory may be rigorously extrapolated to such situations.

The physical-empirical methodology of development of physical models accepted by the authors is based on the correspondence principle formulated by Niels Bohr, according to which a particular theory or procedure must be noncontradictory with a more basic physical theory and with empirical evidence. This condition, however, is often violated in practice, in particular, in stress analysis, where the influence of psychological factors analyzed by Thomas Kuhn - "everybody knows that this is the only correct solution" - is often of major influence [21].

B. Modelling Criteria

As indicated above, modelling criteria are selected according to the degree of complexity of the problem. Evidently, it is impossible to incorporate into the basic physical model (theoretical model) all the known responses of the body of interest to all inputs of various forms of energy. Any model of reality is a simplified model to a lesser or greater degree. A model which is too comprehensive with respect to its purpose becomes impractical - the additional increase in reliability and accuracy of predictions of the associated mathematical models becomes costly. On the other hand, a mathematical model based on a theoretical model which does not contain the necessary components describing all the major coupled physical events is likely to yield either unreliable predictions or predictions whose range of applicability is very narrow. The examples presented below illustrate the above.

The thermoelastic effect, which is readily measurable when the deformation process is close to adiabatic (isoentropic), as seen in Figure 1, can influence

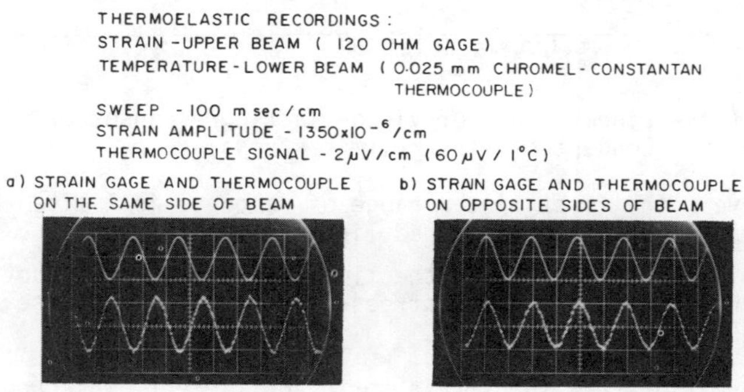

Fig. (1) - Strain analysis of a vibrating beam using the thermoelastic effect

crack propagation and can also be used to collect information on the components of the strain field at the surface of an object. Depending on the circumstances, Hooke's model of deformation may or may not be sufficiently reliable, sometimes it should be replaced by an equation of state containing a thermodynamic term and related material parameters such as the thermal expansion coefficient and the heat capacity coefficient [20].

The actual isochronous mechanical and optical visco-elastic responses of several industrial plastics presented in Figure 2 [14,18] yield various vital pieces of information. Of particular importance is the information that the photoelastic response is not simply related to the mechanical stress/strain response and that the visco-elastic response consists of at least two linear ranges connected by a

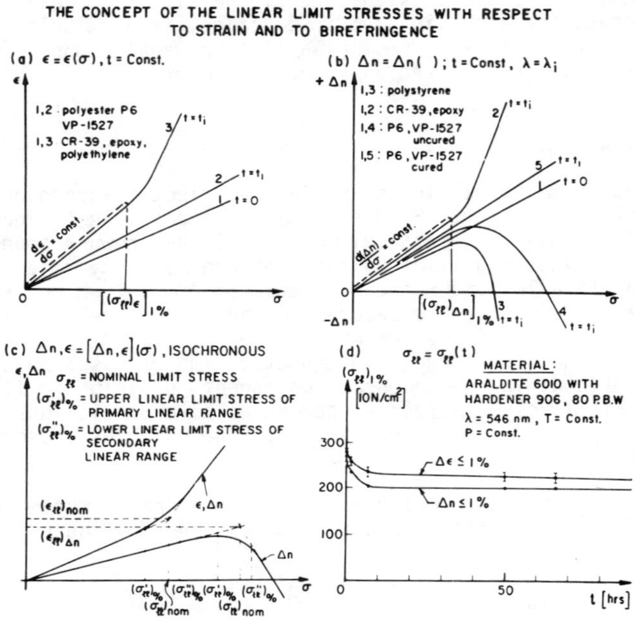

Fig. (2) - Linear range of visco-elastic responses of plastics. Concept of linear limit stress

transition curve: the first linear range of response is limited by the stress value which can be conveniently called linear limit stress, $\sigma_{\ell\ell}$. Furthermore, the linear limit stress depends on the time of loading and is not necessarily the same for the stress-strain and stress-birefringence response, as seen in the last graph in Figure 2 [14,18].

Also, information that the strain optic coefficient, which corresponds to the relaxation modulus, is constant within the first linear range of visco-elastic response indicates that the alteration of the photoelastic coefficient with loading time, called optical creep, which impairs the accuracy of photoelastic experiments, can be by-passed by performing measurements at constant deformation rather than at constant load. It should not be overlooked that the value of the stress-optic coefficient depends significantly on the wavelength of light [1,15,16] and that the nonlinear response and its dependence on the spectral frequency cannot be understood within the framework of a speculative-phenomenological physical model [16].

Figure 3 [14] demonstrates the consequences of exceeding the time-dependent linear limit stress during fatigue tests of homogeneous plastics. Analogous thermal effects may occur during fatigue tests of fiber-reinforced plastics, where the thermodynamic instability may be caused by local stress concentration at the fiber ends.

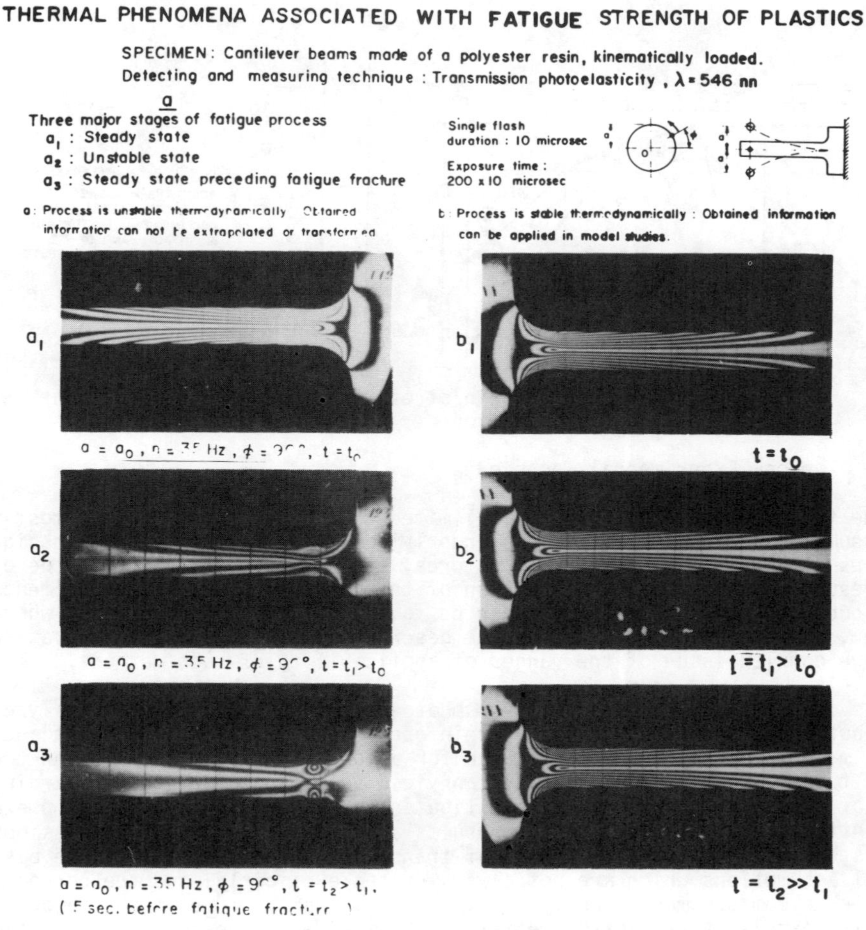

Fig. (3) - Fatigue test of plastics. Consequence of exceeding values of linear limit stress

Figure 4 illustrates that the assumption that light and other forms of energy propagate rectilinearly within a stressed body, where each light ray preserves its identity, is a very approximate assumption which is only acceptable when the optical path is very short [5,6,13,25,27].

Summarizing, one can state that the knowledge of the actual response of materials is indispensable for a rational choice of modelling criteria needed to develop physical and mathematical models of the events of interest that can be taken as foundations for the development of analytical solutions or for designing experimental models. Furthermore, the selection of modelling criteria in engineering research ought to be done on the basis of an overall picture of all the involved processes.

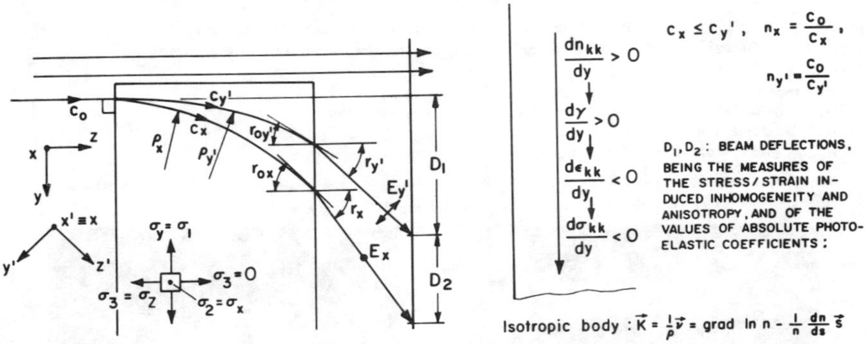

Fig. (4) - Influence of strain/stress gradient on the direction of light propagation

D. Typical Experimental Techniques

The treatment of this subject is limited to a brief review of the most typical assumptions and simplifications underlying various experimental techniques of stress analysis of composite structures, and of their influence on the evaluated results. An extensive discussion of the currently employed experimental techniques in composite mechanics can be found in [2,34] for instance, where, however, the emphasis is placed on the descriptive or applied aspects rather than the more complex issue of the ranges of applicability of such methods.

The use of classical transmission photoelasticity techniques to analyze stress distributions and concentrations in reinforced and composite structures and anisotropic materials has been in existence for a long time [8,10]. Although the methods of transmission photoelasticity can yield valuable information regarding the state of stress in an object of investigation, great caution needs to be exercised in quantitatively interpreting the results and drawing pertinent conclusions. This is because the typical methods of transmission photoelasticity are based on several assumptions which are not always stated explicitly, and the correctness of which is seldom documented analytically or empirically. One of the major assumptions is that light propagates rectilinearly inside a stressed body. In fact, the light paths can be quite curved [13], it can be readily observed that the ordinary and extraordinary rays can separate by about 4 mm over the distance of 100 mm, and can deviate from the rectilinear path by the same amount [27]. This neglected phenomenon, illustrated in Figure 4, influences the reliability and accuracy of all optical methods which employ light to collect information of interest inside a body. Furthermore, photoelastic data obtained by means of transmission photoelasticity are usually analyzed under the assumption that the stress states in plates in regions of notches, crack tips and local contact loads can be analyzed with the methods and solutions of the plane theory of elasticity. This has led to errors which may reach values of up to 30% on the unsafe side [22].

Recent refinements in moire interferometry permit very accurate measurements of surface displacements in composite structures subjected to quasistatic loads [2,34]. The technique has been successfully applied to the determination of displacement fields in the individual plies of multi-directional composites at the free edge, as well as in the top and bottom layers, by numerous researchers in the past. Most recently, quantitative comparison between the results of moire measurements and numerical analysis has been carried out for the free edge problem that has interesting implications regarding the applicability of the classical continuum mechanics approach to the boundary layer problem in inhomogeneous media [7]. However, as the moire method is based on surface deformation measurements, it cannot, in general, be employed to quantitatively deduce the state of deformation in the interior of a body in an independent fashion. Even in situations where a plane state of deformation can be assumed, the free surface will invariably alter this with the result that the surface measurement will include boundary layer effects.

The various photoelastic or moire techniques which use the stress freezing method cannot be reliably applied to stress analysis of composite structures because the stress freezing procedure introduces unknown thermal stresses in structures consisting of thermally different materials.

Theories of measurement based on the method of caustics do not contain terms related to the curvature of light propagating through crack tip regions where large stress gradients occur, nor terms related to the actual three-dimensional stress state in such regions. Consequently, the results of studies utilizing the above techniques, such as those presented in [31], ought to be regarded with caution.

It is customary in dynamic photoelastic model studies to assume that input loads produce harmonic elastic waves of single frequencies, and that the results of dynamic photoelastic studies are transferable to similar structures made of other materials. In fact, any real shock load should be simulated by a system of harmonic waves of various frequencies. Such waves propagate with velocities depending on the wave frequencies and on the initial values of stress components, and are differently attenuated [12]. As a result, the dynamic isochromatics indicate effects whose origin is phenomenologically closer related to the varying group velocities than to the constant phase velocities. Theoretically, such dynamic isochromatics fields are not necessarily related directly to the dynamic stress field in the photoelastic object, and are not directly transferable to the structures made of different materials [12].

The concept of stress is one of the basic concepts in engineering design, including the derived concepts of the stress intensity factors, or, in general, of stress gradients. However, the term "stress" is often used inconsistently. The continuum mechanics concept of "stress at a point" is not directly applicable to real materials, including composite materials. The best, so far, approximation of the stress state in real materials consisting of grains or of fibers and matrix materials is given by the concept of an average stress in a certain region whose size depends on the grain size. It has been shown some time ago that the measurement error in determining the average strain is within 3% when the measurement length is not less than 20 grain sizes [28]. So far, no rational method exists to verify the degree of correspondence between continuum mechanics and micromechanics treatments of stress states in real polycrystalline materials.

At the present time, the ranges of applicability of various experimental methods and related techniques employed in engineering are usually not stated, nor can they be quantified, or ever inferred, because it is not customary to discuss the theoretical basis of these methods and their adequacy. The common substitute of theoretical analysis by reference to various so-called laws obscures the issue. Without multiplying the examples, it is possible to draw the conclusion that there exists a clear need to expand the theoretical bases of various methods of experimental stress analysis by developing more comprehensive, and therefore, more realistic, physical models of the involved phenomena [16]. Of particular importance is the development of sufficiently comprehensive physical models of the structure and responses of materials which satisfy the requirements following from Bohr's correspondence principle. Such models can be taken as the foundation of various safe and efficient analytical and experimental procedures [29,30].

NEW EXPERIMENTAL METHODS

Three new groups of experimental methods of stress analysis are briefly presented below and their efficacy is illustrated by examples: isodyne methods, strain gradient methods and thermoelastic effect method.

All three methods are nondestructive methods. The first two methods can be used to determine stress components or their spatial derivatives in materials which are transparent either in the visible band of electromagnetic radiation or in the infrared or, in principle, microwave band. The third group of methods is based on application of the thermoelastic effect which occurs in all bodies having non-zero values of the thermal expansion coefficient.

A. Isodyne Methods: Plane and Differential Elastic Isodynes; Photoelastic Isodynes; Isodyne Coatings

A1. Plane Elastic Iosdynes

The term "plane elastic isodyne" denotes a geometric locus of points at which the total normal force intensities Δp_n, acting on the characteristic sections Δs between two isodynes, are constant; the characteristic section is any section in the plane stress field parallel to the characteristic direction which characterizes the related field of isodynes.

$$\Delta p_n = \text{const} = S_s m_s \tag{1}$$

where S_s denotes a constant material parameter, m_s denotes a dimensionless parameter called "order of the isodyne", and

$$\Delta p_n = \int_s^{s+\Delta s} \sigma_n ds \tag{2}$$

The plane stress field in a homogeneous, isotropic, and linearly elastic plate is characterized by the two-dimensional equilibrium conditions which are satisfied when the thickness of the plate is infinitesimal but finite, or when the stress state is homogeneous. The hypothetical material of such a plate is characterized by two material parameters only, the modulus of elasticity and the Poisson ratio, and follows Hooke's model of the stress-strain relations. In such a hypothetical plate of an infinitesimal thickness δ, loaded on boundary by force intensities p, all the stress components related to direction z, normal to the plate surface, are equal to zero. The derivatives of all the stress components with respect to z-direction are meaningless.

Plane stress fields induced in such hypothetical elastic plates can be conveniently described by the Airy stress function $\Phi(x,y)$, which yields the known expressions for the components of the stress tensor, with respect to the chosen Cartesian coordinate system (x,y). The isodynes can be expressed in terms of the first derivatives, $\Phi_x(x,y)$ and $\Phi_y(x,y)$ of the Airy stress function with respect to coordinates x and y, and in terms of integration functions $f_x(y)$ and $f_y(x)$ which depend on the boundary conditions, in the following manner:

x-isodynes (isodynes related to x-characteristic direction):

$$p_y(x,y) = \frac{\partial}{\partial x}\Phi(x,y) + f_x(y) = \Phi_x(x,y) + f_x(y) = S_s m_{sx} \tag{3}$$

y-isodynes (isodynes related to y-characteristic direction):

$$p_x(x,y) = \frac{\partial}{\partial y}\Phi(x,y) + f_y(x) = \Phi_y(x,y) + f_y(x) = S_s m_{sy} \tag{4}$$

The functions $p_x(x,y)$ and $p_y(x,y)$ can be understood as isodyne surfaces spanned over the plate surface. With regard to the technique of evaluating the stress components on the basis of relations (3) and (4), it is convenient to introduce isodyne functions which are the cross-sections of isodyne surfaces cut by planes normal to the plate surface and containing the chosen characteristic directions

$$m_{sxx} = m_{sxx}(x,y_o),\ m_{syy} = m_{syy}(x_o,y) \tag{5}$$

where x_o and y_o denote chosen values of ordinates x and y. The normal stress components are thus proportional to the slopes of the corresponding isodyne functions

$$\sigma_{yy} = S_s \frac{\partial}{\partial x} m_{sxx} \text{ and } \sigma_{xx} = S_s \frac{\partial}{\partial y} m_{syy} \tag{6}$$

It can be shown that the shear stress components σ_{xy} and σ_{yx} are proportional to the slopes of the cross-isodyne functions t_{xy} and t_{yx}, which are the cross-sections of the isodyne surfaces cut by planes normal to the plate surface and normal to the characteristic directions

$$m_{sxy} = m_{sxy}(x_o,y),\ m_{syx} = m_{syx}(x,y_o) \tag{7}$$

Summarizing, two isodyne surfaces related to two mutually perpendicular characteristic directions readily yield four independent pieces of information on the values of the plane stress tensor σ_{xx}, σ_{yy}, σ_{xy} and $\sigma_{yx} = \sigma_{xy}$. Isodyne functions directly yield the values of the total normal force components acting on characteristic directions which can be used to check the reliability of the procedure by applying the equilibrium conditions.

A2. Differential Elastic Isodynes

The concept of plane elastic isodynes can be reliably and successfully applied as a basis of various procedures of analysis of stress states in plates which are characterized by relatively low values of the gradients of stress components. High gradients of stress components indicate the presence of local or extended three-dimensional stress states, which cannot be reliably determined by using procedures based on two-dimensional equilibrium equations. Such three-dimensional stress states in plates, where all the stress components are functions of the distance from the middle plane of the plate, occur in regions of crack tips, notches, local contact and thermal loads, and interference zones of elastic waves. The knowledge of the actual stress values in such regions is of utmost importance with regard to the load-carrying capacity of the structure, damage accumulation and the service life of the structure.

In situations where the values of the stress components in planes coplanar with the middle plane of the plate deviate by not more than 30% from the values at the plate surface [22], it is possible to define differential elastic isodynes in a manner analogous to the definition of plane elastic isodynes. Differential isodynes are the geometric loci of points in the characteristic planes of real plates along which the difference of intensities of normal forces acting on the characteristic section between two differential isodynes in the directions within and normal to the characteristic planes are constant. For instance,

$$\Delta p_{y,z}(x,y,z = z_i) = \int_{s}^{s+\Delta s} (\sigma_y-\sigma_z)ds = S_s\Delta m_{s(y,z)}(x,y,z = z_i) = \text{const.} \quad (8)$$

Equation (8) represents a surface defined as a difference between two related surfaces of the x-isodynes.

A3. Photoelastic Isodynes

As shown in [19,21-23,25,26], it is easy to obtain experimentally fields of photoelastic isodynes, which are identical either to the plane elastic or differential elastic isodynes, using a particular scattered-light technique when a set of specified conditions is satisfactorily approximated [17]. The resulting photoelastic isodynes are given by the lines of constant intensities of particular sheets of light scattered from the primary light beam of intensity I_o which propagates in one of the characteristic directions and is scanned in the characteristic plane, e.g.,

$$I^x/I_o = (I^x)_n = \sin^2\pi m_{sx} = \sin^2(\phi_{sx}/2) = 0.5(1 - \cos\phi_{sx}) = \text{const.} \quad (9)$$

where m_{sx} denotes the normalized linear-relative retardation R/λ, and ϕ_{sx} denotes the angular-relative retardation, both of which are related to the secondary principal stresses σ_1^x and σ_3^x acting in the plane normal to the x-direction of the primary beam. When the conditions specified and discussed in the pertinent references (e.g., [14] and [17]) are satisfied, the following expressions describe satisfactorily the physical event of interest

$$\phi_{sx} = 2\pi m_{sx} = 2\pi S_s^{-1} \int (\sigma_1^x-\sigma_3^x)dx, \text{ etc.} \quad (10)$$

where m_{sx} denotes the orders of x-isodynes.

Thus, for instance, for the x-characteristic direction, one obtains

$$m_{sx} = S_s^{-1} \int (\sigma_{yy}-\sigma_{zz})dx = S_s^{-1}(p_y-p_z) = \text{const.} \tag{11}$$

One must note that in the regions where high stress gradients occur, the generally rectilinear path of the primary beam I_o can be locally curved, and that both conjugated light beams, ordinary and extraordinary, can separate. The resulting experimental complications, and, simultaneously, the additional information on the three-dimensional stress fields, are discussed in the pertinent references.

Two basic isodyne measurement techniques have been developed so far: the light-intensity modulation technique and the modulation technique of the spatial frequency of residual isodynes.

A4. Isodyne Coatings

The isodyne coating techniques, both the amplitude and the spatial frequency modulation techniques, have been developed to analyze stress states at the surface of nontransparent objects. Reference [24] contains information on the theory and technique of the isodyne coating method. The spatial frequency modulation is particularly useful in the isodyne coating technique because it gives resolution which is about one order of magnitude higher than the resolution of the amplitude modulation technique. An example of application of isodyne coatings in stress analysis of actual objects is given in Figure 5.

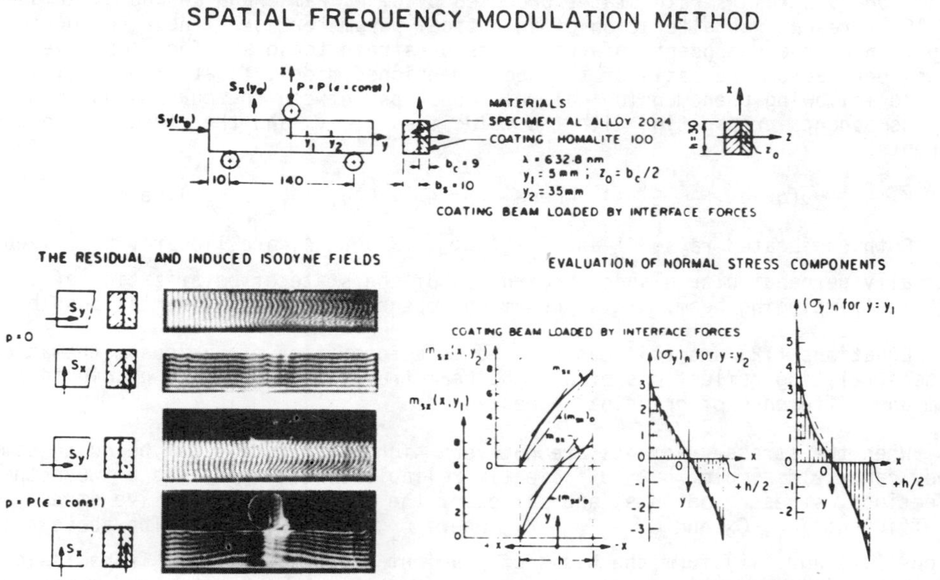

Fig. (5) - Isodyne coatings in **stress** analysis of non-transparent specimens

B. Strain Gradient Methods

The basic phenomenon taken as the basis of the strain gradient methods, which supplies a quantitative basis for the empirical relations presented in [13], is presented in Figure 4. More extensive data on the strain gradient effects, methods and techniques are given in [6,21,25,27]. Only a general outline of the method is given below.

It has been known for a very long time that the path of light propagating through inhomogeneous bodies is not rectilinear; the light path is curved. It has been observed that a noticeable mirage effect may also be produced by stress states in photoelastic objects [13,27]. This effect limits the resolution of photoelastic measurements and obscures isochromatic fields in regions of high birefringence gradients [21]. However, this effect was utilized by Acloque and Guillemet a quarter of a century ago to measure particular residual stress states (see [23] in [25]). Also, it has been observed that the optical anisotropy caused by an inhomogeneous stress state produces a noticeable separation of the light beam: this effect has been used to develop the theories and techniques of methods that can be used to determine the gradients of some linear functions of stress components and the values of absolute stress-optic coefficients [27].

It is common to present the influence of optical inhomogeneity of a body on the path of light propagation by the simplified relation

$$\overline{K} = \frac{1}{\rho}\,\overline{\nu} = \frac{1}{n}\,(\text{grad}n - \frac{dn}{ds}\,\overline{s}) = \text{grad } nn = \frac{1}{n}\frac{dn}{ds}\,\overline{s} \qquad (12)$$

The most general phenomenological-mathematical model of optical anisotropy produced by stress/strain states proposed by G. N. Ramachandran and S. Ramaseshan in 1961 presents an alteration of an optical parameter as a linear homogeneous function of the components of the stress or strain tensors. The simplified relations derived on the basis of the above-mentioned model, together with (12), lead to the following phenomenological relationships between the curvatures of both ray components propagating with the velocities v_1, v_2 and the stress tensor components

$$\overline{K_1} = \frac{1}{\rho_1}\,\overline{\nu}(\text{grad}n_1 - \frac{dn_1}{ds}\,\overline{s}), \text{ where } n_1 = n_0 + C_1\sigma_1 + C_2(\sigma_2+\sigma_3), \text{ etc.} \qquad (13)$$

Both conjugated rays, E_y' and E_x [6,25], Figure 4, are linearly polarized in mutually perpendicular planes, regardless of the state of polarization of the original impinging beam. This phenomenon opens various experimental possibilities.

Equations (12) and (13) can be used as a foundation of various mathematical models relating deflections of a light beam traversing a body to gradients of the sum and difference of principal stresses.

When the stress gradients are not very high, such models can be quite simple; they can yield simple, but sufficiently reliable linear relations between the deflections, stress gradients, and values of the absolute and relative stress-optic coefficients C_1, C_2 and $C_\sigma = C_1 - C_2$, where $C_1, C_2, C_\sigma = C(\lambda, t)$. The above relations (12) and (13) form the basis of gradient photoelasticity. Transmission photoelasticity supplies the relation $C_\sigma = C_1 - C_2$ which can be used as an additional piece of information. An example of application of the strain gradient method in stress analysis is presented in Figure 6. More examples are given in [27].

It must be noted that the curvature and separation of conjugated ordinary and extraordinary rays can be quite noticeable. For instance, as already mentioned,

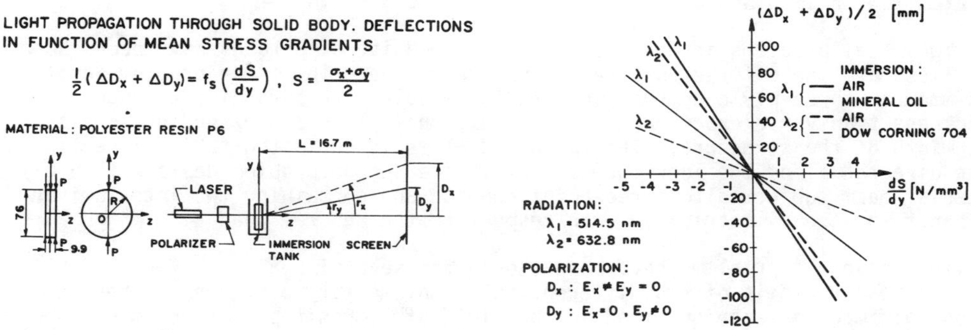

Fig. (6) - Determination of gradients of the sums of principal stresses using strain gradient method

it often occurs that the common photoelastic materials cause deflections and separations of light beams up to 4 mm over the distance of 100 mm. This alters the fields of isochromatics in regions of crack tips, sharp notches and contact loads.

Figure 6 shows that the mean deflections of light beams propagating through a transparent plate are caused by both the curvature of light beams within the plate and the rotation of the faces of the plate. At a shorter wavelength, the effect of curvature prevails over the effect of face rotation. So far, these basic relationships are not considered in theories and techniques of experimental methods which use transmitted light to collect information on the stress state, such as the methods of caustics, transmission photoelasticity, etc.

C. Thermoelastic Methods

Assuming that the isothermal and adiabatic deformation processes in continuous, homogeneous, isotropic and elastic bodies are equivalent with respect to the amount of heat exchanged or produced, one obtains

$$\frac{dT}{T} = -\frac{\alpha}{c_\varepsilon} d(\sigma_{kk}) \tag{14}$$

The above relation can be taken as a basis of various experimental methods of strain/stress analysis and defect detection [20]. One must note that exceeding the linear limit stress leads to irreversible thermodynamic processes, some of which are used as the foundation of defect detection techniques [35].

Since elastic shear waves cannot produce any temperature alteration, the thermoelastic effect can be used to distinguish easily between the dilatational and the shear waves.

The thermocouples used in experiments respond to the two mechanisms of heat energy transport, conduction and radiation, and to the strain induced in the thermocouples themselves. Consequently, various thermoelastic strain gages can be designed.

APPLICATION OF ISODYNES TO PROBLEMS IN COMPOSITE STRUCTURES

Two model problems in the mechanics of composite structures have been chosen to illustrate the efficacy of the isodyne technique. The first problem involves a symmetric three-ply construction with the middle ply containing through-the-thickness transverse crack normal to the external plies and extending across the width of the structure. The laminated plate is subjected to in-plane tension directed along the longitudinal axis. The second problem deals with a bimaterial beam subjected to three-point bending and containing two interlaminar disbonds located symmetrically with respect to the central load.

The above two problems have been chosen for several reasons. First of all, they represent models of two different but complementary problems of great technological importance encountered in the field of composites. Interlaminar and transverse cracks in composite materials and structures are often difficult to avoid and thus accurate characterization of stress fields in the vicinity of such defects is indispensable from the point of view of defining flaw criticality of advanced composites under various loading conditions. Secondly, the functional form of stress distributions in the vicinity of crack tips for the above configurations as predicted by quasi-static plane elasticity solutions is markedly different, thus presumably leading to different failure modes. As at present, there are no generally accepted theories to predict the direction of crack growth and the critical loads for the above cases, the experimental results are of theoretical interest to fracture mechanicians. Thirdly, the possibility of a three-dimensional state of stress in the vicinity of the simulated defects, as has already been demonstrated in the case of homogeneous beams containing surface notches, see [22], raises the issue of the range of applicability of the plane elasticity solutions in problems of this type. Lastly, experimental models utilized in this study, being inherently three-dimensional, can be taken as iconic models of real structures. Consequently, extrapolation of the experimentally obtained stress distributions along various planes within the model to the actual engineering structures is a direct process, a feature of Isodyne Photoelasticity which illustrates both the power and versatility of the developed technique.

The experimental results for the first problem are given in Figure 7 where the maximal shear stress variation τ_{13} in the $y = 0$ plane at different z_i locations above the crack has been determined as a function of the axial distance from the crack tip. The results are presented for two material configurations with different Poisson's ratios for the inner and outer plies: polyester-glass-polyester and polyester-aluminum-polyester systems. The significant influence of the Poisson's ratio on the internal stress distribution in the vicinity of the crack is clearly evident both in the recordings of isodyne fields and the calculated results. It is seen that the loci of the maximum values of the shear stress τ_{13} occur at different axial distances from the crack tip in the two configurations.

For comparison, the predictions of an approximate analytical model [32] are included in Figure 7. As the variation of the normal axial stress across the ply thickness has been neglected in the approximate model, the correlation between the analytical solution and isodyne data is mainly of a qualitative nature at points sufficiently remote from the crack. In the region close to the crack tip, however, the empirical and analytical results differ radically. Whereas the approximate analytical model yields continuously increasing values for τ_{13}, the experimental data indicate local maxima as already mentioned. On the other

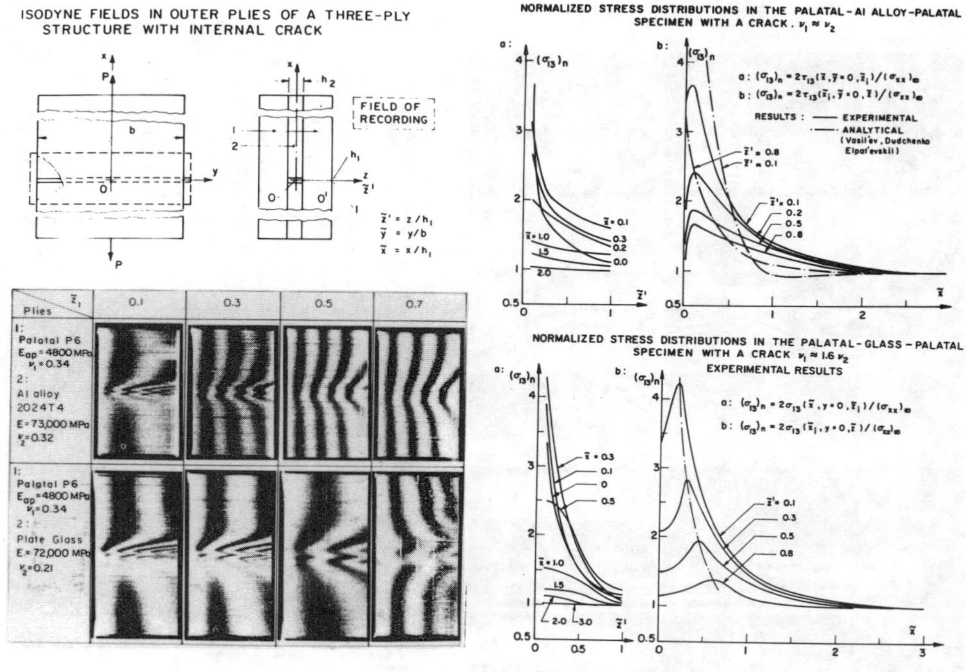

Fig. (7) - Isodyne stress analysis of three-ply structures with internal cracks. Influence of Poisson ratios on the stress field

hand, the local solution for the stress distribution around a crack perpendicular to a bi-material interface developed by Zak and Williams [36] does indeed exhibit such local maxima of τ_{13} near the crack tip [26]. However, the influence of the Poisson's ratio on the localized crack tip stresses predicted by the above plane elasticity solution does not appear to be as significant as indicated by the experimental results.

The experimental and analytical results for the second problem are presented in Figure 8. Figure 8a illustrates the recordings of the x- and y-isodyne fields along with the recordings of the dark and light isochromatics fields included for comparison. The disturbances in the stress fields caused by the presence of the two interlaminar disbonds is clearly visible in the two sets of recordings, which also illustrate the near-perfect symmetry of the stress distributions with respect to the centrally applied load.

The analytical model chosen for comparison with the experimental results is based on the two-dimensional elasticity formulation which has been proposed by Erdogan and Gupta [4] among others. It employs an infinite Fourier Transform technique and thus is strictly applicable to finite-height, laminated plane geometries extending to $\pm\infty$ in the x-direction. Consequently, the experimental-analytical correlation is not expected to be valid in regions close to the free end of

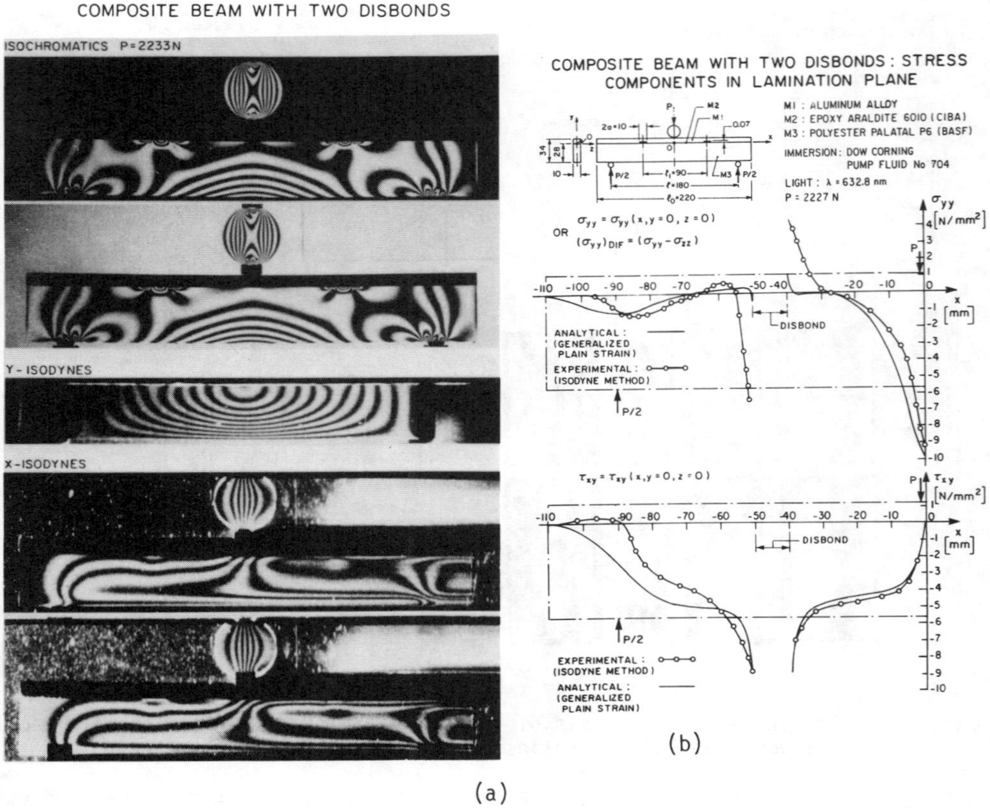

Figs. (8a,8b) - Stress analysis of a composite beam with two local disbonds under three-point loading. Analytical and experimental results

the structure. Although the formulation yields a physically inadmissible solution in the crack-tip region less than 10^{-6} to 10^{-4} of the crack length as discussed by Malyshev and Salganik [11], this limitation is frequently deemed of no practical consequence as all infinitesimal elasticity solutions for this type of problem must break down so close to the crack tip, including those which circumvent the problem of crack-tip material interpenetration [33].

The analytical expressions for the tractions in the plane of lamination can be shown to reduce to the following set of singular integral equations coupled in the u and v components of the displacement discontinuities $\theta_u(x)$ and $\theta_v(x)$ across the crack interfaces when the crack faces are free of tractions:

$$\begin{Bmatrix} -\sigma_{yy}(x) \\ \\ -\tau_{xy}(x) \end{Bmatrix} = \begin{bmatrix} 0 & B_{12} \\ \\ -B_{12} & 0 \end{bmatrix} \begin{Bmatrix} \theta_v(x) \\ \\ \theta_u(x) \end{Bmatrix} + \frac{1}{\pi} \int_{-a}^{+a} \begin{bmatrix} B_{11} & 0 \\ \\ 0 & B_{11} \end{bmatrix} \begin{Bmatrix} \theta_v(x') \\ \\ \theta_u(x') \end{Bmatrix} \frac{dx'}{x'-x}$$

$$+ \frac{1}{\pi} \int_{-a}^{+a} \int_{-\infty}^{+\infty} \begin{bmatrix} A_{11}(s,x,x') A_{12}(x,x,x') \\ A_{21}(s,x,x') A_{22}(s,x,x') \end{bmatrix} \begin{Bmatrix} \theta_v(x') \\ \theta_u(x') \end{Bmatrix} ds dx'$$

$$- \frac{1}{\pi} \int_{-\infty}^{+\infty} \begin{bmatrix} F_{11}(s,x) F_{12}(s,x) \\ F_{21}(s,x) F_{22}(s,x) \end{bmatrix} \begin{Bmatrix} p_y^+ \\ p_y^- \end{Bmatrix} ds \qquad (15)$$

where

$\theta_v(x)$, $\theta_u(x)$ = 0 in the region $|x| > a$, a being half the crack length,

s = Fourier-transform variable,

B_{11}, B_{12} = constant parameters describing the degree of material property mismatch of the upper and lower beam,

$A_{ij}(s,x,x')$,$F_{ij}(s,x)$ = regular kernels obtained from inversion of the Fourier-transformed structural matrix of the composite beam,

p_y^-, p_y^+ = externally applied normal fractions per unit width.

It ought to be mentioned that the applicability of the traction-free condition in the above formulation is verified by experimental data which clearly indicates that the traction components $\tau_{xy}(x)$ and $\sigma_{yy}(x)$ vanish on the crack faces.

The solution of the above system of equations closely follows the general methodology outlined in [3] where a weighted residual technique is employed to reduce the singular integral equations to a set of algebraic equations in unknown coefficients of Jacobi polynomials used to approximate the displacement discontinuities $\theta_v(x)$ and $\theta_u(x)$. Determination of the unknown coefficients, which has been carried out using an 11-term expansion in Jacobi polynomials and 384 Gaussian integration points in the numerical evaluation of the integrals in the above formulation, consequently allows one to obtain the full field interfacial stresses given above.

The experimental-analytical correlation of the shear and normal traction components acting in the plane z = 0 along the interface containing the disbonds is presented in Figure 8b for the generalized plane strain case. The difference between the plane stress and generalized plane strain results for the interlaminar tractions as predicted by the analytical model is not very significant in this case and thus the plane stress results are not shown. Several interesting features of the correlation study are noteworthy. The correlation between the experimentally obtained values of the shear traction component $\tau_{xy}(x)$ and the analytical results is very good as the interlaminar crack is approached from the center of the beam, i.e., in the inner region. The correlation in the outer region, although very similar in form despite the apparent shift close to the outer crack tip, is not as good as the free end of the beam is approached. As mentioned before, this might be explained by the finite-length effects associated with the free end which are not taken into account in the analytical model.

The comparison between the analytical results for $\sigma_{yy}(x)$ and the experimentally obtained differential stress values $(\sigma_{yy})_{diff} = \sigma_{yy} - \sigma_{zz}$ indicates the presence of a significant out-of-plane normal stress component in the polyester beam along the lamination plane in the vicinity of the interlaminar disbonds and the points of external load application. The difference between the externally applied normal tractions and the integral of the difference $\sigma_{yy} - \sigma_{zz}$ along the plane of lamination as determined from the isodyne recordings is approximately 30%. On the other hand, this difference becomes very small in the plane located half way across the height of the polyester beam where the methods of transmission photoelasticity are applicable.

The fracture patterns of the failed bi-material beams are presented in Figure 9. The primary crack appears to have initiated at the outer tip of the disbond

Fig. (9) - Fracture patterns in composite beams with two local disbonds under three-point loadings. Top: beam before test; left: primary, or initial fractures; right: secondary, or dynamic fractures

at an angle of approximately 66° and then apparently changed direction and propagated dynamically at an angle of 59° with respect to the plane of lamination. The subsequent fracture process involved delamination of the inner region and crack path deflection away from the lamination plane into the polyester beam before the inner crack tip of the other symmetrically located disbond was reached. The fracture process was catastrophic for all the specimens tested with no apparently significant initial quasi-static crack growth along the plane of lamination in the outer region.

It is interesting to note that the crack growth initiated at the outer tip where the normal traction component σ_{yy} is compressive as predicted both by the isodyne and analytical calculations. Furthermore, both the experimental data and the analytical model indicate that the shear traction component increases faster near the external crack tip than the internal tip, although the analytical model predicts that this difference is smaller than obtained experimentally. On the other hand, the axial stress component σ_{xx} in the polyester beam along the plane of lamination is very high and has different sign in the vicinity of the outer and inner crack tips: it is tensile as the disbond is approached from the outer region and compressive as the disbond is approached from the inner region. The correlation of the various theories that predict the direction of crack growth and critical loads necessary to initiate crack propagation for the above configuration with the experimental results will be reported at a later time.

CONCLUSIONS

The presented examples show that the isodyne techniques, strain gradient techniques and the auxiliary thermoelastic effect techniques can be applied successfully and in a self-contained manner to solve various stress analysis problems of importance in strength and service life optimization of composite structures and materials. These examples illustrate that the isodyne techniques supply, whole field-wise, data on the two dimensional and three dimensional stress states which cannot be obtained by using other, presently available, experimental methods. In particular, the photoelastic isodyne techniques can directly supply information on:

- total values of the normal and shear forces acting in conveniently chosen cross-sections of a plate; this allows one to check immediately the reliability and accuracy of the performed measurements;

- values of the normal and shear stress components of plane stress fields, which can be checked immediately (four independent pieces of information about three unknown quantities);

- local regions in plates where the stress states are clearly three dimensional;

- all components of the three dimensional stress states that occur in plates in the regions of notches, cracks, local load application, local heat sources, etc.;

- values of stresses in composite structures, in particular, stresses in lamination planes, peel stresses, and stress distributions across the thickness of laminated components;

- values of the stress components in objects of arbitrary shape, by performing measurements in several directions;

- thermal stresses;

- residual stresses induced during assembly or caused by local plastic deformations.

The isodyne techniques can be applied to determine the internal forces and stress components in objects that are transparent to radiation either within the visible band or the near infrared band or the far infrared band, and that exhibit scattering properties similar to those described by the Rayleigh model.

The strain gradient techniques supply data on gradients of the sums and differences of principal stresses, and on rotation of faces of plates in regions of high strain gradient related to strong local three-dimensional stress states in plates.

The thermoelastic effect techniques are of particular interest in the analysis of vibrations, and in the design of fatigue tests for plastics, ceramics, composite materials and other advanced materials. Of particular interest is the efficacy of this technique in studying the propagation and interference of elastic waves. The ease with which this technique allows one to distinguish between longitudinal and shear strain waves is quite spectacular.

There is an urgent need to advance our knowledge and to develop reliable analytical tools in the following areas:

- transition from the stress analysis at the micromechanics (and local) level to the stress analysis at the macro (and global) level;
- applicability (reliability and ranges of acceptable errors) of linear relations in regions of large deformations;
- theory of modelling and choice of modelling criteria;
- physical basis for reliable analytical presentation of the actual creep and relaxation phenomena;
- actual transfer functions of testing and measurement systems and reliability of experimental results (comparison with standard experiments).

ACKNOWLEDGEMENT

The authors gratefully acknowledge the support given by the Natural Sciences and Engineering Research Council of Canada under Grant No. A2939. Also, the authors would like to thank Xin-hua Ji for her assistance.

REFERENCES

[1] Cloud, G. L. and Pindera, J. T., "Techniques in infrared photoelasticity", Experimental Mechanics, 8, pp. 193-201, 1968.

[2] Daniel, I. M., "Experimental methods in applied mechanics", J. Applied Mechanics, 50, pp. 963-976, 1983.

[3] Erdogan, F., "Approximate solutions of systems of singular integral equations", SIAM J. Applied Math., 17, pp. 1041-1059, 1969.

[4] Erdogan, F. and Gupta, G., "Layered composites with an interface flaw", Int. J. Sol. Struc., 7, pp. 1089-1107, 1971.

[5] Experimental Mechanics in Research and Development, edited by J. T. Pindera, H. H. E. Leipholz, F. P. J. Rimrott and D. E. Grierson, Solid Mechanics Division, University of Waterloo, Study No. 9, Waterloo, 1973.

[6] Hecker, F. W. and Pindera, J. T., "Influence of stress gradient on direction of light propagation in photoelastic specimens", VDI-Berichte, 313, pp. 745-754, 1978.

[7] Herakovich, C. I., Post, D., Buczek, M. B. and Czarnek, R., "Free edge strain concentrations in real composite laminates: experimental-theoretical correlation", J. Applied Mechanics, 107, pp. 787-793, 1985.

[8] Herrmann, K. P., "Micromechanical analysis of basic fracture effects in self-stressed fibrous composites", Fracture of Fibrous Composites, edited by C. T. Herakovich, ASME AMD, 74, New York, 1985.

[9] Hirschfeld, T., "Instrumentation in the next decade", Science, 230, pp. 286-291, 1985.

[10] Kemmochi, K. and Uemura, M., "Measurement of stress distribution in sandwich beam under four-point bending", Experimental Mechanics, 20, pp. 80-86, 1980.

[11] Malyshev, B. M. and Salganik, R. L., "The strength of adhesive joints using the theory of fracture", Int. J. Frac. Mech., 1, pp. 114-128, 1965.

[12] Meyer, M. L. and Taylor, D. A. W., "Notes on photoelasticity of impact phenomena", Int. J. Mechanical Sciences, 25, pp. 169-173, 1983.

[13] Pindera, J. T., "Technique of photoelastic studies of plane stress states", (in Polish), Rozprawy Inzynierskie, 3, pp. 109-176, 1955.

[14] Pindera, J. T., "Rheological properties of some polyester resins, part I, II, III", (in Polish), Rosprawy Inzynierskie, Polish Acad. Sciences, 3, pp. 361-411 and 481-540, 1959.

[15] Pindera, J. T. and Cloud, G., "On dispersion of birefringence of photoelastic materials", Experimental Mechanics, 6, pp. 470-480, 1966.

[16] Pindera, J. T., "On physical basis of modern photoelasticity techniques", Beitrage zur Spannungs-und Dehnungsanalyse, Vol. V, Akademie-Verlag, Berlin, pp. 103-130, 1968.

[17] Pindera, J. T. and Straka, P., "Response of the integrated polariscope", J. of Strain Analysis, 8, pp. 65-76, 1973.

[18] Pindera, J. T. and Straka, P., "On physical measures of rheological responses of some materials in wide ranges of temperature and spectral frequency", Rheologica Acta, 13, pp. 338-351, 1974.

[19] Pindera, J. T. and Mazurkiewicz, S. B., "Photoelastic isodynes: a new type of stress modulated light intensity distribution", Mechanics Research Communications, pp. 247-252, 1977.

[20] Pindera, J. T., Straka, P. and Tschinke, M. F., "Actual thermoelastic response of some engineering materials and its applicability in investigations of dynamic response of structures", VDI-Berichte No. 313, VDI-Verlag, Dusseldorf, pp. 579-584, 1978.

[21] Pindera, J. T., "Foundations of experimental mechanics: principles of modelling, observation and experimentation", New Physical Trends in Experimental Mechanics, edited by J. T. Pindera (CISM Courses and Lectures No. 264), Springer-Verlag, Wien, pp. 188-326, 1981.

[22] Pindera, J. T. and Krasnowski, B. R., "Determination of stress intensity factors in thin and thick plates using isodyne photoelasticity", Fracture Problems and Solutions in the Energy Industry, edited by L. A. Simpson, Pergamon Press, pp. 147-156, 1981.

[23] Pindera. J. T. and Mazurkiewicz, S. B., "Studies of contact problems using photoelastic isodynes", Experimental Mechanics, 21, pp. 448-455, 1981.

[24] Pindera, J. T., Issa, S. S. and Krasnowski, B. R., "Isodyne coatings in strain analysis", Proc. of 1981 Spring Meeting, SESA, (Dearborn, Michigan, May 31-June 4, 1981), SEM (Formerly SESA), pp. 111-117, 1981.

[25] Pindera, J. T., "New development in photoelastic studies: isodyne and gradient photoelasticity", Incoherent Optical Techniques in Experimental Mechanics, edited by F. P. Chiang, Optical Engineering, 21, pp. 672-678, 1982.

[26] Pindera, J. T., Krasnowski, B. R. and Pindera, M.-J., "Theory of elastic and photoelastic isodynes. Samples of applications in composite structures", Experimental Mechanics, 25, pp. 272-281, 1985.

[27] Pindera, J. T. and Hecker, F. W., "Basic theories and experimental methods of gradient photoelasticity", Experimental Mechanics, to be published.

[28] Rohrbach, Ch., "Dehnungsmesstreifen mit metallischen Trager als schnell messbereites, feuchtigkeitsunempfindliches Messelement fur Dehnungsmessungen auf Beton", Der Bauingenieur, 33, pp. 265-268, 1958.

[29] Sih, G. C., "The state of affairs near the crack tip", Modelling Problems in Crack Tip Mechanics, edited by J. T. Pindera, Martinus Nijhoff Publishers, Dordrecht, pp. 65-90, 1984.

[30] Sih, G. C., "Mechanics and physics of energy density theory", Theoretical and Applied Fracture Mechanics, 4, pp. 157-173, 1985.

[31] Theocaris, P. S. and Stassinakis, C. A., "Experimental solution of the problem of a curvilinear crack in bonded dissimilar materials", Int. J. Fracture, 13, pp. 13-26, 1977.

[32] Vasil'ev, V. V., Dudchenko, A. A. and Elpatevskii, A. N., "Analysis of the tensile deformation of glass-reinforced plastics", Mekhanika Polimerov, 1, pp. 144-147, 1970.

[33] Wang, S. S. and Choi, I., "The interface crack between dissimilar anisotropic composite materials", J. Applied Mechanics, 50, pp. 169-178, 1983.

[34] Whitney, J. M., Daniel, I. M. and Pipes, R. B., "Experimental mechanics of fiber reinforced composite materials", SEM Mechanics Monograph No. 4, Society for Experimental Mechanics, 1984.

[35] Wilson, D. W. and Charles, J. A., "Thermographic detection of adhesive-bond and interlaminar flaws in composites", Experimental Mechanics, 21, pp. 276-280, 1981.

[36] Zak, A. R. and Williams, M. L., "Crack point stress singularities at a bi-material interface", J. Applied Mechanics, 30, pp. 142-143, 1963.

COMPOSITE MATERIALS AND REPAIR

R. Jones

Department of Defence, Aeronautical Research Laboratories
Melbourne, Australia

ABSTRACT

This paper describes the design guidelines for repairs to fibre composite structural components. The repairs vary from simple injection of resins into small delaminations and disbonds, to a variety of patching procedures for damage to areas up to 100 mm or so. Emphasis will be on repairs that can be carried out under field or depot conditions; however, some of the repairs described will be more appropriate to factory conditions. Many of the repairs for composite faced sandwich panels are similar to those described for metallic honeycomb structures.

INTRODUCTION

Graphite/epoxy composites have many advantages for use as aircraft structural materials, including their high specific strength and stiffness, resistance to damage by fatigue loading and their immunity to corrosion. Thus, extensive use of these composites should reduce the high maintenance costs associated with repair of corrosion damage normally encountered with conventional aluminum alloys, particularly those exposed in a marine environment. Similarly, costs associated with repair of damage due to fatigue should also be substantially reduced, since the composites do not, in general, suffer from the cracking encountered with metallic structures, particularly cracking resulting from fretting around fastener holes or from corrosion pitting.

However, maintenance costs associated with repair of service contact damage is expected in increase, since graphite/epoxy is essentially an unforgiving brittle material - unable to yield plastically under overload. Even quite modest impacts (by metallic standards) can lead to internal damage in the form of delaminations, which may result in a marked strength reduction particularly under compression loading. The impact area may not be apparent from surface examination because of the absence of permanent information.

Other more severe handling and environmental damage will also occur; however, this is also common to metallic structures, particularly those of honeycomb construction.

It is the purpose of this paper to consider the repair of graphite/epoxy aircraft components of various configurations. Repairs described here will vary from simple injection of resins into small delaminations and disbonds, to a variety of patching procedures for damage to areas up to 100 mm or so. Emphasis will be on repairs that can be carried out under field or depot conditions; however, some of the repairs described will be more appropriate to factory conditions. Many of the repairs for composite faced sandwich panels are similar to those described for metallic honeycomb structures. This section will highlight some of the salient factors concerning damage, inspection and repair criteria.

At the moment, repairs to thick monolithic structures with through the thickness damage is done using mechanically fastened doublers. For impact damage, this procedure may also be adopted.

Externally bonded scarf patches now offer a viable alternative, even for monolithic structures. One major disadvantage with mechanically fastened repairs is that they provide a new path for moisture ingress. However, perhaps the most important single requirement is to develop a need to repair criteria. For delamination and impact damage, this approach is simplified since as the size of the damage increases, a stage is reached after which any further increase does not further reduce the residual strength.

DAMAGE NATURE

Graphite/epoxy composites, while having many advantages, are prone to a fairly wide range of defects and damage which may significantly reduce residual strength. These can arise during manufacture, Table 1, or during service, Tables 2 and 3. Of the various types of defect listed, hole elongation, penetration and delaminations are the most severe. Delaminations, i.e., single or multiple internal cracks whose planes are parallel to the surface of a component, which arise in service are probably the most insidious because they can cause severe reductions in compressive strength, and are difficult to detect.

Table 1 - Typical manufacturing defects

Defect	Typical Causes
Voids	Poor process control
Delaminations	Inclusion of release film Poor process control Faulty hole formation procedures
Disbonds (in bonded joints)	Poor fit of parts Inclusion of release film Poor process control
Surface damage	Poor release procedure Bad handling
Misdrilled holes	Faulty jigging

Delaminations may develop during service, due to the presence of excessive interlaminar shear stresses or through-the-thickness tensile stresses at: (a) holes, [1], (b) free edges, [2], (c) regions of section changes, and (d) bonded joints [3]. However, the most important source (because the probability of occurrence is high) is impact, for instance, from dropped tools or from stones thrown up from the run-

Table 2 - Typical service mechanical damage

Defect	Typical Causes
Cuts, scratches	Mishandling
Abrasion	Rain/grit erosion
Delaminations	Impact damage
Disbonds	Impact damage Overload
Hole elongation	Overload/bearing failure
Dents (with delaminations and crushed core)	Impact damage Walk in no-step regions Runway stones
Edge damage	Mishandling of doors and removable parts
Penetration	Battle damage Severe mishandling - e.g., forklift

Table 3 - Typical environmental damage

Defect	Typical Causes
Surface oxidation	Lightning strike Overheat Battle damage (e.g., laser)
Delamination	Freeze/thaw stressing (due to moisture expansion) Thermal spike (causing steam formation)
Disbonds (in honeycomb panels)	As for delaminations
Core corrosion	Moisture penetration into honeycomb
Surface swelling	Use of undesirable solvents, e.g., paint stripper

way. While impact can cause a significant amount of delamination, often the only external indication is a very small surface indentation; thus damage of this type is often referred to as barely visible impact damage (BVID). The problem of BVID is of particular concern because the damage is unlikely to be discovered unless the region is subjected to non-destructive inspection (NDI), generally using ultrasonic procedures, when it is usually readily detectable. Unfortunately, however, most routine NDI is likely to be confined to potential hot-spots such as critical joints; frequent full-scale NDI is costly and time-consuming.

It is highly desirable that procedures are available so that the possible occurrence of delamination-type defects is allowed for in the design and certification of composite aircraft structures and in the development of approaches for through-life support, to provide a rationale for setting inspection intervals, particularly for highly stressed regions, and also to provide repair/reject criteria. For metallic aircraft components, such an approach is available for cracks under tension-dominated loading and is based on linear elastic fracture mechanics (LEFM) and experimental data on crack growth under cyclic loading; these topics form the foundation of damage tolerance design.

THE DAMAGE TOLERANCE APPROACH - APPLICATION TO COMPOSITES

Generally, damage tolerance is concerned with the ability of a structure to contain representative weakening defects under representative loading and environment with suffering excessive reduction in residual strength, for some stipulated period of service. Consideration must be given to both the location and initial size of the defect and to its likely size after a representative period of service, e.g., between inspection intervals, (see [16,23]). The concept is used specifically as a design philosophy by the US Air Force for metallic aircraft, where cracks of a prescribed size (usually just below the size where they would be readily detected by NDI) are assumed to exist at critical regions in the structure, especially at fastener holes. It is necessary, as part of certification of the structure, to demonstrate analytically (using LEFM) and by testing that within a specified interval based on the inspection interval (or a number of lifetimes if inspection is not feasible) the cracks could not propagate to the extent where the strength of the structure would be reduced below safe limits. The related concept of durability (essentially a cost of ownership aspect) is concerned with demonstrating that very small cracks will not grow to a size that would result in functional impairment or require costly repair during the life of the structure.

The development of a damage tolerance approach to composites is greatly complicated in comparison to metals because of the multiplicity of failure models and the sensitivity of composites to out-of-plane stresses. Also, environmental effects associated with moisture absorption by epoxy matrices and temperature excursions (both to high and low levels) need to be considered.

Tensile Loading

Significant defects for composites under tensile loading are holes (including fastener holes), scratches and deep cuts. It is found [4] that, depending on ply configuration, fatigue cycling under tension loading tends to remove stress concentrations by causing delaminations and splitting at the edges of the hole or notch, increasing residual strength - in some cases, to net-section values. Thus, interest centres more on static strength than on residual strength following fatigue cycling. Several LEFM type approaches have been made to determine the strength of composites with holes or slits, generally with some degree of success claimed, [5,6] although these approaches have been criticized; see for instance, [7,8]. Despite these criticisms, and the obvious need for improvements, there is some cause for optimism with respect to the eventual development of a damage tolerance approach for tensile loading.

Compression Loading

The situation concerning damage tolerance under compressive loading is much more complex than for tensile loading, particularly since there is no existing foundation based on studies with metals. Here (besides delaminations), holes, notches and deep scratches produce reductions in static residual strength and, in contrast to tension behaviour, cyclic loading is detrimental. Under cyclic compressive loading, although the degree of stress concentration is initially reduced by fatigue, the loss in stiffness due to the formation and growth of delaminations and general cracking can lead to propagation of damage by local deformation of surrounding material, [9]. Thus the residual strength may fall as a function of the number of cycles. Further, since the behaviour depends on the support offered to the fibres and plies against buckling, the behaviour is matrix-sensitive making the fatigue effect more marked under hot/wet conditions.

Influence of Delaminations

Delaminations are a major concern under compression loading (their influence under tensile loading is generally small); since the delaminated region acts as a low modulus zone, it may produce a loss in residual strength of the order of that produced by an unfilled hole of comparable area. Further, there is concern that in moisturised composites, the delamination may grow due to freeze/thaw cycling (ice has an expansion coefficient of 1.12×10^{-4} °C^{-1} down to -50°C, whereas eopxy has a coefficient of about 6×10^{-5} °C^{-1}) or due to steam formation in thermal spike conditions, as would arise during supersonic flight.

At present [10], the design of composite structures is primarily driven by static strength requirements. The design allowable (B-basis) strains for structural features are normally determined by strength tests on the features in the most environmentally degraded condition expected to be seen in service; for compression members, this is the hot/wet conditions. (The strength test may be preceded by spectrum fatigue loading of the features for a number of lifetimes). Allowable strains determined in this way automatically take account of the effect of representative stress raisers, such as fastener holes, since these are incorporated in the features, and experience to date indicates that these strain levels are such as to provide not only an adequate safe fatigue life but also a reasonable measure of resistance to delamination damage, at least on a small scale. Thus, a crude form of allowance for damage is automatically incorporated into current designs. However, service experience is limited, particularly with respect to severe environmental cycling, and the safe limits on delamination size, allowing, for example, for service growth, are unknown. Further, future applications of composites are likely to be based on much less conservative strain allowables - for instance, by employing strengthened fastener holes or by the use of all bonded structure - to improve on current levels of weight-saving. Consequently, the requirement for a damage tolerance methodology will be quite vital.

DAMAGE ASSESSMENT

Laminates with Holes

Several methods have been successfully developed for predicting the residual strength of a laminate containing a hole or a notch. The point stress and the average stress methods developed in [12-14] are by far the simplest to use.

The average stress failure criterion [12] assumes failure to occur when the average value of σ_y over some fixed length, a_o, ahead of the hole first reaches the unnotched tensile strength of the material σ_o. That is, when

$$\frac{1}{a_o}\int_R^{R+a_o} \sigma_y(x,o)dx = \sigma_o \qquad (1)$$

Using the well-known form for σ_y, where y is in the direction of the load, gives the ratio of the notched to unnotched strength as

$$\frac{\sigma_N}{\sigma_o} = 2(1-\phi)/[2-\phi^2-\phi^4+(K_T-3)(\phi^6-\phi^8)] \qquad (2)$$

where

$$\phi = R/(R+a_o) \qquad (3)$$

and where R is the radius of the hole.

In practice, the quantity a_o is determined experimentally from strength reduction data.

The point stress criterion assumes that failure occurs when the stress σ_y at some fixed distance, d_o, ahead of the hole first reaches the unnotched tensile stress.

$$\sigma_y(x,o)\Big|_{x=R+d_o} = \sigma_o \qquad (4)$$

It was shown in [12] that the point stress and average stress failure criteria are related and that

$$a_o = 4d_o \qquad (5)$$

With the use of modern structural analyses programs, it is relatively easy to apply these failure criteria to more complex problems. For example, recent tests [14] were carried out on various 16-ply graphite/epoxy laminates (AS/3501-5) with holes. The laminates were:

1. $(0/\pm45_2/0/\pm45)_s$

2. $(0_2/\pm45/0_2/90/0)_s$

3. $(0/\pm45/90)_{2s}$

The results are shown in Table 4 and are compared to predicted values using the average stress method with a_o = 2.3 mm. As can be seen from Table 4, the average stress criterion provides accurate estimates of the strength reduction due to the presence of holes. This method is widely used in the aerospace industry and has been applied to estimation of strength reduction due to battle damage [15].

Table 4 - Static strength predictions; from [14]

Hole Size and Placement	Laminate No.	% of Unnotched Strength	
		Test	Average Stress
2-4.8 mm diameter countersunk of Figure 8(a)	1	58.9	53.6
"	2	48.1	51.4
"	3	51.8	53.2
2-4.8 mm diameter countersunk of Figure 8(b)	3	48.7	45.9
2-6.4 mm diameter countersunk of Figure 8(a)	3	53.1	50.3
1-6.4 mm diameter non countersunk	2	54.0	52.6

Delamination and Impact Damage

A review of effects of impact damage on the compressive strength of graphite/epoxy composites is given in [45].

The two failure criteria, i.e., the condition for catastrophic growth of the damage, which offer the most promise are

(1) the energy release rate, used in [11,32-35,39],

(2) the strain energy density hypothesis described in [40] and used in [35, 36,39,48].

(1) Energy Release Rate Approach

When using the energy release rate (i.e., G) approach, it must be remembered that the growth will be mixed mode, i.e., G = G(I,II,III), and non-self-similar. Consequently, in order to calculate G, the delamination shape must be known at each step. In the problems considered in [29,34], which are essentially two-dimensional, G can be evaluated using simple numerical methods, or even by using J integral [41] around the tip of the delamination.

For problems which are not two-dimensional, a line integral J_1^* has been developed [41]. This integral is defined by

$$J_1^* = \lim_{\delta \to 0} \oint \left[\frac{1}{2} \sigma_{ij} \frac{\partial u_i}{\partial x_j} dx_2 - T_i \frac{\partial u_i}{\partial x_1} ds\right] \tag{6}$$

where the integration is along a contour of radius δ normal to the front of the delamination (u_i is the displacement component; x_i, i = 1,3, is a system of Cartesian coordinates with $x_1 = x$). This integral coincides with the classical definition of J for two-dimensional problems but, for three-dimensional problems, returns the "local energy" required for self-similar growth, i.e., circles into concentric circles, and not the true energy release rate. Since delamination growth is often mixed mode and non-self-similar, J_1 is of questionable value.

(2) The Strain Energy Density Approach

This approach may be considered as an extension of the point stress failure criterion to allow for mixed mode failure. Define the strain energy density in the normal fashion:

$$\frac{dW}{dv} = \frac{1}{2}\sigma_{ij}\varepsilon_{ij} \tag{7}$$

Then, for a two dimensional problem, the strain energy density approach has two basic hypotheses which apply for crack extension:

(1) The crack will grow in the direction $\theta = \theta_0$ of maximum potential energy density (viz: minimum strain energy density).

(2) Failure occurs when the stress field at a distance r_c ahead of the crack (or hole) in the direction $\theta = \theta_0$ of the minimum strain energy density, is such that

$$(dW/dV)_{\substack{r=r_c\\ \theta=\theta_0}} = (dW/dv)_c \tag{8}$$

Here, $(dW/dv)_c$ is the value, at failure of the strain energy density of the undamaged laminate.

In this formulation, r_c plays a similar role to d_0, the damage zone size used in the point stress failure formulation. Indeed, as the first approximation, we can take $r_c = d_0$. In this approach, the strain energy density function S as defined by

$$\frac{dW}{dv} = \frac{S}{r} + \text{higher order terms} \tag{9}$$

plays a central role. For Mode I failure of orthotropic material, S is proportional to the stress intensity factor K_I. Indeed, the above hypothesis can be readily expressed in terms of S, viz:

Failure occurs in the direction $\theta = \theta_0$ for which

$$\frac{\partial S}{\partial \theta} = 0 \text{ and } \frac{\partial^2 S}{\partial_0^2} > 0 \tag{10}$$

and when the load is such that

$$S\big|_{r=r_c} = S_c \ (= \tau_c\ (dW/dv)_c) \tag{11}$$

For three dimensional damage, we must first locate at each point along the damage front the local minimum of the strain energy density function S_{min} (= dw/ dv_{min}). Failure then initiates at the point along the front which has the maximum value of S_{min}.

For incremental growth, each point along the damage front advances a distance r which is determined from the relationship

$$\frac{S_1}{r_1} = \frac{S_2}{r_2} = \frac{S_i}{r_i} + [\frac{dW}{dv}]_c \tag{12}$$

Using these failure criteria, it is possible to assess the severity of the damage and determine the loss in residual strength. In the case of impact damage, and edge delaminations, recent experimental and numerical work [23,31,39,43,45, 48] has shown that as the size of the damage increases, a stage is reached after which any further increase in size results in only a small increase in $(S_{min})_{max}$* and G and, hence, only a small decrease in the residual compressive strength. This dramatically simplifies the analysis and reduces the amount of coupon testing required in order to determine the severity of the damage.

REPAIR PROCEDURES

There are two main approaches to repairing damaged components, viz: (1) patch, for major damage and (2) non-patch, for minor damage. These two approaches are summarized in Tables 5 and 6.

Table 5 - Non-patch repair procedures for minor damage

Repair Procedure	Applicable Defect
Resin injection	Connected voids
Potting or filling	Minor depressions Skin damage in honeycomb panels Core replacement in honeycomb panels Fastener hole elongation
Heat treatment	Remove entrapped moisture in honeycomb panels Dry out absorbed moisture
Surface coating	Seal honeycomb panels Restore surface protection

*In the strain energy density approach, failure is assumed to occur when the maximum value of S_{min} along the front of the delamination, defined as $(S_{min})_{max}$, reaches the critical value S_c.

Table 6 - Patch repairs for major damage: all are capable of restoring ultimate strain allowables to the limits of laminate thickness noted

Patch Procedure	Application
Bonded External Patch	
Graphite/epoxy, i) co-cured, ii) pre-cured layers, iii) pre-cured. Titanium alloy foil.	Repairs to skins, particularly on honeycomb panels, up to 16 plies thick. Well suited for field application.
Bonded Flush Patch	
Graphite/epoxy (usually co-cured). Titanium alloy foil scarfed or stepped.	Repairs to skins 16 to 100 plies thick, holes up to 100 mm. May be difficult to employ under field conditions.
Bolted External Patch	
Titanium alloy (usual). Aluminum alloy.	Repairs to monolithic skins 50 to 100 plies, holes up to 100 mm. Suited for field applications. Field repairs to impact damage.

Initial Design Considerations

The majority of the repairs mentioned in Table 6, with the exception of repairs to impact damage, may be idealized as two dimensional problems. In the past several approximate formulae have been developed for these cases. However, given the present availability of finite element packages and their ease of use, it is now simpler to design the repair on the basis of a two-dimensional finite element investigation.

When analyzing a glue line or a laminate, the aspect ratios of the elements often exceed 20:1. In these cases, elements without midside nodes should not be used and the stiffness matrices should be calculated in double precision using reduced integration.

Let us now turn our attention, to field repairs which by necessity are external patches either bonded or bolted to the structure.

Externally Bonded Patches

The external bonded repairs to holes can be modelled as half of an overlap joint. The adhesive must be allowed to yield and allowance must be made for the different coefficients of thermal expansion in the repair and in the parent material. In the case when the patch and the parent material have equal stiffness and thickness, the maximum load carrying capacity of the joint may be estimated, see [46], as

$$P = 2(\tau_p(\frac{1}{2}\,\sigma_e + \sigma_p)E_r t_r)^{1/2} \tag{13}$$

where τ_p is the effective yield stress of the adhesive, σ_e and σ_p are the strain to yield and the plastic strain to failure, respectively, t_a is the adhesive thickness and E_r, t_r are the Young's modulus and thickness of the repair, respectively.

The edges of the patch must be tapered to reduce the peel and shear stresses. The optimum taper can easily be obtained using finite element analysis; however, a ply drop off every 3 mm is a common taper angle and is recommended for initial design purposes.

To increase the fatigue life of the repair, it is desirable that the adhesive behaves elastically throughout its life. This can be achieved by varying the patch and adhesive thicknesses such that the maximum adhesive shear and peel stresses remain within the yield surface. In the absence of local bending, the peak adhesive shear stresses will occur around the hole and, in the elastic regime, may be approximated as

$$\tau_{max} = K_a \sigma_o \pi\lambda / 2E_P \tag{14}$$

where

$$K_a = 1/(t_A/G_A + 3t_R/G_R + 3t_P/8G_P) \tag{15}$$

$$\pi\lambda = [\beta^{-1} E_P t_P \ (1 + t_P E_P / E_R t_R)]^{1/2} \tag{16}$$

$$\sigma_o = E_P t \sigma_P / (E_r t_r + E_P t_P) \tag{17}$$

and

$$\beta = (t_A/G_A + t_r/3G_r + t_P/3G_P)/(t_A/G_A + 3t_r/8G_r + 3t_P/8G_P)^2 \tag{18}$$

Here, G_a, G_r and G_P are the interlaminar shear module of the adhesive, repair and parent material respectively.

To illustrate the effectiveness of this technique, let us consider the repair of 38.1 mm hole in a graphite epoxy laminate of dimensions 635 mm x 508 mm x 2.29 mm subject to an uniaxial stress, see [50]. The hole is repaired using an externally bonded patch with dimensions 152 mm x 508 mm x 0.762 mm applied to both sides of the laminate. The patches to be considered are (i) titanium and (ii) a graphite epoxy laminate with the same ply configuration as the basic laminate. The effectiveness of the repairs schemes are compared in Table 7 where K_{Tu} and K_{TP} are the stress concentration factors before and after patching.

It is interesting to note that, while both repairs significantly reduce the stress concentration factor, the titanium patch is more effective.

Bolted Repairs

Bolted field repair techniques have been developed for laminates up to 12.7 mm thick and have been shown to be capable of restoring strength to components containing damage up to 102 mm in diameter [47]. The design methodology is particularly straightforward and includes the following steps.

Table 7 - Ratio of stress concentration factors K_{TP}/K_{Tu}

Panel Laminate Layup Patch	$(\pm 45/0/90)_{2S}$	$(\pm 45)_{4S}$
(i)	0.33	0.32
(ii)	0.46	0.54

(a) Estimate patch load using the formulae

$$P_p = \varepsilon E_r t_r D \tag{19}$$

where E is the required strain level at ultimate load, E_r and t_r are the modulus and the thickness of the patch respectively and where D is the diameter of the damage.

(b) Estimate number of fasteners required, defined as n, using the formulae

$$n = \frac{P_P}{V} \tag{20}$$

where V maximum allowable shear load of the fasteners. Obtain a second estimate for n based on failure due to exceeding the bearing stress allowables.

$$n = 1.25 \frac{P_p}{F_{BRV}} dt_p \tag{21}$$

Here, t_p is the thickness of the parent material, d is diameter of the fasteners to be used is the maximum of these two estimates.

The size of patch is then dictated by the requirement that the fasteners be separated by a minimum edge distance of 2d.

Impact Damage

The bolted repair concept can be easily extended so as to allow for repairs to impact damage. As mentioned in Section 4, and in [45], as the size of the damage increases, the residual compressive strength asymptotes to a nonzero value σ_R. It is only necessary to increase the residual strength to that of a hole equal in size to diameter of the fasteners used, typically 6 mm's. This now defines the value of ε to be used in equation (19) and as a first estimate, we take D to be the area of the impact damage.

Recent numerical work at ARL has shown that the fasteners do not act as "crack or delamination, stoppers" and that the main role played by the repair is to lower the net section stress. Denoting the required strength as $\hat{\sigma}$ and the strength of the damaged component as σ_R, the minimum patch thickness can now be estimated using the formulae

$$\frac{\hat{\sigma}}{\sigma_R} = (1 + E_R t_R / E_P t_P)$$

THE MOISTURE PROBLEM

Moisture can cause serious problems during repair implementation if it is not removed by an initial heat treatment. Graphite/epoxy laminates may absorb up to 1.5% moisture during exposure in humid environments. The moisture is actually absorbed by the epoxy matrix where (by acting as a plasticizer), it reduces some resin sensitive mechanical properties, such as compressive strength at elevated temperature. Moisture can also be trapped in voids and delaminations where it can cause severe damage due to expansion effects in a thermal spike exposure or in a freeze/thaw cycle, and, in the case of honeycomb panels, where it can cause corrosion of the aluminum alloy core.

During patch application, the moisture may turn to steam which can (i) split the laminate in voided regions, (ii) form voids in the adhesive, and (iii) form voids in matrix of the repair laminate (if it is being cocured). Damage in the matrix can be particularly severe if the heat treatment is performed above its (reduced) glass transition temperature, when the strength of the matrix is quite low. In all cases, the result may be a severe degradation in mechanical properties. The problem of moisture removal is a much more serious problem in a thick laminate (50 plies or more), since days of heating may be required to affect removal. In general, however, it may not be necessary to remove all moisture, since only the surface moisture causes problems (i) and (ii) referred to previously. Rapid surface drying may be achieved by using vacuum assistance.

Thin laminates (16 plies or less) can be dried out fairly rapidly. However, if the laminate forms the face of a honeycomb panel, excessive internal steam formation may result in a blown skin. One approach is to accept the presence of moisture in the laminate and allow for it in terms of reduced design allowables - both for the patch, if cocured, and for the adhesives. Unfortunately, as shown in [49], this results in a significant reduction in the static strength of the repaired structure. Consequently, for thick laminates, the only course may be to rely on mechanically fastened repairs, or to use low temperature cure adhesive.

REFERENCES

[1] Rybicki, E. F. and Schmueser, D. W., "Effect of stacking sequence and lay-up angle on free edge stresses around a hole in a laminated plate under tension", J. Composite Materials, 12, pp. 300-313, 1978.

[2] Pipes, B. R. and Pagano, N. J., "Interlaminar stresses in composite laminates under uniform axial extension", J. Composite Materials, 4, pp. 538-548, 1970.

[3] Hart-Smith, L. F., "Adhesive-bonded double-lap joints", NASA-CR-112235, 1983.

[4] Ramani, S. V. and Williams, D. P., "Fatigue of filamentary materials", ASTM STP 636, 1977.

[5] Dorey, G., "Damage tolerance in advanced composite materials", RAW TR77172, 1977.

[6] Phillips, D. C. and Tetelman, A. S., "The fracture toughness of fibre composites", Composites, 3, pp. 216-223, 1972.

[7] Kanninen, M. F., Rybicki, E. F. and Brenson, H. F., "A critical look at current applications of fracture mechanics to the failure of fibre reinforced composites", Composites, 3, pp. 17-22, 1977.

[8] Wang, A. S. D., "Fracture mechanics of sub-laminate cracks in composite laminates", AGARD Conference on Characterisation, Analysis and Significance Defects in Composite Materials, AGARD-CP-355, 1983.

[9] Saff, C. R., "Compression fatigue life prediction methodology for composite structures", NADC-80235-60, 1980.

[10] Guyett, P. R. and Cardrick, A. W., "The certification of composite airframe structures", Roy. Aeron. Soc. J., pp. 188-203, 1980.

[11] Wilkins, D. J., "The engineering significance of defects in composite structures", AGARD Conference on Characterisation, Analysis and Significance of Defects in Composite Materials, AGARD-CP0355, 1983.

[12] Nuismer, R. J. and Whitney, J. M., "Uniaxial failure of composite laminates containing stress concentrations", ASTM STP 593, pp. 117-142, 1975.

[13] Nuismer, R. J. and Labor, J. D., "Application of the average stress failure criterion: part II - compression", J. Comp. Materials, 13, pp. 49-60, 1979.

[14] Nuismer, R. J. and Labor, J. D., "Application of the average stress failure criterion: part I - tension", J. Comp. Materials, 12, p. 238, 1978.

[15] Husman, G. E., Whitney, J. M. and Halpin, J. C., "Residual strength characterisation of laminated composites subjected to impact loading", ASTM STP 568, pp. 92-113, 1975.

[16] Lauraitis, K. N. and Sandorff, P. E., "Experimental investigation of the interaction of moisture, low temperature and low level impact on graphite/epoxy composites", NADC 79102-60, 1980.

[17] Garrett, R. A., "Effects of defects on aircraft composite structures", AGARD Conference on Characterisation, Analysis and Significance of Defects in Composite Materials, AGARD-CP-355, 1983.

[18] Bishop, S. M. and Dorey, G., "The effect of damage on the tensile and compressive performance of carbon fibre laminates", AGARD Conference on Characterization, Analysis and Significance of Defects in Composite Materials, AGARD-CP-355, 1983.

[19] Potter, R. T., "The significance of defects and damage in composite structures", AGARD Conference on Characterisation, Analysis and Significance of Defects in Composite Materials, AGARD-CP-355, 1983.

[20] Walter, R. W., Johnson, R. W., June, R. R. and McCarty, J. E., "Designing for integrity in long life composite aircraft structures, fatigue of filamentary composite materials, ASTM STP 636, 1977.

[21] Geier, W., Vilismeier, J. and Weiserber, D., "Experimental investigation of delaminations in carbon", AGARD Conference on Characterisation Analysis and Significance of Defects in Composite Materials, AGARD-CP-355, 1983.

[22] Labour, J. D., "Impact damage effects on the strength of advanced composites. Non-destructive evaluation and flaw criticality for composite materials", ASTM STP 969, 1979.

[23] Avva, V. S., "Impact initiated damage in laminated composites", AFOSR-TR-82-1038, 1982.

[24] Wang, A. S. D. and Slomiana, M., "Fracture mechanics of delamination initiation and growth", NADC-79056-60, 1982.

[25] Webster, J. D., "Flaw criticality of circular disbond defects in composite laminates", University of Delaware, NASA-CR-164830, 1981.

[26] Konishi, D. Y. and Honston, W. R., "Fatigue effects on delamination and strength degradation in graphite/epoxy laminates", Composite Materials: Testing and Design, ASTM STP 674, 1979.

[27] Wilkins, D. J., "A comparison of the delamination and environmental resistance of a graphite/epoxy and graphite/bismaleimide", General Dynamics, 1982.

[28] Russell, A. J. and Street, K. N., "Factors affecting the interlaminar fracture energy of graphite/epoxy laminates", Progress in Science and Engineering of Composites, T. Hayashi, K. Kawata and S. Umekawa, eds., ICCM-IV, 1982.

[29] Wang, A. S. D. and Crossman, F. W., "Initiation and growth of transverse cracks and edge delaminations in composite laminates. Part 1. An energy method", J. Composite Materials, 14, pp. 71-87, 1980.

[30] Crossman, F. W., Warren, W. J., Wang, A. S. D. and Law, G. E., "Initiation and growth of transverse cracks and edge delaminations in composite laminates. Part 2. Experimental correlation", J. Composite Materials, 14, pp. 88-108, 1980.

[31] Chou, P. C., Wang, A. S. D. and Miller, H., "Cumulative damage model for advanced composite materials", AFWAL-TR-82-4083, 1982.

[32] Rybicki, E. F., Schmueser, D. W. and Fox, J., "An energy release rate approach for stable crack growth in the free edge delamination problem", J. Composite Materials, 11, pp. 470-487, 1977.

[33] Wang, S. S., "Edge delamination in angle ply composites", NASA-CR-165439, 1981.

[34] Wang, S. S., "Fracture mechanics for delamination problems in composite materials", Progress in Science and Engineering of Composites, T. Hayashi, K. Kawata and S. Umekawa, eds., ICCM-IV, Vol. 1, pp. 287-296, 1982.

[35] Jones, R. and Callinan, R. J., "Analysis of compression failure in composite materials", Progress in Science and Engineering of Composites, T. Hayashi, K, Kawata and S. Umekawa, eds., ICCM-IV, Vol. 1, pp. 447-454, 1982.

[36] Chou, S. C., "Delamination of T300/5208 graphite epoxy laminates", Proc. of the 2nd USA-USSR Conference on Fracture of Composites, G. C. Sih and V. Tamuzh, eds., Martinus Nijhoff Publishers, pp. 247-263, 1982.

[37] Ratwani, M. M. and Kan, H. P., "Compression fatigue analysis of fibre composites", AIAA J. of Aircraft, 18 (6), pp. 458-462, 1980.

[38] Ratwani, M. M. and Kan, H. P., "Compression fatigue analysis of fibre composites", NADC-78049-69, 1979.

[39] Jones, R., Broughton, W., Mousley, R. F. and Potter, R. T., "Compression failures of damaged graphite epoxy laminates", Composite Structures, 3, 2, pp. 167-186, 1985.

[40] Sih, G. C. and Chen, E. P., Mechanics of Fracture 6: "Cracks in composite materials", Martinus Nijhoff Publishers, 1981.

[41] Blackburn, W. S. and Hellen, T. K., "Calculation of stress intensity factors in three dimensions by finite element methods", Int. J. Num. Methods in Engineering, 11, pp. 211-229, 1977.

[42] Stellbrink, K. K. and Aoki, R. M., "Effect of defect on the behaviour of composites", Progress in Science and Engineering of Composites, T. Hayashi, K. Kawata and S. Umekawa, eds., ICCM-IV, Vol. 1, pp. 853-860, 1982.

[43] Lauraitis, K. N., Ryder, T. and Pettit, D. E., "Advanced residual strength degradation rate modelling for advanced composite structures", Vol. II, AFWAL-TR-79-3095, 1981.

[44] Jones, R., Baker, A. A. and Callinan, R. J., "Residual strength of impact damaged composites", Composite Structures, 24, pp. 371-372, 1984.

[45] Baker, A. A., Jones, R. and Callinan, R. J., "Damage tolerance of graphite epoxy composites", 4, pp. 15-44, 1985.

[46] Hart-Smith, L. J., "Analysis and design of advanced composite bonded joints", NASA CR-2218, 1974.

[47] Bohlman, R. E., Renieri, G. D. and Libeskind, M., "Bolted field repair of graphite epoxy wing skin laminates", Proc. International Workshop on Defence Applications of Advanced Repair Technology for Metal and Composite Structures, July 22-24, 1981, Naval Research Labs., Washington; also presented at ASTM Specialist Symposium on Joining of Composite Materials, 1980.

[48] Janardhana, M. N., Brown, K. C. and Jones, R., "Designing for tolerance to impact damage of fastener holes in graphite epoxy laminates under compression", Journal of Theoretical and Applied Fracture Mechanics, Vol. 5, No. 1, pp. 51-55, 1986.

[49] Hart-Smith, L. J., "Design and analysis of bonded repairs for fibrous composite aircraft structures", Proc. Int. Workshop on Defence Applications of Advanced Repair Technology for Metal and Composite Structures, Naval Research Labs., Washington, pp. 261-268, 1981.

[50] Jones, R., Callinan, R. J. and Aggarwal, K. C., "Analysis of bonded repairs to damaged fibre composite structures", Eng. Fracture Mechanics, 17, 1, pp. 37-46, 1983.

RECENT DEVELOPMENTS IN ADVANCED COMPOSITES RESEARCH

T.-W. Chou

Center for Composite Materials and Mechanical Engineering Department
University of Delaware
Newark, Delaware 19716 USA

ABSTRACT

This paper provides an overview of the recent findings of Chou and co-workers in advanced fiber composite materials research conducted at the Center for Composite Materials. Both theoretical and experimental results are reported. In the analysis and characterization work, fibers in discontinuous, continuous, hybrid and woven forms have been examined. Areas of special emphasis include (a) performance of advanced thermoplastic composites, (b) structure-performance maps of 2-D and 3-D textile structural composites, (c) hybrid composites for printed circuit boards, and (d) analytical modeling and experimental evaluation of flexible composites.

INTRODUCTION

Significant advancements have been made in the research of fiber composite materials during the last decade. Major developments can be found in ceramic fibers, advanced thermoplastic matrices, processing technologies and near-net-shape fabrications, hybridization of fibers, multidirectional fiber reinforcement based upon textile composites, and high temperature composites based upon metal and ceramic matrices. The intent of this paper is to summarize some of our recent research findings relevant to the above areas.

SHORT FIBER REINFORCED THERMOPLASTIC COMPOSITES

Short fiber reinforced thermoplastic composites have been gaining increasing importance in consumer and automotive applications because of their unique performance. They also offer the advantages of design flexibility and energy efficiency for high rate productions. In many of the applications of these composites, the matrices are often exposed to extreme temperature environment [1,2]. Consequently, laboratory test should be conducted on these composites to assess their performance under service conditions. A research program is in progress to determine the strength-temperature relationship of three thermoplastics: polyetherimide (PEI), polyethersulphone (PES), and polyetheretherketone (PEEK) as well as their short fiber reinforced composites. These materials were manufactured by injection molding at LNP Corporation. The fiber content of these composites is 30 percent by weight; the average length of short glass fibers is 200 μm.

Tension and compression (using IITRI fixtures) tests of these materials have been performed at -75, 23 and 150°C; bending fatigue tests have been conducted at 23 and 150°C. Tension and compression specimens are straight-sided coupons; flexure fatigue specimen has a triangular shape. All specimens were fabricated so as to have the loading axis transverse to the mold fill direction. Some of the major findings are outlined below [3].

The effect of temperature on composite stress-strain behavior is illustrated in Figure 1. All three types of composites showed higher modulus at -75°C than

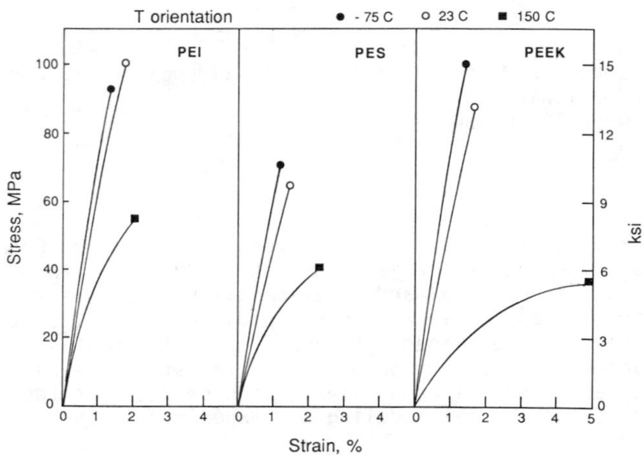

Fig. (1) - Stress-strain behavior of short glass fiber reinforced thermoplastic composites

those at 23°C, the greatest reduction in modulus and strength occurred in PEEK composite at 150°C. A similar stress-strain behavior of matrices was observed at all three temperatures, with the exception that PEEK matrix exhibited more than 200 percent extension without fracture at 150°C. Furthermore, the strengths of all matrices were lower at -75°C than those at 23°C; the reason for this behavior is not clear. Thus, the temperature dependence of the modulus and strength of these composites cannot be totally attributed to the characteristics of their matrices. PEI, PES and PEEK composites also showed, respectively, 39, 68 and 60 percent higher strength at -75°C than those at 23°C under compressive loading.

Flexure fatigue tests were performed on a Krouse sheet bending machine at 30 Hz, and r = -1.0. Fatigue data are presented in the S-N plot (Figure 2); instead of data points, the scatter bands are indicated. The scattering was fairly uniform at both high and low stress levels. It is apparent that fatigue resistance of PEEK composites was considerably higher than that of PEI and PES composites. Because PEI composites have the highest tensile and compressive strengths, we could not find correlation between static strength and fatigue resistance. However, since neither PEI nor PES is as tough as PEEK at room temperature, the test results are not surprising. At 150°C, PEEK composites suffered the greatest degradation in fatigue resistance. This may be explained by its relatively low

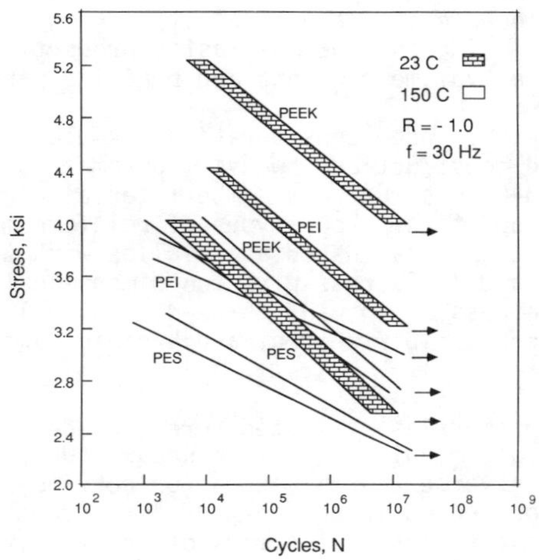

Fig. (2) - Flexural fatigue life (S-N) curves of short glass fiber reinforced thermoplastic composites

glass transition temperature, 140°C, which is 10° lower than the testing temperature; one would expect degradation of material property to occur under this condition. The arrows in Figure 2 indicate no failure (runout) after 30 million cycles. Runout occurred, respectively, at 22 to 32 percent and 37 to 47 percent of composite ultimate tensile strengths at 23 and 150°C. Since runout is a good indicator of relative fatigue resistance at low-stress level, it may be an important consideration in the design of engineering structures.

The effect of testing temperature on the dominant failure mechanisms is also investigated. Examination of fracture surfaces indicating brittle fracture which is characterized by the formation of brittle fracture bands at micro level in all three types of composites and matrices at -75 and 23°C. For composites fracture at 150°C, the ductile failure mode is dominating. The major features on the fracture surfaces are fibrils (> 100 μm), flow structures, and cavities caused by high temperature. There is also evidence of poor glass fiber and matrix adhesion.

The study of crack propagation in short glass fiber composites is underway; study of short carbon fiber reinforced thermoplastic composite has been initiated.

STRUCTURE-PERFORMANCE MAPS OF TEXTILE STRUCTURAL COMPOSITES

Recently, Chou and Yang [4] have constructed the comprehensive "structure-performance maps" to assess the capability of various reinforcement configurations with different fiber and matrix combinations.

Parametric studies have been performed based upon the analytical modeling works developed for various innovative reinforcement concepts such as 2-D and 3-D woven and/or braided structures. In the case of 2-D biaxial woven fabric composites, three analytical models have been developed: mosaic model, crimp model and bridging model to characterize the complex geometrical structures and establish the constitutive relations [5-15]. In the case of 3-D braided composites, two analytical methods have been established: elastic energy approach and fiber inclination model for predicting the thermo-elastic properties as functions of fiber spatial orientation, fiber volume contents and braiding parameters [16-21].

The geometric parameters under consideration include fiber orientation in unidirectional laminated constructions, weaving parameters in 2-D fabrics, and braiding parameters in 3-D constructions. The material parameters are fiber and matrix thermo-elastic properties. Four types of reinforcement forms are under consideration: unidirectional laminated angle-plies with the off-axis angle (θ) ranging from 0° to 90°, a 0°/90° cross-ply, two-dimensional woven fabrics ranging from plain weave to 8-harness satin, and three-dimensional braided composites with the braiding angle between a fiber segment and braiding axis in the range of 15° to 35°.

The thermo-elastic properties evaluated here include composite axial Young's modulus, transverse Young's modulus, in-plane shear modulus, axial thermal expansion coefficient and transverse thermal expansion coefficient. The results of parametric studies are presented in the form of "structure-performance maps". Figures 3 and 4 give the examples of the maps of the longitudinal Young's modulus

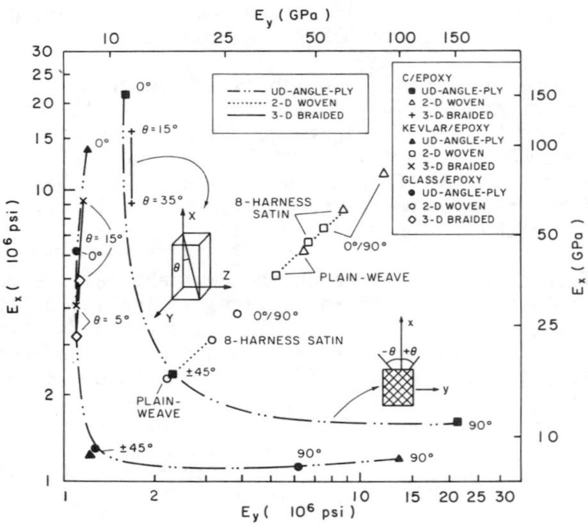

Fig. (3) - E_x versus E_y for epoxy matrix composites [4]

versus the transverse Young's modulus of epoxy and aluminum matrix composites with different reinforcing fibers and reinforcement geometries.

By the construction of the structure-performance maps, the effective composite properties based upon various reinforcement forms, and fiber and matrix combinations can be easily assessed. The uniqueness of various textile structural reinforcements also has been demonstrated. These comprehensive performance maps can

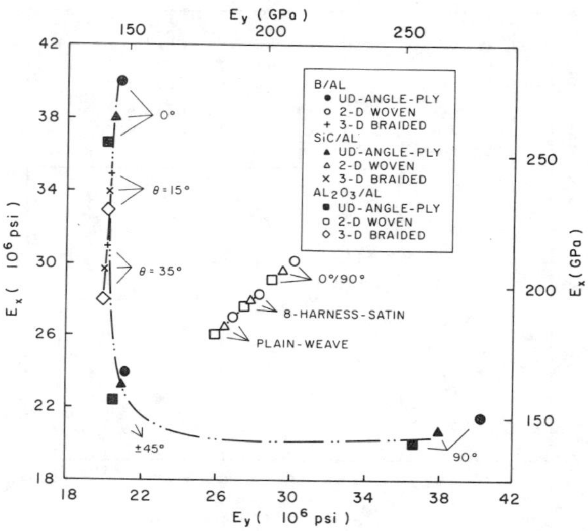

Fig. (4) - E_x versus E_y for aluminum matrix composites [4]

provide the data base necessary for material selections and guidance for future investigations of advanced composites.

HYBRID WOVEN FABRIC COMPOSITES FOR PRINTED CIRCUIT BOARDS

Woven fabric composites have been used extensively as substrate materials for multilayer printed circuit boards (PCB). Conventional PCBs typically use glass fabric reinforced epoxy resin composites as substrate materials. However, the mismatch in coefficients of thermal expansion (CTE) between ceramic chip carriers (CCC) and the PCB will result in stresses on the solder joint, which may crack due to thermal fatigue. In order to meet the requirements of thermal cycling associated with many chip carrier applications, new substrate materials with CTE comparable to those of the CCC must be developed. Various innovative approaches have been examined by researchers in their pursuit of new substrate materials as alternatives to glass fabric reinforced epoxy composites. Kevlar fabrics, quartz fabrics and hybrid fabrics of Kevlar and glass have been adopted for reinforcing various resin materials.

Hybrid fabric constructions can provide a wide variety of material selections and reduce the cost of fabrication. Recently, Chou and Yang [22] have performed an extensive parametric study to evaluate the potential of various hybrid woven fabric composites as a low CTE substrate material for printed circuit applications. The results from the parametric study are presented in the form of "structure-performance map". These comprehensive maps provide a guideline for printed circuit substrate materials selection and design. Figure 5 shows an example of the map of the longitudinal CTE versus the transverse CTE for various plain-weave hybrid fabric constructions in polyimide matrix composites.

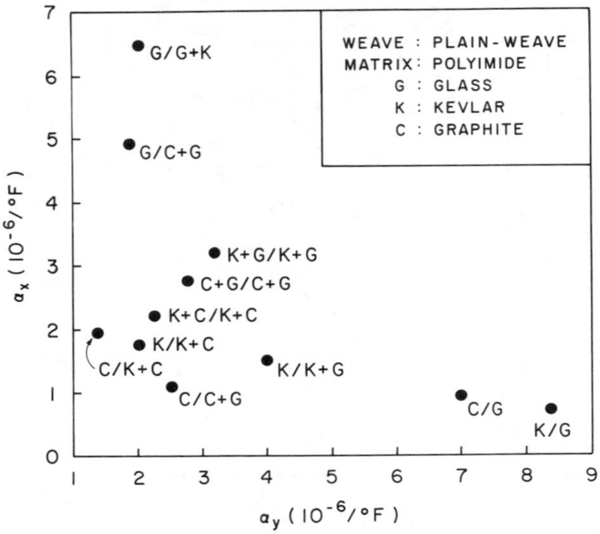

Fig. (5) - α_x versus α_y for various plain-weave hybrid composites in polyimide matrix [22]

These results clearly indicate that hybrid fabric constructions using the combination of different fibers arranged in the fill and/or warp directions can be used as a low CTE substrate material. The volume fractions of the different fibers in the hybrid construction can be adjusted to produce a wide range of CTE values that can match CTE of the CCC. Thus, hybrid fabric composites offer a great potential for tailoring a low CTE substrate material through suitable materials selection, geometrical design, and fiber hybridization.

FLEXIBLE COMPOSITES

The term "flexible composites" is used to describe composite materials for which the usable range of deformation is much larger than that of the conventional thermosetting or thermoplastic polymer based composites. This large nonlinear deformation could be realized through the use of elastomeric matrices and by suitable fiber geometric arrangements. In order to sustain large deformation, the fibers must be able to deform accordingly with the matrix. This can be achieved by: (a) using short fibers, (b) arranging continuous fibers in such an orientation that they are allowed to rotate as the load increases, and (c) using reinforcements in woven, knitted or braided forms.

Chou and co-worker [23] have proposed to arrange continuous fibers in one spatial direction with sinusoidal curvature in elastomeric matrices. There were two types of fiber arrangements, based upon (1) iso-phase model, Figure 6, and (2) random-phase model, Figure 7. The prediction of nonlinear stress-strain responses of the composites is performed by a step-wise incremental analysis. Numerical results of the uniaxial tensile stress-strain relations have been obtained

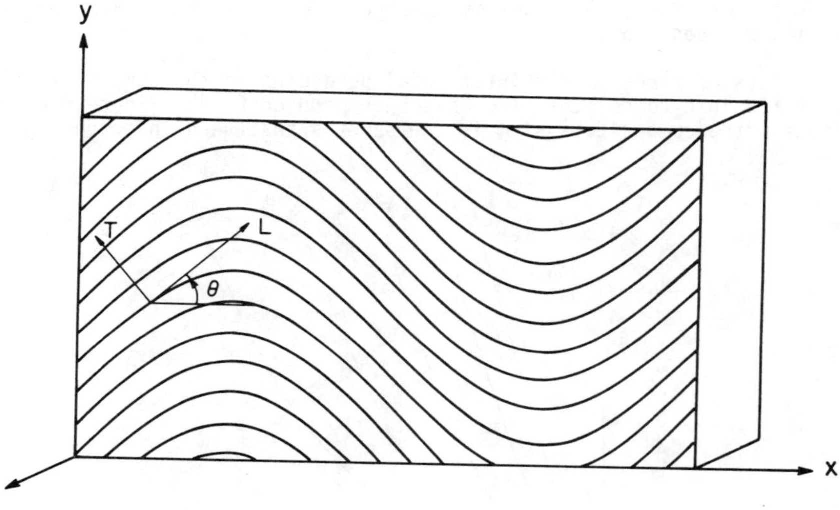

Fig. (6) - Iso-phase model [23]

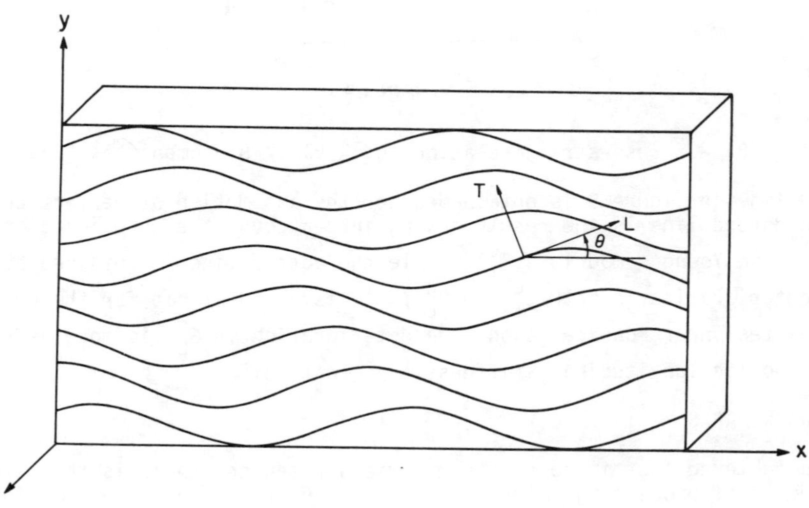

Fig. (7) - Random-phase model [23]

for several types of composites, containing glass or Kevlar fibers in an elastomeric polymer. Figure 8 gives the tensile stress-strain curve of Kevlar/PBT composites at various a/λ values, predicted by the iso-phase model. Here, a and λ

denote, respectively, the amplitude and wavelength of the sinusoidal fiber shape. Rubber elasticity is assumed for the polybutylene matrix. Fiber volume fraction is 50%. Failure of the composites is assumed to occur when the average fiber axial strain reaches 3.5%.

The effects of fiber/matrix interfacial debonding in the stress-strain responses of flexible composites also have been studied [24]. Figure 9 demonstrates such effects; the composite system is the same as that used in producing Figure 8.

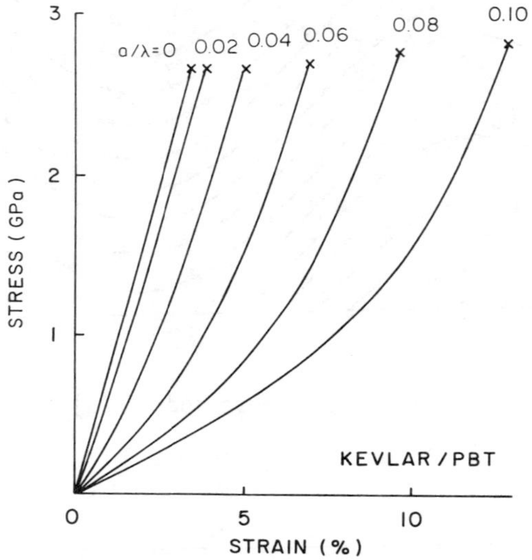

Fig. (8) - Stress-strain relations of Kevlar/PBT composites [23]

The solid line in Figure 9 is obtained under the assumption of perfect bonding. The chain-dotted line is the result taking into account the deterioration only in the transverse Young's modulus (E_T^C), while the dotted line is obtained by considering the deterioration in both E_T^C and G_{LT}^C. It is obvious that for the curved fiber composites under consideration, the deterioration in G_{LT}^C is more detrimental to the composite longitudinal stiffness than that in E_T^C.

CONCLUDING REMARKS

It is believed that in composite materials research, there is the strong interdependence of processing, microstructure, property, stress, deformation, and performance of composites. An integrated and system approach to composites research requires the knowledge in the areas of processing science, materials design and modeling, mechanics, durability, and design science. The research results reviewed in this paper point out some recent developments in these areas. The engineering of better advanced composites can be accomplished only through the continuous optimization and integration of the knowledge gained in these areas.

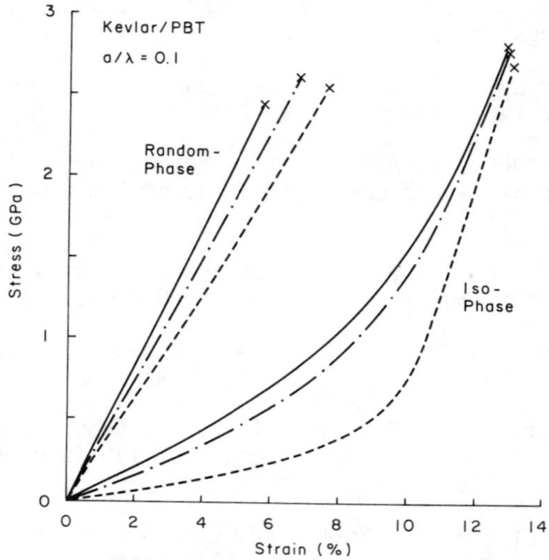

Fig. (9) - Stress-strain relations of Kevlar/PBT flexible composites [24]

REFERENCES

[1] Yau, S. C. and Chou, T. W., "Flexural fatigue of short fiber reinforced high temperature thermoplastics", SAMPE Quarterly, Vol. 16, No. 3, pp. 18-21, 1985.

[2] Hsu, P. L., Yau, S. S. and Chou, T. W., "Stress-corrosion cracking and its propagation in aligned short fiber composites", J. Mater. Sci., in press.

[3] Yau, S. S. and Chou, T. W., "Effects of low temperature on short fiber reinforced thermoplastics", to appear in the Proceedings of the Second Symposium on Test Methods and Design Allowables for Fiber Composites, ASTM, Phoenix, November 3-4, 1986.

[4] Chou, T. W. and Yang, J. M., "Structure-performance maps of polymeric, metal and ceramic matrix composites", to be published in Metallurgical Transaction, 1986.

[5] Ishikawa, T., "Anti-symmetric elastic properties of composite plates of satin weave cloth", Fiber Science and Technology, Vol. 15, pp. 127-145, 1981.

[6] Ishikawa, T. and Chou, T. W., "Stiffness and strength behavior of woven fabric composites", J. of Mat. Sci., Vol. 17, pp. 3211-3220, 1982.

[7] Ishikawa, T. and Chou, T. W., "One-dimensional micromechanical analysis of woven fabric composites", AIAA Journal, Vol. 21, pp. 1714-1721, 1983.

[8] Ishikawa, T. and Chou, T. W., "Stiffness and strength properties of woven fabric composites", Progress in Science and Engineering of Composite Materials, Proceedings of ICCM IV, 1982.

[9] Ishikawa, T. and Chou, T. W., "Elastic behavior of woven hybrid composites", J. of Comp. Mat., Vol. 16, pp. 2-19, 1983.

[10] Ishikawa, T. and Chou, T. W., "In-plane thermal expansion and thermal bending coefficients of fabric composites", J. of Comp. Mat., Vol. 17, pp. 92-104, 1983.

[11] Ishikawa, T. and Chou, T. W., "Thermoelastic analysis of hybrid fabric composites", J. of Mat. Sci., Vol. 18, pp. 2260-2268, 1983.

[12] Ishikawa, T. and Chou, T. W., "Nonlinear behavior of woven fabric composites", J. of Comp. Mat., Vol. 17, pp. 399-413, 1983.

[13] Ishikawa, T., Matsushima, M., Hayashi, Y. and Chou, T. W., "Experimental Confirmation of the theory of elastic moduli of fabric composites", J. of Comp. Mat., 19, p. 443, 1985.

[14] Chou, T. W., "Analysis of textile structural composites: an overview", in Proceedings of the Army Symposium on Solid Mechanics, Advances in Solid Mechanics for Design and Analysis, AMMRC, Watertown, Mass., 1984.

[15] Chou, T. W., Shehata, B. I. and Yang, J. M., "Analytical and experimental studies of textile structural composites", Proceedings of the International Conference on Rotorcraft Basic Research, American Helicopter Society, 1985.

[16] Yang, J. M., Ma, C. L. and Chou, T. W., "Elastic stiffness of biaxial and triaxial woven fabric composites", Proceedings of the 29th SAMPE National Symposium, Reno, 1984.

[17] Ma, C. L., Yang, J. M. and Chou, T. W., "Elastic stiffness of three-dimensional woven fiber composites", ASTM STP 893, Philadelphia, 1986.

[18] Yang, J. M., Ma, C. L. and Chou, T. W., "Fiber inclination model of three-dimensional textile structural composites", to be published in J. of Composite Materials, 1986.

[19] Yang, J. M. and Chou, T. W., "Thermo-elastic analysis of three-dimensional fabric composites", in Advances in Aerospace Science and Engineering, ASME, New York, 1984.

[20] Majidi, A. M., Yang, J. M., Pipes, R. B. and Chou, T. W., "Mechanical behavior of three-dimensional woven fiber composites", Proceedings of the Fifth International Conference on Composite Materials, San Diego, 1985.

[21] Chou, T. W., "Modeling and characterization of textile structural composites", Proceedings of the European Conference on Composite Materials", September 1985.

[22] Chou, T. W. and Yang, J. M., "Performance optimization of woven fabric composites for printed circuit boards", paper presented at the 1986 Materials Research Society Spring Meeting, Palo Alto, Ca., April 1986.

[23] Chou, T. W. and Takahashi, K., "Nonlinear elastic behavior of flexible fiber composites", submitted for publication.

[24] Takahashi, K. and Chou, T. W., "Modeling of the interfacial behavior of flexible composites", Proceedings of the Symposium on "Interface in Composites", 115th AIME Annual Meeting, New Orleans, March 1986.

ACOUSTIC EMISSION AND MICROFRACTURE OF A SOLID

Y.-H. Pao*

Institute of Applied Mechanics, National Taiwan University
Taipei, Taiwan, Republic of China

ABSTRACT

Experiments and theory of acoustic emission (AE) are reviewed, and their applications to investigate microfractures of a steel specimen is discussed in this lecture. Aided by a recently developed calibration procedure and a method for solving integral equations with multiple source parameters (deconvolution), the time function, crack orientation, and mode of failure of each microcrack can all be determined from the AE tests.

INTRODUCTION

Acoustic emission (AE) has been a promising technique for detecting and characterizing microcracks in solids. Early works on the theory, experiments, and applications of AE were reviewed in three articles in 1978 [1-3]. Recent works are reported, aside from many research journals, in the international symposium on acoustic emission held bi-annually [4].

In this lecture, we discuss some of the basic theories and experiments of AE, and their application to investigate the initiation and coalescing of microcracks (less than 1 mm) in a structural steel specimen in fracture toughness tests [5]. A similar investigation was made for an aluminum specimen in fatigue test [6]. These results [5,6] showed clearly that AE technique not only can locate precisely the microcracks relative to the main fractured surface, but also can determine the orientation and mode (tensile or shear failure) of the cracks. It supplies new information about the dynamics of fracture hitherto unavailable from other experimental techniques.

ACOUSTIC EMISSION TECHNIQUE

Acoustic signals emitted by microcracks in a solid propagate in the form of elastic waves which can be detected by piezoelectric transducers, or other sensors, placed on a surface of the solid, Figure 1. The voltage output of the transducer, usually in the order of microvolts, is amplified, sometimes filtered, and recorded for data processing.

---

*On leave of absence from Cornell University, Ithaca, New York 14853, USA.

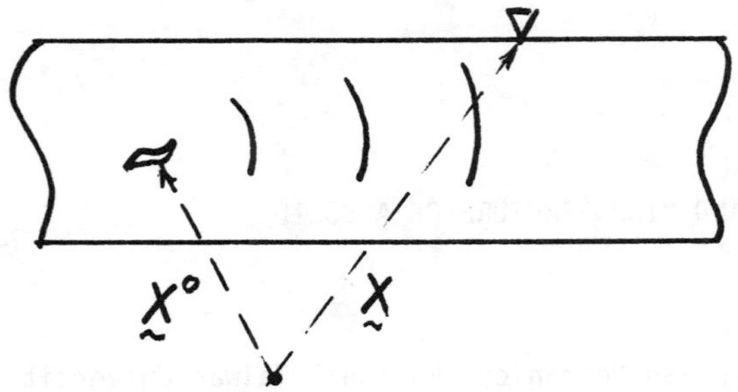

Fig. (1) - Source of emission at x^o and recording transducer at $\underset{\sim}{x}$ of an AE experiment

The analog data of voltage versus time are fed into various instruments for correction analysis, spectral analysis, or ring-down and event counting. The data can also be digitized through an analog to digit converter (A/D) and then stored in a computer for further processing. A typical arrangement of acoustic emission experiment is shown in Figure 2 [2].

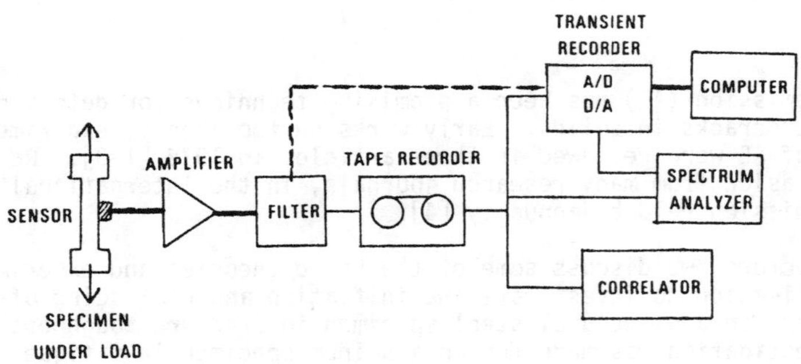

Fig. (2) - A typical arrangement of an AE experiment. (From [2])

In Figure 3, we show a simulated AE signal recorded by a piezoelectric transducer (2 mm diameter) which is placed at one side of a glass plate of 0.95 cm in thickness. The signal is generated by another transducer excited by a single sine pulse of 2 μsec in duration. The source transducer is placed at the other side of the plate, about 20 cm from the receiver. The recorded signal is so complex that it is usually classified as random. The objective of AE is to determine the location and characterization (time history, spatial distribution, orientation, etc.) of the source from such signals.

1. Source Location

If the onset of an AE signal can be distinctly identified, the location of an AE source at (x^o, y^o, z^o) can be determined from the data of four or more receivers

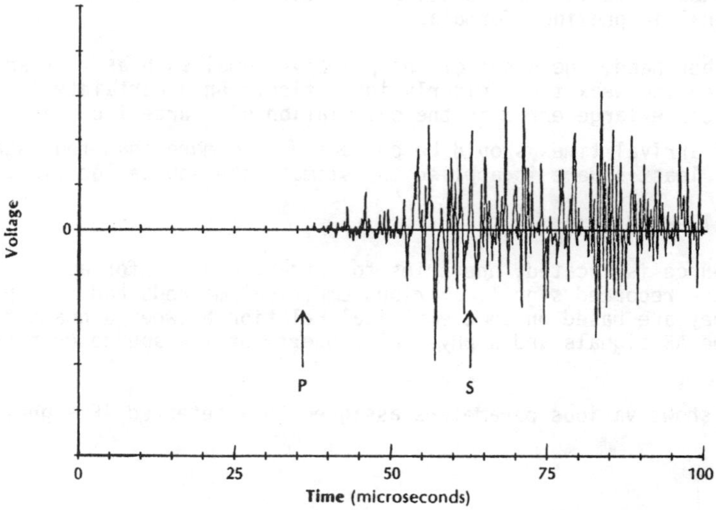

Fig. (3) - Acoustic emission by a simulated AE source in a glass plate using piezoelectric transducers as source (a sine pulse) and receiver (about 15 thickness from the source)

at positions (x_i, y_i, z_i) $(i = 1,2,3,...)$. Let t^o be the time of emission and t_i be the arrival time of the primary signal at each receiver. If the velocity of the primary wave, c_p, is known, the four unknowns x^o, y^o, z^o and t^o may be determined from the following set of equations, Figure 1.

$$[(x_i-x^o)^2 + (y_i-y^o)^2 + (z_i-z^o)^2]^{1/2} = c_p(t_i-t^o) \quad i = 1,2,3,4 \tag{1}$$

In some cases, two distinctive waveforms with arrival times, t_{ip} and t_{is} $(i = 1,2,...)$, are identifiable in all recorded signals. Let the primary signals and secondary signals travel with a constant speed c_p and c_s respectively. The source time t^o is thus eliminated as an unknown and the location is determinable from three equations:

$$[(x_i-x)^2 + (y_i-y)^2 + (z_i-z)^2]^{1/2} = (t_{is}-t_{ip}) \frac{c_p c_s}{c_p-c_s} \quad i = 1,2,3 \tag{2}$$

In most experiments, the c_p is taken as the speed of the longitudinal wave (P-wave, compressional wave) of an elastic solid, and c_s as the speed of the transverse wave (S-wave, shear wave). When the source and receiver are near the same surface, the Rayleigh surface wave which travels with a speed of about 0.92 of that for shear wave is often distinguishable. It can then be treated as the secondary signal in previous formula.

On the other hand, the onset of the primary signal such as that shown in Figure 3, is often too weak to be clearly identified. An uncertainty in the arrival time t_{ip} may cause large error in the calculation of source location. In that case, data of arrival times should be collected from more than four stations and the method of least square is applied to estimate the source location.

2. Signal Processing

Once a source is located, one tries to extract other information about the source from the recorded signal. Various empirical methods had been proposed in the past. They are based on some empirical relation between a quantitative parameter of the AE signals and a physical property of the source or propagating medium.

Figure 4 shows various parameters assigned to a detected AE signal [2]. An

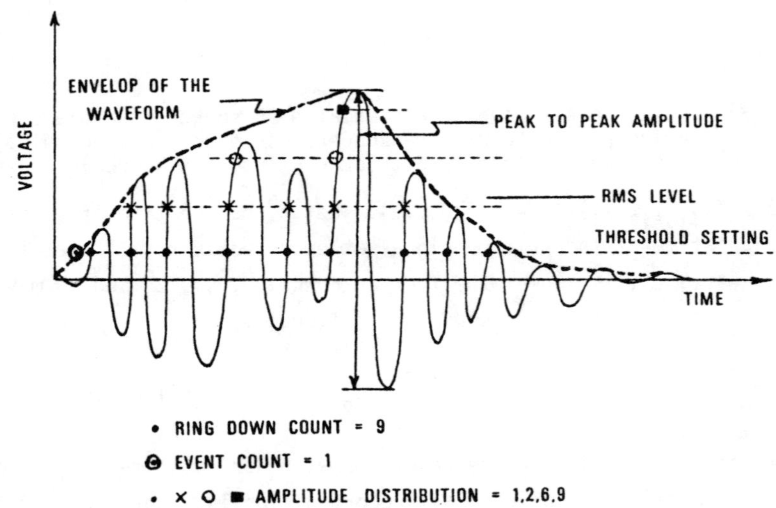

Fig. (4) - Schematic diagram showing various parameters assigned to a detected AE signal by various processing techniques. (From [2]).

event is defined by an envelop for a portion of the signal. Within an event, one can calculate the "peak to peak" amplitude "root mean square value" (RMS), or "amplitude distribution", and count the number of waves the amplitude of which exceed a preset threshold (ring down count). One may also evaluate the frequency spectrum of the event by taking the Fourier transform of the signal.

The event count of AE signals has been related to the stress-strain curve of a specimen in uniaxial test as shown in Figure 5 [5]; or to the S-N curve of a speci-

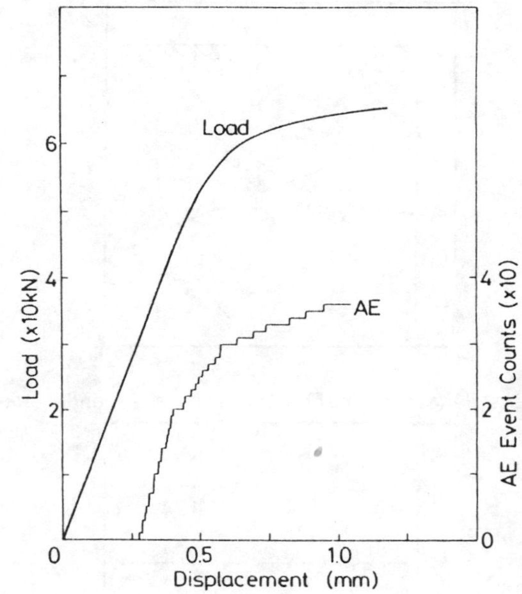

Fig. (5) - Relations between load, AE event count, and displacement. (From [5])

men in fatigue test. Exhibitions of such empirical relations have drawn a lot of attention to the AE technique in early days. Many were disappointed later, however, because the data were unreproducible.

3. Elastic Waves in a Wave-Guide

The difficulty of extracting useful information from an AE signal may be seen from the analytical results of elastic waves in a plate. A center of dilatation which simulates the collapsing of a void or creation of a vacancy generates a P-wave in an elastic solid. Upon impinging a boundary, the P-wave is reflected into two waves, a P-wave and a S-wave as shown in Figure 6. Each of these two waves is reflected into two more by another boundary. The multiple reflections and scattering often distort completely the original wave form of the source.

Shown in Figure 7 are three theoretical wave forms generated by the same source and received at the same location [7]. The first one is for the source in an unbounded medium, the second in a half space while the receiver is at the surface, and the third in an infinite elastic plate bounded by two parallel planes. Although the time function for the source is a single triangular pulse, the third wave form may be construed as one event, two events, three events,..., etc., which renders the event count and ring-down count totally meaningless.

Applications of the theory of elastodynamics and gradual development of analytical methods for solving inverse problems of elastic waves have added a new dimension to the AE technology. In the following, we discuss the theory of radia-

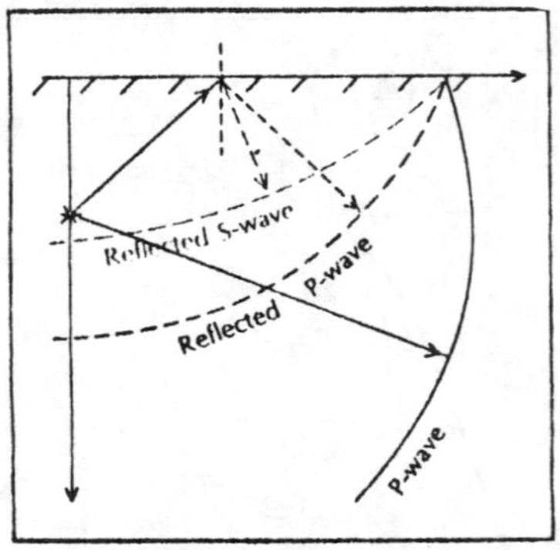

Fig. (6) - An incident P-wave is reflected into a P- and S-wave at a boundary

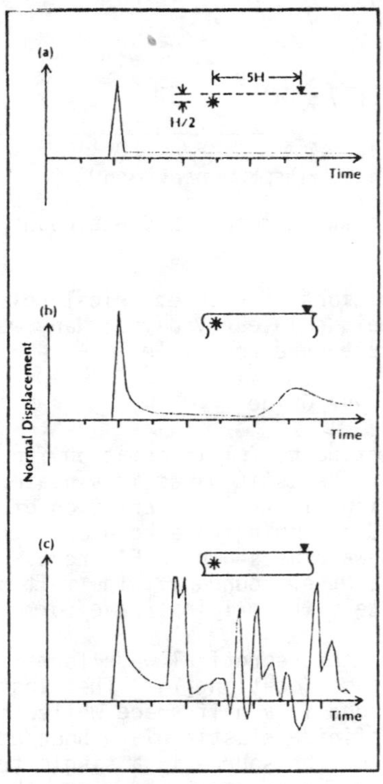

Fig. (7) - Acoustic emission by a point source. (a) in an infinite medium; (b) in a half space with later arrivals of S-wave and the Rayleigh wave; (c) in a large plate (Lamb wave)

tion by a crack and its application to determine the crack initiation in a steel specimen.

RADIATION OF ELASTIC WAVES BY A CRACK

Elastic waves emitted by a fractured surface may be analyzed according to the theory of elastodynamics [8,9]. In the theory, a crack is described by a surface of discontinuity across which the displacement vector $u(x,t)$ is discontinuous. The jump of displacement across a surface $A(x)$ is defined as

$$[|u(x,t)|] \equiv u^{+}(x,t) - u^{-}(x,t) \tag{3}$$

where u^{+} is on the positive side indicated by the positive direction of a unit normal $n(x)$ to A, and u^{-} is on the negative side, Figure 8. Combined with the

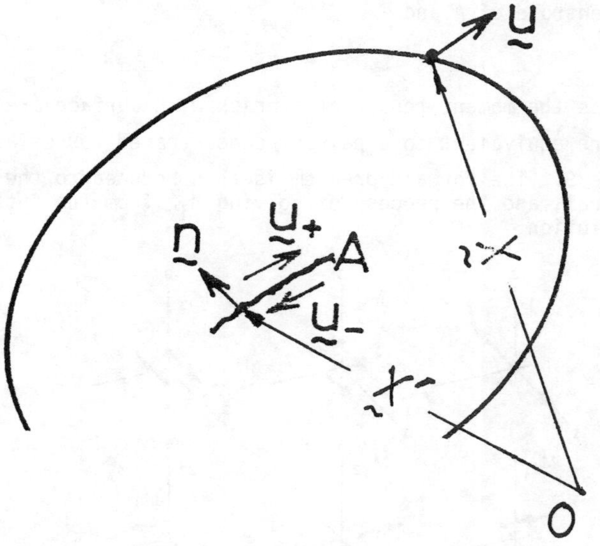

Fig. (8) - A source of displacement jump across a cracked surface A

elastic moduli tensor C of the medium, the unit normal n and jump $[|u|]$ form a second order tensor known as the moment density tensor m,

$$m_{jk}(x,t) = C_{jk\ell m} n_m(x)[|u_\ell(x,t)|] \tag{4}$$

If $G_{ij}(x-x', t-t')$ denotes Green's displacement tensor, which is the displacement at x and t in the direction of x_i due to an impulsive point force applied at x' and t' in the direction of x_j, the waves at (x,t) generated by a crack at x', t' are given by

$$u_i(x,t) = \int_0^t dt' \iint_A m_{jk}(x',t')\partial'_k G_{ij}(x-x',t-t')dS' \tag{5}$$

where $\partial'_k \equiv \partial/\partial x'_k$. Knowing the Green's tensor $\underset{\approx}{G}$, one may evaluate the wave field $\underset{\sim}{u}$ when $\underset{\approx}{m}$ is given. This is called the *forward problem*. Conversely, when $\underset{\sim}{u}$ is given, one may determine the source function $\underset{\approx}{m}$. This is known as the *inverse problem*.

When the distance between the source and receiver $|x-x'|$ is much larger than the characteristic length of the cracked surface, the area integral over A is approximated by its mean value,

$$u_i(\underset{\sim}{x},t) \approx \int_0^t M_{jk}(\underset{\sim}{x}^o,t')\partial_k G_{ij}(\underset{\sim}{x}-\underset{\sim}{x}^o,t-t')dt' \qquad (6)$$

where $\underset{\sim}{x}^o$ is the centroid of A and

$$M_{jk} = m_{jk}A \qquad (7)$$

The M_{jk} is known as the moment tensor of a crack with surface area A. The nine components of $\underset{\approx}{M}$ are equivalent to a pair of concentrated forces acting at a point as shown in Figure 9. The forward problem is then reduced to the evaluation of a convolution integral, and the process of solving M_{jk} from the integral equation is called deconvolution.

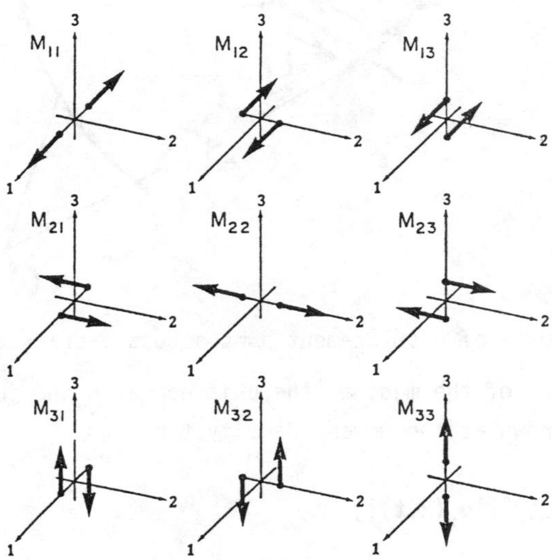

Fig. (9) - Nine components of a moment tensor M_{ij}

Equation (6) forms the basis for determining the AE source characteristics, assuming its location x^o has been determined. The solving of M_{jk} by deconvolution requires precise measurements of u_i at many locations $\underset{\sim}{x}$, and a priori knowledge of Green's tensor $\partial_k G_{ij}$.

The Green's tensors are very difficult to obtain and are reported only for elastic waves in an unbounded medium, a half space, an infinite plate [10], and a sphere. The precise measurement of u_i over a long duration requires careful calibration of transducers.

The solving of M_{jk} in equation (6) is by no means simple. Conventional deconvolutions involve only one unknown function in the integral equation, but equation (6) contains a sum of six unknowns (M_{jk} are symmetric) with six kernel functions. This is simplified by first assuming a common time function S(t) for all six components

$$M_{jk}(\underset{\sim}{x}^o,t) = \hat{M}_{jk}(\underset{\sim}{x}^o)S(t) \tag{8}$$

Equation (6) is then reduced to

$$u^{(i)}(\underset{\sim}{x},t) = \int_o^t \hat{M}_{jk}S(t')\partial_k G_j^{(i)}(\underset{\sim}{x}-\underset{\sim}{x}^o,t-t')dt' \tag{9}$$

The index i is replaced by the superscript (i) so as to emphasize that only one component, a scalar quantity, is involved in the deconvolution. The unknown time function S(t) and six coefficients $\hat{M}_{jk}$ can then be solved by a double iterative procedure [11].

Once the $\hat{M}_{jk}$ are determined, the $\underset{\sim}{n}$ and $[|u|]$ are found by solving a system of algebraic equations (4). The entire process of AE experiments, calibration, and deconvolution has been applied to monitor the fracture of a steel specimen by Ohira and Pao [5]. Their work is described in the following section.

CHARACTERIZING MICROCRACKS BY AE

In the experiment conducted by Ohira and Pao [5], an eight channel AE system was used to record AE signals of a steel specimen during fracture toughness test. The dimension of the specimen (A533B, a medium strength structural steel) and locations of transducers are shown in Figure 10. The transducers are made of thin and long piezoelectric cylinders, each is connected to an amplifier, a transient recorder (digitizer) and a common computer as shown in Figure 2. A pre-crack is generated at the tip of the notch by fatigue test, and statical loads were applied gradually to the specimen. The signals emitted by the microcracks during loading were orded simultaneously by all transducers.

Two distinctly different types of wave forms were recorded. As shown in Figure 11, Type 1 wave is in the form of sharper rising, rapid oscillation, and high frequency in content; Type II wave is in the form of gradual rising, slowly changing, and low-frequency in content. For each event of emission, all transducers recorded the same type of waves.

Since the onset of each wave is clearly identifiable from the record and wave speed, c_p, is known, the location of each source of emission is determined by applying equation (1). Positions of all sources so determined during loading are shown in Figure 12. Because the transient recorder is inactive while transferring

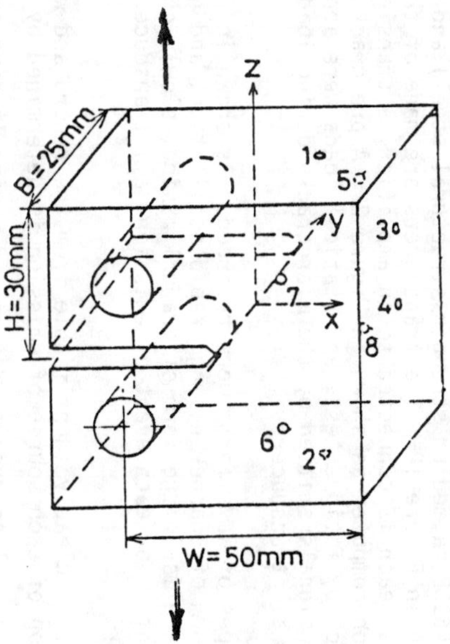

Fig. (10) - Notched specimen and positions of transducers

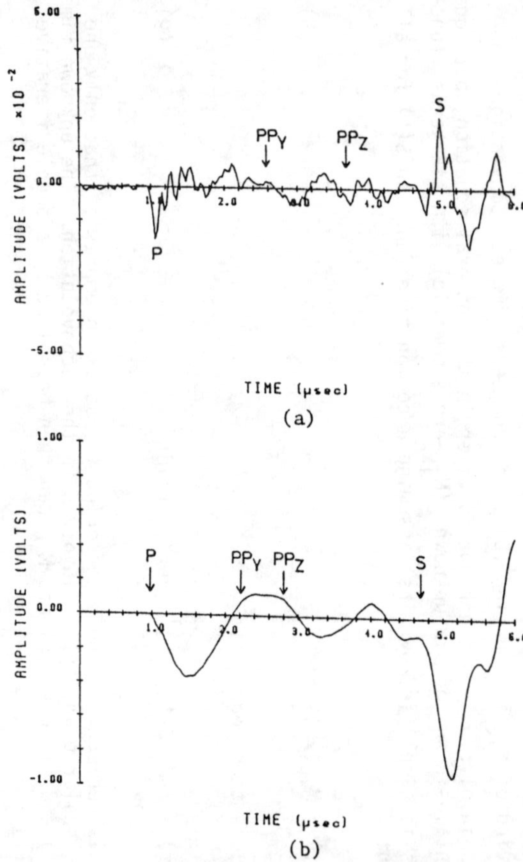

Fig. (11) - Typical wave forms. (a) Type I wave form; (b) Type II wave form. (From [5])

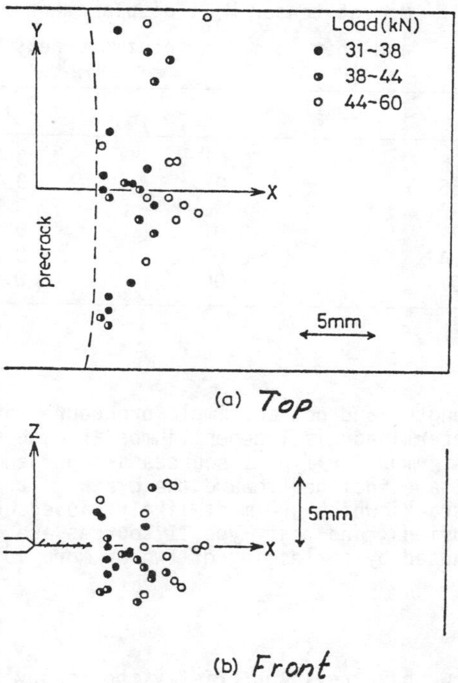

Fig. (12) - Locations of AE sources. (From [5])

data into the computer (a few sec. dumping time), some of the AE events might have been missed in the recording.

We note that all sources of acoustic emission are situated ahead of the edge of the precrack and outside the "yield zone" (not shown) around the edge. Since there exist in the A533B steel microinclusions of MnS, the AE events are generated by microcracks around the MnS inclusions. This is verified partially from the voids around MnS inclusions observed near the surface of total fracture of the specimen after the test.

The voltage output of the transducers are converted to normal displacement signals by transfer functions which are obtained by a sequence of calibration tests done on the specimen. Convolutions of the transfer functions with the wave forms shown in Figure 11 yield $u^{(i)}(x,t)$ in equation (9) at all receivers; theoretical analysis furnishes the Green's functions $\partial_k G^i_j(t)$. The unknown coefficients $\hat{M}_{jk}$ and source time functions S(t) are then determined by a double iterative deconvolution process [11]. Sample results are shown in Table 1 for $\hat{M}_{jk}$. In the table, the components for $[|u|]$ and $\tilde{n}$ in column matrices which are calculated from equations (4) and (7) are also shown.

Table 1 - Moment tensor $\hat{M}_{jk}$, displacement jump u_ℓ, and unit normal n_m of two cracks

$\hat{M}_{jk}$			u_ℓ	n_m
0.59	0.15	0.01	0.10	0.10
0.15	0.66	0.07	0.23	0.23
0.01	0.07	1.00	0.97	0.97
0.46	0.23	-0.01	0.25	-0.10
0.23	0.64	0.07	0.10	0.35
-0.01	0.07	1.00	0.96	0.93

Repeating this lengthy and not so simple procedure for each event recorded, Ohira and Pao [12] determined displacement jumps at each source point and found that the displacement jumps for Type I sources are perpendicular to the crack, and those for Type II are inclined toward the crack. It implies that Type I sources are tensile cracks which are most likely caused by decohesion at the surface of microinclusions, and that Type II sources are closer to shear cracks which are probably caused by coalescing of two or more voids around the microinclusions.

CONCLUSION

In this lecture, we have reviewed briefly the theory and experiments of acoustic emission and discussed how they are applied to determine locations and characteristics of microcracks in a steel specimen during fracture toughness tests. Recent advancement in the analysis of inverse problems of elastodynamics has made it possible to determine the time function, orientation, and displacement discontinuity at each microcrack (a single source) from an AE experiment.

The results reported by Ohira and Pao [5,12] and Scruby [6] showed clearly that the acoustic emission technique is effective in the investigation of the dynamics of fracture. Additional discussions of applying AE techniques to fracture of solids are given in the review article by Scruby [13].

REFERENCES

[1] Spanner, J. C., "Acoustic emission - applications and trends", in Elastic Waves and Non-Destructive Testing of Materials, (ed. Y. H. Pao), AMD, Vol. 29, American Society of Mechanical Engineers, New York, pp. 71-84, 1978.

[2] Hsu, N. N. and Hardy, S. C., "Experiments in acoustic emission waveform analysis for characterization of AE sources, sensors and structures", ibid, pp. 85-106, 1978.

[3] Pao, Y. H., "Theory of acoustic emission", ibid, pp. 107-128, 1978.

[4] Onoe, M., Yamaguchi, K. and Takahashi, H., (eds.), Progress in Acoustic Emission II, Proceedings of International Acoustic Emission Symposium, The Japanese Society of Non-Destructive Inspection, Tokyo, Japan, 1984.

[5] Ohira, T. and Pao, Y. H., "Microcrack initiation and acoustic emission during fracture toughness tests of A533B steel", Metallurgical Transactions, Vol. 17A, pp. 843-851, 1986.

[6] Scruby, C. B., "Defect characterization and monitoring by acoustic emission", Review of Progress in Quantitative Non Destructive Evaluation, Vol. 4B, (eds. D. O. Thompson and D. E. Chimenti), Plenum Press, New York, pp. 643-650, 1985.

[7] Pao, Y. H., Gajweski, R. R. and Ceranoglu, A. N., "Acoustic emission and transient waves in an elastic plate", Journal of Acoustical Society of America, Vol. 65 (1), pp. 96-105, 1979.

[8] Aki, K. and Richards, P. G., Quantitative Seismology, Theory and Methods, Vols. I, II, F. H. Freeman Co., San Francisco, 1980.

[9] Burridge, R. and Knopoff, L., "Body force equivalent for seismic dislocations", Bulletin of Seismological Society of America, Vol. 54, pp. 1875-1888, 1964.

[10] Ceranoglu, A. N. and Pao, Y. H., "Propagation of elastic pulses and acoustic emission in a plate, part I, II, III", Journal of Applied Mechanics, Trans. ASME, Vol. 48, pp. 125-147, 1981.

[11] Michaels, J. E. and Pao, Y. H., "The inverse problems for an oblique force on an elastic plate", Journal of Acoustic Society of America, Vol. 77, pp. 2005-2011, 1985.

[12] Ohira, T. and Pao, Y. H., "Quantitative characterization of microcracking in a steel specimen by acoustic emission", Materials Science Center, Report No. 5863, Cornell University, 1986.

[13] Scruby, C. B., "Quantitative acoustic emission techniques", in Research Techniques in Non-Destructive Testing, Vol. 8, (ed. R. S. Sharpe), Academic Press, New York, pp. 141-210, 1985.

SLIDING ELLIPSOIDAL INHOMOGENEITIES UNDER SHEAR*

T. Mura, R. Furuhashi and T. Mori

Department of Civil Engineering, Northwestern University
Evanston, Illinois 60201 USA

ABSTRACT

It was found by the authors [1] that when an ellipsoidal inclusion (with the same elastic moduli as the matrix) is subjected to a uniform shear-type eigenstrain in the directions of the principal axes of the ellipsoid and cannot sustain any shear traction along the interface, the stress field vanishes everywhere by leaving a slide along the interface.

In this paper, we consider an ellipsoidal inhomogeneity under shear and investigate constitutive equations of composite materials as an application of this unexpected result of the sliding shear-type inclusion.

INTRODUCTION

It was found by Mura and Furuhashi [1] that an ellipsoidal inclusion with a uniform shear-type eigenstrain ε^*_{ij} in its principal directions does not cause any stress field if the inclusion can slide along the interface and cannot sustain any interfacial (tangential) shear stress. This result is contrastive with the perfectly bonded inclusion of Eshelby where the stress is uniform in the inclusion.

Mura and Furuhashi's theory is limited to small deformations and rotations and is subjected to the condition $a_i \neq a_j$, where a_1, a_2 and a_3 are principal half-axes of the ellipsoid. The discussion in this paper, therefore, is subjected to the same limitations. For non-shear-type eigenstrains, the situation is more complicated, as seen from papers by Ghahremani [2], Mura, Jasiuk and Tsuchida [3] and Tsuchida, Mura and Dundurs [4]. It is worthwhile, however, to pursue the subject further on the basis of Mura and Furuhashi's results.

The most interesting application may be found in composite materials. Average mechanical properties of composite materials can be evaluated from mechanical properties of elements composing the composite materials. Many papers are available in this field; however, most of them are based upon the assumption that the inhomogeneities or inclusions are perfectly bonded to the matrix. A few papers

---

*This research was supported by U.S. Army Research Grant DAAG-29-85-K-0134.

have considered the sliding and debonding of interfaces, but none of them have incorporated the important property of the ellipsoidal inhomogeneity mentioned in this papaer.

ELLIPSOIDAL INHOMOGENEITY

Consider an ellipsoidal inhomogeneity Ω with principal half-axes a_1, a_2 and a_3 in an infinite elastic medium. A shear stress σ^o_{ij} is applied at infinity, where the shear directions coincide with the principal axis direction of the ellipsoid. Let the elastic moduli of the matrix D-Ω and the ellipsoid be $C_{ijk\ell}$ and and $C^*_{ijk\ell}$, respectively, (see Figure 1). We assume further that the ellipsoidal inhomogeneity Ω slides along the interface by $\underset{\sim}{b}$ and that the uniform shear-type plastic strain ε^p_{ij} is caused in Ω.

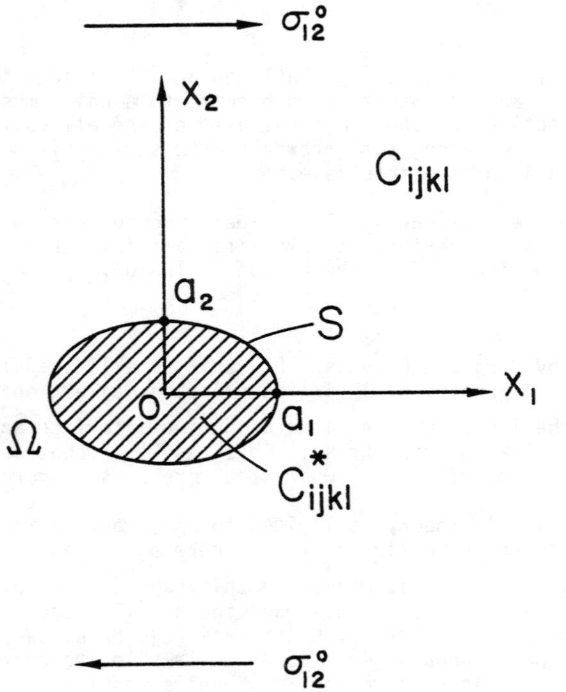

Fig. (1) - Ellipsoidal inhomogeneity under shear

The applied shear stress σ^o_{ij} causes displacement u^o_i when no inhomogeneity exists. However, the displacement field is disturbed by the inhomogeneity of an ellipsoid, plastic deformation in Ω, and slip $\underset{\sim}{b}$ on S. When the disturbed displacement and stress are denoted by u_i and σ_{ij}, we have Hooke's law,

$$\sigma^o_{ij} = C_{ijk\ell} u^o_{k,\ell} \text{ in D} \tag{1}$$

$$\sigma^o_{ij} + \sigma_{ij} = C_{ijk\ell}(u^o_{k,\ell} + u_{k,\ell}) \text{ in } D\text{-}\Omega \tag{2}$$

and

$$\sigma^o_{ij} + \sigma_{ij} = C^*_{ijk\ell}(u^o_{k,\ell} + u_{k,\ell} - \varepsilon^p_{k\ell}) \text{ in } \Omega \tag{3}$$

The surface traction on S must be continuous across the interface,

$$[\sigma_{ij}]n_j = 0 \tag{4}$$

where $[Q] = Q(\text{out}) - Q(\text{in})$ on S, $Q(\text{out})$ is a quantity defined in $D\text{-}\Omega$ on S and $Q(\text{in})$ is that in Ω on S, and n_i is the outward unit vector normal to S.

$\underset{\sim}{b}$ is the discontinuity of tangential displacement on S, namely

$$[u_i] = -b_i \tag{5}$$

and

$$b_i n_i = 0 \tag{6}$$

The last condition (6) is satisfied when $\underset{\sim}{b}$ is chosen as

$$b_i(\underset{\sim}{x}) = \varepsilon^*_{ij}x_j - \omega_{ij}x_j \tag{7}$$

where

$$\omega_{12} = \varepsilon^*_{12}(1/a_1^2 + 1/a_2^2)/(1/a_1^2 - 1/a_2^2) = -\omega_{21}$$

$$\omega_{23} = \varepsilon^*_{23}(1/a_2^2 + 1/a_3^2)/(1/a_2^2 - 1/a_3^2) = -\omega_{32} \tag{8}$$

$$\omega_{31} = \varepsilon^*_{31}(1/a_3^2 + 1/a_1^2)/(1/a_3^2 - 1/a_1^2) = -\omega_{13}$$

In the expressions above, ε^*_{ij} are constant tensor components which must be determined under several physical conditions on S and $\varepsilon^*_{11} = \varepsilon^*_{22} = \varepsilon^*_{33} = 0$. Generally, ε^*_{ij} are not constant and are functions of position and time. In this paper, we restrict ourselves to the case where ε^*_{ij} are constant but can be functions of time; otherwise, we cannot obtain analytical solutions.

Slip $\underset{\sim}{b}$ expressed by (7) with (8) has the interesting property of transforming an ellipsoid

$$x_1^2/a_1^2 + x_2^2/a_2^2 + x_3^2/a_3^2 = 1 \tag{9}$$

into the identical ellipsoid

$$(x_1')^2/a_1^2 + (x_2')^2/a_2^2 + (x_3')^2/a_3^2 = 1 \tag{10}$$

in the framework of the small deformation theory, where $x_i' = x_i + b_i$. Therefore, the deformation of the ellipsoid with slip $\underset{\sim}{b}$ does not cause any misfit relative to the matrix.

Slip $\underset{\sim}{b}$ is regarded as Somigliana's dislocation and the corresponding eigendistortion is expressed by

$$\beta^*_{ji}(\underset{\sim}{x}) = -b_i n_j \delta(S-\underset{\sim}{x}) \tag{11}$$

where $\delta(S-\underset{\sim}{x})$ is the Dirac delta function defined on the interface S and

$$\beta^*_{ji}(\underset{\sim}{x})d\underset{\sim}{x} = -b_i n_j dS \tag{12}$$

where $d\underset{\sim}{x} = dx_1 dx_2 dx_3$. This equation will be used later for the evaluation of the deformation contributed by $\underset{\sim}{b}$.

EQUIVALENT INCLUSION METHOD

Stress (3) is simulated by an ellipsoidal inclusion with uniform eigenstrains ε^{**}_{ij} in a homogeneous infinite body where the inclusion slides by $\underset{\sim}{b}$ given by (7). This equivalent inclusion method requires that in Ω

$$\sigma^o_{ij} + \sigma_{ij} = C^*_{ijk\ell}(u^o_{k,\ell} + u_{k,\ell} - \varepsilon^p_{k\ell}) = C_{ijk\ell}(u^o_{k,\ell} + u_{k,\ell} - \varepsilon^{**}_{k\ell}) \tag{13}$$

The displacement u_i is assumed to be equivalent to that caused by the sliding inclusion with eigenstrain ε^{**}_{ij} in the homogeneous material. u_i is evaluated from Volterra's formula (see Mura [5], pp. 33 and 39),

$$u_i(\underset{\sim}{x}) = -\int_\Omega C_{j\ell mn}\varepsilon^{**}_{mn} G_{ij,\ell}(\underset{\sim}{x}-\underset{\sim}{x}')d\underset{\sim}{x}' + \int_S C_{j\ell mn} b_m(\underset{\sim}{x}')G_{ij,\ell}(\underset{\sim}{x}-\underset{\sim}{x}')n_n dS(\underset{\sim}{x}') \tag{14}$$

where $G_{ij}(\underset{\sim}{x}-\underset{\sim}{x}')$ is Green's function. By integration by parts, (14) is written as

$$u_i(\underset{\sim}{x}) = -\int_\Omega C_{j\ell mn}(\varepsilon^{**}_{mn} - b_{m,n})G_{ij,\ell}(\underset{\sim}{x}-\underset{\sim}{x}')d\underset{\sim}{x}' + b_i(\underset{\sim}{x}) \tag{15}$$

where $b_{m,n} = \varepsilon^*_{mn} - \omega_{mn}$ from (7).

The first integral in (15) was explicitly evaluated by Eshelby [6] for the isotropic medium. It becomes a linear function of $\underset{\sim}{x}$, and the strain in Ω becomes constant,

$$(1/2)(u_{i,j} + u_{j,i}) = S_{ijk\ell}(\varepsilon^{**}_{k\ell} - \varepsilon^*_{k\ell}) + \varepsilon^*_{ij} \tag{16}$$

where $S_{ijk\ell}$ is Eshelby's tensor (see Mura [5], p. 66). When (16) is substituted into (13),

$$\begin{aligned}\sigma^o_{ij} + \sigma_{ij} &= C^*_{ijk\ell}(u^o_{k,\ell} + S_{k\ell mn}(\varepsilon^{**}_{mn} - \varepsilon^*_{mn}) + \varepsilon^*_{k\ell} - \varepsilon^p_{k\ell}) \\ &= C_{ijk\ell}(u^o_{k,\ell} + S_{k\ell mn}(\varepsilon^{**}_{mn} - \varepsilon^*_{mn}) + \varepsilon^*_{k\ell} - \varepsilon^{**}_{k\ell})\end{aligned} \tag{17}$$

When σ^o_{ij}, ε^p_{ij} and ε^*_{ij} are given, ε^{**}_{ij} is determined from (17). The displacement disturbance u_i can be determined from (15) after knowing ε^{**}_{ij}.

Let us discuss (17) for some special cases. When $C^*_{ijk\ell} = C_{ijk\ell}$ and $u^o_{k,\ell} = 0$, we have the case Mura and Furuhashi [1] have considered, where $\varepsilon^{**}_{ij} = \varepsilon^p_{ij}$ holds. Therefore, when ε^*_{ij} is taken as ε^p_{ij}, the stress field vanishes everywhere. This means that the shear-type plastic strain does not cause any stress field except that slip $b_i = \varepsilon^p_{ij}x_j - \omega_{ij}x_j$ is left along the interface. This result is also valid even when $C^*_{ijk\ell} \neq C_{ijk\ell}$. It can be seen from the left-hand side of (17) when $u^o_{k,\ell} = 0$, $\varepsilon^{**}_{mn} = \varepsilon^p_{mn}$ and $\varepsilon^*_{k\ell} = \varepsilon^p_{k\ell}$.

INTERFACIAL CONDITIONS

Slip b_i or ε^*_{ij} given by (7) and (8) is determined by interfacial conditions. The simplest condition is that the interface cannot sustain any tangential shear traction. This condition is expressed as

$$(\sigma^o_{ij} + \sigma_{ij})n_j - (\sigma^o_{jk} + \sigma_{jk})n_j n_k n_i = 0 \tag{18}$$

where n_i is the unit vector normal to the interface S. Since σ^o_{ij} and σ_{ij} are uniform shear in Ω and since n_i is a function of positions, the above condition is satisfied only when $\sigma^o_{ij} + \sigma_{ij} = 0$, i.e., the stress is identically zero in Ω. This means that the inhomogeneity behaves like a void (private discussion with J. Dundurs, 1983). The slip necessary for this complete relaxation is obtained when equation (17) is equal to zero, since the vanishing of stress is equivalent to the vanishing of the elastic strain,

$$\begin{aligned} &\varepsilon^o_{k\ell} + S_{k\ell mn}(\varepsilon^{**}_{mn} - \varepsilon^*_{mn}) + \varepsilon^*_{k\ell} - \varepsilon^p_{k\ell} = 0 \\ &\varepsilon^o_{k\ell} + S_{k\ell mn}(\varepsilon^{**}_{mn} - \varepsilon^*_{mn}) + \varepsilon^*_{k\ell} - \varepsilon^{**}_{k\ell} = 0 \end{aligned} \tag{19}$$

where $\varepsilon^o_{ij} = \frac{1}{2}(u^o_{i,j} + u^o_{j,i})$. In the above equations, $\varepsilon^o_{k\ell}$, $\varepsilon^p_{k\ell}$ and $S_{k\ell mn}$ are known. The solutions for (19) are

$$\varepsilon^{**}_{k\ell} = \varepsilon^p_{k\ell} \tag{20}$$

and

$$\begin{aligned} \varepsilon^*_{12} &= \varepsilon^p_{12} - \varepsilon^o_{12}/(1-2S_{1212}) \\ \varepsilon^*_{23} &= \varepsilon^p_{23} - \varepsilon^o_{23}/(1-2S_{2323}) \\ \varepsilon^*_{31} &= \varepsilon^p_{31} - \varepsilon^o_{31}/(1-2S_{3131}) \end{aligned} \tag{21}$$

where $1 - 2S_{ijk\ell} > 0$.

If no sliding is allowed on S, ε^*_{ij} must be zero and ε^{**}_{ij} is obtained from (17) for given ε^p_{ij} and ε^o_{ij}, i.e., Eshelby's equivalent inclusion method for a perfectly bonded inhomogeneity.

When the inhomogeneous inclusion is allowed to slip under the applied shear stress σ^o_{ij}, the initial state of the inclusion is the perfectly bonded state and the final state is the relaxed state, where the tangential shear traction disappears. At a time t = t between the initial and final states, the slip quantity b_i may be subjected to a relaxation law, for instance,

$$\frac{d}{dt}(b_i) = (b^*_i - b_i)/\tau \tag{22}$$

where b^*_i is given by (21), (7) and (8) and τ is a relaxation time. The solution of (22) is

$$b_i = b^*_i \{1 - \exp(-t/\tau)\} \tag{23}$$

and the corresponding ε^*_{ij} is

$$\varepsilon^*_{ij} = \{\varepsilon^p_{ij} - \varepsilon^o_{ij}/(1 - 2S_{ij})\}\{1 - \exp(-t/\tau)\} \tag{24}$$

where $S_{12} = S_{1212}$, etc.

CONSTITUTIVE EQUATIONS

Let us consider the constitutive equations of a body which contains many ellipsoidal inhomogeneities with the same orientation and is subjected to an applied shear stress σ^o_{ij} in the direction of the principal axis of the ellipsoids. Assume that the inhomogeneity Ω (the sum of the ellipsoidal inhomogeneities) is plastically weak compared with the matrix D-Ω and that the ellipsoids can slide through the law described by (22).

Suppose that the plastic strain ε^p_{ij} is caused instantaneously in Ω by σ^o_{ij} at t = 0. The inhomogeneous inclusions slip by b_i along the interfaces at t = t. The stress disturbance σ_{ij} at time t is caused by the inhomogeneity of Ω, ε^p_{ij} in Ω, and b_i on S. σ_{ij} has the property

$$\int_D \sigma_{ij} d\underset{\sim}{x} = 0 \tag{25}$$

since σ_{ij} is an eigenstress. The average value of the corresponding elastic strain also vanishes,

$$\int_D \{\frac{1}{2}(u_{i,j} + u_{j,i}) - \varepsilon^{**}_{ij} - \frac{1}{2}(\beta^*_{ji} + \beta^*_{ij})\} d\underset{\sim}{x} = 0 \tag{26}$$

Thus, from (12), we have

$$\int_D \frac{1}{2}(u_{i,j} + u_{j,i}) d\underset{\sim}{x} = \int_\Omega \varepsilon^{**}_{ij} d\underset{\sim}{x} - \int_S \frac{1}{2}(b_i n_j + b_j n_i) dS(\underset{\sim}{x}) \tag{27}$$

The surface integral, the last term in (27), is transformed into $\int_\Omega \varepsilon^*_{ij} d\underset{\sim}{x}$. By denoting the volume fraction of Ω by f, the average strain γ_{ij} in the material D is written as the sum of ε^o_{ij} and the value evaluated from (27), i.e.,

$$\gamma_{ij} = \varepsilon^o_{ij} + f(\varepsilon^{**}_{ij} - \varepsilon^*_{ij}) \tag{28}$$

since ε^{**}_{ij} and ε^*_{ij} are uniform in Ω. Another uniform strain $\bar{e}_{ij}$ must be added to (28) when the interaction between inclusions and the back stress is taken into account. According to Tanaka and Mori [7], the back stress $\bar{\sigma}_{ij}$ is uniform and given by

$$\bar{\sigma}_{ij} = C_{ijk\ell}\bar{e}_{k\ell} = -f\sigma^\infty_{ij} \tag{29}$$

where σ^∞_{ij} is σ_{ij} in Ω and is given by (17). Then

$$\bar{e}_{ij} = -f\{S_{ijmn}(\varepsilon^{**}_{mn} - \varepsilon^*_{mn}) + \varepsilon^*_{ij} - \varepsilon^{**}_{ij}\} \tag{30}$$

When (30) is added to (28), we have

$$\gamma_{ij} = \varepsilon^o_{ij} + 2f(\varepsilon^{**}_{ij} - \varepsilon^*_{ij}) - fS_{ijmn}(\varepsilon^{**}_{mn} - \varepsilon^*_{mn}) \tag{31}$$

The relation (constitutive equation) between σ^o_{ij} and γ_{ij} is obtained from (31). Let us consider some examples.

<u>Case 1</u>. $C^*_{ijk\ell} = C_{ijk\ell}$ and $\varepsilon^p_{ij} = 0$.

Non-elastic deformation is caused only by $\underset{\sim}{b}$ on S. $\underset{\sim}{b}$ is caused by σ^o_{ij} under the assumption that the interface is weak and slides according to (23). Since the material is homogeneous, $\varepsilon^{**}_{ij} = 0$. After complete relaxation at $t = \infty$, ε^*_{ij} is given by (21). When the applied stress is σ^o_{12}, we have from (31), (24) and (17),

$$\gamma_{12} = \sigma^o_{12}/2\mu - 2f(1-S_{1212})\varepsilon^*_{12}$$

$$\varepsilon^*_{12} = -\{\sigma^o_{12}/2\mu(1-2S_{1212})\}\{1 - \exp(-t/\tau)\} \tag{32}$$

$$\sigma^o_{12} + \sigma_{12} = 2\mu(\varepsilon^o_{12} - 2S_{1212}\varepsilon^*_{12} + \varepsilon^*_{12})$$

The relation between σ^o_{12} and γ_{12} is shown in Figure 2. Line OA is achieved at $t = 0$. Deformation AB is caused by the sliding on S. The strain at B is achieved at $t = \infty$. The stress $\sigma^o_{12} + \sigma_{12}$ in Ω is less than σ^o_{12}, and it becomes zero at B. When unloaded, point C in Figure 2 is achieved instantaneously and the original virgin state O is recovered at $t = \infty$ after the unloading. At the moment of unloading C, the stress in Ω is $-\sigma^o_{12}$. The stress will be relaxed in the opposite direction. When ε^*_{12} becomes $\varepsilon^o_{12}/(1-2S_{1212})$, the stress in Ω and D-Ω becomes zero.

When the relaxation time τ is very close to zero, the stress-strain curve may may be the one shown in Figure 3.

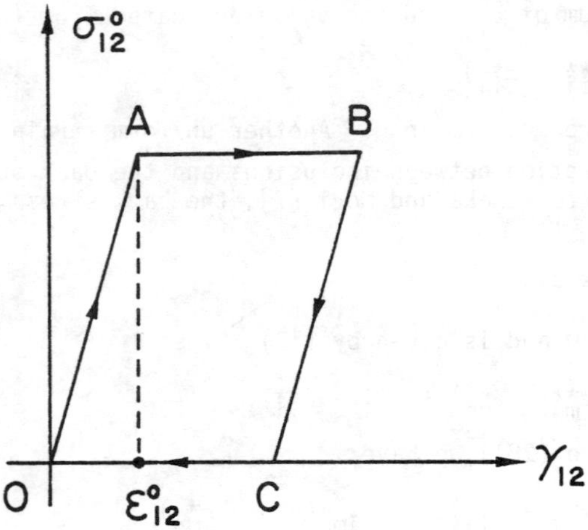

Fig. (2) - Stress-strain relation of the material shown in Fig. (1)

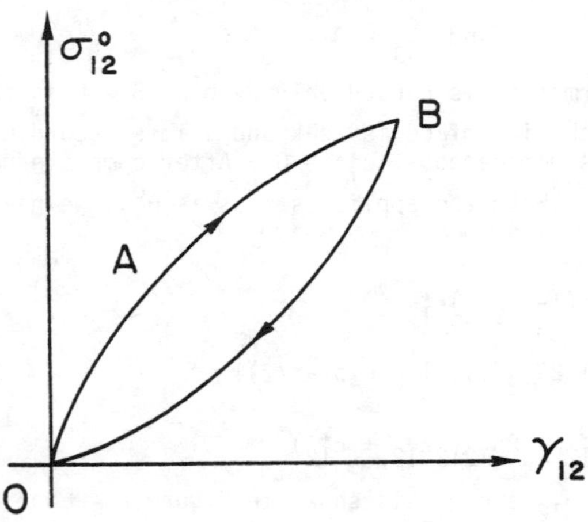

Fig. (3) - Stress-strain curve when the relaxation time is small

Case 2. $C^*_{ijk\ell} \neq C_{ijk\ell}$ and $\varepsilon^p_{ij} = 0$.

Equation (17) gives ($\Delta\mu = \mu^* - \mu$)

$$\varepsilon^{**}_{12} = -\Delta\mu\{\varepsilon^o_{12} + \varepsilon^*_{12}\,(1-2S_{1212})\}/(\mu+2\Delta\mu S_{1212}) \tag{33}$$

when the applied stress is σ^o_{12}. The average strain is given by (31), where ε^*_{12} is obtained from (24), namely

$$\varepsilon^*_{12} = -\varepsilon^o_{12}\{1 - \exp(-t/\tau)\}/(1-2S_{1212}) \tag{34}$$

When (34) is substituted into (33),

$$\varepsilon^{**}_{12} = -\Delta\mu\varepsilon^o_{12}\exp(-t/\tau)/(\mu+2\Delta\mu S_{1212}) \tag{35}$$

It can be seen from (31) that the contribution of ε^{**}_{12} to γ_{ij} is negative when $\Delta\mu > 0$. The nature of stress strain curves, however, is essentially the same as Case 1. It must be pointed out that the stress and strain state at point B is independent of $C^*_{ijk\ell}$.

Case 3. $C^*_{ijk\ell} = C_{ijk\ell}$ and $\varepsilon^p_{ij} \neq 0$.

From (17), we immediately have $\varepsilon^{**}_{ij} = \varepsilon^p_{ij}$. Consider the case of σ^o_{12} and ε^p_{12}. Equation (31) with (24) becomes

$$\gamma_{12} = \varepsilon^o_{12} + 2f(1-S_{1212})[\varepsilon^p_{12} - \{\varepsilon^p_{12} - \varepsilon^o_{12}/(1-2S_{1212})\}\{1 - \exp(-t/\tau)\}] \tag{36}$$

The stress in Ω is always uniform and is given by (17),

$$\sigma^o_{12} + \sigma_{12} = 2\mu\{\varepsilon^o_{12} + (1-2S_{1212})(\varepsilon^*_{12}-\varepsilon^p_{12})\} \tag{37}$$

where

$$\varepsilon^*_{12} = \{\varepsilon^p_{12} - \varepsilon^o_{12}/(1-2S_{1212})\}\{1 - \exp(-t/\tau)\} \tag{38}$$

The stress-strain curve is similar to Figure 2. The strain at $t = \infty$ is the same as that of B for Case 1. The strain at $t = 0$ is

$$\gamma_{12} = \sigma^o_{12}/2\mu + 2f(1-S_{1212})\varepsilon^p_{12} \tag{39}$$

and that at $t = \infty$ is

$$\gamma_{12} = \sigma^o_{12}/2\mu + 2f(1-S_{1212})\sigma^o_{12}/2\mu(1-2S_{1212}) \tag{40}$$

It is interesting to note that γ_{12} at $t = \infty$ is not necessarily larger than γ_{12} at $t = 0$.

STRESS DISCONTINUITY ON INTERFACE

The stress jump across the interface is obtained by a slight modification of the method developed by Hill [8] and Walpole [9].

For a line element, dx_j on S, we have from (5),

$$[u_{i,j}]dx_j = -b_{i,j}dx_j \tag{41}$$

The above relation is satisfied by choosing

$$[u_{i,j}] = -b_{i,j} + \lambda_i n_j \tag{42}$$

since $n_j dx_j = 0$. The constant λ_i is determined from the condition (4). Since

$$[\sigma_{ij}] = C_{ijk\ell}\{[u_{k,\ell}] - [\varepsilon^{**}_{k\ell}]\}$$
$$= C_{ijk\ell}\{[u_{k,\ell}] + \varepsilon^{**}_{k\ell}\} \tag{43}$$

the relation (4) is written as

$$[\sigma_{ij}]n_j = C_{ijk\ell}\{-\varepsilon^{*}_{k\ell} + \lambda_k n_\ell + \varepsilon^{**}_{k\ell}\}n_j = 0 \tag{44}$$

Solving λ_i from (44), we have

$$\lambda_i = -C_{jkmn}(\varepsilon^{**}_{mn}-\varepsilon^{*}_{mn})n_k N_{ij}(\underset{\sim}{n})/D(\underset{\sim}{n}) \tag{45}$$

where $N_{ij}(\underset{\sim}{n})$ and $D(\underset{\sim}{n})$ are the cofactor and the determinant of matrix $C_{ipjq}n_p n_q$.

Then we have

$$\sigma_{ij}(\text{out}) = \sigma_{ij}(\text{in}) + C_{ijk\ell}[-C_{pqmn}(\varepsilon^{**}_{mn}-\varepsilon^{*}_{mn})n_q n_\ell N_{kp}(\underset{\sim}{n})/D(\underset{\sim}{n}) + (\varepsilon^{**}_{k\ell}-\varepsilon^{*}_{k\ell})\}] \tag{46}$$

where $\sigma_{ij}(\text{in})$ is evaluated from (17).

REFERENCES

[1] Mura, T. and Furuhashi, R., "The elastic inclusion with a sliding interface", J. Appl. Mech., 51, p. 308, 1984.

[2] Ghahremani, F., "Effect of grain boundary sliding on an elasticity of polycrystals", Int. J. Solids and Structures, 16, p. 825, 1980.

[3] Mura, T., Jasiuk, I. and Tsuchida, E., "The stress field of a sliding inclusion", Int. J. Solids and Structures, 21, p. 1165, 1985.

[4] Tsuchida, E., Mura, T. and Dundurs, J., "The elastic field of an elliptic inclusion with a slipping interface", J. Appl. Mech., 53, p. 103, 1986.

[5] Mura, T., Micromechanics of Defects in Solids, Martinus Nijhoff, 1982.

[6] Eshelby, J. D., "The determination of the elastic field of an ellipsoidal inclusion, and related problems", Proc. Roy. Soc. London, A241, p. 376, 1957.

[7] Tanaka, K. and Mori, T., "The hardening of crystals by non-deforming particles and fibres", Acta Met., 18, p. 931, 1970.

[8] Hill, R., "Discontinuity relations in mechanics of solids", Progress in Solid Mechanics 2, eds. I. N. Sneddon and R. Hill, North-Holland, p. 245, 1961.

[9] Walpole, L. J., "A coated inclusion in an elastic medium", Math. Proc. Camb. Phil. Soc., 83, p. 495, 1978.

SECTION I

PROCESSING AND MANUFACTURING

GRAPHITE FIBER REINFORCED CERAMICS, TiB_2/TiC/MoC BY SHS PROCESS

S.-E. Hsu, C.-I. Chen, C.-R. Lee and S.-J. Chen

Chung Shen Institute of Science and Technology
P.O. Box 1-26, Lungtan, Taiwan 32500, Republic of China

ABSTRACT

An economical processing sequence was developed involving self-propagation high temperature synthesis of TiB_2/TiC/MoC, containing up to 20% graphite fiber by volume. Graphite fiber, having length of 100 ∿ 1000 μm and diameter of about 8 μm, was performed in random orientation. Formation of porosity in the synthetic ceramic composite is crucial for SHS process. Efforts in this experiment were concentrated on the densification of the final product. It is found that the density of graphite fiber reinforced ceramic composite is a function of composition, particle size, pressure of powder compaction and pressure of environment during reaction. Porosity and crack can be effectively reduced by proper selection of particle size or by addition of moderator such as Ni or Mo powder. The strengthening effect of graphite fiber to SHS TiB_2/TiC/MoC ceramics seems very subtle in comparison with the strength of ceramics.

INTRODUCTION

The self-propagation high-temperature synthesis (SHS) is a simple and economical process to produce refractory materials. Titanium diboride and titanium carbide have been synthesized by this way for a long time [1,2] because of its good properties of hardness, corrosion resistance and high temperature strength. The products of TiB_2 or TiC by SHS process generally include a large fraction of porosity or crack. It is not so dense as the products made by the other ways such as sintering [3], chemical vapor deposition [4] or plasma spray [5]. Although some attempts such as SHS under high pressure [6], or under centrifugal force [7] have been developed to improve density and structure, the results are still unsatisfied.

The reason of creation of pores or cracks accompanied with SHS reaction is speculated [8] that the restricted mobility of remained solid phase during reaction prevents their mutual coalescence to form a more dense product. Meanwhile, the unreacted portion of the green compact during reaction might fail to match the reaction front which is constricted due to coalescence of both reactants and products. Apparently, the binding strength of the molten state in reaction front between reacted and unreacted region is a strong factor to effect density. Unfortunately, synthesis only occurs at reaction front, self-propagated along a wave-propagation front; therefore, wave type laminated crack is generally resulted due to mismatch. In order to eliminate the defects of laminal cracks, this paper uses three approaches to reach the goal. First, put metal additive to improve the wettability of molten state at reacting front to both reacted and unreacted portions. Second, use pitch to pretreat the green compact to restrict the synthesis which occurs only in a small local area. Third, use graphite fiber to hold unreacted green compact of carbon and reacted skeleton product on both sides of reaction front together to prevent large cracks and eventually to strengthen the products. Since graphite fiber is active at high temperature, it is easy to react with reactant and additives. Special protection has to be done to survive the fiber.

EXPERIMENTAL PROCEDURE

A. Raw Material

The titanium powder (Ti), boron carbide powder (B_4C), molybdenum powder (Mo), petroleum pitch and graphite fiber are all commercially available. The characteristics of raw materials are shown in Tables 1, 2 and 3.

Table 1 - The properties of raw material powder

	Particle Size FSS	Surface Area of Particle
Titanium	15μ	0.31 m^2/g
Molybdenum	8μ	0.15 m^2/g
B_4C	7.7μ	0.55 m^2/g

Table 2 - The properties of pitch

Composition		
Ash Content	wt% Max.	0.15
Sulfur Content	wt% Max.	3.0
Benzene Insoluble	% Max.	10.0
Quinoline Insoluble	% Max.	1.0
Physical		
Softening Point	°C	90-130
Density	g/cc	1.20-1.30
Coke Value	wt% Min.	40

Table 3 - The properties of graphite fiber

Diameter	8.5	μm
Tensile Strength	0.62	GPa
Young's Modulus	41	GPa
Density	1.53	

B. Preparation of Specimens

Specimens are prepared by two different ways. One is the conventional method to consolidate the mixtures through cold press and then ignite in graphite powder compact to form products. The flow chart is shown in Figure 1. The other method

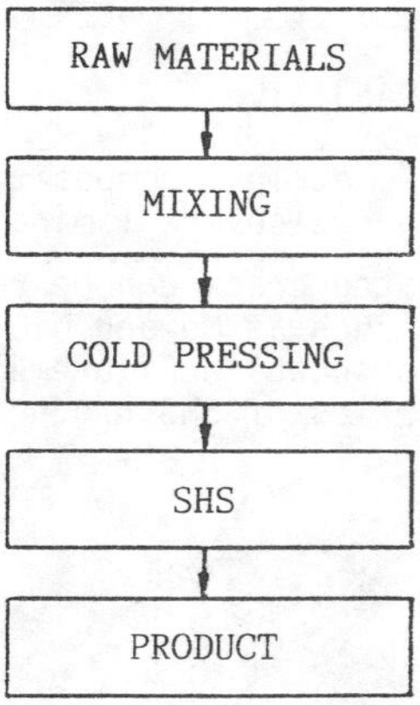

Fig. (1) - Specimen preparation flow chart - I

is to pretreat titanium powder and graphite fiber by pitch. Then the mixture was put in the steel mold and heated up to 350°C in order to melt pitch for press. The pressure is about 200 kg/cm^2 before the cap of mold was tight up to keep the pressure inside the mold. The tightened mold then was removed to the furnace for pyrolysis of the pitch which transfers to coke at 550°C. The specimen was then taken out from mold and machined for further pyrolysis up to 900°C in controlled atmosphere at the heating rate of 5°C/min to reduce the volatile material. The baked specimen is available to be ignited by plasma torch to perform SHS process. The flow chart has been shown in Figure 2.

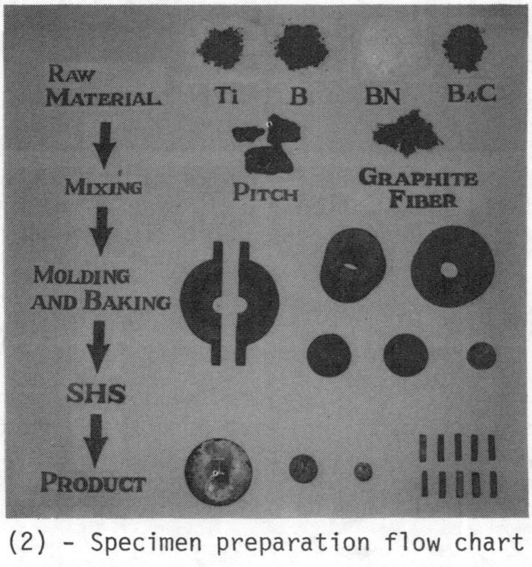

Fig. (2) - Specimen preparation flow chart - II

C. Procedures and Apparatus

The SHS specimen was observed by both optical microscopy and SEM. The X-ray diffraction was used to identify the structure before and after SHS. The properties of MOR, hardness and density are also measured.

RESULTS AND DISCUSSION

A. The SHS of Ti, B_4C and Additives

The SHS reaction of stoichiometrical composition (i.e., 3 mole Ti and 1 mole B_4C) without any additives produces severe laminated crack in the specimen as shown in Figure 3. This laminated crack can be reduced by mixing with some moderator of heavy metallic elements such as Mo and Ni. The photograph of the synthetic specimen with additive of Mo is shown in Figure 4. The influence of the additive metals on the mechanical properties of SHS product is shown in Figure 5.

Fig. (3) - Synthetic specimen ($3Ti + B_4C$) after SHS

Fig. (4) - Mo added $5Ti + B_4C$ specimen after SHS

Figure 5 illustrates that the hardness of the synthetic product decreases rapidly with increasing the amount of additive nickel. While the additive molybdenum which affects the hardness slightly improves the surface crack significantly, indicating that the additive of molybdenum can moderate the reaction rate to such an optimum extent that the molten phase at reaction front has sufficient time to bond the reacted and unreacted portion firmly with the same pace of wave front propagation.

The hardness of the product also depends on the compact pressure. The hardness of SHS product as a function of compact pressure is shown in Figure 6.

The density of the product also have the similar relationship with compact pressure. The final size of specimen might be changed depending on the characteristics of titanium powder. Smaller particle size with less surface area will constrict the specimen to a larger extent. The influence of particle size of Ti-powder on the shrinkage of SHS product is shown in Figure 7. It is speculated that

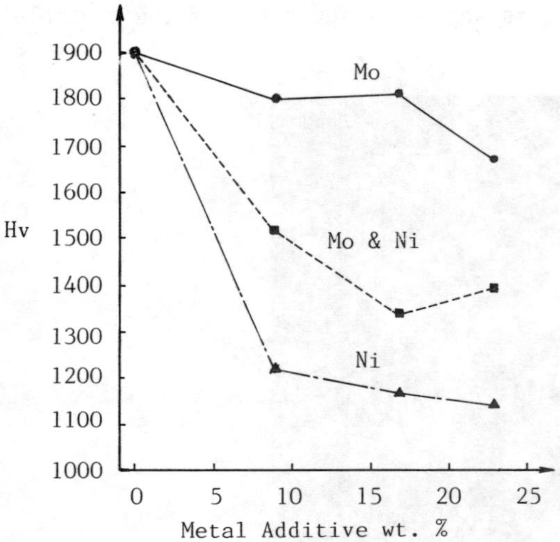

Fig. (5) - The influence of the additive metals on the mechanical properties of SHS product

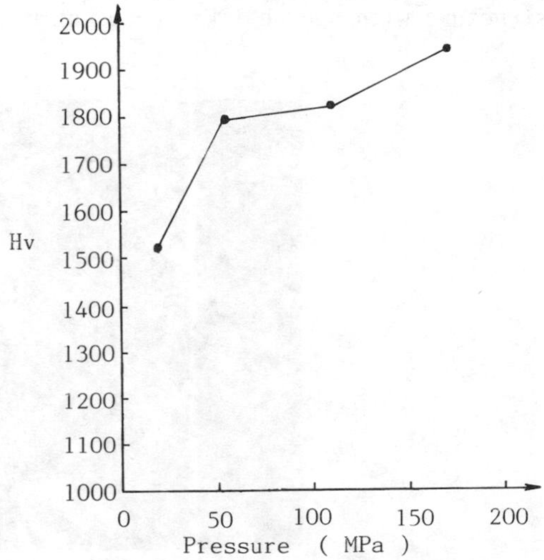

Fig. (6) - The influence of compact pressure on hardness

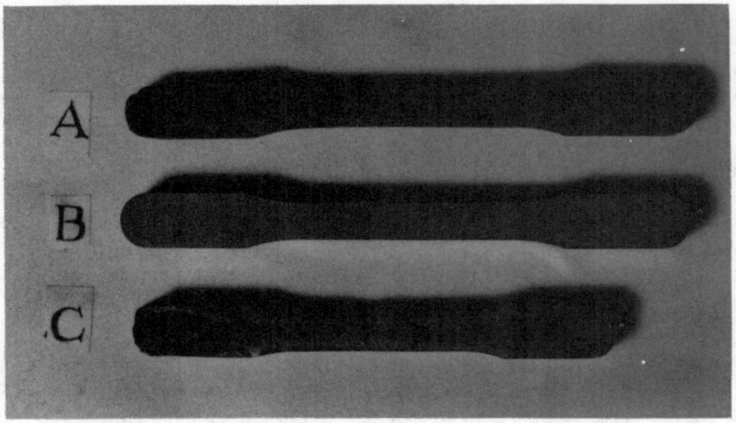

Fig. (7) - The influence of Ti powder particle size on shrinkage of SHS product. (A): Ti: 18μ, 2.23 m^2/g; (B): Green compact before SHS; (C): Ti: 15μ, 0.31 m^2/g

the change of specimen size is a function of contact angle between molten titanium and reactant or product or TiC. Large contact angle, i.e., less wettability, reduces the compactability of molten titanium at the reaction front with reacted and unreacted portion. Consequently, the microcrack will be initiated at reaction front causing a less dense material behind. The residual molten-Ti, if any, might exude out on the surface leaving a group of pores in the material. A SEM microstructure with sponge-like porous structure is shown in Figure 8. If the particle

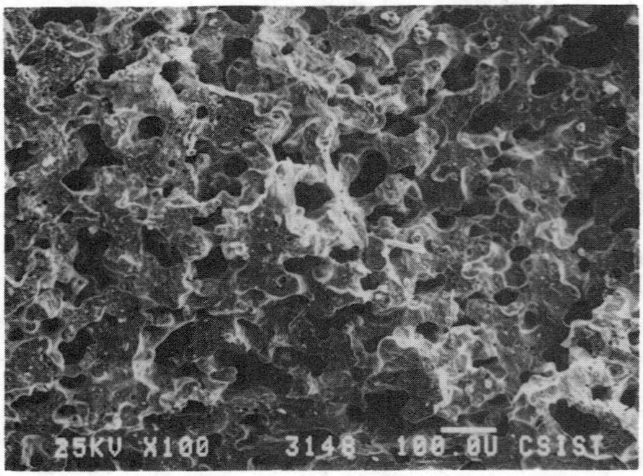

Fig. (8) - SEM microstructure of SHS

size of titanium is sufficiently small, wettability of molten-Ti would not be serious during synthesis reaction since it is easy to diffuse into B_4C, a more dense material with large shrinkage would result upon good coalescence. Tests are conducted to examine the permeability by soaking the specimen in the water. Results indicate that the pores are interconnected quite extensively. The X-ray diffraction shows that the final product is a composite of TiB_2/TiC/MoB_2/MoC as indicated in Figures 9(a) and 9(b).

B. The Influence of Graphite Fiber Additive

Attempts have been tried to reinforce the TiB_2/TiC/MoC composites by addition of graphite fibers. Various volume fraction of graphite fiber mixed with 60 wt% Ti, 23 wt% B_4C, 17 wt% Mo mixture and then ignited in graphite compact. The influence of graphite fibers on modulus of rupture (MOR) is shown in Figure 10. Indicating that mechanical strength is an inverse function of volume fraction of graphite fiber, it is found that graphite fiber almost vanished in microstructure of SHS product. The X-ray diffraction pattern show TiB, MoB, residual graphite and residual titanium coexist with normal structures of TiB_2, TiC, MoB_2 and MoC. This result indicates that the mobility of titanium was restricted by graphite for additive during reaction and the reaction was incomplete. The bonding force of SHS product was so weak that Vicker's hardness could not be tested. Microstructure, as shown in Figure 11, reveals there is only few direct fusion bonding of particles.

C. The SHS Product of Pitch Coated Raw Materials

In order to avoid the direct contact Ti and to save the graphite fibers, raw material of titanium and graphite fibers are coated with pitch. Microstructure

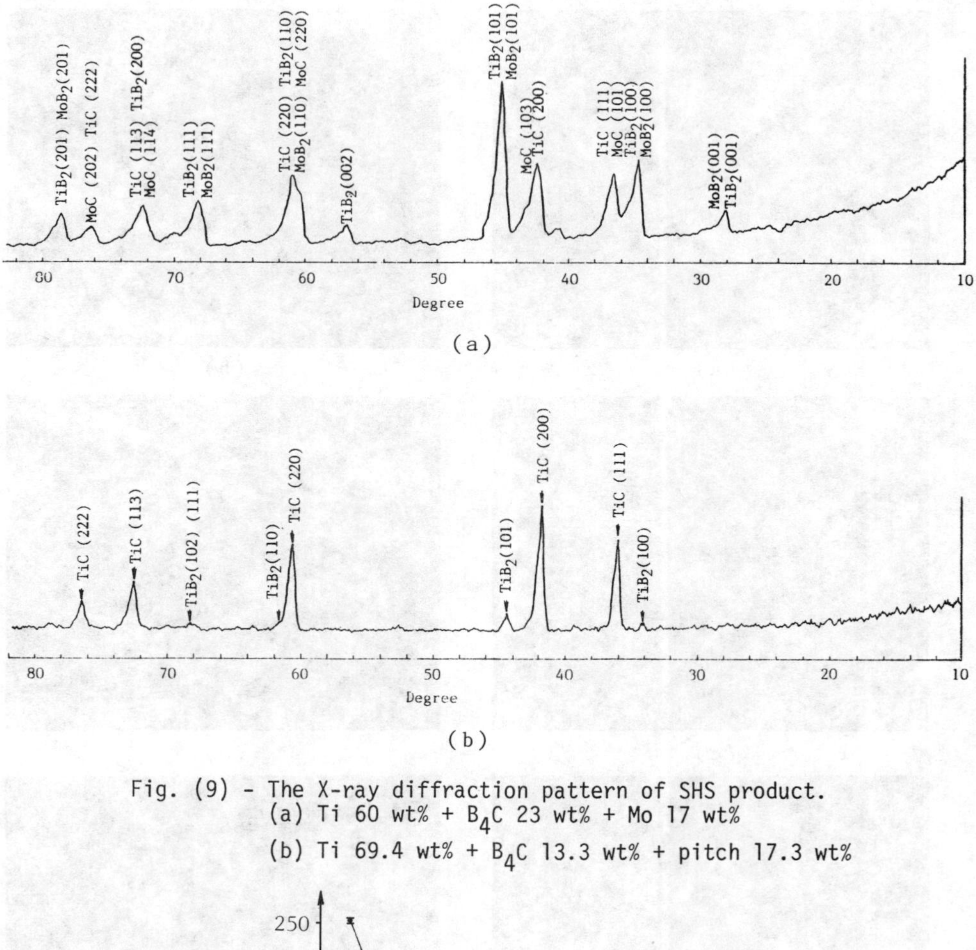

Fig. (9) - The X-ray diffraction pattern of SHS product.
(a) Ti 60 wt% + B_4C 23 wt% + Mo 17 wt%
(b) Ti 69.4 wt% + B_4C 13.3 wt% + pitch 17.3 wt%

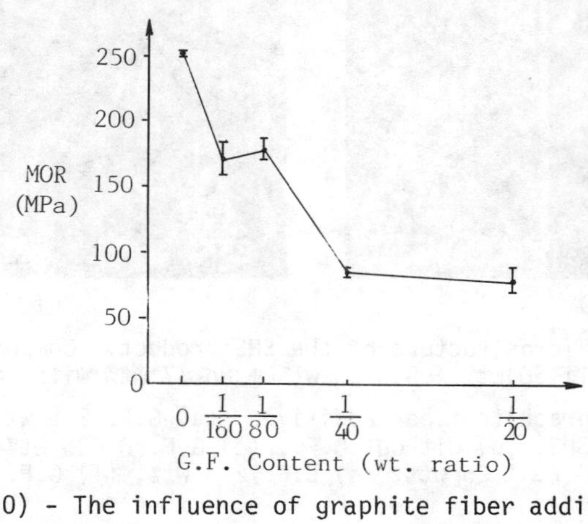

Fig. (10) - The influence of graphite fiber additive on MOR

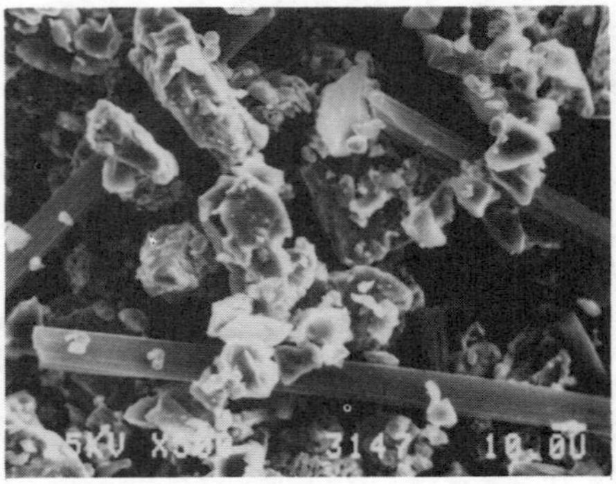

(a)

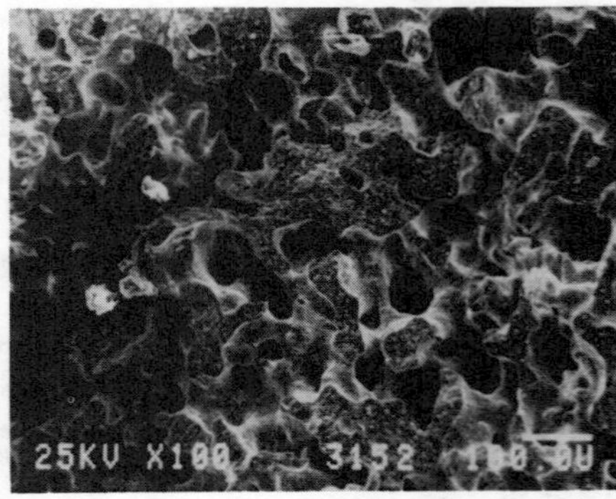

(b)

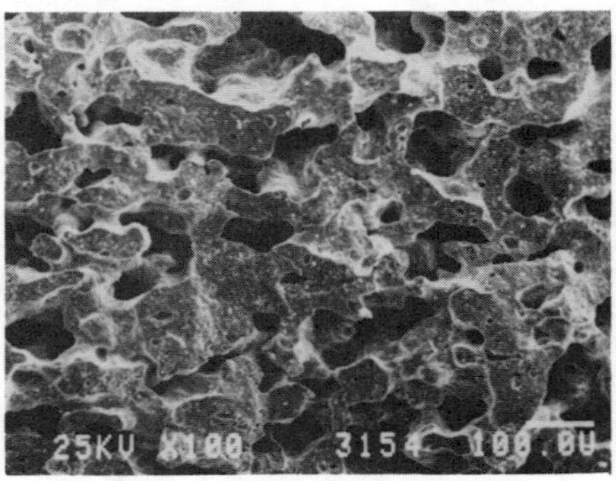

(c)

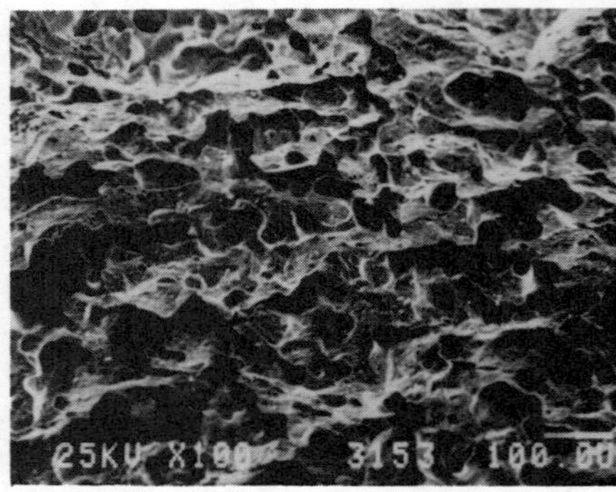

(d)

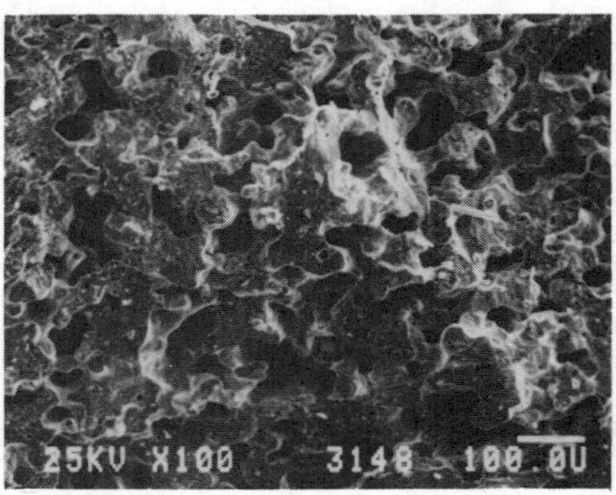

(e)

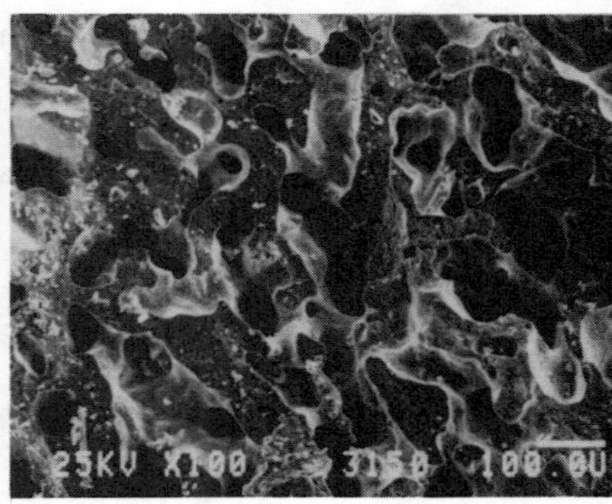

(f)

Fig. (11) - Microstructure of the SHS product. Composition: Ti 60 wt% + B_4C 23 wt% + Mo 17 wt% with various graphite fiber additive. (a) G.F. 2.5 wt% before SHS, (b) without G.F., (c) G.F. 0.625 wt%, (d) G.F. 1.25 wt%, (e) G.F. 2.5 wt%, (f) G.F. 5.0 wt%

of SHS product containing pitch coated graphite fiber is shown in Figure 12. The composition of mixture before SHS is Ti 69.4 wt%, B_4C 13.3 wt%, pitch 17.3 wt%

(a) (b)

Fig. (12) - Microstructure of SHS. Ti powder and graphite fiber coated with pitch. (a) before SHS; (b) after SHS

with or without 2.5 wt% of graphite fiber. Although graphite fibers have survived after SHS reaction, the remaining amount is obviously decreased. This indicated that a number of graphite fibers reacted with titanium directly. The dimension of specimen after SHS kept no change. Although the number of pores seemed to increase, the size of pores in microstructure grew much less than the previous. More uniformity of the structure indicated the product had local fusion during reaction. The SEM microstructure showed that the bonding strength between graphite fiber and matrix is rather weak. Each fiber was pulled out. The comparison of MOR before and after SHS as shown in Table 4 indicates there is no reinforced effect at all upon addition of graphite fiber. The sample without graphite fiber

Table 4 - SHS properties before and after reaction

Composition (wt %)	Condition	Density (g/c.c.)	MOR (MPa)
Ti 69.4 B_4C 13.3	Before SHS	2.92	40.7
Pitch 17.3	After SHS	2.78	33.2
Ti 67.6 B_4C 13.0 G.F. 2.5	Before SHS	2.72	42.5
Pitch 16.9	After SHS	2.67	27.3

additive has slightly higher density, lower reaction rate and lower MOR before SHS. The X-ray pattern, Figure 9(b), shows that the major portion of the final product is TiC and the minor portion of TiB_2. A further comprehensive study is carrying on in this laboratory. The final goal of this study is hopefully to understand the kinetics and mechanisms of SHS process. Eventually, we hope to find the control factor to govern the kinetics.

CONCLUSIONS

1. Metal additive can reduce the laminated crack of Ti-B_4C system; Mo is one of the best additives.
2. The density of Ti-B_4C SHS composite is a function of particle size, particle surface area of titanium and compact pressure.
3. The additive of carbon or graphite reduces the bonding force in Ti-B_4C SHS system.
4. Graphite fiber, exists in Ti-B_4C SHS system, has no evidence of reinforced effect at all at this stage of experiment.

REFERENCES

[1] Merzhanov, A. G. and Borovinskaya, I. P., "Self-propagating high temperature synthesis of refractory inorganic compound", Dokl. Chem. (Eng. Trans.), 204, pp. 429-432, 1972.

[2] Champagne, B. and Dallaire, S., "Structure and properties of TiB_2-based materials produced by the reaction of ferrotitanium and boron powders", Ceram. Eng. Sc. Proc., 5, pp. 702-710, 1984.

[3] Takatsu, S. and Ishimatsu, E., "Sintering and properties of titanium diboride - alloy with refractory metal binder", Int. J. Refract. Hard Met., 1, pp. 75-80, 1982.

[4] Person, H. O. and Randich, E., "Coating of metals with titanium diboride by chemical vapor deposition", Chemical Abstract, 89:10319n, 1978.

[5] Morozumi, S., Kikuchi, M. and Kanazawa, S., "Plasma spray coating of low Z ceramics on molybdenum", J. Nucl. Material, 103, p. 279, 1981.

[6] Zaritsanos, P. and D'Andrea, J., "Synthesis of high density titanium diboride by SHS", DARPA/ARMY Symposium, Daytona Beach, Fl., p. 29, 1985.

[7] Frankhouser, W. L., Brendly, K. W. and Kieszek, M. C., "Gasless combustion synthesis of refractory compound", p. 99, 1985.

[8] Henshow, W. F., Niiler, A. and Leete, T., "Self-propagating high temperature synthesis of ceramics in vacuum", Ceramic Eng. and Sci. Proceedings, 4, p. 634, 1983.

A STUDY ON THE MAGNETIC PROPERTIES OF TWO-COMPONENT MAGNETIC COMPOSITES

S.-E. Hsu, K.-L. Wang, T.-I. Hou and L.-C. Su

Chung Shan Institute of Science and Technology
Lungtan, Taiwan 32500, Republic of China

ABSTRACT

The magnetic properties of two-component magnetic composites are studied theoretically and experimentally. The most commonly used rare-earth hard magnetic materials, such as $SmCo_5$, $Sm(Co,Fe,Cu,Zr)_{7.4}$, $Nd_{15}Fe_{77}B_8$, are utilized as constituents. Good agreements between theoretical calculations and experimental results are generally obtained. There is no obvious difference in phase-distribution between isotropic and anisotropic cases. Due to the fact that intrinsic coercivity is a strong function of particle size, $Nd_{15}Fe_{77}B_8$ powder seems unlikely suitable for being used in a particulate magnetic composite. The explosive compaction has been used to increase the bulk density of composites. About 90% theoretical density has been obtained up to now.

INTRODUCTION

The magnetic composites which consist of two magnetic components have been studied in the previous papers [1,2] theoretically as well as experimentally. The extensively used resin-bonded magnets which consist of only one magnetic component are the extreme case for previously mentioned composites. In those studies, the method of theoretical calculation has been developed which can be used for composites consisting of two magnetic components. A good agreement between theoretical calculations and experimental results has been obtained. In those studies, none of the rare-earth hard magnetic component was used and all the specimens were in isotropic state.

The rare-earth magnets have been more widely used than before and most of the magnets are used in anisotropic state. The purpose of the present study is to further verify the theoretical feasibility in the case of anisotropic composites which consist of at least one rare-earth hard magnetic component.

In this paper, three kinds of the most commonly used rare-earth materials, such as $SmCo_5$, $Sm(Co,Fe,Cu,Zr)_{7.4}$, $Nd_{15}Fe_{77}B_8$, are utilized as constituents. Three systems of two-component magnetic composites, including $SmCo_5$-Ba-Ferrite, $SmCo_5$-$(Co,Fe,Cu,Zr)_{7.4}$, $Nd_{15}Fe_{77}B_8$-$SmCo_5$, are studied. The comparisons of the magnetic properties between theoretical calculations and experimental results were made. The explosive compaction method was used to increase the bulk density of the magnetic composites.

EXPERIMENTAL

The hard magnetic powders, including $SmCo_5$, $Nd_{15}Fe_{77}B_8$, $Sm(Co,Fe,Cu,Zr)_{7.4}$ and Barium Ferrite, are prepared by vibration milling of corresponding sintered magnets. The magnets have been heat-treated to optimize its magnetic properties, which have been measured before the magnets are crushed. Two kinds of powder in different volume fraction are thoroughly mixed in a vibration mill. Then, these powder mixtures and constituent powders are compacted in a cold isostatic press (CIP) at a pressure of 30,000 psi. Before compaction, the powders are aligned in a magnetic field of 40 KOe. In this way, the anisotropic green compacts are obtained. For the purpose of comparison, the isotropic ones are also solely prepared without magnetic alignment. The green compacts are resin-infiltrated. Then, the magnetic properties of these composites are measured by using a hysteresis loop tracer which can produce 32 KOe high magnetic field in a 1/4" air gap of the pole pieces. Because of the rate-earth hard constituent, an 80 KOe high field pulse magnetizer was used to magnetize the samples before the demagnetization curves in the second quadrant are measured. The magnetic properties discussed in this paper, including remanence Br, coercivity Hc, and maximum energy product (B.H)max, can be obtained from these demagnetization curves.

To increase the density of green compacts, the explosive compaction method has been used. After CIP process, the green compact with about 60% of the theoretical density can reach to more than 80% bulk density after further explosive compaction.

RESULTS AND DISCUSSION

A. One Magnetic Component Composite

Before going into the main subject of studying two-component magnetic composite, we first consider the magnetic properties of the constituents used.

The magnetic properties of anisotropic resin-bonded $SmCo_5$ composites as a function of volume fraction of $SmCo_5$ are shown in Figure 1. The maximum volume fraction of $SmCo_5$ is 70% and the maximum magnetic properties are Br = 5600 G, Hc = 5200 Oe, (B.H)max = 7.3 MG.Oe. The curves in Figure 1 are obtained by theoretical calculation. From theoretical consideration, the magnetic properties of composite can be found from the known ones of its constituents. On the other hand, the properties of any constituent can also be calculated if the magnetic properties of both composite and another constituent are known. By using the equations developed in the previous papers, we calculated the constituent properties of 100% $SmCo_5$ from experimental values of composite and those of other nonmagnetic component whose properties are already known. After two constituent properties are found, the magnetic properties of composite in any volume fraction can be easily calculated. The comparison between the two results in Figure 1 shows that the theoretical calculations are in good agreement with the experimental results.

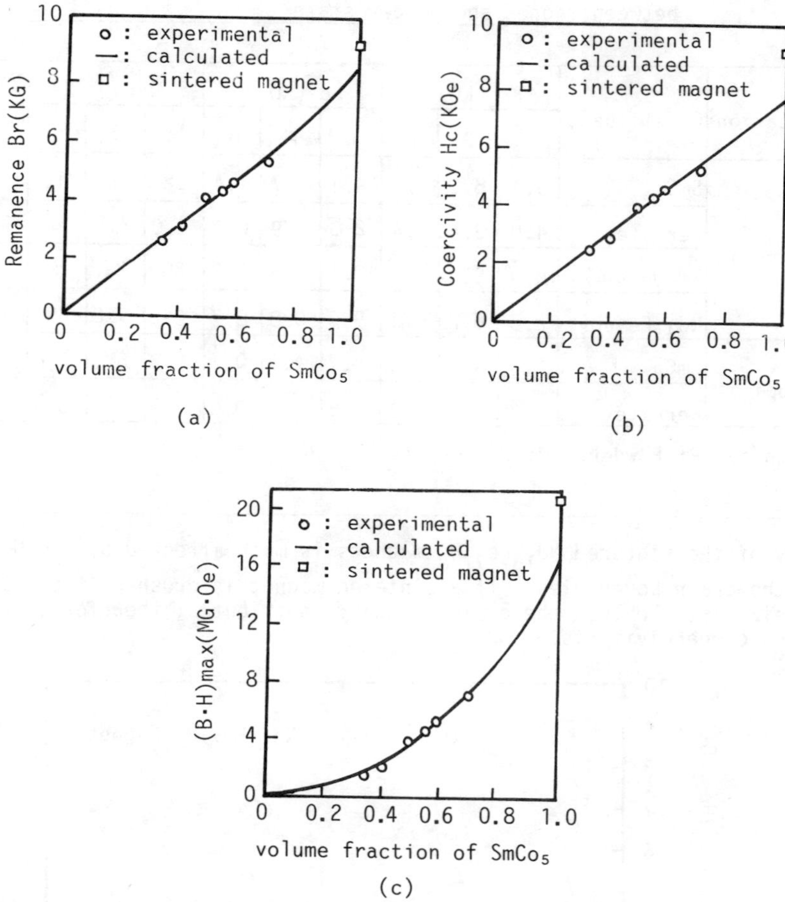

Fig. (1) - The magnetic properties of anisotropic resin-bonded $SmCo_5$ magnets

The constituent properties of 100% $SmCo_5$ are a little lower than those of sintered magnet. This discrepancy may be induced from (1) crystal structure change during vibration milling; (2) less degree of orientation or (3) an additional demagnetizing field applied to the particle due to its smoothless surface. The x-ray diffraction patterns show no evidence of crystal structure change. The degree of orientation should be the same because the green compacts of both resin-bonded and sintered magnets are prepared from the same processes. Thus, surface defects may be the dominant factor which results in this discrepancy.

Table 1 shows the same results observed for other magnetic constituents, For $Nd_{15}Fe_{77}B_8$ powder, the magnetic properties is very particle-size dependent. This is shown in Figure 2. It can be explained as the following: the intrinsic

Table 1 - The comparison of magnetic properties of constituents between magnet and powder state

Orientation	Constituent	Br (KG)			Hc (KOe)			(B·H)max (MG·Oe)			Average Particle Size(μm)
		M	P	D	M	P	D	M	P	D	
Anisotropic	$SmCo_5$	9.4	8.5	0.9	9.1	7.7	1.4	22	18	4	5
	Ferrite	4.0	3.4	0.4	2.3	1.9	0.3	4.0	2.8	1.2	2
	$Sm(Co,Fe,Cu,Zr)_{7.4}$	10.9	9.4	1.1	10.1	7.6	2.2	30	22	8	10
	$Nd_{15}Fe_{77}B_8$	12.5	6.2	6.3	9.5	0.8	8.7	39	9.6	30	5
Isotropic	$SmCo_5$	5.0	4.7	0.3	4.7	4.0	0.7	6.3	5.5	1.8	5
	Ferrite	2.4	2.1	0.3	1.9	1.3	0.6	1.5	1.1	0.4	2

M: Magnet, P: Powder, D: Difference

coercivity of the sintered $Nd_{15}Fe_{77}B_8$ magnets is much affected by the Nd-rich phase on the grain boundaries. If a sintered magnet is crushed, the smaller the particle size is, the less the grain boundary contributes; therefore, the lower the intrinsic coercivity is.

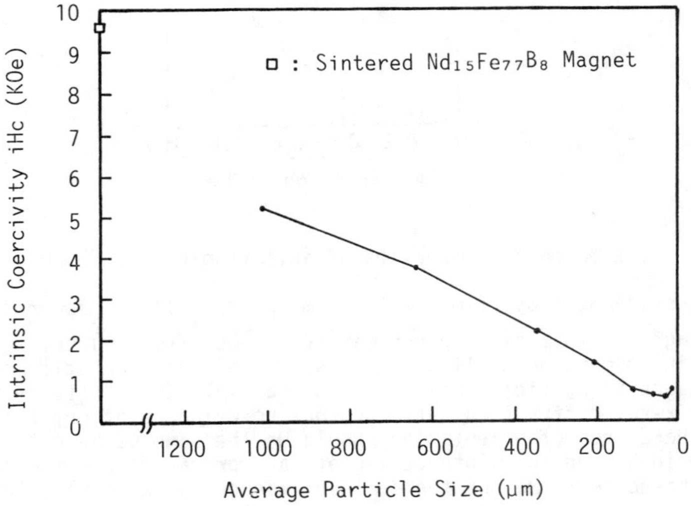

Fig. (2) - The dependence of the intrinsic coercivity iHc on the average particle size of $Nd_{15}Fe_{77}B_8$ powder

B. Two Magnetic Component Composites

$SmCo_5$ - Ba-Ferrite System

Figure 3(a), (b) show the experimental results of the demagnetization B-H curves of composites and its constituents in $SmCo_5$ - Ba-Ferrite system for isotropic and anisotropic cases, respectively. In all samples, the volume fraction

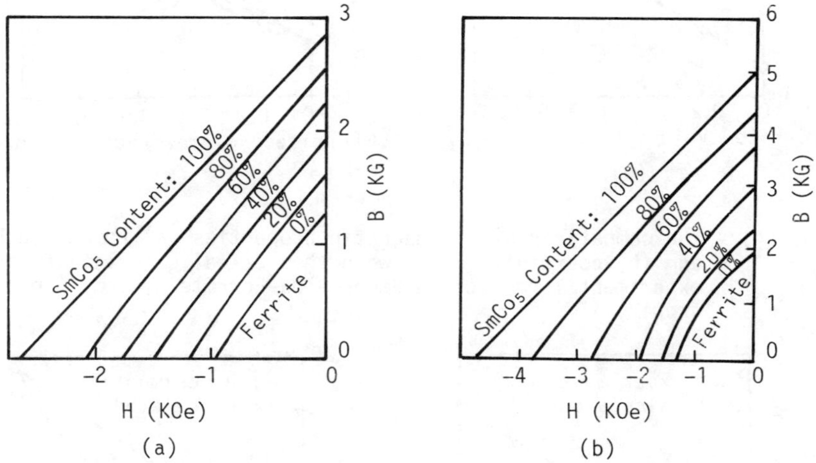

Fig. (3) - The demagnetization curves of composites and its constituents in $SmCo_5$ - Ba-Ferrite system for (a) isotropic and (b) anisotropic cases

of the magnetic component is controlled to 60%. For theoretical calculation, a random phase distribution is assumed. Using the known properties of its constituents, the remanence and coercivity of the composite can be found. The comparison between the two results are shown in Figures 4 and 5 for isotropic and anisotropic, respectively. In both cases, good agreements are obtained. This indicates that the theoretical calculation can be equally applied to both isotropic and anisotropic cases. From the optical microscope photograph shown in Figure 6, we find that there is no obvious difference in phase distribution between these two different cases.

$Sm(Co,Fe,Cu,Zr)_{7.4}$ - $SmCo_5$ System

Figure 7 shows the demagnetization curves of composites and its constituents in $Sm(Co,Fe,Cu,Zr)_{7.4}$ - $SmCo_5$ system. The volume fraction of the magnetic component in all samples is controlled to about 60%. The magnetic properties of $Sm(Co,Fe,Cu,Zr)_{7.4}$ are a little higher than those of $SmCo_5$. The B-H curves of these two constituents are linear and nearly have the same slope (i.e., $m_1 \simeq m_2$).

From theoretical consideration, there exists unique solution for any phase-distribution when $m_1 = m_2$. The remanence and coercivity of composites are vol-

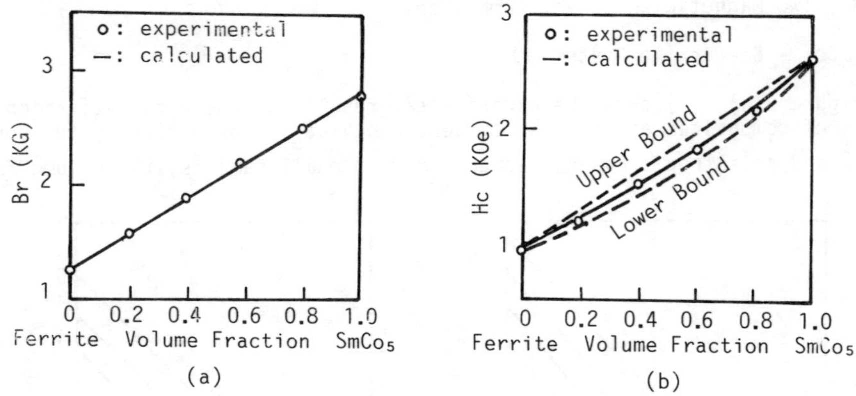

Fig. (4) - The comparison of the magnetic properties (a) remanence Br and (b) coercivity Hc between theoretical calculations and experimental results in $SmCo_5$ - Ba-Ferrite system (isotropic)

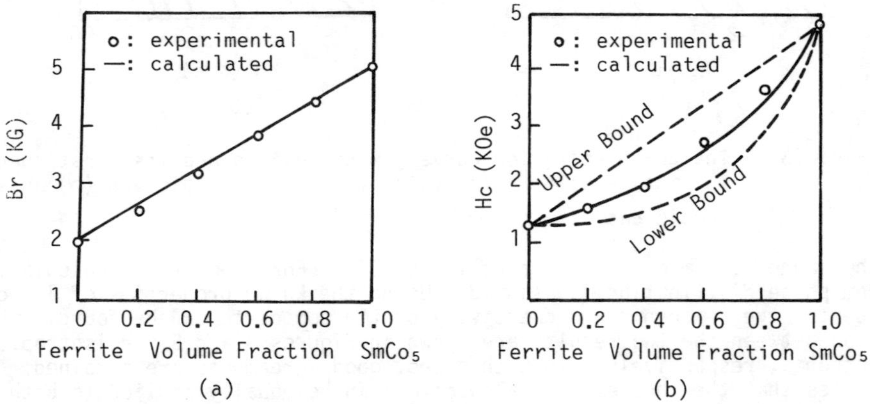

Fig. (5) - The comparison of the magnetic properties (a) remanence Br and (b) coercivity Hc between theoretical calculations and experimental results in $SmCo_5$ - Ba-Ferrite system (anisotropic)

ume average of that of its constituents. Figure 8 shows the comparison of magnetic properties between theoretical calculations and experimental results of this system. Good agreements are obtained. This verifies the theoretical prediction in the case of $m_1 = m_2$.

$Nd_{15}Fe_{77}B_8$ - $SmCo_5$ System

Figure 9(a), (b) show the demagnetization curves of composites and its constituents in $Nd_{15}Fe_{77}B_8$ - $SmCo_5$ system for 5 μm and 630 μm average particle size

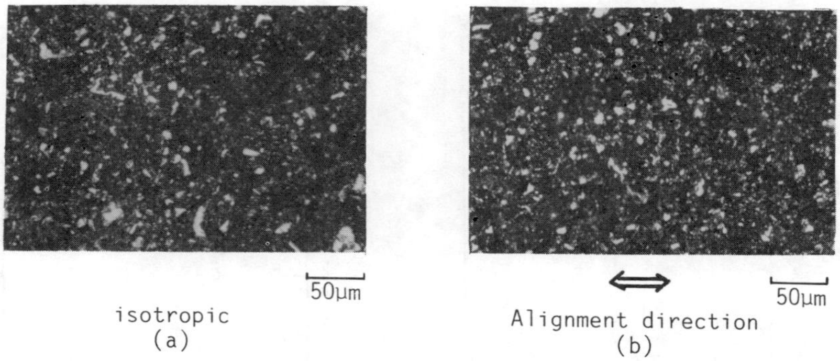

Fig. (6) - The optical microscope photographs of composites in $SmCo_5$ - Ba-Ferrite system for (a) isotropic and (b) anisotropic cases. $SmCo_5$ content 60%

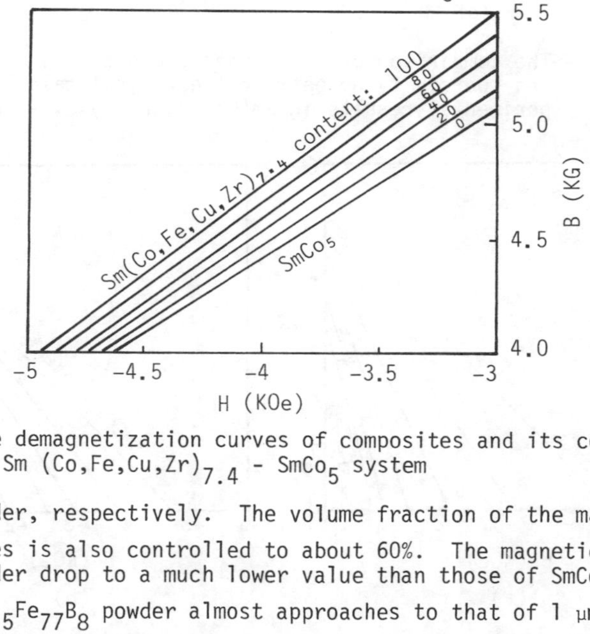

Fig. (7) - The demagnetization curves of composites and its constituents in Sm $(Co,Fe,Cu,Zr)_{7.4}$ - $SmCo_5$ system

of $Nd_{15}Fe_{77}B_8$ powder, respectively. The volume fraction of the magnetic component in all samples is also controlled to about 60%. The magnetic properties of $Nd_{15}Fe_{77}B_8$ powder drop to a much lower value than those of $SmCo_5$. The coercivity of 5 μm $Nd_{15}Fe_{77}B_8$ powder almost approaches to that of 1 μm powder studied by Oesterreicher [3]. The B-H curves of its two constituents are nearly linear but with quite different slopes (i.e., $m_1 >> m_2$). It is noted that all the B-H curves in the same system would pass through one point.

Figures 10 and 11 show the comparison of magnetic properties between theoretical calculations and experimental results in this system for 5 μm and 630 μm of $Nd_{15}Fe_{77}B_8$ powder, respectively. Because $m_1 >> m_2$, the space between upper and lower bounds is broadened. Good agreements are also obtained.

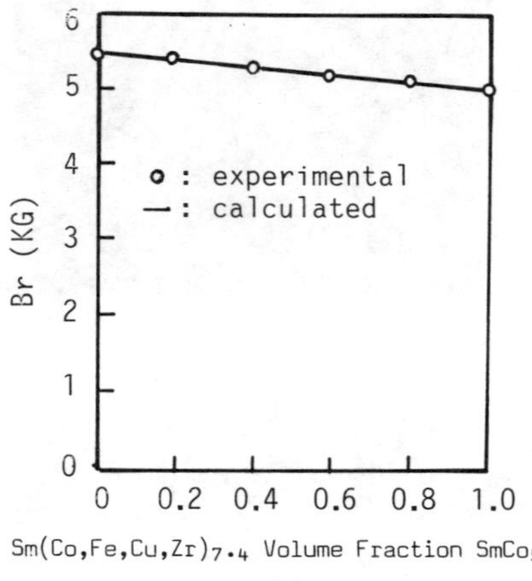

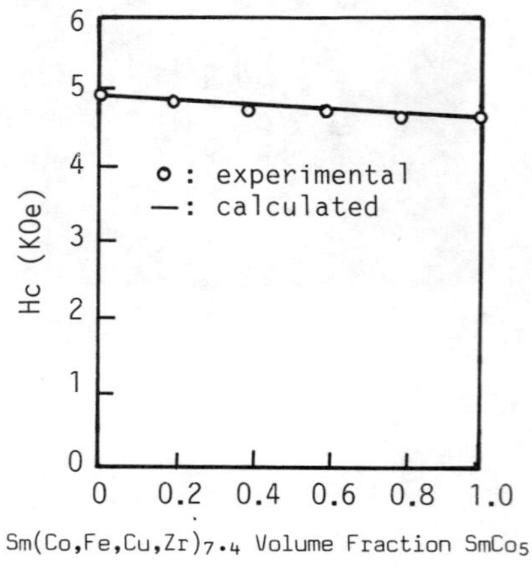

Fig. (8) - The comparison of the magnetic properties (a) remanence Br and coercivity Hc between theoretical calculations and experimental results in $Sm(Co,Fe,Cu,Zr)_{7.4}$ - $SmCo_5$ system

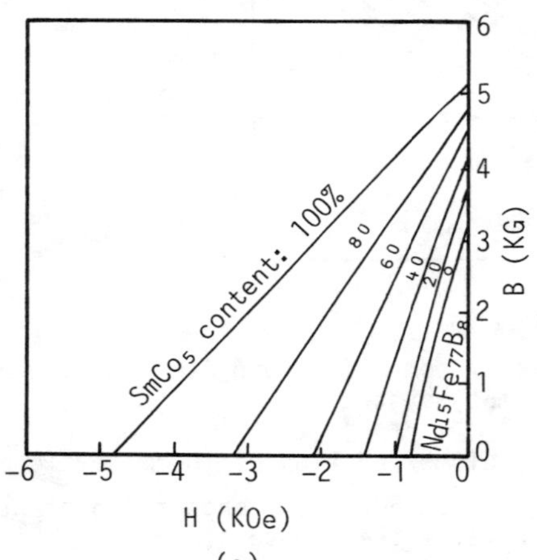

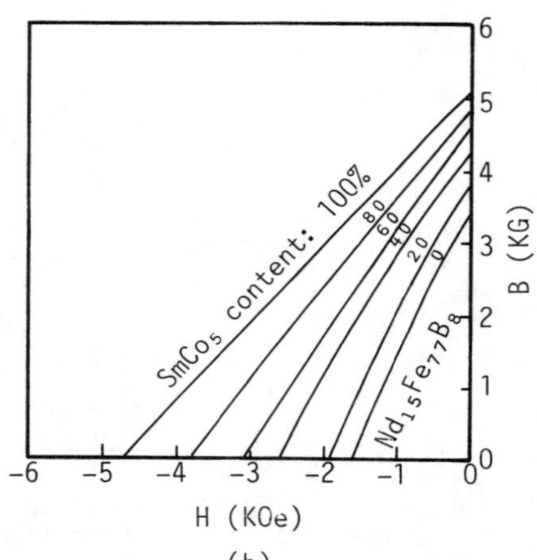

Fig. (9) - The demagnetization curves of composites and its constituents in $Nd_{15}Fe_{77}B_8$ - $SmCo_5$ system. Average particle size of $Nd_{15}Fe_{77}B_8$ powder (a) 5 μm; (b) 630 μm.

C. Increasing the Bulk Density

The relative density of composite can be increased from original 60% theoretical density after CIP process to a higher value followed by an explosive compaction. The results are shown in Figure 12. After explosive compaction, the

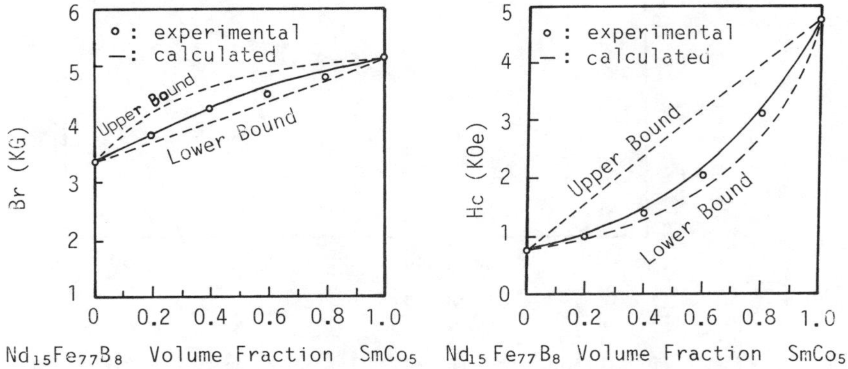

(a) (b)

Fig. (10) - The comparison of the magnetic properties (a) remanence Br and (b) coercivity Hc between theoretical calculations and experimental results in $Nd_{15}Fe_{77}B_8$ - $SmCo_5$ system. (5 μm)

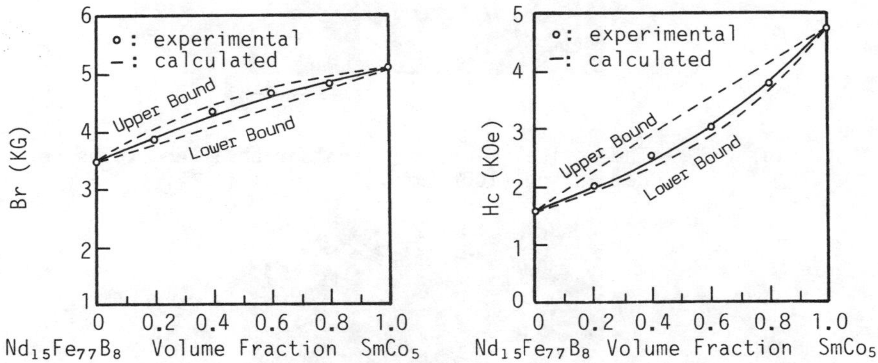

(a) (b)

Fig. (11) - The comparison of the magnetic properties (a) remanence Br and (b) coercivity Hc between theoretical calculations and experimental results in $Nd_{15}Fe_{77}B_8$ - $SmCo_5$ system. (630 μm)

sample is about 17 mm in diameter and 7 mm in height. Because of one-direction compaction, the density is not uniform. For sample exploded by 80g dynamite, the average relative density is about 87%. However, 3 mm of the exploded end reaches 90% relative density. The optical microscope photographs are shown in Figures 13 and 14. It seems that the heat dissipated during the explosive compaction does not cause any damage to the sample. The results above indicate that

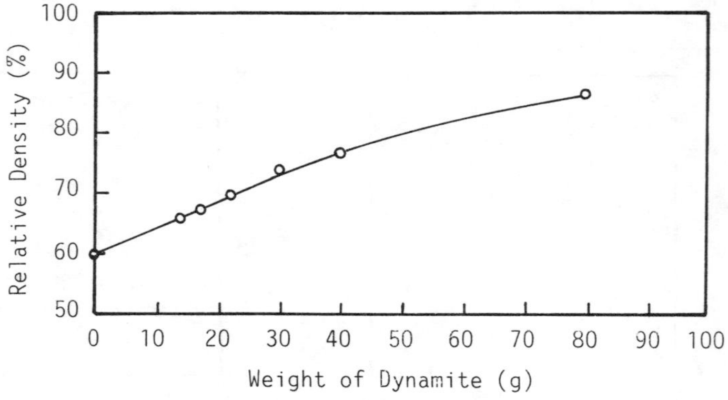

Fig. (12) - The bulk density after explosive compaction (100% $SmCo_5$)

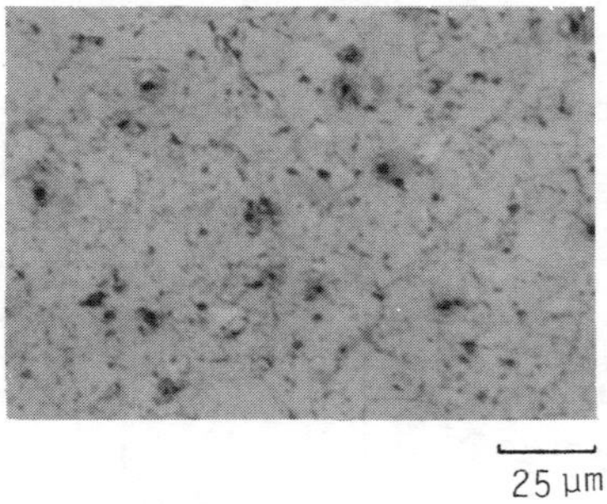

Fig. (13) - The optical microscope photograph after explosive compaction (100% $SmCo_5$)

Fig. (14) - The optical microscope photograph after explosive compaction (40 vol% $SmCo_5$, 60 vol% $Nd_{15}Fe_{77}B_8$)

the explosive compaction method can be used to obtain high bulk density of composites. It is possible that the bulk density of composite would reach that of sintered magnets by using this process. However, compaction method must be improved to obtain uniform density.

The magnetic properties of samples after explosive compaction are shown in Table 2. The remanence after explosive compaction is much lower than those of

Table 2 - The magnetic properties of samples after explosive compaction

Materials	Br (Gauss)	Hc (Oersted)	(B·H)max (MG·Oe)	Relative density (%)	weight of dynamite used (g)
100% $SmCo_5$	6250	3800	6.3	87	80
	(5100)**	(4750)	(6.1)	(60)	0
40 vol% $SmCo_5$ 60 vol% $Nd_{15}Fe_{77}B_8$*	4625	2250	2.3	85	70
	(4100)	(1400)	(1.6)	(60)	0
100% $Nd_{15}Fe_{77}B_8$*	2500	600	0.5	80	60
	(3100)	(700)	0.8	(60)	0

* 5 μm Average particle size.

** The values in brackets indicate the magnetic properties of those before explosive compaction.

sintered magnet. Generally speaking, only a little improvement is obtained compared to that before explosive compaction. In some cases, a little decrease of Br value is even observed. This may be induced from the obvious change of orientation of aligned powder. The coercivity after explosive compaction shows the same results. Further studies must be undertaken to obtain composites which have uniform high density and good degree of orientation.

CONCLUSION

The theoretical calculation can be applied to the magnetic composites which are anisotropic and contain rare-earth hard magnetic component. There is no obvious difference in phase-distribution between isotropic and anisotropic cases.

The magnetic properties of the constituent powder are a little lower than those of corresponding sintered magnet. The magnetic properties of the $Nd_{15}Fe_{77}B_8$ powder are very sensitive to its particle size. Intrinsic coercivity drops very sharply when the particle size is decreased. It is suggested that Nd-Fe-B system with fine particle size is not suitable for being used in a particulate magnetic composite.

When the two-component magnetic composites are used, it is feasible to increase the magnetic properties as well as the bulk density by explosive compaction. Further studies must be undertaken to obtain composites which have uniform high density and good degree of orientation by this process.

REFERENCES

[1] Wang, K. L., Hsu, S. E. and Chen, J., "A study on the magnetic properties of two-phase particulate magnetic composites", J. Magn. Magn. Mat., 30, p. 37, 1982.

[2] Wang, K. L., Hsu, S. E. and Chen, J., "The hysteresis loop of two-phase particulate magnetic composite with non-linear hard magnetic components", J. Magn. Magn. Mat., 37, p. 77, 1983.

[3] Oesterreicher, H., "Potential of $Fe_{14}R_2B$ type materials for magnetic recording", Proceedings 4th International Symposium on Magnetic Anisotropy and Coercivity in Rare Earth-Transition Metal Alloys, Dayton, Ohio, p. 705, 1985.

MULTI-DIRECTIONAL CARBON/CARBON COMPOSITES PREPARED BY HIP AND CVD PROCESSES

K. Yang, T.-M. Wu and C.-P. Chang

Chung Shan Institute of Science and Technology
P.O. Box 1-26-1, Lungtan, Taiwan 32500, Republic of China

ABSTRACT

Carbon/carbon composites consist of prescribed arrays of carbon (graphite) fibers within a carbon (graphite) matrix. The latter may be formed by impregnated with pitch by hot isostatic processing (HIP) or by a chemical vapor deposition (CVD) process.

Both HIP and CVD processes were carried out in this work. 2-D carbon clothes and 3-D woven preforms were used as starting materials. In CVD process, the reaction temperature, pressure, concentration and flow rate of gas, the volume fraction of fibers and location of samples were important factors of the deposition rate. In HIP process, the temperature, pressure and hipping cycles were used as control parameters.

The studies include the measurements of density, mechanical properties, thermal properties and the examination of microstructure of resulting specimens of two processes. Close relationships have been found between the properties of densified specimens and the parameters of processes.

Comparing with the two processes, CVD shows good thermal properties on the surface and HIP shows excellent infiltrating ability. An attempt was made to combine CVD and HIP processes in the densification of carbon/carbon composites. The results show significant improvement in mechanical and thermal properties.

INTRODUCTION

Graphite is one of the best materials for engineering service at elevated temperature. The unique properties that make it so desirable for high temperature application is the high strength and chemical inertness up to 2000°C. In spite of these advantages, bulk graphite has serious problems in certain applications. Since it is a brittle material with low strain to failure, graphite is susceptible to mechanical and thermal stress failure. The low and anisotropic mechanical strength makes it unsuitable in structure application.

The best method to solve these problems is to reinforce the carbon (graphite) matrix with carbon (graphite) fibers to form the composite. Carbon/carbon composite overcomes the basic problem of low strength, anisotropic mechanical characteristics and brittleness that are typical of bulk graphite. Through a proper orientational arrangement of fibers, carbon/carbon composite has the special strength characteristics to meet the design needs.

The basic reinforcement concept used in the development of 3-D composites is shown in Figure 1. Carbonaceous substrates can be formed by proper arrangement

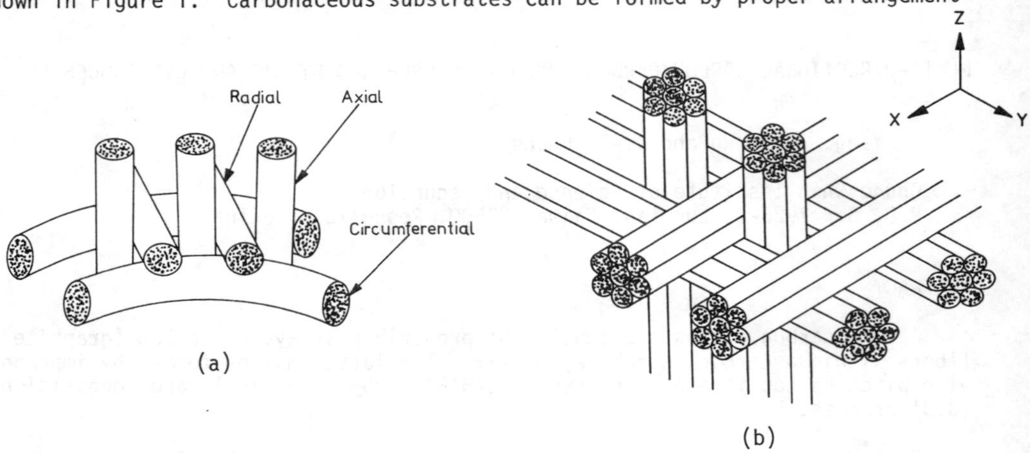

Fig. (1) - Concept of 3-D constructions; (a) block, (b) cylinder

of fibers to attain desired properties. Thermal, mechanical and physical properties of the composites may be enhanced by the appropriate design of weaving parameters such as fiber orientation, volume fraction of fiber in required directions, fiber spacing, weaving density and types of fibers [1].

There are two basic approaches to the formation of the carbon matrix. The first one is to obtain the matrix through the impregnation and carbonization of a carbon-bearing organic material. The second one is through the chemical vapor deposition (CVD) from a hydrocarbon gas.

The most widely used method to introduce carbon matrix into a multi-directional fibrous preform is through the impregnation of the carbonaceous substrate with an organic material such as resin, pitch or tar, and subsequent carbonization or graphitization under inert atmosphere. The conventional impregnation process is conducted at atmosphere pressure and many repeated cycles are needed to reduce the porosity to an acceptable level [2]. The multi-cycles and long process time are main results of: (1) low carbon yield under atmosphere pressure, (2) lack of pressure to an uniformed impregnation and (3) high gas evolution during carbonizing of impregnant [3-5].

A technique was developed to utilize isostatic pressure to increase the efficiency of impregnation and carbonization [6]. Several investigators reported that high carbon yield, high density and good microstructure were obtained as pressure increased [4,5,7]. Fiber Materials, Inc., developed a hot isostatic processing (HIP) equipment that was specially designed to perform the impregnation and carbonization under high pressure and temperature [7]. The main advantages of HIP process are: (1) the isostatic pressure during impregnation maintains a compressive force on the impregnant and prevents running out from pores, (2) high carbon yield and low volatization of carbon-bearing substance under pressure, (3) the pressure improves the uniformity of pore filling as the viscosity of pitch decreases with increasing temperature, and (4) shorter operation time and denser matrix can be obtained.

Another important method to form the matrix of carbon/carbon composites is CVD process. The chemical vapor deposition of carbon from a hydrocarbon gas into a carbon substrate is a complex and difficult technique. The basic process is a pyrolytic reaction of gas followed by the diffusion of active carbon species into the fiber structure. An uniform carbon deposition is usually required in addition to processing the desired properties. Although there are numerous papers concerning the techniques and properties of carbon/carbon composites prepared by CVD process, little is known about the phenomenology and still less about the processes on which it is based.

The isothermal and temperature gradient CVD techniques are two main processes for the formation of carbon composites. In isothermal process, which is emphasized in this paper, the carbon substrate is heated in a reactor with almost uniform temperature. The hydrocarbon gas comes in contact with hot substrate, through a series reaction, carbon is deposited, and hydrogen and by-products are released. The pyrolytic temperature to have carbon deposit on substrate was reported in a wide range of 1050-2000°C [8-11]. Reactors used in CVD process for producing carbon/carbon composite are usually in small size of only several centimeters in diameter to have good temperature profile and gas flow distribution. However, some reactors as large as 72 inches in diameter have been used to do research work on this topic [12,13]. In CVD process, the deposition rate is a function of various factors, such as temperature, pressure, concentration and flow rate of gas, sample configuration and location of samples. Marinkovic and Dimitrijevic [14] reported that infiltration rate increases with temperature and the partial pressure of gas but decreases with reaction time. The shape and size of open pores, and fiber content of substrate have strong influence on the infiltration rate.

The main applications of carbon/carbon composites are expected to be for thermal protection and structural component at high temperature. The application of carbon/carbon composites for the thermal protection of reentry vehicle has been demonstrated [15,16], and the performance of carbon/carbon composites for nosetip application has been reported [16,17]. Advanced propulsion systems have a need for materials which are more erosion resistant with higher dimensional stability and less thermal-structural failure during service. It has been reported [18-23] that carbon/carbon composites have the advantages of simpler design concept, stabler shape and lower ablation rate for nozzle applications. Disk-brake systems made from carbon/carbon composites have undergone extensive development [23,24]. Another application of this material is not-pressing die. The advantages of carbon/carbon for this application are that larger parts can be fabricated, a longer life time can be obtained and higher pressing pressure can be utilized [25].

The object of this work is to produce high performance 2-D and 3-D carbon/carbon composites through HIP and CVD processes. The techniques for substrate and matrix formation are discussed in this paper. Mechanical and thermal properties of multi-directional carbon/carbon composites were examined and a plasma jet testing was used to illustrate the high temperature abrasion and oxidation resistant characteristics of products.

EXPERIMENTAL APPARATUS AND PROCEDURES

A. Substrate Formation

In the preparation of 2-D preforms, two dimensional carbon clothes from HITCO were first dipped in phenolic resin and then cured at 120°C to form prepregs. The substrates were formed by hot pressing the prepregs at 160°C and 710 psi. Post cure at 150°C was carried out after hot pressing and a heat treatment at 900°C was followed to evaporate the volatile content.

The 3-D preforms were woven manually in mutual perpendicular directions with carbon fibers from FMI (density = 1.7 g/cm^3). The woven preforms had a volume fraction of vacancy approximate to 60 percent. The largest preform has been made in this laboratory in cubic and cylinder shape has a dimension of 20 x 20 x 20 cm^3 and 32 cm x 38 cm^3, respectively. Pictures in Figure 2 show the 3-D woven preforms.

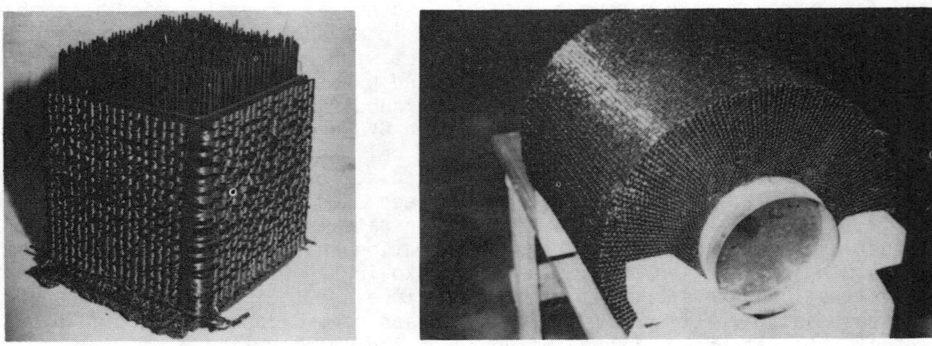

(a) (b)

Fig. (2) - 3-D woven preforms, (a) cubic, (b) cylinder

B. HIP Process

The HIP processes were carried out in this laboratory with a hot isostatic pressing equipment from FMI. The set-up of equipment is shown schematically in Figure 3. Properties and composition of petroleum pitch used as impregnant are shown in Table 1. The preforms were initially vacuum-pressure impregnated using a hot resin transfer system and cured at 250°C. This initial impregnation stage was performed under low pressure (<100 psi) to fix the shape and construction of substrates. After initial impregnation, preforms were submerged in pitch, sealed in a thin wall stainless container and was ready for HIP processing.

In HIP process, the sealed container was placed in furnace and was put in a pressure vessel. The preforms and impregnant were heated at a programmed rate to 650°C, while the pressure was increased and maintained at 13,500 psi. The temperature and pressure profile of HIP process are shown in Figure 4. After HIP process, the composites were carbonized in inert atmosphere at 1000°C and further graphitized at 2200-2600°C. The whole processes of impregnation, carbonization and graphitization are shown in Figure 5. The processes were repeated several cycles in order to attain the desired density of composites.

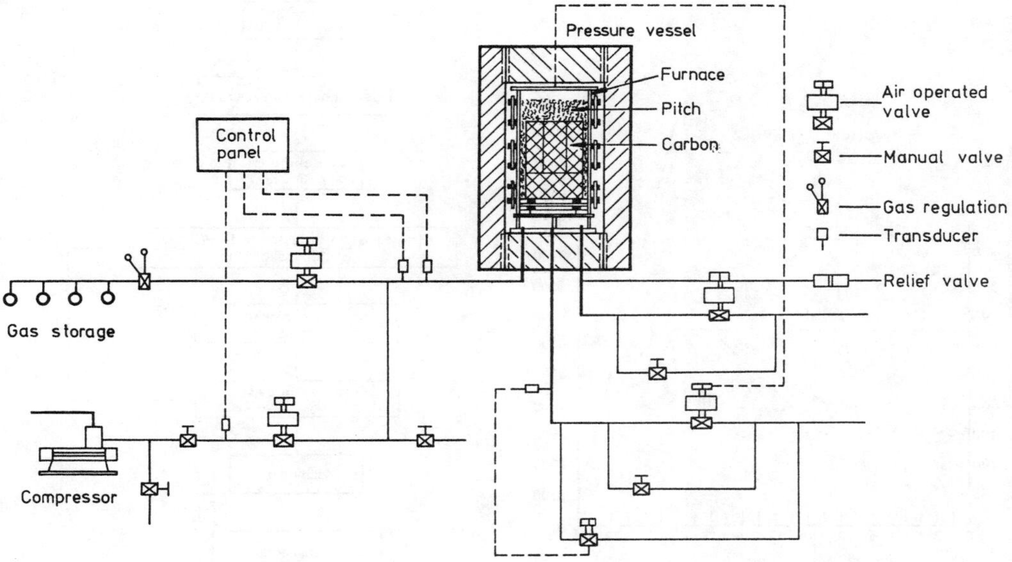

Fig. (3) - Set-up of HIP equipment

Table 1 - Properties and composition of petroleum pitch

Softening point	90 - 130°C
Density	1.20 - 1.30 g/cm^3
Carbon content	>40%
Ash content	<0.15%
Xyline insoluble	<10%
Quinolin insoluble	<1%
Sulfur content	<3%
Viscosity	25 - 30 cps

C. CVD Process

Materials used in CVD process were carbon/carbon composites which were initially hipped to fix the shape and attain certain density. Specimens with a dimension of 1.4 x 2.0 x 11 cm^3 were cleaned by supersonic in acetone solution and then placed in reactor. The set-up of reactor is shown schematically in Figure 6. The heating zone is 120 cm in length with uniform temperature distribution in ±10°C.

Before the CVD process, the reactor was evacuated to 5 torr or below and then purged with nitrogen gas. The heating rate of specimens was controlled at 120°C/hr until 1150°C. The pyrolytic reaction and deposition of carbon were carried out at 1150°C with C_2H_2 gas diluted in 1:1 ratio by N_2 gas as carrier. The flow

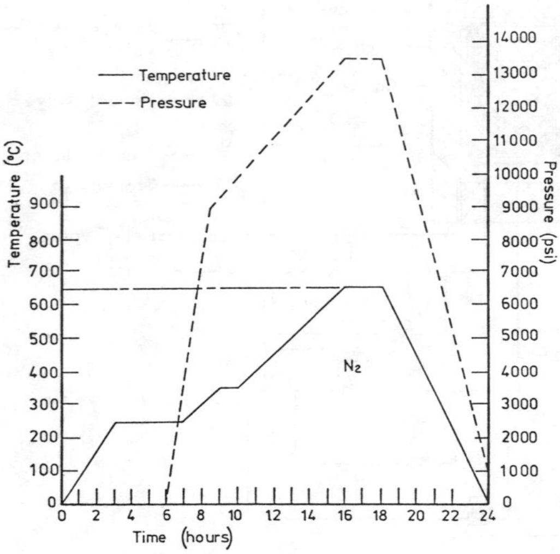

Fig. (4) - The temperature and pressure profile of HIP process

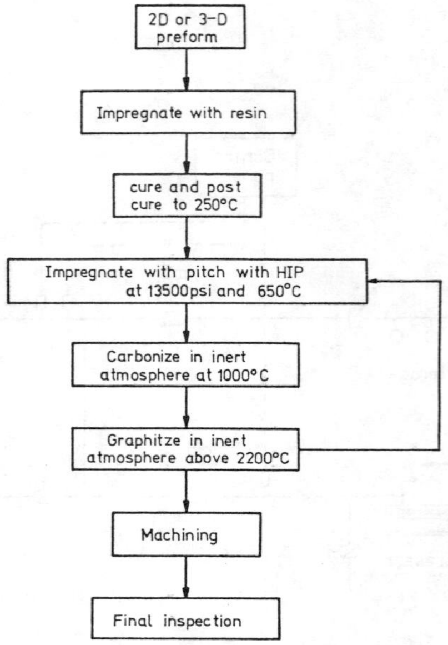

Fig. (5) - The flow chart of HIP process

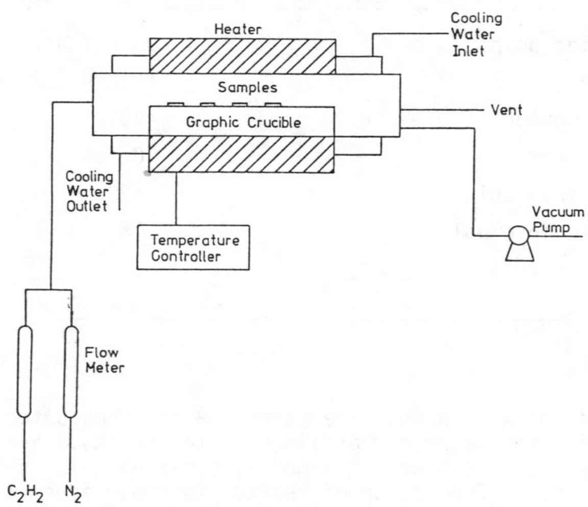

Fig. (6) - Set-up of CVD reactor

rate of gas was 700 cm^3/min. Figure 7 illustrates the process of this operation, Figure 8 shows the reactions taking place in CVD process, and profiles of temperature and gas flow rate are shown in Figure 9. The CVD process was repeated several cycles until a desired deposition depth was obtained.

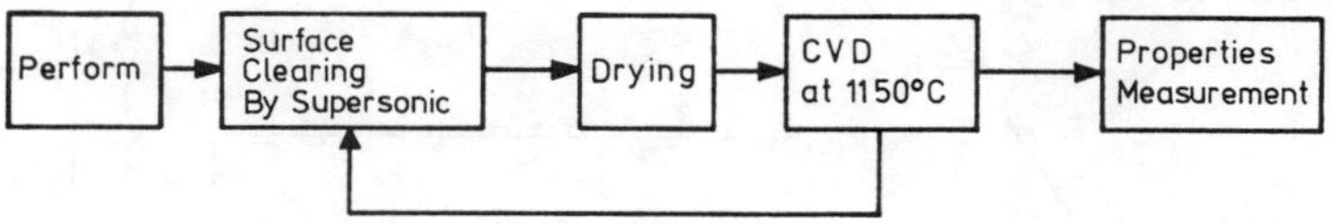

Fig. (7) - Flow chart of CVD process

$$C_2H_2 \xrightarrow{1150°C} A^* \text{ (active species)}$$

$$A^* \xrightarrow{\text{deactive}} C_2H_2$$

$$A^* \xrightarrow{\text{active site}} \text{deposit on substrate}$$

Fig. (8) - Reactions taking place in CVD process

D. Examinations of Physical, Mechanical and Thermal Properties

The carbon/carbon composites prepared by HIP and CVD processes were examined for their densities and flexural strengths. The flexural strengths of composites were measured in a three point flexural testing with 0.25" x 1.25" x 2" test pieces which have a span of 1". Also, the microstructures were examined with JEOL scanning electronic microscope.

The abrasion and oxidation resistant characteristics of composites were investigated with a plasma jet. The set-up of plasma testing is shown in Figure 10. The plasma jet was generated by 110V and 500A with N_2 gas in 50 psi, 100 cfh and H_2 gas in 50 psi, 15 cfh. The temperature of jet was 1800°C or above.

RESULTS AND DISCUSSION

The rate of densification of HIP process, as shown in Figure 11, is dependent on the number of hipping cycle. After six repeated impregnations, a 3-D carbon/

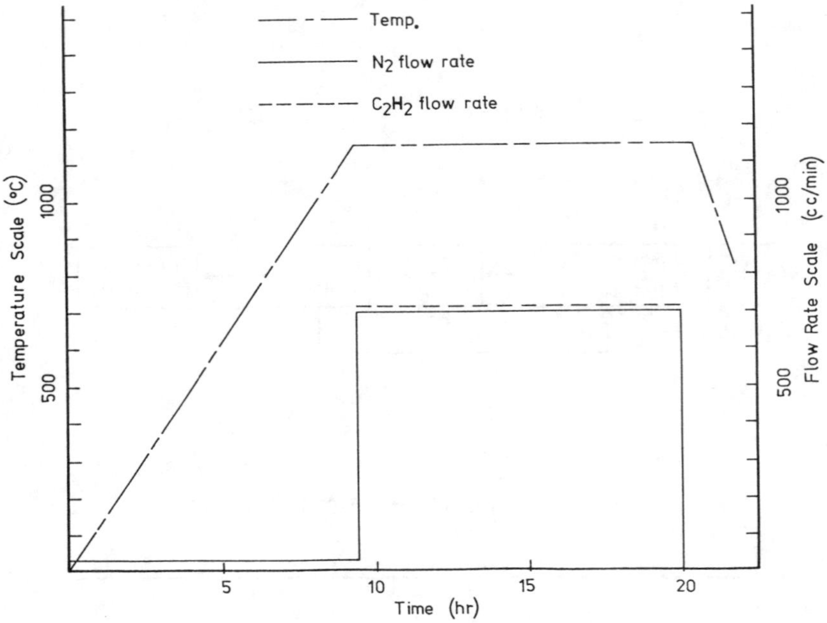

Fig. (9) - Profile of temperature and gas flow in CVD process

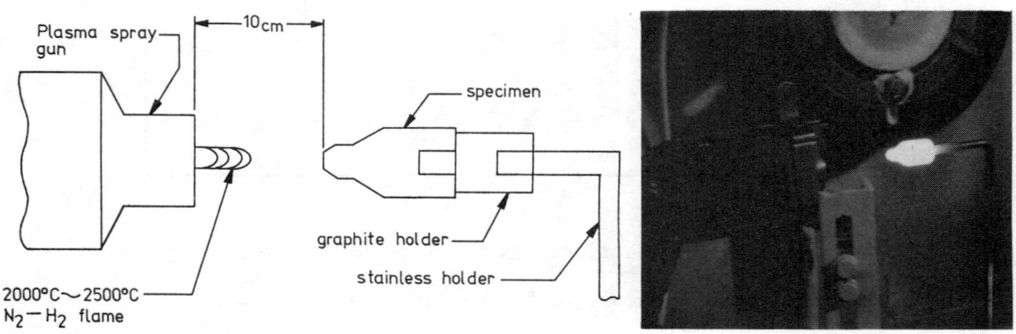

Fig. (10) - Illustration of plasma testing

carbon composite with density of 1.98 g/cm^3 was obtained. A picture of densified billet is shown in Figure 12. The relationship between density and cycles of treatment for 3-D carbon/carbon prepared by CVD process is shown in Figure 13. It was found that the optimum reaction time per cycle is twelve hours, longer reaction time caused carbon crust accumulated on the surface of specimens and prevented further infiltration. In our study, the location of samples has great influence on the deposition rate, the density changed at different positions are plotted in Figure 14.

Microstructures of 3-D carbon/carbon composites prepared by HIP and CVD processes are shown in Figures 15 and 16, respectively. It is clear that voids and pores are gradually closed as density increased with hipping cycles, and CVD process further diminished small voids on the surface. Figure 17 shows that after

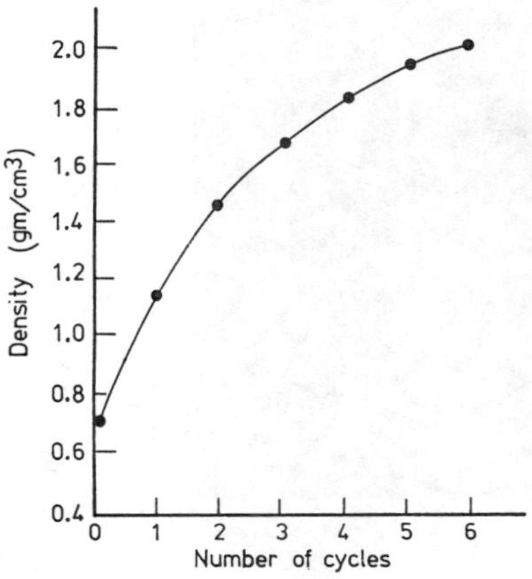

Fig. (11) - Relationship between cycles and densities of 3-D c/c prepared by HIP

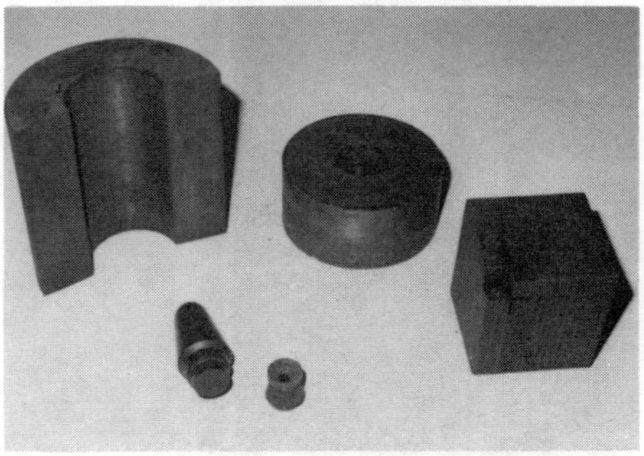

Fig. (12) - Densified 3-D c/c billet

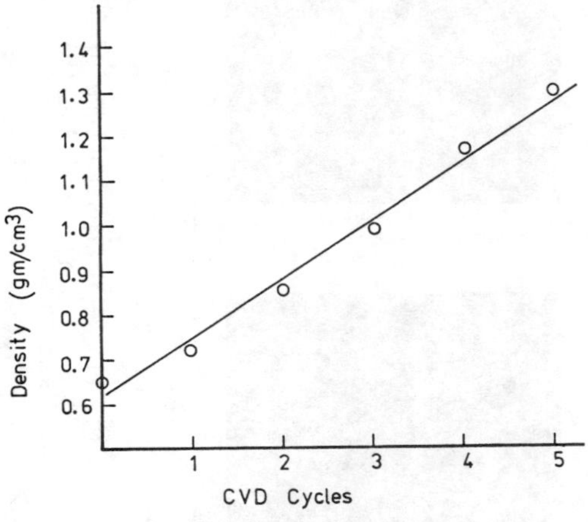

Fig. (13) - Relationship between reaction cycles and densities of 3-D c/c prepared by CVD

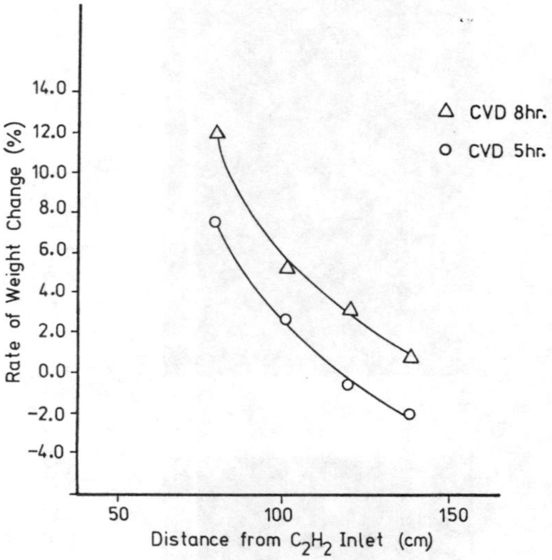

Fig. (14) - Density change at different positions in CVD process

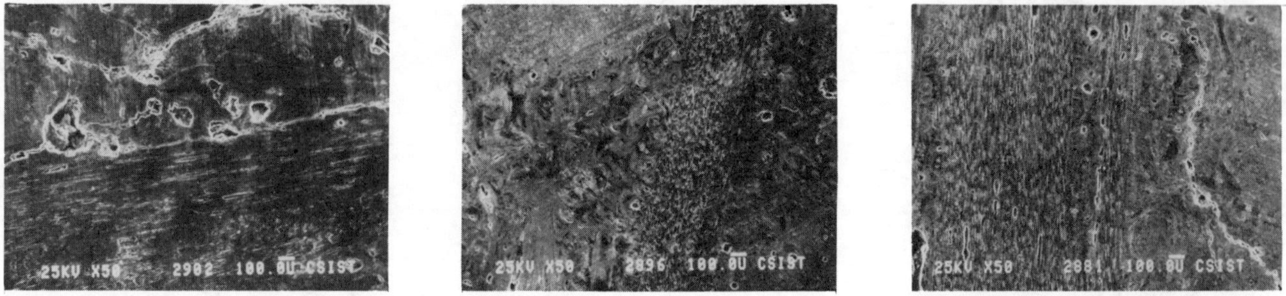

(a) (b) (c)

Fig. (15) - Microstructure of 3-D c/c prepared by HIP. (a) ρ = 1.70, (b) ρ = 1.85, (c) ρ = 1.95

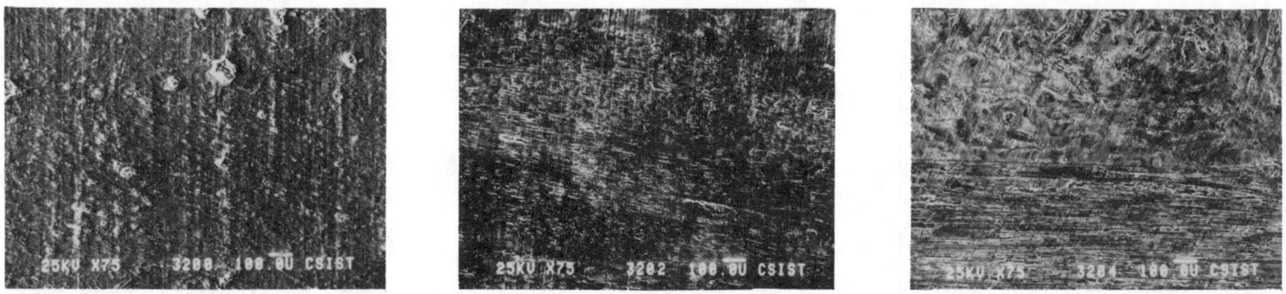

(a) (b) (c)

Fig. (16) - Microstructure of 3-D c/c prepared by HIP + CVD. (a) ρ = 1.75, (b) ρ = 1.86, (c) ρ = 1.98

Fig. (17) - Carbon deposit on the 3-D c/c composite

two cycles of CVD treatment, a 0.03 mm thick infiltration layer on the surface was obtained. Microstructure of 2-D carbon/carbon composites prepared by HIP and CVD are shown in Figures 18 and 19.

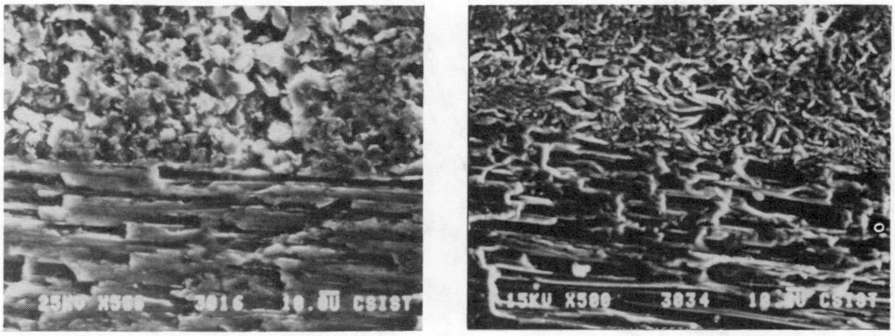

(a) (b)

Fig. (18) - Microstructures of 2-D c/c prepared by HIP, (a) ρ = 1.26, (b) ρ = 1.50

Flexural strengths of carbon/carbon composites prepared by HIP and CVD processes are shown in Figure 20 in comparison with G-90 graphite. The mechanical strength of carbon/carbon composites increases as density increased. The flexural strengths of composites prepared by both HIP and CVD processes are higher than that of the bulk graphite, and a thin layer of CVD infiltration has significant advantage on the mechanical strength of composite.

The abrasion and oxidation resistance of 3-D carbon/carbon composites and graphite were illustrated with plasma jet testings. Figure 21 are pictures of composites and graphite before and after plasma spraying. The data shown in Table 2 indicate that carbon/carbon composites prepared by HIP eroded less than the bulk graphite and a layer of CVD infiltration on hipped composites has excellent erosion resistance at high temperature.

(a)

(b)

Fig. (19) - Microstructures of 2-D c/c prepared by HIP + CVD.
(a) ρ = 1.30, (b) ρ = 1.52

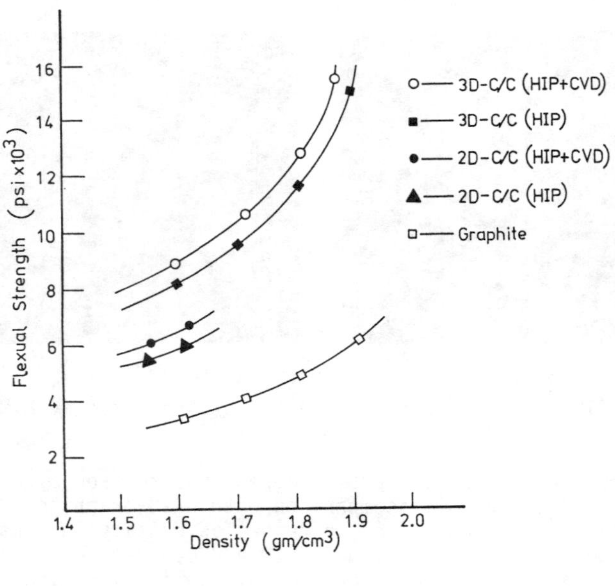

Fig. (20) - Relationships between flexural strength and density
of c/c composites and bulk graphite

c/c by HIP c/c by HIP+CVD graphite
(a)

c/c by HIP c/c by HIP+CVD graphite
(b)

Fig. (21) - Specimens of (a) before plasma spraying,
(b) after plasma spraying

Table 2 - Weight loss of c/c composites and graphite after plasma spraying

Specimens	Weight Loss %	Weight Loss % Ave.
Graphite	5.85	6.01
ρ = 1.90 g/cm^3	6.19	
	6.12	
	5.88	
3-D c/c by HIP	5.14	5.09
ρ = 1.88 g/cm^3	5.12	
	5.06	
	5.02	
3-D c/c by HIP + CVD	4.38	4.52
ρ = 1.88 g/cm^3	4.75	
	4.03	
	4.90	

CONCLUSIONS

Both HIP and CVD processes were carried out successfully in this laboratory for the preparation of multi-directional carbon/carbon composites. HIP process is a very useful technique to produce high density carbon/carbon composites. Through six repeated hipping cycles, a 3-D carbon/carbon composite with a density of 1.98 g/cm^3 was obtained. The chemical vapor deposition of carbon to form composite is a difficult and complex technique. A variety of parameters, such as temperature, pressure, flow rate and location of specimens affect the rate and properties of deposition.

The mechanical strength of carbon/carbon composites, either prepared by HIP and CVD processes, depends directly on their density. The results of experiments showed that the density of composites is increased as cycles of treatment are increasing. A thin layer of carbon deposited on the surface of composite further increases its mechanical strength.

The plasma jet testing showed that carbon/carbon composite has superior abrasion and oxidation resistant characteristics over bulk graphite at elevated temperature, and CVD process on the surface has significant advantages on thermal properties of composite. The combination of HIP and CVD processes in this work resulted in a 3-D carbon/carbon product with high density, high strength and excellent thermal properties at extreme environment.

REFERENCES

[1] Cook, J. L. and Crawford, J. A., "Multidirectional substrates for advanced composites", 76th Annual Meeting of Am. Ceram. Soc., March 1975.

[2] Taverna, A. R. and McAllister, L. E., "The development of high strength three dimensionally reinforced graphite composites", 73rd Annual Meeting of Am. Ceram. Soc., April 1971.

[3] Cook, J. L. and Burns, R. L., "Pressure carbonization of petroleum pitch", Sym. on Petroleum Derived Carbon, Am. Chem. Soc., April 1975.

[4] Forrest, M. A. and Marsh, H., "The effects of pressure on the carbonization of pitch and pitch/carbon fibre composites", J. Mater. Sci., Vol. 18, p. 978, 1983.

[5] Forrest, M. A. and Marsh, H., "The carbonization of blends of pitches and resins to produce anisotropic carbon and the effects of pressure", J. Mater. Sci., Vol. 18, p. 991, 1983.

[6] Cook, J. L. and Lambdin, F., "Fabrication of discontinuous high-fiber content, isotropic carbon/carbon composites", Report Y-1784, Oak Ridge Y-12 Plant, Oak Ridge, Tennessee, 1971.

[7] Lachman, W. L., "Multidirectionally reinforced carbon/carbon composites", Proc. of 1978 Int. Conf. on Composite Materials, p. 1302, 1978.

[8] Branoff, B., Pierson, H. O. and Schuster, D. M., "The effect of chemical vapor-deposition conditions on the properties of carbon-carbon composites", Carbon, Vol. 11, p. 177, 1973.

[9] Bullock, R. E. and Mckague, E. L., Jr., "Radiation effects on mechanical properties of carbon/carbon composites", Carbon, Vol. 11, p. 547, 1973.

[10] Hirakata, H. M. and Ibaraki, T. I., "Carbon-carbon composite material and method for its production", U.S. Pat. 4, 225, 569.

[11] Jasienko, S. and Machikowski, J., "Studies on pyrolytic carbons obtained by acetylene pyrolysis at 1273K-II", Carbon, Vol. 19, p. 205, 1981.

[12] Baver, D. W. and Kotlensky, W. V., Am. Ceram. Soc. 23rd Pacific Coat Region Meeting, San Francisco, Ca., 1970.

[13] Gebhardt, J. J., Yodsnukis, J. J. and Stover, E. R., 4th Nat. SAMPE Tech. Conf., Palo Alto, Ca., 1972.

[14] Marinkovic, S. and Dimitrijevic, S., "Carbon/carbon composites prepared by chemical vapor deposition", Carbon, Vol. 23, No. 6, p. 691, 1985.

[15] Carnahan, K. R. and Kiger, R. W., "Carbon-carbon composites for space shuttle reentry thermal protection", Adv. Mater. Composites and Carbon, Sym. of Am. Ceram. Soc., p. 171, 1971.

[16] Korb, L. J., Morant, C. A., Calland, R. M. and Thatcher, C. S., "The shuttle orbiter thermal protection system", Bull. Am. Ceram. Soc., Vol. 60, No. 11, p. 1188, 1981.

[17] Cristina, V. D., "Hyperthermal ablation performance of carbon/carbon composites", AIAA Paper No. 71-416, AIAA 6th Thermophysics Conf., Tennessee, April 1971.

[18] Bailie, J. A., "Application of composites to missile structure - a review", SAMPE Quarterly, Vol. 12, No. 2, p. 1, 1981.

[19] Kutlensky, W. V., Flock, T. G., Rogers, D. M., Davis, H. O. and Ellis, R. A., "A review of advanced materials for MX booster nozzles", SAMPE Quarterly, Vol. 8, No. 3, p. 1, 1977.

[20] Kutlensky, W. V., Flock, T. G., Rogers, D. M., Davis, H. O. and Ellis, R. A., "A review of advanced materials for MX booster nozzles", SAMPE Quarterly, Vol. 8, No. 4, p. 20, 1977.

[21] Kutlensky, W. V., Flock, T. G., Rogers, D. M., Davis, H. O. and Ellis, R. A., "A review of advanced materials for MX booster nozzles", SAMPE Quarterly, Vol. 8, No. 4, p. 27, 1977.

[22] Laramee, R. C. and Canfield, A., "Carbon/carbon composites solid rocket nozzle material processing, design and testing", Thioko/Wasato, Div., Pub. No. 0371-31382B, 1971.

[23] "Disk-brake system for aircraft is made from carbon composites", Product Engineering, p. 54, July 6, 1970.

[24] Solon, J. N. and Grahan, G. R., U.S. Pat. No. 3,552,533.

[25] Wise, D. C. and Trask, R. B., "Graphite composites for use as hot pressing dies", 73rd Annual Meeting of Am. Ceram. Soc., Chicago, April 1971.

CHARACTERIZATION OF PZT-POLYMER PIEZOELECTRIC COMPOSITES

C.-S. Hao, C.-W. Hsieh, P.-C. Yao, S.-J. Yang and S.-E. Hsu

Chung Shan Institute of Science and Technology
P.O. Box 1-26-6, Lungtan, Taiwan, 32500, Republic of China

ABSTRACT

This paper discusses PZT/polymer composite characteristics for underwater applications. The investigation of connectivity include 0-3 and 3-3 types of composite materials. The composite of 0-3 connectivity is made by uniform embedding sintered PZT powder in a polymer matrix. The particle size of PZT powder, powder distribution, the kinds of polymer and thickness of materials effect on hydrostatic properties of composited piezoelectric materials. Composites with 3-3 connectivity are made by sintered porous PZT ceramic and back-filled with polymer and its porosity, polymers and thickness effect on piezoelectric properties.

INTRODUCTION

Nowadays, PZT [$Pb(Zr,Ti)O_3$] ceramics have been excelled as a piezoelectric transducer material. If applied to underwater as a hydrophone, the PZT ceramics would suffer from several disadvantages, for example, low sensitivity, brittle, etc. In recent years, several types of PZT-polymer composites have been fabricated to improve the acoustic impedance matching with water [1,2]. Newnham et al [3-9] studied on two phases of PZT/polymer composites. Different types of macrosymmetry and interphase connectivity of composites are very different in piezoelectric properties. In all composites, the density (ρ) and permittivity (ε) of solely PZT ceramic are lowered by the introduction of a polymer phase, then increase the elastic compliance and product of $\overline{d}_h$ and $\overline{g}_h$, i.e., figure of merit for hydrophone applications. Therefore, the composite materials are more suitable to underwater applications.

Yamada [2] had fabricated 0-3 type composite by PZT powder embedding in PVDF matrix. Safari et al [6] prepared by pouring the spheres onto a flat surface to form a monolayer and then covering them with a suitable polymer. The methods of fabricating PZT/polymer composites will give flexible composite transducers.

The 3-3 connectivity of piezoelectric composite was made by Skinner [7] using a lost-wax method with coral as a starting material. Through the casting wax, leaching, slip casting of PZT and sintering could obtain porous structure. Shrout et al [8,9] fabricated three-dimensionally interconnected structure by PZT powder and polymethylmethacrylate (PMMA) sphere mixture. Composites with 3-3 connectivity were made by sintered porous PZT ceramic and back-filled with polymer.

This paper has been concerned with new method for fabricated 0-3 type and 3-3 type piezoelectric composites, studies the PZT distribution in two-phase system of composites and effect on piezoelectricity.

EXPERIMENTAL

The composite of 0-3 type was made by uniform embedding sintered PZT powder in a polymer matrix. To prepare the ceramics, 95 wt% of PZT-501A[+] was mixed with 5 wt% of 4% PVA solution. After mixing and drying the powder, 1 inch diameter and 4 ~ 6 mm thick, were pressed at 15000 lbf. The pellets were placed on an alumina crucible and binder was burned out at 550°C for one hour. Sintering was carried out in a sealed alumina crucible and PbO-rich atmosphere was maintained with $PbZrO_3$ powder, then at a heating rate of 200°C per hour with a soak period of 75 minutes at 1250°C. We could smash the sintered PZT and choose the range of particle size, then degas the mixture of PZT powder and epoxy. The 0-3 composites were made by sintered PZT powder which smashed PZT and chooses the range of particle size 44-74 μm, 74-149μm and 149-250μm, respectively, A, B and C. These kinds of polymer choose epoxy 230 and 212, respectively, 1 and 2. The sample fabricated by Doctor-Blade* and punched any shape needed.

For 3-3 type composites, we used soft PU sponge as a starting material, structure of porous material could be obtained by acid leaching. 85 wt% PZT-501A was mixed with 15 wt% of 8% PVA solution, then employed casting-like method. The skeleton of sponge could absorb the mixture. After drying, the porous materials were placed on an alumina crucible, PU and binder were burned out to 360°C in ten hours. Used the condition of sintering at a heating rate of 200°C per hour with a soak period of one hour at 1250°C. After sintering, porous PZT was placed in a box shape container which made by aluminum foil and back-filled with epoxy**.

The epoxy was cured at room temperature for ten hours. Finally, the composites were polished on silicon carbide paper to expose the PZT and to ensure that the faces of the composite were smooth and parallel. Electrodes of air-dried silver paste were applied. The composites in 100°C oil bath for a few minutes ensured thermal equilibrium with surrounding, poling was done at a field of 20 to 125 Kv/cm for 30 minutes. As temperature was cooling to 60°C, cut down the powder source. Piezoelectric properties were measured after 24 hours.

The dielectric constants of all the composites were measured using an automated capacitance bridge§. The piezoelectric coefficient $\overline{d}_{33}$ along the poling direction was measured using a d_{33} meter[++]. The hydrostatic piezoelectric coeffi-

---

[+] Ultrasonic Powder, Inc.

[*] Tam Ceramics, Inc., M-164

[**] CIBA-GEIGY, Inc., Aradite 230, 212

[§] Hewlett Packard, Inc., M-4192A

[++] Channel Products, Inc.

cient $\overline{d}_h$ was measured by changing pressure in an oil chamber at the rate of 200 psi/sec and collecting the charge with a charge amplifier[+++] and recorder. Figure 1 shows the apparatus[§§] used to determine the $\overline{d}_h$ value. The piezoelectric voltage coefficients $g_{33} = d_{33}/\varepsilon_o K_{33}$ were calculated from the measured values of $\overline{d}_{33}$, $\overline{d}_h$ and $\overline{K}_{33}$.

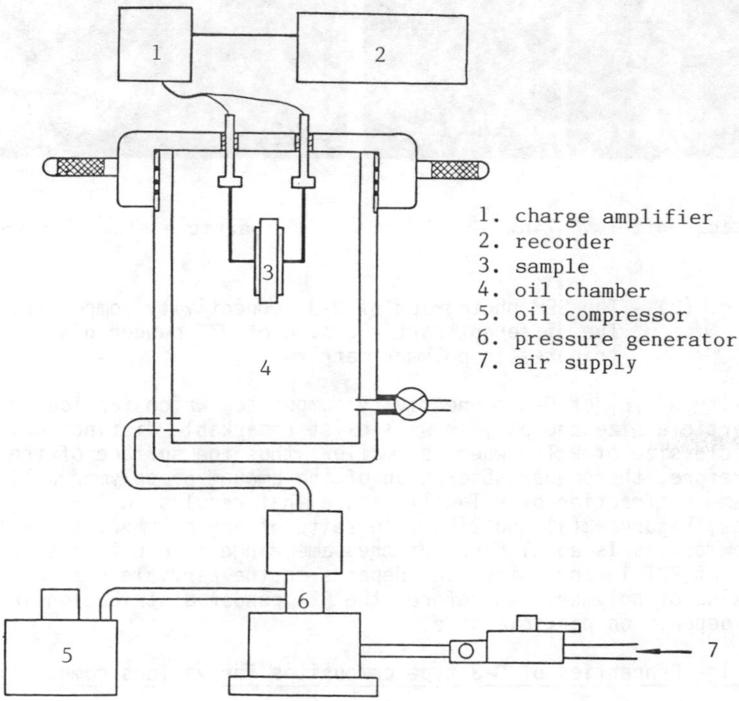

Fig. (1) - The scheme of measured $\overline{d}_h$

RESULT AND DISCUSSION

From Figure 2, the 0-3 composite with different PZT particle size, we can discover the PZT powder is discontinuous distribution in polymer matrix that the composite corresponds with the definition of 0-3 type composite.

[+++]Bruel and Kjer, Inc., Type 2650

[§§] Yamchen Industrial Co., Ltd.

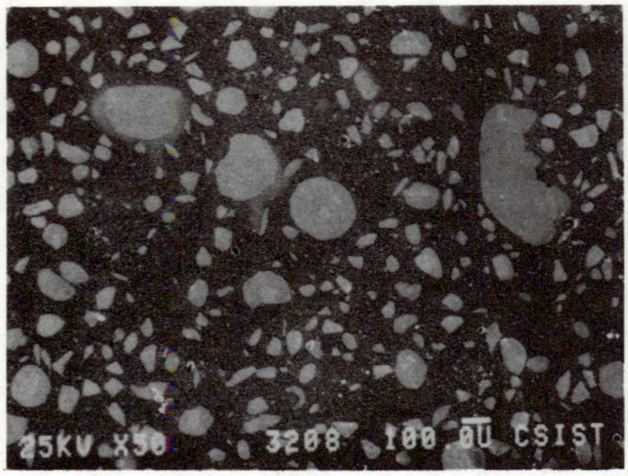

(a)

particle size: 44~74μm

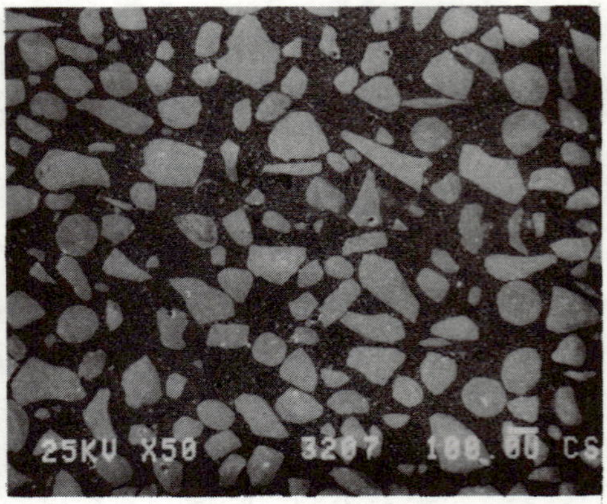

(b)

particle size: 74~149μm

Fig. (2) - The SEM photograph of 0-3 connectivity composite. The different particle size of PZT powder distributes in polymer matrix.

The density of (ρ) of 0-3 connectivity composites which fabricated with the different particle size and polymer will exist remarkable distinction. From Table 1, the particle size of PZT powder is smaller, thus the surface of the powder is larger, therefore, the powder absorption of the amount of polymer will increase and the volume of fraction of PZT will reduce that results in lowering the density of composites, Figures 2(a) and 2(b). In spite of any polymer, the volume fraction of PZT in composites is equal based on the same range of particle size. The volume fraction of PZT in the composites depends on the particle size of PZT powder, not on the kind of polymer. Therefore, the PZT powder distribution in a polymer matrix only depends on particle size.

Table 1 - Properties of 0-3 type composites for various compositions

	1		2	
	ρ	q	ρ	q
A	3.19	0.3	3.26	0.31
B	3.34	0.33	3.42	0.33
C	3.82	0.4	-	-

ρ: density of composites (g/cm^3); q: volume fraction of PZT. The permittivity and density of polymer 1 is 9.36 and 1.17, respectively. The permittivity and density of polymer 2 is 10.46 and 1.2, respectively.

The strength of electric field is a driving force which can reorient the dipole moment at the period of poling. During poling with a period of 30 minutes at 100°C, the various strength of electric field are applied to the same composition of composites, the piezoelectric coefficient $\overline{d}_{33}$ increases with field but the dielectric constant ($\overline{K}_{33}$) relieves disorder, Tables 2 and 3.

For particle size effect, the value of $\overline{d}_{33}$ increases with particle size at the same field, Tables 2 and 3. Because the particle size of powder increases with the volume fraction of piezoelectric material, Table 1, the driving force of poling will reduce dissipation in the polymer phase of PZT powder surface. The driving force of poling may easily focalize on the domain of PZT crystal which dipole moment occurs reorientation arrangement. For different thickness, 1.0, 1.2 and 1.5 mm, piezoelectric coefficient $\overline{d}_{33}$ will present a maximum value at 1.2 mm, Table 2. It may influence each other by two contrary factors which the contents of PZT and polymer increase at the same time.

Table 2 - The piezoelectric properties of 0-3 type composites using polymer 1 in different composition and poling field (thickness: 1.0 mm)

$\overline{K}_{33}$ = dielectric constant, $\overline{d}_{33}$ = piezoelectric coefficient (pC/N), $\overline{g}_{33}$ = voltage coefficient (mV-m/N)

	25 Kv/cm			50 Kv/cm			75 Kv/cm			100 Kv/cm			125 Kv/cm		
	$\overline{K}_{33}$	$\overline{d}_{33}$	$\overline{g}_{33}$	$\overline{K}_{33}$	$\overline{d}_{33}$	$\overline{g}_{33}$	$\overline{K}_{33}$	$\overline{d}_{33}$	$\overline{g}_{33}$	$\overline{K}_{33}$	$\overline{d}_{33}$	$\overline{g}_{33}$	$\overline{K}_{33}$	$\overline{d}_{33}$	$\overline{g}_{33}$
	45.9	5	12.3	37.66	8	24.0	34.59	9	29.4	36.05	11	34.5	41.55	12	32.6
A	-	-	-	-	-	-	-	-	-	38.03^a	14^a	41.6^a	-	-	-
	-	-	-	-	-	-	-	-	-	49.5^b	12^b	27.4^b	-	-	-
	38.6	11	32.2	37.88	12	35.8	39.37	14	40.2	39.29	15	43.1	40.26	15	42.1
B	-	-	-	-	-	-	-	-	-	37.87^a	17^a	50.7^a	-	-	-
	-	-	-	-	-	-	-	-	-	56.79^b	9^b	17.9^b	-	-	-
C	51.84	15	32.7	56.98	18	35.7	49.4	24	54.9	-	-	-	-	-	-

[a]the thickness of composite is 1.2 mm

[b]the thickness of composite is 1.5 mm

Table 3 - The piezoelectric properties of 0-3 type composite using polymer 2 in different composition and poling field (thickness = 1.0 mm)

$\overline{K}_{33}$ = dielectric constant, $\overline{d}_{33}$ = piezoelectric coefficient (pC/N), $\overline{g}_{33}$ = voltage coefficient (mV-m/N)

	25 Kv/cm			50 Kv/cm			75 Kv/cm			100 Kv/cm			125 Kv/cm		
	$\overline{K}_{33}$	$\overline{d}_{33}$	$\overline{g}_{33}$	$\overline{K}_{33}$	$\overline{d}_{33}$	$\overline{g}_{33}$	$\overline{K}_{33}$	$\overline{d}_{33}$	$\overline{g}_{33}$	$\overline{K}_{33}$	$\overline{d}_{33}$	$\overline{g}_{33}$	$\overline{K}_{33}$	$\overline{d}_{33}$	$\overline{g}_{33}$
A	35.47	9	28.7	36.11	9	28.2	37.11	11	35.5	33.70	12	40.2	34.01	13	43.2
B	40.26	12	33.7	37.24	12	36.4	37.19	14	42.5	40.03	16	45.2	38.98	16	46.4

The composites are made by piezoelectric material PZT and polymer. Therefore, the piezoelectric properties of composites depends on electric properties of polymer. In the experiment, the breakdown voltages of polymer 1 and 2 are above 450 kv/cm and 290 kv/cm, polymer 2 (epoxy 212) can sustain the lower field. Due to the smaller resistance to electric field, the driving force of poling may reduce dissipation in the polymer phase, then increase the $\overline{d}_{33}$ value of composites, (Tables 2,3) and easily poles at the same field, Figure 3.

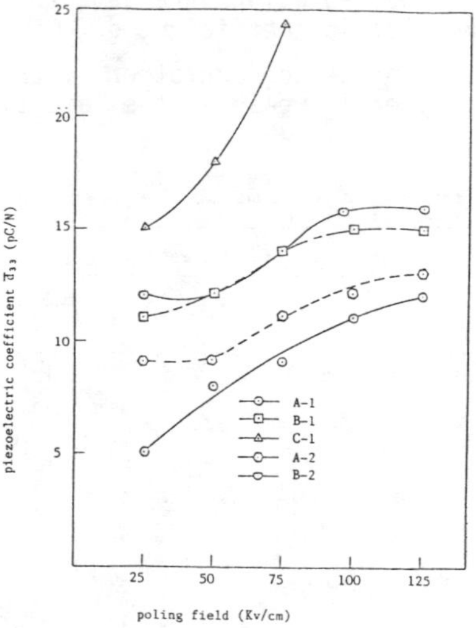

Fig. (3) - Relative $\overline{d}_{33}$ versus poling field, the 0-3 connectivity composite in various composition (particle size - polymer). Thickness: 1.0 mm

For 3-3 connectivity of composites, we can suitably choose different extent of foaming of PU sponge as starting material. We obtain various pore size and porosity (p) of starting materials. Thus, we can control the pore size or porosity (p) of PZT ceramic, at the same time, the density of composites will be controlled. The pore sizes, 1.5 mm and 1.0 mm, are measured in Figure 4. The composite has larger pore size or high porosity (p) that has much more polymer phase, then lowers the density of composite, Table 4. High porosity of composite, the $\overline{d}_{33}$ value is lowering as the content of piezoelectric ceramic decreases, Table 4.

For the polymer effect, the 3-3 connectivity composites are just the same as 0-3 type composites, the composites have higher $\overline{d}_{33}$ value by using polymer 2 as a polymer matrix, Table 4. Compared with 3.0 mm and 1.0 mm of thickness, 3-3 type composite is three-dimensional connectivity, but the thickness of composite may affect the mechanism of pole, the thinner composite has larger $\overline{d}_{33}$ value.

To the figure of merit for underwater application, we can discover from Table 5. The piezoelectric material of composite with polymer has better the figure of merit for underwater application than the solely PZT at the same polihg condition,

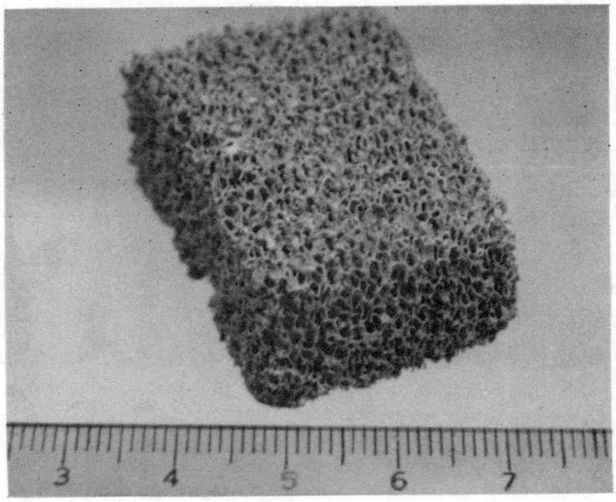

(a)
pore size: 1.5mm

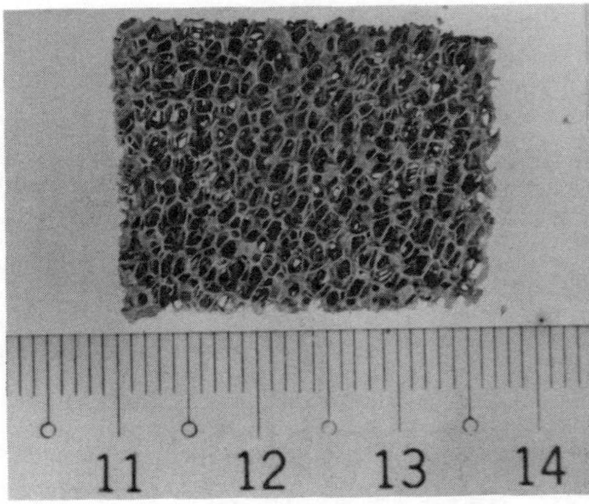

(b)
pore size: 1.5mm

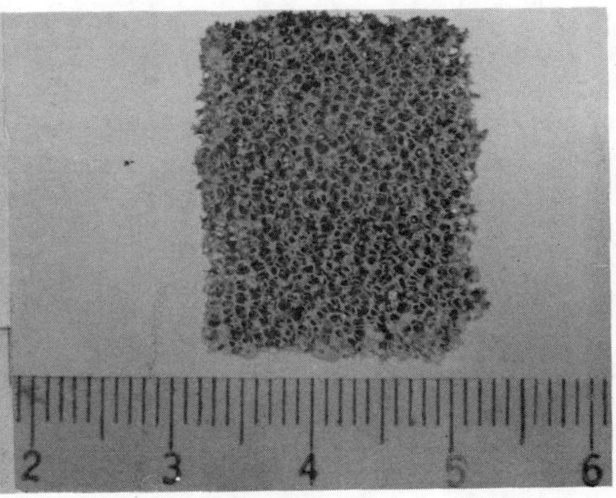

(c)
pore size: 1.0mm

Fig. (4) - The photograph of porous sintered PZT ceramic (adsorption on PU skeleton). (a) and (b): large pore; (c): small size

although in the hydrostatic pressure 1500 psi corresponding to 1050 m water depth. The larger particle size of PZT powder is worse figure of merit. The 0-3 connectivity composite which compose of larger particle size of PZT powder that contain smaller amount of polymer has lower figure of merit than the smaller one. The 3-3 connectivity composite which has skeleton of porous PZT ceramic is three-dimensional connectivity and low density, its figure of merit is better than 0-3 connectivity composite. From Figure 5, the curve decreases, the hydrostatic pressure affect on the hydrostatic piezoelectric coefficient $\overline{d}_h$ of various materials. Because the increasing rate of charge is smaller than the pressure, the $\overline{d}_h$ value will decrease with hydrostatic pressure increasing.

Table 4 - The physical and piezoelectric properties of 3-3 type composite in different composition (poled at 20 Kv/cm, 100°C, 30 min)

$\overline{K}_{33}$ = dielectric constant, $\overline{d}_{33}$ = piezoelectric coefficient (pC/N), $\overline{g}_{33}$ = voltage coefficient (mV-m/N)

Polymer	1								2				
Thickness	3.0 mm			1.0 mm					1.0 mm				
Property	$\overline{K}_{33}$	$\overline{d}_{33}$	$\overline{g}_{33}$	ρ	P	$\overline{K}_{33}$	$\overline{d}_{33}$	$\overline{g}_{33}$	ρ	P	$\overline{K}_{33}$	$\overline{d}_{33}$	$\overline{g}_{33}$
Larger Pore	192.6	60	35.2	2.55	0.81	415	90	24.5	2.59	0.81	282	115	46.1
Smaller Pore	-	-	-	2.87	0.74	229	120	59.2	-	-	-	-	-

The pore size of larger pore is 1.5 mm and the smaller one is 1.0 mm.

Table 5 - Piezoelectric properties of PZT-polymer composites (poled at 20 Kv/cm, 100°C, 30 min)

$\overline{K}_{33}$ = dielectric constant, $\overline{d}_h$ = hydrostatic piezoelectric coefficient (pC/N)

$\overline{g}_h$ = hydrostatic voltage coefficient (mV - m/N)

$\overline{d}_h \times \overline{g}_h$ = hydrostatic figure of merit (x 10^{-15} m^2/N), all measurements made at 25°C, 500 psi

	$\overline{K}_{33}$	$\overline{d}_h$	$\overline{g}_h$	$\overline{d}_h \times \overline{g}_h$
PZT-501A	1359	21	1.75	37
3-3 Connectivity[a]	189	35	21	735
0-3 Connectivity[b] (250 ∿ 420 μm)	64.6	10.7	18.7	200
0-3 Connectivity[b] (149 ∿ 250 μm)	53.9	11.4	23.9	272
0-3 Connectivity[b] (74 ∿ 149 μm)	48.4	15	35	525

[a]The composite uses the larger pore size of porous ceramic.

[b]The composite is made by polymer 2 (epoxy 212) and different particle size.

CONCLUSION

The larger particle size of 0-3 connectivity composite or larger strength of electric field can increase the $\overline{d}_{33}$ value. The PZT powder distribution in a polymer matrix only depends on particle size. The pore size or porosity of 3-3 type composites can affect on the piezoelectric properties. At the same time, the composite is composed of PZT and polymer, so the physical properties of polymer must influence the piezoelectric properties of composite. The composite is better than the solely PZT applied to underwater. For 0-3 composite, the one of smaller particle size which has larger amount of polymer can increase the function of underwater application. We can realize the hydrostatic pressure effect on the hydrostatic piezoelectric coefficient $\overline{d}_h$, the $\overline{d}_h$ value will decrease with hydrostatic pressure increasing.

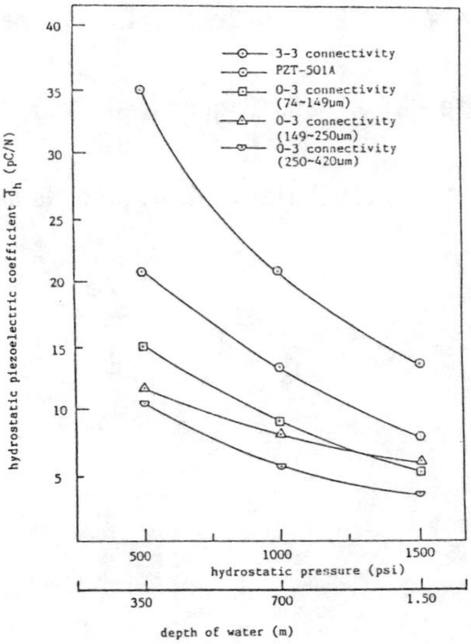

Fig. (5) - Effect of pressure on the hydrostatic piezoelectric properties of composite sample

This paper has been concerned with the forming method of composites. The 0-3 type composite is formed by Doctor-Blade method. It can be more easily fabricated large area and any shape-needed which fits with the punch. The 3-3 type composite employs the PU sponge as a starting material. PU sponge is easily obtained and is one of raw materials in industry, we can control the pore size through the foaming process. So PU sponge is an excellent starting material.

ACKNOWLEDGEMENT

We wish to thank Mr. Chung-Tang Lin, Mr. Jau-Zen Jang and our other colleagues at the Solid State Materials Laboratory.

REFERENCES

[1] Rund, J. and Galgoci, E. C., Mat. Res. Bull. 19, 253, 1984.

[2] Yamada, T., Ueda, T. and Kitayama, T., J. Appl. Phys., 53, 4328, 1982.

[3] Klicker, K. A., Biggers, J. V. and Newnham, R. E., J. Am. Ceram. Soc. 64, 5, 1981.

[4] Safari, A., Newnham, R. E., Cross, L. E. and Schulze, W. A., Ferroelectrics, 41, 197, 1982.

[5] Ting, R. Y., Ferroelectrics, 49, 251, 1983.

[6] Safari, A., Halliyal, A., Bowen, L. J. and Newnham, R. E., J. Am. Ceram. Soc., 65, 207, 1982.

[7] Skinner, D. P., Newnham, R. E. and Cross, L. E., Mat. Res. Bull. 13, 599, 1978.

[8] Shrout, T. R., Schulze, W. A. and Biggers, J. V., Mat. Res. Bull. 14, 1533, 1979.

[9] Rittenmyer, K., Shrout, T., Schulze, W. A. and Newnham, R. E., Ferroelectrics, 4, 189, 1982.

GAS-PERMEABILITY IN GRAPHITE FIBER COMPOSITE AS A RESULT OF CVD PROCESSING

Y.-H. Lee

College of Engineering, National Taiwan University
Taipei, Taiwan, Republic of China

ABSTRACT

Gas-permeability measurements in vacuum and at low pressures have been performed on artificial graphites and graphite fiber composites. Results showed that the measurements carried out under vacuum and at pressures up to 1 atm exhibited various gas-permeability curves having a least permeability point close to the vacuum part. In a second experiment, a CVD pyrolytic carbon process was utilized to coat the artificial graphites and graphite fiber composites, and gas-permeability measurements were also carried out on these as-deposited samples. Due to the different coating temperatures in CVD process, sensitive analysis data were obtained. Results show that low propane gas concentrations are beneficial to a low infiltration condition. A good linear relationship has also been found between the CVD coating time and the infiltration condition with the gas-permeability measurement.

INTRODUCTION

There has been a pronounced improvement in shear strength of composites with the deposition of carbon on the surface of graphite fibers, with only minor loss of tensile strength of fibers. Carbon-carbon systems formed by the controlled pyrolysis of carbon fabric reinforced plastic laminates constitute a class of materials useful in hyperthermal environments. Among the advantages by these carbon-carbon composites are high emittance, elastic, non-brittle behavior, high strength to weight ratio and outstanding ablative characteristics.

The mechanical and physical properties of carbon/graphite composites has been mentioned by Delmonte [1] in his book. He said that many of the properties characterization methods for composites have been extrapolated from the metal industry. In consequence, methods of analysis have been developed which reflect the homogeneity of the metals. These methods cannot anticipate the numerous variables in the composite constituents or in the structure of the composite variables which have significant influence on the behavior of the composite structure. Here, the same conception stimulates us to consider what a microstructure, like porosity can have a significant influence on that of the composite structure, especially, after a chemical vapor deposition (CVD) processing. The author [2,3] has conducted an experiment to prove that on the pyrolytic carbon CVD processing, there are two crucial rate-determining mechanisms existing. They are the chemical reaction process and the diffusion reaction process [4]. The former is directly related to the application of CVD densification of microporous substances and the latter is associated to a performance of manufacturing a sound film for the protection from abrasion. As for the applications, it can be found in many excellent papers [5-7]. Here, we are interested in the microstructure of the carbon/carbon or carbon/graphite composites. For example, work was performed to improve high quality carbon-carbon composite materials for re-entry vehicle thermal protection applications. High strength carbon-carbon composites can be achieved by post treatment through porosity control. However, porosity control must begin at the fabric resin impregnation process, then re-pyrolysis, chemical vapor infiltration and heat treatment procedures. Control of porosity to optimize the matrix pore spectra for subsequent post pyrolysis impregnation and/or CVD treatments is highly desirable since it should directly improve the matrix-binder-reinforcement relationship. Up to date, it is believed that the basic process that resulted in the desired matrix binder-reinforcement relationship would be adaptable for all carbon-carbon systems. The dimension measurements and weight loss percentage were made for each performance after pyrolysis. An advanced investigation method is necessary to be developed to match the more advanced research needs. Gas-permeability measurement in near vacuum or at low pressures is of such an elaborated method. Details of this measurement are to be described.

GAS-PERMEABILITY MEASUREMENT OF CARBON/CARBON COMPOSITES

A. Chemical Reaction Process and Diffusion Reaction Process

Mechanism of pyrolytic carbon formation and the kinetics of the thermal decomposition of raw material gases have been reviewed by Kotlensky [4]. As intended in this paper, description of actual phenomena of such CVD on the carbon/carbon or the carbon/graphite composites will be given with our experimental data.

Experimental system has been described previously [2,3]. Here, we put point on the explanation of how to select an optimum coating or infiltration condition. For example, at 1 atm (760 mmHg), the reactant gases were fed into the reaction chamber with a flow rate of 38 ml/min (0.05 cm/sec). It is found that their deposition rates depended on their deposition temperatures, as shown in Figure 1. Between the deposition rate and the propane gas concentration, a nearly linear relationship could be observed. Reflection point was found under higher temperature conditions. It is meant that a change of conversion rate from propane gas to pyrolytic carbon film occurs at higher temperature regions with higher concentration conditions. These data were depicted in Figure 2 to show a result similar to Arrhenius plot. A characteristic deposition temperature about 1350°C existed, despite of varying reactant gas concentration. Deposition reactions above this characteristic temperature were controlled by the rate of diffusion reaction and below it were determined by the rate of chemical reaction. This indicates that the region above 1350°C subjects to a diffusion reaction process and below 1350°C, it undergoes a chemical reaction process. The microstruction can be seen in Photograph 1.

B. Gas-Permeability Measurement of Carbon/Carbon Composites

Gas permeability measurement has been utilized to evaluate porous solid media from long ago. Hewitt [8] has reviewed this measurement within graphite that promote a good understanding of these process. However, owing to the advanced high technique of manufacturing processes of graphite and composite, the gas-permeabili-

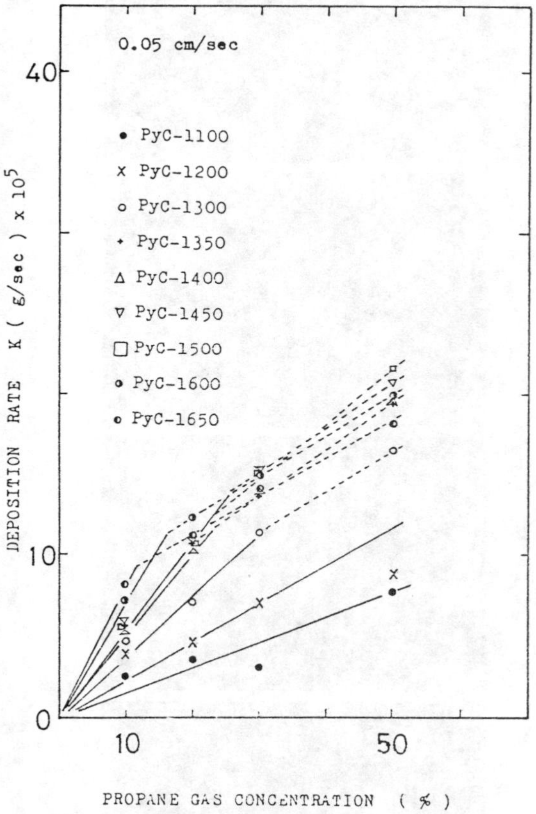

Fig. (1) - Relationship between deposition rates and propane gas concentrations

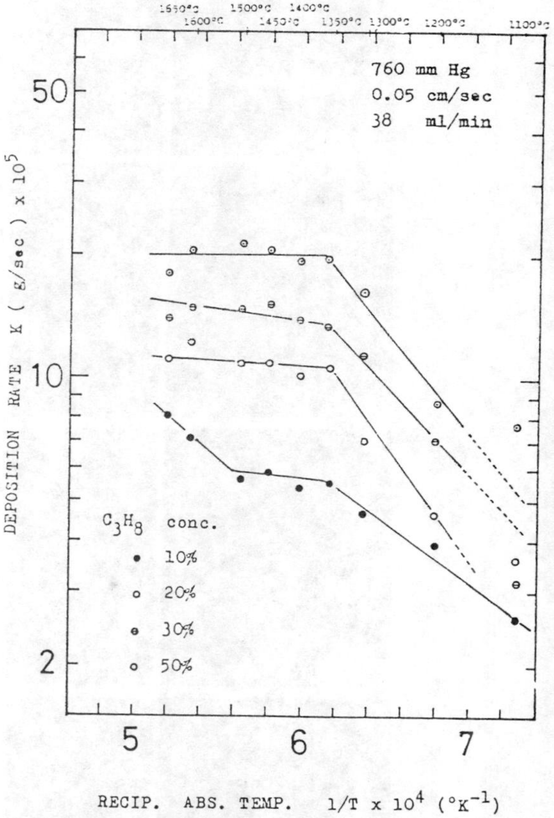

Fig. (2) - Deposition rates are dependent on their deposition temperatures

ty measurement face a variant in its physical meaning. The author has conducted an experiment [9] with this method but treated these data with another procedure.

Gas-permeability apparatus and its measuring procedure are similar to that already reported [2,9]. Here, it is worthy of emphasizing that the gas-permeability was referred to a ratio of "calculated pore radius" in place of the usual one. The "calculated pore radius" was computed with a semiempirical equation of Creutz and Zumwalt [10] to infer a minimum value by computer. Such a minimum value suggests that at this pressure the pore of solid media is near to its mean free path [11]. Experimental data shown in Figure 3 indicate that a minimum existed on every measured curve, even in the case of chemical vapor deposition at higher temperature. All the pyrolytic carbon deposited samples, whose data shown in Figures 1 and 2 can be calculated and depicted in Figure 4. It is evident that the deposition temperature near 1350°C (the characteristic temperature) exhibit a dense

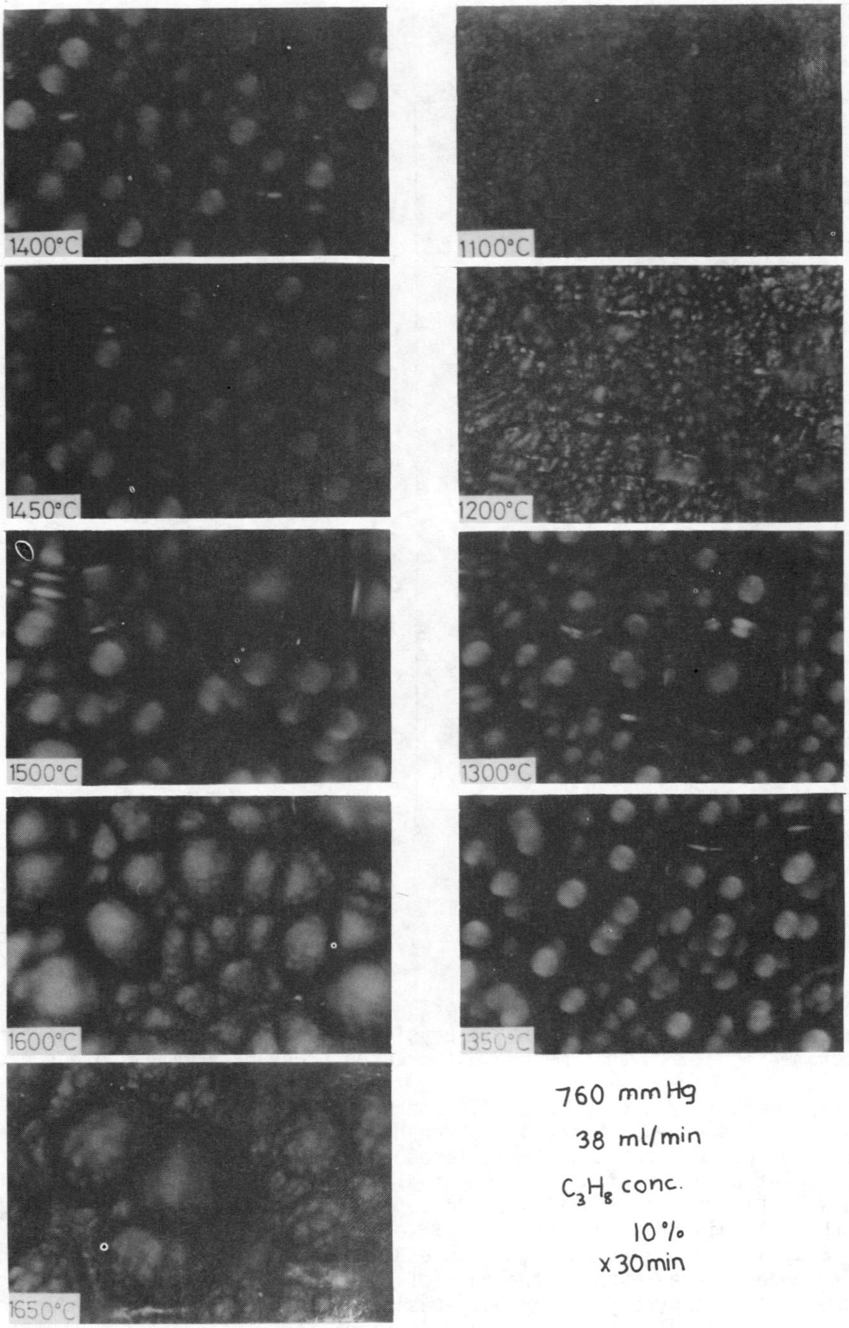

Photograph (1)

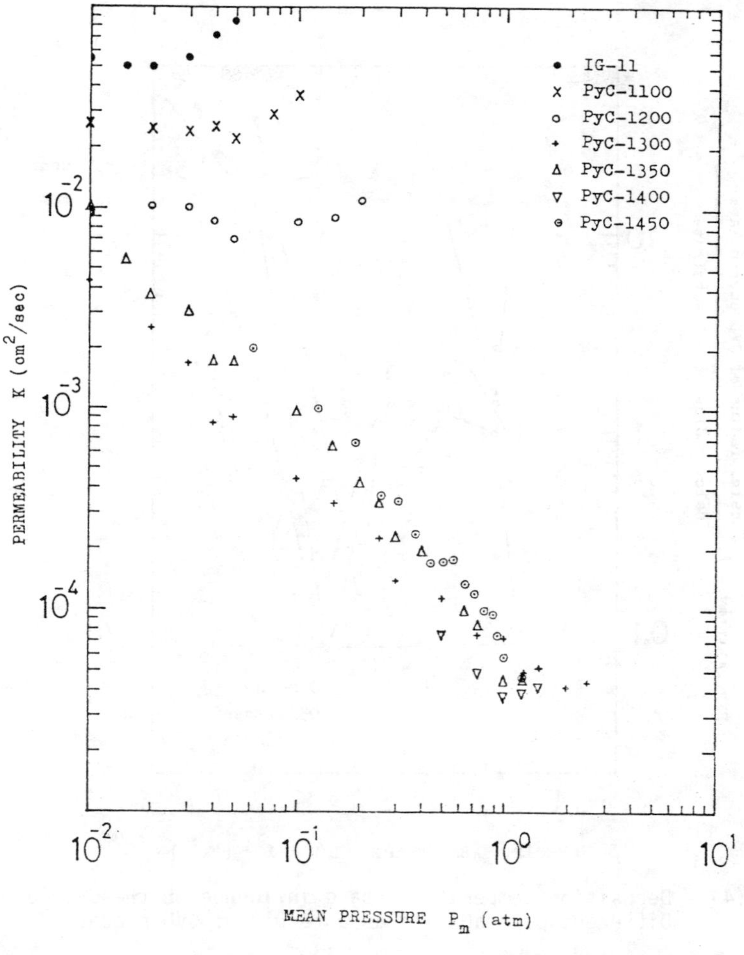

Fig. (3) - Relationship between permeability and mean pressure

film coated. The lower the concentration of reactant gas, the denser the as-deposited film becomes.

GAS-PERMEABILITY MEASUREMENT OF CARBON/GRAPHITE FIBER COMPOSITES

A. Impregnation Percentage and Geometric Factor

From the experimental data mentioned above, a conclusion was deduced that the CVD pyrolytic carbon deposited at temperature higher than the characteristic value of 1350°C subjects to a diffusion reaction process and below that of the characteristic temperature undergoes a chemical reaction process. The latter is directly related to the application of CVD densification of microporous substances, or say, infiltration, impregnation. That is, the reaction region we intend to investigate for the carbon/graphite fiber composites application.

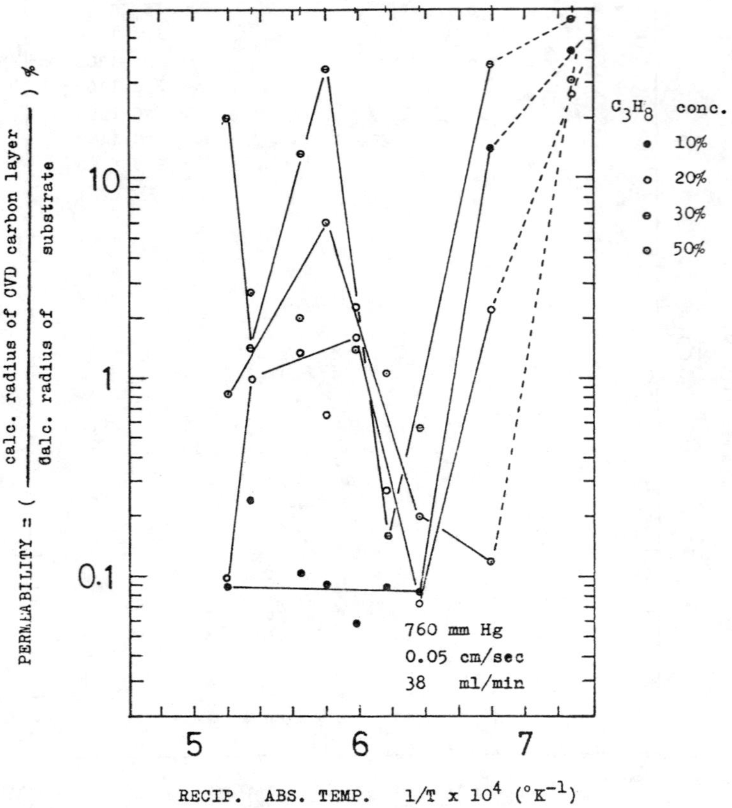

Fig. (4) - Deposition temperatures have influence on the CVD carbons. Different permeability as a result of CVD process

In this experiment, test specimens were cut from an exit cone into 3 cm diameter discs. Its substrate composition with carbon fiber structures bonded with resin or pitch char systems. Dimension of these specimens all have been indicated in Table (not shown here).

In spite of the specimens being cut in different shape, the weight of coated film increased with the increasing of specimens' volume. Results were shown in Figure 5. The isothermal CVD infiltration process had conducted under the condition of having 0.05 cm/sec gas rate at 20 torr pressure. The concentration of propane gas was 20% and its compositions were propane gas in 8 ml/min and carrier gas (95% Ar + 5% H_2) in 30 ml/min. In order to evaluate the merit of this CVD infiltration process, a calculation method was derived from assuming that the density of deposited carbon is of 2 g/cm^3 and the density of the three dimensions carbon/graphite fiber composite is also assumed of 2 g/cm^2 post infiltration. Through a procedure of having the coated weight of the surface carbon film sub-

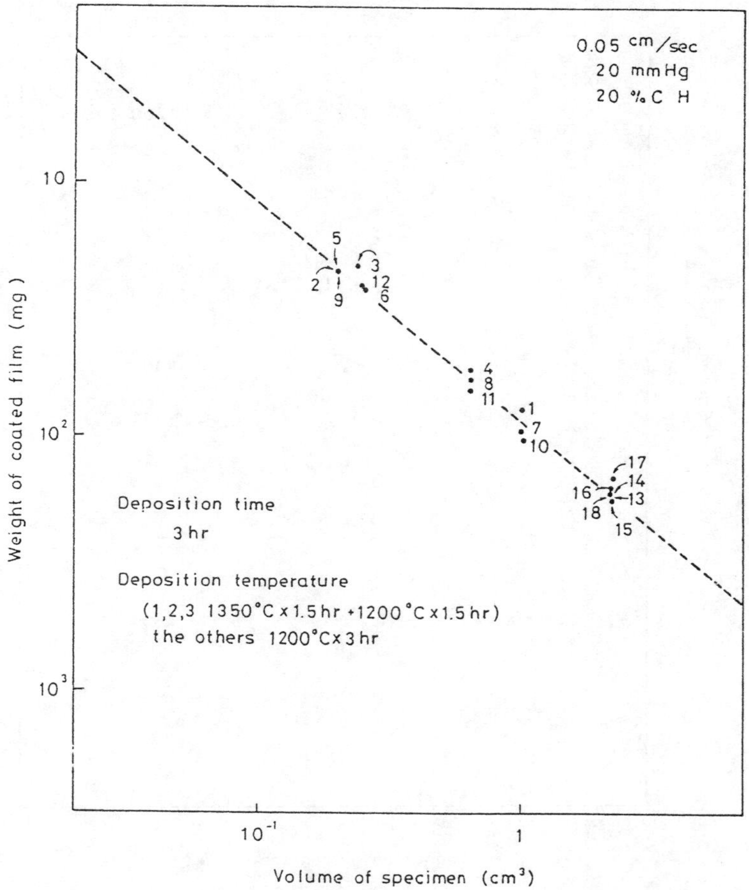

Fig. (5) - The weight of coated film increased with the increasing of specimens' volume

tracted from the total deposited weight, the remainder is impregnated carbon weight. The equation is shown as:

$$\text{Impregnation } (\%) = \frac{(\text{amount of impregnated carbon})}{(2\text{-density of substrate}) \times (\text{volume of substrate})} \times 100\%$$

Figure 6 exhibits no clear relationship between impregnation and volume of specimen. However, it is evident that the temperature condition of the CVD infiltration process influence its impregnation very much. At 1200°C, the infiltration prevailed over that of 1350°C.

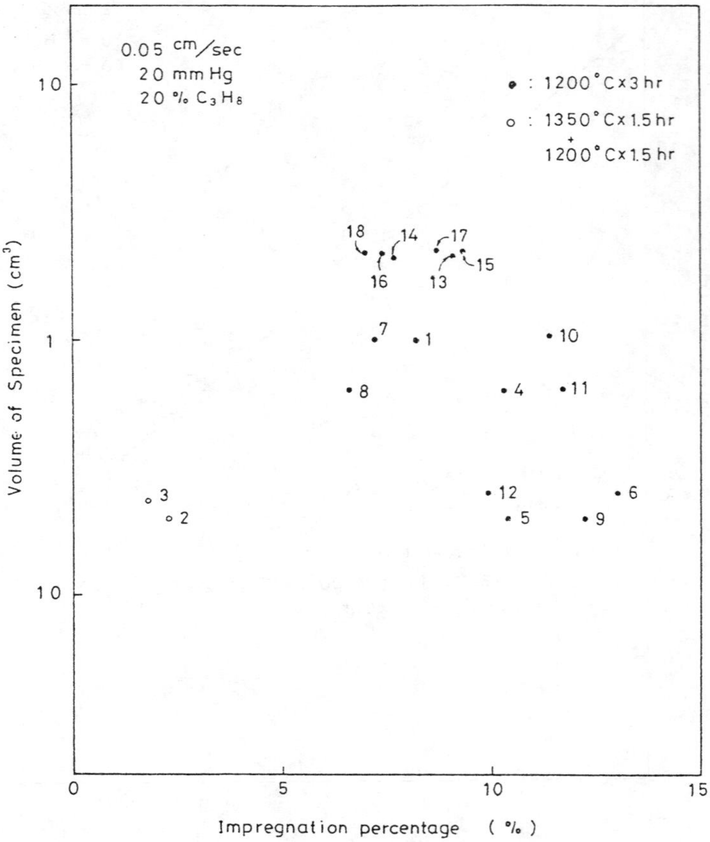

Fig. (6) - Impregnation percentage with respect to volume of specimen

B. Impregnation Percentage and Concentration Temperature

In previous section, results showed that the lower temperature was beneficial to the CVD infiltration process. Here, another experiment was conducted at concentration temperature 1200°C and 1150°C, respectively. The results confirmed it as indicated by Figure 7. The time factor seemingly failed to interpret any good response between deposition rate and impregnation.

C. Gas Pressure of the CVD Infiltration Process

Atmosphere in the reaction chamber at lower pressure can control the deposition rate and influence their impregnation state. In Figure 8, it exhibits a nonlinear result between gas pressure and deposition rate or impregnation percentage. The ceramic pedestal which supported the suscepter was found that it has some advan-

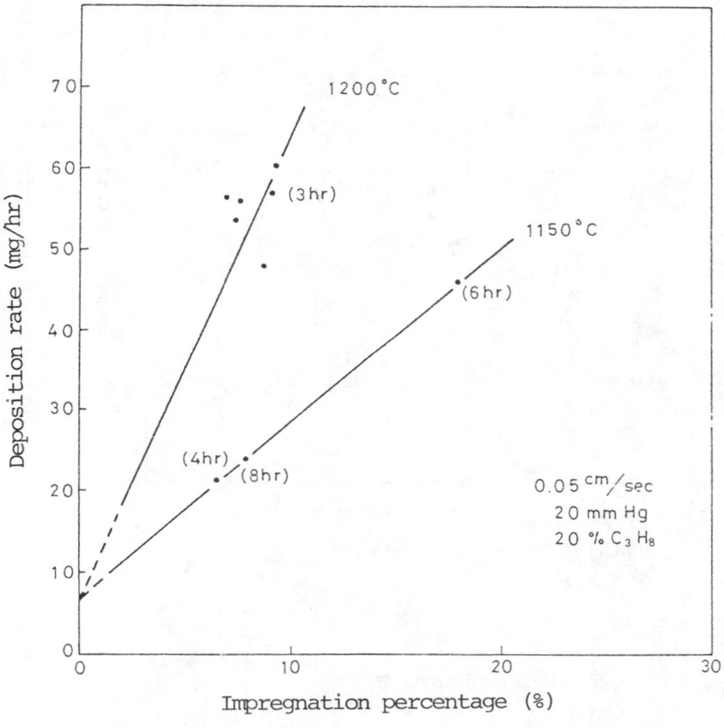

Fig. (7) - Deposition rate influences impregnation effect

tages for the CVD process comparing with the use of a graphite pedestal, at least against its deposition rate.

D. Activation Temperature with Respect to the CVD Process

Activation temperature was taken into consideration that through the pre-treatment at such an optimum temperature, the graphite fiber composite substrate became more activated to be susceptible to the isothermal CVD infiltration process. Figure 9 was illustrated to prove such a phenomenon. It is evident that the characteristic temperature 1350°C render an account of good sensitivity for the impregnation. However, it is difficult to interpret this effect with its deposition rates as shown in Figure 10.

E. Gas-Permeability Measurement of Carbon/Graphite Fiber Composites

The deposition temperature is very important to the impregnation of the carbon/graphite fiber substrate. It is mentioned above in the previous sections. Now, from another point of view, an experiment was conducted under the CVD conditions as shown in Figure 11 and investigated these specimens with the gas-permeability measurement. Results are also indicated in Figure 11. There is a good response

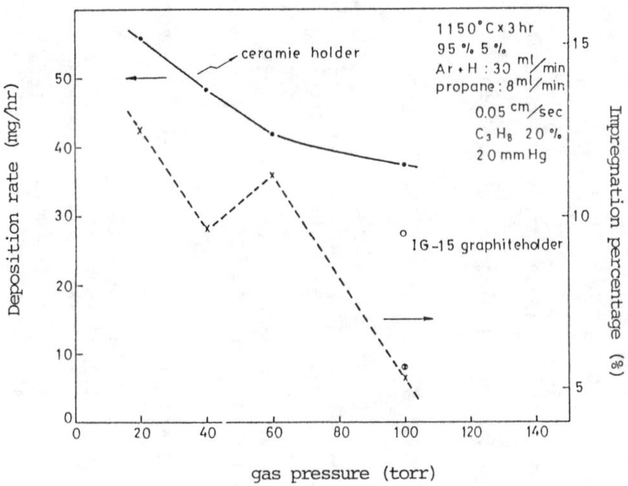

Fig. (8) - Gas pressure influences deposition rate and impregnation percentage

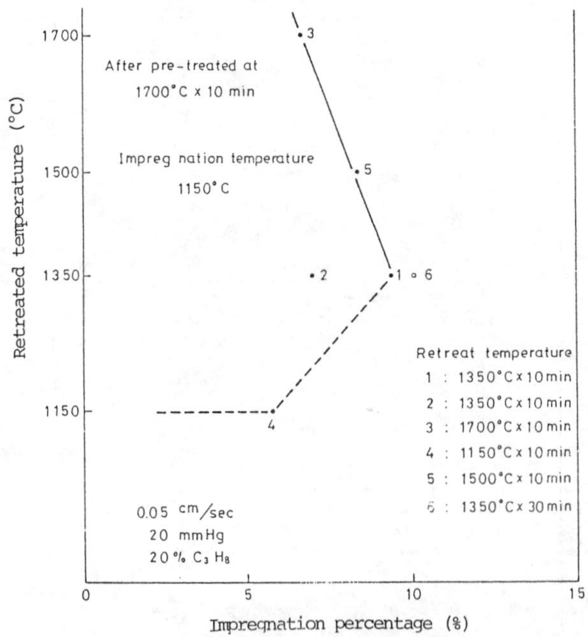

Fig. (9) - Influence of the different retreated temperatures to the impregnation at 1150°C for 3 hours

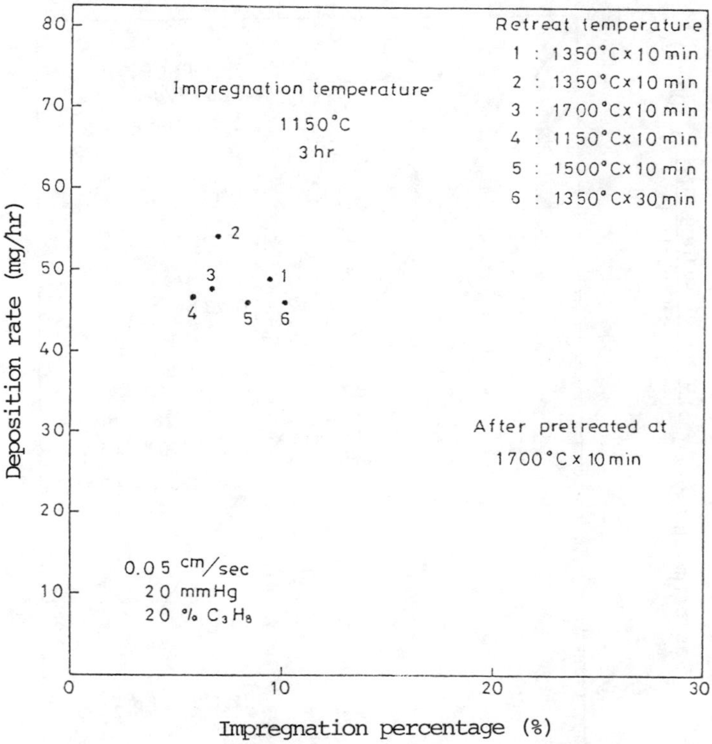

Fig. (10) - Deposition rates exhibit no linear relationship with respect to the impregnation

between gas-permeability and its pyrolytic carbon coating time or say infiltration time. It is worthy of recommending investigators to utilize this procedure to determine how much time needed to reach this goal.

CONCLUSIONS

Pyrolytic carbon microstructures have been shown to be a function of processing conditions. A characteristic temperature was found at 1350°C in a series of chemical vapor deposition processes. Experimental data are helpful in determining the detailed mechanisms of a chemical reaction rate-controlling and a diffusion reaction rate-controlling.

Gas-permeability measurement was utilized to investigate the pyrolytic carbon layers to nitrogen gas as a result of CFD process. New calculation method was developed to evaluate the microstructure. It is relatively sensitive to prove that the lower concentration of reactant gas is beneficial to a dense coating.

Carbon/graphite fiber composites through an isothermal CVD processing at 1200°C and 1150°C proved itself an excellent impregnation process. Retreated tem-

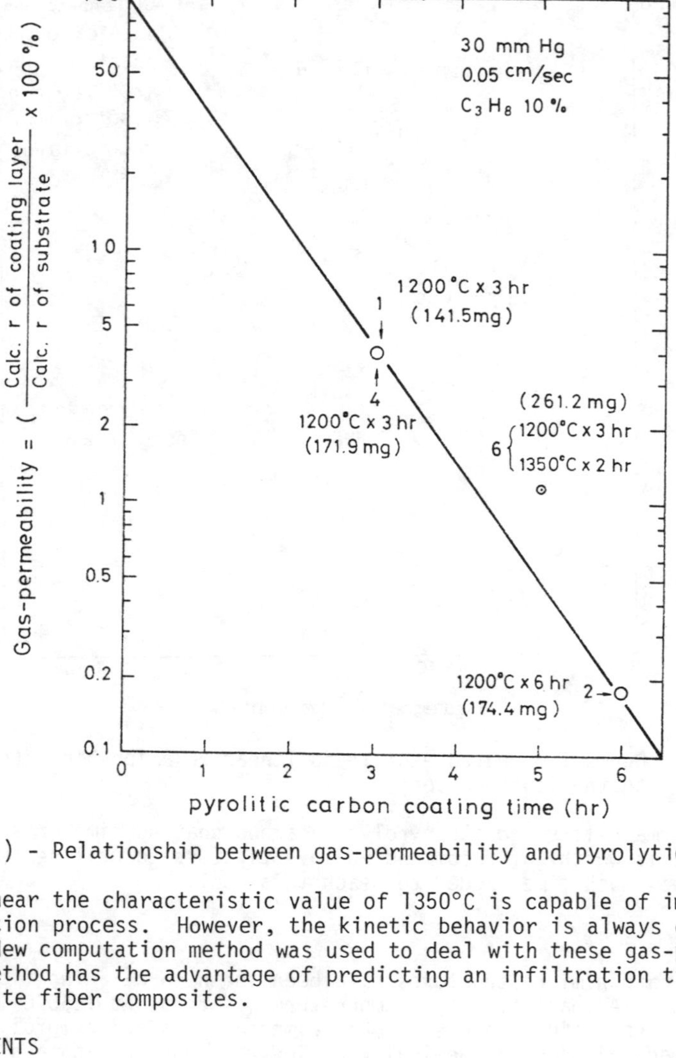

Fig. (11) - Relationship between gas-permeability and pyrolytic carbon

perature at near the characteristic value of 1350°C is capable of improvement in the impregnation process. However, the kinetic behavior is always difficult to interpret. New computation method was used to deal with these gas-permeability data. The method has the advantage of predicting an infiltration time for the carbon/graphite fiber composites.

ACKNOWLEDGEMENTS

This work has been made with the help of Dr. Fujii at Japanese Government Industries Research Institute in Osaka. Nissan Motor Company, Ltd., and Toyo Tanso Company, Ltd., are greatly appreciated for offering specimens.

REFERENCES

[1] Delmonte, J., Technology of Carbon and Graphite Fiber Composites, Van Nostrand Reinhold Company, 1981.

[2] Lee, Y. H., Fujii, R. and Matsuo, K., 47th Annual Science Lecture Meeting of Japan Chemical Society, Kyoto, 1983.

[3] Lee, Y.-H., to be published.

[4] Kotlensky, W. V., "Deposition of pyrolytic carbon in porous solids", Chemistry and Physics of Carbon, Vol. 9, pp. 174-255, 1981.

[5] Kotlensky, W. V., "A review of CVD carbon infiltration of porous substrates", 16th National SAMPE Symposium, California, 1971.

[6] Bauer, D. W. and Kotlensky, W. V., "Relationship between structure and strength for CVD carbon infiltrated substrates", 23rd Pacific Coast Regional Meeting, American Ceramic Society, California, 1970.

[7] Warren, J. W. and Williams, R. M., "Isothermal CVD processing", 4th National SAMPE Technical Conference, California, 1972.

[8] Hewitt, G. F., "Gaseous mass transport within graphite", Chemistry and Physics of Carbon, Vol. 1, pp. 72-120, 1979.

[9] Lee, Y. H., "Gas-permeability within isotropic high-density artificial graphite", Chinese Journal of Materials Science, 17A, No. 2, p. 34, 1985.

[10] Creutz, E. S. and Zumwalt, L. R., J. Appl. Phys., 33, p. 2883, 1962.

[11] Pollard, W. G. and Present, R. D., Phys. Rev., 23, p. 762, 1948.

PROPERTY-TEMPERATURE RELATIONSHIP OF TSTR POLYIMIDE/GRAPHITE FIBER COMPOSITES

S.-W. Leu, W.-Y. Chen, S.-E. Hsu

Chung Shan Institute of Science and Technology
P.O. Box 1-26-2, Lungtan, Taiwan 32500, Republic of China

and

L.-C. Hsu

Macromolecular Material, Inc.
P.O. Box 16190, Rocky River, Ohio 44116 USA

ABSTRACT

Toho ST-3 Graphite fiber reinforced Triaryl-S-Triazine Ring (TSTR) cross-linked composites were fabricated from aromatic nitrile terminated imide oligomers. Their forced torsional viscoelastic properties were determined up to 428°C and flexural strengths up to 600°C. Evaluation of these TSTR polyimide composites for industrial uses as well as for military applications is underway.

INTRODUCTION

Highly heat resistant polymeric composites are needed for aeronautical and space applications as well as for the development of advanced tactical missiles and fighters.

The polyimide class of polymers is known to possess a high degree of thermoxidative stability. But processing difficulty has limited their use as matrices for fabricating resin/fiber composites. The problem has been solved by using the combination of both "Catalytic Trimerization of Aromatic Nitriles" and "In Situ Polymerization of Monomeric Reactants" technologies. Via these technologies, a new group of polyimides containing Triaryl-S-Triazine Rings (TSTR) with improved thermoxidative stability may be prepared [1].

The purpose of our study is to seek for TSTR polyimide composites having good mechanical properties at elevated temperatures.

EXPERIMENT

A. Materials

(1) 3,3',4,4' - Benzophenone tetracarboxylic dianhydride (BTDA).

(2) 4,4' - Methylene dianiline (MDA).

(3) p-Aminobenzonitrile (PABN).

(4) p-Toluenesulfonic acid (PTSA).

All the above chemicals were of reagent grade and used as received.

(5) Toho-ST-3 Graphite fibers were supplied by Japanese Toho Company.

B. Schematic Reaction Equations for the Preparation of TSTR Polyimide Resin

(1) Esterification

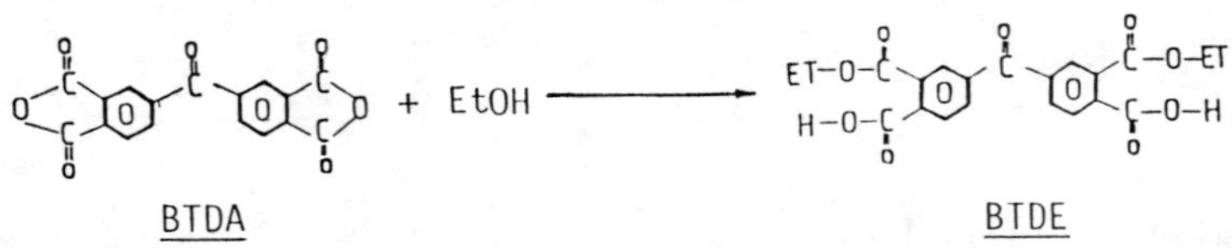

(2) Imidization

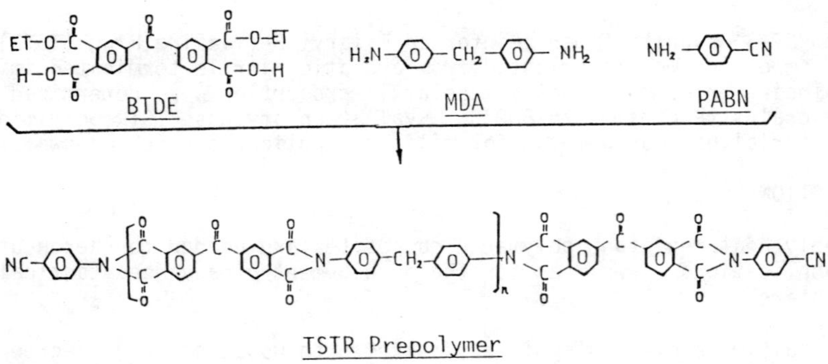

(3) Trimerization

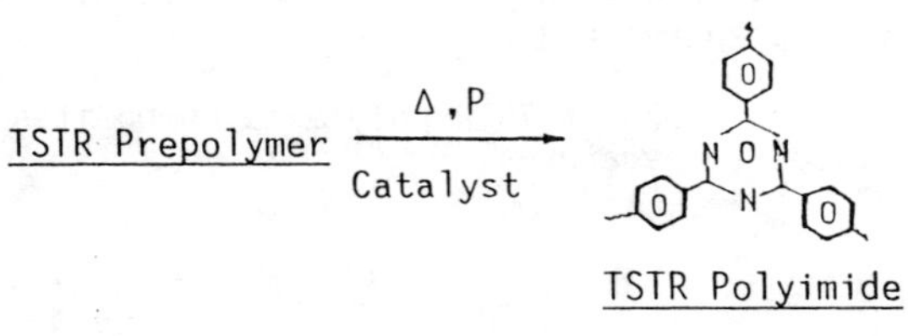

C. General Procedure for the Fabrication of TSTR Composites

Japan's Toho ST-3 graphite fibers were impregnated with an ethanol solution of TSTR prepolymer and wound on a mandrel of filament winding machine to form prepregs. The prepregs were heated to reduce their solvent content to have suitable tack and drape properties. They were then stacked in layers and heated for imidization. The completely imidized prepregs were transferred to a hot molding press and cured under pressure to form composites with resin content of 40% by weight. The composites may be further post cured in an oven if needed.

D. Determination of the Rheological Properties of Composites

Forced torsion rheological properties of TSTR/ST-3 composites via Rheometrics Inc. Rheometric Dynamic Spectrometer RDS-7700 are shown in Figure 1 and Table 1. Those of PMR-15/ST-3 and Epoxy (MY720+DDS)/ST-3 composites are also shown in Figure 2 and Figure 3 for comparison.

E. Determination of Flexure Properties of Composites

Via a Universal Instron 1125, the flexure properties of TSTR/ST-3 composites at various temperature up to 600°C are shown in Table 2, Figure 4 and Figure 5. Those of PMR-15/ST-3 and Epoxy (MY720+DDS)/ST-3 composites are also shown for comparison. Ultimate flexural strength and modulus of elasticity are determined in accordance with ASTM-D-790 except modified as follows: The flexural test specimen consists of a straight-sided rectangular cross section having the specified thickness. Span/thickness ratio is maintained at 20 to 1. Nominal specimen length is 50 mm but varies with thickness around 1.9 mm. The specimen width remains constant at 12.7 mm. The flexural specimens are loaded at 38.1 mm span points in a three point flexure. Ultimate flexural strength is calculated with equation (1). The modulus of elasticity is calculated with equation (2). In the environmental chamber, the samples are soaked at the testing temperatures for 3 minutes before conducting the test.

$$F_b = \frac{3PS}{4bt^2} \tag{1}$$

$$E_b = \frac{1}{4}\frac{S^3}{bt^3}\, m \tag{2}$$

where:

F_b: stress in the outer fiber throughout the load span

P: load at a given point on the load deflection curve

S: support span

b: width of specimen

t: thickness of specimen

E_b: modulus of elasticity in bending

m: slope of the tangent to the initial straight-line portion of the load-deflection curve.

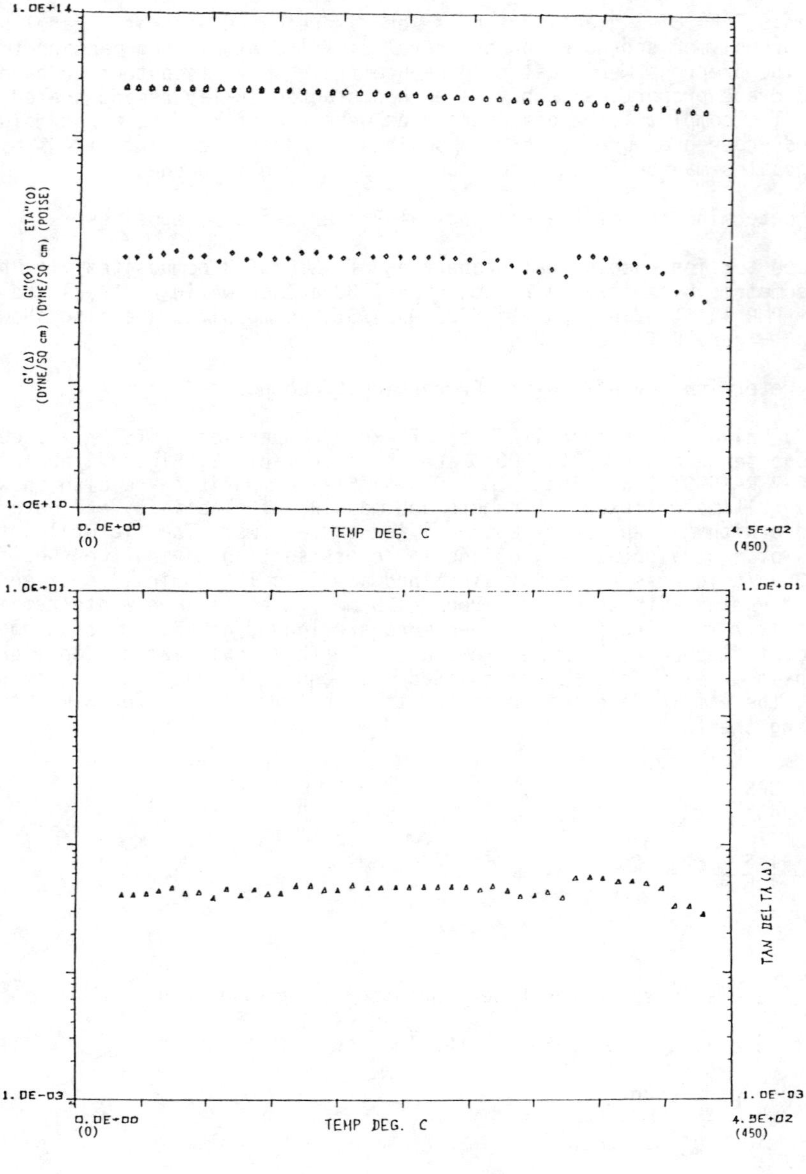

Fig. (1) - RDS-7700 forced torsion test of TSTR/ST-3 composite. Temperature sweep 10°C/step, rate: 1 rad/sec, strain: 0.15%, soak: 1 min/step

Table 1 - RDS-7700 forced torsion test of TSTR/ST-3 composite

Test Mode: Temperature sweep. 10°C/step
Rate: 1 rad/sec. Strain: 0.15% Soak: 1 min/step

G' DYNE/SQ CM	G" DYNE/SQ CM	ETA* POISE	TANDEL	TIME MINUTES	TEMP DEG.C
1.362E+13	9.192E+11	1.365E+13	6.750E-02	1.8	25
1.341E+13	8.132E+11	1.343E+13	6.067E-02	4.3	34
1.331E+13	7.636E+11	1.333E+13	5.740E-02	6.8	44
1.329E+13	6.948E+11	1.330E+13	5.230E-02	9.2	53
1.334E+13	6.784E+11	1.336E+13	5.086E-02	11.6	63
1.347E+13	6.824E+11	1.349E+13	5.067E-02	14.2	73
1.361E+13	7.147E+11	1.363E+13	5.249E-02	16.6	83
1.380E+13	7.628E+11	1.382E+13	5.528E-02	19.1	93
1.400E+13	7.807E+11	1.402E+13	5.576E-02	21.6	102
1.416E+13	8.153E+11	1.418E+13	5.759E-02	24.2	112
1.436E+13	9.374E+11	1.439E+13	6.529E-02	26.6	121
1.454E+13	1.056E+12	1.458E+13	7.261E-02	29.1	131
1.477E+13	1.037E+12	1.481E+13	7.020E-02	31.6	141
1.479E+13	1.066E+12	1.482E+13	7.212E-02	34.0	151
1.482E+13	1.079E+12	1.486E+13	7.280E-02	36.4	161
1.489E+13	1.035E+12	1.492E+13	6.953E-02	38.9	172
1.517E+13	9.350E+11	1.520E+13	6.163E-02	41.3	182
1.514E+13	9.100E+11	1.517E+13	6.009E-02	43.7	192
1.519E+13	9.303E+11	1.522E+13	6.124E-02	46.2	202
1.421E+13	1.050E+12	1.425E+13	7.386E-02	48.5	211
1.470E+13	8.478E+11	1.472E+13	5.769E-02	51.2	221
1.479E+13	9.243E+11	1.482E+13	6.249E-02	53.6	231
1.415E+13	8.050E+11	1.417E+13	5.691E-02	56.0	241
1.450E+13	9.034E+11	1.453E+13	6.230E-02	58.4	251
1.348E+13	8.685E+11	1.351E+13	6.442E-02	60.8	261
1.374E+13	8.705E+11	1.377E+13	6.336E-02	63.1	270
1.451E+13	8.746E+11	1.453E+13	6.028E-02	65.5	280
1.292E+13	9.121E+11	1.296E+13	7.059E-02	67.9	291
1.357E+13	1.017E+12	1.361E+13	7.492E-02	70.3	300
1.448E+13	7.125E+11	1.449E+13	4.922E-02	72.7	310
1.379E+13	9.292E+11	1.382E+13	6.740E-02	75.1	320
1.429E+13	7.278E+11	1.430E+13	5.095E-02	77.5	329
1.419E+13	7.336E+11	1.420E+13	5.172E-02	79.9	340
1.423E+13	7.496E+11	1.425E+13	5.269E-02	82.3	349
1.429E+13	7.654E+11	1.431E+13	5.355E-02	84.7	359
1.413E+13	7.839E+11	1.415E+13	5.547E-02	87.0	369
1.412E+13	7.808E+11	1.415E+13	5.528E-02	89.4	379
1.415E+13	8.038E+11	1.417E+13	5.682E-02	91.8	389
1.373E+13	7.496E+11	1.375E+13	5.461E-02	94.2	399
1.385E+13	7.869E+11	1.387E+13	5.682E-02	96.6	409
1.398E+13	7.928E+11	1.400E+13	5.672E-02	99.0	419
1.384E+13	7.850E+11	1.386E+13	5.672E-02	101.3	428

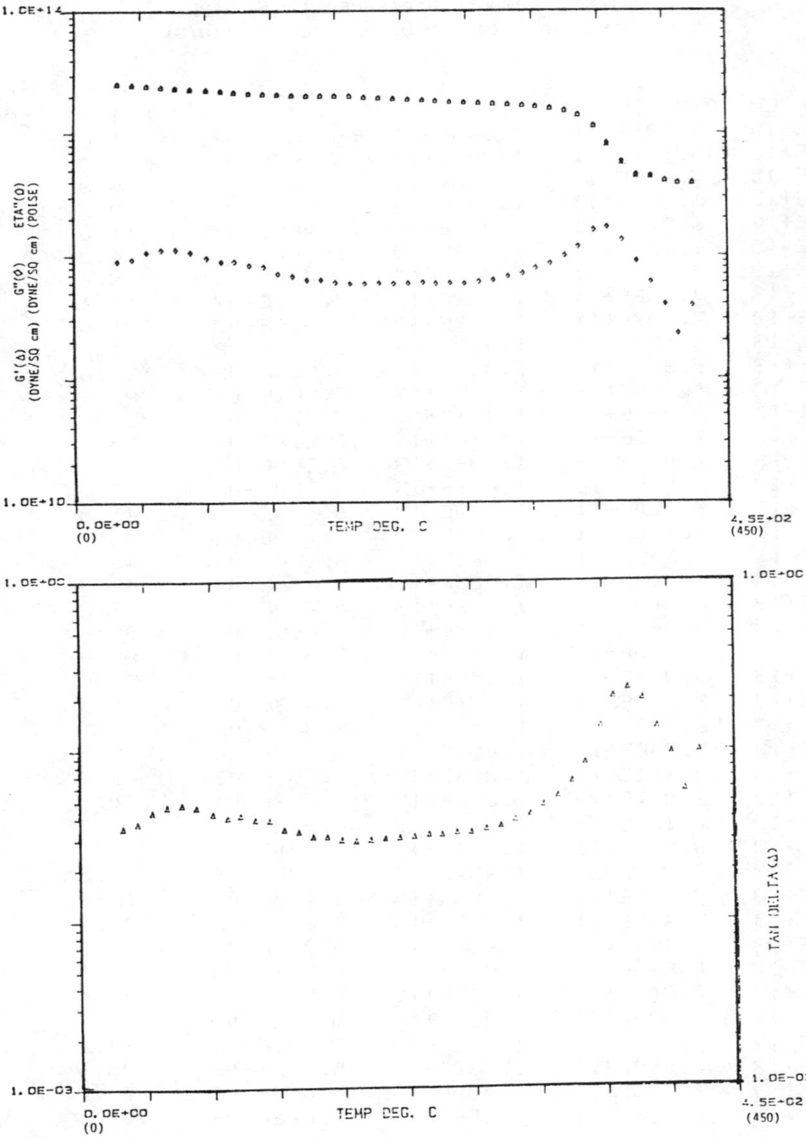

Fig. (2) - RDS-7700 forced torsion test of PMR-15/ST-3 composite. Temperature sweep 10°C/step, rate: 1 rad/sec, strain: 0.15%, soak: 1 min/step

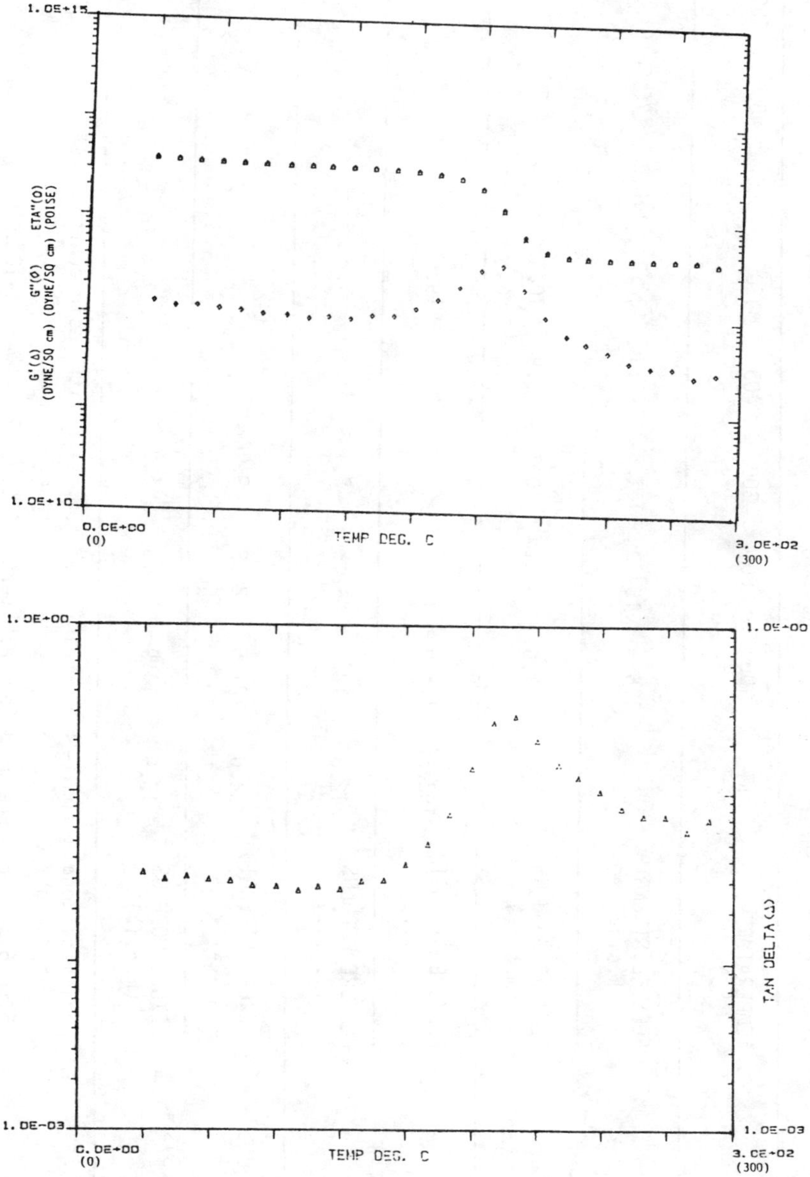

Fig. (3) - RDS-7700 forced torsion test of EPOXY (MY720+DDS)/ST-3 composite. Temperature sweep 10°C/step, rate: 1 rad/sec, strain: 0.15%, soak: 1 min/step

Table 2 - Flexure strength versus temperature of TSTR/ST-3, PMR-15/ST-3, EPOXY (MY720+DDS)/ST-3 composites

SPECIMEN $(0°)_8$	TEMPERATURE (°C)	RT	100	200	300	400	450	500	550	600
TSTR/ST-3	FLEXURE STRENGTH (k.s.i.)	209.5	176.1	157.0	151.8	141.3	132.3	110.3	76.7	61.8
	FLEXURE MODULUS (m.s.i.)	12.4	11.6	12.1	10.7	9.9	10.2	9.4	8.2	6.7
PMR-15/ST-3	FLEXURE STRENGTH (k.s.i.)	192.8	175.0	136.0	66.4	19.6	19.8			
	FLEXURE MODULUS (m.s.i.)	12.8	12.3	12.3	8.1	1.8	1.8			
EPOXY (MY720+DDS)ST-3	FLEXURE STRENGTH (k.s.i.)	184.0	121.5	27.2	9.76					
	FLEXURE MODULUS (m.s.i.)	15.9	14.9	5.3	1.9					

* unidirectional graphite fiber reinforced composite with resin content of 40% by weight .

* 5 to 6 samples were used for each test. C.O.V. is between 4-10%.

* Toho-ST-3 graphite fibers were used.

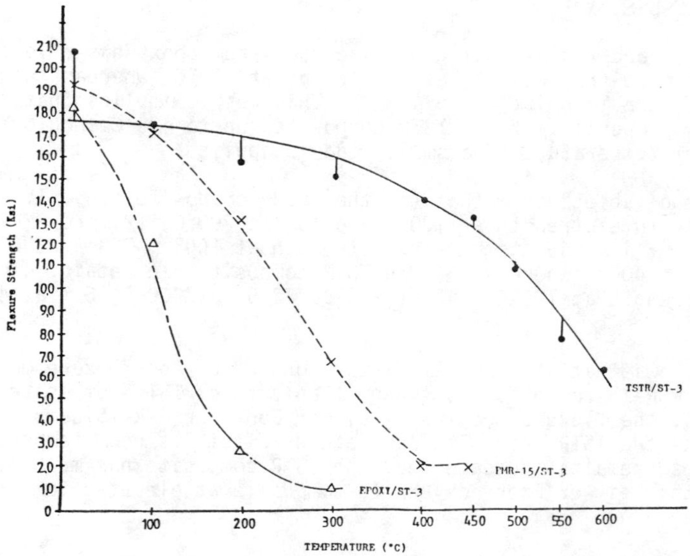

Fig. (4) - Flexure strength versus temperature of TSTR/ST-3, PMR-15/ST-3, EPOXY(MY720+DDS)/ST-3, U.D. composites

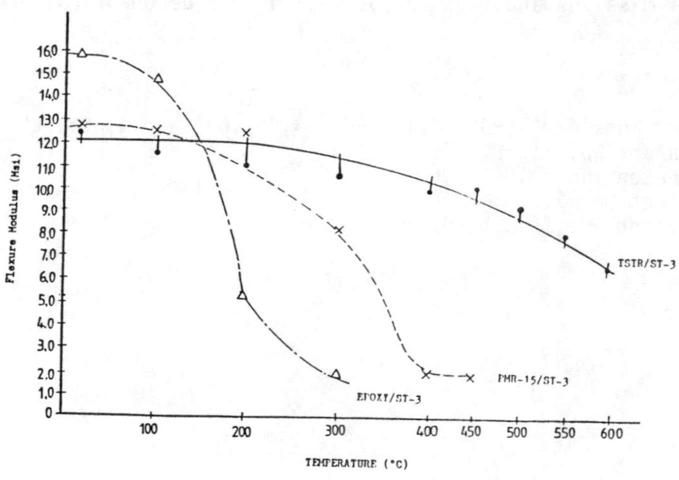

Fig. (5) - Flexure modulus versus temperature of TSTR/ST-3, PMR-15/ST-3, EPOXY(MY720+DDS)/ST-3, U.D. composites

RESULTS AND DISCUSSION

Figures 1, 2 and 3 show that composite made from Epoxy has a Tg point at 220°C and that made from PMR-15 has a Tg point at 360°C, whereas the TSTR composite does not have a Tg point up to 428°C. This result implies that, in general, the mechanical properties of the TSTR composites up to 428°C should be nearly the same as at room temperature if a small load is applied.

Figure 4 and Table 2 show that (1) the Epoxy composite can only retain 66% of its room temperature strength at 100°C and 15% at 200°C, (2) the PMR-15 composite can retain 91% of its room temperature strength at 100°C, 71% at 200°C, 34% at 300°C and 10% at 400°C whereas (3) the TSTR composite can retain 84% of its room temperature strength at 100°C, 74% at 200°C, 72.5% at 300°C, 67% at 400°C and 37% at 550°C.

Figure 5 shows that at room temperature or 100°C, the flexure modulus of either Epoxy or PMR-15 composite is higher than that of TSTR composite. However, at about 160°C, the flexure modulus of either Epoxy or PMR-15 composite drops sharply whereas the TSTR composite can retain 76% of its room temperature modulus at 500°C. These results indicate that the TSTR composite has much higher mechanical strength than either Epoxy or PMR-15 composite at elevated temperatures.

CONCLUSION

TSTR polyimide composites can be easily prepared by using the combination of "Catalytic Trimerization of Aromatic Nitriles" and "In Situ Polymerization of Monomeric Reactants" technologies.

TSTR polyimide composites possess extremely high thermoxidative stability and are the most promising structural material known today for the development of advanced tactical missiles and fighters as well as for aeronautical and space applications.

REFERENCE

[1] U.S. Patent Nos. 4,061,856; 4,159,262; 4,458.041 and 4,555.565.
British Patent No. 1,518,386.
Canadian Patent No. 1,068,696.
French Patent No. 2,287,448.
Swedish Patent No. 7510867-0.

EFFECT OF THE MICROSTRUCTURES AND PROCESSINGS ON THE STRENGTH OF CARBON FIBER

S. C. Yu, Y. H. Yen, W. F. Chen

Cheng Kung University
Tainan 70101, Taiwan, Republic of China

and

H. Y. Lin

Industrial Technology Research Institute
Hsinchu, Taiwan, Republic of China

ABSTRACT

A study on the correlation of microstructures and properties of both commercial and domestically processed carbon fibers was made. A fluctuation in the microstructure distribution in the transverse direction for the carbon fiber processed in the present study was observed with the use of electron microscopy techniques. Measured mechanical properties of these materials are reflected by the variation in the degree of carbonization and the growth of the graphite single crystal in the carbon fibers.

The trace element analysis for carbon fiber indicates that some impurity substances act as the catalytic agents in the carbonization process. However, the presence of the impurity elements may also introduce defect substructures into the carbon fiber when the temperature of heat treatment is in the excess of 1200°C.

INTRODUCTION

The ideal crystal structure of graphite is composed of regularly stacked (ABAB...) hexagonal carbon sheets (carbon basal planes). Within the basal plane, carbon atoms are linked together by strong covalent bonds while these carbon sheets are held together by relatively weak Van der Waals force. Carbon-carbon distance is 1.42A within the sheet and the separation between two adjacent carbon sheets is 3.35A. Therefore, the physical properties of graphite is highly anisotropic. For instance, the theoretical tensile modulus of elasticity is 100 GPa [1] in the direction parallel to the basal plane, yet the same parameter is only 35 GPa in the direction perpendicular to the basal plane.

The basic structural unit in carbon fibers is "turbostratic graphite". The essential structural features in turbostratic graphite are similar to those in the ideally perfect graphite crystal. The main difference is that the stackings of the carbon sheets in turbostratic graphite are not as regular as in the perfect graphite crystal but displaced one with respect to another as shown in Figure 1.

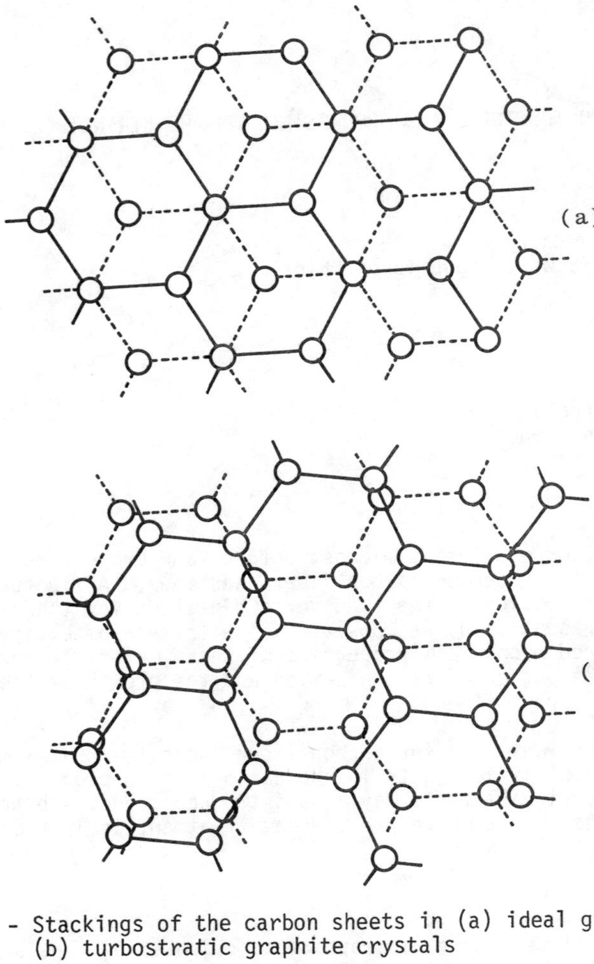

(a)

(b)

Fig. (1) - Stackings of the carbon sheets in (a) ideal graphite and (b) turbostratic graphite crystals

Because of its unique characteristic in specific strength, carbon fibers have become one of the most important materials in today's technological application. Carbon fibers were first manufactured in early 1960, and the demand for the carbon fibers as composite material for sporting goods has been increasing ever since late 1970. In view of its significant application in automobile industry and aeronautics and space technology, carbon fibers and graphite fibers have been extensively studied by various laboratories throughout the world.

At present, the consumption rate for carbon fibers in this country is about 400 tons a year and all are imported from foreign countries, largely from Japan and the US. In consideration of the potential advancement and development of the carbon fiber technology in this country, it is urgently required to establish the research capability and to carry out fundamental study on carbon fibers. A basic research work on the structure, microstructure and property characterization for carbon fibers was initiated three years ago. This is a paper to present our latest results on the analysis of the PAN-based carbon fibers, both imported and domestically processed materials.

EXPERIMENTAL PROCEDURES

A. Chemical Composition Analysis

The concentrations of carbon, hydrogen and nitrogen in carbon fibers were determined with the use of a CHN 600 analyzer manufactured by LECO company.

Trace elements in carbon fibers were analyzed with atomic absorption spectroscopy (Varian 100), ultraviolet spectroscopy (Shimazu UV 240) and X-ray fluorescence technique (Rigaku's 360M XRF). The semi-quantitative composition data obtained by the X-ray fluorescence method are mainly used to check the analysis results obtained from the AA technique. The trace elements contents reported here are based on AA and UV analyses.

B. Substructure Examination

Carbon fiber samples used in the microstructural observation were prepared with the technique of ultramicrotomy in which some 500A thick film of the carbon fiber specimens can be obtained. The unit we used is a ULTRACUT E microtome made by Reichert-Jang company. The suitable thin carbon fiber flakes can also be prepared by grinding the fibers in a mortar. Transmission electron microscopy analysis was carried out with two electron microscopes, one manufactured by Joel company (JEM-200 CX STEM) and the other by Hitachi company (H-700H STEM). Scanning electron microscopy work was made with JSM-35 SEM manufactured by Joel company.

EXPERIMENTAL RESULTS AND DISCUSSION

A. C, H, N Contents

The concentrations of carbon, hydrogen and nitrogen elements for seven commercial carbon fibers were determined and the results are listed in Table 1. With respect to the contents of carbon and nitrogen, it is evident from this analysis that the seven commercial fibers can be divided into two groups. The first group involves two different brands of carbon fibers, labeled EHMS and HM-35. The second group includes the rest of the five brands of commercial fibers, labeled HTA-7, EXAS, ST-1, ST-3 and T-300. Carbon fibers of the first group have the carbon concentration of nearly 100% while the fibers in the second group contain only about 95% of carbon and approximately 4% of nitrogen impurity. The variation in the carbon concentration correlates very well with the pattern of the tensile modulus for these fibers. In other words, carbon fibers with high modulus (360 and 350 GPa respectively for EHMS and HM-35) are almost pure carbon in composition while those five fibers of low modulus (between 235-240 GPa) contain less carbon and more nitrogen impurity. With PAN fibers as precursor materials, it is more difficult to remove nitrogen than hydrogen in the carbon fiber manufacturing process. Denitrogenation process started at around 600°C and there are still about 4-5% (by weight) nitrogen left as the temperature increased to as high as 1300°C.

Table 1 - C, H, N contents in seven commercial carbon fibers (wt%)

	EHMS	HM-35	HTA-7	EXAS	ST-1	ST-3	T-300
C	100.27	99.45	95.52	95.00	95.89	94.76	93.72
H	0.02	0.03	0.09	0.05	0.10	0.13	0.10
N	0.06	0.11	4.00	3.88	3.98	4.55	5.37
Trace Elements	0.14	0.12	0.35	0.50	0.60	0.87	0.46
Total	100.49	99.71	99.96	99.43	100.57	100.31	99.65
Strength (GPa)	2.62	2.45	3.30	3.10	3.70	4.40	2.24
Modulus (GPa)	360	350	240	235	240	240	235

This indicates that the fibers of the first group were heat treated at much higher temperature than those of the second group.

Figure 2 is the temperature dependence of the tensile modulus for carbon fibers as reported by Johnson [2]. From Johnson's plot, the temperatures of heat treatment for the high modulus fibers (EHMS and HM-35) and the low modulus ones (HTA-7, EXAS, ST-1, ST-3 and T-300) are estimated to be roughly 2300°C and 1300°C respectively, and this is in good agreement with the result we obtained based on the C, H, N analysis.

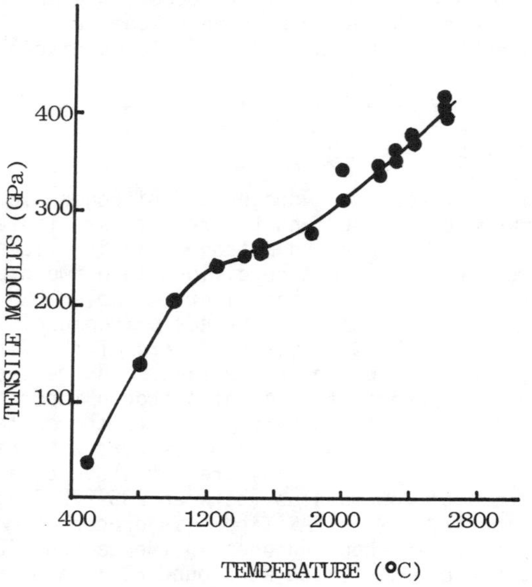

Fig. (2) - Elastic modulus as a function of temperature [2]

B. Trace Element Concentrations

Table 2 lists the analysis result of the trace elements content in a variety of commercial carbon fibers. It appears that the high modulus and low strength fibers, EHMS and HM-35, are relatively free from trace elements impurity. On the other hand, the high strength fibers such as ST-3 (4.4 GPa) and ST-1 (3.7 GPa) are rich in the trace elements concentration.

Table 2 - Trace element concentrations of commercial carbon fibers (ppm)

	EHMS	HM-35	HTA-7	EXAS	ST-1	ST-3	T-300	Nikkiso
Na	5	6	27	1210	21	19	22	279
Mg	17	24	103	25	66	95	78	49
Fe	51	36	36	310	83	217	115	67
Zn	--	--	102	11	306	265	8	37
Mn	--	--	7	16	20	164	7	13
Ni	--	--	--	--	--	--	--	--
Si	1360	1020	3040	3280	4540	3850	3900	24--
Co	--	--	--	--	--	--	--	--
K	--	16	26	58	25	32	47	16
Cu	--	--	--	--	--	--	--	--
Cr	--	--	--	--	--	--	--	--
Pb	--	--	--	--	--	--	--	--
Ca	10	60	160	60	900	4070	380	100
Total	1443	1162	3501	4970	5961	8712	4557	2971
Strength (GPa)	2.62	2.45	3.30	3.10	3.70	4.40	2.24	--
Modulus (GPa)	360	350	240	235	240	240	235	--

As reported by Oya and Marsh [3], Fishbach [4] and Marsh et al [5] that certain elements, alloys and compounds possess catalytic effect in the graphitization (carbonization) process of precursor fibers. Elements with catalytic effect are circled out in Table 3. It is required only trace amount of these catalytic elements dispersed evenly throughout the precursor material, so that the carbonization or graphitization can be achieved at lower temperature. For instance, Marsh

Table 3 - Circled elements are those with catalytic effect in graphitization [3]

Li	**Be**											**B**	C	N	**O**
Na	**Mg**											**Al**	**Si**	P	**S**
K	**Ca**	Sc	**Ti**	**V**	**Cr**	**Mn**	**Fe**	**Co**	**Ni**	**Cu**	Zn	Ga	**Ge**	As	Se
Rb	Sr	Y	**Zr**	Nb	**Mo**	Tc	Ru	Rh	Pd	**Ag**	Cd	In	Sn	Sb	Te
Cs	**Ba**	La	Hf	**Ta**	**W**	**Re**	Os	Ir	**Pa**	**Au**	Hg	Tl	Pb	Bi	Po

pointed out that with iron as catalytic material, turbostratic graphite can be formed at a temperature as low as 600°C.

Table 4 lists the contents of trace elements with catalytic effect for various brands of carbon fibers. It is interesting to note that, except T-300, there

Table 4 - Concentrations of trace elements with catalytic effect for eight different commercial carbon fibers (ppm)

	EHMS	HM-35	HTA-7	EXAS	ST-1	ST-3	T-300	Nikkiso
Mg	17	24	103	25	66	95	78	49
Fe	51	36	36	310	83	217	115	67
Mn	--	--	7	16	20	164	7	13
Si	1360	1020	3040	3280	4540	3850	3900	2400
Ca	10	60	160	60	900	4070	380	110
Total	1438	1140	3346	3691	5609	8396	4480	2639
Strength (GPa)	2.62	2.45	3.30	3.10	3.70	4.40	2.24	
Modulus (GPa)	360	350	240	235	240	240	235	

exists a trend between the tensile strength and the total amount of catalytic trace elements. Among them, high strength carbon fiber ST-3 (4.4 GPa) leads the way in the catalytic element concentration. Figure 3 plots the tensile strength versus total catalytic trace element contents for fibers HM-35, HTA-7, ST-1 and ST-3, all produced by one single manufacturer, Toho company. In this plot, the mechanical strength of carbon fibers appears to be correlated to the total amount of catalytic elements in the fibers. However, the low strength HM-35 fiber could be originated from the fact that it had been heat treated at relatively high temperature (estimated to be about 2300°C). The trace element impurities had melted and evaporated out of the fibers at this temperature and left voids, microvoids

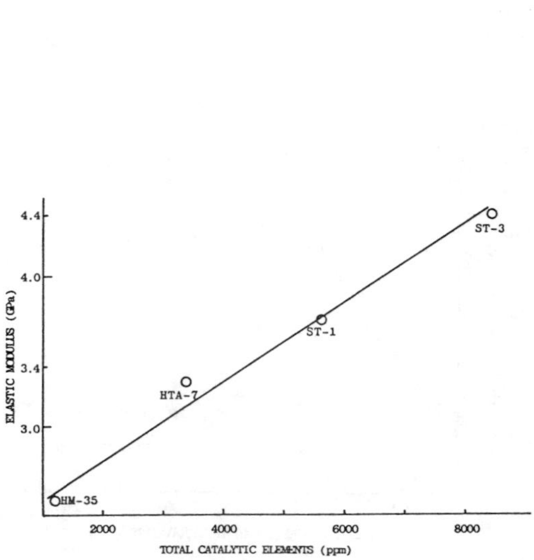

Fig. (3) - Tensile strength versus total catalytic elements for Toho's carbon fibers

or other types of defects in the fibers. These structural defects resulted from the melt or evaporation of impurity elements may have contributed to the low strength (2.4 GPa) for HM-35 fibers.

According to Johnson [2], the tensile strength of carbon fibers increases with increasing temperature of heat treatment as shown in Figure 4, until about 1200°C when the strength begins to decrease with further increase in temperature.

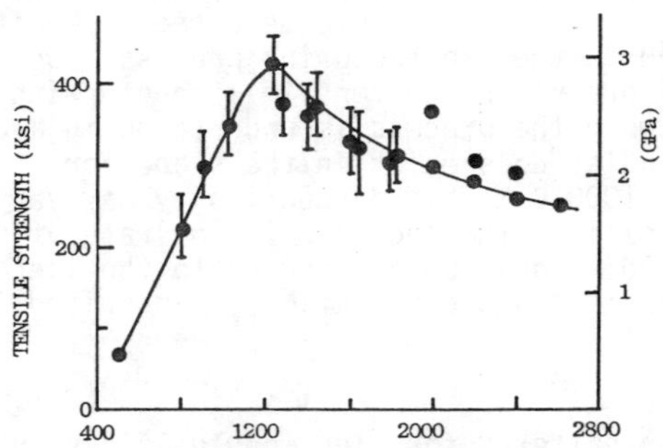

Fig. (4) - Tensile strength as function of temperature [2]

Moreton and Watt [6] reported a continuous increase in the tensile strength of the fibers with increasing temperature, Figure 5, when heat treatment is processed in a clean room (number of particles with dimension greater than 0.5 μm in cubic feet less than 100). A gauge length dependence of the tensile strength for carbon fibers was observed by Moreton [7] that the strength of the fibers decreases with increasing gauge length. This indicates that the fiber strength is very sensitive to the amount of defects in the fibers, the longer the gauge length, the higher the total amount of flaws and thus the lower the strength. Defects such as voids, cavities and inclusions in the fibers may have inherited from the precursor fibers [8,9]. But the drop of the tensile strength processed at tem-

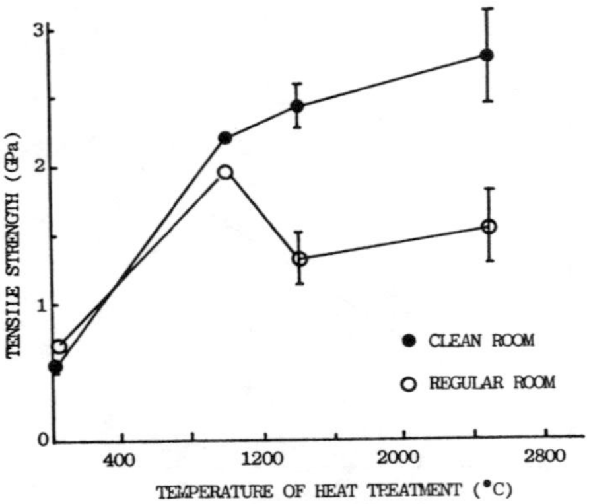

Fig. (5) - Tensile strength versus temperature plot in different environments

perature higher than 1200°C can also be attributed to the flaws of voids or cavities in the fibers resulted from the evaporation of the impurities at high temperature condition [10,11]. Mrozowski cracks originated from the anisotropic contraction of the fibers in the cooling process were regarded by some authors to be intimately related to the drop of strength of the fibers beyond 1200°C [12].

The above discussion suggests that the catalytic impurities have a mixed effect on the strength of carbon fibers depending upon the temperature of heat treatment. Impurities in the fibers could be either inherited from the precursor material or contaminated in the manufacturing process. Despite the fact that there is a continuous increase in the tensile strength with increasing temperature of heat treatment when the process is undertaken in a clean room as shown in Figure 5, a discontinuity does exist in the slope for the strength versus temperature curve at about 1000°C. This discontinuity may very well imply that regardless of the contamination, the decrease in the rate of strength increase with temperature at about 1000°C could be attributed to the increase in the defect content derived from the melt and evaporation of inherited impurities.

C. STEM Examination

Electron diffraction patterns for high modulus fiber EHMS, high strength fiber ST-3 and domestically processed fibers marked C3-5-1 and T-1-2 are shown in Figure 6. Figure 7 is a schematic electron diffraction pattern in which $\Delta\phi$ and $\Delta\theta$ represent respectively the degree of preferred orientation and the dimension of L_c, the stacking height of the carbon sheets in the crystallographic c direction. Based on the rather diffuse (002) peak observed for C3-5-1 and T-1-2 in comparison with less diffuse (002) peak for EHMS and ST-3, it is inferred that the fibers C3-5-1 and T-1-2 are low in the extent of graphitization. The large variation in $\Delta\phi$ observed for the different grains of the same sample C3-3-1 (Figure 8) as well as other domestically manufactured fibers indicates a fluctuation in the degree of preferred orientation in the fibers. This fluctuation may suggest that the fibers could have sheath and core structure or that the fibers are not homogeneous in the transverse direction. This suggested inhomogeneity of the

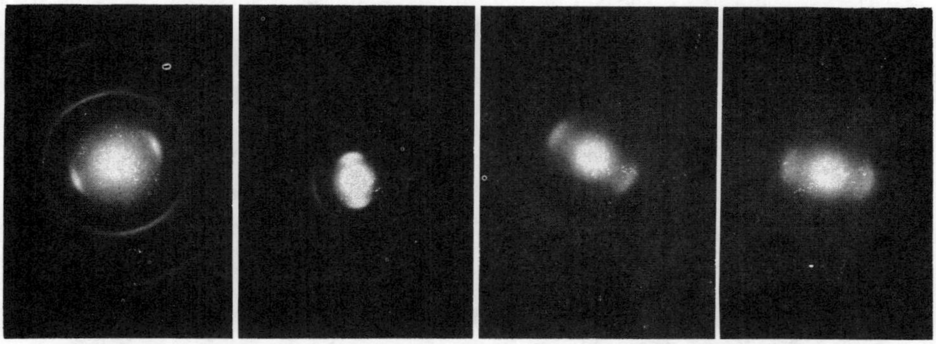

(a) EHMS (b) ST-3 (c) C3-5-1 (d) T-1-2

Fig. (6) - Electron diffraction patterns for (a) EHMS, (b) ST-3, (c) C3-5-1, (d) T-1-2

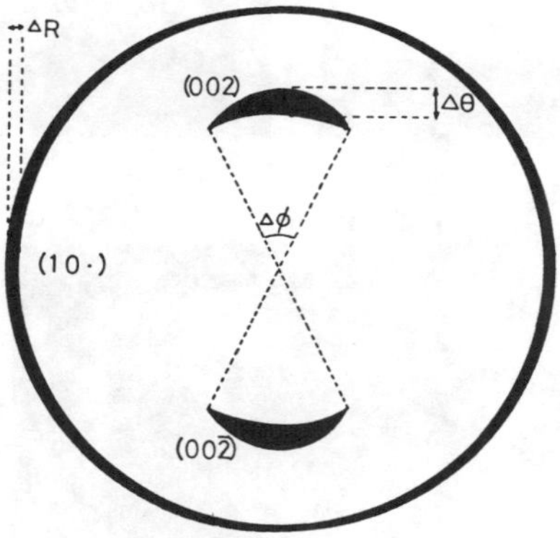

Fig. (7) - Schematic electron diffraction pattern showing Δϕ and Δθ

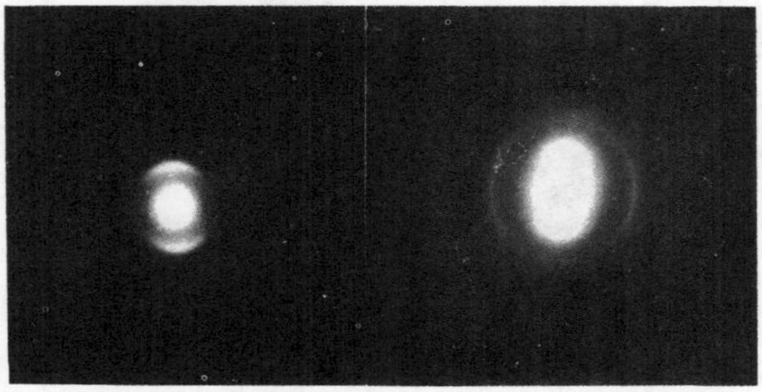

Fig. (8) - Diffraction (002) peaks for two different grains of the same sample

fiber is also supported by the variation in color in the transverse section of the fiber observed in the optical microscopy work.

Transverse sections observed with SEM technique for the fibers C3-1-1, C3-2-1 and C3-5-1 are shown in Figure 9. Fibrillar texture is predominant in

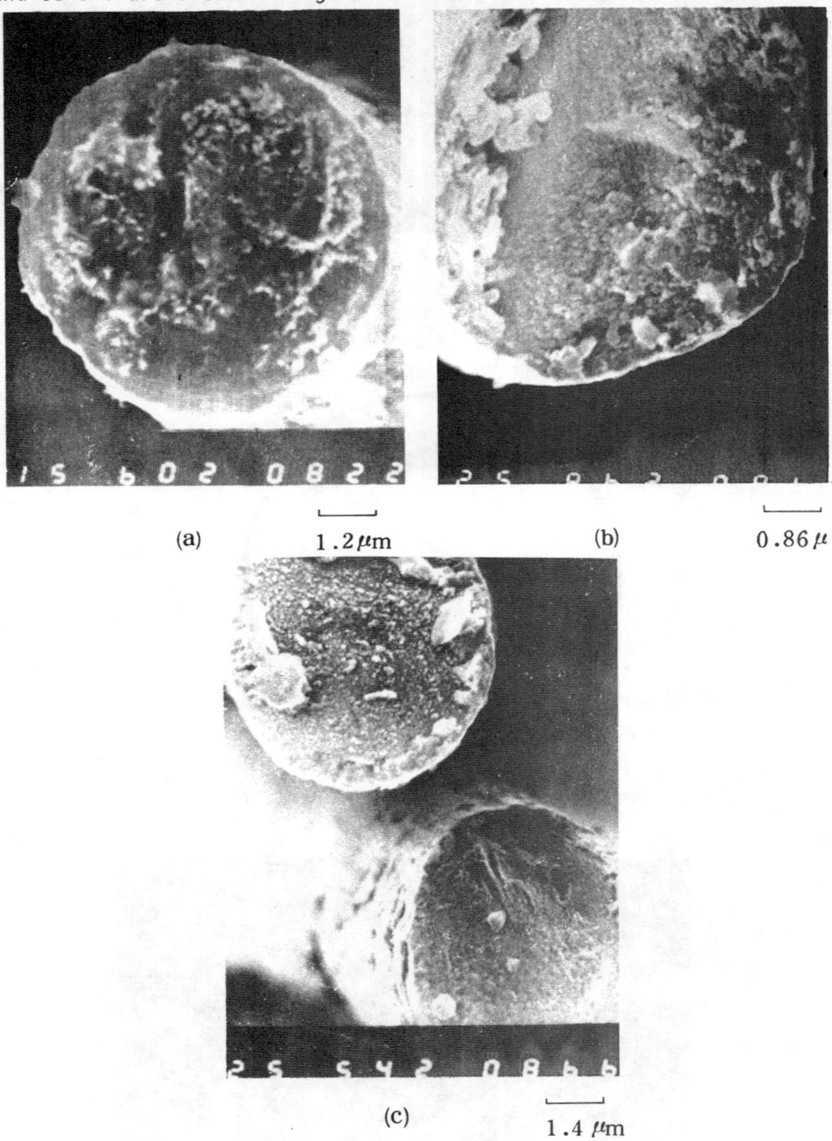

Fig. (9) - SEM photographs of the transverse sections for (a) C3-1-1, (b) C3-2-1, (c) C3-5-1 showing fibrillar textures

all of the three fibers. The details of the imperfection structures for C3-1-1, C3-2-1 and C3-5-1 are depicted in Figure 10 and Figure 11. The structural flaws

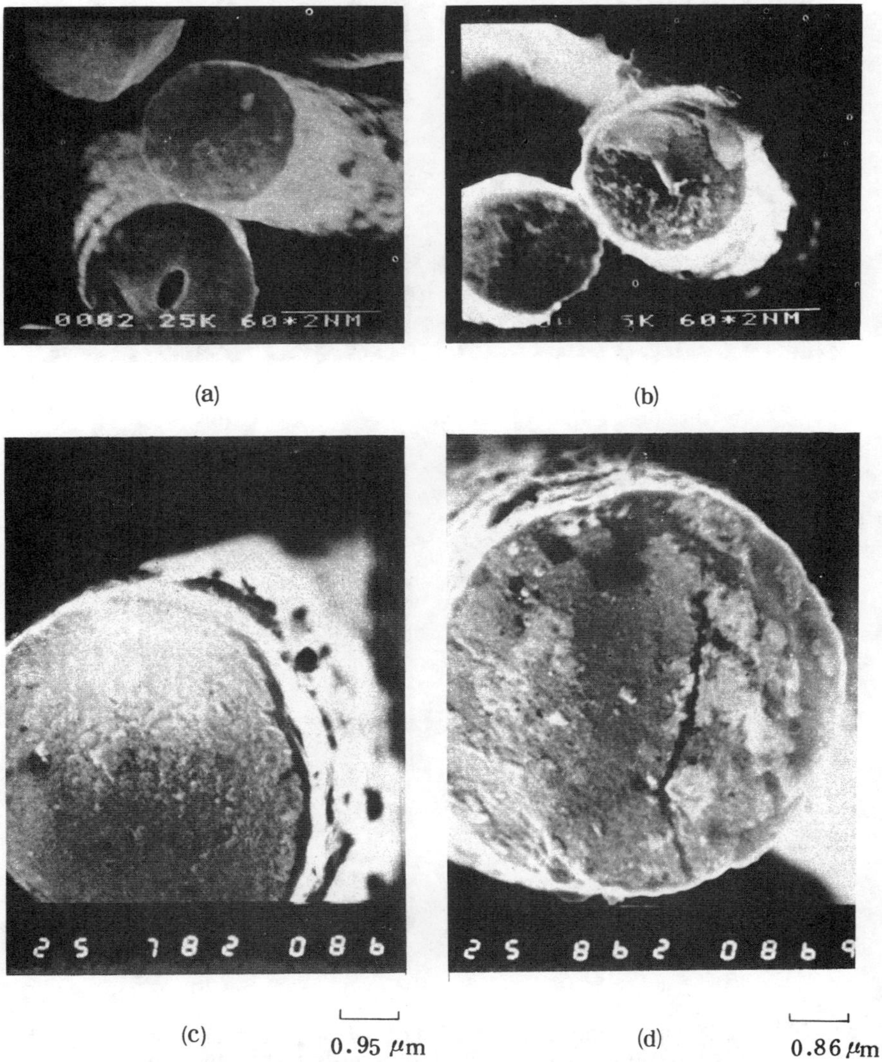

Fig. (10) - SEM photographs of the transverse sections for (a),(b) C3-1-1, (b) C3-2-1, (d) C3-5-1, depicting various defect structures

involve cracks, voids, holes, impurity aggregates and regular or irregular surface defects. The inferior characteristics in strength for the domestically processed carbon fibers (Table 5) are largely related to the abundance of the internal defects and the surface flaws as well.

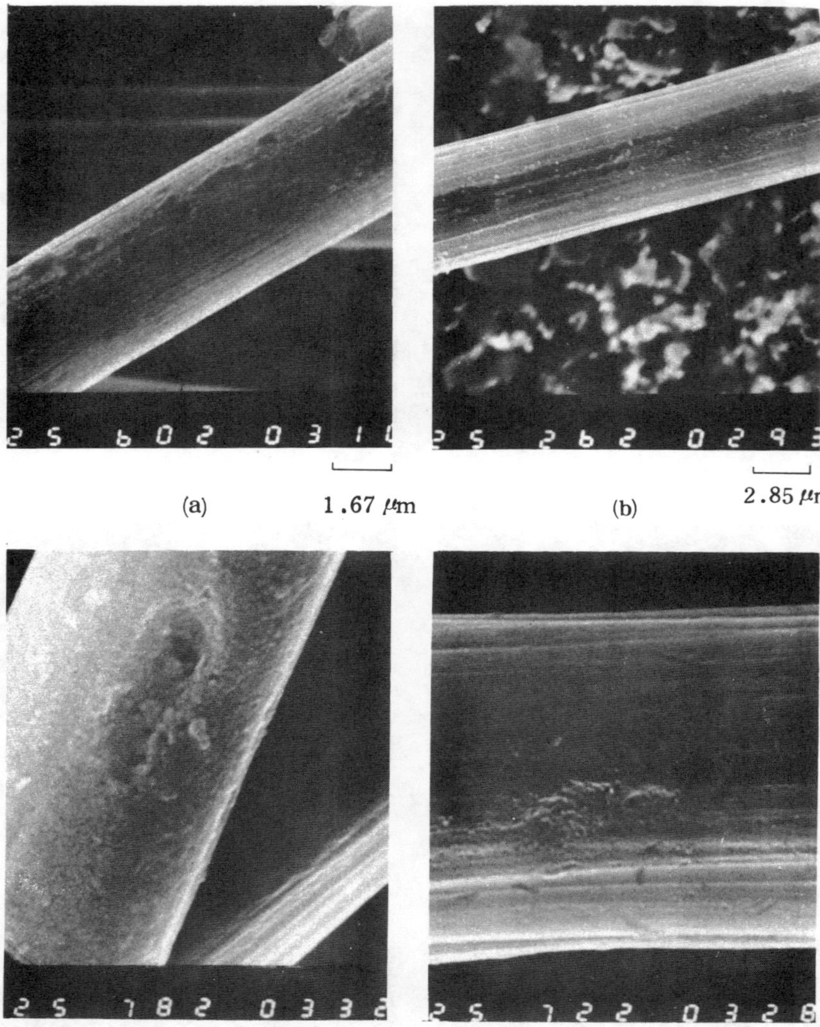

(a) (b) (c) (d)

Fig. (11) - SEM photographs of the longitudinal sections for (a) C3-5-1, (b) C3-2-1, (c),(d) C3-3-1, showing the surface flaws

Table 5 - Tensile strengths of domestically processed carbon fibers

C3-1-1	0.44GPa
C3-2-1	1.73
C3-3-1	2.11
C3-5-1	3.63
T-1-1	2.01
T-1-2	3.28
T-1-3	2.50

REFERENCES

[1] Blakslee, O. L., Proctor, D. G., Seldin, E. J., Spence, G. B. and Weng, T., "Elastic constants of compression-annealed pyrolytic graphite", J. Appl. Phys., 41, p. 3373, 1970.

[2] Johnson, J. W., "Factors affecting the tensile strength of carbon-graphite fibers", Appl. Polym. Symp., No. 9, p. 229, 1969.

[3] Oya, A. and Marsh, H., "Phenomena of catalytic graphitization", J. Mater. Sci., 17, p. 309, 1982.

[4] Fishbach, D. B., "The kinetics and mechanism of graphitization", Chem. and Phys. of Carbon, 7, p. 83, 1971.

[5] Marsh, H., Crawford, D. and Taylor, D. W., "Catalytic graphitization by iron of isotropic carbon from polyfurfuryl alcohol, 725-1090K. A high resolution electron microscope study", Carbon, 21, p. 81, 1983.

[6] Moreton, R. and Watt, W., "Tensile strength of carbon fibers", Nature, 247, p. 360, 1974.

[7] Moreton, R., "The effect of gauge length on the tensile strength of R.A.E. carbon fibers", Fibre Sci. and Tech., 1, p. 273, 1969.

[8] Johnson, J. W. and Thorne, D. J., "Effect of internal polymer flaws of strength of carbon fibres prepared from an acrylic precursor", Carbon, 7, p. 659, 1969.

[9] Thorne, D. J., "Distribution of internal flaws in acrylic fibers", J. Appl. Polym. Sci., 14, p. 103, 1970.

[10] Burnay, S. G. and Sharp, J. V., "Defect structure of PAN-based carbon fibres", J. Microscopy, 97, p. 153, 1973.

[11] Sharp, J. V., Burnay, S. G., Matthews, J. R. and Harper, E. A., "Defect limitation to the strength of carbon fibres", Int. Conf. on Carbon Fibres and their Composites, The Plastic Institute, London, Paper 5, 1974.

[12] Jones, B. F. and Duncan, R. G., "The effect of fibre diameter on the mechanical properties of graphite fibres manufactured from polyacrylonitrile and rayon", J. Mater. Sci., 6, p. 289, 1971.

ROLLING AND FORGING CHARACTERISTICS OF ALUMINUM MATRIX FILAMENTARY COMPOSITES

W. Wang

Department of Mechanical Engineering, Northeastern University
Boston, Massachusetts 02115 USA

and

T. Erturk

Department of Mechanical Engineering, Massachusetts Institute of Technology
Cambridge, Massachusetts 02139 USA

ABSTRACT

The rolling formability and the effects of secondary processing on mechanical properties of fiber reinforced metal matrix composites are reviewed. Duraluminum reinforced with various metallic fibers is capable of undergoing extensive rolling reduction in the direction of fiber alignment without significant damage to the fibers or interfaces, provided that the ratio of the flow strength of the reinforcement to the flow strength of the matrix is equal to or less than 3. Metal matrix composites with brittle fibers, on the other hand, cannot be rolled without fiber damage in the fiber direction. Rolling or forging of metal matrix composites in the transverse direction (perpendicular to the fiber) is possible even with brittle fibers. A proper amount of cold rolling in transverse direction can be beneficial to fiber or whisker reinforced Al matrix composites. Beneficial effects of transverse cold rolling or forging can be attributed to the relief of residual stress, rotation and alignment of the reinforcement, work hardening of the matrix, and dislocation network substructure buildup in the matrix. However, the tensile strength will deteriorate as a result of fiber fracturing and misalignment caused by increasing cold rolling. Hot rolling may decrease the strength of the matrix while increasing the ductility. This, in turn, may enhance the post cracking energy absorption capacity (strain to fracture) of the composite.

INTRODUCTION

Metal matrix composites have the potential of exceeding conventional materials by significant improvements in the strength-to-weight and modulus-to-weight ratios, and in high temperature capability. Incorporating metal, ceramic, or graphite fibers in metal matrix composites makes possible to tailor mechanical and other properties for specific applications. For example, metal matrix composites with discontinuous fibers can exhibit excellent resistance to creep, and continuous fiber reinforcement can, in principle, eliminate creep deformation at elevated temperatures in the fiber direction, provided that the fibers remain elastic and intact. The high specific stiffness and strength, coupled with high thermal conductivity and low thermal expansion in the fiber direction make metal matrix composites very attractive for high temperature applications. Different operational requirements in load bearing structures can be met by selecting combinations of matrices, fibers, and tailoring their interfaces to yield composites with high specific stiffness and strength, coupled with adequate fracture toughness and fatigue resistance.

Processing metal matrix composites into near net shapes economically, however, is problematic. The conventional consolidation techniques of vacuum hot press/diffusion bonding, powder metallurgical processing, along with fiber tow infiltration, are expensive. Solidification processing of metal matrix composites, on the other hand, appears to provide the route to produce near net shape products in a cost-effective manner [1]. This is mainly due to the low viscosity of metals during pressure infiltration.

Deformation processing of the composites consolidated using these primary techniques, including pressure casting, might be necessary to complete shaping and sizing, and to improve mechanical properties. The latter may be affected through rolling and forging by modifying the residual stress patterns, pores and other defects developed during primary processing, by hardening the matrix, by improving the bond between the fibers and the matrix, and so forth. The intent of this paper is to review the limited number of research papers published in the open literature on the rolling and forging characteristics of Al-alloy filamentary composites. Formability criteria for metal matrix composites containing ductile fibers in longitudinal rolling along with effects of rolling on mechanical properties will be discussed in some detail.

ROLLING

Several experimental results were reported on the effects of cold rolling [2-6] and hot rolling [7] on the microstructure and properties of Al-matrix composites. Only little amount of longitudinal cold rolling (parallel to the direction of the fibers or the major axis of the whiskers) could be applied to the nonmetal fiber or whisker reinforced Al-matrix composites because of extensive edge cracking and breakage of fibers [6]. However, significant amounts of transverse cold rolling (perpendicular to the direction of the fibers or the major axis of the whiskers) were found to be feasible for Al-B fiber composites [6] and Al-Al_3Ni whisker composites [5].

The criteria for the onset of plastic instability in the fiber in metal matrix composites having ductile fibers in longitudinal rolling were established by Moore and co-workers [2]. They studied several metal matrix composite systems reinforced with a variety of strong metallic reinforcements (fibers or sheets). It was assumed that the reinforcement was well bonded to the matrix to allow loading of the reinforcement up to its maximum strength. Upon rolling, local necking of the reinforcement occurs, although the external cross section of the composite remains effectively uniform. If the flow strength σ_m of the matrix is negligibly small compared with the flow strength σ_r of the reinforcement, necking of the reinforcement will start when

$$(d\sigma_r)/(d\varepsilon_r) = \sigma_r \qquad (1)$$

where ε_r is the natural strain in the reinforcement and therefore in the composite for plane strain deformation. Equation (1) is analogous to the criterion for tensile instability for an unrestrained uniaxial tensile test. Necking or localized deformation begins when the rate of strain hardening equals to the instantaneous stress. If the flow strength of the matrix is not negligible, it will impose a constraint to necking of the reinforcement and raise the effective flow strength of the reinforcement. In this case the instability criterion can be written as

$$(d\sigma_r)/(d\varepsilon_r) = \sigma_r - \sigma_t \qquad (2)$$

where σ_t is a function of the current flow strength of the matrix and acting in a direction perpendicular to the reinforcement in the necked region. When a plane strain deformation is imposed on the matrix surrounding the neck, σ_t is approximately equal to 0.5 σ_m. Thus, the matrix will be important in delaying the onset of plastic instability only when σ_m is an appreciable fraction of σ_r.

In rolling a homogeneous metal, the dominant stress system is compressive. The tensile instability does not impose limitations on the extent of deformation. The application of a front tensile stress σ_f in the rolling direction will reduce the required rolling pressure σ_p [8]. Nevertheless, for the case of filamentary or laminar composites, an induced tensile stress in the rolling direction will be developed in the reinforcement by longitudinal rolling. The stresses involved in rolling a filamentary composite are illustrated in Figure 1. During longitudinal

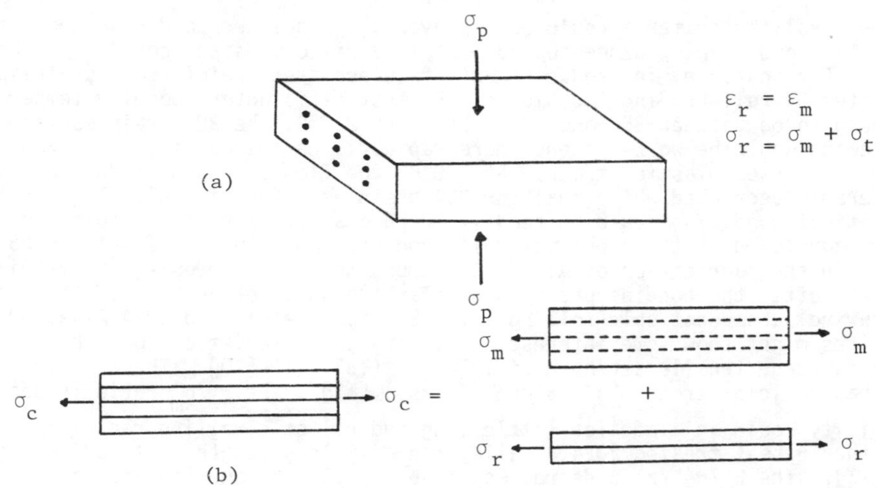

Fig. (1) - The stresses involved in rolling a filamentary composite

rolling, the composite is in plane strain transversely and the strain levels in both the reinforcement and the matrix are approximately the same. Under these conditions, the apparent flow stress of the composite σ_c can be estimated by the law of mixture [9]:

$$\sigma_c = \sigma_r V_f + \sigma_m(1-V_f) \tag{3}$$

where V_f is the volume fraction of the reinforcement. The apparent composite flow strength σ_c is larger than the flow strength of the matrix σ_m while smaller than that of the reinforcement σ_r. Thus, the apparent rolling pressure corresponding to σ_c will be insufficient to maintain a comparable amount of strain in the strong reinforcement. A stress concentration will be established in the reinforcement in the rolling direction. It functions like an applied front tensile stress σ_f on the strong reinforcement to promote the strain. When the flow strength of the matrix σ_m is much smaller than that of the reinforcements σ_r, σ_c will be appreciably smaller than σ_r. The apparent rolling pressure σ_p will be small and the induced front tensile stress σ_t could be a significant fraction of σ_r. In such a system, tensile instability can develop in the strong reinforcement at high strains when the work hardening rate falls to less than the instantaneous fiber flow stress. The plastic instability is therefore sensitive to the flow strength ratio σ_r/σ_m. Experimental results showed that necking would be a serious problem in rolling when the σ_r/σ_m ratio is larger than 3. At the same time, reinforcements which exhibit a sustained rate of work hardening will delay the onset of plastic instability for the same σ_r/σ_m ratio [2].

These rolling criteria could be employed as guidelines to design composite fabrication and shaping processes for metal matrix composites containing ductile fibers. For instance, in producing 304 stainless steel reinforced duraluminum composites by roll-bonding techniques, the most satisfactory bonding temperature was found to be between 250 and 300°C [4]. At 200°C, the 304 stainless steel reinforcement would be work-hardened more rapidly, the instability condition could be met at a lower plastic strain. At 400°C, the flow strength of duraluminum is considerably decreased while that for 304 stainless steel is only slightly lowered. The critical condition could be readily reached at low strains. Moore and coworkers concluded that the optimum roll-bonding temperatures are between 250 and 300°C. In the fabrication of Al-matrix composites, age hardening the matrix immediately after the bonding pass would delay the onset of necking of the reinforcement during subsequent cold rolling. The Be fiber reinforced 2024 Al-alloy matrix composites might show some interesting rolling characteristics too. Be fibers have an average tensile strength of 1274 MPA (185 ksi) [10]. The annealed 2024 Al alloy has a yield strength of 75 MPa (11 ksi) [11]. The σ_r/σ_m ratio is about 17. Instability would initiate for little longitudinal cold rolling reduction. However, the T6 heat treated 2024 Al alloy has a yield strength up to 395 MPa (57 ksi) [11]. The σ_r/σ_m ratio decreases to about 3.2. A moderate amount of cold rolling might be feasible.

In the longitudinal rolling of a Cu fiber reinforced pure Al-matrix composite, the instantaneous fiber flow stress after heavy cold work should be equal to the tensile strength of heavily cold worked Cu wires, i.e., 413 MPa (60 ksi) [10]. The instantaneous matrix flow stress could be estimated as the yield strength of a commercially pure 1100 Al after more than 80% cold reduction as 180 MPa (26 ksi) [11]. With a ratio of σ_r to σ_m about 2.3, a significant amount of cold reduction in the direction of fiber alignment was obtained [2].

Although the above criteria were established for metal-metal composites, it could also serve as guidelines for the composites reinforced with non-metallic fibers despite the fact that local necking is rare for brittle non-metallic fibers. For instance, little cold rolling could be imposed to the non-metallic boron fiber reinforced Al composites. The average tensile strength of the boron filaments is about 2691 MPa (390 ksi) [6]. However, the effective strength for B fibers in Al matrix was calculated by Kreider and Leverant as 1932 MPa (280 ksi) [12]. Before any cold rolling, the matrix flow stress of annealed commercially pure 1100 aluminum is about 90 MPa (13 ksi) [11]. The ratio σ_r/σ_m has a value of about 21, which is much larger than 3. On the other hand, thickness reductions greater than 90% were obtained by transverse cold rolling [6]. In transverse rolling under plane strain conditions, there is little or no induced tensile stress in the fibers, and tensile instability will not occur. The failure of composites will then be controlled by fiber splitting or interfacial separation. Although the matrix can flow readily around the fibers, this produces secondary tensile stresses at the horizontal poles of the fibers, leading to the formation of lenticular voids around them. Thus, transverse rolling formability is closely related to the interfacial bond stress, which allows a significant reduction for Al-B composites.

EFFECTS OF ROLLING ON MECHANICAL PROPERTIES

Only a limited number of studies on the effects of rolling on mechanical properties of fiber or whisker reinforced metal matrix composites have been reported. The following is a brief summary of obtained results.

The transverse cold rolling characteristics of uni-directionally reinforced Al-B fiber composites were investigated by rolling several composites with fiber volume fractions of 16, 30 and 44.5%, respectively [6]. The results indicated that transverse cold rolling has a beneficial effect in raising the tensile strength of the composites. Compared with the as-annealed samples, a 138 MPa (20 ksi) increase in longitudinal tensile strength was observed in the 20% cold-worked 44.5% boron composite. The same effect, to a lesser extent, was also observed with lower volume fraction composites.

Salkind and co-workers [5] have demonstrated that Al_3Ni whisker reinforced Al matrix composite could be readily cold rolled transversely. Rolling in the whisker direction, however, was not feasible without damage to the whiskers. Thickness reductions of more than 98% were achieved without cracking in transverse rolling while extensive edge cracking was noted after only 15% longitudinal rolling. The transverse tensile strength of the composite continuously increased with increasing cold work up to 98% thickness reduction. The longitudinal strength was also increased by transverse cold rolling until a peak strength was reached. The initial volume fraction of the unbroken fibers was essentially unaffected until the peak strength has been reached. Afterwards, more whiskers are broken and their contributions to the composite strength become smaller because the length to diameter aspect ratios are smaller. A 50% cold reduction increased the transverse tensile strength by 55.2 MPa (8 ksi) for a 10% Al_3Ni whisker reinforced aluminum, whereas the longitudinal tensile strength was increased nearly three times this amount. The enhancement of the transverse tensile strength resulted mainly from the strain hardening effect of cold working the matrix. The tensile strength of commercial pure 1100 aluminum could be increased by 48.3 MPa (8 ksi) at 50% cold work [11]. The large increase in longitudinal tensile strength could not be achieved by cold working the matrix alone. Any increase in the whisker strength by cold work was unlikely because of the inherent brittle nature of the material. It was suggested that the marked gain was related to the residual stress relieving action and the increase of elastic limit of the matrix [5]. The added increment in longitudinal tensile strength might also be attributable to a higher strain hardening effect in the matrix due to the presence of reinforcement. The presence of whiskers could restrict the free flow of the matrix material, and the strain in the matrix was greater than the apparent global strain.

In contrast, longitudinal hot rolling has only a little effect on the composite strength. The principal effect of longitudinal rolling is to improve, while not dramatic, the strain to fracture [7]. A total of 80% thickness reduction was given to a SiC whisker reinforced 2024 Al-alloy matrix composite by longitudinal hot rolling at about 650°C followed by a T6 heat treatment. This resulted in decreases in longitudinal and transverse yield strengths by 1 MPa (1.6 ksi) and 43 MPa (6.2 ksi) respectively, corresponding to 2.5% and 10% decreases. The negligible and somewhat negative effects of hot rolling followed by T6 heat treatment on mechanical strength were attributed to the following: (1) No strain hardening in the matrix could be credited for hot rolling followed by T6 heat treatment. (2) The dislocation network substructure buildup during hot rolling, if any, would be annealed in the subsequent T6 heat treatment. (3) Any residual stress buildup during the fabrication process would also be relieved during the hot rolling and the subsequent T6 heat treatment processes. The improvement of fracture strain is a consequence of the more uniform distribution of the reinforcements throughout the matrix produced by hot rolling.

FORGING

Due to the anisotropic nature of filamentary composites, forging can be carried out only in the transverse direction. Deformation occurs nominally in plane strain as the fibers prevent the flow of the material in the fiber direction. An analysis of transverse forging, including friction and geometry effects, of 2024 Al-alloy matrix reinforced with 25% stainless steel fibers was carried out by Erturk and co-workers [13,14]. The main damage was found to be lenticular void formation at the horizontal poles of the fibers. They showed that stresses developed in the composite during forging was a function of friction prevailing at the die/workpiece interface and the specimen geometry, i.e., the length to thickness ratio. Friction increased the hydrostatic stress level in the composite for a given specimen geometry, and thereby was beneficial in preventing void formation. At high friction levels and large aspect ratios, deviation from plane strain occurred due to the tensile stresses developed in the fibers. In the transverse rolling of fiber reinforced metal matrix complsites, plane strain conditions are caused by mainly the constraining effects of the material outside the deformation zone, whereas in transverse compression, mainly fibers provide the constraint. Thus, substantial tensile stresses can develop in compression experiments and in transverse forging. It is clear that quality of interfaces will play a major role in determining the amount of reduction possible in the forging of metal matrix composites, provided that proper process design (combinations of workpiece geometry and friction conditions) is exercised. Forming limit criteria can be established for different composite systems as has been demonstrated for the aluminum/stainless steel systems by Erturk and co-workers [13,14].

CONCLUSIONS

Rolling metal matrix composites in the fiber direction without damage to the material is possible only for composites having ductile fibers and restricted to systems with a ratio of fiber strength to matrix strength equal to or less than 3. Large deformations, however, are achievable in the transverse direction. Transverse cold rolling or forging of metal matrix composites may provide both the necessary shaping and sizing operations and improve their mechanical properties. Quality of the interfaces will play the major role in obtaining large deformations. Forging operations can be designed to produce defect free products using forming limit criteria. Hot rolling or forging, lacking the strengthening effects of the matrix is less efficient in property improvement.

REFERENCES

[1] Cornie, J. A., Chiang, Y. M., Uhlmann, D. R., Mortensen, A. and Collins, J. M., Cer. Bull., 65, p. 293, 1986.

[2] Moore, J. J., Wilson, D. V. and Roberts, W. T., Mat. Sci. Eng., 48, p. 107, 1981.

[3] Moore, J. J., Wilson, D. V. and Roberts, W. T., Mat. Sci. Eng., 48, p. 113, 1981.

[4] Moore, J. J., Wilson, D. V. and Roberts, W. T., Mat. Sci. Eng., 48, p. 123, 1981.

[5] Salkind, M., George, F. and Tice, W., Met. Trans. AIME, 245, p. 2339, 1969.

[6] Getten, J. R. and Ebert, L. J., Trans. ASM, 62, p. 869, 1969.

[7] Nieh, T. G. and Karlak, R. J., J. Mat. Sci. Let., 2, p. 119, 1983.

[8] Meyers, M. A. and Chawla, K. K., Mechanical Metallurgy, Prentice-Hall, p. 112, 1984.

[9] McDanels, D. L., Jech, R. W. and Werton, J. W., Met. Trans. AIME, 233, p. 636, 1965.

[10] Chou, T. W., Kelly, A. and Okura, A., Composites, p. 187, 1985.

[11] ASM Metals Handbook, Desk ed., pp. 6-38, 1985.

[12] Kreider, K. and Leverant, G., "Boron fiber metal-matrix composites by plasma spraying", Tech. Rept. AFML-TR-66-219.

[13] Erturk, T., Kuhn, H. A. and Lawley, A., Met. Trans. 5A, p. 2295, 1974.

[14] Erturk, T. and Kuhn, H. A., J. Eng. Mat. Tech., 101, p. 3, 1979.

THERMAL ANNEALING STUDY OF COLD POWDER METALLURGY PROCESSED Cu-Ni COMPOSITES

Y. D. Yao

Institute of Physics, Academia Sinica
Taipei, Taiwan, 11529, Republic of China

ABSTRACT

Cu-Ni composite samples were prepared by swaging and drawing techniques. The temperature-dependent electrical resistivity was studied between 4K and 1,100K. The isothermal irreversible electrical resistivity changes due to thermal annealing was measured at 300°C, 500°C and 700°C. The saturation magnetization of these Cu-Ni composites is decreased after thermal annealing. It is evident from our experimental results that the diffusion between Cu and Ni is very slow below 570°K; and the alloying effect is not completed even annealing at 700°C for two weeks.

INTRODUCTION

The outstanding mechanical properties of composite materials has led not only to extensive research on the mechanical properties of composite materials, but also to a highly developed technology [1]. In comparison, relatively little effort has been devoted to the other physical properties of composite materials; such as the electrical and magnetic properties, etc. The different behaviors of the electrical and magnetic properties after thermal treatments are very important probes to understand the special characteristic of the composites. There is not only a need to improve and optimize these properties but also the possibility of devising materials showing novel electrical and magnetic behaviors [2,3]. In this investigation, we give experimental information on the influence of the electrical and magnetic properties of Cu-Ni composite thin wires after the thermal annealing treatment. Our experiment results and their significance are briefly presented below.

EXPERIMENTAL CONSIDERATION

The Cu-Ni composite samples were fabricated following the same procedure that developed for metal-metal composite materials [4,5]. Figure 1 illustrates the procedure of the cold powder technique which has been used to produce Cu-Ni composite samples. Using roughly 100 μm copper and nickel powders, we mixed these powders together with 50 percent by weight. Copper tubes with roughly 12.7 mm outer diameter and roughly 11.1 mm inner diameter were used as an external jacket to contain these powders. The mixed powders were tapped into the Cu tube, the tube was evacuated, and then the compact was swaged and wire drawn to final size. Swaging techniques were used for the preliminary elongation, and the remaining de-

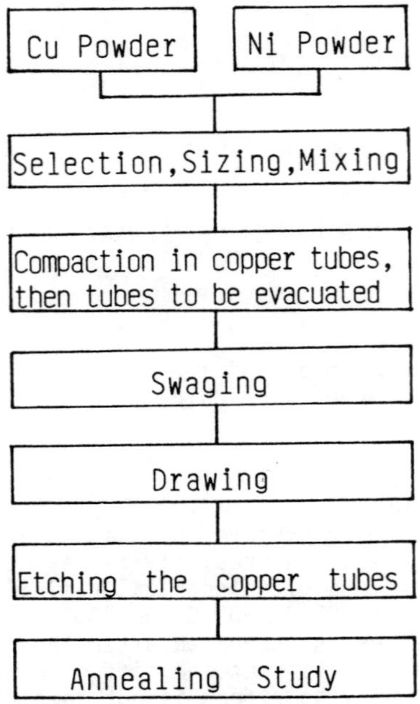

Fig. (1) - Cold powder processes

formation to a wire with a final outer diameter of roughly 0.25 mm was made by wire drawing. Nominal areal reduction ratio of roughly 2500 were obtained without any intermediate anneals.

The electrical resistivity of these Cu-Ni composite thin wires was determined using the conventional four-probe technique. A D.C. current was maintained constant to about one part in 10^5 and the voltage was measured with an accuracy to a few nanovolts. Temperatures between 4K and 300K were achieved in a Displex closed-cycle refrigeration system (Model CS-308, Air Products). Temperatures between 300K and 1,100K were produced by a Marshall vacuum furnace. The furnace could be either evacuated or filled with helium gas. Magnetic measurements were made with a vibrating sample magnetometer at room temperature with the field along the fiber axis.

RESULTS AND DISCUSSION

Figure 2 shows the temperature dependence of the electrical resistivity ρ of the Cu-Ni composite sample in the temperature range between 4K and 1,100K. The rate of the temperature change was roughly about 1K/min. One low temperature run (4K - 300K) and three high temperature runs (300K - 1,100K) have been taken. The whole measuring procedures are indicated by numerically labelled arrows in Figure 2. The data of the low temperature run (4K - 300K) were completely reproducible. The first high temperature run was started after the low temperature run, and the electrical resistivity were measured with increasing temperature up to about 725K. Below roughly 570K the slope of the ρ versus T curve is quite stable, however, between 570K and 725K the slope increases rapidly; this suggests that the alloying

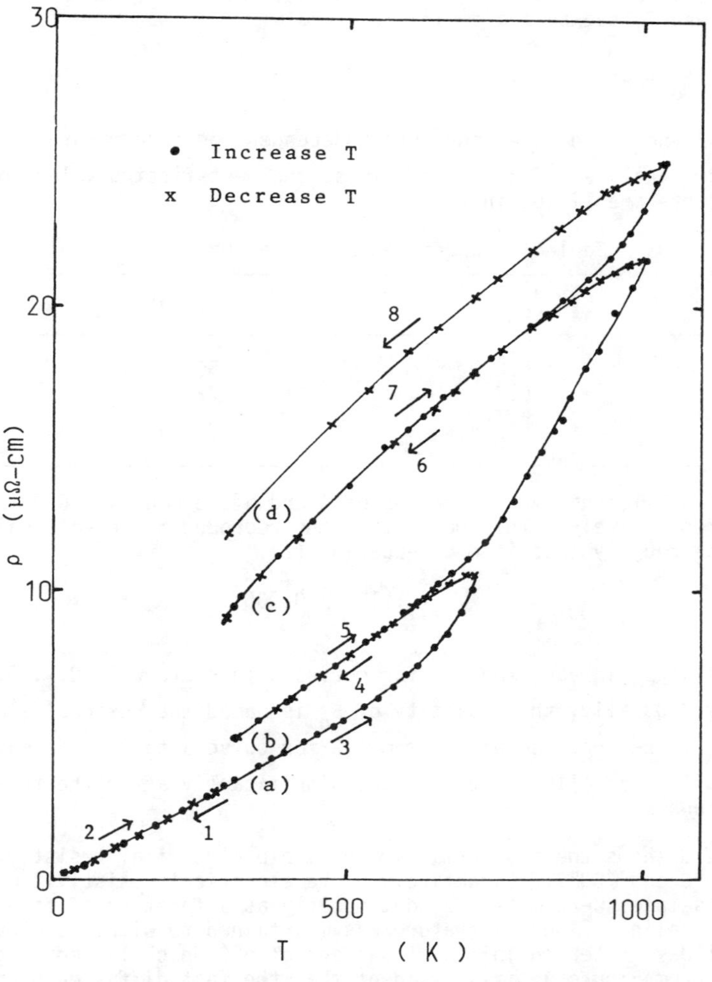

Fig. (2) - Electrical resistivity of Cu-50wt.%Ni composite as a function of absolute temperature between 4K and 1,100K

effect begins roughly around 570K. After reaching approximately 725K, temperatures were decreased down to about 300K. It is obvious that there is a large shift between the ρ versus T curves of increasing and decreasing temperatures. This means that the diffusion happens between Cu and Ni; and some part of the elements becomes CuNi alloys. The second and third high temperature runs were measured with the same situation as the first high temperature run, except that the highest temperatures achieved were roughly 1,000K and 1,050K for the second and third high temperature runs, respectively.

The four ρ versus T curves, named by a, b, c and d in Figure 2, represent the reproducible data points. The electrical resistivity data of each curve (a to d) were fitted to the empirical formula:

$$\rho(T) = \rho_o + AT^n \qquad (1)$$

where ρ_o, A and n are some constants independent on temperature. According to the computer analyses of our data points, the best fitting values of ρ_o, A and n for each curve are listed in Table 1.

Table 1 - Experimental values of ρ_o, A and n

Curve	ρ_o	A (x 10^{-4})	n
a	0.2	57	1.103
b	0.7	72	1.108
c	5.7	84	1.107
d	9.0	82	1.101

From Table 1, the average values of A and n are 0.0074 ± 0.0017 and 1.105 ± 0.005, respectively. This means that the reproducible electrical resistivity data can be roughly described by equation (2).

$$\rho(T) = \rho_o + (0.0074 \pm 0.0017)T^{1.105 \pm 0.005} \qquad (2)$$

The values of ρ_o in equation (2) for curves a to d are 0.2, 0.7, 5.6 and 9.0, respectively. Usually, this quantity of ρ_o is named the residual electrical resistivity [6]. The increase of ρ_o from 0.2 for curve a to 9.0 for curve d suggests that the effect of alloying take place significantly after the three high temperature runs.

Figure 3 shows the isothermal irreversible electrical resistivity study at 300°C, 500°C and 700°C, respectively. The electrical resistivities of the Cu-Ni composite samples were measured continuously as a function of time during isothermal annealing. The temperature was maintained to within 2°C over a period of several days after an initial transient of 5°C which lasted roughly about 10 minutes. From Figure 3, it is evident that the fast diffusion happens between Cu and Ni at 700°C for roughly 2 x 10^4 minutes (i.e., roughly two weeks), and the alloying effect of the Cu-Ni composite is not completed even annealing at 700°C for two weeks.

Figure 4 shows the characteristic of the room-temperature electrical resistivity of the Cu-Ni composite as a function of the annealing time at 300°C, 500°C and 700°C. If we compare it with the room temperature electrical resistivity of Cu-Ni alloys [7], it seems that the Cu-Ni composites gradually change to the Cu-Ni alloys with roughly fifty percent Ni concentration after annealing at 700°C for two weeks.

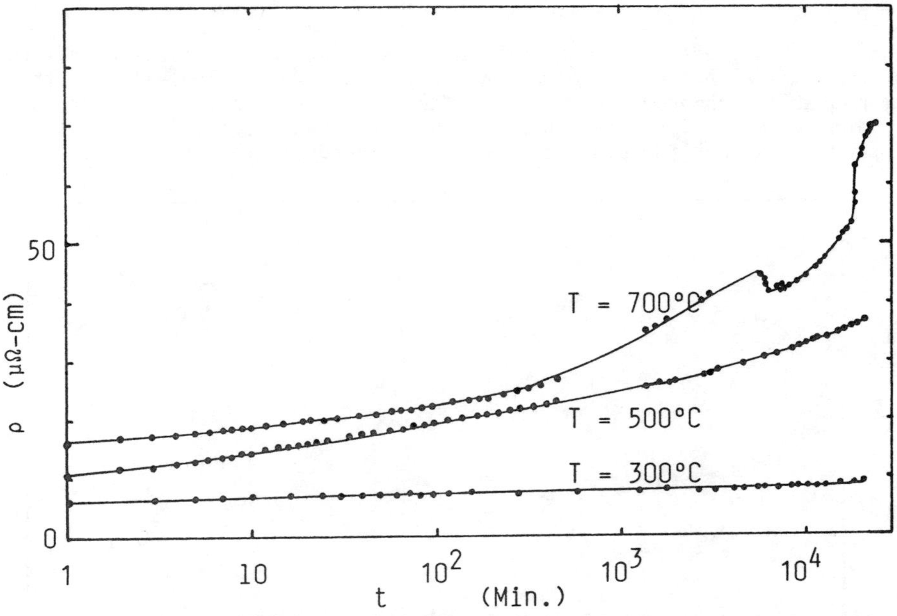

Fig. (3) - Electrical resistivity of Cu-50wt.%Ni composite as a function of annealing time at 300°C, 500°C and 700°C

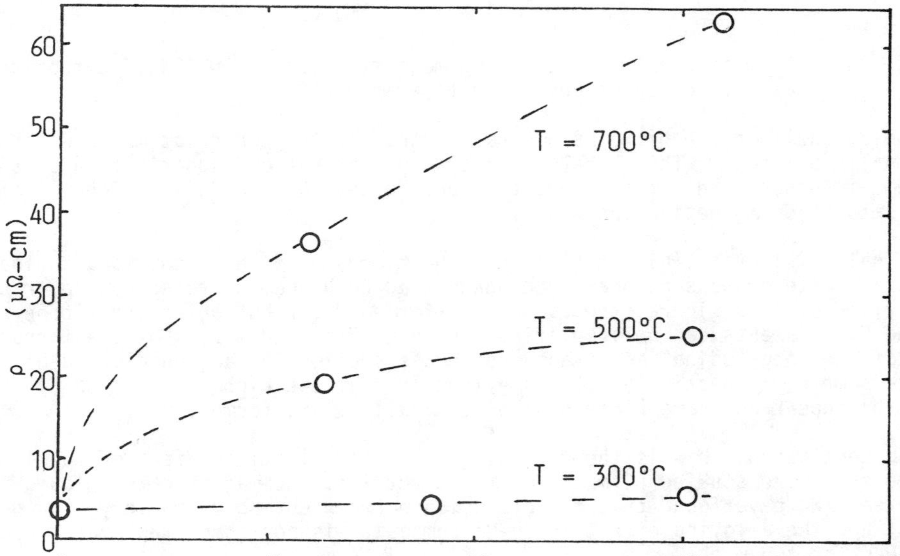

Fig. (4) - Room-temperature electrical resistivity of Cu-50wt.% composite as a function of annealing time at 300°C, 500°C and 700°C

Figure 5 represents the room temperature saturation magnetization of the Cu-Ni composite as a function of annealing time at 700°C. The value of the saturation magnetization drops very fast at the beginning; and this decrease of the saturation magnetization after thermal annealing means that some electrons of the copper atoms seek the place of the lowest energy in the system and find it

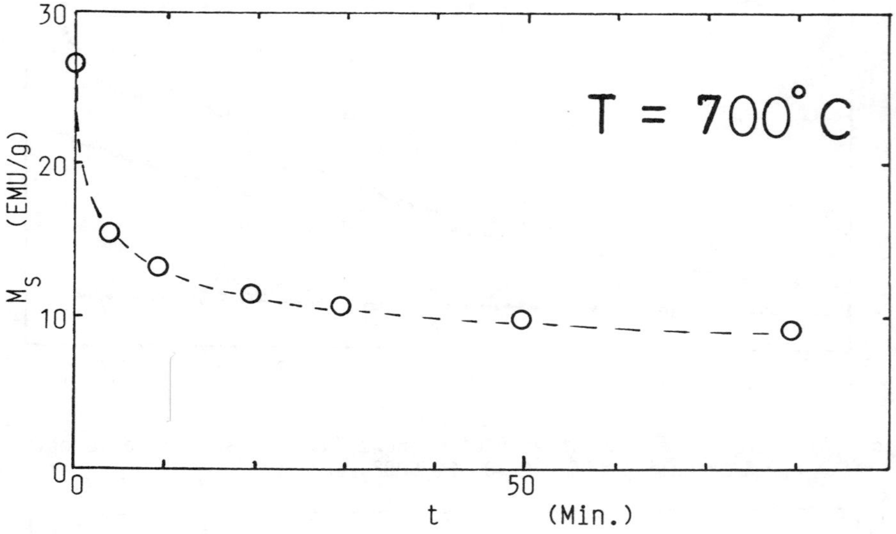

Fig. (5) - Room-temperature saturation magnetization of Cu-50wt.% composite as a function of annealing time at 700°C

in the 3d shell of the nickel atoms rather than in the copper atoms to which they originally belonged. This behavior lowers the saturation magnetization of the system; in other words, the alloying effect between Ni and Cu is evident from this saturation magnetization study.

Finally, Figures 6(a) and 6(b) show the microstructure of the longitudinal and transverse cross sections of an unannealed Cu-Ni composite thin wire, respectively. From this figure, the roughly uniformed distribution of the ribbon-shaped Cu and Ni filaments can be identified. Figures 7(a) and 7(b) show the microstructure of the longitudinal and transverse cross sections of an annealed Cu-Ni composite sample, respectively. It is evidently that the ribbon-shaped Cu-Ni filaments disappear and some kinds of the Cu-Ni alloys are formed.

In conclusion, this is the first time that the alloying effect of Cu-Ni composite thin wire studied by means of the electrical and magnetic techniques is reported. We have found that the diffusion between Cu and Ni is very slow below 570°C; and the alloying effect in Cu-Ni composite is not completed even annealing at 700°C for two weeks.

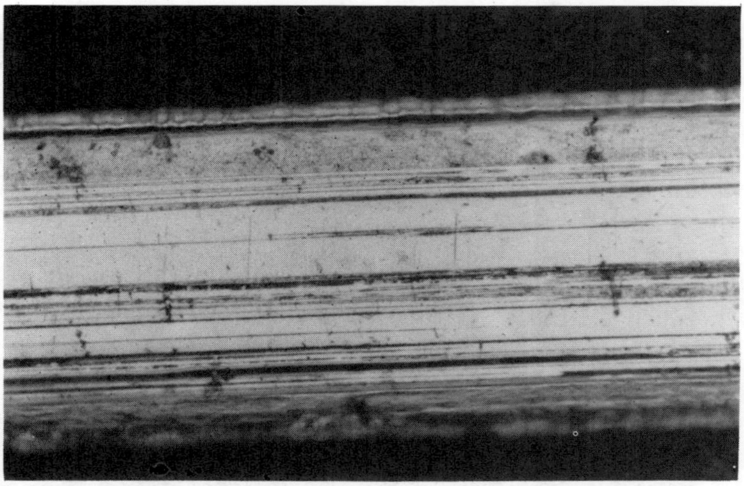

(a) longitudinal cross section

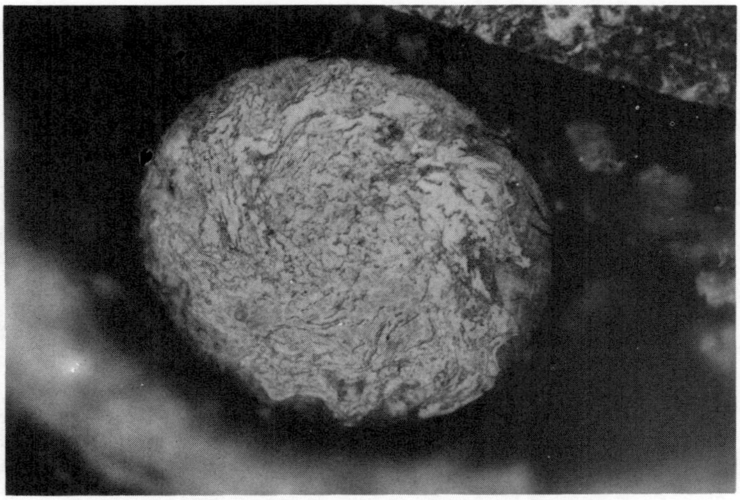

(b) transverse cross section

Fig. (6) - Micrograph of polished cross sections (a: longitudinal and b: transverse) of an unannealed Cu-50wt.%Ni composite with the magnification of 250

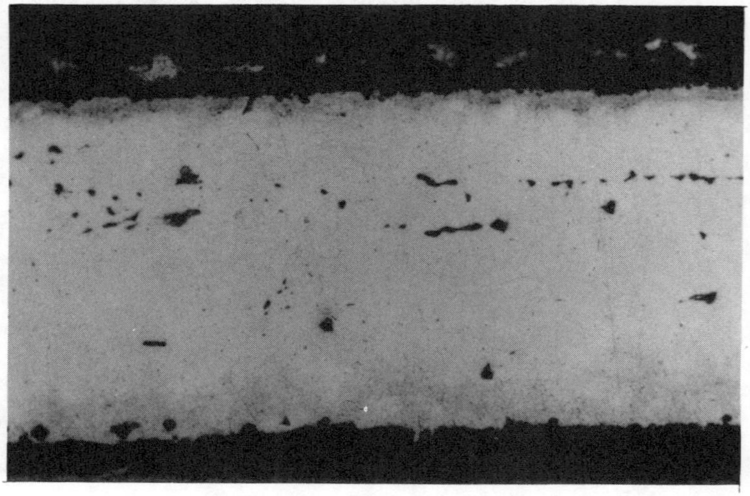

(a) longitudinal cross section

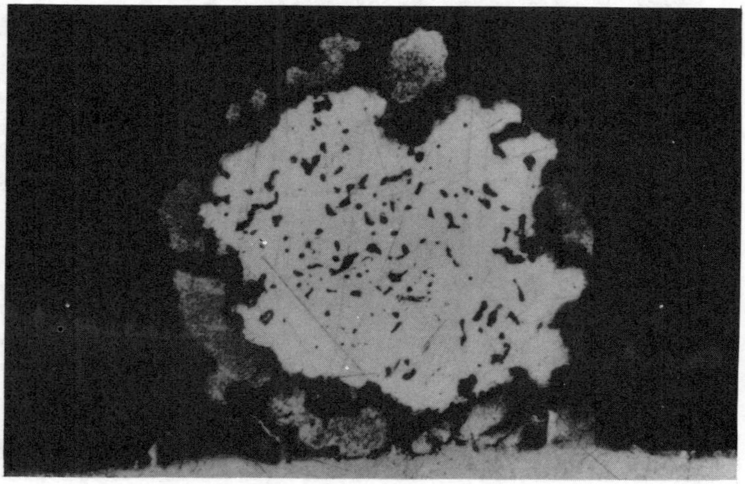

(b) transverse cross section

Fig. (7) - Micrograph of polished cross sections (a: longitudinal and b: transverse) of an annealed Cu-50wt.%Ni composite with the magnification of 250

ACKNOWLEDGEMENTS

The author is grateful to the National Science Council of the Republic of China for the financial support of this work under Contract No. NSC75-0208-M001-09.

REFERENCES

[1] Christensen, R. M., "Mechanics of composite materials", John Wiley & Sons, Inc., 1979.

[2] Bhushan, B., Kashyap, S. C. and Chopra, K. L., "Novel nonohmic binary composite", Appl. Phys. Lett., 38, p. 160, 1981.

[3] Abukay, D., Rao, K. V., Arajs, S. and Yao, Y. D., "Electrical resistivity of aluminium-boron composites between 78K and 400K", Fib. Sci. Technology, 10, p. 313, 1977.

[4] Yao, Y. D. and Foner, S., "Ferromagnetic multifilamentary Fe and Ni wires with high coercive fields produced by powder metallurgy processing", Appl. Phys. Lett., 43, p. 697, 1983.

[5] Chen, I. S. and Yao, Y. D., "Annealing study of multifilamentary Cu-Zn thin wires", Chin. J. Mat. Sci., 18A, 1986.

[6] Yao, Y. D., "Temperature-dependent resistivity in Ni-rich alloys", Phys. Stat. Sol., B94, p. K137, 1979.

[7] Yao, Y. D. and Tsai, J. H., "Magnetic phase transition in nickel-rich nickel-copper alloys", Chin. J. Phys. 16, p. 189, 1978.

EFFECT OF FORMALDEHYDE/PHENOL RATIO, VISCOSITY AND CATALYST TYPE ON THE CHARACTERISTICS OF CASTING PHENOLIC RESINS

S.-L. Jwo and W.-Y. Chen

Chung Shan Institute of Science and Technology
P.O. Box 1-26-2, Lungtan, Taiwan 32500, Republic of China

ABSTRACT

This study presents experimental results on the preparation and characteristics of sodium carbonate, barium hydroxide, sodium and potassium hydroxide-catalyzed casting type phenolic resins polymerized in two-liter glass reactors over a broad range of formaldehyde/phenol (F/P) molar ratios. The effects of catalyst type, viscosity and F/P molar ratio on its compressive strength were investigated, and the resulting resins were explained by using electronic microscopy technique.

An optimum strength was obtained for the sodium carbonate or sodium hydroxide-catalyzed resin system at a certain F/P molar ratio and viscosity. The relation between the density and strength of the cast phenolic resins was also studied.

INTRODUCTION

Phenol-formaldehyde resins are divided into two main classes - novolacs and resoles. Novolacs are almost invariably prepared by heating phenol and formaldehyde with acid catalysis at a F/P molar ratio of less than 1. These resins are usually dehydrated and cooled to give glassy products which are pulverized, compounded with fillers, hexamethylenetetramine, flow promoters, etc., and employed as industrial molding powders.

Resoles are base-catalyzed phenol-formaldehyde resins generally prepared with a F/P molar ratio of greater than 1. Resoles are prepared by refluxing phenol, formaldehyde, catalyst and water for a variable period of time, resulting in a water-or alcohol-soluble varnish. The finished resin is useful for coating or impregnating paper, cloth, glass fabric, etc., which is subsequently used to prepare industrial or decorative laminates.

The first step in the manufacture of a phenolic casting resin is similar to that of a resole. The casting resins are prepared using a formaldehyde excess (F/P molar ratio = 1.2 - 2.5) with alkaline catalyst. After completed reaction, the mixture is neutralized with a weak organic acid such as lactic acid, and the water is then removed as far as possible. The resin is poured or cast into open molds and hardened at 60-80°C for 24-48 hours [1].

Casting phenolic resins are widely used in many applications such as decorative articles, bowling balls, aircraft tools and mechanical parts because of its easy processability, high performance of the products made from it and relatively low cost. To obtain products with good properties, the type of catalyst and F/P molar ratio as well as the reaction conditions are important.

A wide range of resin types is obtained if the type and amount of catalyst and F/P molar ratio as well as the reaction conditions are varied. The effect of F/P ratio [2-12] and catalyst type [10-13] on the properties of pure phenolic resin has been represented and investigated. However, very few papers have been published for the casting type phenolic resin, especially, on the relation between the strength and the viscosity of the resin.

This study was undertaken to discuss the properties of several series of basic-catalyzed casting phenolic resins as a function of F/P molar ratio, type of catalyst and reflux time (or viscosity), primarily with a view to developing an optimum reaction conditions for the preparation of high-compressive cast phenolic resin. The relationship between the resulting resins and the compressive strength of the products was also explained by using electronic microscopy technique.

EXPERIMENT

A. Preparation of Phenolic Casting Resin

A number of syntheses were carried out in which the molar ratios were: formaldehyde, 1.2, 1.4, 1.5, 1.6, 1.8, 2.1, 2.3, 2.5, moles; phenol 1.0 mole.

The conditions of experimentation are given in Tables 1-4. The required quantities of reagent-grade phenol, 37% aqueous formaline, and catalyst (catalyst/phenol equivalent ratio = 0.047) were charged in a 2-L three-necked flask fitted with stirrer, thermometer and reflux condenser. The complete apparatus was then lowered into a heating mautle, refluxing for three hours at 64-68°C, 1 hour at 74-78°C, and 1 hour at 82-84°C, cooled, and neutralized with a weak organic acid such as 50% lactic acid, which was equivalent to the amount of catalyst (e.g., NaOH, etc.) added. The mixture was dehydrated under 20 mmHg and 55°C until the water is removed as far as possible, 0.2 phr. Glycerol then was added. The resin was heated at 80-85°C to obtain a lower viscosity resin for casting easily, the resin was poured or cast into thin stainless steel open molds, and vacuumed for 1 hour at 80-85°C to drive the remained water and bubbles, and then hardened at 80°C for 24 hours, 90°C for 1 hour, 120°C for 3 hours and 150°C for 5 hours.

Similarly, a series of casting phenolic resins were prepared at a constant F/P = 1.5 molar ratio using variable reflux time and temperature, with sodium hydroxide catalyst. The conditions of experimentation are given in Table 5.

B. Characterization of Cast Phenolic Resin

The viscosity of phenolic casting resin was determined by using a Brookfield viscometer (HBT model) at 25°C. The density was determined in accordance with the method specified in ASTM standard D792-66. For determination of compressive strength, the loading surfaces of the specimen were made parallel and the load was applied perpendicular to the face in accordance with the method specified in ASTM D695-80. The test was carried out on an Instron Machine (Model 1125). The results of the compressive strength and density of various specimens of casting resin are shown in Tables 1-5.

Table 1 - Casting phenolic resin properties as a function of F/P mole ratio for sodium hydroxide catalyst

	N1	N2	N3	N4	N5	N6	N7
Phenol, g.	650	600	585	575	550	500	450
CH_2O(37%), g.	673	724	757	794	854	906	970
NaOH(s), g.*	13	12	11.7	11.5	11	10	9
F/P Mole Ratio	1.2	1.4	1.5	1.6	1.8	2.1	2.5
Density (g/cm^3)	1.245	1.267	1.271	1.276	1.278	1.282	-
Compressive Strength (ksi)	28.2±0.3	31.3±0.8	33.3±0.8	31.5±0.8	30.3±0.7	25.6±0.7	crack

*NaOH(s) is dissolved in 10 ml of water before this is added to the reaction mixture.

Table 2 - Casting phenolic resin properties as a function of F/P mole ratio for potassium hydroxide catalyst

	K1	K2	K3	K4	K5	K6	K7
Phenol, g.	650	600	585	575	550	500	450
CH_2O(37%), g.	673	724	757	794	854	906	970
KOH(s), g.*	18.2	16.8	16.4	16.1	15.4	14.0	12.6
F/P Mole Ratio	1.2	1.4	1.5	1.6	1.8	2.1	2.5
Density (g/cm^3)	1.249	1.267	1.268	1.270	1.275	1.278	1.297
Compressive Strength (ksi)	27.3±0.6	28.0±0.6	31.5±0.6	30.5±0.9	28.0±0.9	22.9±0.7	21.3±0.9

*KOH(s) is dissolved in 10 ml of water before this is added to the reaction mixture.

Table 3 - Casting phenolic resin properties as a function of F/P mole ratio for $Ba(OH)_2 \cdot 8H_2O$ catalyst

	B1	B2	B3	B4	B5	B6	B7	B8
Phenol, g.	650	600	585	575	550	500	475	450
CH_2O(37%), g.	673	724	757	794	854	906	942	970
$Ba(OH)_2 \cdot 8H_2O(s)^*$, g.	51.3	47.4	46.1	45.4	43.4	39.4	37.5	35.5
F/P Mole Ratio	1.2	1.4	1.5	1.6	1.8	2.1	2.3	2.5
Density (g/cm^3)	1.270	1.279	1.284	1.289	1.293	1.296	1.307	-
Compressive Strength (ksi)	29.5±0.8	31.2±0.8	32.2±0.8	31.3±0.6	29.6±0.8	27.3±0.6	26.2±0.6	crack

*$Ba(OH)_2 \cdot 8H_2O(s)$ is dissolved in 60 ml of hot water (50°C) before this is added to the reaction mixture.

Table 4 - Casting phenolic resin properties as a function of F/P mole ratio for sodium carbonate catalyst

	C1	C2	C3	C4	C5	C5	C6
Phenol, g.	650	600	585	575	550	500	475
CH_2O(37%), g.	673	724	757	794	854	906	942
$Na_2CO_3(s)$, g.*	17.2	16.0	15.5	15.2	14.6	13.3	12.6
F/P Mole Ratio	1.2	1.4	1.5	1.6	1.8	2.1	2.3
Density (g/cm^3)	1.274	1.275	1.277	1.279	1.284	1.299	-
Compressive Strength (ksi)	31.0±0.5	31.2±0.6	33.1±0.5	31.1±0.7	30.0±0.4	25.7±0.5	crack

*$Na_2CO_3(s)$ is dissolved in 30 ml of hot water (40°C) before this is added to the reaction mixture.

Table 5 - Casting phenolic resin properties as a function of resin viscosity for NaOH catalyst and F/P = 1.5

	A	B	C	D	E	F	G	H	I
Refluxing Temperature and time (hr.): 66±1(°C)	4 (hr.)	6	8	1	2	3	4	5	5
76±1(°C)	0	0	0	1	1	1	1	1	2
82±1(°C)	0	0	0	1	1	1	1	1	2
Viscosity, 25°C(cps.)$\times 10^3$	14	16	85	122	144	188	550	840	1170
Compressive Strength (ksi)	*	27.5 ±0.9	28.0 ±1.0	29.8 ±0.6	31.2 ±0.6	33.3 ±0.8	32.0 ±0.8	31.1 ±0.7	30.5 ±0.2

*Bubbles spread over the specimen like a foam.

C. Scanning Electron Microscopy

We made the surface of specimen developed from the bending fractured. Five sets of electron micrographs of fracture surface of casting phenolic resin systems are shown in Figures 4-8.

RESULTS AND DISCUSSION

A. Effect of Catalyst

The effect of catalyst types on the compressive strength of casting resin is shown in Figure 1. In comparison with potassium hydroxide, there is little effect

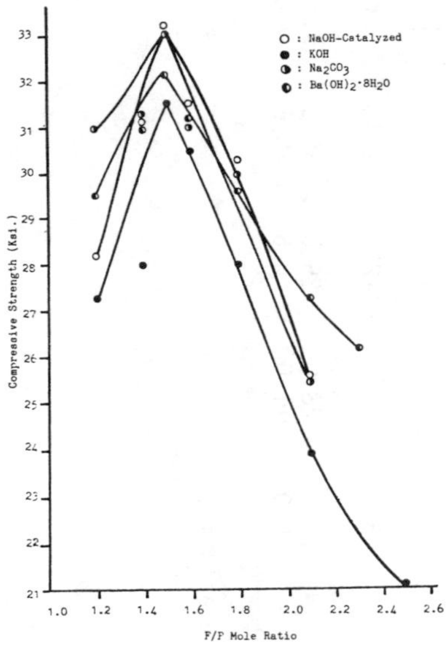

Fig. (1) - Compressive strength as a function of F/P mole ratio for four catalyst systems

of sodium hydroxide-octahydrate, sodium hydroxide and sodium carbonate on the compressive strength of casting phenolic resins. Strictly speaking, it may be noted, however, that the compressive strength of casting resin is slightly higher than the others for using sodium hydroxide or sodium carbonate as catalyst, especially at a 1.5:1 F/P molar ratio.

B. Effect of F/P Molar Ratio

Graphic comparisons of compressive strength and density of the cast phenolic resin as a function of F/P molar ratio are presented in Figure 1 and Figure 2. Figure 1 shows that an optimum compressive strength is obtained at a 1.5:1 F/P molar ratio, falling gradually at lower or higher ratios, in spite of using any

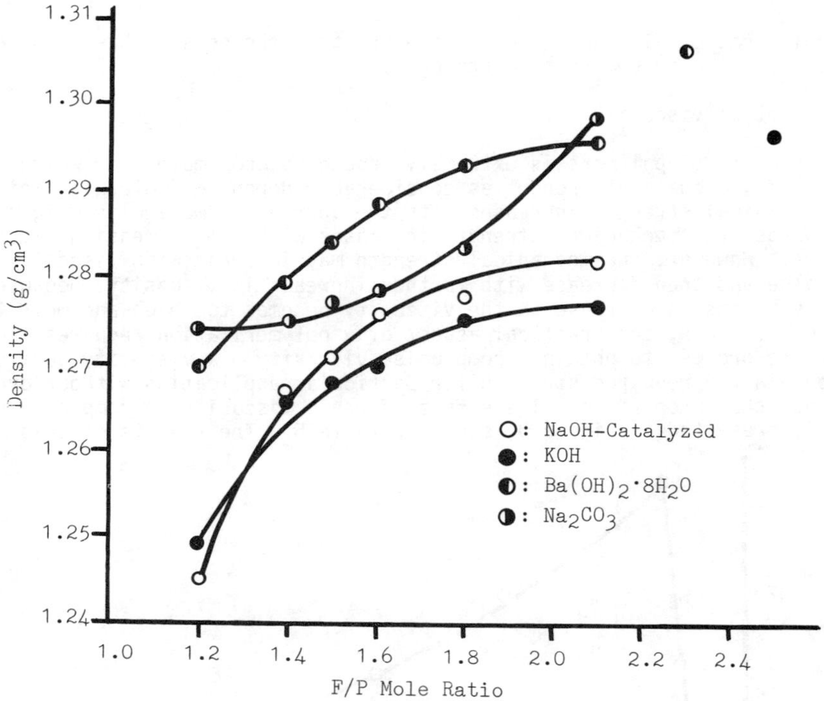

Fig. (2) - Density as a function of F/P mole ratio for four catalyst systems

one type of catalyst. Theoretically, 1.5 mol of formaldehyde is needed for the complete three-dimensional cross-linking of 1 mol of phenol, see product (1), [14]. The more complete three-dimension cross-linking is, the higher mechanical

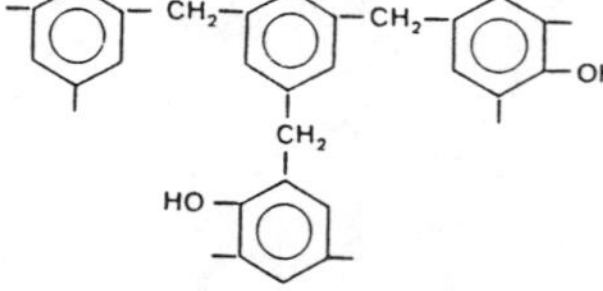

(1)

strength it possesses. The experimental results indicate that the reality and theory are in close agreement. The possible reason is that it will remain some free monomers (formaldehyde or phenol) and very low molecular weight polymers (self-condensation products, etc.) after completed reaction as F/P molar ratio is more than 1.5 or less than 1.5. If the polymer contains a large fraction of very low molecular weight material mixed with high molecular weight material, the polymer will be extremely brittle and will have lower strength than normal one [15], and the diminution of mechanical strength depends mostly upon the amount of low molecular weight polymers. The more the F/P molar ratio is larger than 1.5, the more low molecular weight polymers remain; therefore, the compressive strength of casting resin is less. Figure 2 shows that the density increases with increasing F/P molar ratio. It can be seen, however, that the density has no direct proportion relation to compressive strength.

C. Effect of Viscosity

The viscosity of polymers is extremely dependent upon molecular weight [16]. Most important, mechanical properties considerably depend on molecular weight and the mechanical strength increases with the increasing molecular weight [17]. In other words, the mechanical strength increases with the increasing viscosity of polymers. However, the mechanical strength may increase with viscosity to a maximum value and then decrease with further increase in viscosity, because processability begins to decrease as the viscosity becomes too high and melt flow too difficult. Thus, the practical aspect of a polymerization requires one to carry out the process to obtain a compromise viscosity - a viscosity sufficiently high to obtain the required strength for particular application without overly sacrificing other properties. The effect of resin viscosity on compressive strength is presented in Figure 3, see also Table 5. The results of this study

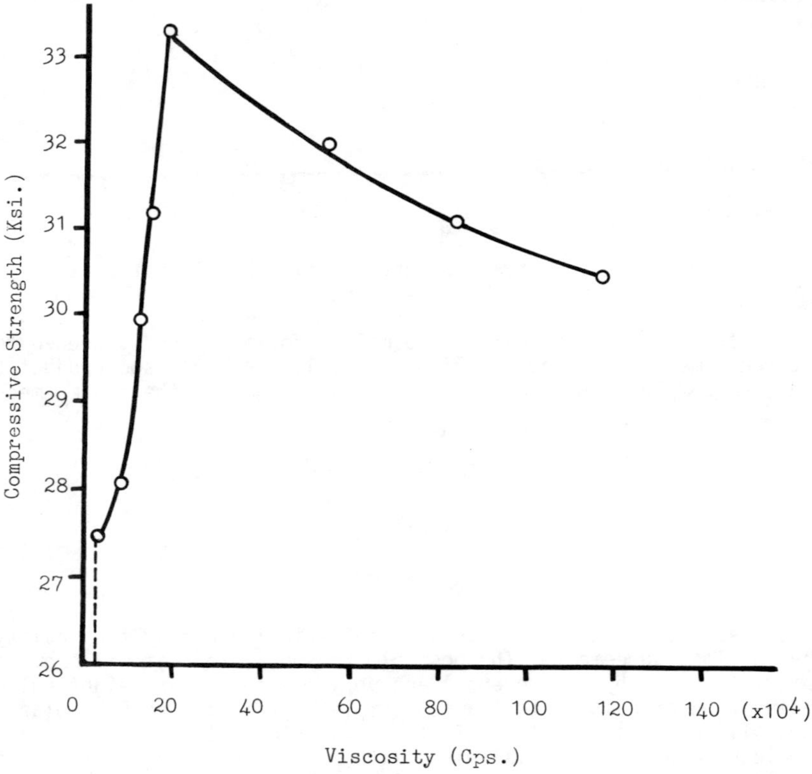

Fig. (3) - Compressive strength versus viscosity for casting phenolic resins (F/P mole ratio = 1.5)

show that a better viscosity of casting resin, in the range of 180,000 - 500,000 centi-poises, is necessary to obtain a higher compressive strength of casting phenolic resins.

D. Scanning Electron Microscopy

Four sets of scanning electron micrographs of fracture surface of casting phenolic resins are shown in Figures 4-7. These specimens are B1, B3, B7; C1, C3, C6; K1, K3, K7; N1, N3 and N6, respectively. The fracture surface of N and K specimens which is smooth, B and C specimens is rough, and microvoids are dispersed throughout the matrix. The size of the microvoids in the specimen is shown as follows:

Figure No.	4(N1)	4(N3)	4(N6)	5(K1)	5(K3)
Microvoid size, μ	5-9	1-1.5	2-3	1-12	1-4
Figure No.	6(B1)	6(B3)	7(C1)	7(C3)	
Microvoid size, μ	1-4	1-2	0.5-1.0	0.2-0.5	

Moreover, the compressive strength of each specimen is N3>N1, K3>K1, B3>B1 and C3>C1. The results show that the more the size of microvoids is, the less the compressive strength of casting resin is. On the other hand, the fracture surface of C6 and K7 specimen presents a form of mountain ridge, the B7 specimen presents many hollows on the fracture surface, and the fracture surface of N6 specimen presents many hollows and microvoids (2-3μ), the compressive strength of these specimens is less.

A set of scanning electron micrographs of the fracture surface of casting phenolic resins for different viscosity is also shown in Figure 8. These specimens are B, F and I specimen, respectively. The fracture surface of specimens is smooth; similarly, microvoids are dispersed throughout the matrix. The size of microvoids of B (4-8μ) and I (1-2.5μ) is larger than F (1-1.5μ), respectively, and the compressive strength of each specimen is F>I>B. The results confirmed that the compressive strength increases with the increasing viscosity of polymers, however, the strength will decrease with further increase in viscosity, because processability begins to decrease as the viscosity becomes too high.

CONCLUSIONS

Several series of basic-catalyzed casting phenolic resins were prepared as a function of F/P molar ratios of 1.2 to 2.5, type of catalyst and viscosity. An optimum compressive strength was obtained for using sodium hydroxide or sodium carbonate-catalyzed resin system, especially at a 1.5:1 F/P molar ratio, and in the range of 180,000 - 500,000 centi-poises viscosity of phenolic casting resin. Generally speaking, the cast phenolic resin density increases with increasing F/P molar ratio; however, the density has no direct proportion relation to compressive strength.

ACKNOWLEDGEMENT

The authors express their appreciation to Mr. Shin-Yeh Chung for his help in preparation of samples, to Mr. Torng Shan and Yeo-Fu Wang for their work on microscopy. We also thank Mr. Ter-Shoung Lai for assistance on compression test.

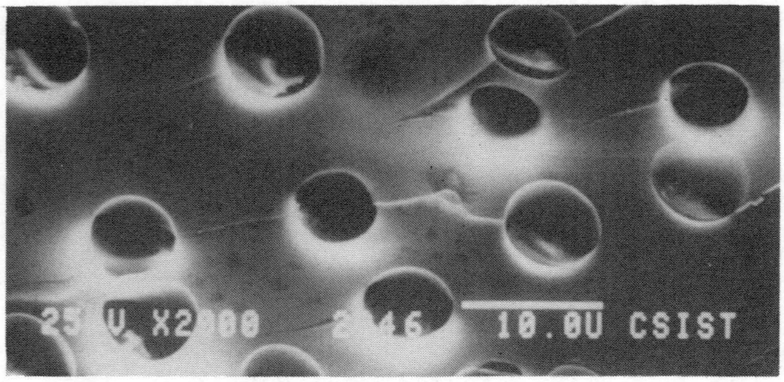

(N1)

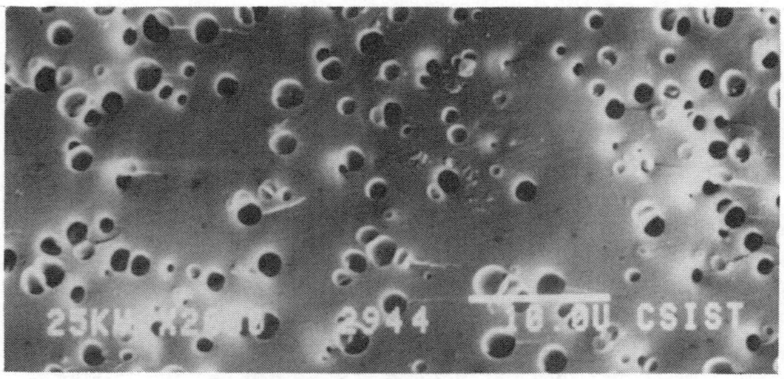

(N3)

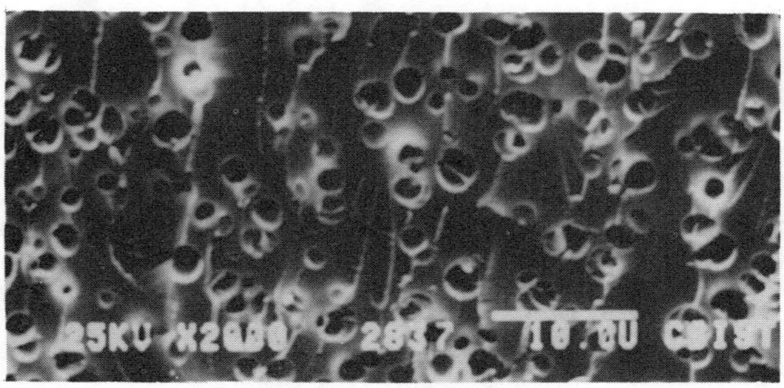

(N6)

Fig. (4) - Scanning electron micrograph of fracture surface of NaOH-catalyzed cast phenolic resin: (N1) F/P = 1.2; (N3) F/P = 1.5; (N6) F/P = 2.3

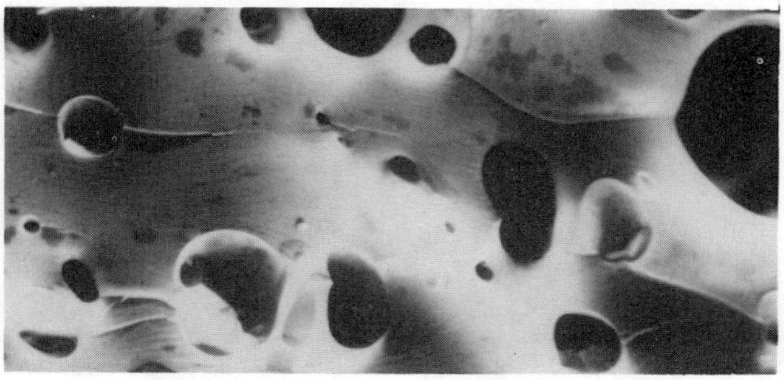

(K1)

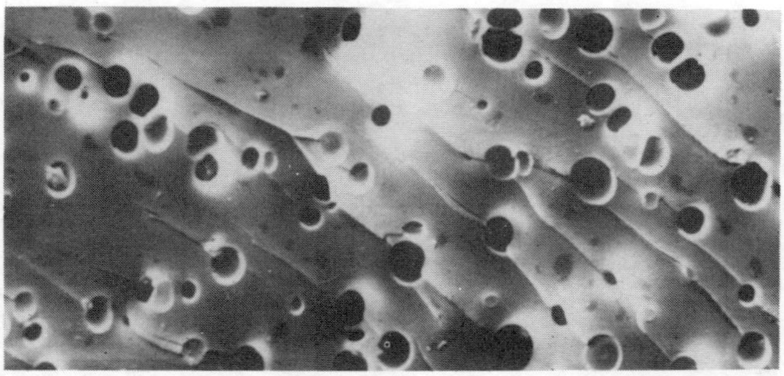

(K3)

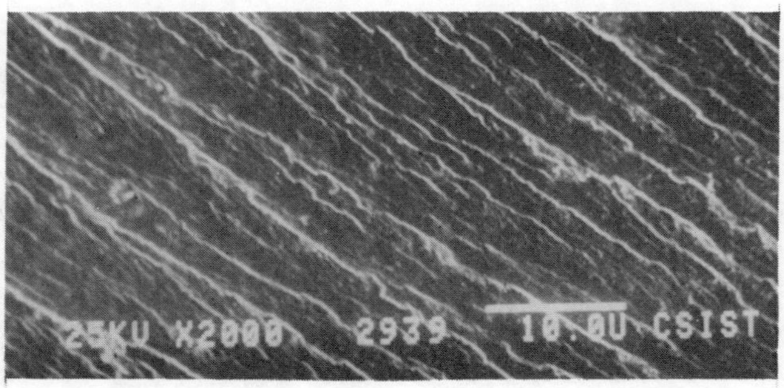

(K7)

Fig. (5) - Scanning electron micrograph of fracture surface of KOH-catalyzed cast phenolic resin: (K1) F/P = 1.2; (K3) F/P = 1.5; (K7) F/P = 2.3

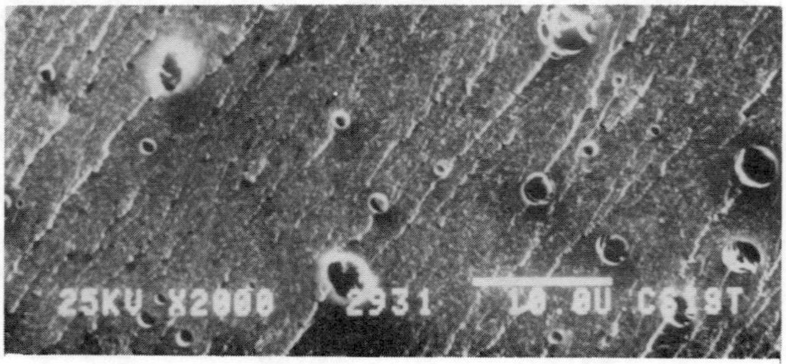

(B1)

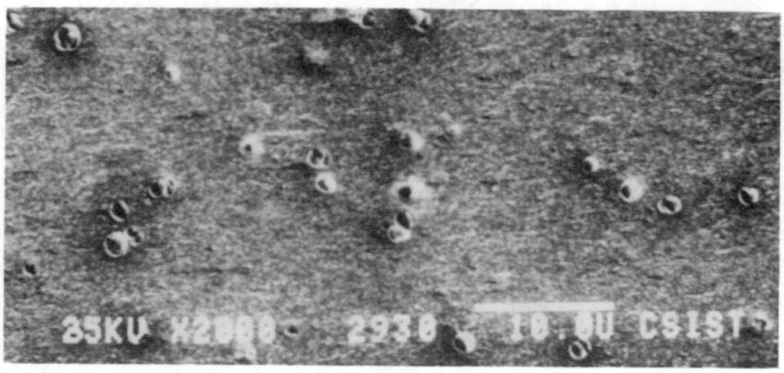

(B3)

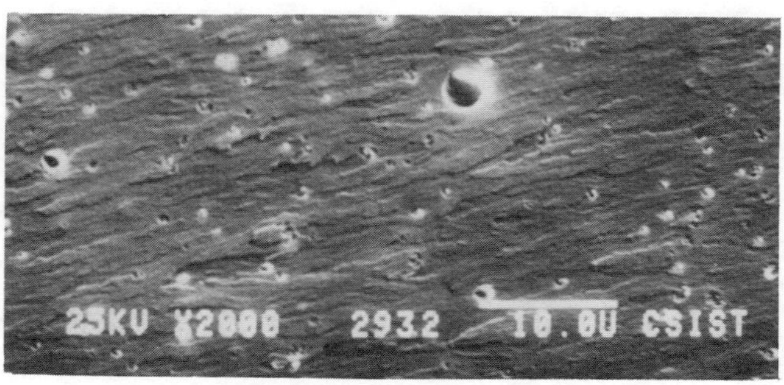

(B7)

Fig. (6) - Scanning electron micrograph of fracture surface of $Ba(OH)_2 \cdot 8H_2O$-catalyzed cast phenolic resin: (B1) F/P = 1.2; (B3) F/P = 1.5; (B7) F/P = 2.3

(C1)

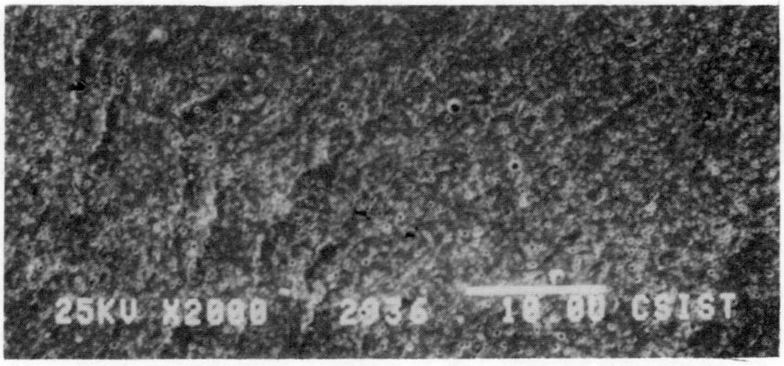

(C3)

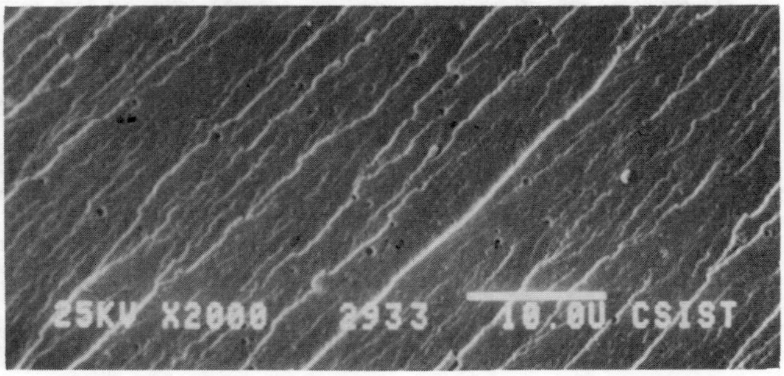

(C6)

Fig. (7) - Scanning electron micrograph of fracture surface of Na_2CO_3-catalyzed cast phenolic resin: (C1) F/P = 1.2; (C3) F/P = 1.5; (C6) F/P = 2.1

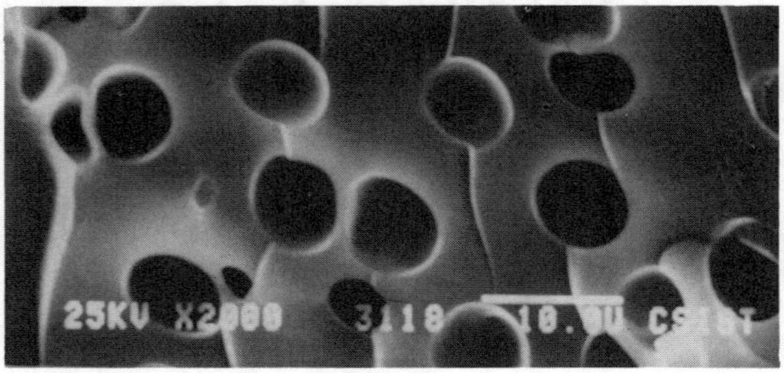

(B)

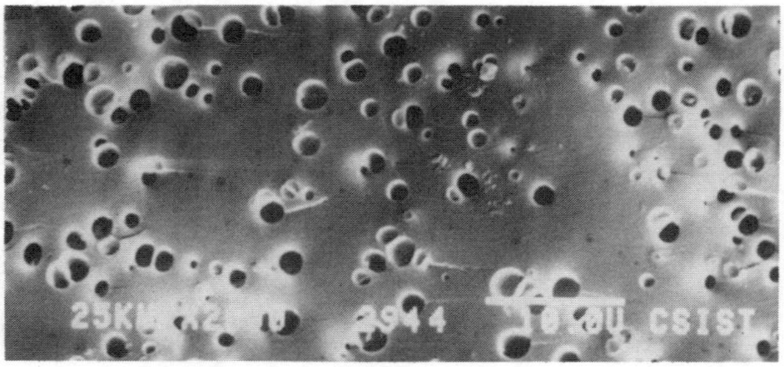

(F)

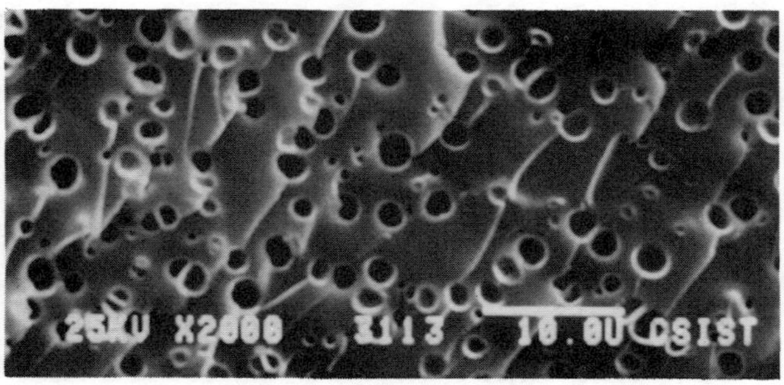

(I)

Fig. (8) - Scanning electron micrograph of fracture surface of NaOH-catalyzed cast phenolic resin for different viscosity (F/P mole ratio = 1.5): (b) 14,000 cps; (F) 188,000 cps; (I) 1,170,000 cps

REFERENCES

[1] Murayama, S., Phenolics, Nikang Ind. News Co., p. 392, 1970 (in Japanese)

[2] Kenney, G. F., Engineering Properties and Applications of Plastics, Wiley, New York, p. 95, 1964.

[3] Katovic, Z., J. Appl. Polym. Sci., 11, p. 85, 1967.

[4] Learmonth, G. S. and Searle, D. P., J. Appl. Polym. Sci., 13, p. 437, 1969.

[5] Mitchell, S. J., Pickering, R. S. and Thomas C. R., J. Appl. Polym. Sci., 14, p. 175, 1970.

[6] Duval, M., Bloch, B. and Kohn, S., J. Appl. Polym. Sci., 16, p. 1585, 1972.

[7] Chow, S. and Steiner, P. R., J. Appl. Polym. Sci., 23, p. 1973, 1979.

[8] Singh, D. and Ohri, S., J. Appl. Polym. Sci., 27, p. 1191, 1982.

[9] Christiansen, A. W., J. Appl. Polym. Sci., 30, p. 2279, 1985.

[10] Lochte, H. W., Strauss, E. L. and Conley, R. T., J. Appl. Polym. Sci., 9, p. 2799, 1965.

[11] King, P. W., Mitchell, R. H. and Westwood, A. R., J. Appl. Polym. Sci., 18, p. 1117, 1974.

[12] Meier, J. F. and Frank, P. P., J. Appl. Polym. Sci., 21, p. 1383, 1977.

[13] Peer, H. G., Rec. Trav. Chem., 78, p. 851, 1959.

[14] Knop, A. and Scheib, W., Chemistry and Application of Phenolic Resins, Springer Verlag, New York, p. 28, 1979.

[15] Merz, E. H., Nielsen, L. E. and Buchdahl, R., Industrial and Engineering Chemistry, 43, p. 1396, 1951.

[16] Drexler, L. H., J. Appl. Polym. Sci., 14, p. 1857, 1970.

[17] Odian, G., Principles of Polymerization, John Wiley, New York, p. 20, 1981.

SECTION II

THERMAL AND MOISTURE EFFECTS

MECHANICAL BEHAVIOR OF SiC/2014 Al COMPOSITE AT HIGH TEMPERATURE

S.-C. Chou and J. L. Green

Army Materials Technology Laboratory
Watertown, Massachusetts 02172-0001 USA

ABSTRACT

The composite material consists of 25 volume percent of SiC whiskers in 2014-T4 aluminum matrix. Specimens were taken from hot-isostatically-pressed cylindrical billets and tested at high temperatures reached at high heating rates. Tensile tests only were performed at a strain rate of approximately 5 sec^{-1}. Comparisons were made with those obtained at room temperature, and also with those from 2014-T4 aluminum tests to show the effect of whisker reinforcement.

INTRODUCTION

Many occasions in space application, a structure must withstand rapid temperature increases to a high temperature for a short period of time. Designing a structure, which must satisfy the requirement of having high specific stiffness and is subjected to such a hostile environment, requires the utilization of a material to its utmost effectiveness. This necessitates the need of determination of mechanical properties of materials in that environment. To accomplish this, a facility was developed which has the capability of varying temperatures and heating rates. Heating rate was accomplished by utilizing the resistance of the specimen in combination with controlling the energy output. A strain measuring device was developed that would be able to withstand high temperature, have fast response to dynamic loadings and measure strain to fracture.

In this paper, the attention is focused on the behavior of SiC/2014-T4 aluminum at high temperatures reached at high heating rate. The blended powder of 2014 aluminum and SiC whisker was hot-isostatically pressed into various sizes of cylindrical billets. Specimens were machined from the billets and tested in a pneumatic machine at high strain rates and high temperature in tension. Comparisons were made with those obtained at room temperature, and also with those from 2014-T4 aluminum test to show the effect of whisker reinforcement.

MATERIAL DESCRIPTION AND SPECIMEN CONFIGURATION

The metal-matrix composite material investigated in this study was 25 volume percent silicon carbide whiskers, SiC in a 2014 Al alloy matrix. The mixture was hot-isostatically-pressed (HIP) into a form of various sizes of cylinders.

Before specimens were taken from these cylindrical billets, the HIP material was given the conventional heat treatment which brought 2014 Al to T4 condition. The whiskers were 0.2 to 0.5 μm in diameter and had an initial length-to-diameter ratio between 50 and 80. The ratio after HIP is unknown. It was expected that the whiskers were randomly oriented and dispersed in the matrix so that the mechanical properties would be isotropic. The isotropy of the HIP cylindrical billets was verified by testing specimens taken from those billets in both axial and radial directions. Specimens were tested in tension. The geometry and dimensions of these specimens are shown in Figure 1.

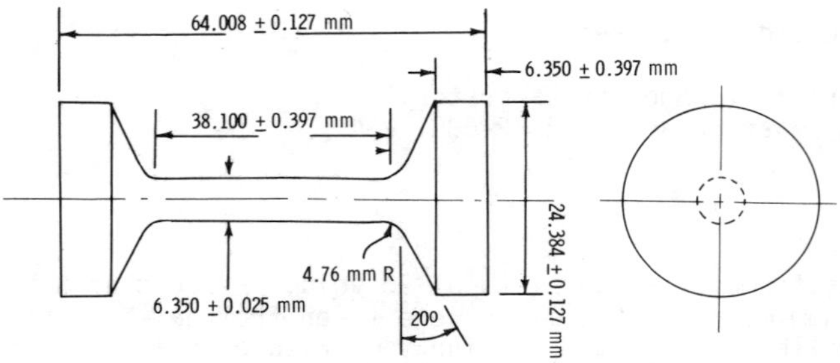

Fig. (1) - Dimensions of test specimens

EXPERIMENTAL ASSEMBLY

The heating apparatus utilizes the resistance heating technique. This method involves passing current through the specimen using the specimen's resistance to convert the current to heat. The apparatus consists of an autotransformer which is used to control the power output of apparatus, a step down transformer for converting low current high voltage to high current low voltage, and a high voltage relay which is located between the autotransformer and computer. The system is actuated by a command in the test program to trigger the relay and allow the current flowing to the specimen. Specimen heating rate is proportional to the energy dissipation rate. Assuming an ideal situation, in which all energy is converted to heat, the heating rate may be estimated by

$$AMS \frac{\Delta T}{\Delta t} = \frac{V^2}{R}$$

where A is the conversion factor, M is the mass of specimen, S is the specific heat of specimen, T is the temperature, t is the time, V is the applied voltage to specimen and R is the resistance of the specimen.

Since no real system is an ideal one, the estimated heating rate just provides a starting point from which the desired heating rate for a particular specimen geometry and material can be obtained by adjusting the voltage and measuring the temperature and time history with an appropriate thermocouple attached to the specimen. The setting on the autotransformer is then used on all subsequent tests for that particular heating rate.

Because A.C. current is passed through specimens, the possibility of skin effects must be examined. Skin effects are that when current is alternating, there are internal effects which tend to force the current to flow mostly in the outer portion of the conductor [1]. The equation, which determines the (diameter) thickness beyond which the skin effect is significant, is

$$\delta = \frac{1}{\sqrt{F\mu\sigma\pi}}$$

where δ is the thickness of conductor (cm), σ is the conductivity (ohm-cm), μ is the permeability (Henrys/meter) and F is the frequency (Hz).

For aluminum, the resistivity = 2.828×10^{-6} ohms-cm, so the conductivity

$$\frac{1}{\text{resistivity}} = 1/283 \times 10^{-6} = 0.35 \times 10^{6}\ (\text{ohm-cm})^{-1}$$

σ is 35×10^{6} $(\text{ohm-m})^{-1}$, μ is 1.26×10^{-6} Henrys/meters, F is 60 Hz and $\delta = 1.1$ cm.

The value of diameter at which skin effects must be considered is greater than the specimen's diameter (0.635 cm). This ensures that the specimen has a uniform temperature in the gage section. It is assumed that this conclusion is also applicable to SiC/Al specimens.

Special attention must be paid to the design of the grip system, because it requires being isolated electrically from the specimen and also strong enough to sustain the load. Tapered aluminum oxide inserts were used to transmit load and provide insulation both thermally and electrically. The complete grip assembly is shown in Figure 2. The inserts are machined to fit into split collar bushings and conform to the specimen's geometry. The specimen is inserted into the split collars first, before the two split collars are assembled and threaded into the top and bottom cylindrical couplers. Copper electrodes are used as conductors which are spring loaded to maintain contact with both ends of the specimen at all times. Copper electrodes and springs are insulated from the cylindrical couplers with nylon bushings.

Specimens are shielded from the atmosphere to prevent the oxidation by an environment chamber which has openings for strain and temperature monitors. The chamber is overpressurized with inert gas.

OPTICAL EXTENSOMETER

One of the difficult problems encountered in the high temperature mechanical testing is the measurement of strains, particularly at dynamic conditions (high strain rates such as those tests performed in this study). The maximum temperature which can be sustained by the wire strain gages is around 270°C. The high temperature gage made by plasma spray can be used up to 550°C ∿ 830°C but the cost is too prohibitive to be used in the destructive testing. Clip gages and extensometers constitute alternate measuring techniques, but they are mechanically fragile and must be removed prior to specimen failure. Furthermore, the maximum response time of ordinary extensometers is about 10 Hz which is much too slow for the high strain rate test. Laser has been utilized in the optical tracking

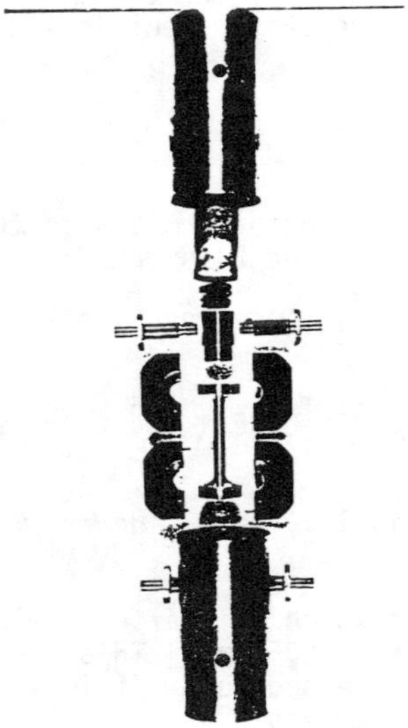

Fig. (2) - Grip assembly for high heating rate tests

method [2] which requires attaching two flags to the specimen as targets to be tracked. The disadvantage is that at large plastic deformations, the targets could be out of range of the viewport.

Since there were no extensometers available that could meet the requirements of being able to withstand high temperatures and fast response time to measure strains at strain rates up to 50/sec., a special optical extensometer was designed and built. The principle of operation of this device is based on the detection of changes in light intensity. The light intensity is varied by the relative displacement of two contiguous evenly spaced diffraction gratings on two flat quartz lens which contact intimately with each other. The gratings were composed of vacuum deposited chrome which was later chemically etched to produce a grid system of alternating lines and spaces of equal widths. The grating used in this study was 1968.5 lines per cm which gives a resolution of 0.00051 cm/fringe. The width of spacings has a deviation of less than 1%. Lenses are mounted in metal sliders with the two grid surfaces contacting each other. The metal sliders are in turn spring loaded to the specimen with a wedge to ensure the intimate contact during the test. The relative displacement of the two gratings produces a diffraction pattern. The diffraction pattern was monitored by a reflective optical switch which consists of an infrared light emitting diode and an optically matched photo-transistor in the same package. The reflective optical switch is excited by a 5 volt D.C. battery to eliminate any A.C. fluctuations. Reflected light intensity is modulated by lens displacement, and the variation is sensed by photo resistor and converted into an electrical impulse, which, in turn, is amplified and recorded. The optical switch has a rise time of 6 μsec which is equivalent to 66 kHz. Based on the frequency response of optical switch, the allowable maximum strain rate within the capability of the switch can be determined. Since the velocity or displacement rate, $\dot{d}$, is given by

$$\dot{d} = (\text{frequency}) \times (\text{wave length})$$

where wave length is the grid line spacing, the strain rate $\dot{\varepsilon}$ is

$$\dot{\varepsilon} = \dot{d}/(\text{gage length}) = \frac{(\text{frequency}) \times (\text{wave length})}{\text{gage length}}$$

Therefore, for a grid spacing of 0.00051 cm (5000 lines/inch) and the gage length of 2.86 cm, the strain rate is

$$\dot{\varepsilon} = \frac{66{,}000 \times 0.00051}{2.86} = 11.8 \text{ sec}^{-1}$$

which is larger than the strain rate of approximately 5 sec^{-1} used in this study.

The lenses are mounted in sliders with ceramic knife edges, Figure 3. To maintain proper alignment and contact, a second set of sliders is mounted oppo-

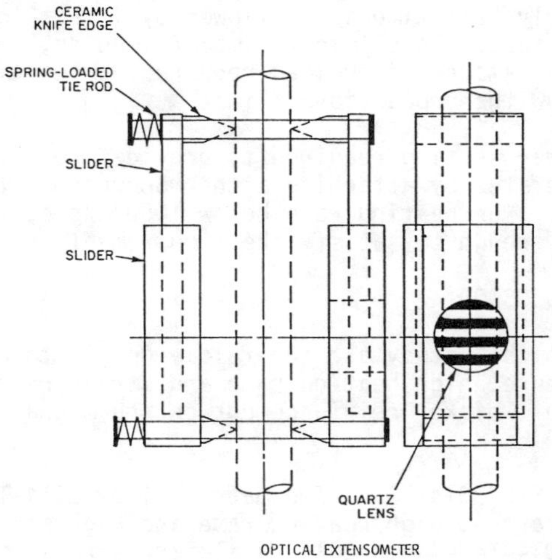

Fig. (3) - Lens assembly of optical extensometer

site the first set. Spring loaded tie bolts connect opposite sliders and provide clamping action. Ceramic knife edges secure the sliders to the specimen and also provide thermal and electric isolation for the gage. The gage length is represented by the initial distance between knife edges. At fracture, the sliders separate and each half remains attached to its portion of the specimen. Reassembly requires simply slipping the two halves together.

MECHANICAL TEST MACHINE AND PROCEDURE

As it was mentioned previously, the effect of heating rate on mechanical properties is the primary interest of this study. In order to observe such effect, the loading rate must also be high; in other words, the test time must be very short in comparison with the total heating time to the prescribed temperature. To perform such a test, a test machine with a capability of pneumatic operation is used in this study. Nitrogen is used as a medium for driving the piston. Two fast acting valves are used to evacuate either top or bottom of the piston creating a pressure differential which causes the piston to move. Control over the piston's speed is accomplished by varying orifice sizes in the fast acting valves, the pressure of nitrogen, and the length of travel by the piston. In this manner, a nominal strain rate will be achieved which, of course, also depends on the ductility of the material.

Temperature was measured by either thermocouple or a non-contact optical pyrometer. The signals from load cell, optical extensometer, and thermocouple/pyrometer are recorded by a transient recorder, such as Nichole Oscilloscope, Lacroy, etc., which has a sample rate of 2 MHz. The data are stored in a memory first, then played back and analyzed after the test.

During heating, specimen will be elongated which depends on the coefficient of thermal expansion of the material. This occurrence will cause unloading such that a specimen loses contact with the grips. Under this circumstance, free run will occur in the early test time and followed by impact loading when specimen and grips come in contact. To eliminate impact loading, the specimen is preloaded prior to heating. The amount of preload required varies with test temperature, but must be kept below the proportional limit.

The autotransformer setting required to provide the voltage for the prescribed heating rate is calibrated by attaching a thermocouple on the specimen and recording its time history. For heating rate below 1000°C/sec, the response of the thermocouple is fast enough to satisfy the requirement.

TEST RESULTS AND DISCUSSION

The objectives of this study are two-fold. One is to characterize the material at high temperature, high heating rate and strain rate, while the other is to study the effect of whisker reinforcement on the mechanical properties at such conditions.

Tensile tests of SiC/2014-T4 Al and matrix alloy 2014-T4 aluminum were performed at high temperature, high heating rate and high strain rates. The test conditions for SiC/2014-T4 Al and 2014-T4 aluminum are listed in Table 1. Test results are shown in Figures 4 through 7.

One point which should be mentioned here is that the optical extensometer used in this study only gives the average strain over the gage length. The average strain can give an obscure view of the real deformation of the specimen, particularly if a specimen exhibits necking before fracture takes place. In this situation, the recorded maximum average strain by the optical extensometer is not the true strain at failure cross section. A correction was made in the analysis of test results. The diameter of the necked section where fracture occurred was measured after the test, and the ultimate strength was calculated based on the

Table 1 - Test conditions for SiC/2014-T4 and 2014-T4 aluminum

Material	Temperature °C	Heating Rate °C/sec	Holding Time sec	Strain Rate sec^{-1}
SiC/2014-T4	RT	N/A	N/A	≈5
	540	250	None	≈5
	540	28	30	≈5
2014-T4	RT	N/A	N/A	≈5
	260	110	None	≈5
	288	14	30	≈5
	540	250	None	≈5

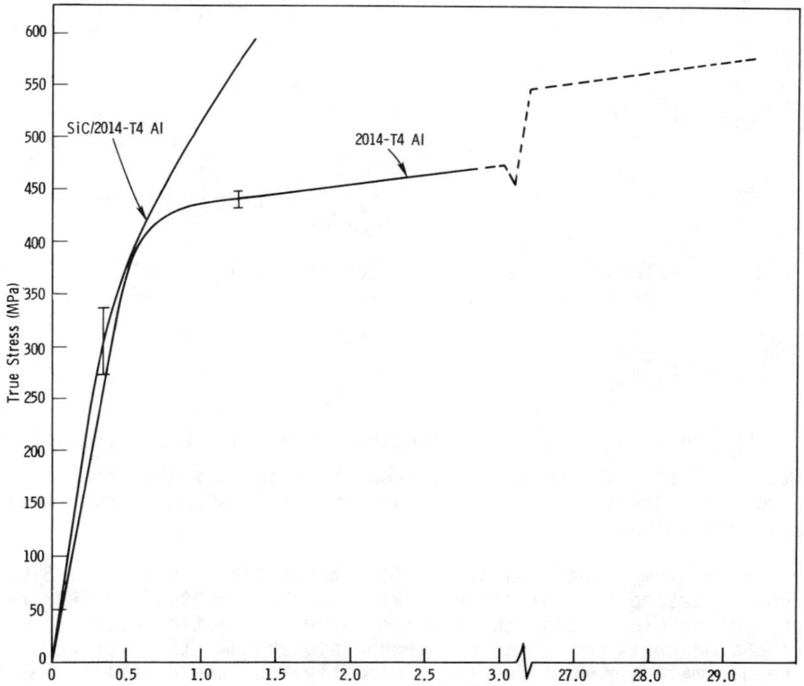

Fig. (4) - Tension test of 2014-T4 aluminum and SiC/2014-T4 aluminum at room temperature and a strain rate of 10^{-4} sec^{-1}

measured necked area rather than the original area. The strain at failure was calculated by assuming that the volume of material remains constant during the plastic deformation which yields the expression for the true strain, e,

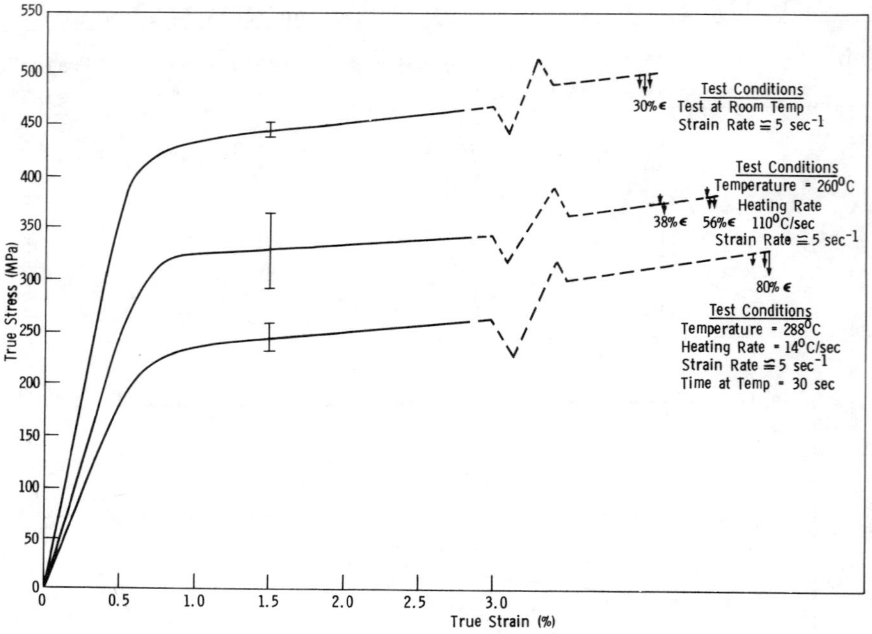

Fig. (5) - Tensile stress-strain curves of 2014-T4 aluminum tested at various conditions

$$e = \ell n \left(\frac{L}{L_o}\right) = \ell n \left(\frac{A_o}{A}\right)$$

where A_o and L_o are original area and length, while A and L are those in the deformed state. The stress-strain curves shown here include the interpretation discussed above and the axes are labeled as "true stress" and "true strain" to reflect this correction.

Figure 4 shows the comparison of tensile stress-strain curves of SiC/Al and 2014-T4 aluminum tested at ambient temperature. The effect of whisker reinforcement is very noticeable in that the average value of elastic modulus for SiC/2014 Al is 120.7 GPa which is more than 60 percent higher than that for the matrix material. The ultimate strength increases slightly, but the ductility has been drastically reduced in comparison with those of matrix alloys.

Heating rate and holding time effect were demonstrated in Figures 5 and 7. The results in Figure 5 show that the modulus of 2014-T4 Al has been reduced from 71.7 GPa to 49.0 GPa at 260°C with 110°C/sec heating rate, but it is further reduced to 42.7 GPa when the heating rate is 14°C/sec and the temperature was held for 30 seconds. Meantime, the average ductility was increased from 30 percent to 50 percent, and further to 80 percent, if the temperature was held for 30 seconds.

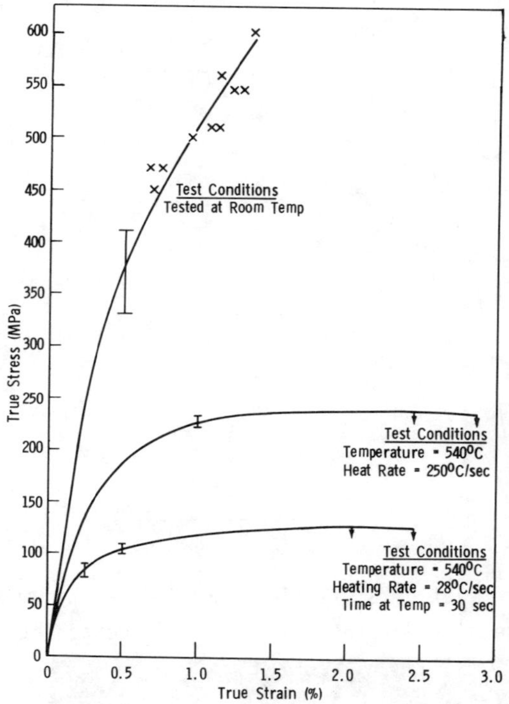

Fig. (6) - Tensile stress-strain curves of SiC/2014-T4 aluminum tested at various conditions

One should notice that the high temperature, high heating rate and short holding time do reduce the modulus. However, the modulus is much higher than that listed in the handbook because the values in the handbook are obtained after the specimen has been soaked in the temperature for a long time.

Results from tests of SiC/2014 Al at the temperature of 540°C and two different heating rates are shown in Figure 6. For the heating rate of 250°C/sec and no holding time, the modulus is 80 GPa and the ductility is approximately 2.5 percent. When the heating rate is 28°C/sec and the temperature is held for 30 seconds, the modulus is reduced to 53.8 GPa and the ductility is reduced somewhat rather than increased. This point will be discussed later.

Finally, Figure 7 shows the comparison of stress-strain curves of SiC/Al and 2014-T4 Al at temperature of 540°C heated up at 250°C/sec. The modulus of 2014-T4 Al is about 22 GPa, but with whisker reinforcement, the modulus is 80 GPa, which is almost the same as that of the matrix alloy at ambient temperature. It is interesting to notice that at 540°C, the ductility of 2014-T4 Al is less than 1 percent, due to the oxidation and decohesion at the grain boundaries.

To answer some of the questions raised by observing the mechanical (tensile) behavior, the fracture surfaces of tested specimen were examined under Scanning Electronic Microscope (SEM). Figure 8 shows the fracture surface of an SiC/Al

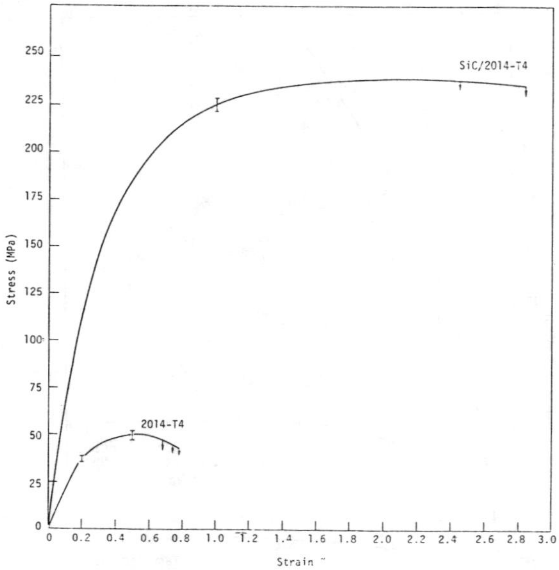

Fig. (7) - Tensile test of SiC/2014-T4, 2014-T4 aluminum at strain rate of 5 sec^{-1} and temperature of 540°C heated at 250°C/sec

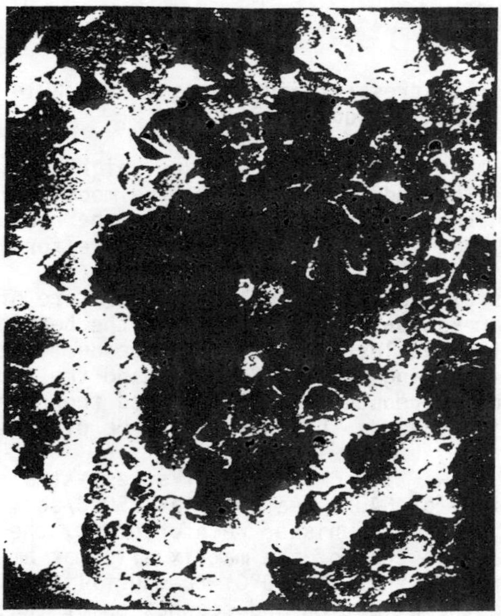

Fig. (8) - Typical fracture surface of SiC/aluminum tested at room temperature (magnification: 5000X)

Fig. (9) - Fracture surface at the center of a SiC/aluminum specimen tested at temperature of 540°C (magnification: 5000X)

Fig. (10) - Fracture surface at the edge of a SiC/aluminum specimen tested at temperature of 540°C (magnification: 5000X)

specimen tested at ambient temperature. The silicon carbide whiskers are visible and many dimples or shear lips indicate the local ductile fracture in the matrix alloy.

The fracture surfaces of a SiC/Al specimen tested at 540°C are also examined. The specimen clearly shows a primary crack initiation starting from the center of the specimen. At 5000 magnification, Figure 9 shows fiber pull-out at the center of the specimen, an indication of interfacial debonding. The short shear lips in the matrix alloy with rounded edges are indications of localized melting and oxidation which is over all the fracture surface. The periodic striation at the shear lips indicates that fiber pull-out are random localized micro-occurrences. At the edge of the specimen, Figure 10, matrix cleavage occurs where there is practically no whisker reinforcement. This indicates a non-uniform mixing of the 2014 powder with whiskers and also the decohesion at grain boundaries which is caused by heating and oxidation at the grain boundaries. It is known that the 2014-aluminum is an Al-cu alloy with the eutectic temperature equal to 540°C, at which the specimens were tested in this study.

CONCLUSION

A high temperature, high heating rate and strain rate facility was assembled and an optical extensometer measuring the strain at high temperature was developed. The 2014-T4 aluminum reinforced with 25 volume percent of SiC whiskers and the matrix alloy were heated up to high temperatures with the maximum temperature at the eutectic temperature of the matrix alloy, and tested at high strain rate. Test results indicate that at the same strain rate, 2014-T4 aluminum exhibits decreasing modulus with increasing temperature, decreasing heating rate and increasing holding time, while ductility increases under the same condition until the eutectic temperature 540°C was reached, at which point, the grain boundary decohesion and oxidation take place and ductility was drastically reduced.

For SiC/2014 Al, the whisker reinforcement has increased the modulus at ambient temperature 120.7 GPa which is 60 percent higher than that of matrix alloy, but drastically reduced the ductility to 1.5 percent. At high temperature, the modulus decreases but retains a large portion of it even at the eutectic temperature of the matrix alloy. For instance, the modulus at 540°C heated up at 250°C/sec is 80 GPa which is slightly higher than the modulus of the matrix alloy at room temperature. Based on these results, it seems that it is advantageous to use SiC/Al instead of aluminum in a situation where the temperature is too high for the aluminum alloy to sustain.

REFERENCES

[1] Smythe, W. R., Static and Dynamic Electricity, Third Edition, McGraw-Hill, New York, 1968.

[2] Marion, R. H., "A new method of high-temperature strain measurement", Experimental Mechanics, Vol. 18, No. 4, pp. 134-140, April 1978.

FLAMMABILITY OF EPOXY

J.-F. Chang and W.-Y. Chen

Chung Shan Institute of Science and Technology
P.O. Box 1-26-2, Lung-Tan, Taiwan, Republic of China

ABSTRACT

Several commonly used flame retardants such as antimony oxide, aluminum hydroxide, halogen-based compounds, phosphorus-based compounds had been used to improve the fire resistance of epoxy.

Horizontal flammability tests, vertical flammability tests and ignition tests had been done to determine the flammability of flame retardant added epoxy. Accelerated aging tests, water absorption, uncured resin viscosity, yield char ratio, heat distortion temperature and composite flexural strength had also been tested to help determine the suitable flame retardant.

If the water absorption is negligible, solid phosphorus-based compounds will be the best flame retardant of epoxy.

INTRODUCTION

The basic mechanism of burning is the same for epoxy as for other plastics. Plastics burn when they are subjected to sufficient heat to break down the polymer structure into volatile gases. The gases burn and radiate heat from the flame back to the plastic, causing further pyrolysis and evolution of flammable gases.

The more common approach to obtain fire resistance is to add one or more of the commonly used fire-retardant elements to a flammable polymer system. This may be done either with an inert additive or by copolymerizing a moderate amount of a monomer containing the requisite element. The elements are antimony, chlorine, bromine, phosphorus and mixtures of these into polymer system.

Phosphorus-based flame retardants alter the chemistry of the thermal degradation process so that more nonvolatile char is formed and reduce the heat flow from the flame to the solid plastic to decrease the rate of pyrolysis and the production of flammable gases.

Aluminum hydroxide absorbs heat through endothermic reactions to reduce the amount of heat available to degrade the polymer and generate nonflammable gases which reduce the concentration of flammable gases.

Halogen and halogen with antimony oxide generate chemicals that act as flame "snuffers" through a direct effect on the combustion reaction.

Achieving good fire resistance often requires 5-20 Wt% of the fire-retardant element in the polymer. At such high level, the effect on polymer properties is substantial and must be considered.

EXPERIMENT

The resin system is a conventional bisphenol A-type epoxy resin, which is a blend of diglycidyl ether of biphenol A-type resin (epon 828) modified with a cycle-aliphatic epoxy resin reactive diluent (ERL 4206), an amine-type curing agent composed of metaphenylene diamine and 4, 4' - methylene dianiline (Tonox 60/40). This resin system was added several kinds and amounts of flame retardants including antimony oxide, aluminum hydroxide with 35% water. Dechlorane Plus (65%Cl), tris (2-chlorethyl) phosphate (EP.: 10.8%P, 36.7%Cl), Der 542 (Brominated epoxy, EEW = 305-355, 44-48Br%) $[(NH_4)_3PO_3]_n$ to do the following tests.

A. Composite Flexural Strength and Modulus

Composite flexural strength and modulus tests were conducted using ASTM D790.

B. Resin Viscosity Test

Uncured resin viscosity tests were conducted using ASTM D2393.

C. Water Absorption Tests

Water absorption tests were conducted using ASTM D570. The specimens dried in oven at 60°C for 48 hours were placed in a container of distilled water maintained at a temperature of 90°C for 48 hours, and then reconditioned for the same time and temperature as used in the original drying period.

D. Heat Distortion Temperature

Heat distortion temperature was conducted using ASTM D648.

E. Horizontal Flammability Test

The installation of horizontal flammability test and procedures were the same as ASTM D635. However, we determined the weight loss in burning and burning time after removing the burner instead for the sake of comparing the flammability of epoxy easily. Besides, we watch the smoke density when the specimen was burning.

F. Vertical Flammability Test

The installation of vertical flammability test is shown in Figure 1. The test procedures were the same as those of horizontal flammability test. The combustion time from removing the burner to flame proceeding to the mark (t1) or to ceasing (t2) and the weight loss had been determined.

G. Accelerated Aging Test

Horizontal flammability test was also used in accelerated aging test. Two kinds of accelerated aging tests had been done. (1) The specimens were immersed

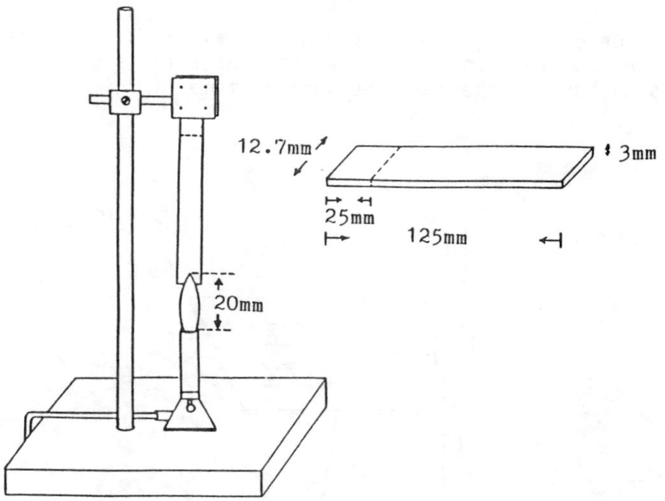

Fig. (1) - Vertical flammability test

in distilled water at 90°C for 240 hours and then in oven at 100°C for 240 hours. (2) Each side of the specimens were shined with 500W medium pressure mercury lamp (UV) for 500 hours. Its installation is shown in Figure 2.

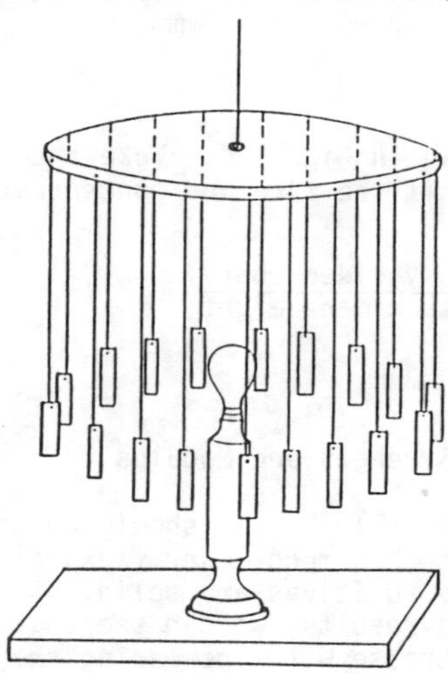

Fig. (2) - UV shining accelerated aging test. Light source: 500W medium pressure mercury lamp. Distance from lamp to specimen is 16 cm

H. Ignition Test

The installation of ignition test is shown in Figure 3. Take the burner under the specimen, and turn on the stop watch. Stop the stop watch when the specimen begins to burn. Take down the ignited time.

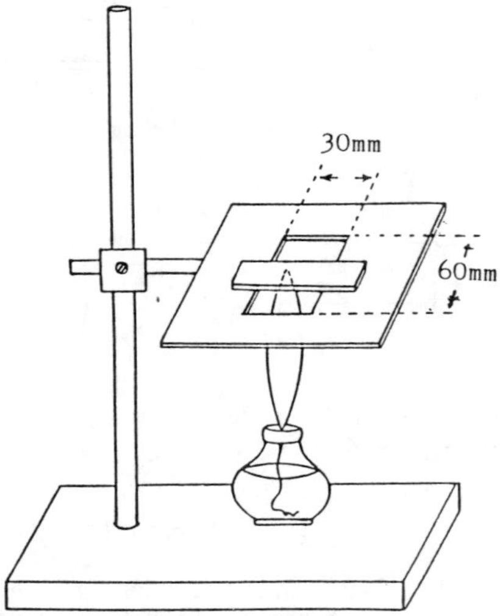

Fig. (3) - Ignition test (specimen dimension 60 x 12.7 x 3 mm)

I. Yielded Char Ratio

Its installation is shown in Figure 4. Take the specimen in the pot when the pot is burned to red-hot until the fire and concentrated smoke disappear.

$$\text{Yielded char ratio } R = \frac{\text{yielded char}}{\text{specimen weight}}$$

RESULTS AND DISCUSSION

A. Composite Flexural Strength and Modulus

The results are shown in Table 1. It should be noted that liquid flame retardant - EP. added composite has much lower flexural strength, but its modulus has little effect. When the additives are solid, flexural strength differences are small. According to the results, we can see that the modulus and strength increase a little and then decrease with increasing the additives.

B. Resin Viscosity Test

The results are given in Table 1. Uncured resin viscosity is greatly related to its processing-filament winding and hand lay up processing. For example, its

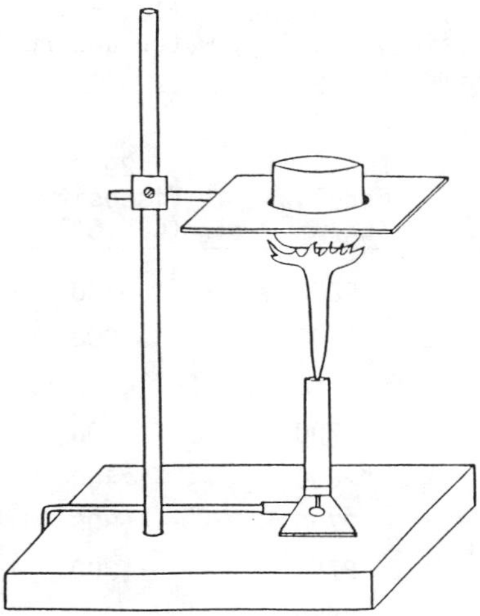

Fig. (4) - Yield char ratio (specimen dimension 12.7 x 12.7 x 3 mm)

viscosity must be controlled. The viscosity increased with the increasing additive, except EP. EP. is a low viscosity liquid, so it can dilute the resin. Der 542 made the viscosity largely increased with its added amount because of its higher molecular weight and melting in resin system.

C. Water Absorption Test

The percentage of water absorbed results are shown in Table 1. Resin dominated properties of composite were significantly affected by its water absorption. Phosphorus-based compound made the resin have much higher water absorption, but Halogen and aluminum hydroxide decreased it a little. It may be because phosphorus-based compounds have higher water absorption or solubility, and it seems reasonable to use halogen substituted or higher molecular weight oligomer.

D. Heat Distortion Temperature

As shown in Table 1, all flame retardants except EP., added resin had not considered significantly different. We can see that unreactive liquid filler largely decreases HDT of resin, and solid fillers seldom affect it.

E. Horizontal Flammability Test

The results are shown in Table 2. One way to reduce fire hazard is to use polymer especially tailored to resist burning. By making a polymer rich in ring structures and low in readily oxidizable side groups, low flammability may be achieved. Because of its aromatic ring structure, epoxy is self-extinguished in horizontal flammability test which will be classed UL-94HB.

When these additives were used alone, considering the combustion time, we can see that the effect of fire resistance was phosphorus - based compounds, bromin $>$ based compound chlorine $>$ based compound. The resin with 2% phosphorus had the

Table 1 - Additives versus composite flexural strength and modulus, uncured resin viscosity, water absorption and heat distortion temperature

(Resin)+ Additives	Flexural Strength kg/mm^2	Flexural Modulus kg/mm^2	29°C Viscosity (C.P.S.)	Water Absorption (%)	HDT
Neat Resin	42.0	2200	700	2.10	149
P:7%	37.5	2700	3,000	7.10	149
P:5%	39.0	2750	1,500		
P:2%	45.0	2700	800		
P:2% Cl:6%	41.0	2600	1,400		148
P:2% Cl:10%	44.0	2700	1,800		150
P:2% Br:5%	45.5	2750	1,300		
P:2% Br:10%	41.5	2550			149
P':2% Cl':7%	31.0	2200	400	7.00	99
Cl:30%	40.5	2600	5,000	1.05	
Cl:20%	39.0	2550	2,500		
Cl:10%			1,100		
Sb:4% Cl:4%	43.0	2750	800		
Sb:8% Cl:8%	42.0	2650	1,200		
Sb:6% Cl:10%			1,600		151
Sb:9% Cl:13%	41.0	2300	2,100	1.55	
Br:20%	42.0	2700	6,000	1.40	
Br:15%	42.5	2700	2,800		
Br:10%	43.5	2700	1,600		
Sb:2% Br:4%	45.5	2700	950		
Sb:4% Br:8%	45.5	2750	1,200		
A:17%	43.5	2400			
A:30%	40.5	2400	900		149
A:38%					147
A:45%			2,700	1.50	148
Sb:7%				2.05	

P : Phosphorus content from $[(NH_4)_3PO_3]n$
Cl: Chlorine content from Dechlorane Plus
Br: Bromine content from Der 542
A : Aluminum oxide content
P' : Phosphorus content from EP
Cl': Chlorine content from EP
Sb : Antimony content from Sb_2O_3

Table 2 - Additives versus horizontal flammability test

(Resin)+ Additives	t1 (sec)	W1 (g)	Smoke Density	t2 (sec)	W2 (g)	t3 (sec)	W3 (g)
Neat Resin	100	1.25	X	70	0.80	60	0.80
P:7%	0	0.10	0	0	0.15		
P:5%	0	0.10	0				
P:2%	10	0.10	0				
P:2% Cl:6%	5	0.15	0			5	0.15
P:2% Cl:10%	5	0.15	0	0	0.15	5	0.15
P:2% Br:5%	0	0.15	0				
P:2% Br:10%				0	0.15		
P':2% Cl':7%	10	0.20	0	10	0.20	10	0.25
Cl:30%	5	0.20	Δ				
Cl:20%	10	0.30	Δ	5	0.30	5	0.35
Cl:10%	25		X			15	0.40
Sb:4% Cl:4%	30	0.50					
Sb:8% Cl:8%	10	0.30	X			5	0.30
Sb:6% Cl:10%	5	0.25	Δ			5	0.30
Sb:9% Cl:13%	0	0.25	Δ	0	0.20		
Br:20%	5	0.35	Δ			0	0.35
Br:15%	10	0.35	X	5	0.35		
Br:10%	25	0.50	X			15	0.45
Sb:2% Br:4%	35	0.35				25	0.55
Sb:4% Br:8%	5	0.40	X			5	0.45
A:17%						75	0.70
A:30%	70	0.50	Δ				
A:38%							
A:45%	95	0.50	0				
Sb:7%	60	0.65	X				
Sb:3%	70	0.70	X				
P':1.9% Cl:7.7% Cl':6.3% Sb:7.7%	65	1.10	X				

Combustion time after removed the burner: t1 for unaged specimen, t2 for water and over stored specimen, t3 for UV shined specimen. Weight loss: W1 for unaged specimen, W2 for water and oven stored specimen, W3 for UV shined specimen. Smoke density: X for concentrated smoke, Δ for medium smoke, 0 for light smoke.

same effect of it with 15% bromine or with 20% chlorine. The resin exhibits good fire resistance with 5% phosphorus incorporated. The fire ceased immediately when removing the burner. This result is better than those contained chlorine 30% or bromine 20%. The addition of antimony oxide can largely reduce bromine or chlorine added amount (synergistic effect). Aluminum hydroxide, even large amount added, has little effect on combustion time. The greater fire resistance of bromine over chlorine compounds may be due to more rapid elimination of hydrogen bromide to yield a high concentration of hydrogen bromide. The hydrogen halide may alter the mechanism of the termination of flaming.

Phosphorus - based compounds much lower its smoke density, aluminum hydroxide had some effect, but antimony oxide or halogen - based materials had little effect. It seemed reasonable to postulate that smoke density is related to its flammable volatile material amount when aromatic compound is burned. Phosphorus - based compound is a condense phase flame retardant, H_3PO_4 was generated in burning which is a catalyst for plastics to carbonate, and lower the volatile materials. Aluminum hydroxide decreased the heat back to the condense phase, so it decreased the speed of pyrolysis, and yielded H_2O to dilute the flammable volatile materials. Halogen and antimony oxide had no major change in general exothermic decomposition process, so its smoke density is still high. Because phosphorus - based compound is a condense phase flame retardant, it makes the weight loss of added resin much lower. Antimony with halogen makes the specimen have good self-extinguished property, but it has little effect on burning rate when the burner is under the specimen, so its weight loss is a little higher than phosphorus - based compound. Aluminum hydroxide decreased the heat back to the polymer, so it made the weight loss of resin a little lower. Antimony oxide is used almost exclusively with heat-labile halogen compounds. Without halogen, antimony oxide is an inert filler, so it has little effect on combustion time, weight loss and smoke density. No satisfactory theory has yet been suggested to explain the synergism between halogen and antimony oxide compounds in conferring fire retardancy on polymer compositions. Antimony trihalide or antimony oxyhalide has been suggested as the active inhibitor produced under conditions of burning. This assumption was supported by the fact that 75% of antimony vaporized during burning process. However, as our yield char ratio results, antimony-halogen fire retardant compositions produce a considerable amount of carbonaceous residue when burned.

F. Vertical Flammability Test

We show the results in Table 3. The results are similar to horizontal flammability tests but the flammability difference of specimens is higher than the former. General epoxy resin is self-extinguished in horizontal flammability test, but it is flammable in vertical flammability test. Phosphorus-based compound is more effective than halogen.

Flame retardant added epoxy can be classed UL-94V-0 when it contained 5% phosphorus, 20% chlorine or 15% bromine. When added halogen-based compound to epoxy contained phosphorus-based compound, we can get better fire resistance. When we added antimony oxide to epoxy with halogen, the fire resistance of epoxy was greatly improved. When t2 is zero, the specimen weight loss is phosphorus added epoxy, both halogen and phosphorus added epoxy, both halogen and antimony oxide added epoxy. Therefore, it seems reasonable to index that when the burner is under the specimen, its burning rate is phosphorus added epoxy, both halogen and phosphorus added epoxy, both halogen and antimony oxide added epoxy, halogen added epoxy, and the more the halogen is added, the more rapidly it burned.

It should be noted that when the resin contained phosphorus, halogen and antimony oxide, it is surprisingly flammable. Both phosphorus and antimony oxide with halogen inhibition effect were reduced (antagonistic effect). It seems reasonable to postulate that phosphorus is being converted to an ineffectual form, specifically $SbPO_4$ which is known to readily produce from antimony oxide and H_3PO_4.

G. Accelerated Aging Test

The results are shown in Table 2. There is no significant difference in horizontal flammability after accelerated aging. Although our storage time is not enough, it tells us these flame retardants have some durability at least.

H. Ignition Test

Ignition test is important when the material must resist fire for a very short period. We show the results in Table 3. Aluminum hydroxide had very good effect to make the specimen ignite slowly. It is surprising to see that halogen made the specimen ignite easily. It seems that halogen made the polymer decompose easily, so the polymer under the burner will ignite quickly. In $Al(OH)_3$ added epoxy, $Al(OH)_3$ decomposes first when the burner is under the specimen. It is an endothermic reaction to the specimen needs more energy to ignite, and ignite slowly. Phosphorus-based compounds have little effect on ignited time.

I. Yielded Char Ratio

We show the results in Table 3. All flame retardants made the char ratio increase, and the effect was phosphorus-base compound > both halogen-based compound and antimony oxide or phosphorus-base compound > halogen-based compound > aluminum hydroxide. When the specimen was heated at high temperature, phosphorus-based compound added specimen produced nonvolatile phosphorus acids, halogen-based compound added epoxy produced HX. Both of them can be improved to produce carbonaceous residue when burned. Aluminum hydroxide produces aluminum oxide at high temperature, and which is remained in residue.

CONCLUSIONS

The same in other polymers, epoxy contained antimony oxide, halogen and phosphorus had antagonism effect. Except Ep., all flame retardants had little effect on flexural strength and HDT, and some increased in flexural modulus and viscosity. Accelerated aging test seldom affected the horizontal flammability test of flame retardants added specimens. Aluminum hydroxide had little effect on self-extinguished property, good effect on ignition test, lowered smoke density, burning rate, and made a little increasing char yield. Halogen or halogen with antimony oxide had good effect on self-extinguished property, some increasing char yield, bad effect on ignition test, and little effect on smoke density and burning rate when the burner was under the specimen. Both aluminum hydroxide and halogen lowered the water absorption of the specimen.

Phosphorus-based compound had good effect on self-extinguishment and on reducing burning rate and smoke density. Comparing the weight loss when the burner was under the specimen, phosphorus added epoxy < chlorine added epoxy < bromine added epoxy. Antimony oxide has no effect is halogen is absent, but if it is present, self-extinguished property is good.

Table 3 - Additives versus vertical flammability test, ignition test and yield char ratio

(Resin)+ Additives	tl (sec)	t2 (sec)	Weight Loss	Ignition Time (sec)	Yield Char Ratio (%)
Neat Resin	25			9	5
P:7%		0	0.10	11	
P:5%		0	0.10	9	26
P:2%		15	0.40	8	
P:2% Cl:6%		5	0.15	8	
P:2% Cl:10%		0	0.15	8	23
P:2% Br:5%		0	0.20		24
P:2% Br:10%		0	0.25	6	27
P':2% Cl':7%				6	
Cl:30%				4	
Cl:20%		10	0.75		20
Cl:10%					
Sb:4% Cl:4%		30	0.70	5	
Sb:8% Cl:8%		0	0.45		23
Sb:6% Cl:10%		0	0.55	4	
Sb:9%Cl:13%		0	0.45		24
Br:20%		0	1.65		16
Br:15%		0	0.65		
Br:10%		15	1.70	5	
Sb:2% Br:4%		15	1.60		11
Sb:4% Br:8%		0	0.75		
A:17%					16
A:30%		65	0.90		
A:45%		110	1.00	29	
Sb:3%	20				
Sb:7%	20				
P':1.9% Cl:7.7% Cl':6.3% Sb:7.7%	30				

tl: the combustion time from removing the burner to flame proceeding to the mark
t2: the combustion time from removing the burner to ceasing of the flame

Without considering water absorption of epoxy resin, phosphorus must be the best fire retardant because of (1) being more effective, (2) reducing smoke density, (3) reducing burning rate and (4) without bad effect on ignition.

REFERENCES

[1] Hilade, C. J., Flammability Handbook for Plastics, 3rd ed., p. 31, 1982.

[2] Seymour, R. B. and Carraher, C. E., Polymer Chemistry, p. 383, 1981.

[3] Schwartz, S. S. and Goodman, S. H., Plastics Materials and Processes, p. 32, 1982.

[4] Beckwith, S. W. and Wallace, B. D., Kevlar/Epoxy Composite Case Material Aging and Environmental Behavior, CPIA Proceedings, 1982.

[5] Standard for Tests for Flammability of Plastic Materials for Parts in Device and Appliances, UL-94, 1980.

[6] Gottlieb, J. M., A theory of flame retardant finishes", Textile Res. J., 26, p. 156, 1956.

[7] Tang, W. K. and Neill, W. K., "Effect of flame retardants on pyrolysis and combustion of α-cellulose", J. Polymer Sci., Vol. 6, p. 65, 1964.

[8] Modern Plastics Encyclopedia, McGraw-Hill Co., Inc., New York, p. 470, 1966.

[9] Fenimore, C. P., Flammability Characteristics of Polymeric Materials, Polymer Conference Series, Wayne State University, June 1966.

[10] Associated Lead Manufactures Ltd., Antimony Phosphate, British Pat. 792, 997, 1958.

[11] Hovde, P. J., "Combustion properties of glass fiber reinforce polyester (GRP) and the effect of some additives and fillers", J. Fire Flammability, Vol. 11, p. 184, 1980.

WATER AND SOLVENT EFFECTS ON CARBON FIBER/EPOXY COMPOSITES

K. J. Chang and R. Y. Wu

Materials Research Laboratory, Industrial Technology Research Institute
Hsin-Chu, Taiwan, Republic of China

ABSTRACT

The effect of water and solvents absorption of graphite-epoxy composites immersed in water and in various solvents were investigated. The moisture and solvents content as a function of time at room temperature were measured. The lowering of the glass transition temperatures of the saturated specimens were measured by Rheometric Mechanical Spectrometer (RMS) and the changes of mechanical properties were studied by flexural test.

It was concluded that both the water and solvents have the plasticizing effect on the resin which not only reduces the glass transition temperature of the resin but also the resin modulus and strength.

INTRODUCTION

Owing to their excellent performance characteristics, i.e., light and strong, composite materials have been gaining more applications. One of the applications, we are looking into at MRL, is to use the carbon fiber/epoxy composite material to build the structural components for the robots used in automatic painting for the automobile industry. For optimum and safe design of the structure, it is definitely necessary to know the long-term material performances under the hostile conditions where many solvents exist.

A considerable number of papers have been published in recent years discussing the long-term behavior of composite materials. However, most of the previous investigations have been concerned with moisture absorption of graphite-epoxy composites exposed to humid air. Very few data are available on the effect of solvents to the composite materials [1,2]. The objective of the present investigation was to determine the physical changes of graphite-epoxy composites submerged in water and various type of solvents: distilled water, salt water, gasoline, diesel fuel, toluene, rosin, sulfuric acid (20 wt%) and sodium hydroxide (10 wt%) solution when the composites were saturated with the above solvents.

EXPERIMENTAL

The fiber was supplied by Hysol/Grafil (XA-S) with 6000 filaments per bundle. The epoxy resin was formulated in MRL. All the specimens were cut into sizes suitable for later tests. The solvent contents were measured by weighing the soaked specimens by an electronic precision balance and diagrams of liquid content versus the square root of time were drawn to determine the saturated liquid contents and diffusivities of materials [3].

The glass transition temperature was obtained by using a Rheometrics Mechanical Spectrometer after the composites saturated with liquids, flexural tests were performed, because it was found that the percentage reduction in the flexural strength is, in general, more severe than in other properties [4]. All the measurements were taken after the specimens were saturated and the tests were conducted at a temperature of 23 ±1°C.

RESULTS AND DISCUSSION

The liquid contents as a function of time for composite immersed in liquids are shown in Figures 1 through 5. The saturated liquid contents and diffusivities

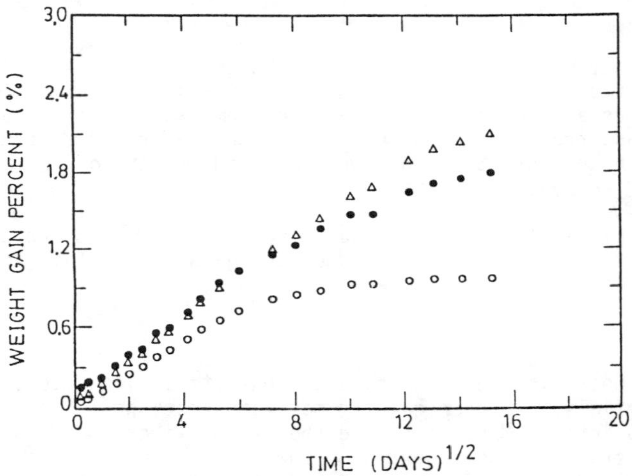

Fig. (1) - Moisture contents as a function of time for composite immersed in humid air, salt water and distilled water. o - 80% RH; ● - salt water; Δ - distilled water

of materials are presented in Table 1. From Figure 1, the diffusion curves in the salt water and distilled water are very similar at the initial step, but the water content in distilled water becomes more in the later stage. In Figure 2, if the whole diffusion curve in the diesel is vertically shifted down, we find the two curves are very similar. The amount of diesel fuel at the start is probably due to the diesel fuel absorbed on the surface of specimen which cannot evaporate for a long time after the specimen was withdrawn from the diesel bath. This shifting value is about 0.37%. We must subtract from the apparent saturated fuel content. Then, the actual saturated fuel content is lower than that one in gasoline as shown in Table 1. Figure 4 shows the diffusion behavior in dichloromethane. Initially, the solvent contents gradually increase with time, then the contents suddenly increase to its saturation point (about 12%).

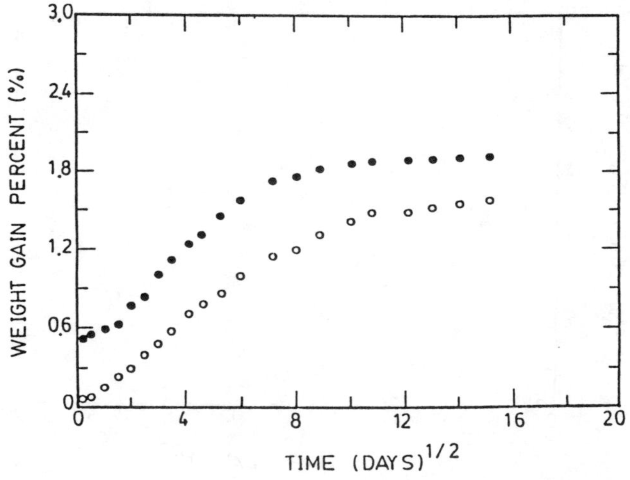

Fig. (2) - Liquid contents as a function of time for composite immersed in gasoline and diesel. o - gasoline; ● - diesel

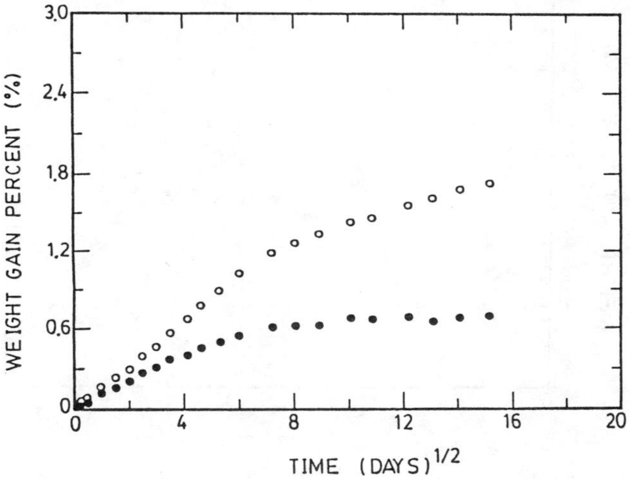

Fig. (3) - Liquid contents as a function of time for composite immersed in toluene and rosin. o - toluene; ● - rosin

During the glass transition temperature (T_g) measurements in the RMS, if the heating rate is slow, then significant amount of solvents may evaporate before T_g can be taken and thus affect the accuracy of the results. On the other hand, if the heating rate is fast, the materials will not have time to reach its equilibrium

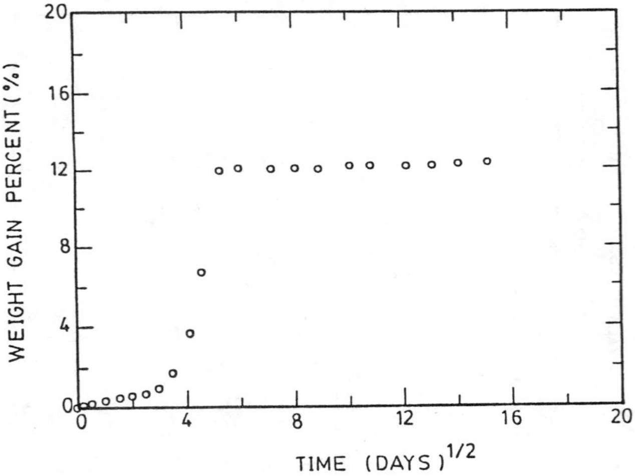

Fig. (4) - Liquid contents as a function of time for composite immersed in dichloromethane

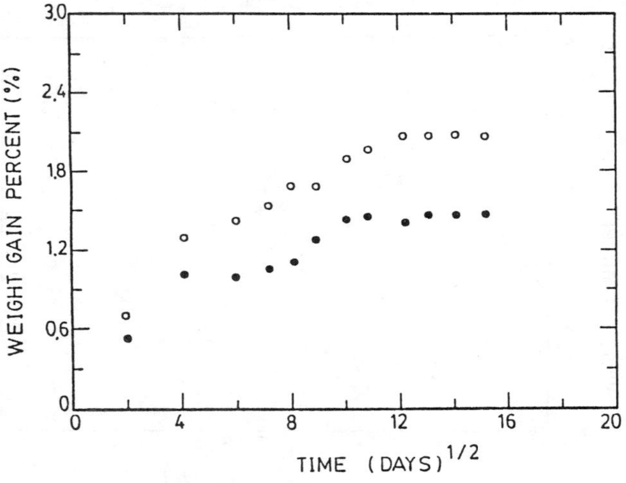

Fig. (5) - Water contents as a function of time for composite immersed in H_2SO_4 (aq) and NaOH (aq). o - 20 wt% H_2SO_4; ● - 10 wt% NaOH

temperature with the heating chamber, therefore, there will be an overestimation of T_g. In this study, we used two heating rate, the results are shown in Table 2. It is clearly shown in Table 2 that both water and solvents have more or less plasticizing effects on the composite. Dichloromethane is the best solvent, makes the composite into "rubbery state" and no T_g can be found from 40°C to 250°C in RMS spectrum.

Table 1 - Saturated liquid contents and diffusivities of composite laminates

Conditions / Diffusion Constants	80% R.H.	Salt Water	Distilled Water	Gaso-line	Diesel	Toluene	Rosin	Dichloro-methane	20 wt% H_2SO_4	10 wt% N_aOH
Saturated Contents (%)	0.98	1.80	2.10	1.58	1.55	1.72	0.70	12.34	2.07	1.47
Diffusivities x 10^{-3} mm^2/day	1.82	0.92	0.67	1.35	1.81	1.03	1.92	-	-	-

Table 2 - Glass transition temperature (T_g) of composite laminates

Heating Rate	Conditions / Prop-erties	Dry	80% R.H.	Salt Water	Distilled Water	Gaso-line	Diesel	Toluene	Rosin	Dichloro-methane	20 wt% H_2SO_4	10 wt% N_aOH
5	T_g (°C)	143	106	98	95	98	101	95	116	-	109	103
(°C/min)	Lower in T_g (°C)	-	37	45	48	45	42	48	27	-	34	40
20	T_g (°C)	164	122	110	110	110	117	103	131	-	123	111
(°C/min)	Lower in T_g (°C)	-	42	54	54	54	47	61	33	-	41	53
Liquid Contents (%)		-	0.91	1.53	1.74	1.52	1.89	1.43	0.69	12.2	2.05	1.39

Table 3 - Flexural properties of composite laminates

Conditions / Flexural Properties	Dry	80% R.H.	Salt Water	Dis-tilled Water	Gaso-line	Diesel	Toluene	Rosin	Dichloro-methane	20 wt% H_2SO_4	10 wt% N_aOH
Flexural Strength (Ksi)	249.1	195.9	194.4	191.2	196.6	205.9	198.2	221.1	50.1	230.9	209.6
S.D.	9.9	9.4	11.7	8.5	6.0	11.6	6.3	13.3	4.5	6.2	9.9
C.V.%	3.97	4.82	6.01	4.05	3.06	5.65	3.16	6.03	8.93	2.68	4.72
Lower in Flexural Strength (%)	-	21.4	22.0	23.2	21.1	17.3	20.4	11.2	79.8	7.3	15.9
Liquid Contents (%)	-	0.91	1.53	1.74	1.52	1.89	1.43	0.69	12.2	2.05	1.39

Table 3 shows the flexural strength changes when the graphite-epoxy composites were saturated with liquids. It was found that the lowering of flexural strengths were in direct proportion to the lowering to T_g.

CONCLUSIONS

Both the saturated liquid contents and diffusivities of materials were estimated and calculated. The effect of solvents and water on the properties of graphite-epoxy composite were also investigated. It was found that all the liquids showed more or less detrimental effects on the composite, the flexural strength drop is, in general, more than 10%. It is especially worth while to mention that the residual flexural strength is only 20% of original strength and the T_g of the composite drop below the room temperature in the dichloromethane. We should be especially careful in designing structural components made with such a composite which may be in contact with solvents.

REFERENCES

[1] Loos, A. C. and Springer, G. S., "Moisture absorption of graphite-epoxy composites immersed in liquids and in humid air", J. Composite Materials, 13, p. 131, 1979.

[2] Loos, A. C. and Springer, G. S., "Moisture absorption of polyester E glass composites", J. Composite Materials, 14, p. 142, 1980.

[3] Shen, C. H. and Springer, G. S., "Moisture absorption and desorption of composite materials", J. Composite Materials, 10, p. 2, 1976.

[4] Rege, S. K. and Lakkad, S. C., "Effect of salt water on mechanical properties of fibre reinforced plastics", Fibre Science and Technology, 19, p. 317, 1983.

THERMAL AND MECHANICAL ANALYSIS OF ACID ANHYDRIDES CURED EPOXY RESIN SYSTEMS

H.-T. Chiu and C.-B. Chen

National Taiwan Institute of Technology
Taipei, Taiwan, Republic of China

ABSTRACT

This paper presents the results of a study in which thermal and mechanical analysis were used to optimize curing process and formulation for acid anhydrides cured epoxy resin systems. Using DuPont 9900 differential scanning calorimetry (DSC) and Fourier transform infrared (FT-IR), cure times and conversion were determined. The mechanical properties such as tensile, flexural and impact property were also examined. These results are discussed with respect to the kinetic data which obtained from DSC and suggest the optimum formulation of epoxy/anhydride system.

INTRODUCTION

In recent years, epoxy resins have been extensively studied as matrix materials for composite structures as well as adhesives for aerospace applications. In order to make optimum use of epoxies as structural materials, it is important to know what the curing process is, to what extent it proceeds, the mechanical properties of the cured material and how these are influenced by temperature. The curing process has been monitored by several techniques, among them d.c. conductivity [1], dielectric analysis [2], Fourier transform infrared absorption (FT-IR) [3] and differential scanning calorimetry (DSC) [4].

In this report, a study of the epoxy-anhydride polymerization kinetics with benzyladimethylamine (BDMA) as promoter is described. Data analysis for the activation energy (E), pre-exponential factor (Z), heat of reaction (ΔH) and order of reaction (n) were carried out using DuPont DSC 9900 thermal analyzer. The extent of cured materials were determined by FT-IR and compared with those obtained from DSC measurement.

The present study, one part of an investigation on the curing and mechanical properties of various epoxide/hardener systems are also discussed. The mechanical properties such as heat distortion temperature (HDT), tensile strength, flexural strength and impact strength are examined as a function of composition and curing process.

EXPERIMENTAL

Materials

The mixtures of the epoxy monomer, the promoter and three anhydrides (see Table 1) were used as reactive systems in this investigation. These materials were used without further purification.

Table 1 - Material systems

Name, Code and Formula	Source and Properties
Diglycidyl ether of bisphenol A (DGEBA) Araldite 6010 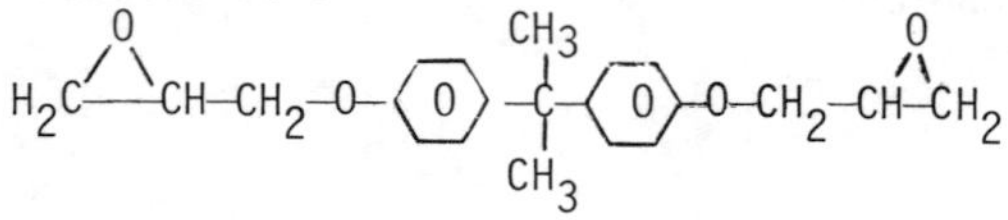	Ciba-Geigy, epoxy equivalent: 190
Nadic Methyl Anhydride (NMA) 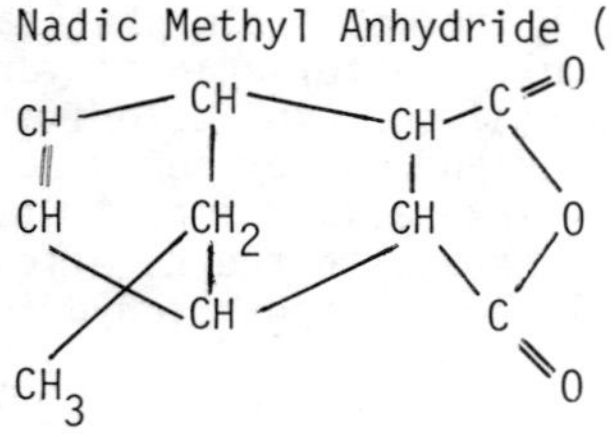	M.W. = 178 M.P. = below 12°C
Hexahydrophthalic Anhydride (HHPA) 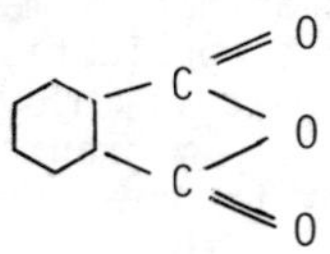	M.W. = 154 M.P. = 35°C
Tetrahydrophthalic Anhydride (THPA) 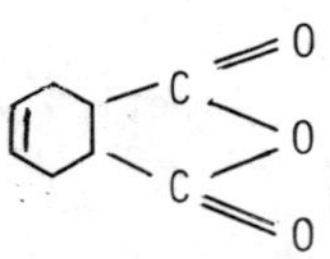	M.W. = 152 M.P. = 99°C
Benzyl Dimethylamine (BDMA) 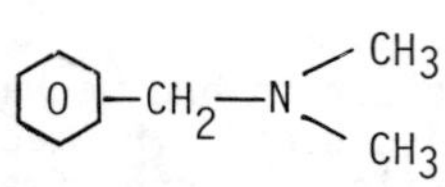	

Sample Preparation

(a) Mixing: Stoichiometric amount or non stoichiometric ratio of epoxy-anhydride-promoter compound was mixed. Prior to mixing, the HHPA and THPA were melted at 50°C and 100°C respectively. Then, adding the promoter, stir thoroughly to ensure uniform dispersion.

(b) Vacuum degassing: It is difficult to prevent air bubbles from forming in the mixture during the mixing operation, therefore, vacuum degassing was required to obtain bubble-free specimens.

(c) Casting: Degassed mixtures were poured into a silicone rubber container to a depth of a few millimeters and allowed to cure in a forced-draft oven.

Measurement of DSC

DuPont Model DSC-9900 thermal analyzer was employed to measure the curing behavior of epoxy resin. The heating rate was 10°C/min, and the specimen size was 5-13 mg. These were dynamic scans from 30°C to 300°C for specimens. The DSC-9900 system to provide rapid automatic calculation of reaction order, activation energy, pre-exponential factor and heat of reaction. These values are then used in the software data analysis program to provide a series of predictive thermal curves to assess the curing process in terms of conversion, time and temperature. This data is of particular value in process development for predicting optimum curing temperatures. It is also useful in process control for optimization of cure advancement or conversion.

FT-IR Spectra

The extent of cure proceeding was determined by Bomen Model DA 3002 Fourier transform infrared spectroscopy (FT-IR). The absorbance at 910 cm^{-1} have been assigned to the epoxide peak, and the extent of reaction can be monitored by the decrease in peak height as a function of time. Variation in epoxide absorbance caused by differences in specimen thickness can be eliminated by normalizing the epoxide peak height to an internal reference peak. Absorbance at 1610 cm^{-1} is due to the phenyl groups in the epoxies, and this peak was used as the intemal standard, [5,6]. In this study, spectra resolution is 2 cm^{-1} and using 100 scans for each sample.

Measurements of Mechanical Properties

(a) Heat Distorsion Temperature (HDT). HDT test were used by Cheast 6510/517 tester, according to ASTM D-648, using 2°C/min heat rate and a stress of 264 psi is applied to all specimens with 12.7 cm span length, 0.62 cm thickness and 1.27 cm width.

(b) Tensile, Flexural and Impact Properties. The tensile and flexural properties were performed by Instron tester (Adamel Lhomargy Division d'Instruments Model Dy-18) at room temperature, according to ASTM D-638-82a type I and ASTM D-790 method I respectively. Cross head speed for tensile test was 5 mm/min and using 2.8 mm/min for flexural test while impact test was conducted according to ASTM D-256-78, method B notched Charpy type. (Frank Pendulum Impact Tester).

RESULTS AND DISCUSSION

Curing Process and DSC Kinetics

The cure exotherm may be used to study the degree of cure and the kinetics of cure of a thermoset system. To study the kinetics of cure, the method of Borchardt and Daniels [7] was used in this study. The method of Borchardt and Daniels allows the calculation of activation energy (E), pre-exponential factor (Z), heat of reaction (ΔH) and order of reaction (n) from a single DSC scan. The method assumes that the reaction follows nth order kinetics and that the temperature dependence of the reaction rate follows the Arrhenius expression. The DSC exotherm is used to measure the two basic parameters required to solve these equations. Firstly, the reaction rate is obtained by determining the peak height to peak area ratio at temperature T. Secondly, the fractional conversion is obtained by calculating the partial area to total area ratios at temperature T. In this report, we use this method as employed with a micro-computer to determine the kinetic parameters of an exothermic cure reaction.

Typical dynamic thermograms of epoxy/anhydride system are shown in Figures 1 to 3. The shift of peak exotherm temperature (T_p) presents the information about reaction of cure. Higher T_p indicate a longer cure schedule for materials. Detail discussions were described following DSC kinetic data analysis.

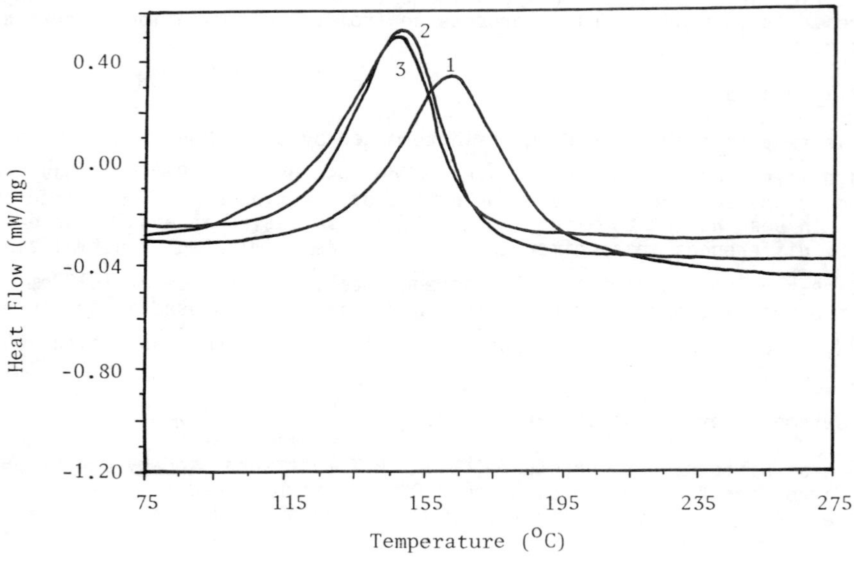

Fig. (1) - Dynamic thermograms of hardener species effect for Araldite 6010 with BDMA (1.5 phr) promoter. Curve 1: NMA; curve 2: HHPa; curve 3: THPA

Data analysis using Arrhenius equation is illustrated by the plot of lnK versus 1/T for runs at different isothermal temperatures gives the activation energy. Figures 4 to 6 show the Arrhenius plots for the influence hardener species, hardener concentration and promoter concentration on epoxy resin system. The numerical data of various activation energy are listed in Tables 2 to 5.

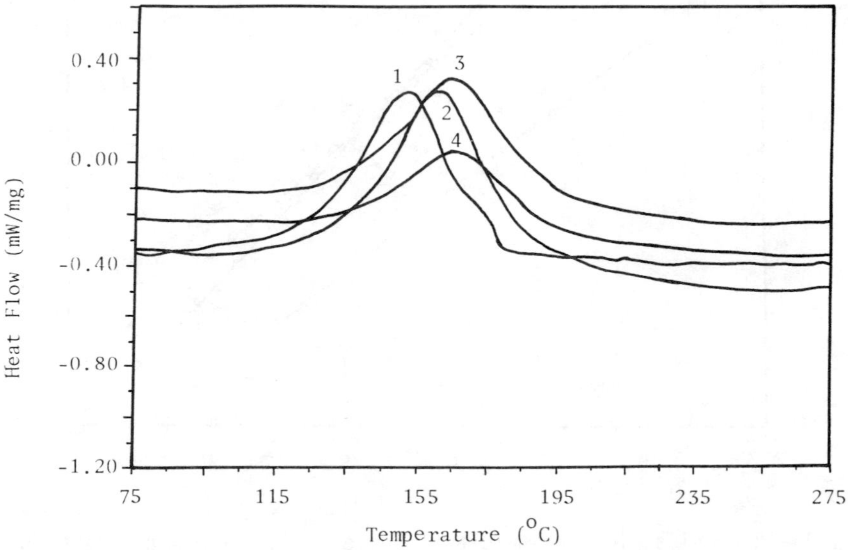

Fig. (2) - Dynamic thermograms of hardener concentration effect for Araldite 6010/NMA/BDMA system. Curve 1: 40 phr; curve 2: 80 phr; curve 3: 110 phr; curve 4: 140 phr

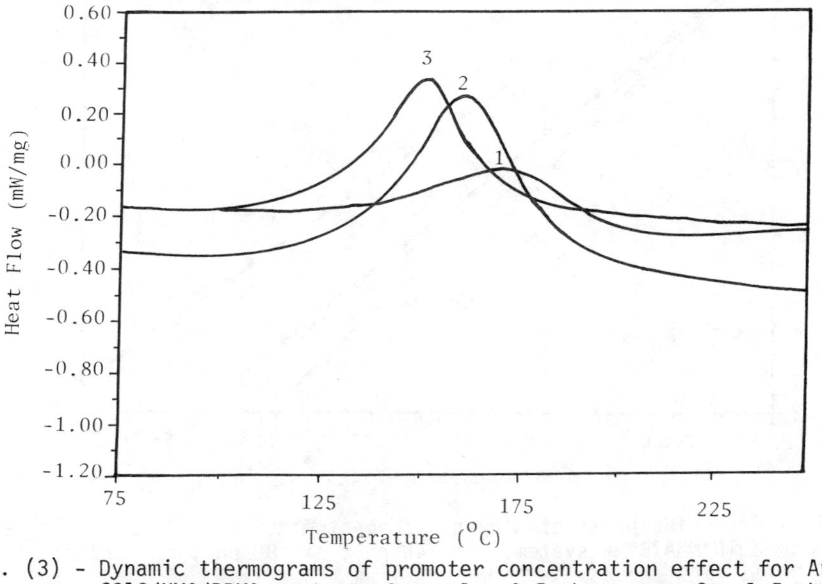

Fig. (3) - Dynamic thermograms of promoter concentration effect for Araldite 6010/NMA/BDMA system. Curve 1: 0.5 phr; curve 2: 1.5 phr; curve 3: 3.0 phr

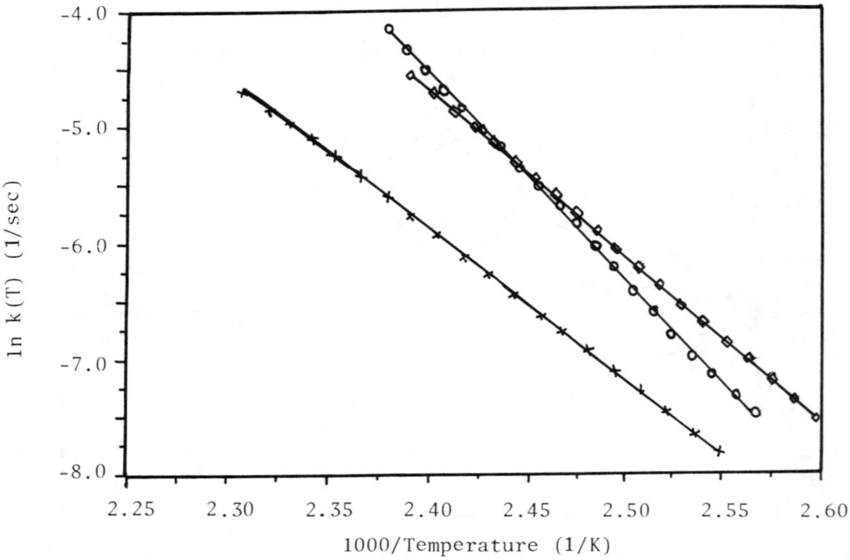

Fig. (4) - Arrhenius plots of hardener species effect for Araldite 6010 cured with BDMA. o: HHPa, +: NMA, ◊: THPA

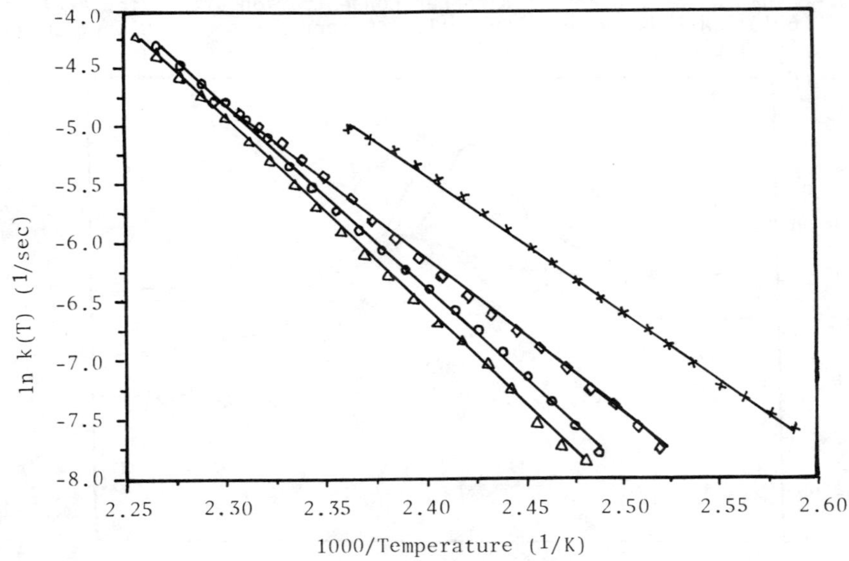

Fig. (5) - Arrhenius plots of hardener concentration effect for Araldite 6010/NMA/BDMA system. +: 40 phr; ◊: 80 phr; o: 110 phr; Δ: 140 phr

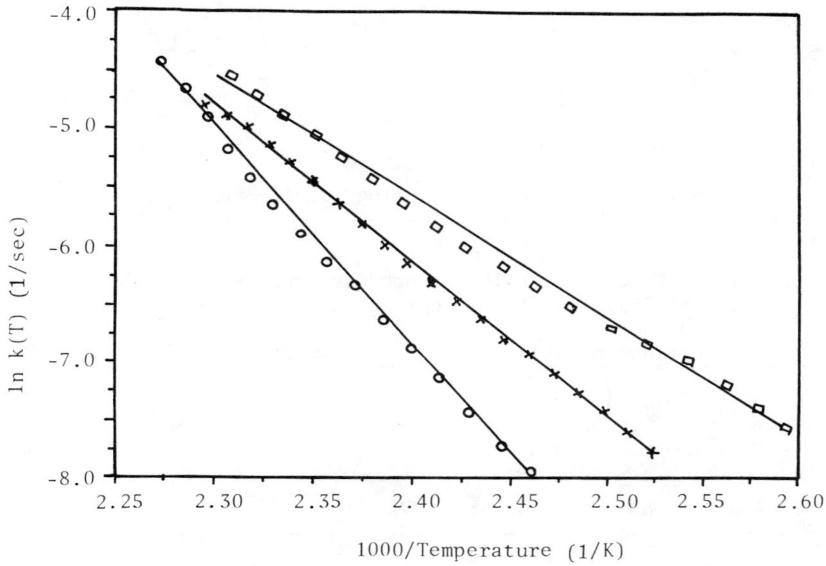

Fig. (6) - Arrhenius plots of promoter concentration effect for Araldite 6010/NMA/BDMA system. o: 0.5 phr; +: 1.5 phr; ◇: 3.0 phr

Table 2 - The influence of hardener effect on DSC kinetics

Resin	NMA Concentration	A/E*	n	E(KJ/mole)	$\log Z(min^{-1})$	ΔH(J/g)
Araldite 6010	40 phr	0.43/1	0.43	97.1	11.60	136.2
	80 phr	0.85/1	1.00	110.3	12.95	177.8
	110 phr	1.17/1	1.66	128.5	15.11	132.6
	140 phr	1.47/1	1.99	136.3	16.02	92.3

*A/E = Anhydride/epoxy ratio.

Table 3 - The influence of promoter on DSC kinetics

BDMA Concentration*	n	E(KJ/mole)	$\log Z(min^{-1})$	ΔH(J/g)
0.5 phr	2.01	154.3	18.15	52.5
1.5 phr	1.00	110.3	12.95	177.8
3.0 phr	0.54	81.7	9.57	148.9

*Formulation: Araldite 6010 + 80 phr NMA + BDMA (varying concentration); A/E = 0.85/1.

Table 4 - The influence of hardener species on DSC kinetics

Curing Agent*	n	E(KJ/mole)	$\log Z(\min^{-1})$	ΔH(J/g)
NMA	1.21	108.1	12.76	175.6
HHTA	1.49	146.5	18.14	155.9
THPA	1.01	117.5	14.45	139.8

*Formulation: (A/E = 1:1), epoxy resin + curing agent + BDMA 1.5 phr

Table 5 - The rate of conversion of epoxide group examined by DSC and FT-IR for different curing condition

No.	Curing Condition*	DSC	FT-IR
A	120°C, 170 min	85%	83.1%
B	120°C, 200 min	92%	86.2%
C	120°C, 335 min	100%	94.8%

*Formulation: Araldite 6010/NMA (80 phr)/BDMA (1.5 phr)

Kinetic results obtained for epoxy/anhydride systems are tabulated in Tables 2 to 4. Table 2 shows the influence of hardener concentration on DSC kinetics. The reaction order and activation energy increased with the concentration of NMA. The higher activation energy was found to have lower rate of cure reaction for the system. However, the influence of promoter on DSC kinetics shows the reaction order and activation energy decreased with the amount of BDMA. This information suggests that increasing the promoter concentration causes a marked decrease of cure time (see Table 3). Table 4 shows the influence of hardener species on DSC kinetics. It was found that HHPA and THPA exhibit low heat of cure (ΔH).

The Extent of Cure Proceeding

The extent of cure proceeding for various formulations of epoxy/anhydride resins were measured by FT-IR and DSC. The isothermal plot and isoconversion plot (see Figures 7 and 8) obtained from DuPont DSC-9900 system may be used to optimize formulations and cure cycles for epoxy resins. These data are of particular value in process development for predicting optimum cure temperatures. It is also useful in process control for optimization of cure advancement or conversion.

The actual extent of cure obtained from DSC data analysis were confirmed by the FT-IR experiment. Comparing Figure 7 with Figure 9, it was found that the extent of cure characterized by DSC and FT-IR existed very coincidence. For example, the extent of cure for Araldite 6010/NMA (80 phr)/BDMA (1.5 phr) cured under 120°C for 170 minutes has 85% conversion rate observed from DSC, while estimated by FT-IR, it is about 83%. The results of conversion obtained from DSC and FT-IR were tabulated in Table 5.

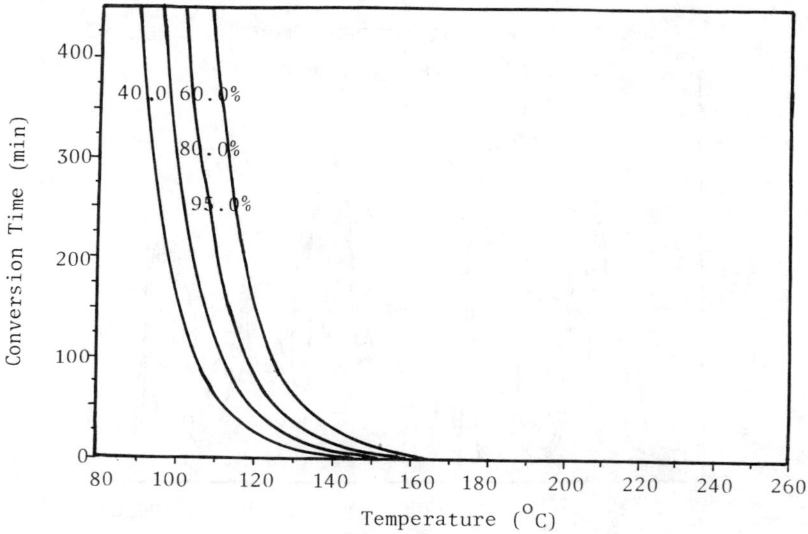

Fig. (7) - Isoconversion plots of Araldite 6010/NMA (80 phr)/BDMA (1.5 phr)

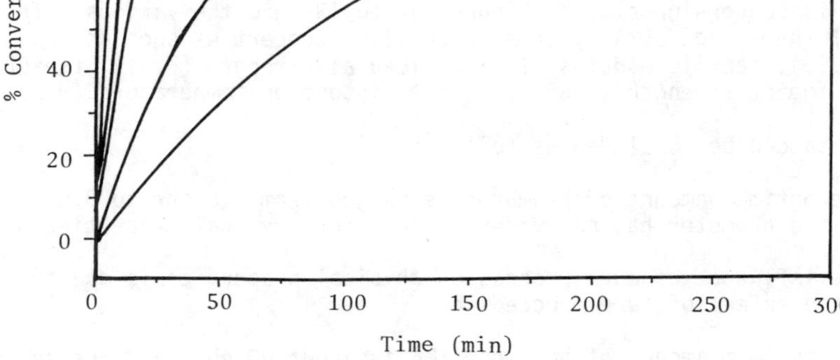

Fig. (8) - Isothermal plots of Araldite 6010/NMA (80 phr)/BDMA (1.5 phr)

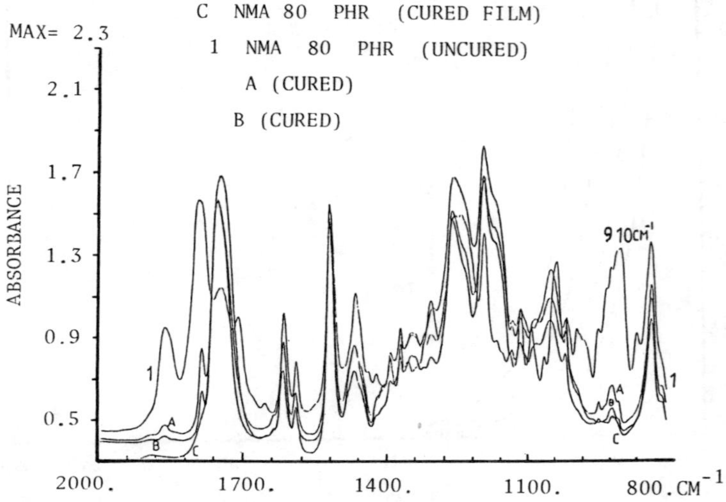

Fig. (9) - FT-IR spectra of various curing condition for epoxy/NMA/BDMA system. A: 120°C x 170 min; B: 120°C x 200 min; C: 120°C x 335 min

Physical Properties of Cured Materials

The cure process and the extent of cure proceeding were discussed above. In order to make optimum use of epoxy as structural materials, the physical properties of the cured materials should be examined. Since the physical properties of epoxy depend on the degree of crosslinking. Hence, to evaluate the influence of hardener concentration, species, and promoter concentration, etc., on physical properties of epoxy systems, constant degree of cure would be considered. Fortunately, from above described, we can control the degree of cure from the information of DSC isoconversion plots. Figures 10 to 13 show the various effects for Araldite 6010/Anhydride/BDMA systems on physical properties such as tensile strength (T.S.), tensile modulus (T.M.), flexural strength (F.S.), flexural modulus (F.M.), impact strength (I.S.) and heat distortion temperature (H.D.T.).

These data can be concluded as follows:

(1) The optimum amount of promoter is ranged from 1.5 phr to 3.0 phr. In this range, the promoter has no evident effect for physical properties of epoxy.

(2) The influence of cure process on physical properties is due to the response for the extent of cure proceeding.

(3) The optimum amount of hardener NMA is about 80 phr, and the concentration of NMA in epoxy have important effect for physical properties of epoxy.

(4) NMA have better tensile and H.D.T. properties than HHPA and THPA, while the impact and flexural properties NMA is bad.

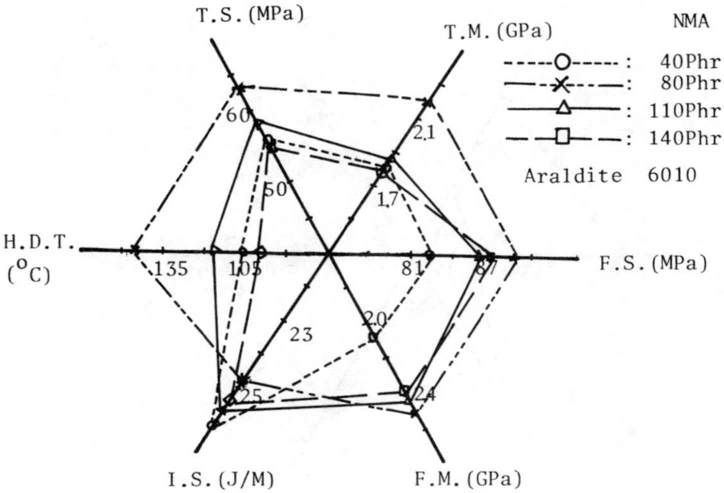

Fig. (10) - Physical properties of hardener concentration effect for Araldite 6010/NMA/BDMA (1.5 phr) system

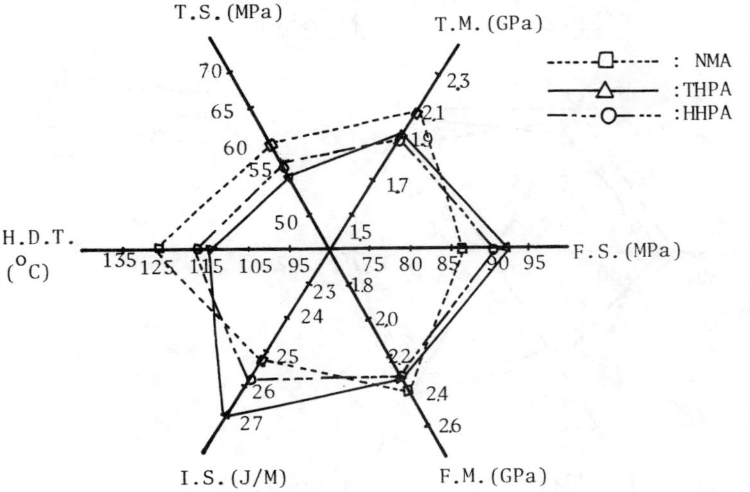

Fig. (11) - Physical properties of hardener species effect for Araldite 6010/Anhydride/BDMA system

ACKNOWLEDGEMENT

The authors wish to thank Mr. C. H. Huang and Mr. K. N. Wang of ER BRIGHT Co., for their assistance in measuring DSC.

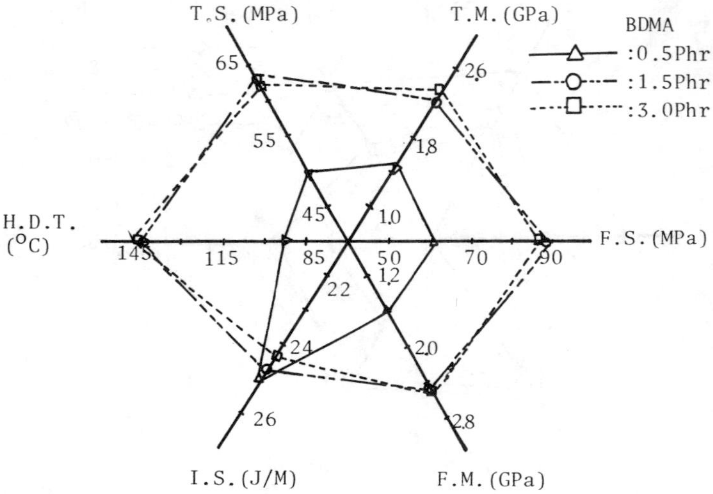

Fig. (12) - Physical properties of promoter effect for Araldite 6010/NMA/BDMA system

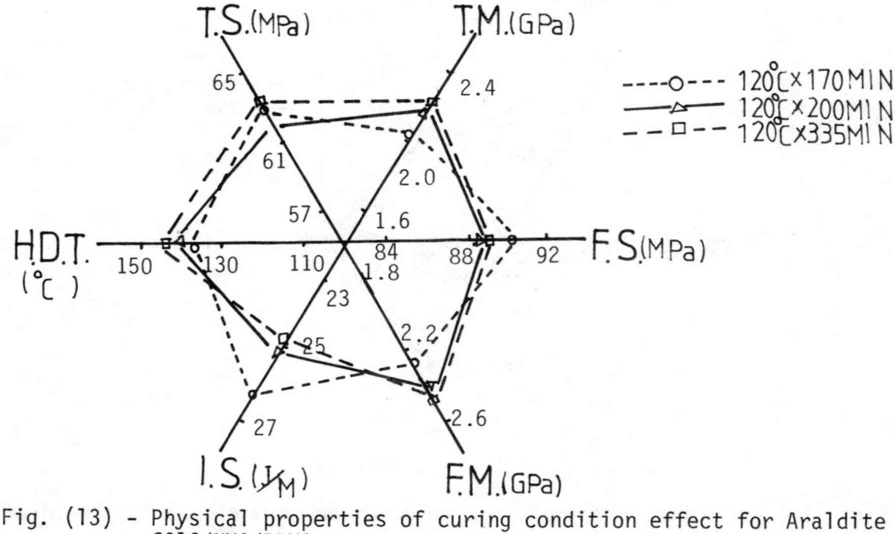

Fig. (13) - Physical properties of curing condition effect for Araldite 6010/NMA/BDMA system

REFERENCES

[1] Warfield, R. W., Testing of Polymers, Volume 1, J. V. Schmitz, ed., Interscience, New York, p. 271, 1965.

[2] Moy, P. and Karasz, F. E., Polymer Engineering and Science, 20, 4, p. 315, 1980.

[3] Dannenburg, H. and Harp, W. R., Anal. Chem., 32, p. 365, 1960.

[4] Leckenby, J. N., The British Polymer Journal, 15, 3, p. 45, 1983.

[5] Buckley, L., Polymer Engineering and Science, 22, 3, p. 166, 1982.

[6] Chang, T. D., Carr, S. H. and Britaing, J. O., Polymer Engineering and Science, 22, 18, p. 1213, 1982.

[7] Borchardt, H. J. et al, J. Am. Chem. Soc., 79, p. 41, 1957.

VIBRATIONS OF HYGROTHERMAL ELASTIC COMPOSITE PLATES

L.-W. Chen

Department of Mechanical Engineering, National Cheng Kung University
Tainan, Taiwan, Republic of China

and

Y. M. Chen

Chung Shan Institute of Science and Technology
Lung Tan, Taiwan, Republic of China

ABSTRACT

A finite element model is performed for the free and forced vibrations of the laminated rectangular composite plate exposed to a steady state hygrothermal environment. The Mindlin plate elements which include the effects of transverse shear and rotatory inertia are used to study the vibration problems of thick composite plates with hygrothermal effects.

A parametric study is done for a symmetrically $[0°,+45°,-45°,90°]_{2S}$ laminated, T300/5208 graphite-epoxy plate with both simply supported and clamped boundary conditions. The results show that the steady state hygrothermal environment has degradational effects on the natural frequencies of a laminated composite plate.

INTRODUCTION

Advanced composite materials are widely used for primary load-bearing structures on the aircraft and space craft, accurate design and analysis methods are becoming more critical. Moisture diffusion in anisotropic heterogeneous composites have been shown to result in significant changes in their physical properties [1,2] and dimensional stability [3,4] which are critically important in the design and analysis of composite structures.

Shen and Springer showed that the physical change due to temperature and moisture effects in the resin results in decreasing the tensile properties [5] and reducing the transverse and shear modulus [6]. Flaggs and Crossman did the linear viscoelastic analysis of the response of the composite laminates during hygrothermal exposure [7]. They showed that the viscoelastic model can predict the hygrothermal response accurately. Sih and Shih studied the hygrothermal stress in a plate subjected to antisymmetric time dependent moisture and temperature boundary conditions [8]. The effects of coupling between moisture and temperature were investigated. Tobing, Shephard and Sternstein used the finite element to study the moisture effects in graphite-epoxy composites [9]. Permanent damage at the interface between fiber and epoxy may be possible due to high humidity in the composite. Wang and Choi worked on the problems of boundary-layer hygroscopic stresses in angle-ply composite laminates [10].

The plate and shell are major structural components in the aerospace industry. The hygrothermal analysis of the laminated composite plates and shells are important to the structural designers. Pipes, Vinson and Chou investigated the hygrothermal response of the laminated system [11]. Flaggs and Vinson studied the hygrothermal effects on the buckling of laminated composite plates [12]. Snead and Palazotto worked on the stability problems of cylindrical composite panels with moisture and temperature effects [13]. The buckling loads will degrade with increasing moisture concentrations and temperatures for both the laminated plates and shells.

The vibration problems of composite plates are very important to the structural designers. Srinivas, Rao and Rao did an exact analysis for vibration of the simply supported homogeneous and laminated thick rectangular plates [14]. Bert and Chen investigated the effect of shear deformation on vibration of the antisymmetric angle-ply laminated rectangular plates [15]. Reddy and Kuppusamy used three dimensional elasticity theorem approach to study the dynamic problems of anisotropic laminated plates [16].

Reismann had discussed the dynamic response of simply supported rectangular plates with uniform distributed loading [20,21]. Rock and Hinton used the finite element method to solve the transient problems of laminated composite plates [22].

The present study uses Mindlin finite element model to solve the free and forced vibration problems for the hygrothermal elastic thick composite plates. This model includes the effects of transverse shear and normal deformation. The effects of moisture and temperature on the natural frequencies of the simply supported and clamped rectangular composite plates are also presented.

HYGROTHERMAL ELASTIC THICK COMPOSITE PLATES

The Mindlin plate theory is used here to describe the behavior of the thick composite plate with hygrothermal effect. The transverse shear deformation and normal deformation are permitted. The displacement field of the Mindlin plate theory is associated to be

$$\bar{u}(x,y,z,t) = u(x,y,t) + zF_x(x,y,t) \qquad (1)$$

$$\bar{v}(x,y,z,t) = v(x,y,t) + zF_y(x,y,t) \qquad (2)$$

$$\bar{w}(x,y,z,t) = w(x,y,t) + f_1F_z(z,t) \qquad (3)$$

where $\bar{u}$, $\bar{v}$ and $\bar{w}$ are final displacements, u, v and w are the midplane (z=0) displacements, and F_x and F_y are the rotations of the normal lines of the undeformed plate in the x-z and y-z planes, respectively. These normal lines are not necessary perpendicular to the midplane after deformation. f_1 is the tracing constant (f_1 = 1) to determine the effect of transverse normal strain due to hygrothermal effect.

The Green's strain tensor is given as

$$\{\varepsilon\} = \begin{bmatrix} \varepsilon_x \\ \varepsilon_y \\ \varepsilon_z \\ \gamma_{yz} \\ \gamma_{xz} \\ \gamma_{xy} \end{bmatrix} = \begin{bmatrix} \frac{\partial \bar{u}}{\partial x} \\ \frac{\partial \bar{v}}{\partial y} \\ \frac{\partial \bar{w}}{\partial z} \\ \frac{\partial \bar{v}}{\partial z} + \frac{\partial \bar{w}}{\partial y} \\ \frac{\partial \bar{u}}{\partial z} + \frac{\partial \bar{w}}{\partial x} \\ \frac{\partial \bar{u}}{\partial y} + \frac{\partial \bar{v}}{\partial x} \end{bmatrix} + \begin{bmatrix} \frac{1}{2}(\frac{\partial \bar{u}}{\partial x})^2 + \frac{1}{2}(\frac{\partial \bar{v}}{\partial x})^2 + \frac{1}{2}(\frac{\partial \bar{w}}{\partial x})^2 \\ \frac{1}{2}(\frac{\partial \bar{u}}{\partial y})^2 + \frac{1}{2}(\frac{\partial \bar{v}}{\partial y})^2 + \frac{1}{2}(\frac{\partial \bar{w}}{\partial y})^2 \\ \frac{1}{2}(\frac{\partial \bar{u}}{\partial z})^2 + \frac{1}{2}(\frac{\partial \bar{v}}{\partial z})^2 + \frac{1}{2}(\frac{\partial \bar{w}}{\partial z})^2 \\ \frac{\partial \bar{u}}{\partial y}\frac{\partial \bar{u}}{\partial z} + \frac{\partial \bar{v}}{\partial y}\frac{\partial \bar{v}}{\partial z} + \frac{\partial \bar{w}}{\partial y}\frac{\partial \bar{w}}{\partial z} \\ \frac{\partial \bar{u}}{\partial x}\frac{\partial \bar{u}}{\partial z} + \frac{\partial \bar{v}}{\partial x}\frac{\partial \bar{v}}{\partial z} + \frac{\partial \bar{w}}{\partial x}\frac{\partial \bar{w}}{\partial z} \\ \frac{\partial \bar{u}}{\partial x}\frac{\partial \bar{u}}{\partial y} + \frac{\partial \bar{v}}{\partial x}\frac{\partial \bar{v}}{\partial y} + \frac{\partial \bar{w}}{\partial x}\frac{\partial \bar{w}}{\partial y} \end{bmatrix} \tag{4}$$

or

$$\{\varepsilon\} = \{e\} + \{\alpha\} \tag{5}$$

The final state of stress of the Kth lamina can be given by the form of

$$\begin{bmatrix} \bar{\sigma}_{xx} \\ \bar{\sigma}_{yy} \\ \bar{\sigma}_{zz} \\ \bar{\sigma}_{yz} \\ \bar{\sigma}_{xz} \\ \bar{\sigma}_{xy} \end{bmatrix}^K = \begin{bmatrix} \sigma_{xx} \\ \sigma_{yy} \\ \sigma_{zz} \\ \sigma_{yz} \\ \sigma_{xz} \\ \sigma_{xy} \end{bmatrix}^K + \begin{bmatrix} \sigma_x^h \\ \sigma_y^h \\ 0 \\ 0 \\ 0 \\ \sigma_{xy}^h \end{bmatrix}^K \tag{6}$$

or

$$\{\bar{\sigma}\}^K = \{\sigma\}^K + \{\sigma^h\}^K \tag{7}$$

where $\bar{\sigma}_{ij}$ is the final stress, σ_{ij} is the incremental stress and σ_{ij}^h is the hygrothermal stress. The stress-strain relationship for the Kth lamina can be written as [17]

$$\{\sigma_i\}^K = [\bar{Q}_{ij}]^K\{\varepsilon_j\}^K \tag{8}$$

The conventional contracted notations for composite materials are used here. Where $\sigma_{ii} = \sigma_i$ for i = 1, 2, 3, $\sigma_{23} = \sigma_4$, $\sigma_{13} = \sigma_5$ and $\sigma_{12} = \sigma_6$. $[\bar{Q}_{ij}]^K$ is the reduced stiffness for the Kth lamina.

The hygrothermal stress is expressed as

$$\sigma_i^{h(K)} = -\bar{Q}_{ij}^{(K)}(\alpha_j\theta(z) + \beta_j M(z)) \quad (i = 1,2,6,\ j = 1,2,3,6) \tag{9}$$

and $\theta(z) = T(z) - T_o(z)$, $M(z) = \bar{M}(z) - M_o(z)$ where $T(z)$ is the temperature distribution and $\bar{M}(z)$ is the moisture distribution within the plate. $T_o(z)$ and $M_o(z)$ are the stress free temperature and moisture distribution, respectively. α_j and β_j are the thermal expansion coefficient and the swelling coefficient, respectively.

The strain energy per unit volume of the Kth lamina is given as

$$W^K = \frac{1}{2}(\{\sigma\}^K)^T\{\varepsilon\} + (\{\sigma^h\}^K)^T\{\varepsilon\} \tag{10}$$

Substitute equations (4)-(9) into equation (10) and neglect higher order terms in displacement gradients, we obtain the following expression

$$W^K = \frac{1}{2}\{e\}^T[\bar{Q}_{ij}]\{e\} + (\{\sigma^h\}^K)^T(\{e\} + \{\alpha\}) \tag{11}$$

The incremental potential energy is written as

$$\Delta W^K = W^K - (\{\sigma^h\}^K)^T\{e\} = \frac{1}{2}\{e\}^T[\bar{Q}_{ij}^T]^K\{e\} + (\{\sigma^h\}^K)^T\{\alpha\} \tag{12}$$

Substitute equation (1) into equation (12) and integrate over the thickness of the plate, then the incremental potential energy per unit area is obtained.

$$\Delta U = \int_{-h/2}^{h/2} \Delta W dz = \frac{1}{2}\{\bar{e}\}^T[\bar{D}]\{\bar{e}\} + \frac{1}{2}\sum_{K=1}^{N}\iint_A \int_{h_K}^{h_{K+1}} f_1\{e'\}^T \frac{\partial F_z}{\partial z}\{Q'\}dzdA$$

$$+ \frac{1}{2}\left[\frac{\partial u}{\partial x}\ \frac{\partial u}{\partial y}\right][N^h]\begin{Bmatrix}\frac{\partial u}{\partial x}\\ \\ \frac{\partial u}{\partial y}\end{Bmatrix} + \frac{1}{2}\left[\frac{\partial u}{\partial x}\ \frac{\partial u}{\partial y}\right][M^h]\begin{Bmatrix}\frac{\partial F_x}{\partial x}\\ \\ \frac{\partial F_x}{\partial y}\end{Bmatrix}$$

$$+ \frac{1}{2}\left[\frac{\partial u}{\partial x}\ \frac{\partial u}{\partial y}\right][N^h]\begin{Bmatrix}\frac{\partial u}{\partial x}\\ \\ \frac{\partial u}{\partial y}\end{Bmatrix} + \frac{1}{2}\left[\frac{\partial F_x}{\partial x}\ \frac{\partial F_x}{\partial y}\right][M^{h*}]\begin{Bmatrix}\frac{\partial F_x}{\partial x}\\ \\ \frac{\partial F_x}{\partial y}\end{Bmatrix}$$

$$+ \frac{1}{2}\left[\frac{\partial v}{\partial x}\ \frac{\partial v}{\partial y}\right][N^h]\begin{Bmatrix}\frac{\partial v}{\partial x}\\ \\ \frac{\partial v}{\partial y}\end{Bmatrix} + \frac{1}{2}\left[\frac{\partial F_y}{\partial x}\ \frac{\partial F_y}{\partial y}\right][M^{h*}]\begin{Bmatrix}\frac{\partial F_y}{\partial x}\\ \\ \frac{\partial F_y}{\partial y}\end{Bmatrix}$$

$$+ \frac{1}{2}\left[\frac{\partial w}{\partial x}\ \frac{\partial w}{\partial y}\right][N^h]\begin{Bmatrix}\frac{\partial w}{\partial x}\\ \\ \frac{\partial w}{\partial y}\end{Bmatrix} \tag{13}$$

where

$$\{\bar{e}\}^T = \left[\frac{\partial u}{\partial x}, \frac{\partial v}{\partial y}, \frac{\partial v}{\partial x} + \frac{\partial u}{\partial y}, \frac{\partial w}{\partial y} + F_y, \frac{\partial w}{\partial x} + F_x, \frac{\partial F_x}{\partial x}, \frac{\partial F_y}{\partial y}, \frac{\partial F_y}{\partial x} + \frac{\partial F_x}{\partial y}\right] \tag{14}$$

$$\{e'\}^T = \left[2\left(\frac{\partial u}{\partial x} + z\,\frac{\partial F_x}{\partial x}\right), 2\left(\frac{\partial v}{\partial y} + z\,\frac{\partial F_y}{\partial y}\right), \frac{\partial F_z}{\partial z}, \left(\left(\frac{\partial u}{\partial y} + \frac{\partial v}{\partial x}\right) + z\left(\frac{\partial F_x}{\partial y} + \frac{\partial F_y}{\partial x}\right)\right)\right] \tag{15}$$

$$\{Q'\}^T = [\bar{Q}^K_{13}, \bar{Q}^K_{23}, \bar{Q}^K_{33}, \bar{Q}^K_{36}] \tag{16}$$

$$(A_{ij}, B_{ij}, D_{ij}) = \int_{-h/2}^{h/2} (i, z, z^2)\bar{Q}^K_{ij}dz \tag{17}$$

$$([N^h], [M^h], [M^*]) = \int_{-h/2}^{h/2} (1, z, z^2)\begin{bmatrix}\sigma^h_x & \sigma^h_{xy}\\ \sigma^h_{xy} & \sigma^h_y\end{bmatrix}^K dz \tag{18}$$

$$[D] = \begin{bmatrix}[A'] & [B']\\ [B]^T & [D']\end{bmatrix} \tag{19}$$

$$[A] = \begin{bmatrix}A'_{11} & A'_{12} & A'_{16} & 0 & 0\\ A'_{12} & A'_{22} & A'_{26} & 0 & 0\\ A'_{16} & A'_{26} & A'_{66} & 0 & 0\\ 0 & 0 & 0 & A'_{44} & A'_{45}\\ 0 & 0 & 0 & A'_{45} & A'_{55}\end{bmatrix}$$

$$[B] = \begin{bmatrix} B'_{11} & B'_{12} & B'_{16} \\ B'_{12} & B'_{22} & B'_{26} \\ B'_{16} & B'_{26} & B'_{66} \\ 0 & 0 & 0 \\ 0 & 0 & 0 \end{bmatrix}$$

$$[D] = \begin{bmatrix} D'_{11} & D'_{12} & D'_{16} \\ D'_{12} & D'_{22} & D'_{26} \\ D'_{16} & D'_{26} & D'_{66} \end{bmatrix}$$

and

$$A'_{ij} = A_{ij} - \sum_{K=1}^{N} \frac{\bar{Q}^K_{i3}\bar{Q}^K_{j3}}{\bar{Q}^K_{33}} (h_{K+1} - h_K)$$

$$B'_{ij} = B_{ij} - \frac{1}{2} \sum_{K=1}^{N} \frac{\bar{Q}^K_{i3}\bar{Q}^K_{j3}}{\bar{Q}^K_{33}} (h^2_{K+1} - h^2_K)$$

$$D'_{ij} = D_{ij} - \frac{1}{3} \sum_{K=1}^{N} \frac{\bar{Q}^K_{i3}\bar{Q}^K_{j3}}{\bar{Q}^K_{33}} (h^3_{K+1} - h^3_K)$$

$$\frac{\partial F_z}{\partial z} = \frac{f_1}{\bar{Q}^K_{33}} [- \bar{Q}^K_{13} \frac{\partial F_x}{\partial x} - \bar{Q}^K_{23} \frac{\partial F_y}{\partial y} - \bar{Q}^K_{36} (\frac{\partial F_x}{\partial y} + \frac{\partial F_y}{\partial x})] + \frac{1}{\bar{Q}^K_{33}} [- \bar{Q}^K_{13} \frac{\partial u}{\partial x}$$

$$- \bar{Q}^K_{23} \frac{\partial v}{\partial y} - \bar{Q}^K_{36} (\frac{\partial u}{\partial y} + \frac{\partial v}{\partial x})] \qquad (20)$$

The derivation of equation (20) can be seen in [18].

The kinetic energy K and the work of external forces We are defined as follows

$$K = \frac{1}{2} \int_V \rho(\dot{u}^2 + \dot{v}^2 + \dot{w}^2)dv \qquad (21)$$

and

$$We = \int_s (P_1\bar{u} + P_2\bar{v} + P_3\bar{w})ds \tag{22}$$

where p_i is the surface force per unit area.

The Hamilton's principle will be used to formulate the finite element model for the free and forced vibration problems. It can be expressed as

$$\delta \int_{t_0}^{t_1} (U-K-We)dt = 0 \tag{23}$$

where K is the kinetic energy, U is the strain energy and We is the work of external forces.

Making use of equation (2) and performing the necessary variational procedure, equation (23) becomes

$$\int_{t_1}^{t_2} [\delta U + \int_v \rho(\ddot{u}\delta u + \ddot{v}\delta v + \ddot{w}\delta w + z^2\ddot{F}_x\delta F_x + z^2\ddot{F}_y\delta F_y)dv - \int_s (P_1\delta\bar{u} + P_2\delta\bar{v}$$

$$+ P_3\delta\bar{w})ds]dt = 0 \tag{24}$$

FINITE ELEMENT MODEL

The nine-noded Lagrange-Mindlin element is used to formulate the finite element model. Each node has five degrees of freedom u, v, w, F_x and F_y. Let u, v, w, F_x and F_y be interpolated over the element by

$$u(x,y,t) = \sum_{i=1}^{9} u_i(t)\phi_i(x,y)$$

$$v(x,y,t) = \sum_{i=1}^{9} v_i(t)\phi_i(x,y)$$

$$w(x,y,t) = \sum_{i=1}^{9} w_i(t)\phi_i(x,y) \tag{25}$$

$$F_x(x,y,t) = \sum_{i=1}^{9} F_{x_i}(t)\phi_i(x,y)$$

$$F_y(x,y,t) = \sum_{i=1}^{9} F_{y_i}(t)\phi_i(x,y)$$

where the $\phi_{i's}(x,y)$ are the shape functions of the nine-noded isoparametric element and $u_i(t)$, $v_i(t)$, $w_i(t)$, $F_{x_i}(t)$ and $F_{y_i}(t)$ are the values of u(t), v(t), w(t), $F_x(t)$ and $F_y(t)$ at node i and time t, respectively [19].

Substitution of equation (25) into equation (24) gives

$$\frac{1}{2}\sum_{i=1}^{9}\sum_{j=1}^{9}\{\delta D_i\}^T[K_{ij}]\{D_j\} + \{\delta D_i\}^T[K^h_{ij}]\{D_j\} + \{\delta D_i\}^T[M_{ij}]\{\ddot{D}_j\} = \sum_{i=1}^{9}\{\delta D_i\}^T\{F_i\} \qquad (26)$$

Performing the Rayleigh-Ritz procedure, equation (26) becomes

$$\left\{\begin{bmatrix} [K_{11}] & [K_{12}] & \cdots & [K_{19}] \\ & [K_{22}] & \cdots & [K_{29}] \\ \text{Symm.} & & \ddots & \vdots \\ & & & [K_{99}] \end{bmatrix} + \begin{bmatrix} [K^h_{11}] & [K^h_{12}] & \cdots & [K^h_{19}] \\ & [K^h_{22}] & \cdots & [K^h_{29}] \\ \text{Symm.} & & \ddots & \vdots \\ & & & [K_{99}] \end{bmatrix}\right\} \begin{bmatrix} \{D_1\} \\ \{D_2\} \\ \vdots \\ \{D_9\} \end{bmatrix} + \begin{bmatrix} [M_{11}] & [M_{12}] & \cdots & [M_{19}] \\ & [M_{22}] & \cdots & [M_{29}] \\ \text{Symm.} & & \ddots & \vdots \\ & & & [M_{99}] \end{bmatrix} \begin{bmatrix} \{\ddot{D}_1\} \\ \{\ddot{D}_2\} \\ \vdots \\ \{\ddot{D}_9\} \end{bmatrix} = \begin{bmatrix} \{F_1\} \\ \{F_2\} \\ \vdots \\ \{F_9\} \end{bmatrix} \qquad (27)$$

where $[K_{ij}]$ is the stiffness coefficient matrix, $[K^h_{ij}]$ is the hygrothermal effect geometry stiffness coefficient matrix and $[M_{ij}]$ is the mass coefficient matrix associated with the vector $\{D_i\}$ in the jth equation, and

$\{D_i\}^T = [u_i, v_i, w_i, F_{x_i}, F_{y_i}]$ is the total nodal value of the ith point. The forced vibration problem can be studied by using equation (27).

To investigate the free vibration problem, the external force $\{F_i\}$ will be set to be zero. Assume the motions are harmonic; then the motion equations can be expressed as

$$
\begin{aligned}
u_i(t) &= u_{oi}e^{i\omega t} \\
v_i(t) &= v_{oi}e^{i\omega t} \\
w_i(t) &= w_{oi}e^{i\omega t} \\
F_{x_i}(t) &= F_{x_{oi}}e^{i\omega t} \\
F_{y_i}(t) &= F_{y_{oi}}e^{i\omega t}
\end{aligned}
\tag{28}
$$

and ω denotes the circular frequency of natural vibration. Substitution of equations (28) into equations (27) gives the generalized eigenvalue problem

$$([K] + [K^h] - \omega^2[M])\{\Delta\} = \{0\} \tag{29}$$

where

$$\{\Delta\}^T = [\{\Delta_1\}, \{\Delta_2\},\ldots,\{\Delta_9\}]$$

and

$$\{\Delta_i\}^T = [u_{oi}, v_{oi}, w_{oi}, F_{x_{oi}}, F_{y_{oi}}]$$

The global equation should be assembled in the usual finite element manner from the element equation (29) and the boundary conditions are imposed before solving the eigenvalue problem.

The Newmark direct integration method is used to solve the forced vibration equation

$$([K] + [K^h])\{D\} + [M]\{\ddot{D}\} = \{F\} \tag{30}$$

The results shall be shown to have good accuracy.

NUMERICAL RESULTS

The simply supported and clamped plates are considered. Both changes of temperature and moisture are functions of thickness only. The biaxial symmetric plate is exploited to model only one quadrant of the plate. The nine-noded Lagrange Mindlin plate element is used here and the element mesh can be seen in Figure 1(a). The free and forced vibration problems of the hygrothermal composite plates are studied. To validate this generally laminated plate element and to de-

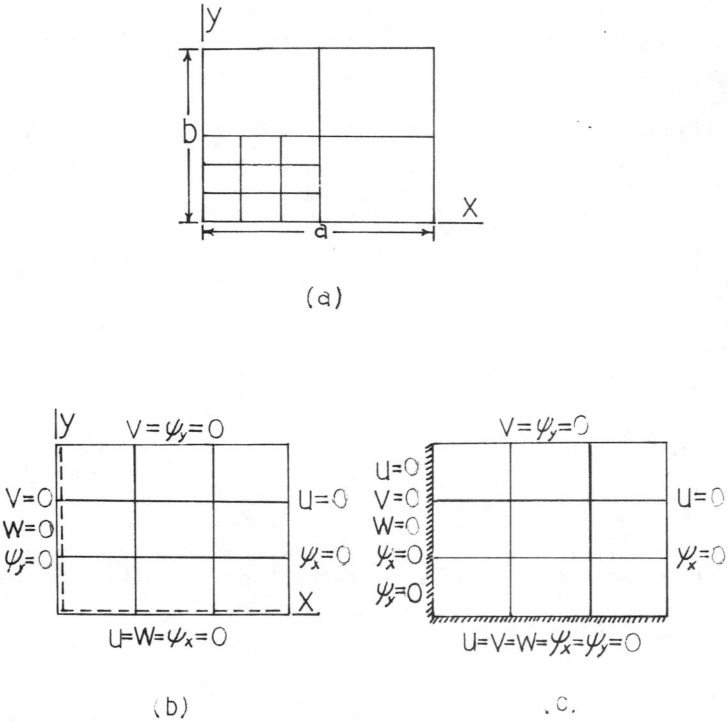

Fig. (1) - (a) Finite element mesh and boundary conditions; (b) simply supported; (c) clamped

termine the discretization error, we use the present model to analyze the free vibration problem of a three-layered isotropic, square laminated plate. The lowest nondimensional natural frequency obtained is 0.09305. By comparing with the exact solution 0.09315 of [14], the present solution is shown to be very accurate.

First, we consider a $[0°,45°,-45°,90°]_{2S}$ T300/5208 graphite-epoxy laminated composite plate with simply supported and clamped boundary conditions. The natural frequencies of the composite plate with steady state hygrothermal conditions are calculated. The lamina properties at 70°F and 350°F are found in [11]. It is assumed that the lamina properties vary linearly between 70°F and 350°F for those which are given as temperature dependent.

The simply supported and clamped boundary conditions are considered in Figure 1(b) and Figure 1(c), respectively. The side lengths of the square plate are a = b = 6 in (15.24 cm) and the thickness is h = 0.088 in (0.22 cm). The plots of natural frequencies versus temperature are plotted in Figures 2 and 3. We can see that the increases of temperature and moisture content shall decrease the natural frequency. The clamped plate can be subjected to a more severe hygrothermal change. The zero frequency point means that the hygrothermal buckling will occur at that point.

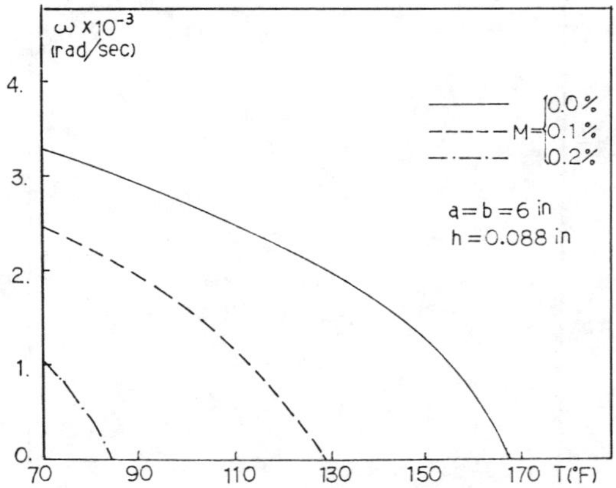

Fig. (2) - Frequencies ω versus temperature with various moisture contents for a simply supported plate

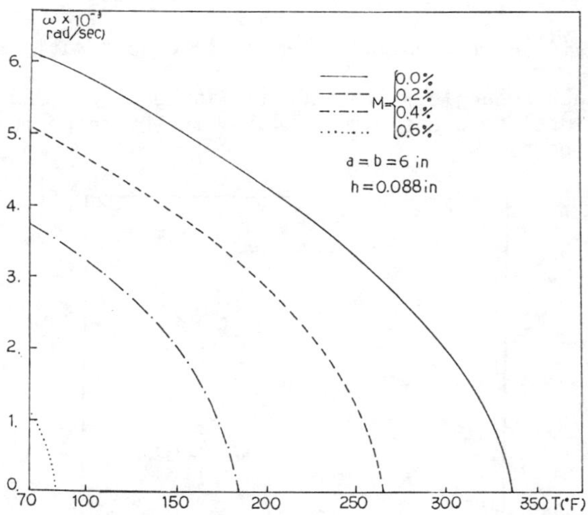

Fig. (3) - Frequencies ω versus temperature with various moisture contents for a clamped plate

The plots of frequencies ω versus the thickness ratio a/h are shown in Figure 4. Two curves with different temperature and moisture conditions are shown. It can be seen again that high temperature and moisture contents will soften the plate. The thicker plate has the stronger stiffness and naturally it has the higher natural frequency as seen in the plots.

Next, we consider an antisymmetric stacking of a simply supported plate. The stacking sequence is $[\theta,-\theta]_{4T}$. The θ changes from 0° to 45°. The plots of fre-

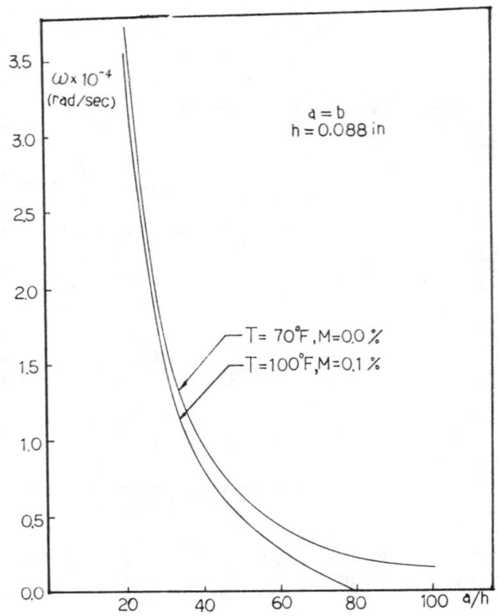

Fig. (4) - Frequencies ω versus thickness ratio a/h

quencies versus stacking angle θ are made in Figure 5. We can see that the plate has the highest natural frequency at θ = 45°. The hygrothermal effect decreases as the value of θ increases.

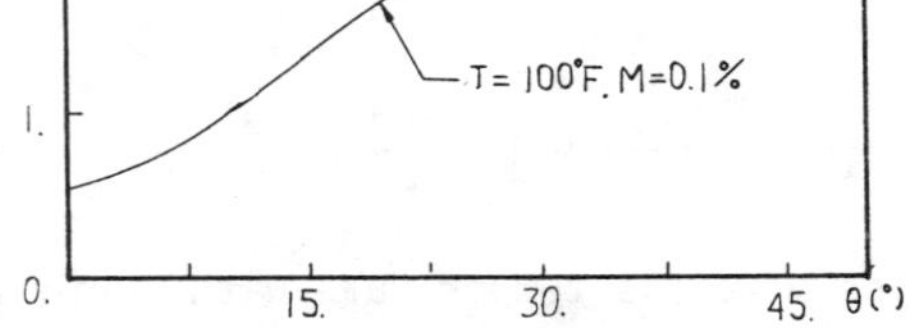

Fig. (5) - The effect of antisymmetric stacking angle θ

Finally, we consider the dynamic problem of a laminated plate subjected to a uniform distribution impulse loading as shown in Figure 6. The Newmark constant-average acceleration method is used to solve the time-dependent problem. First, an isotropic square simply supported plate subjected to the impulse loading q(t) = q_o = 10 N/cm^2 is studied. The present solution of the center deflection of the

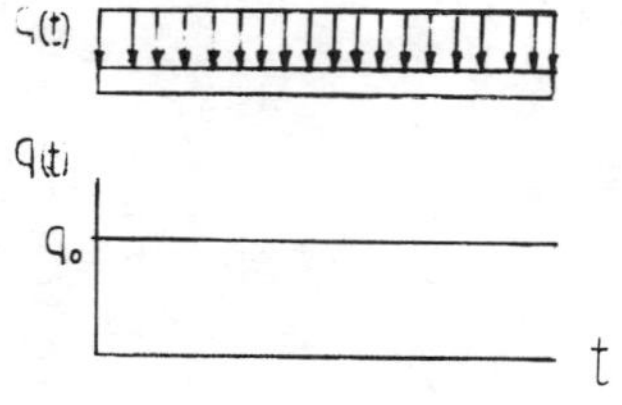

Fig. (6) - The dynamic uniform impulse loading

plate is compared with the Akay's mixed finite element solution [25] and classical solution [26]. It can be seen in Figure 7 that the present finite element solution has very good results.

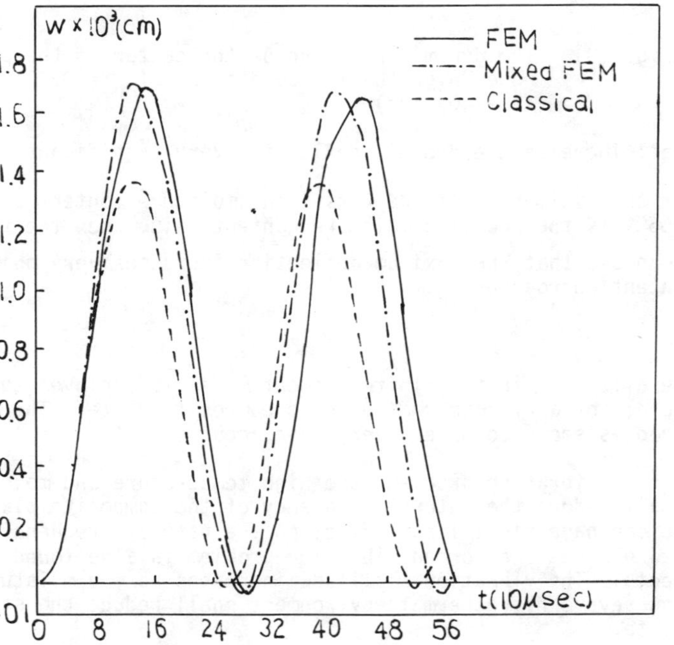

Fig. (7) - The present FEM solution compared with the Akay's mixed FEM solution. ($E = 2.1 \times 10^6$ N/cm², a = b = 25 cm, h = 5 cm, $\rho = 8 \times 10^{-6}$ Nsec²/cm⁴, $\nu = 0.25$, $q_o = 10$ N/cm², 3 x 3 mesh, $\Delta t = 5\mu$ sec)

A $[0°,45°,-45°,90°]_{2S}$ laminated T300/5208 composite plate under hygrothermal condition is considered. The temperature is fixed at 150°F and the moisture constant is varied from 0% to 0.4%. It can be seen in Figure 8 that the high moisture

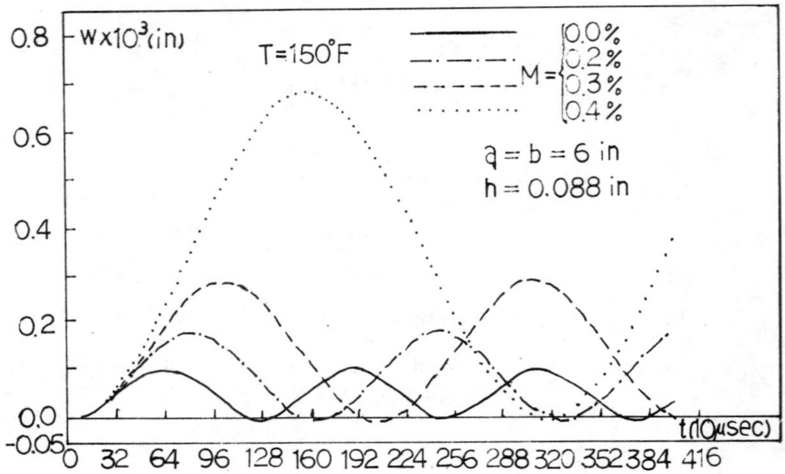

Fig. (8) - The dynamic response of the center of the plate under hygrothermal condition. (q_o = 0.011 bf/in^2, Δt = 40μ sec)

constant shall increase the dynamic deflection very significantly.

Plots of the maximum amplitude versus the moisture content are made in Figure 9. M_B = 0.568% is the buckling moisture content value when the temperature is 100°F. We can see that the maximum deflection increases very quickly when the moisture content approaches M_B.

CONCLUSIONS

The generalized laminated finite element model is employed to solve the vibration problems of a hygrothermal elastic composite plate. The present finite element method is shown to have a very good accuracy.

For the free vibration problem, the high temperature and moisture content environment shall reduce the natural frequency of the composite plate. The change of ply angle can have significant effect on the natural frequency of the angle-ply laminated plate. The forced vibration problem is also found to have hygrothermal effects. The vibration amplitude increases as the moisture content increases. The severe hygrothermal environment shall reduce the stiffness of the composite plate.

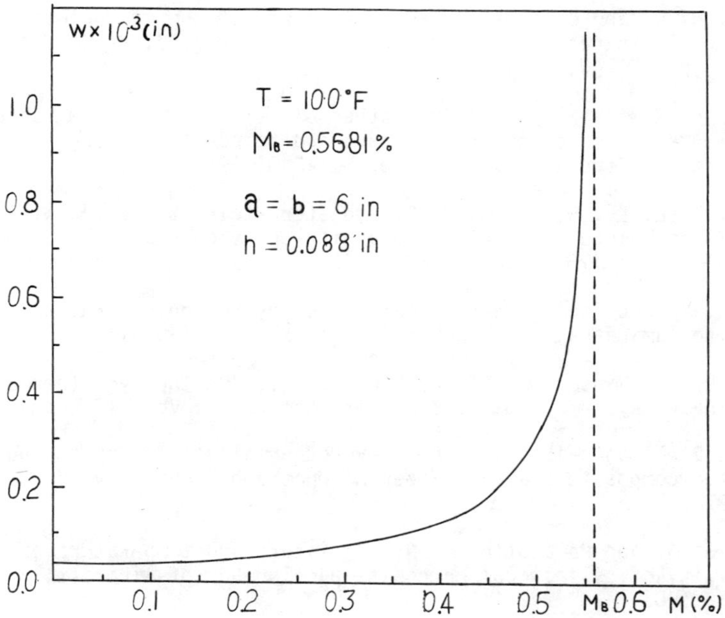

Fig. (9) - The maximum amplitude versus moisture content

REFERENCES

[1] Vinson, J. R. and Walker, W., "The influence of relative humidity and elevated temperature on composite materials and structures", Tech. Rept. AFOSR-TR-77-0030, 1976.

[2] Crossman, F. W., Mauri, R. E. and Warren, W. J., "Moisture altered viscoelastic response of graphite-epoxy composites", Advanced Composite Materials-Environmental Effects, ASTM STP 658, edited by J. R. Vinson, Philadelphia, Pa., pp. 205-220, 1978.

[3] Hahn, H. T. and Kim, R. Y., "Swelling of composite laminate", Advanced Composite Materials-Environmental Effects, ASTM STP 658, edited by J. R. Vinson, Philadelphia, Pa., pp. 98-120, 1978.

[4] Cohen, O. and Maron, G., "A proposal for a coefficient of hygroelasticity", Polymer Engineering and Science, Vol. 18, No. 13, pp. 1001-1005, 1978.

[5] Shen, C. H. and Springer, G. S., "Effects of moisture and temperature on the tensile strength of composite materials", J. of Composite Materials, Vol. 11, 1977.

[6] Shen, C. H. and Springer, G. S., "Environmental effects on the elastic moduli of composite materials", J. of Composite Materials, Vol. 11, 1977.

[7] Flaggs, D. L. and Crossman, F. W., "Analysis of the viscoelastic response of composite laminates during hygrothermal exposure", J. of Composite Materials, Vol. 15, pp. 21-40, 1981.

[8] Sih, G. C. and Shih, M. T., "Hygrothermal stress in a plate subjected to antisymmetric time-dependent moisture and temperature boundary conditions", J. of Thermal Stresses, Vol. 3, pp. 321-340, 1980.

[9] Tobing, F. E. L., Shephard, M. S. and Sternstein, S. S., "Finite element analysis of moisture effects in graphite-epoxy composites", Computers and Structures, Vol. 16, pp. 457-469, 1983.

[10] Wang, S. S. and Choi, I., "Boundary-layer hygroscopic stresses in angle-ply composite laminates", AIAA Journal, Vol. 20, pp. 1592-1598, 1982.

[11] Pipes, R. B., Vinson, J. R. and Chou, T. W., "On the hygrothermal response of laminated systems", J. of Composite Materials, Vol. 10, pp. 129-148, 1976.

[12] Flaggs, D. L. and Vinson, J. R., "Hygrothermal effects on the buckling of laminated composite plates", Fiber Science and Technology, Vol. 11, pp. 353-366, 1978.

[13] Snead, J. M. and Palazotto, A. N., "Moisture and temperature effects on the instability of cylindrical composite panels", J. of Aircraft, Vol. 20, pp. 777-783, 1983.

[14] Srinivas, S., Rao, C. V. and Rao, A. K., "An exact analysis for vibration of simply supported homogeneous and laminated thick rectangular plates", J. of Sound and Vibration, Vol. 12, p. 187, 1970.

[15] Bert, C. W. and Chen, T. L. C., "Effect of shear deformation on vibration of antisymmetric angle-ply laminated rectangular plates", Int. J. of Solids and Structures, Vol. 14, p. 465, 1978.

[16] Reddy, J. N. and Kuppusamy, T., "Natural vibrations of laminated anisotropic plates", J. of Sound and Vibration, Vol. 94, p. 63, 1984.

[17] Tsai, S. W. and Hahn, H. T., Introduction to Composite Materials, Technomic Publishing Co., 1980.

[18] Wilson, D. W. and Vinson, J. R., "Viscoelastic analysis of laminated plate buckling", AIAA Journal, Vol. 22, p. 982, 1984.

[19] Becher, E. B., Carey, G. F. and Oden, J. T., Finite Element, An Introduction, Prentice-Hall, 1981.

[20] Reismann, H., "Forced motion of elastic plates", J. of Applied Mechanics, Vol. 35, pp. 510-515, 1968.

[21] Reismann, H., "Dynamics of rectangular plates", Int. J. of Engineering Science, Vol. 7, pp. 93-113, 1969.

[22] Rock, T. and Hinton, E., "Free vibration and transient response of thick and thin plates using the finite element method", Earthquake Engineering and Structural Dynamics, Vol. 3, pp. 51-63, 1974.

[23] Reddy, J. N., "On the solutions to forced motions of rectangular composite plates", J. of Applied Mechanics, Vol. 49, pp. 403-408, 1982.

[24] Reddy, J. N., "Dynamic (transient) analysis of layered anisotropic composite-material plates", Int. J. for Numerical Methods in Engineering, Vol. 19, pp. 237-255, 1983.

[25] Akay, H. U., "Dynamic large deflection analysis of plates using mixed finite element", Computer and Structure, Vol. 11, pp. 1-11, 1980.

[26] Voltera, E. and Eachmanoglou, E. C., Dynamics of Vibrations, Merril, Columbus, Ohio, 1965.

EFFECTS OF FIBER LENGTH DISTRIBUTION ON ELASTIC MODULI OF SHORT FIBER REINFORCED THERMOPLASTIC

W. K. Chin, H. T. Liu and Y. D. Lee

Department of Chemical Engineering, National Tsing Hua University
Hsin-Chu, Taiwan, Republic of China

ABSTRACT

Fiber breakage during processing will limit the reinforcement effect of fibers in thermoplastic. However, most analytical methods which predict the elastic properties of unidirectional short fiber composites do not put this factor into consideration. Here, we discuss the effects of fiber length distribution represented by two statistical distribution functions, and the ways to apply above fiber length distributions into analytical estimation in order to modify the prediction value of Young's modulus of unidirectional short fiber reinforced composite.

INTRODUCTION

Short fiber reinforced thermoplastics (FRTP) are material in which fibers are embedded in thermoplastic matrices. Through the reinforcement of fibers, mechanical properties, dimensional stability and heat distortion temperature of thermoplastic can be improved.

To predict the elastic modulus of a fiber reinforced thermoplastic, the simulated laminated-plate model is often used. FRTP is treated as unidirectional plies or laminates stacked up in a particular arrangement of orientation. With known elastic properties of constitutional materials, many equations have been developed to estimate the elastic moduli of unidirectional ply [1-4], while Halpin-Tsai equation [1,2] is the most popular one. In Halpin-Tsai equation, fibers are assumed to have uniform length. However, the shear forces of screw and ram during extrusion compounding or injection molding will break fibers which results a length distribution of fibers [5-8]. Since the elastic modulus of composite is not a linear function of fiber length, a uniform fiber length assumption might bring greater prediction error of elastic modulus of composite than expected.

In this paper, we discussed fiber length distribution, their statistical distribution functions and how to apply these distribution functions into calculation in order to obtain a better prediction of the elastic modulus of unidirectional composites.

FIBER LENGTH DISTRIBUTION (FLD)

In FRTP, fiber length can vary from less than one fiber diameter to hundreds of fiber diameter. During extrusion and injection molding processes, the shear forces of screw and ram will break fibers and result a length distribution which is independent of the original length of chopped strand fibers [5]. Figure 1 shows fiber length histograms of glass fiber-reinforced polypropylene which are compounded by twin-screw extruder and followed by injection molding. The proc-

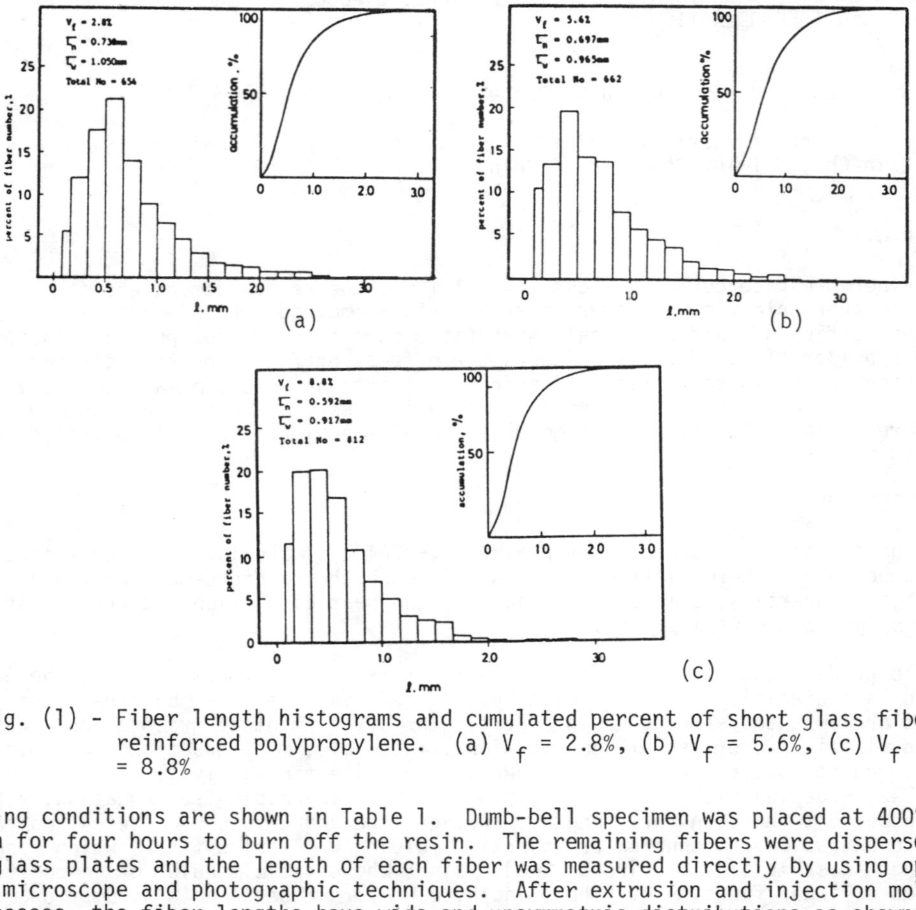

Fig. (1) - Fiber length histograms and cumulated percent of short glass fiber reinforced polypropylene. (a) V_f = 2.8%, (b) V_f = 5.6%, (c) V_f = 8.8%

essing conditions are shown in Table 1. Dumb-bell specimen was placed at 400°C oven for four hours to burn off the resin. The remaining fibers were dispersed on glass plates and the length of each fiber was measured directly by using optical microscope and photographic techniques. After extrusion and injection molding processes, the fiber lengths have wide and unsymmetric distributions as shown in Figure 1. The mean fiber length is only about one tenth of the original length. Statistically, we can choose a suitable probability function to describe this kind of distribution. Since the length distribution is unsymmetric with a tail at the long fiber end, the selected distribution function also needs to meet this requirement. Pennington [9] had used log-normal distribution function to describe this kind of length distribution. In addition, two parameters Weibull distribution was also an unsymmetric distribution function and had been used frequently [8,10,

Table 1 - Processing conditions of glass-fiber reinforced polypropylene

(a) Twin-Screw Extrusion

Cylinder Temperature, °C					Screw (rpm)
1	2	3	4	5	
220	240	230	230	240	75-90

(b) Injection Molding

Injection Pressure	1098 Kg/cm^2
Back Pressure	5 Kg/cm^2
Screw rpm	80 rpm
Cylinder Temperature (3 sections)	260, 260, 250 °C
Mold Temperature	60°C
Injection Time	12 seconds
Mold Cooling Time	35 seconds

11]. In this section, we will discuss these two distribution functions and their characteristics to represent the fiber length distribution.

A. Weibull Distribution

Let us define the probability functions $f(\ell)d\ell$, and $F(\ell)$ as the probability density that the length of fiber is between ℓ and $\ell+d\ell$ and the probability that the length of fiber is less than or equal to ℓ, respectively. The relation of $f(\ell)$ and $F(\ell)$ is

$$F(\ell) = \int_0^\ell f(x)dx \text{ and } \int_0^\ell f(x)dx = 1 \quad (1)$$

If $f(\ell)$ and $F(\ell)$ are characterized by two-parameters Weibull distribution [10], then

$$f(\ell) = \frac{c}{b}\left(\frac{\ell}{b}\right)^{c-1} e^{-\left(\frac{\ell}{b}\right)^c} \quad (2)$$

$$F(\ell) = 1 - e^{-\left(\frac{\ell}{b}\right)^c} \quad (3)$$

where b and c are shape parameters, and ℓ is fiber length. The expected fiber length is

$$L_{exp} = \int_0^\infty xf(x)dx = b\Gamma\left(\frac{1}{c}+1\right) \quad (4)$$

with a standard deviation of

$$S_d = [\int_0^\infty x^2 f(x)dx - (\int_0^\infty xf(x)dx)^2]^{1/2} = b[\Gamma(\frac{2}{c} + 1) - \Gamma^2(\frac{1}{c} + 1)]^{1/2} \quad (5)$$

where $\Gamma(x)$ is the gamma function. The mean fiber length $\overline{L}$ is equal to the expected value of function,

$$\overline{L} = L_{exp} = b\Gamma(\frac{1}{c} + 1) \quad (6)$$

From equation (6), we can see that parameter b can be alternatively expressed as the function of the other parameter c and the mean fiber length. Figure 2 shows Weibull density curves while the mean fiber length is assumed to be 0.6 mm. It

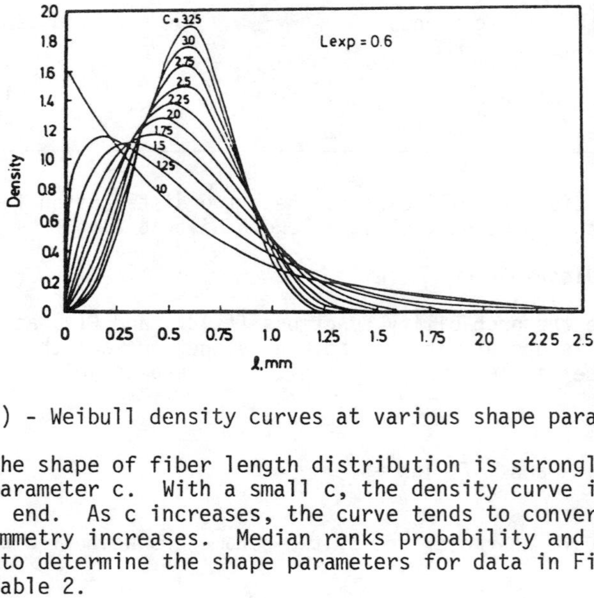

Fig. (2) - Weibull density curves at various shape parameters

indicates that the shape of fiber length distribution is strongly depended on the value of shape parameter c. With a small c, the density curve is skewed with a tail at the long end. As c increases, the curve tends to converge to the mean value and the symmetry increases. Median ranks probability and maximum likelihood method are used to determine the shape parameters for data in Figure 1, the result being shown in Table 2.

B. Log-Normal Distribution

The definitions of log-normal distribution are as follows:

$$f(\ell) = \frac{1}{\sqrt{2\pi} s \ell} e^{-(\ell n \ell - \mu)^2 / 2s^2} \quad (7)$$

$$F(\ell) = \frac{1}{\sqrt{2\pi} s} \int_0^\ell \frac{1}{x} e^{-(\ell n x - \mu)^2 / 2s^2} dx \quad (8)$$

Table 2 - Shape parameters of Weibull and log-normal distributions while simulated Figure 1 data

(a) Weibull Distribution				(b) Log-Normal Distribution				
V_f,%	$\overline{L}_n$,mm	b,mm	c	V_f,%	L_n,mm	L_{mod},mm	μ	s
2.8	0.783	0.732	1.735	2.8	0.783	0.583	-0.382	0.396
5.6	0.697	0.772	1.649	5.6	0.697	0.417	-0.532	0.585
8.8	0.592	0.665	1.761	8.8	0.592	0.333	-0.716	0.619

where s and μ are parameters of distribution function, and ℓ is fiber length. The expected value is

$$L_{exp} = e^{\mu+(s^2/2)} \tag{9}$$

with a standard deviation of

$$S_d = [e^{2\mu+s^2}(e^{s^2}-1)]^{1/2} \tag{10}$$

The mean fiber length, $\overline{L}$, and the most probable length (mode length), L_{mod}, are

$$\overline{L} = L_{exp} = e^{\mu+(s^2/2)} \tag{11}$$

$$L_{mod} = e^{\mu-s^2} \tag{12}$$

Rearranging equations (11) and (12), parameters s and μ can be expressed as

$$S = \sqrt{\frac{2}{3}(\ell n\overline{L}-\ell nL_{mod})} \tag{13}$$

$$\mu = \frac{2\ell n\overline{L}+\ell nL_{mod}}{3} \tag{14}$$

Therefore, given $\overline{L}$, s and μ depends on L_{mod}. Figure 3 shows the density curves of log-normal distribution. As L_{mod} is far less than $\overline{L}$, the distribution is skewed and highly unsymmetric. As L_{mod} approaches to $\overline{L}$, the curve converges to $\overline{L}$ and the symmetry increases. Simulated Figure 1 data by log-normal distribution, the corresponding shape parameters are also shown in Table 2 while $\overline{L}$ and L_{mod} (midpoint of maximum fiber fraction) are obtained directly from Figure 1.

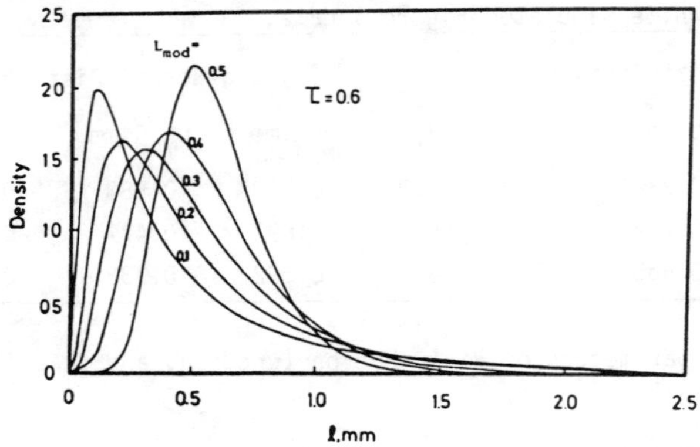

Fig. (3) - Log-normal density curves at various shape parameters

ELASTIC MODULUS OF UNIDIRECTIONAL COMPOSITE

The elastic modulus of unidirectional short fiber composite can be predicted by Halpin-Tsai equation [1,2] which is shown below:

$$\frac{M}{M_m} = \frac{1+\zeta\eta V_f}{1-\eta V_f} \tag{15}$$

where

$$\eta = (\frac{M_f}{M_m} - 1)/(\frac{M_f}{M_m} + \zeta)$$

$$\zeta_{E_1} = 2(\frac{\ell}{d})$$

$$\zeta_{E_2} = 2$$

$$\zeta_{G_{12}} = 1$$

and M_f, M_m, M are elastic moduli of fiber, matrix, and composite respectively, V_f, V_m are volume fraction of fiber and matrix respectively, and ℓ, d are length and diameter of fiber respectively.

In Halpin-Tsai equation, fibers are assumed to have uniform length. Since the elastic modulus is not linear function of fiber length, the predicted elastic modulus using mean fiber length is not the same as the mean elastic modulus. For example, the true mean value of Young's modulus is $\overline{E_1(L)}$ which is defined as

$$\overline{E_1(L)} = \frac{\int_0^\infty E_1(\ell)f(\ell)d\ell}{\int_0^\infty f(\ell)d\ell} \tag{16}$$

where $f(\ell)$ is length distribution density function and $E_1(\ell)$ is the Young's modulus at length ℓ calculated from equation (15), and the Young's modulus with the mean fiber length is $E_1(\bar{L})$. The difference between $E_1(\bar{L})$ and $\overline{E_1(L)}$ is

$$\varepsilon,\% = \frac{E_1(\bar{L})-\overline{E_1(L)}}{\overline{E_1(L)}} \times 100\% \tag{17}$$

In the previous section, Weibull distribution or log-normal distribution had been discussed as suitable probability functions to represent the fiber length distribution. Therefore, equation (2) or (7) can be substituted into equation (16) to calculate $\overline{E_1(L)}$. In Figure 4, A is the Weibull density curve where c and

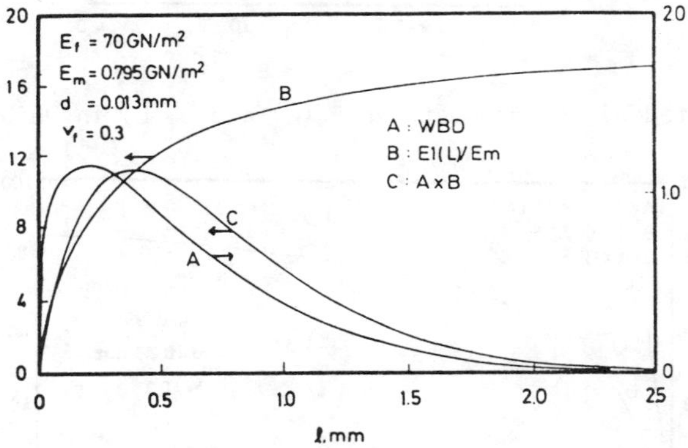

Fig. (4) - Relation between Young's modulus and fiber length; A: Weibull density; B: E_1/E_m in Halpin-Tsai equation; C: AxB

$\bar{L}$ are 1.25 and 0.6 mm, respectively, B is the ratio of E_1/E_m in Halpin-Tsai equation, and C is the product of A and B. Since E_1/E_m is not a linear function of fiber length, long fiber will contribute more reinforcement than short fiber does. Therefore, C curve moves slightly to the right side of A curve. $E_1(\bar{L})$ is the value of B curve at ℓ equal to $\bar{L}$ and $\overline{E_1(L)}$ is the area of C curve divided by the area of A curve. The difference between these two shown in Figure 4 is 13.5%. Figure 5 shows the difference of $E_1(\bar{L})$ and $\overline{E_1(L)}$ versus shape parameter C. Wide and unsymmetric length distribution gives big difference between $E_1(\bar{L})$ and $\overline{E_1(L)}$. Generally, the shape parameter c of fiber length distribution is around 1.5 to 2.5, while the difference between $E_1(\bar{L})$ and $\overline{E_1(L)}$ is around 10 to 5%. Similarly, using log-normal distribution to represent the fiber length distribution, result is shown in Figure 6. In Figure 6, A curve represents the density curve of log-normal dis-

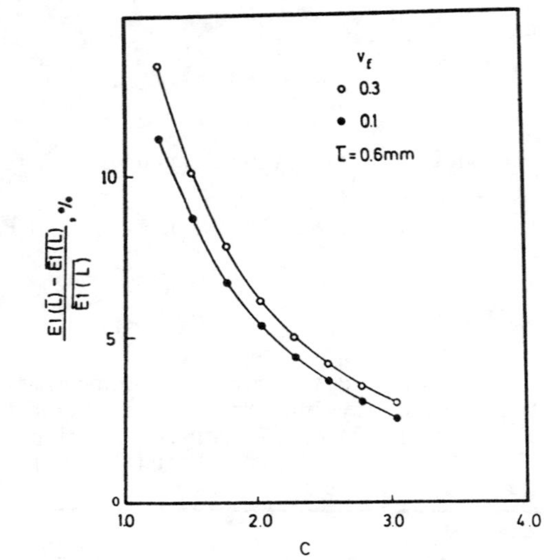

Fig. (5) - Estimated difference between $E_1(\bar{L})$ and $\overline{E_1(L)}$ in Weibull distribution

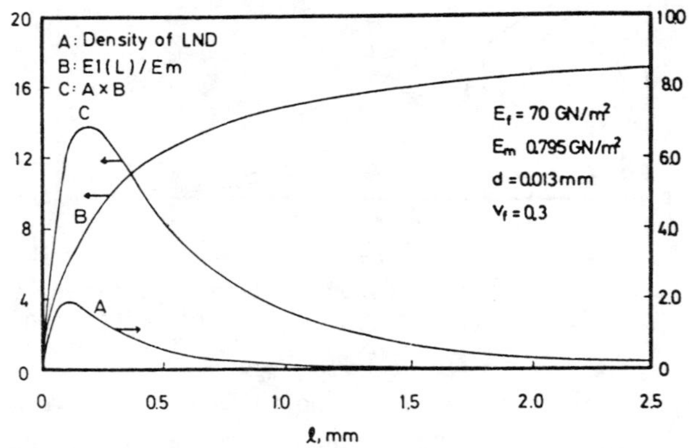

Fig. (6) - Relation between Young's modulus and fiber length; A: log-normal density; B: E_1/E_m in Halpin-Tsai equation; C: AxB

tribution ($\bar{L}$ = 0.6 mm, L_{mod} = 0.1 mm), B curve is the ratio of E_1/E_m in Halpin-Tsai equation, and C curve is the product of A and B. Since the log-normal density curve shown is highly unsymmetric, the difference between $E_1(\bar{L})$ and $\overline{E_1(L)}$ is up to 22.7%. Figure 7 shows the relation of the difference of $E_1(\bar{L})$ and $\overline{E_1(L)}$ versus L_{mod} where the mean fiber length is fixed at 0.6 mm. As L_{mod} approaching $\bar{L}$ gives narrow and symmetry distribution, so less difference between $E_1(\bar{L})$ and $\overline{E_1(L)}$ is observed.

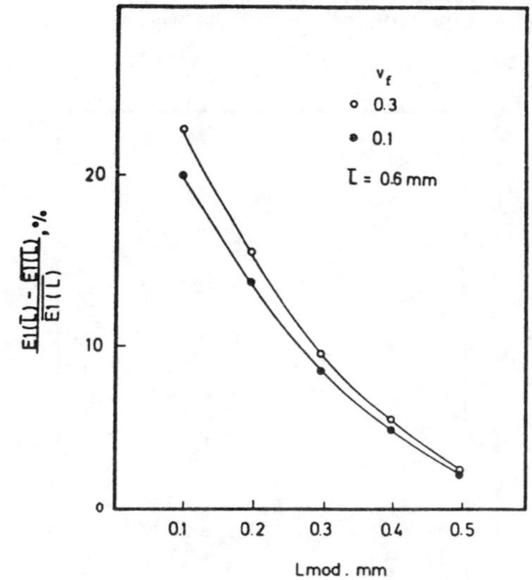

Fig. (7) - Estimated difference between $E_1(\overline{L})$ and $\overline{E_1(L)}$ in log-normal distribution

In some cases, $\overline{L}_w/\overline{L}_n$ value is used as a guide of distribution where $\overline{L}_n$ and $\overline{L}_w$ are number average length and weight average length, respectively. The number average length $\overline{L}_n$ is equal to the expected value, L_{exp}, as shown in equations (6) and (11). The weight average length is defined as

$$\overline{L}_w = \int_0^\infty x^2 f(x)dx/\int_0^\infty xf(x)dx \tag{18}$$

and

$$\overline{L}_n \cdot \overline{L}_w = \int_0^\infty x^2 f(x)dx \tag{19}$$

The standard deviation can be rewritten as

$$S_d^2 = \int_0^\infty x^2 f(x)dx - (\int_0^\infty xf(x)dx)^2 = \overline{L}_n \cdot \overline{L}_w - \overline{L}_n^2 = \overline{L}_n^2 (\frac{L_w}{L_n} - 1) \tag{20}$$

Further arrangement of equation (20) gives

$$\frac{\overline{L}_w}{\overline{L}_n} = \frac{S_d^2}{L_{exp}} + 1 \tag{21}$$

Since expected value and standard deviation are function of shape parameters, therefore $\overline{L}_w/\overline{L}_n$ is also a function of shape parameters. The difference between $E_1(\overline{L})$ and $\overline{E_1(L)}$ is plotted against distribution index $\overline{L}_w/\overline{L}_n$ as shown in Figure 8.

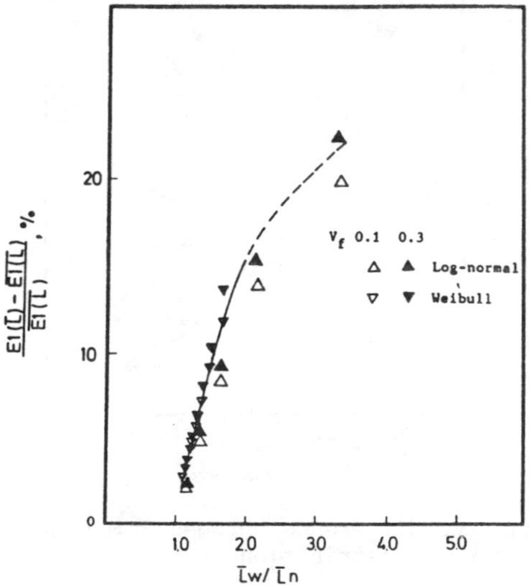

Fig. (8) - Relation of $E_1(\overline{L})$-$\overline{E_1(L)}$ and $\overline{L}_w/\overline{L}_n$

The difference between $E_1(\overline{L})$ and $\overline{E_1(L)}$ is found to be insensitive to the fiber volume content as shown in Figure 8 as well as independent of the type of distribution used (Weibull or log-normal). Therefore, we can use one master curve to represent the relationship between the difference of $E_1(\overline{L})$ and $\overline{E_1(L)}$ and $\overline{L}_w/\overline{L}_n$ of FRTP. A sample $\overline{L}_w/\overline{L}_n$ is easy to obtain. If the fiber length distribution of FRTP at various processing conditions can be represented properly such as $\overline{L}_w/\overline{L}_n$, then we can predict the Young's modulus of unidirectional composite quickly and accurately using the results shown in Figure 8.

CONCLUSIONS

Fiber breakage during extrusion and injection molding processes results a length distribution of fibers within the thermoplastic. Statistical distribution function such as Weibull or log-normal can be used to describe this kind of length distribution. Since the elastic modulus of composite is not a linear function of fiber length, a uniform fiber length assumption might bring more prediction error of elastic modulus of composite than expected. In this paper, we compare the difference between $E_1(\overline{L})$ and $\overline{E_1(L)}$ of unidirectional short-fiber reinforced thermoplastic where the fiber is assumed to have only one type of length distribution. In actual FRTP products, the length distribution may vary through thickness direction. In the complex case, using bulk length distribution probably is the simple way to deal with the problem.

REFERENCES

[1] Halpin, J. C. and Tsai, S. W., "Effects of environmental factors on composite materials", AFML-TR67-423, 1967.

[2] Halpin, J. C., J. Composite Materials, 3, p. 732, 1967.

[3] Lewis, T. B. and Nielsen, L. E., J. Appl. Polym. Sci., 14, p. 1447, 1970.

[4] Chamis, C. C. and Sendeckyj, G. P., J. Composite Materials, 2, p. 332, 1968.

[5] Bernado, A., SPEJ, 26, 1976.

[6] Hull, D., An Introduction to Composite Materials, Cambridge University Press, 1981.

[7] Turkovich, R. and Erwin, L., Polym. Eng. Sci., 23, p. 13, 1983.

[8] Hegler, R. P. and Menning, G., Polym. Eng. Sci., 25, p. 7, 1985.

[9] Pennington, W., Ph.D. Thesis, University of Liverpool, U.K., 1979.

[10] Weibull, W., J. Appl. Mech., 18, p. 2957, 1951.

[11] Coleman, B. D., J. Mech. and Phy. Solids, 7, p. 60, 1958.

CHARACTERIZATION OF MODE I DELAMINATION OF COMPOSITE LAMINATES AT ELEVATED TEMPERATURES

S.-C. Yen and M. El-Zein

Department of Civil Engineering and Mechanics, Southern Illinois University
Carbondale, Illinois 62901 USA

ABSTRACT

The onset of Mode I delamination crack growth of a composite laminate at elevated temperatures was investigated. The composite laminate used in this study contains 24 layers of a Graphite/Epoxy composite, forming a $[0/90/90/0]_{3s}$ stacking sequence. A precrack was embedded in the middle plane of each specimen. When the specimen was loaded, it behaved as if it were two double cantilever beams (DCB). Therefore, a mechanics of material type of beam analysis was used to describe the load-deflection response of the composite specimens. To characterize crack growth onset, the concept of a critical strain energy release rate was applied. An Instron testing machine was used to perform constant cross head rate testing of composite specimens. It was found that the critical strain energy release rate increases with an increase in temperature.

INTRODUCTION

It is known that the ultimate failure of fiber reinforced composite laminates is preceded by a sequence of microstructural events, such as microvoid growth, transverse matrix cracking, fiber-matrix debonding, interlaminar cracking, edge delamination, and fiber fracture, which are all loosely termed damage [1]. It has been found that, of all the damage states mentioned above, delamination is the most life-limiting mechanism in composite laminates [2,3]. It is therefore necessary to investigate the mechanics of delamination growth as part of an overall effort to develop a method for estimating the durability and damage tolerances of composite materials and structures.

Delamination in composite laminates is defined as the separation of surfaces between adjacent layers. The initiation of the delamination zone, in composite materials, may result from local incomplete curing, machining, impact damage, etc. As a result, when a delaminated composite laminate is subjected to an applied load, the delamination crack may expand. Such delamination crack growth is usually attributed to out of plane (plane of laminate) stress components near the vicinity of the crack tip. In particular, delamination crack growth is often caused by an interlaminar normal stress (Mode I). Hence, the objectives of this paper are to study the characteristics of Mode I delamination crack growth in composite laminates.

Currently, fracture resistance type tests are being applied to investigate the characteristics of delamination crack growth. For example, a double cantilever beam (DCB) type of specimen has been utilized to determine the critical strain energy release rate of Mode I delamination [4,5]. This is due to the fact that, for displacement controlled loading, a DCB specimen gives rise to stable crack growth, which makes it well suited for energy release rate measurements.

Much work has been done in the past on Mode I delamination at room temperature [4-8]. However, it is known that polymer based composites usually exhibit temperature dependent material properties. Hence, it is necessary to study the effect of temperature on delamination characteristics.

ANALYTICAL PROCEDURE

The theoretical background of this study is primarily procured from previous literature [4-8]. The basic mechanics of materials and linear elastic fracture mechanics approaches were combined to analyze the onset of delamination crack growth of composite laminates. In particular, composite specimens with precracks (Figure 1), referred to as double cantilever beam (DCB) specimens, were used to measure Mode I delamination characteristics. The parameter used to describe the onset of delamination crack growth is the critical strain energy release rate [8].

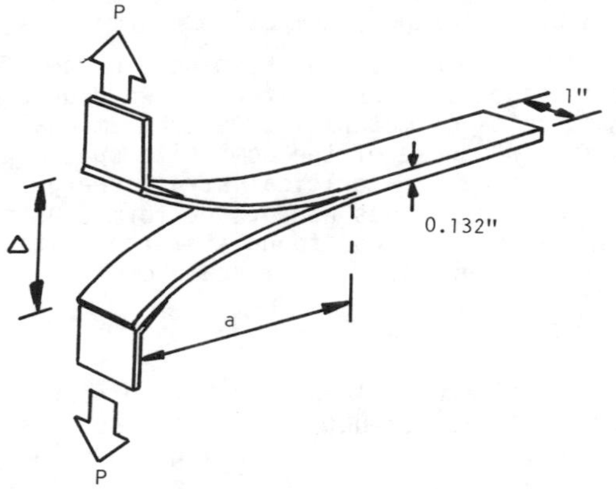

Fig. (1) - The double cantilever beam specimen

As is shown in Figure 1, a DCB specimen, when subjected to a tensile load, behaves as if it were two cantilever beams. The load-deflection relationship, which can be derived from the mechanics of materials approach, is:

$$\Delta = \frac{2Pa^3}{3EI} \tag{1}$$

where P = applied load; E = bending modulus of half of the composite laminate; I = moment of inertia of half of the composite laminate; a = crack length and Δ = crack opening displacement between the two loading points. The strain energy release rate (G) associated with crack growth in a DCB specimen can also be calculated, using:

$$G = \frac{\partial U}{\partial a} = \frac{P^2 a^2}{EI} \tag{2}$$

where U = strain energy of the double cantilever beam. It has been found [8] that when the onset of delamination occurs, the strain energy release rate reaches a critical value. This critical value is a constant and is defined as the critical strain energy release rate, G_{IC}, which can be calculated, using:

$$G_{IC} = \frac{P^2_{cr} a^2}{EI} = \frac{3P_{cr}\Delta_{cr}}{2a} \tag{3}$$

where P_{cr} = load that causes the onset of delamination crack growth and Δ_{cr} = corresponding crack opening displacement (referred to as the critical crack opening displacement). It is clear that based on the approach described above, only critical load and critical crack opening displacement need to be determined experimentally in order to characterize the onset of delamination crack growth.

MANUFACTURE OF COMPOSITE MATERIAL

One of the goals of this research was to establish our capability of manufacturing composite materials and specimens in house. Thus, the cost of specimens can be greatly reduced. On the other hand, the manufacturing history of a particular testing sample can be traced when certain experimental findings need to be explained. Two systems, the pressure vessel system and the hot press system, have been developed to manufacture composite materials. A hot press system was used to fabricate testing specimens for this research.

The hot press system (Figure 2), used in our laboratory, is a small loading frame, consisting of two flat aluminum plates and a manually driven hydraulic jack. The aluminum plates contain heating elements capable of applying high temperatures (up to 400°F). The hydraulic jack, which is attached to the lower aluminum plate, is capable of providing a maximum load of 40,000 lb. An assembly (Figure 3), containing the composite material, a picture frame, nylon and polyester fabrics, etc., is placed between the two flat aluminum plates. The nylon and polyester fabrics are placed on the top and bottom of the composite material. The purpose of the picture frame and the nylon and the polyester fabrics is to restrain resin flow during curing of the composite. Thus, fiber orientation of the composite material can be controlled. After the assembly is placed between the aluminum plates, the lower plate is raised using the hydraulic jack, thus applying pressure. Meanwhile, the heating elements are turned on, providing the heat needed for curing the polymer resin of the composite.

Precise control of temperature and pressure is required in manufacturing composite materials. In particular, temperature must be increased and/or decreased at specific rates and must be maintained at a recommended level for a certain duration. On the other hand, the pressure must also be maintained at a given level. These parameters usually vary for different manufacturing systems. Therefore, a trial and error method has been used to determine the procedures for manufacturing composite material using our system. Typical curing procedures used in our laboratory for making Graphite/Epoxy (T300/934) composites are shown in Figure 4.

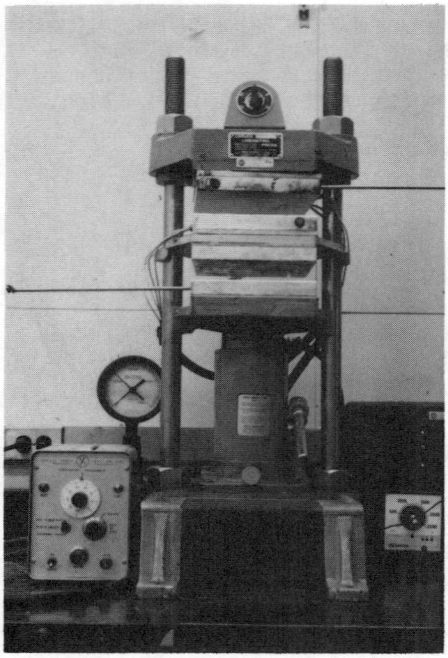

Fig. (2) - The hot press system

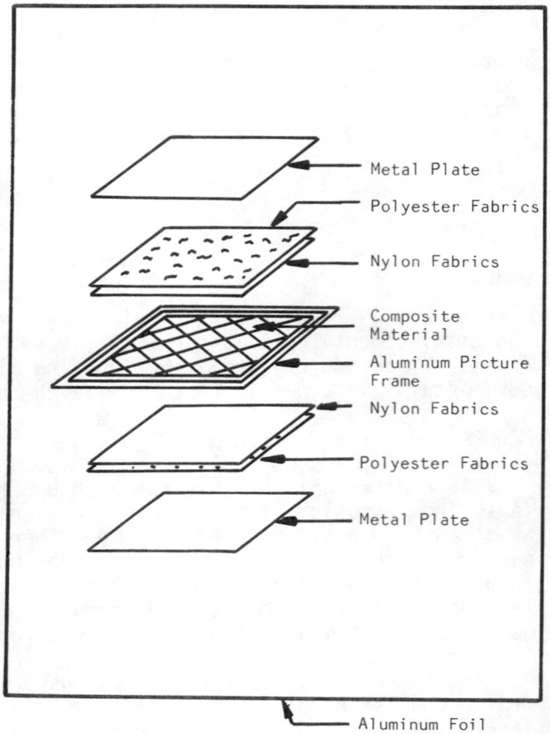

Fig. (3) - Typical arrangement for making composite material

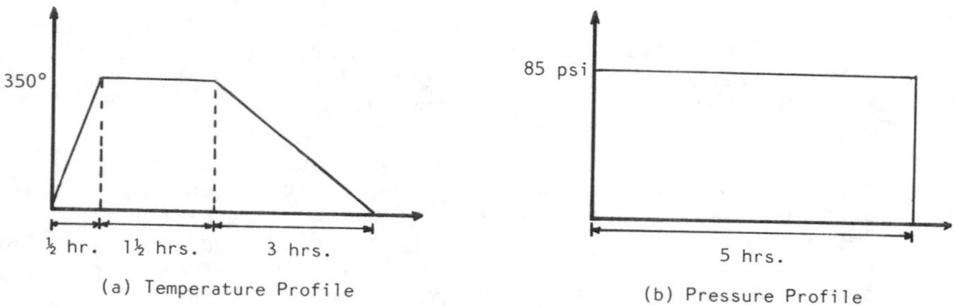

Fig. (4) - Typical manufacturing process for Graphite/Epoxy composite material

MATERIAL AND SPECIMEN

The composite material used in this study was Graphite/Epoxy (T300/934) and was obtained from the vendor (Fiberite Corporation) in a prepeg tape form. To manufacture composite laminates, pieces of prepeg tape are cut and stacked according to a desired orientation sequence. The stacking sequence used in this study was $[0/90/90/0]_{3S}$ (24 plies). To fabricate DCB specimens, a precrack is introduced, by placing a folded plastic film strip, along one edge of the middle plane of the composite laminate. The composite is then placed in the hot press system for curing. After curing, the composite laminate is cut into specimens that are 1 in. wide, 8.5 in. long and have a 2 in. precrack. Galvanized aluminum L-shaped end tabs are bonded to each specimen (as shown in Figure 1) with a high temperature adhesive. Thus, one end of the L-shaped tab is attached to the testing specimen while the other end of the tab is attached to the mechanical grip of a material testing machine.

EXPERIMENTS AND RESULTS

Tensile tests of DCB specimens were conducted using an Instron machine. The machine has a load cell which meets the small load testing requirements (less than 50 lb.) of DCB specimens. In order to test the DCB specimens at elevated temperatures, a furnace was built and attached to the Instron machine (Figure 5). Holes

Fig. (5) - The Instron testing machine

were drilled in the top and bottom of the furnace for the loading rods to go through. The furnace is capable of providing heat up to 400°F. The variation of temperature was about 2°F. To anticipate the possibility of thermal effects on its reading, the load cell was calibrated at elevated temperatures. It was found that temperature has no apparent effect on the load cell's readings.

Specimens were tested at six temperature levels: 73°F, 118°F, 205°F, 255°F, 284°F, and 309°F. Before being tested, a specimen was soaked at the desired temperature for at least 2 hours. One to three specimens were tested at each of the selected temperature levels. Specimens tested at the same temperature level were selected from different panels of composite laminates manufactured at different times. Thus, whether the experimental results from specimen to specimen were reproducible was checked.

Each DCB specimen was subjected to a repeated loading-unloading procedure. Load was applied using a constant cross head rate of 0.2 in./min. Figure 6 shows a DCB specimen under loading. The specimen was loaded in tension until delamina-

Fig. (6) - The DCB specimen under loading

tion crack growth began. The delamination crack was then allowed to advance 0.3 to 0.5 in. before unloading was begun. Once the load returned to zero, the specimen was untightened and then retightened to eliminate possible residual stress in the specimen. After retightening, the specimen was slightly shaken, to check that there was no preload being applied. The Instron machine was then re-zeroed and balanced. This procedure was repeated for every loading-unloading cycle. Typically, each specimen was subjected to more than one cycle. As a result, several data sets were gathered for each specimen.

In each test, several types of data were taken: the load versus crack opening displacement data, the load and the crack opening displacement when delamination crack growth began, and the new delamination crack length before unloading was begun. Load data was recorded from the load cell of the Instron machine. The crack opening displacement was determined using a ruler. The edges of the

DCB specimen were painted with correction fluid and then marked with a pencil at 0.1 in. intervals. Darker pencil tick marks were used at each 1 in. length. As a result, the crack length was measured by taking readings from the ruler-like specimen edges with a magnifying lens.

Figure 7 shows two typical load versus crack opening displacement curves for data obtained from the tensile tests of Graphite/Epoxy DCB specimens. It was

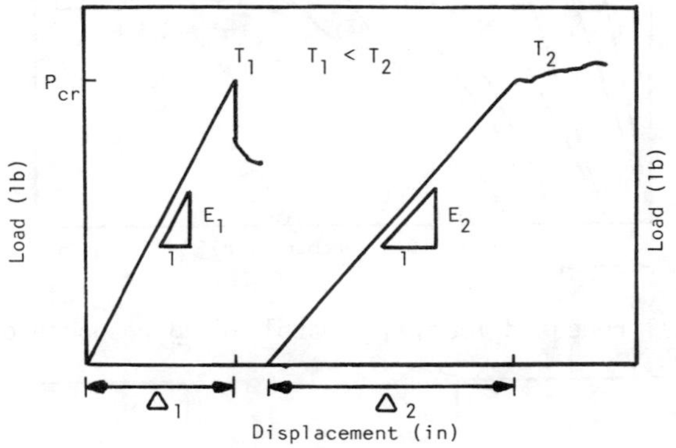

Fig. (7) - Typical load-displacement curve for DCB specimen at different temperature

found that the load-crack opening displacement relationships are linear up to the onset of delamination crack growth. It was also found that at room temperature conditions, delamination crack growth starts abruptly. Hence, a sharp drop in the applied load occurs (Figure 7). On the other hand, at elevated temperature conditions, crack growth appears to be a slow and stable process once delamination has occurred. It was also found that the crack opening displacements appear to be larger at higher temperatures, for the same amount of applied load. Figure 8 shows the typical repeated loading-unloading response of a DCB specimen. It is clear that the compliance of the specimen decreases with an increase in crack length.

Figures 9 through 11 show the relationship between critical load and crack length for three different temperature levels (73°F, 205°F, and 284°F). In each figure, the number of different symbols represent the number of specimens tested for a particular temperature level. The data presented (P_{cr} versus a) was collected from specimens with crack lengths ranging from 2.0 to 6.0 in. The data recorded for specimens with crack lengths less than 2.0 in. was affected by rotation of the L-shaped tabs. Hence, it is not presented in this paper. The solid line in each figure represents the least square straight line fit of the experimental data. For most of the data shown, deviation from the least square line is approximately 7%. Critical loads for specimens with equal crack lengths, but tested at different temperatures, are given in Figure 12. It was found that the critical load is independent of temperature.

Figures 13 through 15 show the relationship between critical crack opening displacement (Δ_{cr}) and crack length (a) for three selected temperature levels.

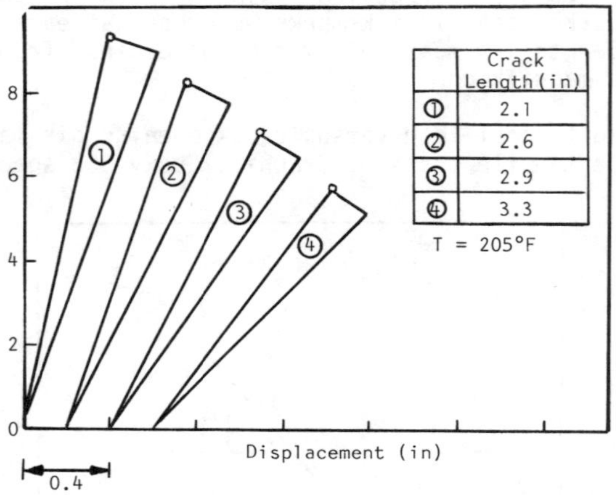

Fig. (8) - Typical repeated loading and unloading response of a DCB specimen

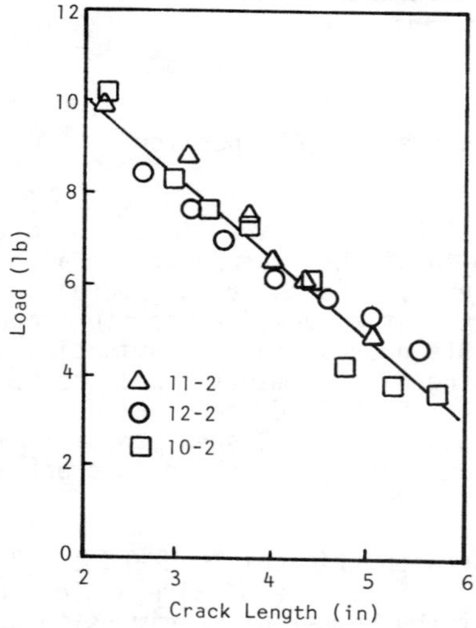

Fig. (9) - The relations between the critical load and the crack length at 73°F

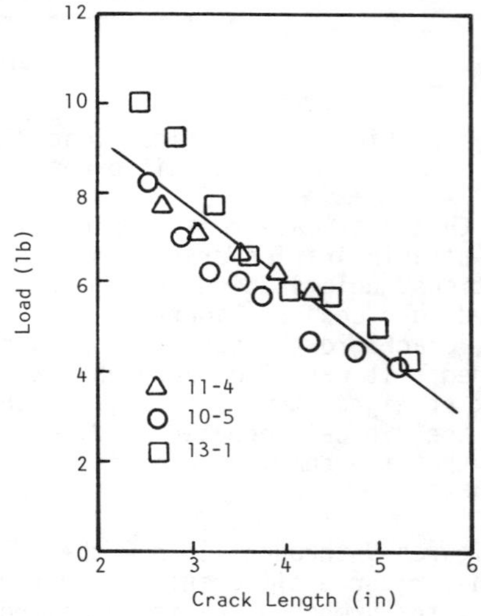

Fig. (10) - The relations between the critical load and the crack length at 205°F

Again, in each figure, the number of different symbols represent the number of different specimens tested. A least line fit was also applied to data collected for each temperature level. The average deviation of experimental data from the best fit line is approximately 5%. The relationship between critical crack opening displacement (Δ_{cr}) and temperature for specimens with equal crack lengths is shown in Figure 16. It appears that the critical crack opening displacements increase with an increase in temperature. By omitting the bad data taken at 205°F, the critical crack opening displacement versus temperature data can be approximated by a smooth curve.

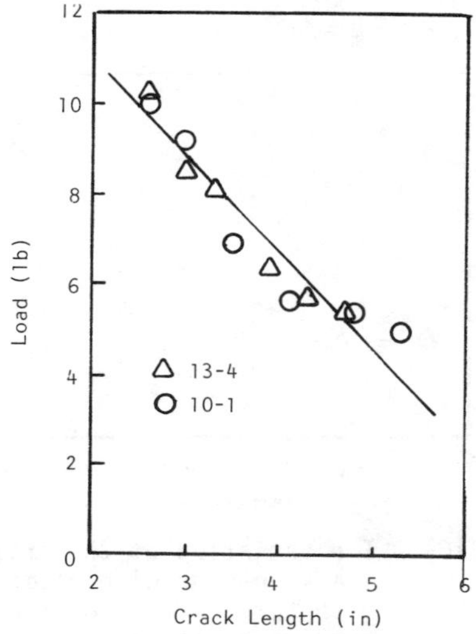

Fig. (11) - The relations between the critical load and the crack length at 284°F

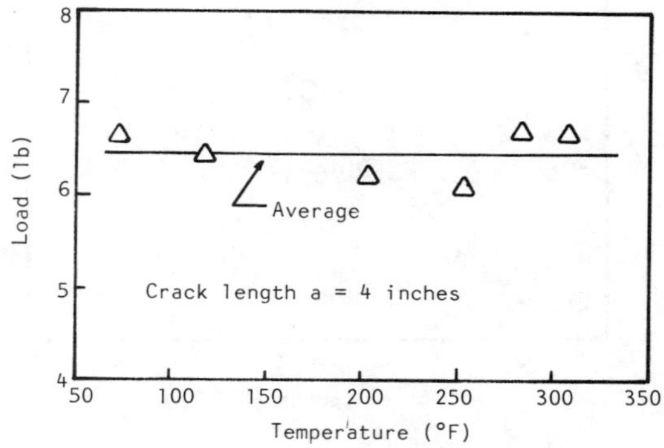

Fig. (12) - The variation of the critical load with temperature for the specimens with the same initial crack length

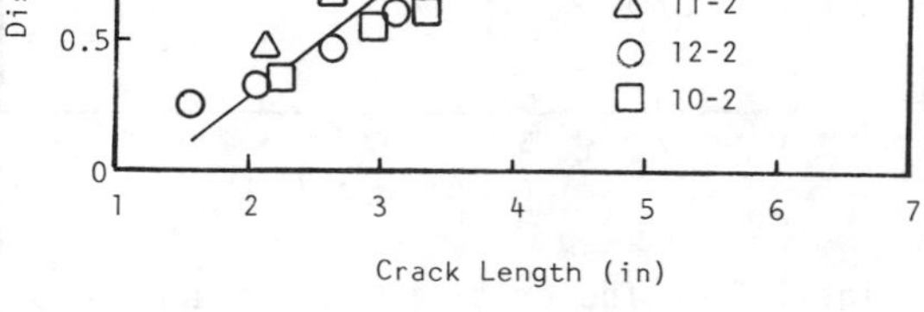

Fig. (13) - The relation between the critical displacement and the initial crack length at 73°F

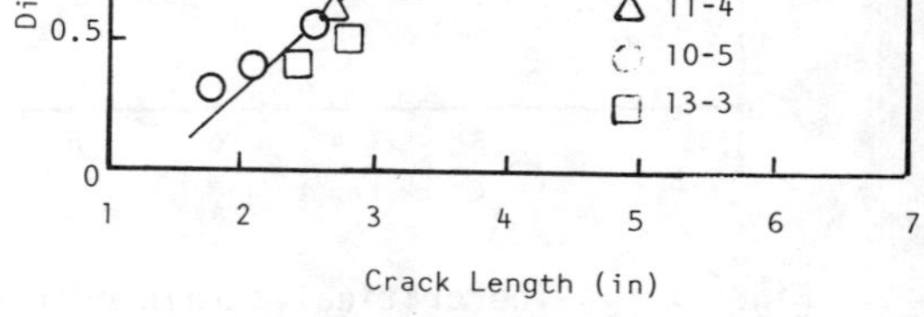

Fig. (14) - The relation between the critical displacement and the initial crack length at 205°F

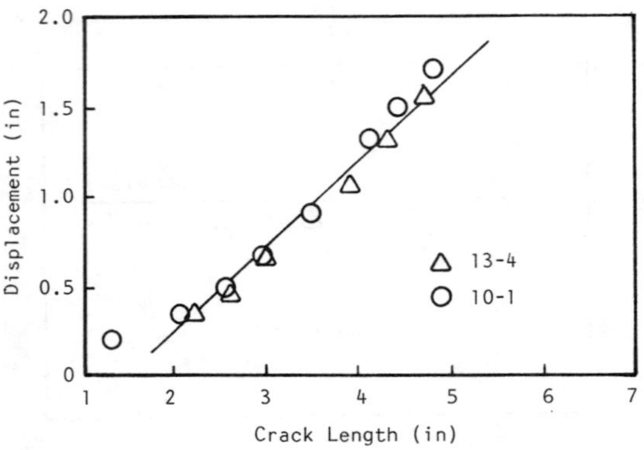

Fig. (15) - The relation between the critical displacement and the initial crack length at 284°F

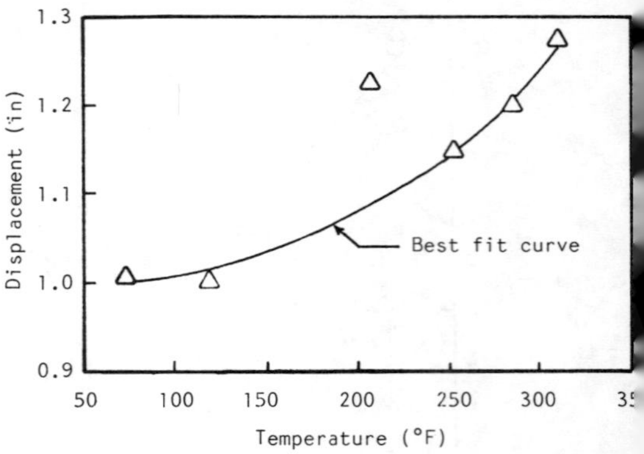

Fig. (16) - The variation of the critical crack opening displacement with temperature for the specimens with the same initial crack length

Critical strain energy release rates, G_{IC}, were calculated using equation (3). Values of G_{IC} for different crack lengths, at each temperature level were plotted in one graph (Figures 17 through 19). It was found that values of the critical strain energy release rate for a given temperature level is practically

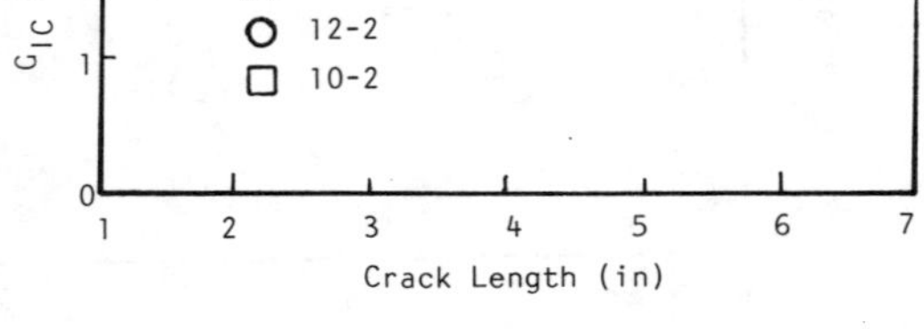

Fig. (17) - The critical strain energy release rate data at 73°F

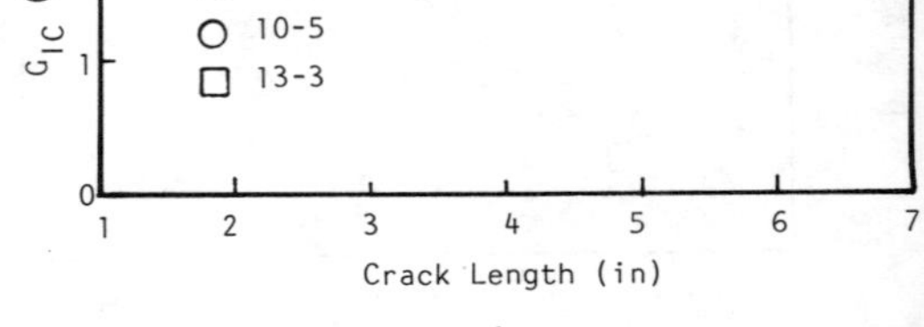

Fig. (18) - The critical strain energy release rate data at 205°F

constant. The average value of G_{IC} for each temperature level was calculated using all data presented in Figures 17 through 19. The average value of G_{IC} for each temperature level was then plotted on one graph (Figure 20). It was found that val-

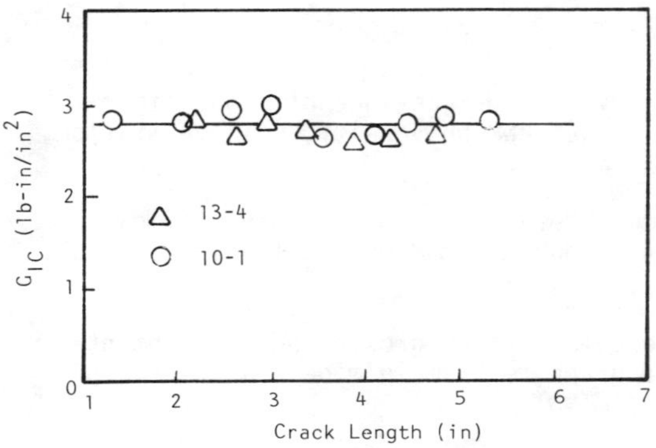

Fig. (19) - The critical strain energy release rate data at 284°F

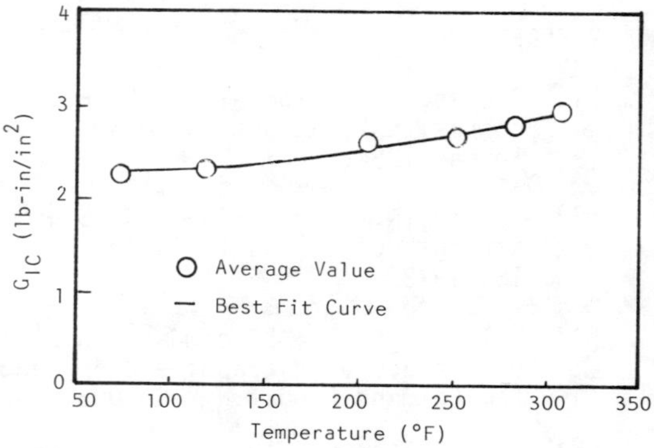

Fig. (20) - Variation of the critical strain energy release rate with temperature

ues of the critical strain energy release rate increase with an increase in temperature. Also, the variation of G_{IC} with temperature can be approximated by a smooth curve. In fact, these findings can be explained by the data presented in Figure 16, where the values of the critical crack opening displacement were larger at higher temperatures, for equal crack lengths. Since the critical strain energy release rate are directly proportional to the critical crack opening displacement (equation (3)), therefore a higher value for the critical strain energy release rate at higher temperature can be concluded. Another point is that the critical strain energy release rate is inversely proportional to the bending modulus of a composite laminate. An increase of G_{IC} with temperature reveals to a decrease in the bending modulus with temperature. This is also coherent to the type of material behavior that has been reported in the literature.

CONCLUSIONS

The effect of temperature on the onset of delamination crack growth in a composite laminate was experimentally studied. Basic mechanics of materials and linear elastic fracture mechanics approaches were used to analyze experimental data. Research findings can be summarized as follows:

1. The composite materials manufactured in our laboratory are of good quality, because repeatable experimental results were obtained from specimens made at different times.

2. The critical load at the onset of delamination crack growth for a given crack length is insensitive to temperature.

3. The critical crack opening displacement at the onset of delamination crack growth, for a given crack length, increases with an increase in temperature.

4. The critical strain energy release rate is a constant for a given temperature level and increases with an increase in temperature.

REFERENCES

[1] Groves, S. E. and Allen, D. H., "A survey of damage in continuous fiber composites", Report No. MM5023-84-6, Texas A&M University, College Station, Texas, 1984.

[2] Pagano, N. J. and Pipes, R. B., "Some observations on the inter-laminar strength of composite laminates", Int. Journal of Mechanical Sciences, Vol. 15, 1973.

[3] Wilkins, D. J., "A comparison of the delamination growth and environmental resistance of a graphite-epoxy and a graphite-bismaleimide", Naval Air System Command Report NAV-GD-0037, August 1981.

[4] Browing, C. E., Adams, F. and Whitney, J. M., "A four-point shear test for graphite/epoxy composites", Composite Materials: Quality Assurance Processing, ASTM STP 797, 1982.

[5] O'Brien, T. K., "Characterization of delamination onset and growth in a composite laminate", Damage in Composite Materials: Basic Mechanics, Accumulation, Tolerance, and Characterization, ASTM STP 775, 1982.

[6] Wang, S. S., "Delamination of crack growth in unidirectional fiber-reinforced composite under static and cyclic loading", Composite Materials: Testing and Design, (5th Conference), ASTM STP 674, 1982.

[7] de Charentenary, F. X. and Benzeggagh, M., "Fracture mechanics of Mode I delamination in composite materials", Proceedings of ICCM3, Paris, August 1980.

[8] Wilkins, D. J., Eisenmann, J. R., Camin, R. A., Margolis, W. S. and Benson, R. A., "Characterizing delamination growth in graphite-epoxy", Damage in Composite Materials: Basic Mechanisms, Accumulation, Tolerance, and Characterization, ASTM STP 775, 1982.

THE EFFECT OF MOISTURE ABSORPTION AND PHYSICAL AGING ON THE PHYSICAL AND MECHANICAL PROPERTIES OF COMPOSITES

M.-F. Sheu, C.-S. Yeh, Y.-P. Bian and C.-L. Ong

Aero Material Department, MRDC, P.O. Box 8676-1-10
Taichung, Taiwan, Republic of China

ABSTRACT

The moisture absorption and physical aging of composite laminates at different temperatures for various periods of exposure can have very significant effects on the physical and mechanical properties of composites, such as the increase or decrease of Tg, the toughening or embrittlement of laminates. In this program, the material that has been studied is epoxy/graphite prepreg. Two experiments have been conducted.

1. Moisture absorption: Cured laminates are conditioned in salt spray chamber and tested to determine the Tg, and mechanical properties at different moisture levels.

2. Physical aging: Cured laminates are annealed or quenched at different temperatures (below Tg or at room temperature) or at some temperature for different periods of time.

The resultant specimens are tested thermally and mechanically to determine the effect of moisture content and physical aging on composites.

INTRODUCTION

Epoxy resin is a thermosetting material. After curing, the low molecular weight epoxy resin becomes highly cross-linking polymer; the density of the cross-linking of resin depends upon the kind and weight percent used hardener and the degree of reaction [1,2]. The molecular structure of a hardener affects the physical and chemical properties of cured epoxy resin. Generally, the aliphatic amine-type hardeners are more active and reacts with epoxy resin at room temperature. The amine cured epoxy resin have the properties of (1) more flexibility, (2) more free volume, (3) higher hydrophilicity. On the other hand, the amide-type hardeners are less reactive and react with resin at higher temperature. The carbonyl group structure of the amide-type hardener can reduce the hydrophilicity of resins. HY-1076E prepreg were prepared with N.N.N',N'-Tetraglycidyl-Methylenedianiline (TGMDA)-DDS resin system. DDS having aromatic structure and rigid property, reacts with TGMDA make the cured composite have high glass transition temperature.

Carbon fiber is a hydrophobic material. The moisture absorption of composites takes place from the cured epoxy resin and the voids in composite. The voids are formed from the solvent and water of prepreg during curing processes. Voids affect the mechanical properties of composite enormously. One percent void content decreases the mechanical properties of composite about 7 to 8 percent. The presence of voids are usually observed in using the hot-press to cure carbon fiber/epoxy resin prepreg. The method used to prepare the composite laminate is vacuum bagged and cured in autoclave and therefore, the void content can be neglected, so we can suppose that the moisture absorption results from epoxy resin.

The difference between thermoset resin and thermoplastic resin is that the cross-linking part of the cured resin is fixed. The only possibility is the side-chain or uncured groups [3]. This molecular behavior of thermoset resin is similar to that of crystalline and amorphous structure of thermoplastic resin. The crystalline lattice of thermoplastic resin breaks while heating, but rearrange to crystalline state while cooling. However, when the thermoset resin are heated above soften temperature, the cross-linking chains decompose. In addition, the physical aging effect of thermoset resin, i.e., heating below the soften temperature, can be described with Figure 1: the molecular structure of the cross-linking areas have no change and only the side chains rearrange to more order structure.

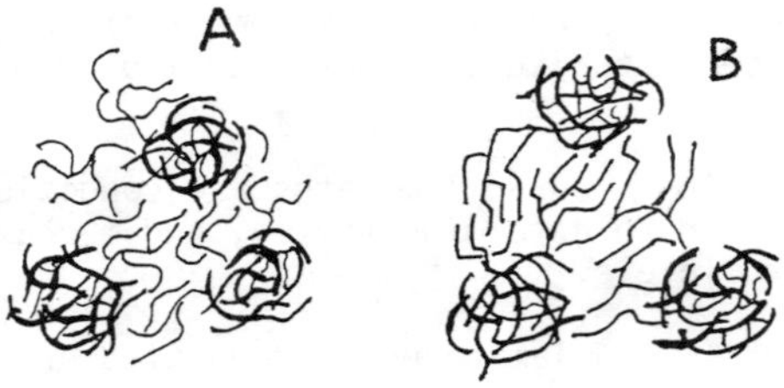

Fig. (1) - The molecular morphology of thermoset resin (A) before aging, (B) after aging

Of polymer material, the more crystalline their structure are, the more rigid or stiff they will be. On the other hand, the less crystalline polymer material will be more flexible or ductile [4]. In carbon fiber/epoxy resin prepreg, the carbon fiber is a high crystalline and oriented material. For the cured epoxy resin, its cross-linking part is regarded as the crystalline state. Comparing with thermoplastic, the thermosetting resin is more rigid and more stable in dimension. The range of molecular motion depends on the residual free volume of the composite at temperature above Tg. The studies of using physical aging to change the molecular morphology of polymer materials are greatly in thermoplastic field, but lacking in thermosetting. Until present, Mijovic et al [5-10] published some papers about the physical aging affecting the thermal properties of composite. In this experiment, three different temperatures (100°, 140°, 160°C) and various aging time (1, 3, 5, 7 hr.) are selected to study the physical aging effect on the mechanical properties of HY-1076E composite.

EXPERIMENT

1. Material and Cure Process

The materials of our experiment are HY-1076E prepreg. The lay-up operation and cure cycle of prepreg are shown as Figures 2 and 3.

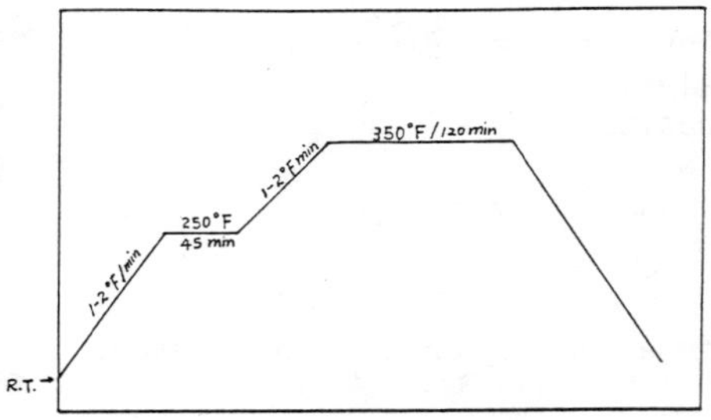

Fig. (2) - Cure cycle of HY-1076E prepreg

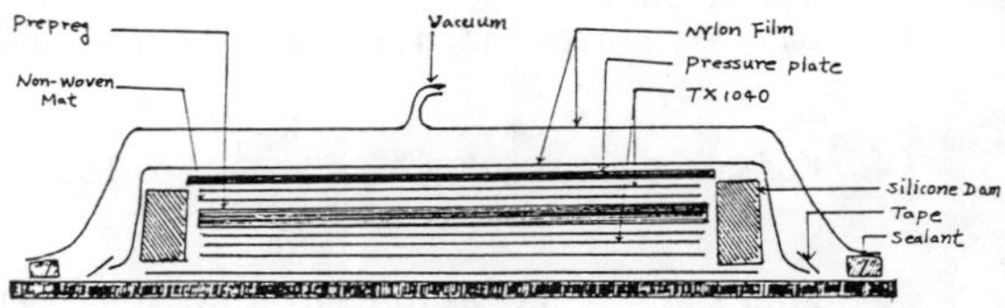

Fig. (3) - Autoclave molding method of HY-1076E prepreg

2. The Moisture Absorption Test

Two methods are used to test the moisture absorption of HY-1076E composite in our experiment.

(a) Salt spray chamber test

The equipment of the salt spray chamber and specimen holder are shown in Figure 4. The specimens are put into the chamber keeping at room temperature.

Fig. (4) - Apparatus of salt spray chamber test

At specific times, we take out the specimens and test the moisture content, interlaminal shear strength (ILSS) of specimens by moisture evolution analyzer (MEA) and Instron.

(b) Cooking test

Using the cooking method to test the moisture absorption of HY-1076E composite, the procedure is as follows:

1) Heat the distilled water to boiling in Heake NB-22 heater.
2) Put the specimens into the boiling water for some days.
3) Take out the specimens and use dry cloth to remove the surface water of specimens.
4) Put the specimens on a table for one hour at room temperature.
5) Weigh the specimens to 0.1 mg with analytic balance.
6) Calculate the moisture content of composite.

$$\text{Moisture content (\%)} = \frac{W_f - W_i}{W_i} \times 100\%$$

where W_i: the weight of dry specimen before cooking and W_f: the weight of moisture absorption specimen.

(c) Physical aging test

The procedure of physical aging of HY-1076E composite is as follows:

1) Put the specimens into oven and dry at 40°C for one day.
2) Take out the specimens and put into the desiccator for 30 min.
3) Aging dry specimens according to the following conditions:

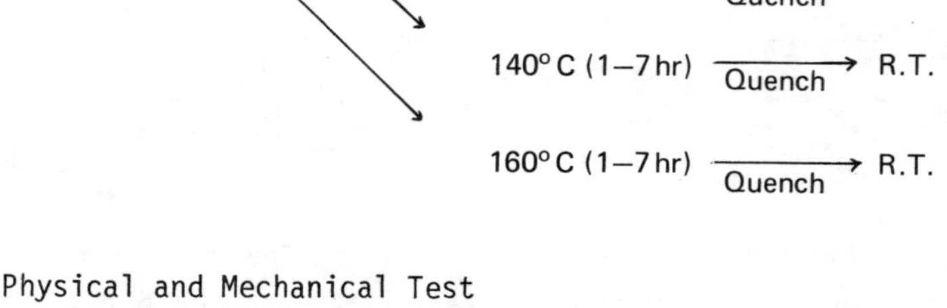

3. The Physical and Mechanical Test

a) Tg and the thermal property. The Tgs of HY-1076E composites were tested with Perkin-Elmer DSC-4 with scan rate 40 degree/min.

b) The mechanical property. All of the mechanical properties of HY-1076E composites are determined according to the ASTM method. The tests are carried out at room temperature and test machine is Instron model 1125.

c) Shear modulus. The shear modulus of HY-1076E composite is determined with RMS-model 605. The temperature range is from 30° to 300°C.

RESULTS AND DISCUSSION

1. The Moisture Content and the Interlaminar Shear Strength (ILSS)

HY-1076E composite are put in the salt spray chamber for a long time. The relationship between moisture content ILSS and spray time of composite is shown as Figure 5.

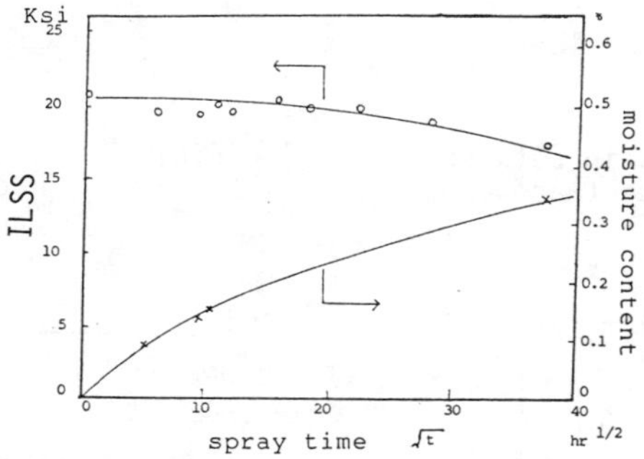

Fig. (5) - The relationship between moisture content, ILSS and spray time

The specimens are put for 500 hours; the values of moisture content and ILSS are 0.25% and 19.5 ksi. The decrease of ILSS is small. But for 1496 hours, the moisture content is 0.36% and ILSS decreases from 20.9 ksi to 17.56 ksi.

The void content is small and may be neglected as described in Introduction. Therefore, the moisture absorption of composite comes from the interaction of epoxy resin and water [11-14]. Because of the high cross-linking density of composite, the water molecule is difficult to permeate the interior of composite, even dipping in water for a long time. Water, in the interior of composite, acts as a plasticizer and causes the expansion of composite [15]. The expansion effect of water can degrade the bonding of the interlayers of composite. Therefore, the increase of the moisture content causes the decrease of ILSS of the composite. Figure 6 shows the relationship between cooking time and moisture content of HY-1076E composite. The Tg and the flexural strength, ILSS of the laminate are decreasing as moisture content is increasing, see Figures 7 to 10.

2. The Change of Tg by Physical Aging

The glass transition temperature of an epoxy resin increases with the increase of the degree of cure of the resin. However, as the curing has arrived at a high enough degree, the Tg keeps at a constant value even for a prolonged cure time. HY-1076E prepreg has been cured at 350°F for two hours in vacuum, then physical aged at 100°, 140° and 160°C for various periods of time. The Tgs of the specimens do not change as probed by DSC testing, Figure 11. The reasons can be explained as follows: (1) in the cure cycle, the HY-1076E composite has reached high degree of cure; (2) when the composite are physical aged below 160°C, the postcure reaction does not happen, but the local molecular motion of the ends of those molecules or side chains do occur.

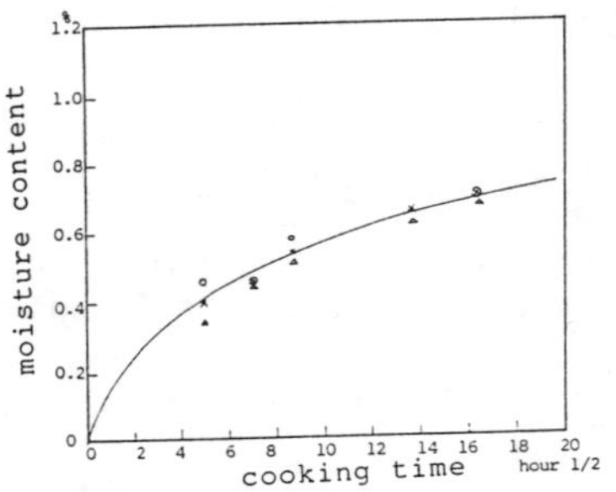

Fig. (6) - The relationship between moisture content and cooking time

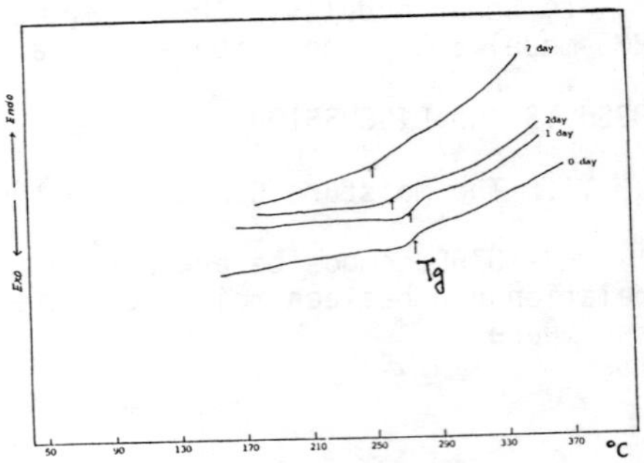

Fig. (7) - DSC curve of HY-1076E composite at various cooking time

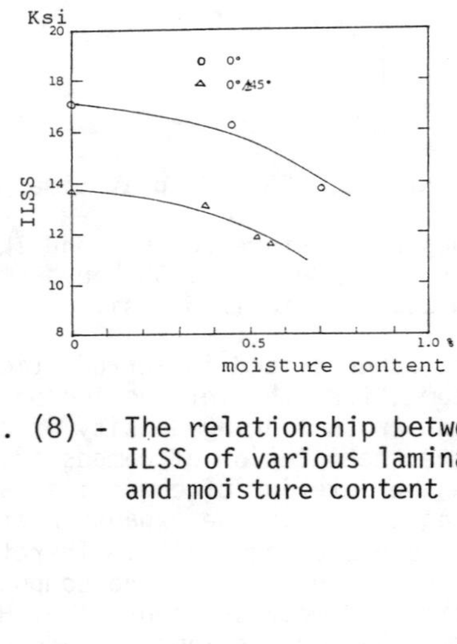

Fig. (8) - The relationship between ILSS of various laminates and moisture content

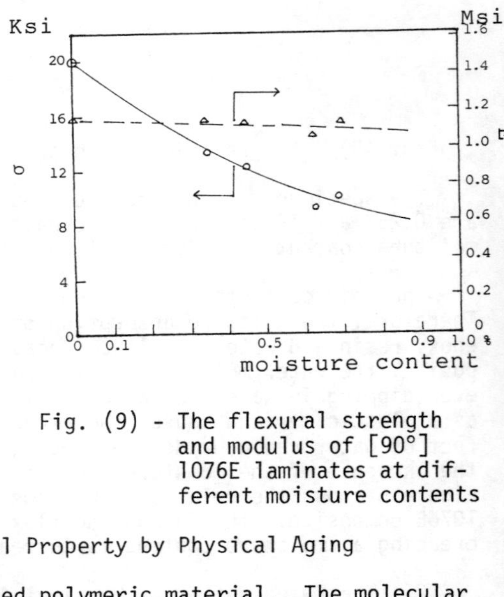

Fig. (9) - The flexural strength and modulus of [90°] 1076E laminates at different moisture contents

3. The Change of the Dynamic Mechanical Property by Physical Aging

HY-1076E composite is a high cross-linked polymeric material. The molecular structure is very rigid, so it has high Tg (257°C). Therefore, the epoxy molecules can move only at high temperature. At low temperature, the energy is only enough to do the segmental motion of the end group or side chain of resin. The aging temperatures used in this experiment is 160°C or below. For unaged composites, the temperature sweeps for the shear modulus with RMS-605 have been made and two damping peaks found, one at 130°C (α_1) and the other at 257°C (α_2) (see Figure 12).

After aging at 140°C for various periods of time, the peak at 130°C has disappeared. This phenomenon can be explained that the physical aging has changed the molecular morphology of epoxy resin. Since the aging at 140°C provides enough

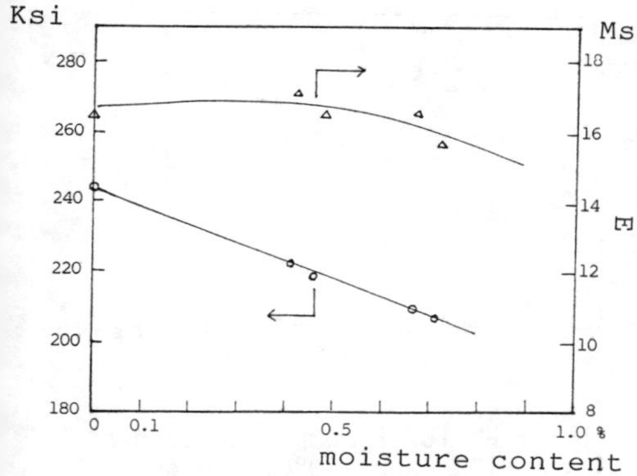

Fig. (10) - The relationship between moisture content and the flexural properties of [0°] laminate of HY-1076E composite

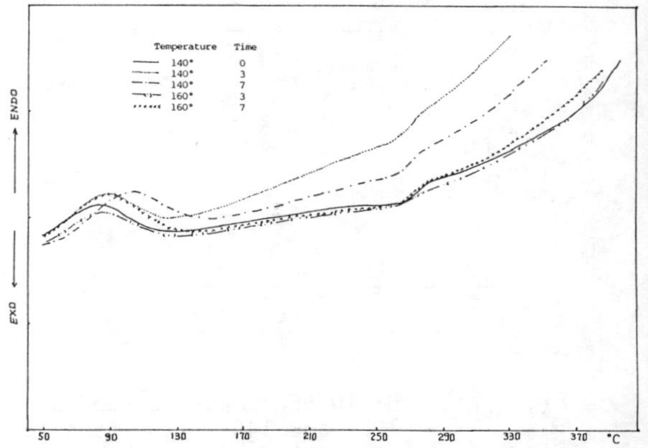

Fig. (11) - DSC curve of HY-1076E composite at different aging temperatures for different periods of time

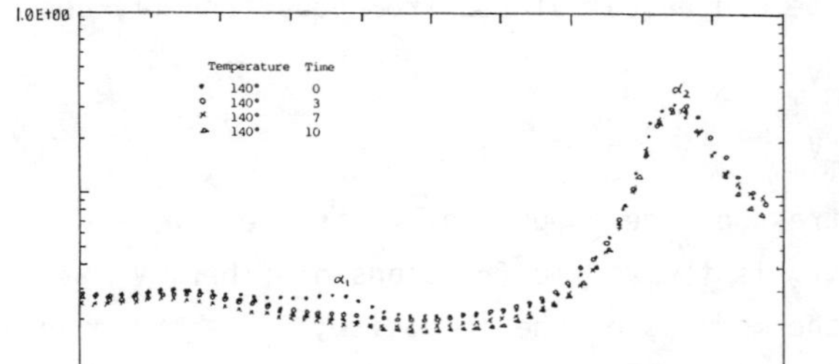

Fig. (12) - The damping curves of HY-1076E composite of 140°C aging

energy for end groups of molecule or side chain to rearrange and get rid of the residual stress of molecular chain, the peak at 130°C disappears.

4. The Aging Temperature/Time and Tensile Properties

During aging of polymer material, the uncured molecular chains take a suitable rearrangement. The energy of rearrangement comes from the aging temperature. Therefore, the higher the aging temperature is, the more active segmental motion and the more rearrangement is. In addition, this rearrangement is more complete for a long time and the composite gets more rigid. From Figures 13 and 14, we can see that the tensile strength and modulus are higher for those composites aged at

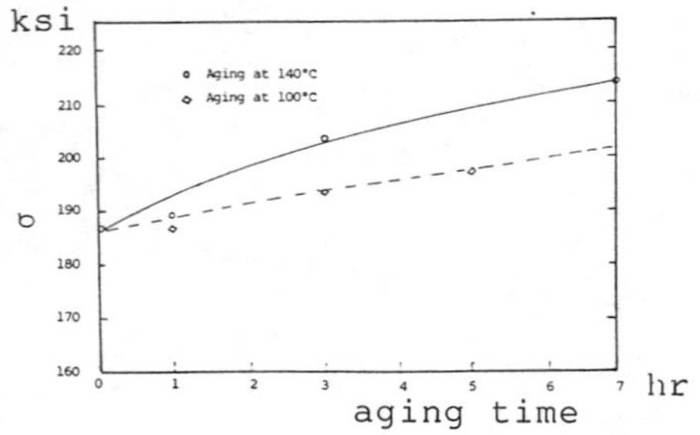

Fig. (13) - HY-1076E composite aged at 100° and 140°C, the tensile strength varies with aging time

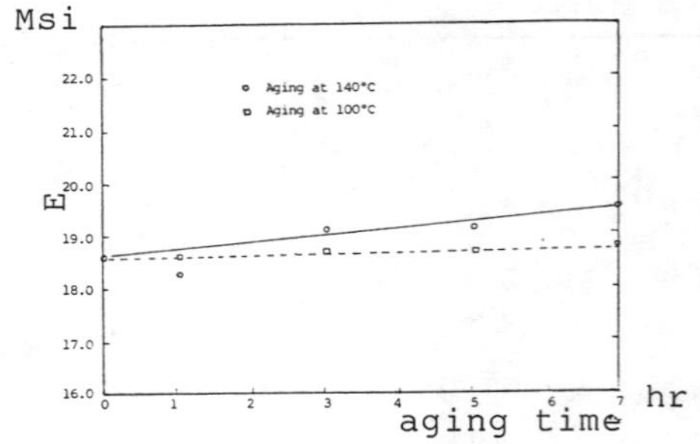

Fig. (14) - HY-1076E composite aged at 100° and 140°C, the curve of tensile modulus and aging time

higher temperature and longer time. But, the increment for the tensile strength is larger than that of the tensile modulus.

According to the mixing rules of composites, the tensile strength and tensile modulus of composite can be calculated from equations (1) and (2)

$$\sigma_c = \sigma_f V_f + \sigma_m V_m \qquad (1)$$

$$E_c = E_f V_f + E_m V_m \qquad (2)$$

where σ_c is the stress of the composite; σ_f is the stress of fiber; σ_m is the stress of matrix; V_f is the volume fractions of fiber; V_m is the volume fractions of matrix; E_c is the modulus of the composite; E_f is the modulus of fiber and E_m is the modulus of matrix.

The values of σ_f and E_f are constant, but σ_m and E_m will be different for different molecular structures. For the comparison of the casting epoxy resin and carbon fiber, the typical values of σ and E are listed in Table 1.

Table 1 - [2]

Sample	Tensile Strength (ksi)	Modulus (ksi)
Net epoxy resin	8-18.9	406-609
Carbon fiber	350-450	33000-49000

These values in Table 1 show that for the composite material, the effect of σ_m on σ_c is larger than the E_m on E_c. The increment of σ_c with the aging temperature at 140°C is larger than that at 100°C. But the change of the tensile modulus of the composite with aging at 100°C is very small.

5. The Physical Aging and the Flexural Properties and ILSS

Figures 15 and 16 show that the physical aging affect the flexural strength and modulus of HY-1076E composite. Although the increase of flexural strength is negligible as aged at 100°C, it is obvious at 140°C or higher. In addition, flexural modulus does not change at 140°C, but it increases at 160°C.

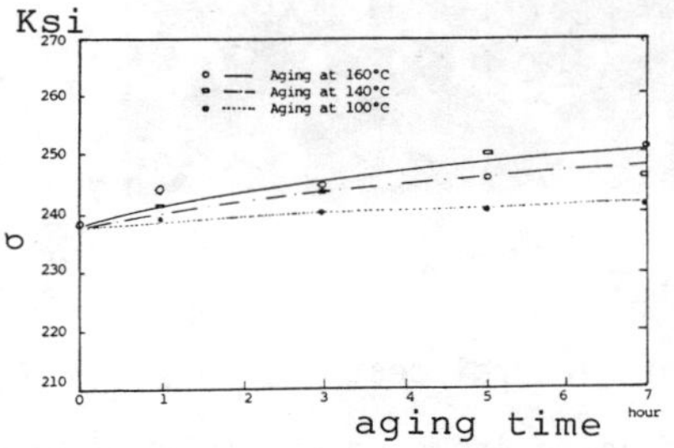

Fig. (15) - The relationship between flexural strength and aging time at various aging temperatures

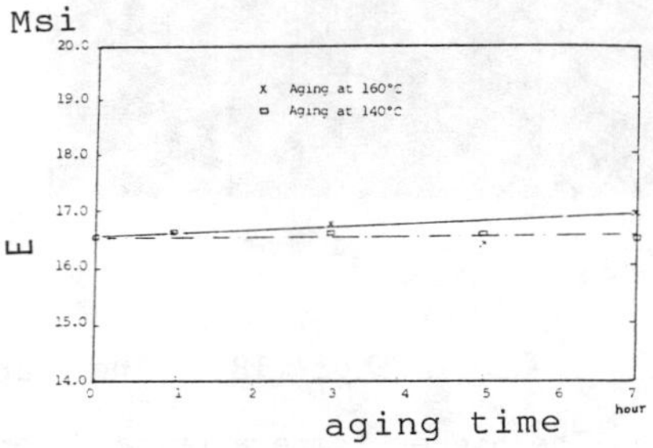

Fig. (16) - The relationship between flexural modulus and aging time at 140° and 160°C

Since physical aging enhances the interlaminar bonding of the laminates, the ILSS of aged composite increases. The effect of rising ILSS is larger for aging at 160°C than that at 140°C, see Figure 17.

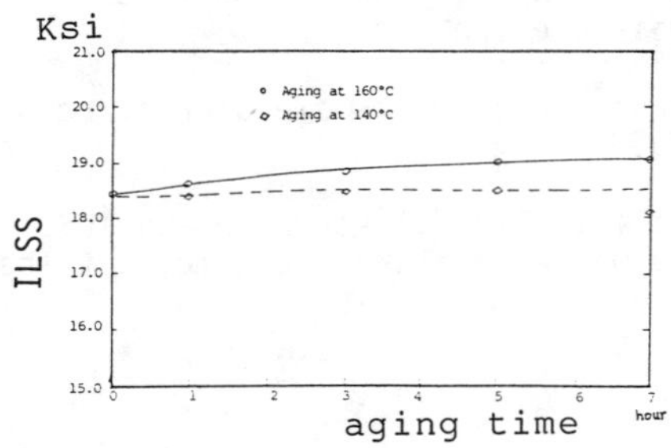

Fig. (17) - The ILSS of HY-1076E laminates at various aging conditions

6. The Physical Aging and the Crack of Flexural Test

Figure 18(A) shows the back of crack pattern of flexural test specimen. At aging temperature 100°C, the crack of specimens are similar to that of quench specimen, and Figure 18(B) at aging temperature 140°C, the crack pattern is the same as that of annealing. Two reasons to explain those two different crack patterns.

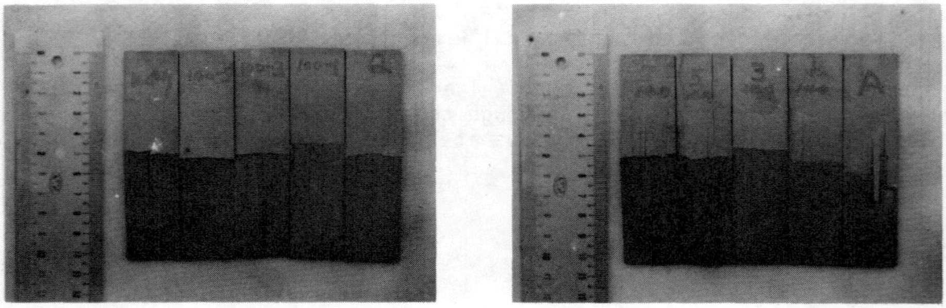

A B

Fig. (18) - The crack pattern of flexural test

a) The higher temperature and longer time of aging increase the interlaminar bonding of composite. Force applied can be completely distributed in specimen, and the cracks may occur at every part of the specimen.

b) For unaged composites, the unrearranged part of molecular networks can absorb energy to relax the stress. Load is borne by rigid carbon fiber until the crack occurs at the center of specimen.

CONCLUSIONS

In this work, the following conclusive observations are listed:

1. During the moisture absorption, the glass transition temperature and mechanical properties (flexural strength, modulus and ILSS) of HY-1076E composite reduce as increasing their moisture content.

2. HY-1076E composites rearrange their molecular structure with physical aging at temperature below Tg.

3. After aging, the tensile, flexural strength and ILSS of HY-1076E composite increase as increasing the aging temperature and time, but the increment of modulus is small.

REFERENCES

[1] Lee, H. and Neville, K., Handbook of Epoxy Resins, McGraw-Hill, New York, 1967.

[2] Lubin, G., ed., Handbook of Composites, Van Nostrand, New York, 1982.

[3] Struik, L. C. E., Physical Aging in Amorphous Polymers and other Materials, Elsevier, Amsterdam, 1978.

[4] Nielsen, L. E., Mechanical Properties of Polymers and Composites, Marcel Dekker, Inc., New York, 1974.

[5] Mazzio, V. F. and Huber, G., SAMPE, Journal, 20, 14, 1984.

[6] Mijovic, J., J. Composite Materials, 19, 178, 1985.

[7] Mijovic, J., J. Appl. Polym. Sci., 27, 1149, 1982.

[8] Mijovic, J., ibid, 27, 2919, 1982.

[9] Kong, E. S. W., Adamson, M. and Mealler, L., Composite Technology Review, 6, 170, 1984.

[10] Kong, E. S. W., ibid, 4, 97, 1982.

[11] Clarke, J. A., Polymer Communications 26, 113, 1985.

[12] Browning, C. E. and Hartmess, J. T., Composite Materials Testing and Design, 3rd Conference, ASTM STP 546, March 1973.

[13] Browning, E. E., Husman, G. E. and Whitney, J. M., Composite Materials Testing and Design, 4th Conference, ASTM STP 617, May 1976.

[14] Kalnin, I. L., AFML-TR-75-51, Vol. 1, Moisture Effect, 1975.

[15] Springer, G. S., Environmental Effects on Composite Materials, Vol. 2, 1984.

[6] [illegible]

[7] [illegible], J. Appl. Polym. Sci., 29, [illegible] 1983.

[8] [illegible]

[9] [illegible] and [illegible], Composite Technology Review, [illegible]

[10] [illegible]

[11] [illegible], Composites, [illegible] 1985.

[12] [illegible], Composite Materials: Testing and Design, ASTM STP [illegible]

[13] [illegible], Composite Materials: Testing and Design ([illegible] Conference), ASTM STP [illegible], May 1976.

[14] [illegible]

[15] [illegible], Composite Materials, [illegible]

INFLUENCE OF MOISTURE ON THE MECHANICAL PROPERTIES OF GRAPHITE EPOXY SYSTEM

A. Ankara

Middle East Technical University
Ankara, Turkey

D. Weisgerber and J. Vilsmeir

Messerschmitt-Bolkow-Blohm
Munich, Germany

ABSTRACT

The absorption of water by the CFRP and the effect of this on static and dynamic behaviour of composite especially at elevated temperature is a problem area in the aerospace applications. The time dependent phenomenon of water absorption into the bulk is considered to be a diffusion process. In this investigation, the influence of thermal spikes on composites containing moisture, on further water absorption was also studied.

Static and fatigue tests were carried out on those specimens exposed to various environmental conditions. The results show clear deteriorations in both static and dynamic properties due to moisture pickup.

WATER ABSORPTION

Advanced resin composites absorb moisture directly from its environment resulting changes in its mechanical properties and dimensional stability. Considerable amount of work has been devoted to understand the mechanism of moisture diffusion into fibre reinforced resin composites to develop means of predicting the behaviour of composite structures under various stress and environmental conditions.

The absorption of moisture in CFRP may be through resin, fiber-matrix interface, and cracks or voids. It is known that the carbon fibers do not absorb moisture. However, the primary mechanism for moisture pickup is surface absorption and diffusion through the matrix. Diffusion normal to the surface is described by Fick's second law which treats the composite as an isotropic material.

This study was concerned with water absorption of unidirectional carbon-epoxy laminates 914C/T300 and 914C/XAS at laboratory test conditions which is not actually met at service life. Water uptake would be more rapid in small test coupon, as used in this work, than most real components.

The moisture gain in 914C/T300 is plotted in Figure 1 for a temperature of 70°C at 85% relative humidity. It is evident that there is a linear relationship

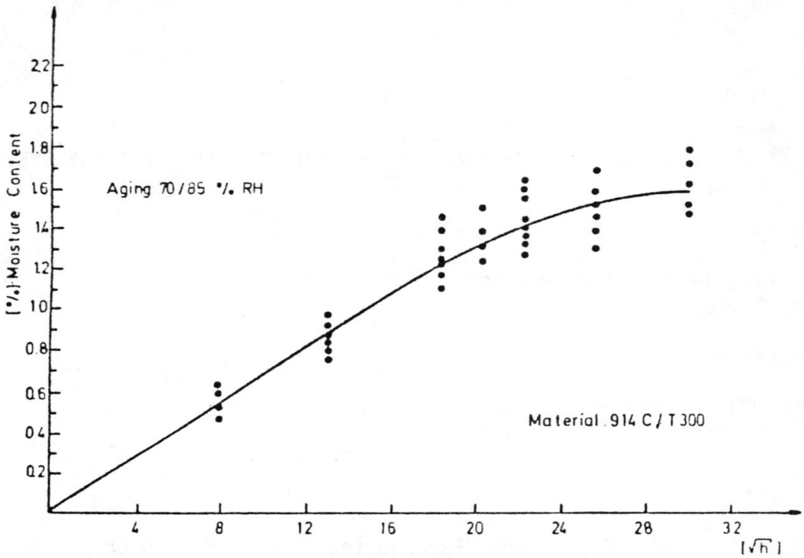

Fig. (1) - Moisture absorption rate of Fibredux 914C/T300

between the weight of water absorbed and square root of time up to a high proportion of water content. Using the initial slope of the curve, the calculated diffusion coefficient was

$$D = 2 \times 10^{-7}\ mm^2 S^{-1}$$

The average saturation moisture gains at different relative humidities are given in Table 1. The equilibrium moisture content is related to the relative humidity.

Table 1 - Moisture saturation values

Moisture Conditioning at 70°C	Saturation T300	Moisture Content % XAS
65%RH	1.19	1.08
75%RH	1.33	-
85%RH	1.71	1.67
95%RH	1.96	1.77

This saturation value of the laminate can also be predicted by using empirical equations similar as suggested by Whiteside et al [1]

$$M = 1.4 \times 10^{-3}\ (R \cdot H)^{1.6}$$

The calculated values by using the above equation were very close to the experimental data.

It is interesting to note that the degree of cross-linking obtained by different post-cure treatments in Fibredux 914/T300 composites had no effect on water absorption kinetics. The investigated range of temperature was between 190 to 210°C and up to 10 hrs. Thus it is believed that after laminate production cycle, a high level of cross-linking density is obtained in Fibredux 914C and a further post-cure would thus have a very minor effect.

INFLUENCE OF THERMAL SPIKING ON THE PROPERTIES OF CFRP

Composite materials technology has advanced to introduce various secondary and primary structural parts made from CFRP for aeronautical and space applications [2]. In supersonic flights, (of about Mach 2), recovery skin temperatures as high as 130°C can be reached and relative humidities for short durations close to 100% are expected, although in Middle Europe, this value would be below 85%RH. Work on water absorption in epoxy resins show that the sorption phenomenon can be influenced by thermal excursions to which the resin may be subjected. The thermal excursions known as thermal spiking has been found to depend mainly on spiking temperature and may cause both the amount and rate of moisture absorption to increase. It is well established that moisture uptake results in a loss of strength of composite at higher temperatures and that the magnitude of reduction depends on the resin formulation, moisture content and testing temperature [3-5]. In real aircraft life, thermal spikes occur during short supersonic flights due to aerodynamic heating and so it is important that the loss of composite strength has to be evaluated. The work presented here have been done on those composite properties that are controlled by the matrix.

Specimens were prepared from Fibredux 914C resin and two different carbon fibers, Toray T300 and Courtaulds XAS, in an autoclave. To study the effects of thermal spikes on the properties of wet Fibredux 914C/T300 and 914C/XAS unidirectional composites, samples were saturated at 70°C from 65 to 95% relative humidities prior to thermal spiking.

For thermal spiking, the samples were removed from hot/wet environmental chamber, cooled to room temperature, then placed into ovens heated to spiking temperature for about 5 minutes followed by air cooling to 70°C. After reaching to 70°C, the samples were placed back to moisture chamber for further moisture conditioning. A schematic representation of the thermal spike cycle is given in Figure 2. Samples were subjected to weekly spike cycles. Table 2 shows the spiking conditions used.

Thermal spiking effects was defined as an increase in moisture saturation concentration and degradation of any mechanical properties. A slight increase in water absorption with thermal spiking between 110°-130°C was observed. The reason of the increase is not clear and requires further investigation.

The transverse tensile strength of 914C/T300 and 914C/XAS were found not to depend on the spiking conditions. This is shown in Figure 3. The results of compressive test are given in Figure 4. These results are similar to transverse test data where no influence of spiking on compressive strength is observed. The interlaminar shear strength at room temperature measured after 10 spike cycles showed no difference between spiked and non-spiked wet samples.

Figure 5 shows the ILSS values of samples tested at 120°C. From the data, it is seen that there is a little drop in the ILSS from 50 MPa to 45 MPa if spiking temperature was raised from 110°C to 130°C.

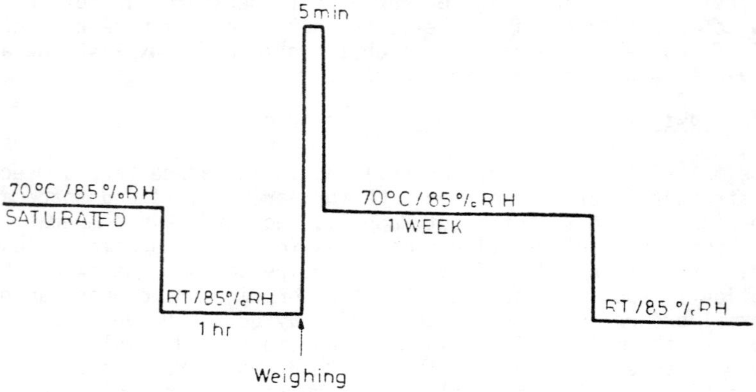

Fig. (2) - Spike cycle

Table 2 - Spiking conditions used

Spike No.	Spike Temperature (C)	Spike Duration (minutes)
1	100	7
2	110	5 and 6
3	120	5
4	125	5
5	130	4 and 5

The results of the glass transition temperature tests are shown in Figure 6. As contrast to dry specimens, the results of torsion pendulum tests indicated that thermal-spiking have resulted in a decrease in the softening point as observed from dynamic shear modulus values. There appeared also a shift of damping peak to lower temperature. If, however, the measurements were carried out on moisture saturated but unspiked specimens, so difference could be detected between spiked and non-spiked wet specimens.

Based on these data, the following trends can be concluded. The effect of thermal spiking for the epoxy systems investigated is of negligible importance to: i) compressive strength, ii) transverse tensile strength, iii) glass transition temperature, iv) strain to failure, v) room temperature interlaminar shear strength. A minor effect of spiking on interlaminar shear strength at 120°C was observed.

THERMOMECHANICAL BEHAVIOUR

As was mentioned earlier, the matrix resins when absorbing moisture, will have a deleterious effect on the resin itself. With increasing moisture and temperature, the T_G of the resin is lowered, this means that a transition from a glassy to a rubbery state takes place [6-8].

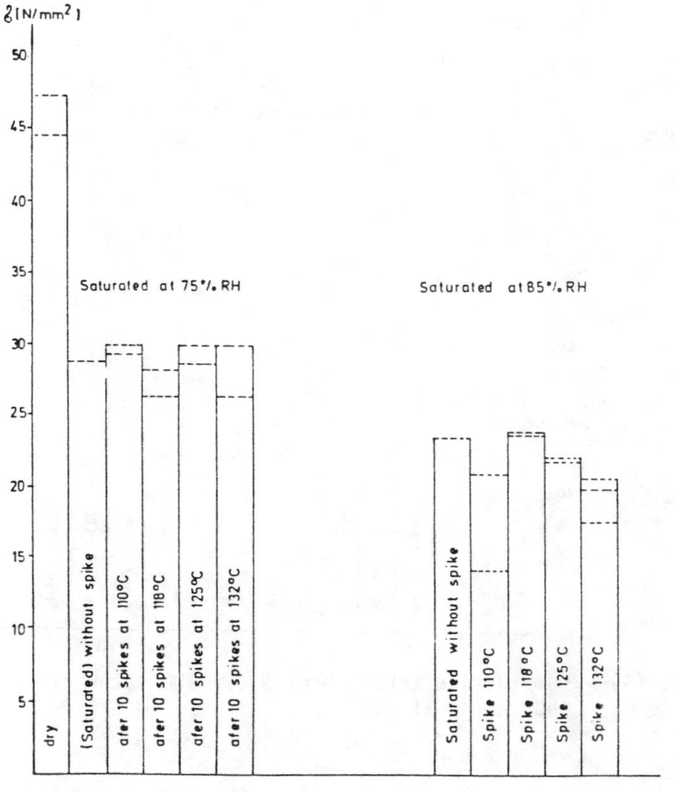

Fig. (3) - Transverse tensile strength of 914C/T300 (all saturated at 70°C) and tested at 120°C

A lot of test techniques are used to measure this material property. The most important methods used are the following:

- DMA
- Torsion pendulum
- HDT
- ISO R 75 (modified method)

All methods give good results in distinguishing between dry and wet materials. For the dry material condition, it is possible to determine the dry T_G value with all test procedures. The temperature regime for the glass transition for the dry material is only small and therefore T_G-values for dry materials are, in many cases, nearly the same measured with different methods. The DMA and torsion pendulum test methods show a sharp peak for the mechanical dissipation factor. This peak, where the material shows the maximum damping effect, is a good characterization of the material in the dry state.

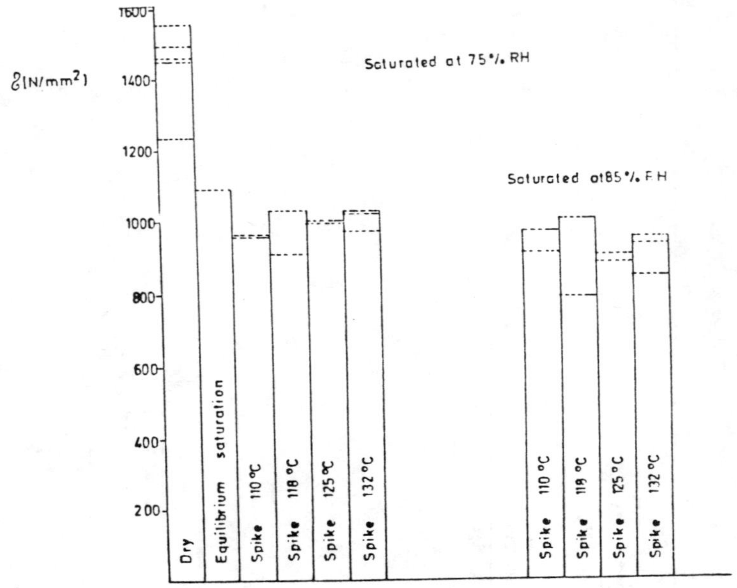

Fig. (4) - Compressive strength of 914C/T300 saturated at 70°C and tested at 120°C

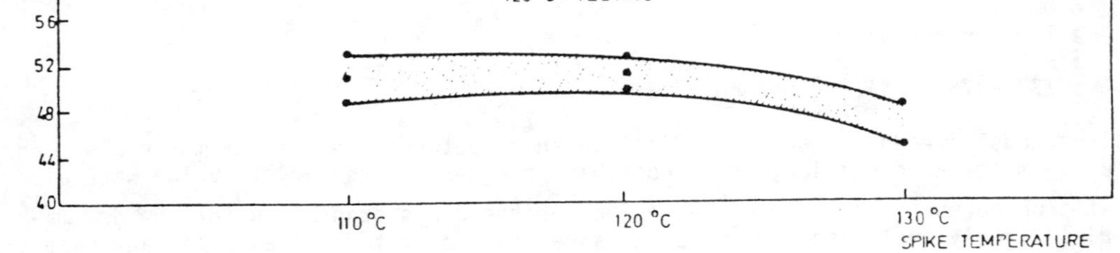

Fig. (5) - Effect of thermal spiking on ILSS values

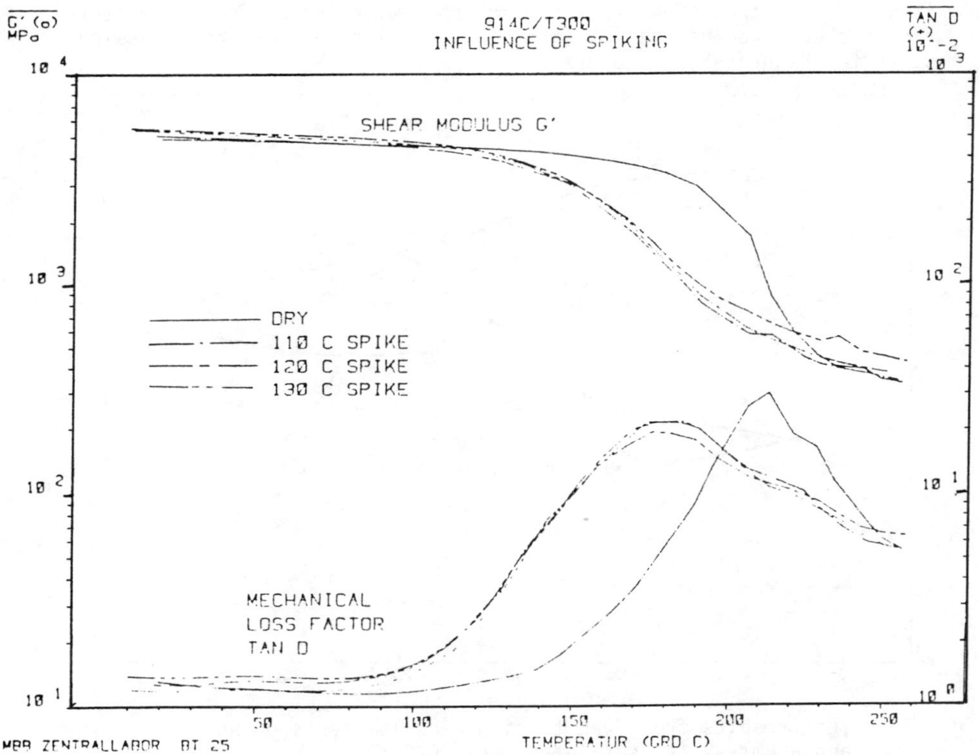

Fig. (6) - Influence of spiking on dynamic behaviour of Fibredux 914C/T300

For wet materials, it is no longer possible to define a small temperature range for glass transition. In torsion pendulum tests, the damping peak is getting very wide, this means that a "classical T_G" defined as the maximum of damping cannot be defined any longer. However, some general statements can be made:

- the shear storage modulus and the mechanical dissipation factor are a function of moisture content, that means that with increasing moisture, the glass transition regime is at a lower temperature

- for all wet materials (epoxies, BMI's) tested at MBB with the torsion pendulum technique, it was found that the tanδ is starting to increase at a temperature between 90 and 110°C and the shear modulus is starting to decrease at similar (Figure 7) temperatures

- the starting temperature for softening being nearly the same for a large number of materials tested in the wet condition, the curves for the shear modulus G' and tanδ may vary considerably with different materials saturated with moisture at a given humidity level, Figure 7

- it was found, that CFRP-systems which show with increasing temperature higher shear modulus than other systems, will have higher compression strength, Figures 8(a) and 8(b).

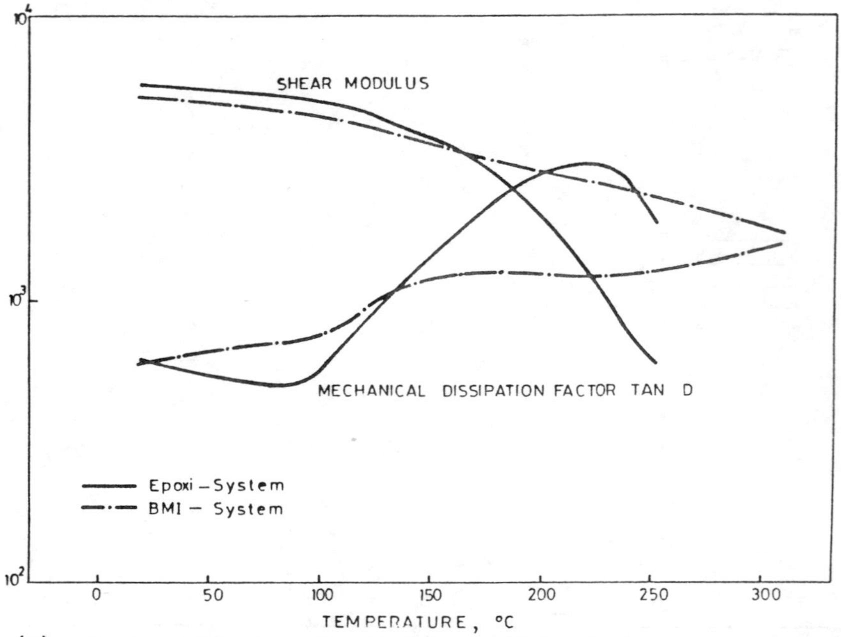

Fig. (7) - Test results from free torsion pendulum tests on an epoxi-system and a BMI-system saturated at 70°C/85%RH

All methods in use to evaluate the thermomechanical behaviour of CFRP-materials are able to give information on the behaviour of a material compared to another system, tested with the same method and in the same condition. It is only possible to qualify which material will have better hot/wet mechanical strength. It is not possible to quantify the improvement by performing only thermomechanical tests. It was impossible too, to find a correlation between thermomechanical analysis and thermal spike behaviour; this means, a temperature at which the thermal spike influence will start (degradation of the material due to thermal excursions) cannot be defined as a general rule applicable for all materials.

STATIC MECHANICAL PROPERTIES

A. Unidirectional Data

There are almost no applications possible for unidirectional CFRP-materials in industrial design. The unique material properties of CFRP are given by the possibility to design strength and modulus of the material by changing the direction of plies in a sheet. In practical design, the 0, ±45, 90 degree laminate family is used very often. Because it is too time-consuming to test the whole variety of laminates which are theoretically possible, it is common practice to evaluate unidirectional material properties and use these together with failure criteria

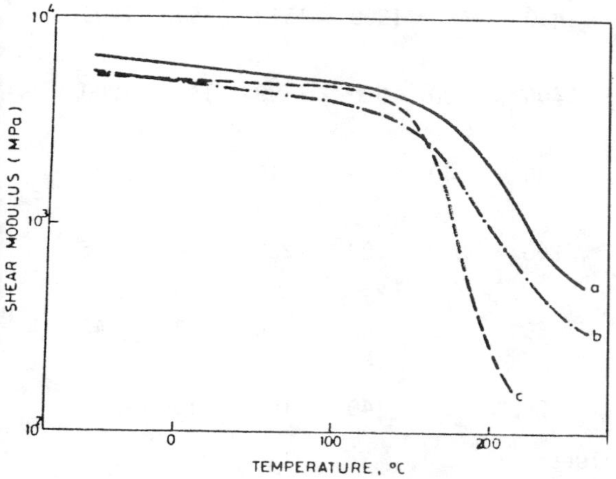

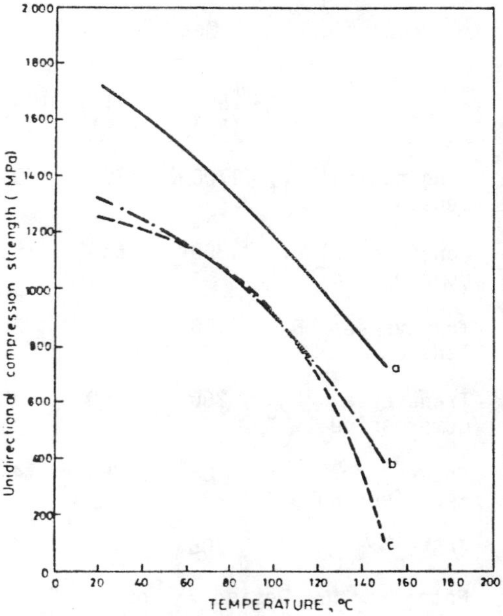

Fig. 8(a) - Free torsion pendulum test results on three different CFRP-systems saturated at 70°C/85%RH

Fig. 8(b) - Unidirectional compression strength of three CFRP-systems saturated at 70°C/85% RH

to calculate the strength, modulus and strain of multidirectional laminates. Only some checks on multidirectional properties are necessary to prove that the theoretical prediction is true.

The multidirectional material properties which are required for calculation are the following:

- longitudinal tension
- transverse tension
- longitudinal compression
- transverse compression
- in-plane shear
- Poisson's ratio

For all these properties, not only the strength has to be known, but Young's or shear modulus and the failure strains are of importance also. A very large number of unidirectional data is available from literature [9].

In Tables 3 and 4, unidirectional properties are given for the CFRP-system Fibredux 914C/T300. The data is given for the room temperature, as received, and for hot/wet condition. The wet specimens had been saturated at 70°C and 80%RH and then tested at elevated temperature. It can be seen that longitudinal tension properties are not greatly affected by moisture and temperature.

Table 3 - Typical strength properties of Fibredux 914C/T300

Properties		Mean Test Values N/mm^2			Design Values[1] N/mm^2			Design Values[2] N/mm^2		
		-55°C (AR)	RT (AR)	100°C (HE)	-55°C (AR)	RT (AR)	100°C (HE)	-55°C (AR)	RT (AR)	100°C (HE)
Longitudinal Tension	F_{1T}	1700	1700	1500	1450	1450	1300	1540	1540	1400
Longitudinal Compression	F_{1T}	1800	1650	1150	1400	1300	920	1590	1450	1000
Transverse Tension	F_{2T}	50	65	30	35	40	25	42	50	27
Transverse Compression	F_{2C}	250	250	120	200	270	100	220	220	106
Shear ±45°-Tension	F_{12}	85	85	45	65	65	40	70	70	42
ILS τ_o		105	105	45	90	90	40	100	100	42

Note: 1 : Design values are "A" values
2 : Design values are "B" values
(AR): Material in condition as received
(HE): Humidity exposed condition; saturated at 70°C/85%RH

Table 4 - Typical modulus properties of Fibredux 914C/T300

Properties	Mean Test Values			Design Values[1]		Design Values[2]	
	-55°C (AR)	RT (AR)	100°C (HE)	RT (AR)	100°C (HE)	RT (AR)	100°C (HE)
Longitudinal Modulus, GN/m^2 E_1	140	140	140	125	125	133	133
Transverse Tension Modulus, GN/m^2 E_2	9	9.5	6		4.0		4.8
Shear Modulus ±45°-Tension, GN/m^2 G_{12}		4.5	1.8				
Poisson's Ratio ν_{12}		0.3	0.3				

Note: 1 : Design allowables are 99% values
2 : Design allowables are 90% values
(AR): Material in condition as received
(HE): Humidity exposed condition; saturated at 70°C/85%RH

Longitudinal compression, transverse tension and compression and in-plane shear strength show a decrease in strength with increasing moisture and temperature. The longitudinal compression modulus is only slightly changed by moisture and temperature. Transverse tension, compression modulus and in-plane shear modulus are significantly reduced by moisture and temperature.

In comparing different resin/fibre-systems, we have found that a material demonstrates a good hot/wet compression strength when matrix dependent moduli show high values in the hot/wet condition.

The two multidirectional properties which are dominating most of the design strength of multidirectional laminates are longitudinal compression and transverse tension.

The stiffness of a laminate, which is very often the most important point in composite application, is governed by the modulus in fibre direction. Therefore, the new, intermediate modulus fibres, which have a modulus 15 to 20% higher than standard fibres, are of great interest in composite design.

MULTIDIRECTIONAL PROPERTIES

Plane multidirectional properties can easily be calculated with unidirectional data and failure criteria. Some other multidirectional properties, like bearing strength, filled hole compression and open hole tension, which have to be known for designing composite structures, are much more difficult to calculate. It is possible to predict these multidirectional properties, but checks by tests work are necessary.

Another important design feature is the compression after impact strength. This is due to the requirement that Barely Visible Impact Damage (BVID) may not reduce the design strength (static and fatigue) of the composite structure. As composite materials are brittle and impact damage can therefore easily produce non-visible delamination between the plies of the composite, the compression after impact behaviour has become a major field of research.

INFLUENCE OF MOISTURE AND TEMPERATURE ON FATIGUE AND RESIDUAL STRENGTH

The static strength and stiffness of CFC is much more effected by service environments like temperature and moisture than metallic materials. In order to investigate effects of temperature and humidity on the fatigue properties specimens as well as structural components, samples were manufactured by MBB using Fibredux 914C/T300 and XAS material.

The following discussion is dealing with results of tests conducted by MBB and LBF applying flight-by-flight loading representative of fighter aircraft [10,11]. All specimens and components were NDT inspected. For the tests, only specimens without defects were used.

Before testing, the specimens were conditioned at 70°C/85% relative humidity or 70°C/95%/RH, respectively. The tests have been performed under the same hot/wet conditions as mentioned above. The ultimate strength tests and the residual strength tests were done at 120°C. The moisture content has been controlled by traveller specimens.

A. Durability Tests on Unnotched and Bolted Specimens

The unnotched specimens as well as the bolted joint specimens (dimensions and laminates are shown in Figures 9 and 11), have been conditioned at 70°C/80% RH before testing. The specimens were then fatigue tested between 50°C and 130°C applying the FALSTAFF loading programme. FALSTAFF is a standardized flight-by-flight loading programme representative of a fighter A/C wing. The unnotched specimens have been loaded according to an upper wing skin (mainly compression), whereas the loading of the bolted joint specimens has been according to a lower skin (mainly tension). The maximum spectrum stress was approximately $\sigma_{ult(wet)}/1.5$. The tests were discontinued at 16,000 flights (4 x life). The moisture content has been kept constant at approximately 1,3% during the test.

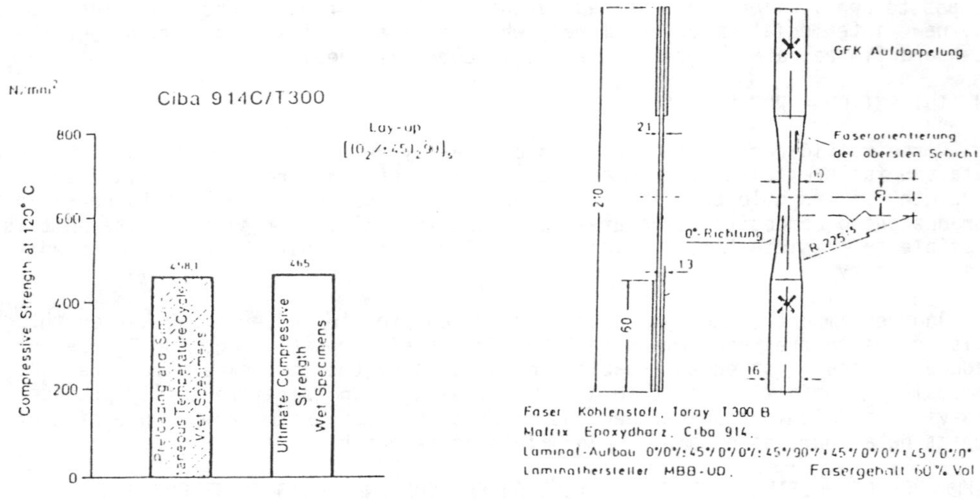

Fig. (9) - Comparison between the residual strength of preloaded (FALSTAFF with simultaneous temperature cycles) CFRP-specimens and the ultimate compressive strength

Fatigue failures could not be observed. The degradation of stiffness during the test was neglectable. After durability testing, the residual strength at 120°C (compression or tension, respectively) has been determined. The results are shown in Figures 9, 10 and 12. No significant difference can be found between the strength of specimens undergoing fatigue loading and those without any preloading. The reduction of residual stiffness was also neglectable.

B. Durability Testing on CFC Components [1]

The durability tests on components have been performed in cooperation between MBB and BAE. The components tested are shown schematically in Figure 13. The components have been preconditioned to approximately 1% moisture content. The parts have then been tested using typical taileron (flight-by-flight) loading programmes and representative loads. The temperature was varied between +80°C and

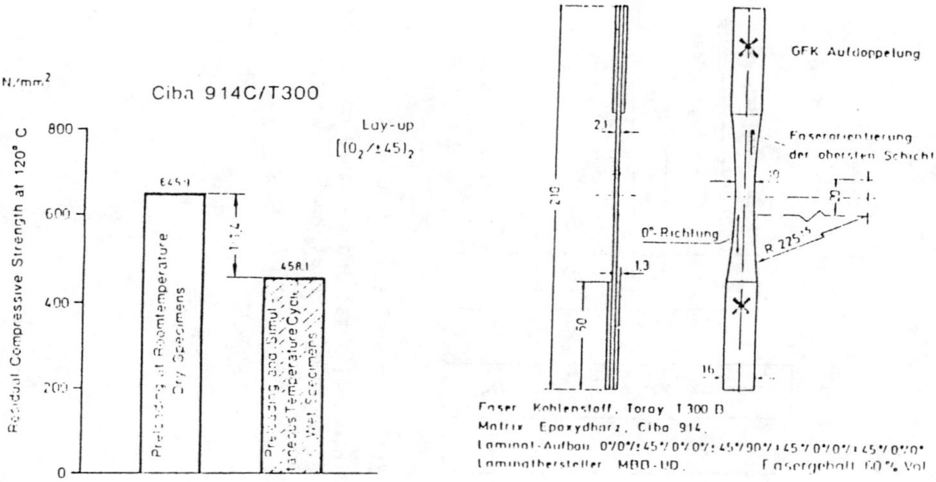

Fig. (10) - Influence of flight-by-flight temperature cycles and moisture on the residual compressive strength of FALSTAFF-preloaded CFRP specimens

-24°C. The moisture content was kept constant at approximately 1% during the durability test and also the subsequent residual strength test.

During the durability test, no failure could be observed. The results of the residual strength tests at 120°C are shown in Figure 14. No significant difference between ultimate strength and residual strength (after durability test) could be found.

Based on the aforementioned test results, the trends as summarized below can be concluded:

- the specimens have not shown any fatigue failure during the durability tests under condition representative for fighter aircraft
- no significant differences have been found between ultimate strength and residual strength, both tested under hot/wet conditions
- on the basis of the present investigations, it can be concluded that unnotched structures are critical to compression, whereas joints are critical to tension
- the residual strength (tension) for joints was almost equal for both conditions, namely for RT/as received as well as for hot/wet
- the ultimate strength (compression) of unnotched specimens under hot/wet is considerably lower than for specimens under RT/as received conditions.

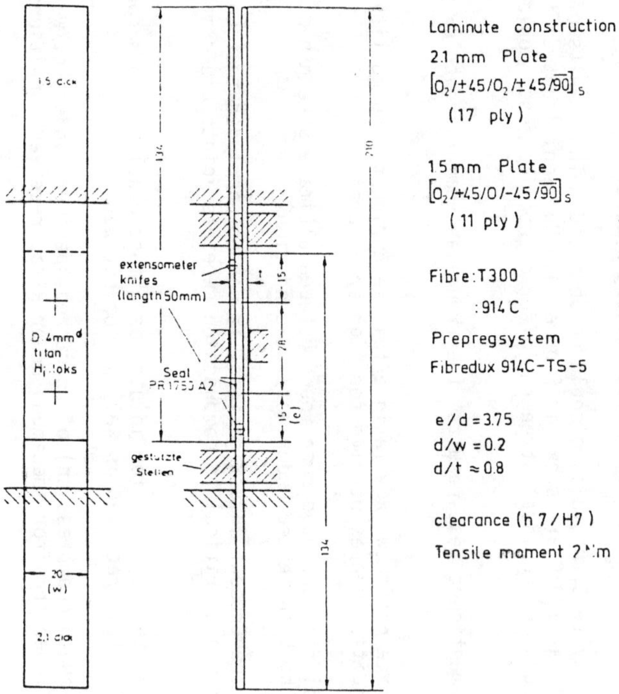

Fig. (11) - Specimen used

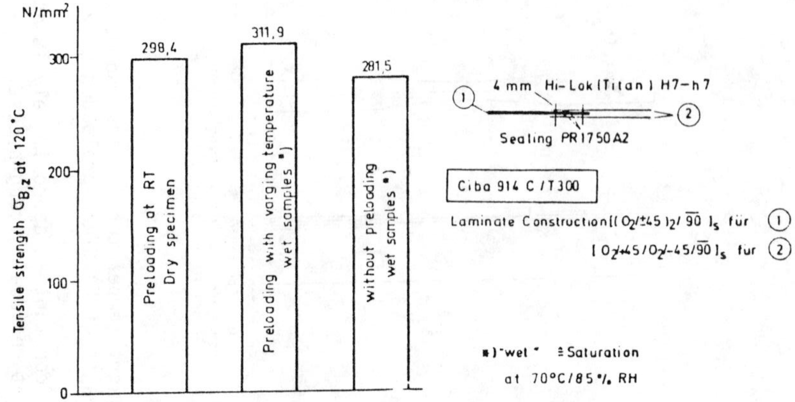

Fig. (12) - Effect of preloading (FALSTAFF - σ_0 = 190 N/mm, 16,000 flights) and moisture content on tensile strength of CFRP at 120°C

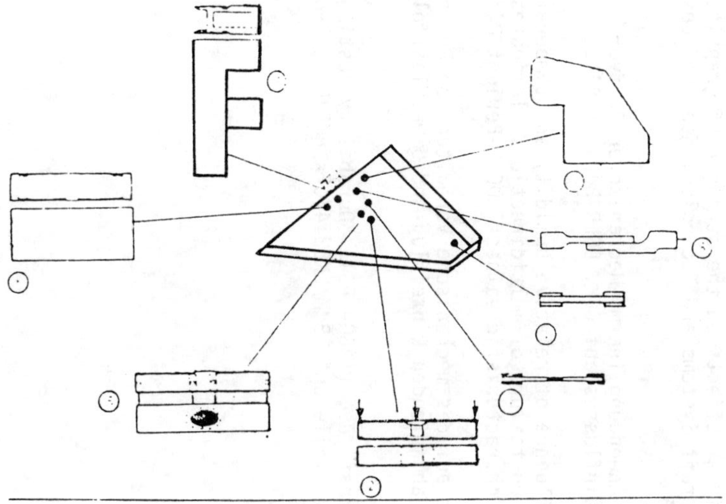

① GRADUATED SKIN THICKNESS SPECIMEN
② CORE JUNCTION SPECIMEN
③ PERFORATED FILLED HONEYCOMB CORE SPECIMEN
④ BOLTED ROOT RIB SKIN SUB-ASSEMBLY
⑤ ACCESS HOLE SUB-ASSEMBLY
⑥ BOUNDARY MEMBER SPECIMEN
⑦ I-BEAM FRONT SPAR SPECIMEN
⑧ BOLTED JOINT SPECIMEN

Fig. (13) - Component durability test programme

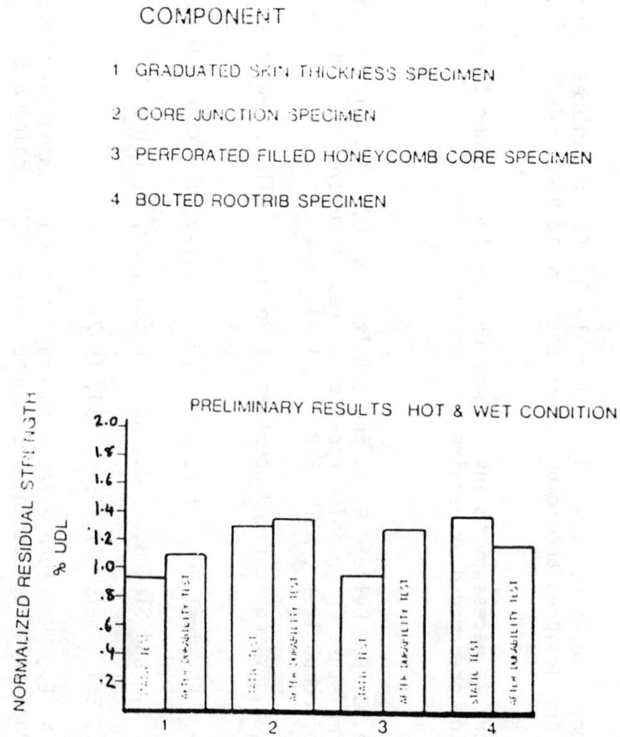

Fig. (14) - Test results from component durability testing

All the above statements are only applicable for CFC structures without predamage, which are designed according to presently valid standards for fighter aircraft.

Further work is necessary to investigate whether these general results can also be used for predamaged or repaired structures.

REFERENCES

[1] Whiteside, J. B., Delsai, R. J. and Schultel, R. L., "The distribution of absorbed moisture in graphite-epoxy after real time cycling long-term behaviour of composites", ASTM, STP 813, 1983.

[2] Ferry, J. D., Viscoelastic Properties of Polymers, Wiley, New York, London, 1961.

[3] Bueche, F., Physical Properties of Polymers, Wiley, New York, London, 1962.

[4] Vieth, W. R. and Sladed, K. J., Colloid Sci., 20, p. 1014, 1965.

[5] Kelly, A., Sci. Amer. 217, 3, p. 160, 1967.

[6] Browing, N. E., "Moisture effects in graphite/epoxy composites", paper given at Internationalem wehrtechnischem Symposium at WIM, Erding 8.5.84 - 10.5.84.

[7] Verette, R. M., "Temperature/humidity effects on the strength of graphite/ epoxy laminates", A/AA 1375, Aircraft Systems and Technology Meeting, Los Angeles, Calif., August 4-7, 1975.

[8] Brandt, J. and Warnecke, J., Kie Anwendung Thermo-Mechanischer Prufverfahren bei der Untersuchung des Feuchteeinflusses auf CFK, DGLR-Bericht 84-02.

[9] Gadke, M., "Evaluation of the influence of relative humidity and temperature on the mechanical properties and on failure of multidirectional laminates on the basis of measured single ply characteristic values", DGLR-Bericht 84/02.

[10] Untersuchung des Umwelteinflusses auf das mechanische Verhalten von Proben aus CFK-Verbunden unter simultan ablaufenden Einzelflug-und Umweltlastfolgen; LBF-Bericht vom 29.03.1985.

[11] Keerl, P. N., Hartmann, W. and Ernst, J., (IABG-TFS), "Durability testing of the TORNADO carbon-fibre-composite Taileron", BMM/LKE214/CFK/PUB/005, 1984.

SECTION III

FAILURE AND FATIGUE ANALYSIS

STABILITY OF A CRACK INTERACTING WITH FIBERS

C. K. Chao

Institute of Fracture and Solid Mechanics, Lehigh University
Bethlehem, Pennsylvania 18015 USA

ABSTRACT

The interaction of a crack with two circular elastic fibers or inclusions is considered. Applied is the strain energy density criterion for determining the stability of the composite system. This is accomplished by finding a length parameter "ℓ" that represents the distance between $[(dW/dV)_{min}^{max}]_g$ and $[(dW/dV)_{min}^{max}]_\ell$. They correspond to the maximum of the minimum strain energy density referred to the fixed global coordinates and local coordinates and referred to the points G and L, respectively. Depending on the relative moduli ratio E_2/E_1 of the fiber and matrix, the location G will change while L always coincides with the crack tip. When the crack and inclusions lie on the axis of load symmetry, G lies in between the crack tip and inclusion boundary.

As the inclusions are displaced slightly away from the axis of load symmetry with an eccentricity e, G can be either above or below the crack plane. Again, it is governed by the ratio of E_2/E_1. In general, instability tends to increase as L and G drift apart. What is significant is that the combined effect of load, geometry and material of a composite can be expressed in terms of a length parameter for describing the stability behavior of a system.

INTRODUCTION

One of the advantages of composites is that failure can occur locally without seriously affecting the load carrying capacity of the system. This includes the breaking of fibers, cracking of the matrix, delamination of interfaces, etc. Defects and flaws are unavoidable when fabricating composites. A common defect is that of a crack in the matrix that can alter the neighboring load transfer characteristics particularly when it is near the fibers whose moduli are many times higher than that of the matrix material. The simple model of a crack between two fibers considered in [1,2] will be investigated by application of the strain energy density criterion for determining the composite system stability.

Past studies [3-6] on the interaction of a crack with circular inclusions involved infinite boundaries. These results are not useful as the point G would lie at infinity such that global instability prevails immediately after failure initiation. This, of course, would defeat the very purpose of constructing composite materials, i.e., to enhance local failure while maintaining the load carrying capacity of the system.

The concept of local and global stationary values of the strain energy density function was first applied to study the problem of a foreign inclusion embedded in an elastic matrix [7]. The locations of $[(dW/dV)^{max}_{min}]_g$ and $[(dW/dV)^{max}_{min}]_\ell$ were found and discussed in terms of the stability of the system. The same idea is applied in this work to evaluate the influence of fiber inhomogeneity on the cracked matrix material. The ratio E_2/E_1 is varied such that different cases of instability are obtained in addition to changes in the relative position of the crack with the fibers.

FINITE ELEMENT ANALYSIS

For simplicity, the composite system in Figure 1 is subjected to an uniform stress of magnitude $\sigma_o = 6.9 \times 10^2$ MPa. The parameters that describe the plate,

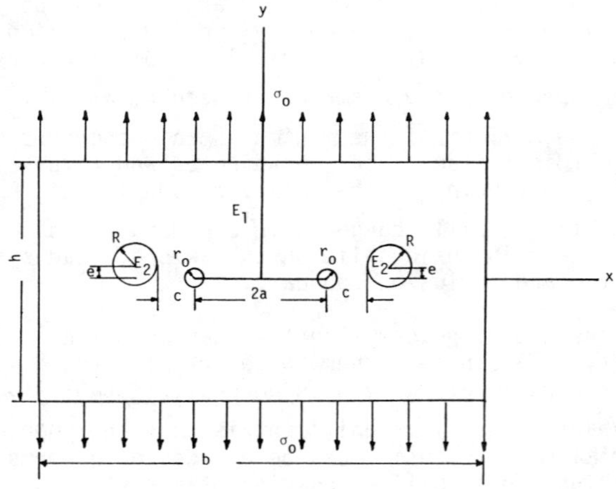

Fig. (1) - Elastic circular inclusions with eccentricity embedded in finite matrix containing a crack under uniform tensile load

crack and fiber are given by 2a = 25.4 cm, b = 101.6 cm, h = 50.8 cm, R/a = 0.6 and c/a = 1. A small circular region of radius r_o = 0.1 cm is selected such that all computations are made outside of this region. The material properties of the matrix are $E_1 = 2.07 \times 10^5$ MPa and $\nu_1 = \nu_2 = \nu = 0.3$. The fiber properties are varied by changing E_2/E_1.

Shown in Figure 2 is the finite element grid pattern. Refinements are introduced near the crack tip. The problem possesses one-half symmetry and is analyzed by considering a total of 56 elements and 320 nodes. The two dimensional and axisymmetric computer program (APES) is applied which employs the twelve node iso-

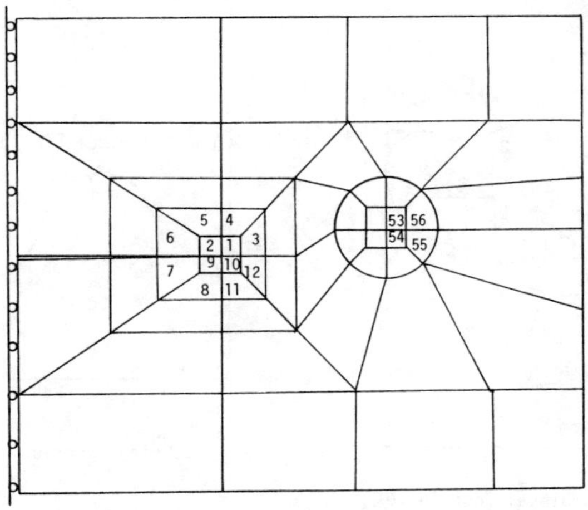

Fig. (2) - Finite element grid pattern with one-half symmetry

parametric elements. A frontal solution technique is chosen for solving the system of linear algebraic equations. Accurate evaluation of $[(dW/dV)^{max}_{min}]_g$ from the nodal values of dW/dV is achieved by applying a smoothing technique based on a least square estimate [8].

FAILURE CRITERION: GLOBAL AND LOCAL CONSIDERATIONS

Failure usually initiates from sites of geometric defects that give rise to oscillation in the energy density. It is desirable to tailor-make composites such that subcritical damage of the material can occur over a large region of the system. Energy can thus be dissipated slowly without having it accumulated and concentrated locally to cause sudden release causing catastrophic fracture.

To be considered are the variations of the strain energy density function dW/dV at each point 0_1, 0_2, etc., in the system as illustrated in Figure 3(a). At a fixed distance r_o from these origins, there prevails a local maximum $[(dW/dV)_{max}]_\ell$ and minimum $[(dW/dV)_{min}]_\ell$ with reference to the angle θ_j ($j = 1, 2,\ldots n$), Figure 3(b). The maximum of these quantities denoted by $[(dW/dV)^{max}_{min}]_\ell$ and $[(dW/dV)^{max}_{max}]_\ell$ are assumed to coincide with the initiation sites of fracture and yielding, respectively. Similarly, it is also possible to obtain stationary values of dW/dV with respect to R and θ in the global system. The maximum of $[(dW/dV)_{min}]_g$ and $[(dW/dV)_{max}]_g$ can be obtained from the contours of constant dW/dV. The troughs of constant dW/dV indicate the quantity $[(dW/dV)_{min}]_g$ while the peaks of constant dW/dV indicate the quantity $[(dW/dV)_{max}]_g$. The quantity $[(dW/dV)^{max}_{min}]_g$ can thus be found from the maximum of the troughs and the quantity $[(dW/dV)^{max}_{max}]_g$ can be obtained from the maximum of the peaks. The pair

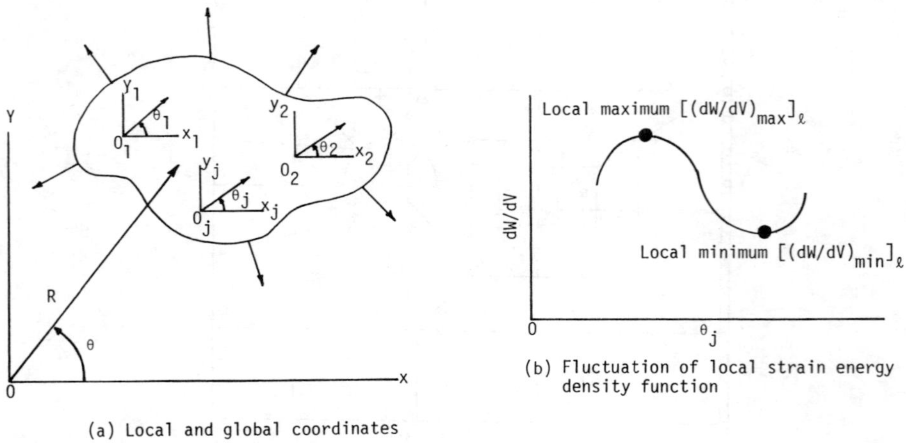

Fig. (3) - Local variations of dW/dV in a continuum

$[(dW/dV)^{max}_{max}]_g$ and $[(dW/dV)^{max}_{max}]_\ell$ will not be discussed here as they pertain to yield prediction. The distance "ℓ" between $[(dW/dV)^{max}_{min}]_\ell$ or L and $[(dW/dV)^{max}_{min}]_g$ or G can serve as a measure of the stability or instability of the system. As the load is increased or decreased monotonically in increments, there arise two possibilities. The first is [7]

$$\ell_1 < \ell_2 < \dots < \ell_j < \dots < \ell_c \qquad (1)$$

where the system becomes progressively unstable and reaches global instability at ℓ_c. It is also possible to have

$$\ell_1 > \ell_2 > \dots > \ell_j > \dots > \ell_o \qquad (2)$$

A stable situation such that damage will eventually be reached arrests at ℓ_o.

DISCUSSION OF RESULTS

As mentioned earlier, the location of $[(dW/dV)^{max}_{min}]_\ell$ or L occurs at the crack tip in most cases. Its magnitude depends on the material and geometric parameters. The normalized values of $[(dW/dV)^{max}_{min}]_\ell$ are plotted against c/2R in Figure 4. Two opposite trends are found and the effects of the fiber are most pronounced when c is small. The maximum of the minimum local strain energy density increases with c for $E_2 > E_1$, i.e., a fiber having a higher modulus than the matrix. This is the usual case. The reverse is obtained when $E_2 < E_1$. For a fixed distance c/2R = 0.6, Figure 5 shows that $[(dW/dV)^{max}_{min}]_\ell$ decreases with increasing E_2/E_1, a result that is expected.

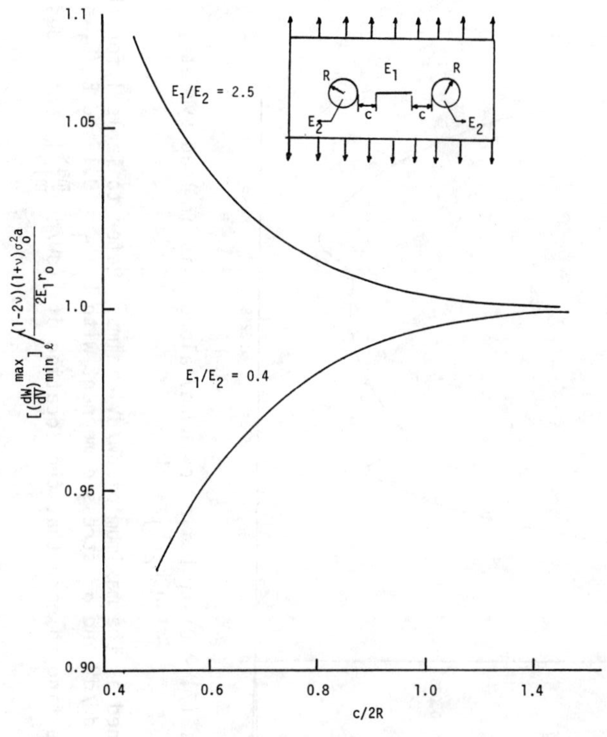

Fig. (4) - Variations of $[(dW/dV)^{max}_{min}]_{\ell}$ with respect to the distance c between a crack and inclusion

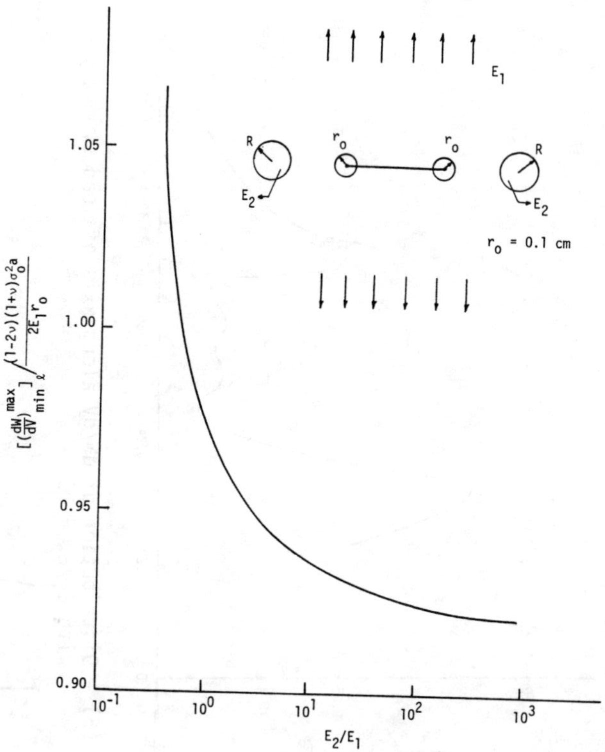

Fig. (5) - Variations of $[(dW/dV)^{max}_{min}]_{\ell}$ with the ratio of inclusion stiffness to matrix stiffness

Figures 6 and 7 display the distribution of strain energy density along the axis of load symmetry for $E_1/E_2 = 10$ and 0.1, respectively. The point G can be

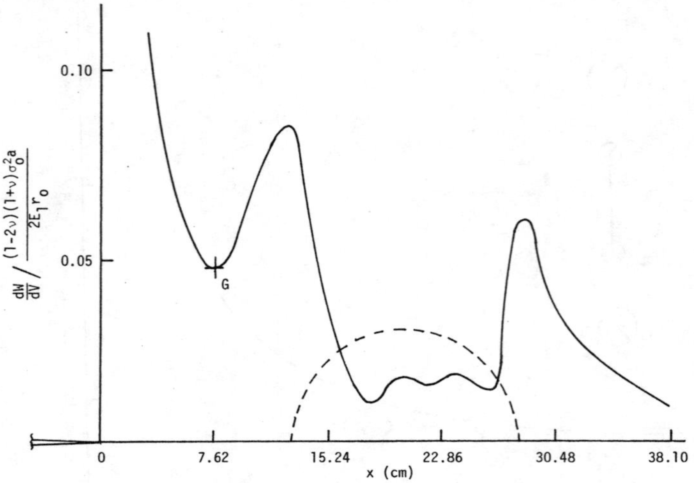

Fig. (6) - Distribution of dW/dV along axis of load symmetry with $E_1/E_2 = 10$

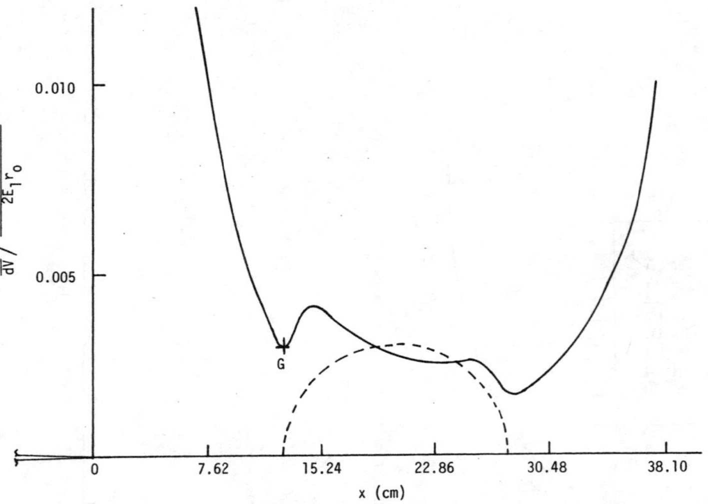

Fig. (7) - Distribution of dW/dV along axis of load symmetry with $E_2/E_1 = 10$

readily obtained as the maximum of dW/dV minima. Refer to Table 1 for the numerical values of dW/dV and σ_y stress component with $E_2/E_1 = 0.1$ and $E_2/E_1 = 10$. It is also interesting to note that the location of $[(dW/dV)_{min}^{max}]_g$ coincides with

Table 1 - Variations of strain energy density dW/dV and normal stress in y-direction along x-axis of symmetry with c/a = 1 and R/a = 0.6

(a) $E_1/E_2 = 0.1$

x (cm)-distance from a crack tip	dW/dV (MPa)	σ_y (GPa)	x (cm)-distance from a crack tip	dW/dV (MPa)	σ_y (GPa)
0.14	65.010	5.593	17.78	0.203	1.030
0.56	35.860	3.490	18.63	0.190	0.984
1.27	12.340	2.300	19.46	0.170	0.934
2.17	4.782	1.620	20.32	0.160	0.900
2.96	3.264	1.345	21.17	0.154	0.885
3.81	2.570	1.200	22.00	0.149	0.878
5.08	2.042	1.070	22.86	0.145	0.880
6.35	1.670	0.960	24.54	0.146	0.844
7.62	1.370	0.863	26.25	0.149	0.837
9.31	0.860	0.673	27.94	0.097	0.484
11.01	0.430	0.490	31.33	0.357	0.386
12.70	0.190	0.340	34.71	0.710	0.554
14.39	0.245	1.130	38.10	0.841	0.638
16.09	0.224	1.074			

(b) $E_1/E_2 = 10$

x (cm)-distance from a crack tip	dW/dV (MPa)	σ_y (GPa)	x (cm)-distance from a crack tip	dW/dV (MPa)	σ_y (GPa)
0.14	76.535	6.024	18.63	0.842	0.207
0.56	42.021	3.767	19.46	0.869	0.210
1.27	14.628	2.499	20.32	0.902	0.213
2.17	5.979	1.794	21.17	0.898	0.212
2.96	4.367	1.543	22.00	0.900	0.210
3.81	3.629	1.414	22.86	0.914	0.209
5.08	3.083	1.309	24.54	0.840	0.203
6.35	2.829	1.255	26.25	0.789	0.196
7.62	2.795	1.244	27.94	3.531	0.979
9.31	3.306	1.354	31.33	2.579	1.135
11.01	5.008	1.603	34.71	0.916	0.678
12.70	5.049	1.115	38.10	0.485	0.395
14.39	0.952	0.224			
16.09	0.891	0.216			
17.78	0.823	0.205			

those at which σ_y attains a local minimum. Figure 8 further shows that the length "ℓ" increases as E_2/E_1 increases from zero to one. A sudden jump then occurs for $E_2/E_1 > 1$ before ℓ becomes constant. Failure tends to be most localized

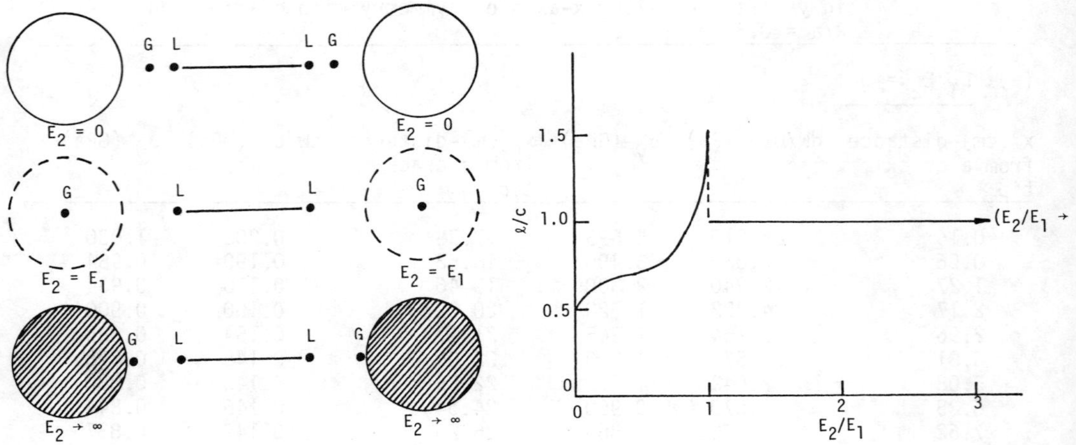

Fig. (8) - Variations of ℓ/c with E_2/E_1

for the case of circular hole (i.e., $E_2 = 0$) where ℓ is the smallest. The most unstable situation corresponds to the absence of fibers, i.e., $E_1 = E_2$. Generally speaking, system stability can be said to be proportional to the size ℓ. This is shown by the results in Table 2 for $c/a = 1$ and $R/a = 0.6$. The length ℓ increases as E_2/E_1 is increased quickly reaching a limit of $E_2/E_1 \simeq 1.0$ beyond which ℓ remains at 12.7 cm.

Table 2 - Distances between $[(dW/dV)^{max}_{min}]_\ell$ and $[(dW/dV)^{max}_{min}]_g$ for different ratios of matrix stiffness to inclusions with $c/a = 1$, $R/a = 0.6$

E_2/E_1	ℓ (cm)
0	6.350
0.1	7.620
0.3	8.573
0.5	9.049
0.7	10.636
1.0	19.050
1.5	12.700
5.0	12.700
∞	12.700

If the fibers are displaced slightly away from the axis of load symmetry, G will change its position depending on the amount of eccentricity e. Shown in Figures 9(a) and 9(b) are the cases of two circular holes and two rigid fibers with $e = R/2$. Note that the crack tends to travel along a path of least resistance as dotted in Figure 9(a) for $E_2 = 0$. When $E_2 = \infty$, however, the failure path is predicted to circle around the fiber even though it is the longer path.

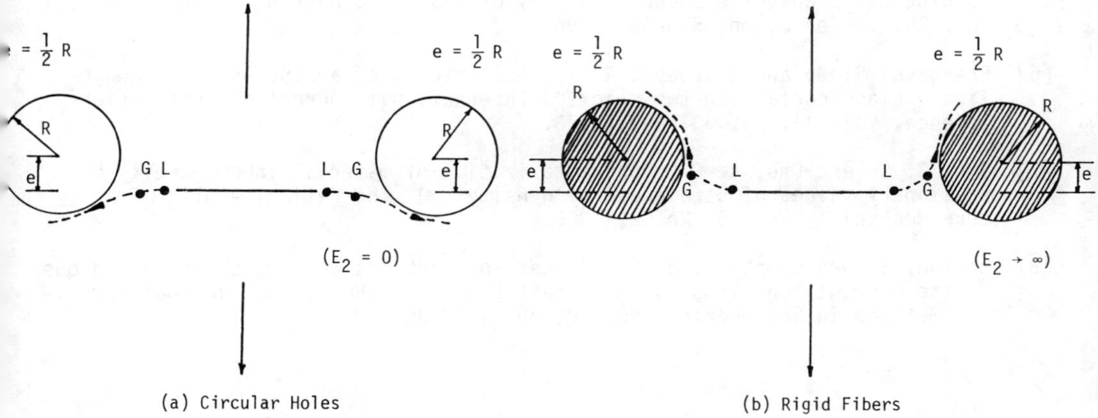

Fig. (9) - Fracture trajectory for both circular hole and rigid inclusion with eccentricity e/R = 0.5

Although the composite system considered in this work is rather simplified, it nevertheless demonstrates the utility of the ℓ-concept used in conjunction with strain energy density criterion. Since composites would possess several modes of failure, it would be impractical to include all of them in an analysis without considering the load-time history. They are a class of material to be tailor-made to satisfy certain performance requirements. The parameter ℓ serves as a useful means of system stability as it considers the combined effect of loading, geometry and microstructure.

ACKNOWLEDGEMENT

The author wishes to thank Professor G. C. Sih for providing bright ideas and the encouragement to complete this work.

REFERENCES

[1] Sih, G. C., Hilton, P. D., Badaliance, R., Shenberger, P. S. and Villarreal, G., "Fracture mechanics for fibrous composites", ASTM STP 521, pp. 98-132, 1973.

[2] Sih, G. C., Chen, E. P. and Huang, S. L., "Fracture mechanics of plastic-fiber composites", Engineering Fracture Mechanics, Vol. 6, pp. 343-359, 1974.

[3] Erdogan, F., Gupta, G. D. and Ratwani, M., "Interaction between a circular inclusion and an arbitrarily oriented crack", Journal of Applied Mechanics, December 1974.

[4] Atkinson, C., "The interaction between a crack and an inclusion", International Journal of Engineering Science, Vol. 10, pp. 127-136, 1972.

[5] Erdogan, F. and Sih, G. C., "On the crack extension in plates under plane loading and transverse shear", Journal of Basic Engineering, Trans. ASME, Vol. 85, Series D, pp. 519-526, 1963.

[6] Bhargava, R. D. and Bhargava, R. R., "Elastic circular inclusion in an infinite plane containing two cracks", International Journal of Engineering Science, Vol. 11, pp. 437-449, 1973.

[7] Sih, G. C. and Chu, R. C., "Characterization of material inhomogeneity by stationary values of strain energy density", J. Theoretical Applied Fracture Mechanics, Vol. 5, No. 2, 1986.

[8] Hinton, E. and Campbell, J. S., "Local and global smoothing of discontinuous finite element functions using a least squares method", Int. Journal Numerical Methods in Engineering, Vol. 8, pp. 461-480, 1974.

LOCAL AND GLOBAL INSTABILITY IN FRACTURE MECHANICS

A. Carpinteri

Istituto di Scienza delle Costruzioni, University of Bologna
40136 Bologna, Italy

ABSTRACT

The Strain Energy Density Theory is applied to beams in flexure. Hardening or softening constitutive laws are utilized in the stress and failure analysis to predict different fail modes in metals or concrete-like materials respectively. The finite element method is applied consistently for each increment of loading so that the rate of change of the strain energy density factor dS with respect to the crack growth rate da results to be constant for each loading increment and specimen size scale. The straight line relationship S versus a rotates in a counterclockwise direction around a common point as the loading increment is increased. The straight line translates as the size of geometrically similar specimens is increased. Brittle behaviours are predicted for large loading increments and/or large sizes. Above an upper size-threshold, the stable crack growth ceases to occur and the brittle failure is achieved prematurely when the global behaviour is still linear. On the other hand, below a lower size-threshold, the unstable crack growth ceases to occur (even in the softening stage) and the ductile failure develops up to the complete specimen separation. For intermediate specimen sizes, the stable crack growth is followed by the unstable crack propagation.

Ductility appears therefore as a mechanical property depending on the size-scale of the structure as well as on the fracture toughness of the material.

INTRODUCTION

The failure of structural elements may occur in different ways. The ultimate loading condition may be accompanied by plastic flow of the element, slow crack growth and/or fast crack propagation, depending on material strength, ductility and toughness as well as on structural size-scale. Initially, uncracked elements of brittle material may reach their maximum loading capacity with a sudden crack formation and uncontrollable crack propagation, whereas cracked elements of ductile material may reach their ultimate condition with the progressive formation of a plastic hinge at the ligament. On the other hand, slow crack growth and plastic deformation can terminate with a rapid crack propagation.

Plastic Limit Analysis and Linear Elastic Fracture Mechanics reproduce the behaviour of very ductile and very brittle materials respectively. For materials with intermediate properties, they are both unable to describe the behaviour under loading and the slow crack growth. Moreover, remarkable scale effects are present in fracture toughness testing. The mechanical specimen behaviour ranges from ductile to brittle only increasing the size-scale and keeping the geometrical shape unchanged. Large components fail due to rapid crack propagation, when the global behaviour is still linear elastic and the slow crack growth and plastic zones have not developed yet. On the contrary, small specimens fail in a ductile manner with a slow crack growth, so that the hardening and softening behaviours may be revealed up to the complete load relaxation and specimen separation.

In the present paper, the Strain Energy Density Theory of Sih [1,2] is applied to analyze the integrity of structural members. Hardening or softening constitutive laws are used in the stress analysis, while the finite element method is applied consistently in both stress and failure analysis for each increment of loading.

As a consequence of the Strain Energy Density Theory, the rate of change of the strain energy density factor dS with respect to the crack growth rate da remained as a constant for each loading increment and specimen size-scale, i.e., dS/da = constant. The straight line relationship S versus a rotates in a counterclockwise direction around a common point as the loading increment is increased. The straight line translates as the size of geometrically similar specimens is increased.

Numerous examples are provided to illustrate how the Strain Energy Density Theory can be applied to predict different fail modes in metals (isotropic or kinematic hardening) and concretes (softening), ranging from plastic collapse to brittle fracture.

STRAIN ENERGY DENSITY FRACTURE CRITERION

In order to evaluate the crack growth increment at each loading step, the Strain Energy Density Theory will be applied, as proposed by Sih [1,2]. It is based on the following fundamental assumptions: (1) the stress field in the vicinity of the crack tip cannot be described in analytical terms because of the relative heterogeneity of the material. A minimum distance r_o does exist below which it is a non-sense to study the mechanical behaviour of the material from a "continuum-mechanics" point of view and to consider macroscopic crack growth increments; (2) outside such a core region of radius r_o, the strain energy density field can always be described by means of the following general relationship:

$$\left(\frac{dW}{dV}\right) = \frac{S}{r} \tag{1}$$

where the strain energy density factor S is generally a function of the three space coordinates; (3) according to Beltrami's criterion, all the material elements in front of the crack tip, where the strain energy density is higher than the critical value $(dW/dV)_c^*$, fail, Figure 1; (4) when the following condition holds:

$$\Delta a = r_o^* = S_o/(dW/dV)_c^* \tag{2}$$

the crack may be considered as arrested, at least from a macroscopical point of view. On the other hand, when the crack growth increment is:

$$\Delta a = r_c^* = S_c/(dW/dV)_c^* \tag{3}$$

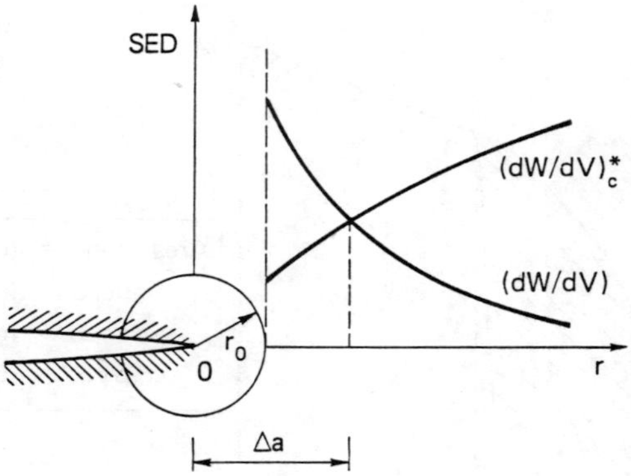

Fig. (1) - Crack growth increment Δa according to the Strain Energy Density Theory

the unstable crack propagation takes place. S_c is a material constant and represents the strength of the material against rapid and uncontrollable crack propagation. S_c is connected with the critical value of the stress-intensity factor K_{IC} through the following equation (plane strain condition) [1]:

$$S_c = \frac{(1+\nu)(1-2\nu)}{2\pi E} K_{IC}^2 \tag{4}$$

STRAIN-HARDENING MATERIALS

A. Material Behaviour

The material behaviour is assumed to follow a multilinear hardening stress-strain relationship, Figure 2. The stress and strain increase up to the failure point c. Once the stress state surpasses the yield point e, permanent deformation or residual stress prevails: in fact, if the load is relaxed at point p, unloading occurs along the line pp'. The new constitutive relation then follows the path p'pc and irreversibility is thus considered.

For an undamaged-material element, the critical value $(dW/dV)_c$ corresponding to failure is the area oecb in Figure 2. At p, the material element is damaged, the area oepp' being the dissipated strain energy density function, $(dW/dV)_d$. Therefore, the available energy density is:

$$\left(\frac{dW}{dV}\right)^*_c = \left(\frac{dW}{dV}\right)_c - \left(\frac{dW}{dV}\right)_d \tag{5}$$

According to the incremental flow theory of plasticity [3], the center of the yield surface can move under plastic deformation and this motion is directed along the vector describing the current deviatoric stress state. The magnitude of the

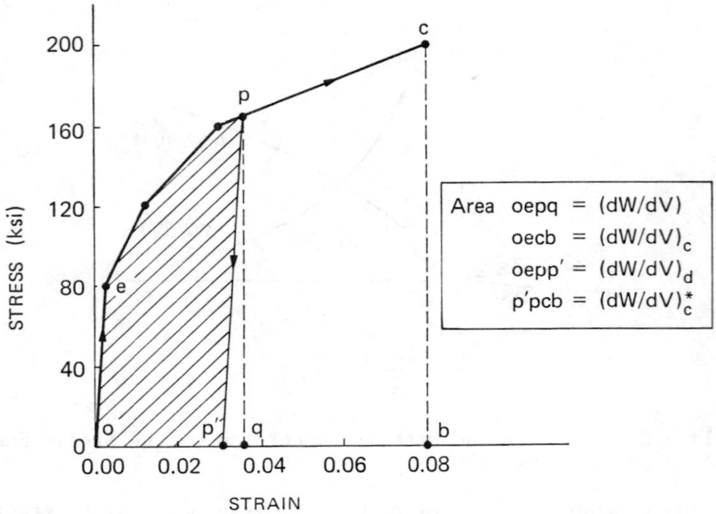

Fig. (2) - Multilinear hardening stress and strain relationship

yield surface center rate is given by the scalar product of the stress rate vector and the unit outward normal vector. Moreover, the rate of the yield surface center is proportional to the coefficient β: *isotropic hardening* is achieved for β=0 and *kinematic hardening* results for β=1. Combined hardening models can be described with values $0<\beta<1$. These possibilities are indicated in Figure 3 which shows the variations of the yield surface for different hardening models.

The failure of material elements ahead of the crack occurs when:

$$\left(\frac{dW}{dV}\right) = \left(\frac{dW}{dV}\right)_c^* \tag{6}$$

For the material considered, Figure 2, this relationship is verified for ε = 0.0475 and σ = 174 ksi when $(dW/dV) = (dW/dV)_c^* = 6.55$ ksi.

Consider a three-point bending specimen, Figure 4, with width b = 5 in, length ℓ = 20 in and initial crack length a_o = 1 in, in a condition of plane strain. The specimen material follows the stress-strain relationship given in Figure 2. Because of symmetry about the load line, only one-half of the specimen geometry needs to be considered. The finite element grid pattern is displayed in Figure 5. A total of 115 nodes and 18 elements are employed. Twelve nodes quadrilateral isoparametric elements and cubic displacement shape functions are used. The r^{-1} singularity in the strain energy density field is developed by means of the 1/9 to 4/9 nodal spacing of the elements adjacent to the crack tip.

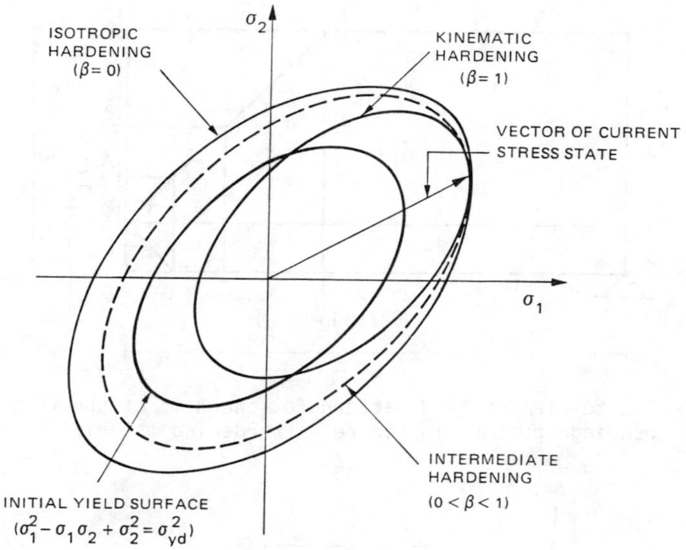

Fig. (3) - Comparison of subsequent yield surfaces for various hardening models

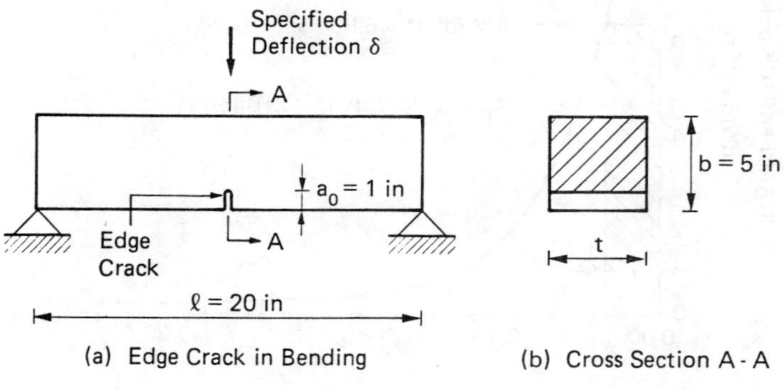

Fig. (4) - Three point bending specimen with an edge crack, subjected to specified deflection δ

Stress and energy state are computed by application of the Plastic Axisymmetric/Planar Structures (PAPST) computer program [4], which incorporates von Mises yield criterion. The crack length increment and the strain energy density factor are computed by equating the strain energy density function dW/dV to the critical value $(dW/dV)_c^* = 6.55$ ksi, Figures 6 and 7. The abscissa of the intersection

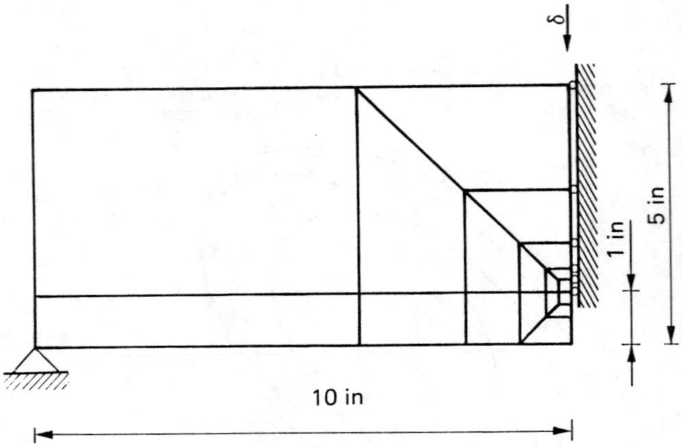

Fig. (5) - Finite element grid pattern for one-half of the three point bending specimen in Figure 4 (hardening material)

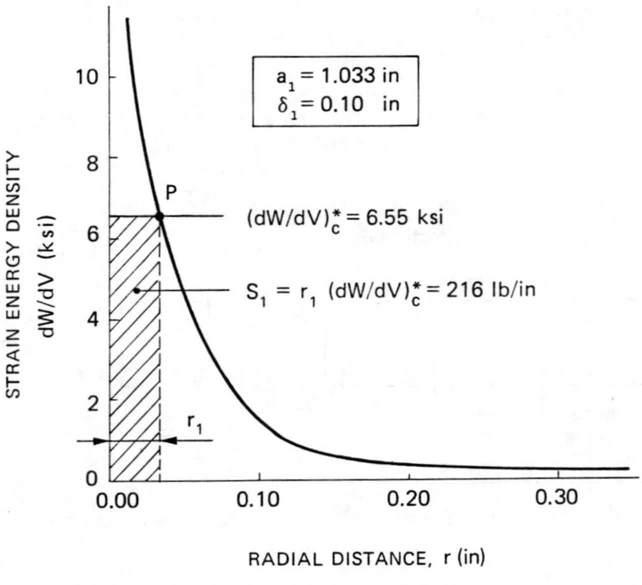

Fig. (6) - Variation of strain energy density function with radial distance for first increment of specified deflection (β=0 and $\Delta\delta$ = 0.05 in)

point P represents the crack length increment while the shaded area corresponds to the strain energy density factor.

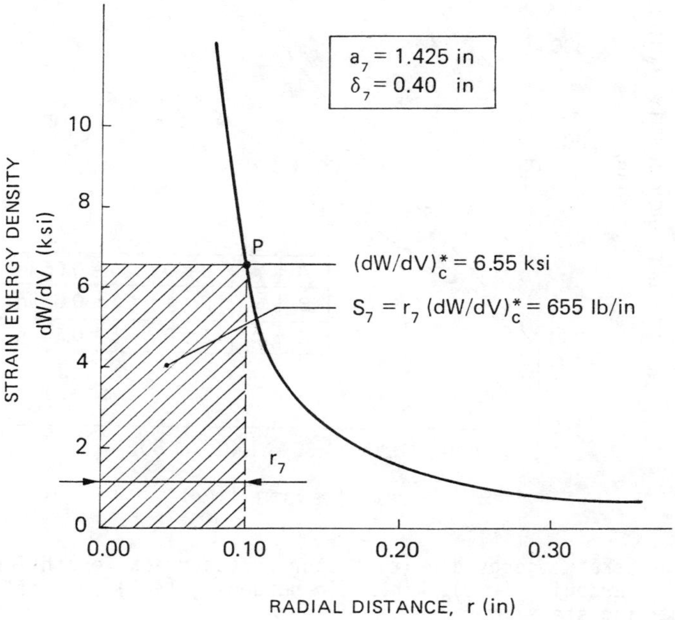

Fig. (7) - Variation of strain energy density function with radial distance for last increment of specified deflection ($\beta=0$ and $\Delta\delta = 0.05$ in)

As was previously stated, unstable crack growth is governed by S_c while local material damage depends on $(dW/dV)_c^*$. Energy dissipation and crack growth very depending on the combination of material properties, loading step and specimen size [5-8]. The results obtained for the three-point bending specimen are useful to understand how the individual variables affect the behaviour of materials that follow a multilinear hardening stress-strain relationship, Figure 2.

B. Isotropic versus Kinematic Hardening

Two different hardening models are considered. For Case A, the rate vector of the yield surface center is zero, whereas for Case B, it is maximum. Even though the process of energy dissipation is a nonlinear phenomenon, the S versus a diagram is linear, Figure 8, i.e., the conditions dS/da = constant prevails. These straight line relations correspond to the situations when both crack growth and structure behaviour are stable. The curve of Case B has a larger slope than the curve of Case A: this means that, for a given material with S_c = constant, a smaller subcritical crack length can be sustained in the case of kinematic hardening.

C. Effect of Loading Step

The effect of loading step is analyzed for a material with isotropic hardening. This effect can be clearly shown by plotting the S versus a diagrams, Fig-

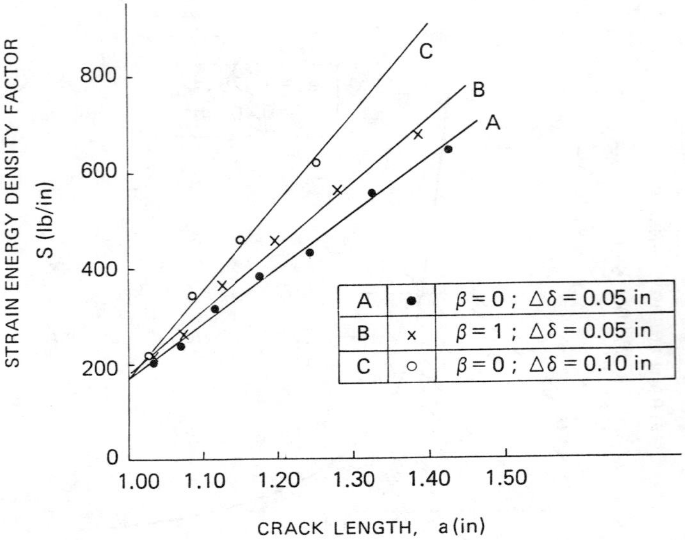

Fig. (8) - Strain energy density factor versus crack length for isotropic hardening (β=0), kinematic hardening (β=1) and different loading steps

ure 8. The two straight lines A and C correspond to $\Delta\delta$ = 0.05 and 0.10 in, respectively. The slope dS/da increases as $\Delta\delta$ is increased: this implies that, for a given value of S_c, the subcritical crack length decreases if larger loading steps are chosen. Such a result is consistent with the experimental observations.

STRAIN-SOFTENING MATERIALS

A. Material Behaviour

Damage of the material at the crack tip and crack growth increments will be computed on the basis of an uniaxial bilinear elastic-softening stress-strain relation, Figure 9. Stress may increase up to the ultimate strength (point U in Figure 9), while strain increases proportionally. Then, only strain may increase, while stress decreases linearly down to zero (point F in Figure 9). If the loading is relaxed when the representative point is in A, the unloading is assumed to occur along the line AO, so that the new bilinear constitutive relation is the line OAF. No permanent deformation is allowed by such a model, but only the degradation of the elastic modulus. In fact, the slope of line AO becomes lower when point A approaches point F. When point A is eventually coincident with F, the effective modulus E^* vanishes, as well as the stress, and fracture or complete separation of the material occurs. On the other hand, when point A is on the elastic branch OU, unloading occurs along the same line and no degradation of the material takes place.

The present model simulates the mechanical damage by decreasing the elastic modulus, E, and the strain energy density which can be absorbed by a material element [6,7,9]. In fact, while for a non-damaged material element, the critical value

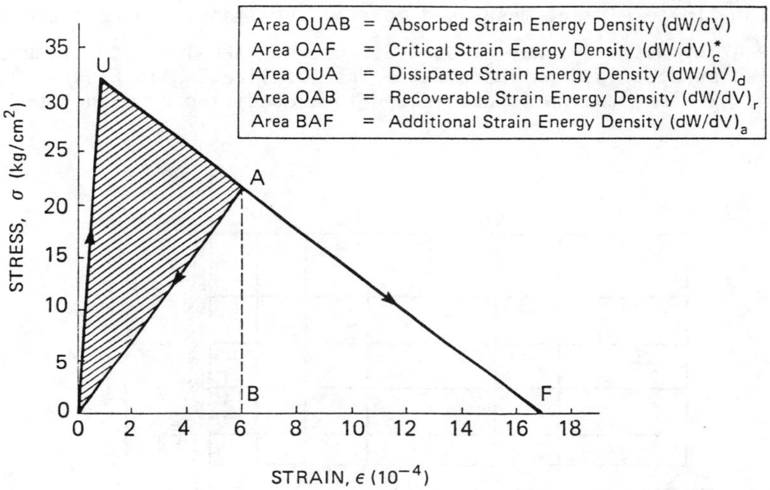

Fig. (9) - Bilinear elastic-softening stress and strain relationship

of the strain energy density, $(dW/dV)_c$, is equal to the area OUF, Figure 9, for a damaged material element with representative point in A, the decreased critical value of the strain energy density, $(dW/dV)_c^*$, is equal to the area OAF. In addition, as is shown in Figure 9, the area OUA represents the dissipated strain energy density, $(dW/dV)_d$, OAB the recoverable strain energy density, $(dW/dV)_r$, and BAF the additional strain energy density, $(dW/dV)_a$. In this way, the decreased critical strain energy density, $(dW/dV)_c^*$, can be expressed as follows:

$$\left(\frac{dW}{dV}\right)_c^* = \left(\frac{dW}{dV}\right)_c - \left(\frac{dW}{dV}\right)_d = \left(\frac{dW}{dV}\right)_r + \left(\frac{dW}{dV}\right)_a \tag{7}$$

The above described model will be extended to the three-dimensional stress conditions, using the current value of the absorbed strain energy density, (dW/dV), as a measure of damage. In other words, the effective elastic modulus, E^*, and the decreased critical value of strain energy density, $(dW/dV)_c^*$, will be considered as functions of the absorbed strain energy density, $(dW/dV) = (dW/dV)_r + (dW/dV)_d$. Stress and strain in the softening condition A, Figure 9, can be expressed in terms of stress and strain in the ultimate and fracture conditions:

$$\sigma = E^*\varepsilon = \frac{\sigma_u \varepsilon_f}{(\varepsilon_f - \varepsilon_u) + \frac{\sigma_u}{E^*}} \tag{8}$$

The three point bending specimen in Figure 4 is reconsidered for the above-described softening material with the following sizes: $b = t = 15$ cm, $\ell = 4b$, $a_o = b/3$. The finite element mesh used for the specimen is represented in Figure 10. The Axisymmetric/Planar Elastic Structures (APES) finite element program [4] is applied at each loading increment. The idealization in Figure 10 utilizes 309 nodes and 52 elements and is considered in a condition of plane strain.

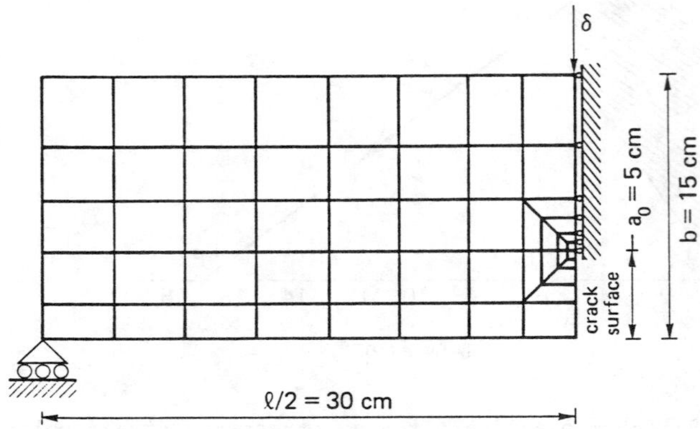

Fig. (10) - Finite element grid pattern for one-half of the three point bending specimen in Figure 4 (softening material)

The damage-crack growth model is applied to the three point bending test specimen, considering a loading process for which the driving (input) parameter is the deflection δ, Figure 11. In this way, it is possible to follow even the softening stage of the load-displacement curve, as in a strain-controlled loading test. In Figure 11, the P-δ relationship is represented for Material 1, Figure 12, and for a constant deflection increment $\Delta\delta = 4 \times 10^{-3}$ cm. At the first step, the stiffness is only 6.21% lower than the original one. Such a decrease is mostly due to the material damage at the crack tip. At the second step, the departure from linearity becomes significant, whereas at the third step, the P-δ curve bends decidely to the right and achieves its maximum. The segments AD and DT marked at the maximum, Figure 11, represent the decreases in the secant stiffness due to material damage and crack growth respectively. At the fourth step, the load P decreases and then the tangent stiffness results to be negative. This is due to the wide damage zone and to the extended crack, which make the specimen more and more flexible. At this stage, the contributions of the material damage and crack growth to the secant stiffness decrease are almost the same. At the fifth and sixth steps, the load continues to decrease while the influence of crack growth increases sharply. When $\delta = 24 \times 10^{-3}$ cm, the load bearable by the specimen is only one fifth of the maximum, while the crack achieves the two thirds of the specimen width.

The distributions of the strain energy density, (dW/dV), and of the effective critical strain energy density, $(dW/dV)_c^*$, over the whole ligament are represented

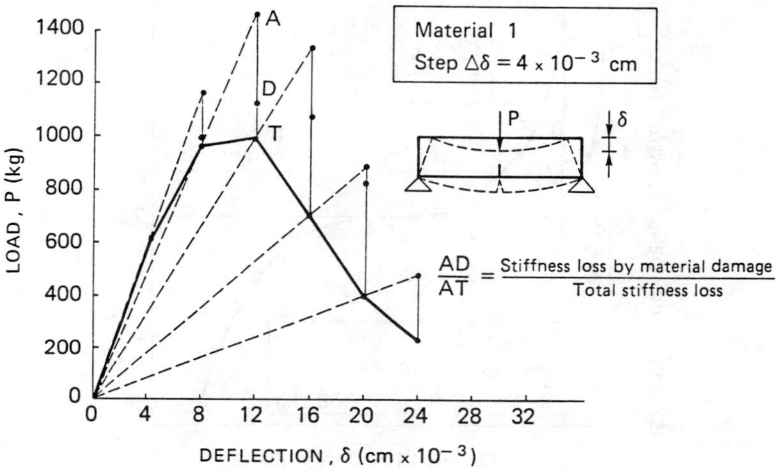

Fig. (11) - Load versus deflection relationship of three point bending specimen

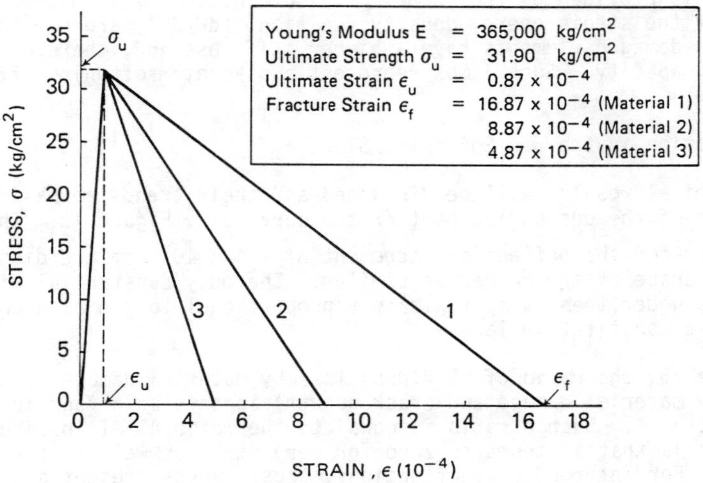

Fig. (12) - Stress-strain curves for three different materials

in Figure 13. While (dW/dV) decreases going away from the crack tip only inside each finite element, $(dW/dV)_c^*$ is a monotonic increasing function. The crack growth

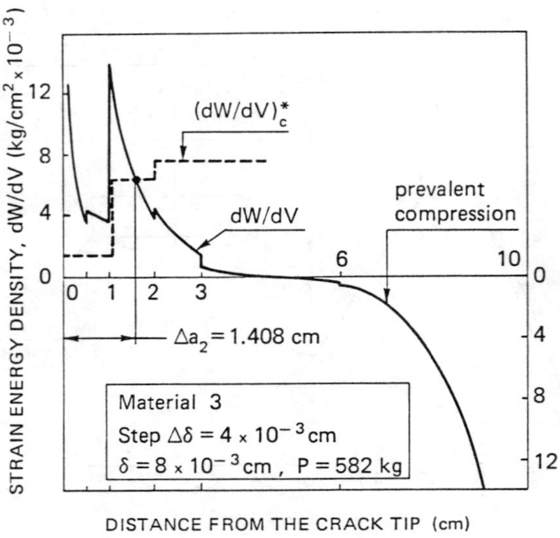

Fig. (13) - Distributions of the strain energy density function, (dW/dV), and of the effective critical strain energy density function, $(dW/dV)^*_c$, over the whole ligament

increment Δa is provided by the intersection of the two distributions. The positive jumps of the strain energy density function, (dW/dV), are due to the fact that the less damaged elements have a higher stiffness and, therefore, a higher load-bearing capacity. Such jumps represent a discrete softening effect ahead of the crack tip.

B. Variation in the σ-ε Softening Slope

The numerical results will be discussed and their trends emphasized when varying the slope of the descending part of the curve σ-ε, Figure 12. The load-deflection responses for the deflection increment $\Delta\delta = 4 \times 10^{-3}$ cm are displayed in Figure 14. The shape of the curves is similar. The only substantial difference is that the area under them is approximately proportional to the area under the corresponding σ-ε constitutive law.

In Figure 15, the ratio of stiffness loss by material damage to total stiffness loss (by material damage and crack growth) appears as a function of the imposed deflection δ. Such a ratio is equal to the ratio AD/AT in Figure 11. The general trend is that it tends to zero for very small as well as for very large deflections. For intermediate deflections, these curves present a maximum which increases the critical strain energy density of the material, $(dW/dV)_c$. In fact, for $\varepsilon_f \to \infty$, the material becomes elastic-perfectly plastic (without permanent deformation) and only material damage occurs, since $(dW/dV)_c$ and $(dW/dV)^*_c$ tend to infinity.

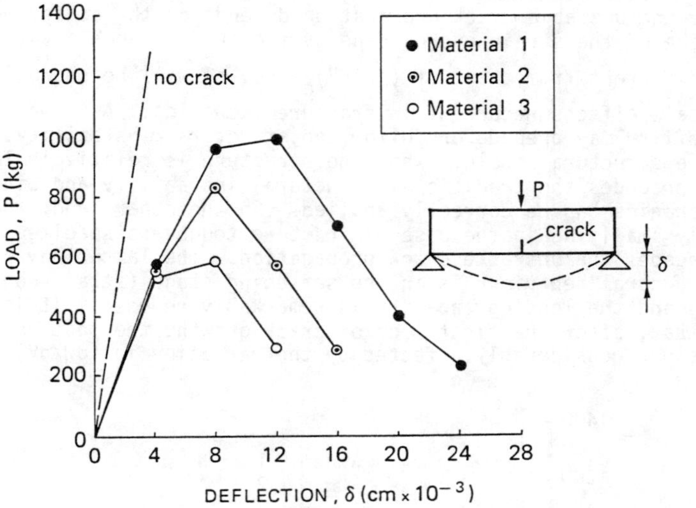

Fig. (14) - Load-deflection diagrams for three different materials (see Figure 12) and $\Delta\delta = 4 \times 10^{-3}$ cm

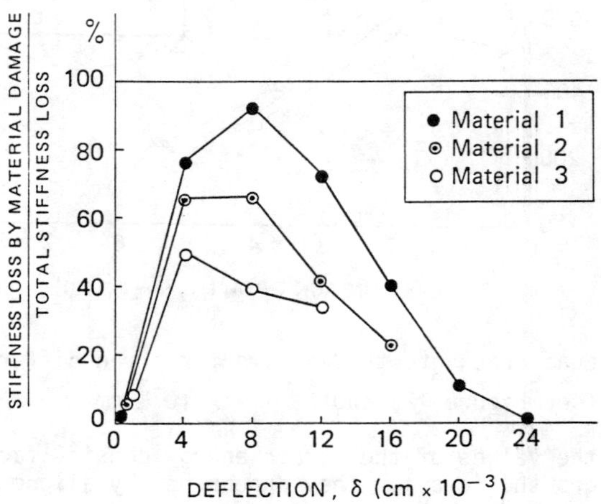

Fig. (15) - Ratio of stiffness loss by material damage to total stiffness loss as a function of the imposed deflection, for three different materials (see Figure 12) and $\Delta\delta = 4 \times 10^{-3}$ cm

In Figure 16, the load P is reported against the crack growth, $(a-a_0)$. At the first steps, the load increases while the crack grows. Then, after reaching a maximum, the load decreases. The maximum represents the transition between stable and unstable structural behaviour. On the other hand, the transition between stable and unstable crack propagation depends on the achievement of the critical value of the strain energy density factor S_c. Such a value has physical dimensions different from those of $(dW/dV)_c$ and, as will be shown later, this produces the scale effects recurrent in fracture mechanics. As a matter of fact, the crack instability may precede or follow the structural instability. This depends mainly on the structural scale. When the structure is relatively large, the crack instability precedes the traditional structural instability and Linear Elastic Fracture Mechanics ban be correctly applied. On the other hand, when the structure is relatively small, as in the case of fracture toughness specimens, the structural collapse precedes the unstable crack propagation. The latter may occur only later, when the structural behaviour is in the softening stage (strain-controlled loading process) and the loading capacity is remarkably reduced. It is interesting to observe that, after the first 2 cm of crack growth, the post-collapse structural behaviour is not considerably affected by the variation in $(dW/dV)_c$, Figure 16.

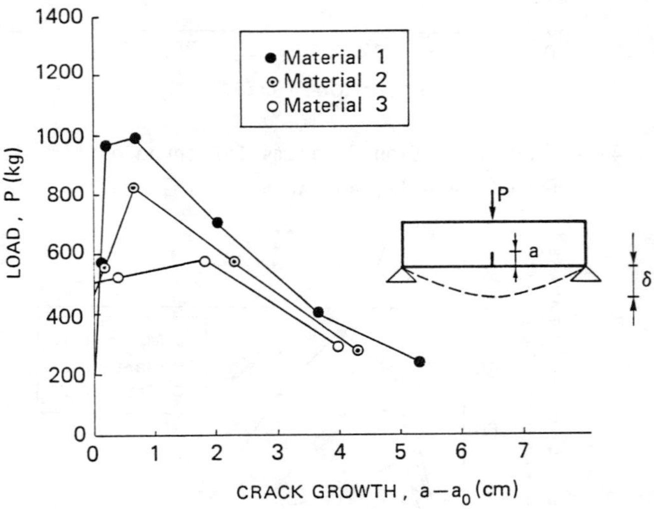

Fig. (16) - Load-crack growth diagrams for three different materials (see Figure 12) and $\Delta\delta = 4 \times 10^{-3}$ cm

In Figure 17, the values of the strain energy density factor, S, are reported against the crack growth, $(a-a_0)$. They appear nearly aligned. The results reported in Figure 17 relate only to small crack growth increments, when the structure is still in a stable P-δ ascending condition. Observe that the strain energy density factor rate, dS/da, has the same physical dimensions of the strain energy density and that it is always lower than the corresponding $(dW/dV)_c$. The more important the damage is, compared with the stable crack growth, the smaller the ratio $(dS/da)/(dW/dV)_c$. We may expect this ratio to tend to zero when $\varepsilon_f \to \infty$ (elas-

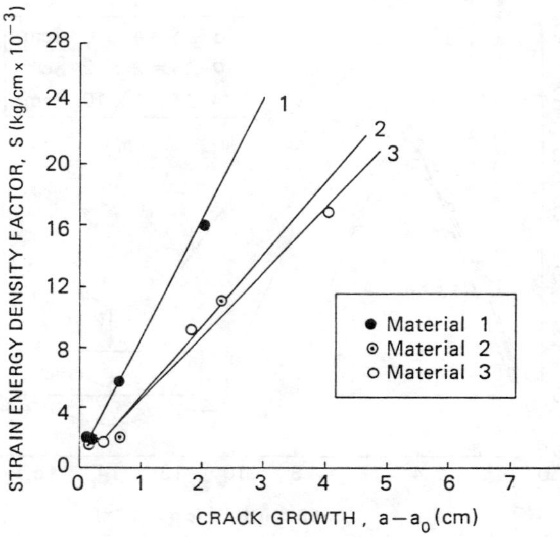

Fig. (17) - Strain energy density factor versus crack growth for three different materials (see Figure 12) and $\Delta\delta = 4 \times 10^{-3}$ cm

tic-perfectly plastic material) and to unity when $\varepsilon_f \to \varepsilon_u$ (elastic-perfectly brittle material).

C. Effect of Loading Step

In Figure 18, the load-deflection curves are displayed for material 3 and three different deflection increments: $\Delta\delta = 4 \times 10^{-3}$ cm, 2×10^{-3} cm, 1×10^{-3} cm. The maximum load, P_{max}, increases by increasing the deflection increment.

In Figure 19, the ratio of stiffness loss by material damage to total stiffness loss is reported as a function of the deflection, δ. It presents the same trends as discussed in section B, i.e., damage prevails over crack growth only for intermediate crack lengths. Moreover, the stiffness loss by crack growth tends to prevail over the stiffness loss by material damage when the load step, $\Delta\delta$, decreases.

In Figure 20, the load P is plotted against the crack growth, $(a-a_o)$. As was already discussed in section B, the maximum of these curves represents the transition between stable and unstable structural behaviour and not necessarily between stable and unstable crack propagation. Also, the load-crack growth diagram is not very strongly affected by the load step, $\Delta\delta$, in the post-collapse condition.

In Figure 21, the values of the strain energy density factor, S, are reported versus the crack growth. As already shown in Figure 17, they appear aligned when

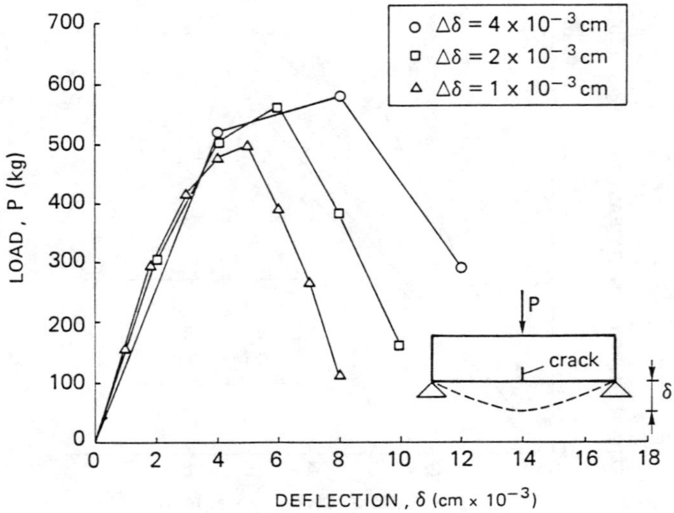

Fig. (18) - Load-deflection diagrams for Material 3 (see Figure 12) and three different loading increments

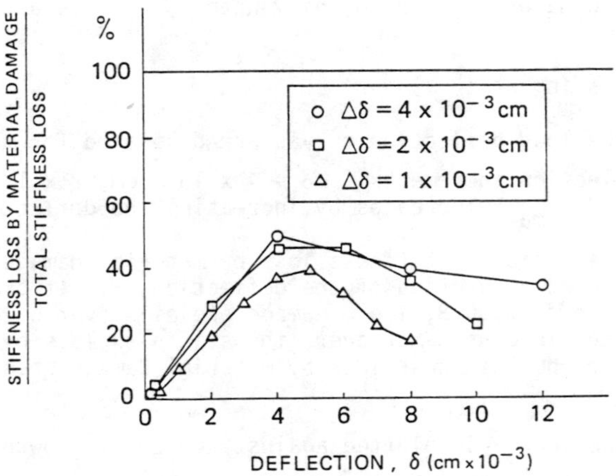

Fig. (19) - Ratio of stiffness loss by material damage to total stiffness loss as a function of the imposed deflection and for three different loading increments (Material 3 in Figure 12)

both crack growth and structural behaviour are stable. The higher the deflection increment, $\Delta\delta$, is, the steeper the linear S-a variation appears to be. This means

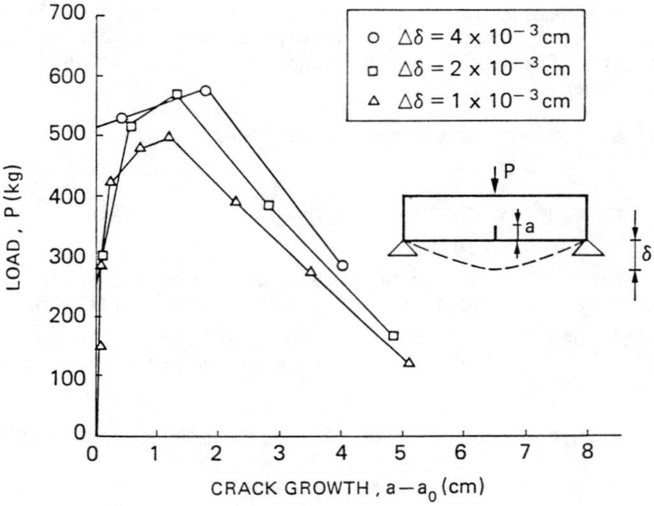

Fig. (20) - Load-crack growth diagrams for three different loading increments and Material 3 (see Figure 12)

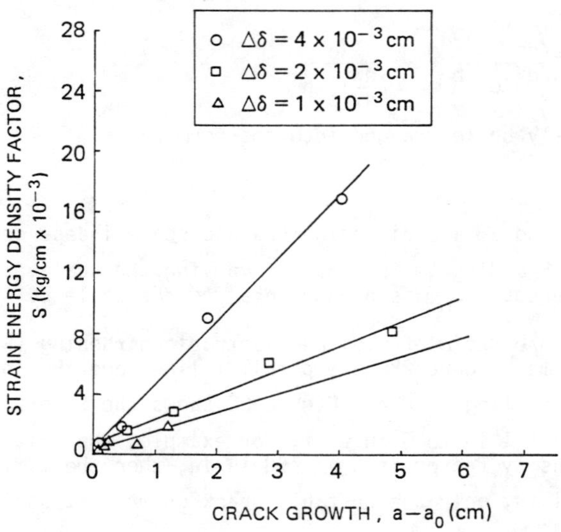

Fig. (21) - Strain energy density factor versus crack growth for three different loading increments and Material 3 (see Figure 12)

that, the critical value of the strain energy density factor, S_c, being constant, with larger load steps the crack instability is achieved for smaller crack lengths.

SIZE EFFECTS ON STRENGTH AND DUCTILITY

Because material damage and crack growth occur in a non self-similar fashion for each step of loading, specimens of different sizes appear to behave differently. This well-known size effect will be analyzed by the results obtained on the three-point bending specimen.

A dimensionless load parameter may be defined:

$$\frac{P}{(\frac{dW}{dV})_c\ b^2} = \pi[\frac{\delta}{b}, \frac{E}{(\frac{dW}{dV})_c}, \frac{\sigma_u}{(\frac{dW}{dV})_c}, \nu, \frac{l}{b}, \frac{t}{b}, \frac{a_o}{b}] \tag{9}$$

as in the Buckingham's theorem for physical similitude and scale modelling. In equation (9), material toughness, $(dW/dV)_c$ and specimen width, b, have been used as the fundamental quantities. The dimensionless load parameter may be regarded as a function of the dimensionless deflection parameter δ/b only, if all others are kept constant.

In the same way, it is possible to define a dimensionless strain energy density factor:

$$\frac{S}{(\frac{dW}{dV})_c\ b} = \Sigma[\frac{a}{b}, \frac{E}{(\frac{dW}{dV})_c}, \frac{\sigma_u}{(\frac{dW}{dV})_c}, \nu, \frac{l}{b}, \frac{t}{b}, \frac{a_o}{b}] \tag{10}$$

Function Σ can be regarded as linear in a/b [5-8]:

$$\frac{S}{(\frac{dW}{dV})_c\ b} = \frac{dS/da}{(dW/dV)_c}\frac{a-a_o}{b} + \frac{S_o}{(\frac{dW}{dV})_c\ b} \tag{11}$$

which may obviously be rearranged into the form [7]:

$$\tilde{S} = \tilde{S}'(\xi-\xi_o) + \tilde{S}_o \tag{12}$$

The constants $\tilde{S}'$ and $\tilde{S}_o$ are dimensionless and scale independent. It follows that the slope of the S-a diagram is constant varying the scale, i.e., dS/da = constant, and that the intercept, S_o, is proportional to the scale b.

For a fixed $\Delta\delta/b$ ratio of 0.01 and isotropic hardening (β=0), all the S versus $(a-a_o)$ diagrams, Figure 22, are parallel lines and the intercept S_o is proportional to the scaling size b. Figure 22 shows the S versus $(a-a_o)$ shifting with b varying from 1 in to 5 in: if, for example, the critical value of the strain energy density factor is S_c = 150 lb/in, then the limiting size is b = 441 in. For b > 4.41 in, only the unstable crack growth is possible.

Figure 23 relates to the softening Material 1, Figure 12, and gives the straight line plots of S versus $(a-a_o)$ with b varying from 15 cm upwards. The critical crack growth decreases with increasing specimen size. With the critical value S_c = 8 x 10^{-3} kg/cm, the limiting size is b = 120 cm. Beyond this size, stable crack

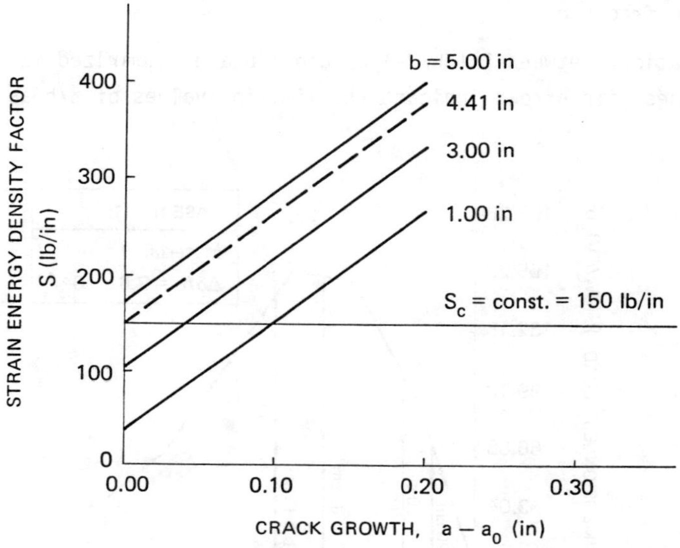

Fig. (22) - Strain energy density factor versus crack growth by varying the specimen size-scale. Isotropic hardening material, $\Delta\delta/b = 0.01$

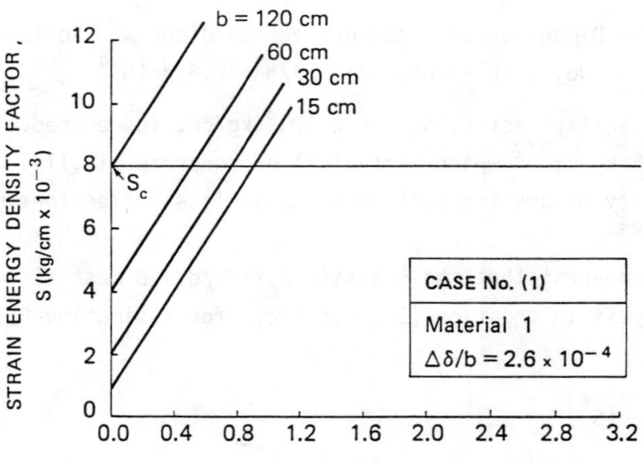

Fig. (23) - Strain energy density factor versus crack growth by varying the specimen size-scale. Softening material (No. 1 in Figure 12), $\Delta\delta/b = 2.6 \times 10^{-4}$

growth ceases to occur and failure corresponds to unstable crack propagation or catastrophic fracture.

The relations between $P/(dW/dV)_c b^2$ and δ/b are summarized in Figure 24. The vertical lines with arrows indicate the limiting values of δ/b as the critical

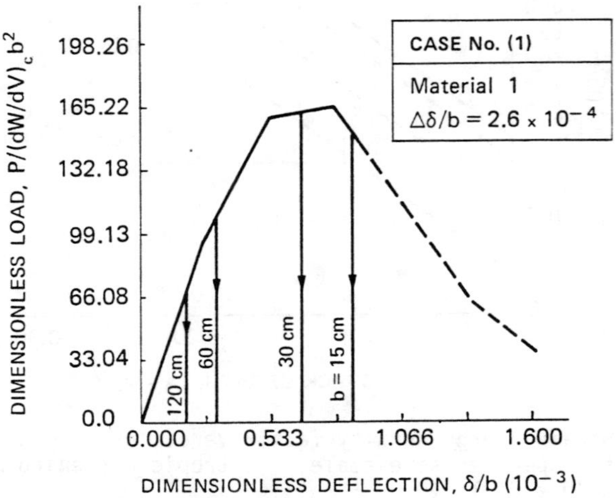

Fig. (24) - Dimensionless load-deflection diagram. Softening material (No. 1 in Figure 12), $\Delta\delta/b = 2.6 \times 10^{-4}$

strain energy density factor, $S_c = 8 \times 10^{-3}$ kg/cm, is reached. This corresponds to $K_{IC} = 144.35$ kg/cm$^{3/2}$ which is typical of concrete [10,11]. It is evident that crack instability occurs for smaller dimensionless deflections of the specimen as size b increases.

It is now apparent that the quantity $S_c/(dW/dV)_c b$ must also enter into the dimensional analysis in equation (9). In fact, for estimating P_{max}, it suffices to consider:

$$\frac{P_{max}}{(\frac{dW}{dV})_c \; b^2} = \pi(S^*) \tag{13}$$

in which S^* is a dimensionless quantity:

$$S^* = \frac{S_c}{(\frac{dW}{dV})_c \; b} = \frac{r_c}{b} \tag{14}$$

Hence, all geometrically similar structures can be regarded as governed by S^*.

This dimensionless quantity can be used to predict the load versus deflection behaviour for all specimen sizes [12].

In conclusion, and recalling equation (12), it is possible to state that above the size, Figure 25

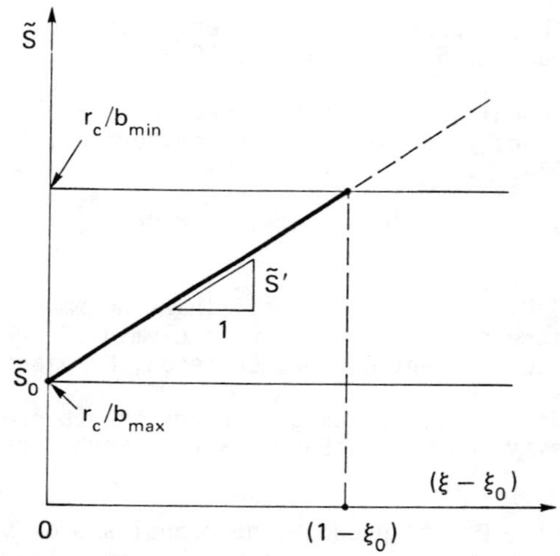

Fig. (25) - Dimensionless strain energy density factor versus crack growth

$$b_{max} = \frac{S_c}{(\frac{dW}{dV})_c \tilde{S}_o} = \frac{r_c}{\tilde{S}_o} \quad (15)$$

stable crack growth ceases to occur and the brittle failure is achieved when the P-δ curve is still in its linear initial course, Figure 24, whereas below the size, Figure 25:

$$b_{min} = \frac{S_c}{(\frac{dW}{dV})_c[\tilde{S}_o+\tilde{S}'(1-\xi_o)]} = \frac{r_c}{\tilde{S}_o+\tilde{S}'(1-\xi_o)} \quad (16)$$

unstable crack growth ceases to occur (even in the softening stage) and the progressive slow crack growth develops up to the complete specimen separation. On the other hand, for $b_{min} < b < b_{max}$, stable (or slow) crack growth is followed by unstable (or fast) crack propagation. Ductility appears therefore as a mechanical property depending on the size-scale rather than on the geometrical shape of the structure.

ACKNOWLEDGEMENTS

The author gratefully acknowledges the "Ministero Pubblica Istruzione" for the financial support received in the development of the present research. Thanks are also due to Professor G. C. Sih, Director of the Institute of Fracture and Solid Mechanics, Lehigh University, Bethlehem, Pennsylvania USA for the very stimulating and fruitful discussions the author had with him during his sabbatical time at Lehigh University.

REFERENCES

[1] Sih, G. C., "Some basic problems in fracture mechanics and new concepts", Eng. Fracture Mechanics, 5, pp. 365-377, 1973.

[2] Sih, G. C. and Macdonald, B., "Fracture mechanics applied to engineering problems - strain energy density fracture criterion", Eng. Fracture Mechanics, 6, pp. 361-386, 1974.

[3] Martin, J. B., Plasticity: Fundamentals and General Results, MIT Press, Cambridge, 1975.

[4] Gifford, L. N. and Hilton, P. D., "Preliminary documentation of PAPST nonlinear fracture and stress analysis by finite elements", David W. Taylor Naval Ship Research and Development Center, Bethesda, Maryland, 1981.

[5] Sih, G. C. and Madenci, E., "Crack growth resistance characterized by the strain energy density function", Eng. Fracture Mechanics, 18, pp. 1159-1171, 1983.

[6] Sih, G. C. and Matic, P., "A pseudo-linear analysis of yielding and crack growth: strain energy density criterion", Defects, Fracture and Fatigue, edited by G. C. Sih and J. W. Provan, Martinus Nijhoff Publishers, pp. 223-232, 1983.

[7] Carpinteri, A. and Sih, G. C., "Damage accumulation and crack growth in bilinear materials with softening", Theoretical and Applied Fracture Mechanics, 1, pp. 145-159, 1984.

[8] Carpinteri, A., "Crack growth resistance in non-perfect plasticity: isotropic versus kinematic hardening", Theoretical and Applied Fracture Mechanics, 4, pp. 117-122, 1985.

[9] Carpinteri, A., "Interpretation of the Griffith instability as a bifurcation of the global equilibrium", Application of Fracture Mechanics to Cementitious Composites, NATO Advanced Research Workshop, September 4-7, 1984, Northwestern University, edited by S. P. Shah, Martinus Nijhoff Publishers, pp. 287-316, 1985.

[10] Carpinteri, A., "Static and energetic fracture parameters for rocks and concretes", Materials and Structures, RILEM, 14, pp. 151-162, 1981.

[11] Carpinteri, A., "Application of fracture mechanics to concrete structures", J. of the Structural Division, ASCE, 108, pp. 833-848, 1982.

[12] Carpinteri, A., "Notch sensitivity in fracture testing of aggregative materials", Eng. Fracture Mechanics, 16, pp. 467-481, 1982.

FAILURE TRAJECTORY INITIATING FROM AN ELLIPTICAL INCLUSION

R. C. Chu

Aeronautic Research Laboratory, AIDC/CSIST
Taiwan, Republic of China

ABSTRACT

This work is concerned with the assessment of material inhomogeneity, on failure behavior of a solid. An inclusion of dissimilar material embedded in a matrix is depicted as the analytical model.

The stress analysis of an elastic elliptical inclusion is considered here with the objective of casting the solution to a form that is amendable to failure prediction based on the strain energy density theory. A single inclusion is studied and some insights into the effect of material inhomogeneity are provided for the failure of the matrix material subjected to static load.

In this paper, the influences of load angle with different degrees of material inhomogeneity are reviewed. The locations of failure initiating from an inclusion are also seen predicted and it shows the same result which has been observed by Kipp and Sih. The fracture trajectories for elements that are outside of the interface layer are projected. All fracture paths for several angles of loading are seen to initiate in an almost normal to the inclusion boundary. This is the same trend as the notch case which has been observed by Sih.

INTRODUCTION

Fracture analysis of inclusions can be found only in a few special cases [1, 2]. Because of the complexity of material nonhomogeneity, work in this area is limited. Although the general solution for an elliptical inclusion is already known [3-5], and the determination of stresses around elastic inclusions has been discussed by many previous investigators [6-8], very little is known with regard to their failure behavior, particularly when the inclusion geometry becomes slender resulting in the high elevation of local stresses [1]. The stress analysis of an elastic elliptical inclusion is reconsidered here with the objective of casting the solution to a form that is amendable to failure prediction based on the strain energy density theory. The knowledge gained for the case of a single inclusion can provide some insights into the effect of material inhomogeneity.

STRESS DISTRIBUTION AROUND AN ELLIPTICAL INCLUSION

For bodies that are nonhomogeneous in nature, it is no longer clear whether fracture will always occur in one phase of the system. As in the case of composite materials, failure will depend on the nature of loading. Static and dynamic loading can lead to very different failure modes as the interaction of material inhomogeneity and energy transfer rate is complex. In order to gain some fundamental understanding of this effect, the present study will be limited to a single inclusion while attention will be first given to the failure of the matrix material subjected to static load.

The geometry to be considered is that of an elastic elliptical inclusion with semi-axes a and b in an infinite medium subjected to tension T that is inclined at an angle γ to the x-axis as shown in Figure 1. The Lame's constants for the medium are λ_2, μ_2 and for the inclusion are λ_1, μ_1. The same Poisson's ratio $\nu_1 = \nu_2 = \nu$ is assumed. By application of the minimum energy theorem [9] and complex variable method, the stress combination in the medium can be expressed in terms of two complex functions $\phi_2(z)$ and $\psi_2(z)$ of the complex variable z = x + iy:

$$(\sigma_x)_2 + (\sigma_y)_2 = 4\mathrm{Re}[\phi_2'(z)]$$
$$(\sigma_y)_2 - (\sigma_x)_2 + 2i(\sigma_{xy})_2 = 2[\bar{z}\phi_2''(z) + \psi_2'(z)] \tag{1}$$

in which the subscript 2 refers to the medium while 1 refers to the inclusion. The complex functions $\phi_2(z)$ and $\psi_2(z)$ can be expressed as

$$\phi_2(z) = \phi_{2a}(z) + \phi_{2b}(z)$$
$$\psi_2(z) = \psi_{2a}(z) + \psi_{2b}(z) \tag{2}$$

The complex functions $\phi_{2a}(z)$, $\psi_{2a}(z)$ correspond to the medium due to the disturbance of the inclusion and they are given by

$$\phi_{2a}(z) = \frac{2RT}{kc^2}[z - (z^2-c^2)^{1/2}](B + Di)$$
$$\psi_{2a}(z) = \frac{2RT}{c^2}\{[\frac{mc^2+4R^2}{2k}]\frac{B+Di}{(z^2-c^2)^{1/2}} + [(A + \frac{4R^2}{kc^2}B) \tag{3}$$
$$+ i(\frac{4R^2}{kc^2}D - C)](z^2-c^2)^{1/2} + [-(A + \frac{4R^2}{kc^2}B) + i(C - \frac{4R^2}{kc^2}D)](z)\}$$

in which R, m and c are defined as:

$$R = \frac{a+b}{2};\ m = \frac{a-b}{a+b};\ c = (a^2-b^2)^{1/2} \tag{4}$$

The quantities

$$\phi_{2b}(z) = \frac{T}{4}\{\frac{4R^2}{c^2}\cos2\gamma[z - (z^2-c^2)^{1/2}] + (z^2-c^2)^{1/2}$$
$$- i[\frac{4R^2}{c^2}\sin2\gamma(z^2-c^2)^{1/2}]\}$$

$$\psi_{2b}(z) = -\frac{T}{4}\{[a^2 + b^2 - c^2\cos 2\gamma]\cosh^{-1}(\frac{z}{c}) + \frac{2R^2}{c^2}$$

$$\times [\frac{a^2+b^2}{c^2}(2z^2-c^2) - \frac{4ab}{c^2}\cos 2\gamma z(z^2-c^2)^{1/2}]$$

$$+ i\frac{2R^2}{c^2}[\frac{2ab}{c^2}\sin 2\gamma(2z^2-c^2) - \frac{2(a^2+b^2)}{c^2}\sin 2\gamma z(z^2-c^2)^{1/2}]\} \quad (5)$$

account for the influence of the notch, i.e., the medium without inclusion. The parameters A, B,...,D are the elastic constants of the medium and inclusion and they are given as

$$A = -\frac{(k+1)(a+b)}{4P}\{[\mu_2\lambda_1(b^2+a^2) + \mu_1\mu_2(2a^2+2b^2-2ab) + 2abk(\mu_1^2+\mu_1\lambda_1)]$$

$$+ \cos 2\gamma[\mu_2\lambda_1(b^2-a^2) + \mu_1\mu_2(2b^2-2a^2)]\}$$

$$B = -\frac{k(k+1)}{4P}\{(b-a)[\mu_2\lambda_1(a^2+b^2) + \mu_1\mu_2(2a^2+2b^2+2ab)$$

$$+ \lambda_1\mu_1(2ab) + \mu_1^2(wab)] + (a+b)\cos 2\gamma[\mu_2\lambda_1(b-a)^2$$

$$+ \mu_1\mu_2(2b^2+2a^2) + (4ab)(\lambda_1\mu_1+\mu_1^2)]\}$$

$$C = \frac{1}{4Q}\mu_1(a-b)(a+b)^2(k+1)\sin 2\gamma$$

$$D = -\frac{1}{4Q}\mu_1(a+b)^3k(k+1)\sin 2\gamma$$

with

$$P = 4ab[\mu_2^2 + \mu_1k(\lambda_1+\mu_1)] + \lambda_1\mu_2[(a+b)^2 + (a-b)^2k] + 2\mu_1\mu_2(k+1)(a^2+b^2)$$

$$Q = 4\mu_2ab + \mu_1[(a-b)^2 + k(a+b)^2]$$

In order to find the stresses ahead of the inclusion, ϕ_2 and ψ_2 will be expanded asymptotically. An arbitrary point $A(r,\theta)$ ahead of the inclusion will be considered, where (r,θ) is measured from the focal point of the ellipse, Figure 2. The variable z can be expanded in series form as

$$z = a(1 + \frac{\alpha\rho}{a} + \frac{\beta\rho}{a}i) \quad (6)$$

where α, β are the nondimensional parameters that define the position of A and $\rho = (b^2/a)$ is the radius of curvature at the end of the major axis with 2c being the focal length of the ellipse. For a sharp ellipse, i.e., $(\rho/a) << 1$, z^2 can be approximated as

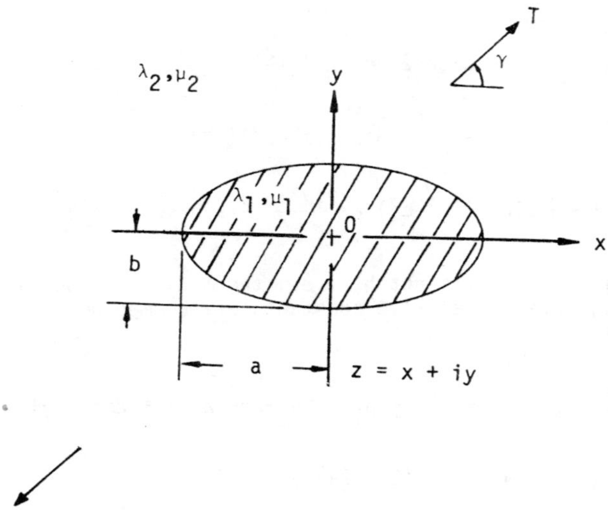

Fig. (1) - Elastic elliptical inclusion in a stressed medium

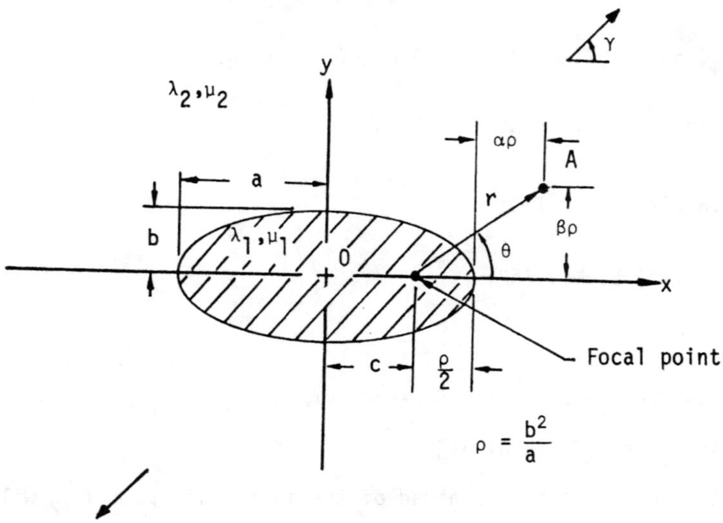

Fig. (2) - Polar coordinates with center at the focal point of ellipse

$$z^2 \simeq a^2(1 + 2\alpha \frac{\rho}{a} + 2\beta \frac{\rho}{a} i) \tag{7}$$

and

$$z^2 - c^2 = 2a\rho \left(\frac{1}{2} + \alpha + i\beta\right) = 2a^2 e^{i\theta} \frac{r}{a} \tag{8}$$

Furthermore, by letting $g = (\rho/a)$, it can be shown that

$$\frac{R^2}{c^2} = \frac{1+g^{1/2}}{4(1-g^{1/2})} = \frac{1}{4}(1 + g^{1/2})(1 + g^{1/2} + g + g^{3/2} + g^2 + \ldots)$$

$$\simeq \frac{1}{4}(1 + 2g^{1/2} + 2g) \tag{9}$$

where the high order terms can be neglected for a narrow elliptical inclusion. The following stress combinations are thus obtained by taking the derivatives of these complex functions $\phi_{2a}(z)$ and $\psi_{2a}(z)$:

$$\begin{aligned}
(\sigma_x)_{2a} = {} & \frac{2}{(a+b)}\left(-\frac{T\sqrt{a}}{\sqrt{2r}}\right)\{\cos\frac{\theta}{2}[(1 + \frac{2b}{a} + \frac{r}{a}\cos\theta)(\frac{3}{2}\frac{B}{k} + \frac{A}{2}) \\
& - (\frac{r}{a}\sin\theta)(\frac{3}{2}\frac{D}{k} - \frac{C}{2}) + \frac{B}{k}\frac{b}{a}] + \sin\frac{\theta}{2}[(1 + \frac{2b}{a} + \frac{r}{a}\cos\theta) \\
& \times (\frac{3}{2}\frac{D}{k} - \frac{C}{2}) + (\frac{r}{a}\sin\theta)(\frac{3}{2}\frac{B}{k} + \frac{A}{2}) + \frac{D}{k}\frac{b}{a}]\} \\
& + \frac{1}{2k(a+b)}\left(-\frac{T\sqrt{a}}{\sqrt{2r}}\right)\{(B\cos\frac{3\theta}{2} + D\sin\frac{3\theta}{2})[(\frac{ab}{a+b})\frac{1}{r} \\
& + \frac{b}{r}(\frac{b-a}{a+b}) + (\frac{b}{a+b})\cos\theta] + (\frac{2a+b}{a+b})(D\sin\theta\cos\frac{3\theta}{2} - B\sin\theta\sin\frac{3\theta}{2})\} \\[2ex]
(\sigma_y)_{2a} = {} & \frac{2}{(a+b)}\left(-\frac{T\sqrt{a}}{\sqrt{2r}}\right)\{\cos\frac{\theta}{2}[(1 + \frac{2b}{a} + \frac{r}{a}\cos\theta)(\frac{B}{2k} - \frac{A}{2}) \\
& - (\frac{r}{a}\sin\theta)(\frac{D}{2k} + \frac{C}{2}) - \frac{B}{k}\frac{b}{a}] + \sin\frac{\theta}{2}[(1 + \frac{2b}{a} + \frac{r}{a}\cos\theta) \\
& \times (\frac{D}{2k} + \frac{C}{2}) + (\frac{r}{a}\sin\theta)(\frac{B}{2k} - \frac{A}{2}) - \frac{D}{k}\frac{b}{a}]\} \\
& - \frac{1}{2k(a+b)}\left(-\frac{T\sqrt{a}}{\sqrt{2r}}\right)\{(B\cos\frac{3\theta}{2} + D\sin\frac{3\theta}{2})[(\frac{ab}{a+b})\frac{1}{r} \\
& + \frac{b}{r}(\frac{b-a}{a+b}) + (\frac{b}{a+b})\cos\theta] + (\frac{2a+b}{a+b})(D\sin\theta\cos\frac{3\theta}{2} - B\sin\theta\sin\frac{3\theta}{2})\} \\[2ex]
(\sigma_{xy})_{2a} = {} & \frac{1}{2k(a+b)}\left(-\frac{T\sqrt{a}}{\sqrt{2r}}\right)\{(B\sin\frac{3\theta}{2} - D\cos\frac{3\theta}{2})[(\frac{ab}{a+b})\frac{1}{r} \\
& + (\frac{b-a}{a+b})\frac{b}{r} + (\frac{b}{a+b})\cos\theta] + (\frac{2a+b}{a+b})(B\sin\theta\cos\frac{3\theta}{2} \\
& + D\sin\theta\sin\frac{3\theta}{2})\} + \frac{1}{(a+b)}\left(-\frac{T\sqrt{a}}{\sqrt{2r}}\right)\{-\cos\frac{\theta}{2}[(\frac{D}{k} - C) \\
& \times (1 + \frac{2b}{a} + \frac{r}{a}\cos\theta) + \frac{r}{a}\sin\theta(A + \frac{B}{k}) + 2\frac{D}{k}\frac{b}{a}] \\
& + \sin\frac{\theta}{2}[(A + \frac{B}{k})(1 + \frac{2b}{a} + \frac{r}{a}\cos\theta) - \frac{r}{a}\sin\theta(\frac{D}{k} - C) + 2\frac{B}{k}\frac{b}{a}]\}
\end{aligned} \tag{10}$$

The stresses are singular of the order $1/\sqrt{r}$ which is the same as that for a notch [10]. The angular variations are affected by both the geometry and material of the inclusion through the constants a, b, A, B, C and D.

In a similar fashion, the asymptotic expansion of the stresses corresponding to $\phi_{2b}(z)$ and $\psi_{2b}(z)$ can be found:

$$
\begin{aligned}
(\sigma_x)_{2b} = {} & \sin^2\gamma \left(\frac{T\sqrt{a}}{\sqrt{2r}}\right) \Big[\Big(\cos\frac{\theta}{2} - \sin\frac{\theta}{2}\cos\frac{\theta}{2}\sin\frac{3\theta}{2} - \frac{1}{2}\frac{b^2}{ar}\cos\frac{3\theta}{2}\Big) \\
& + g^{1/2}\Big(\cos\theta\cos\frac{3\theta}{2} + \frac{1}{2}\sin\theta\sin\frac{3\theta}{2} + \cos\frac{\theta}{2}\Big) \\
& + g\Big(\frac{ra}{b^2}\sin\theta\sin\frac{\theta}{2} + \frac{ra}{b^2}\cos\theta\cos\frac{\theta}{2} + \frac{3}{2}\cos\theta\cos\frac{3\theta}{2} \\
& + \frac{3}{2}\sin\theta\sin\frac{3\theta}{2} + \frac{1}{2}\cos\frac{\theta}{2}\Big)\Big] + \sin\gamma\cos\gamma\left(\frac{T\sqrt{a}}{\sqrt{2r}}\right)\Big[\Big(-2\sin\frac{\theta}{2} \\
& - \sin\frac{\theta}{2}\cos\frac{\theta}{2}\cos\frac{3\theta}{2} + \frac{1}{2}\frac{b^2}{ra}\sin\frac{3\theta}{2}\Big) + g^{1/2}\Big(-2\sin\frac{\theta}{2} \\
& + \sin\theta\cos\frac{3\theta}{2} - 2\cos\theta\sin\frac{3\theta}{2}\Big) + g\Big(\frac{ra}{b^2}\cos\frac{\theta}{2}\sin\theta \\
& - \frac{ra}{b^2}\cos\theta\sin\frac{\theta}{2} - \frac{3}{2}\sin\frac{\theta}{2} + \frac{7}{2}\sin\theta\cos\frac{3\theta}{2} - 4\cos\theta\sin\frac{3\theta}{2}\Big]
\end{aligned}
$$

$$
\begin{aligned}
(\sigma_y)_{2b} = {} & \sin^2\gamma \left(\frac{T\sqrt{a}}{\sqrt{2r}}\right) \Big[\Big(\cos\frac{\theta}{2} + \sin\frac{\theta}{2}\cos\frac{\theta}{2}\sin\frac{3\theta}{2} + \frac{1}{2}\frac{b^2}{ra}\cos\frac{3\theta}{2}\Big) \\
& + g^{1/2}\Big(-\cos\theta\cos\frac{3\theta}{2} - \frac{1}{2}\sin\theta\sin\frac{3\theta}{2} + \cos\frac{\theta}{2}\Big) \\
& + g\Big(\frac{ra}{b^2}\sin\theta\sin\frac{\theta}{2} + \frac{ra}{b^2}\cos\theta\cos\frac{\theta}{2} - \frac{3}{2}\cos\theta\cos\frac{3\theta}{2} \\
& - \frac{3}{2}\sin\theta\sin\frac{3\theta}{2} + \frac{1}{2}\cos\frac{\theta}{2}\Big)\Big] + \sin\gamma\cos\gamma\left(\frac{T\sqrt{a}}{\sqrt{2r}}\right)\Big[\Big(\sin\frac{\theta}{2} \\
& \times \cos\frac{\theta}{2}\cos\frac{3\theta}{2} - \frac{1}{2}\frac{b^2}{ra}\sin\frac{3\theta}{2}\Big) + g^{1/2}\Big(-2\sin\frac{\theta}{2} - \sin\theta\cos\frac{3\theta}{2} \\
& + 2\cos\theta\sin\frac{3\theta}{2}\Big) + g\Big(\frac{ra}{b^2}\sin\theta\cos\frac{\theta}{2} - \frac{ra}{b^2}\cos\theta\sin\frac{\theta}{2} - \frac{3}{2}\sin\frac{\theta}{2} \\
& - \frac{7}{2}\sin\theta\cos\frac{3\theta}{2} - 4\cos\theta\cos\frac{3\theta}{2}\Big)\Big]
\end{aligned}
$$

$$(\sigma_{xy})_{2b} = \sin^2\gamma \left(\frac{T\sqrt{a}}{\sqrt{2r}}\right) [(\cos\frac{3\theta}{2} \sin\frac{\theta}{2} \cos\frac{\theta}{2} - \frac{1}{2}\frac{b^2}{ra} \sin\frac{3\theta}{2})$$

$$+ g^{1/2} (\cos\theta\sin\frac{3\theta}{2} - \frac{1}{2}\sin\theta\cos\frac{3\theta}{2}) + g (\frac{3}{2}\cos\theta\sin\frac{3\theta}{2}$$

$$- \frac{3}{2}\sin\theta\cos\frac{3\theta}{2})] + \sin\gamma\cos\gamma \left(\frac{T\sqrt{a}}{\sqrt{2r}}\right)[(\cos\frac{\theta}{2} - \sin\frac{\theta}{2}$$

$$\times \cos\frac{\theta}{2} \sin\frac{3\theta}{2} - \frac{1}{2}\frac{b^2}{ra}\cos\frac{3\theta}{2}) + g^{1/2} (2\cos\theta + \sin\theta\sin\frac{3\theta}{2})$$

$$+ g (4\cos\theta + \frac{7}{2}\sin\theta)] \qquad (11)$$

In accordance with equations (2), the total near field stresses are obtained by adding the results in equations (10) and (11). Several special cases of this general solution prevail. When $\lambda_1 = 0$ and $\mu_1 = 0$, the solution for an elliptical hole with a very small radius of curvature $\rho \simeq 0$ can be recovered. In this case, all the coefficients A, B, C and D in equations (10) are zero and $(\sigma_x)_{2a}$, $(\sigma_y)_{2a}$ and $(\sigma_{xy})_{2a}$ vanish. The only contribution arises from equations (11). Moreover, the terms containing $g^{1/2}$, g and (b^2/ra) in equations (11) can be dropped as $(b/a \to 0)$ for a line crack problem. These results are the same as those formed in [10] in which a different approach was used. Another special solution of equations (10) and (11) is that of rigid line inclusion with $\lambda_1 = \mu_1 = \infty$ and $b = 0$. This case is given in [1].

FAILURE EMANATING FROM INTERFACE LAYER

When the defect contour is smooth and loading does not coincide with any geometric axes of symmetry, the location of failure initiation is no longer known as a priori*. In such a case, an additional condition is required for determining the failure points on the boundary in addition to those in the interior of the solid. To this end, Sih [11] has proposed the introduction of a finite surface boundary layer**. In consistent with the strain energy density criterion, the energy per unit area, say γ_e, in a layer δ can be computed:

$$\gamma_e = \left(\frac{dW}{dV}\right) \qquad (12)$$

in which (dW/dV) is the strain energy density function, and it can be calculated from the stresses and strains in accordance with equations (10) and (11). The location at which γ_e attains a maximum is assumed to coincide with failure initiation. Referring to Figure 3, the same concept applies to failure initiating

---

* In the case of a sharp crack, failure is assumed to be connected with the crack tip whose location is known.

** Surface is an inherent part of the solid. In practice, it is necessary to associate the behavior of the solid near the surface by averaging over a finite layer distance.

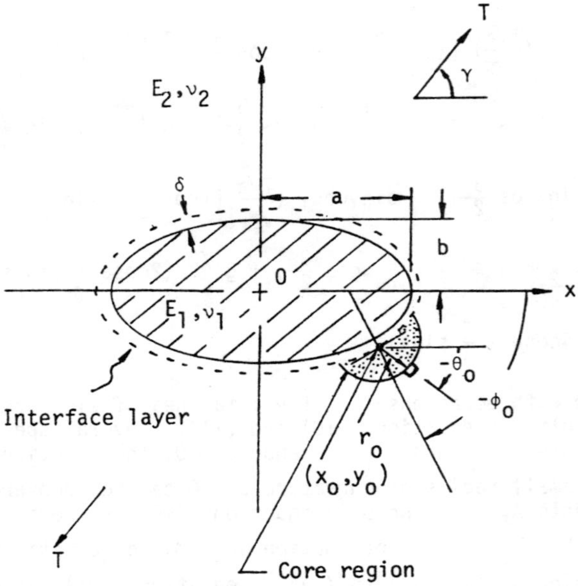

Fig. (3) - Interface layer and core region for an elastic elliptical inclusion in infinite medium under inclined uniform tensile load

from the interface between the inclusion and its surrounding material. An interface layer δ is considered such that failure is assumed to start at some point (x_o, y_o) corresponding to γ_e maximum and connected to an element outside the core region with radius r_o. A critical value $(dW/dV)_c$ is assumed to exist for the failure of the interface layer and the nearest neighboring element.

A. Influence of Load Angle

Figure 4 displays the variations of normalized maximum strain energy density function, $(4\mu/T^2)/(dW/dV)_{max}$, with the angle of loading, γ, for $(b/a) = 0.1$ and $(E_1/E_2) = 0.1, 0.2, \ldots, 0.5$ as shown. All curves exhibit a maximum at $\gamma = 70°$ except for $(E_1/E_2) = 0.1$ where $(dW/dV)_{max}$ occurred at $\gamma = 90°$. This conclusion also holds for the case of a notch [11]. Figure 5 illustrates the influence of decreasing (b/a) by an order of magnitude, i.e., $(b/a) = 0.01$. The peak values of (dW/dV) shifted to $\gamma = 60°$ for all the values of (E_1/E_2) calculated. The effect of (b/a) is illustrated in Figure 6 for $(E_1/E_2) = 0$ and $(E_1/E_2) = 0.1$. All the curves rised monotonically to a maximum at $\gamma = 90°$. The values of (dW/dV) are generally decreased for the inclusion $(E_1/E_2) \neq 0$ as compared to those for the notch with $(E_1/E_2) = 0$.

When the load is compressive, (dW/dV) tends to rise more sharply to a maximum and then decreases with the load angle γ as shown in Figure 7 for different (E_1/E_2) values and $(b/a) = 0.1$. In this case, $(dW/dV)_{max}$ occurs between $\gamma \simeq 40°$

as 50° as (E_1/E_2) is varied from 0.1 to 0.5.

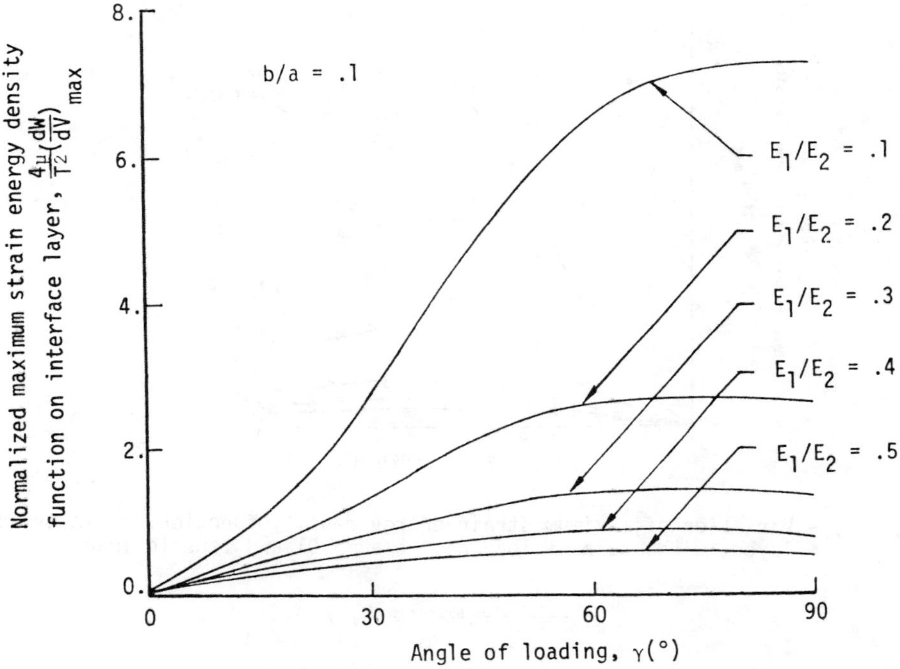

Fig. (4) - Variation of maximum strain energy density function, on interface layer, with angle of loading: b/a = .1 and tensile load

The ratio (E_1/E_2) is seen to affect the fluctuation of (dW/dV) within the medium that contains the inclusion regardless of whether the load is tensile or compressive. This influence is seen to depend on the slenderness of the elliptical inclusion and becomes more pronounced as (E_1/E_2) is decreased approaching towards a narrow cavity or crack.

B. Failure Initiation Angle

Referring to Figure 3, the location of failure initiation is described by the angle ϕ which will change as the direction of T is altered by γ. The sign of ϕ will depend on whether T is positive or negative and can be determined from the stationary values of (dW/dV) as discussed earlier in connection with γ_e in equation (12). More precisely, it is the location where (dW/dV) attains its maximum that is assumed to coincide with failure initiation.

For (b/a) = 0.1, Figure 8 gives a plot of $-\phi$ with γ for T positive and (E_1/E_2) = 0, 0.1, etc. The angle $-\phi$ is seen to decrease as γ is increased from 0° to 90°. The decrease is most severe as (E_1/E_2) approaches zero for the case of a cavity.

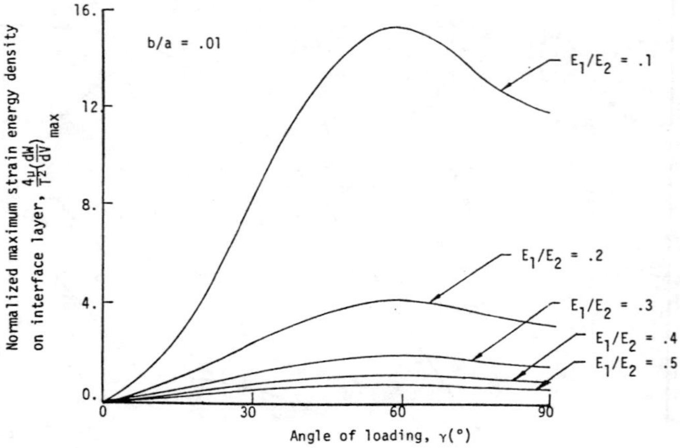

Fig. (5) - Variation of maximum strain energy density function, on interface layer, with angle of loading: $b/a = .01$ and tensile load

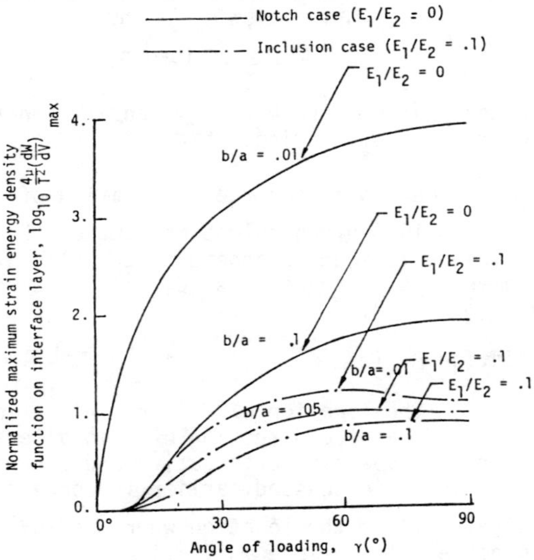

Fig. (6) - Comparison of $\frac{4\mu}{T^2}$ $(dW/dV)_{max}$ on interface layer between inclusion case and notch case: tensile load

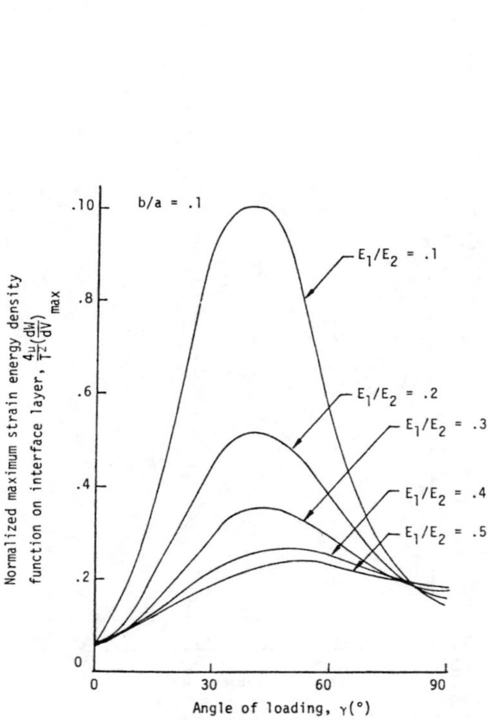

Fig. (7) - Variation of maximum strain energy density function, on interface layer, with angle of loading: b/a = .1 and compressive load

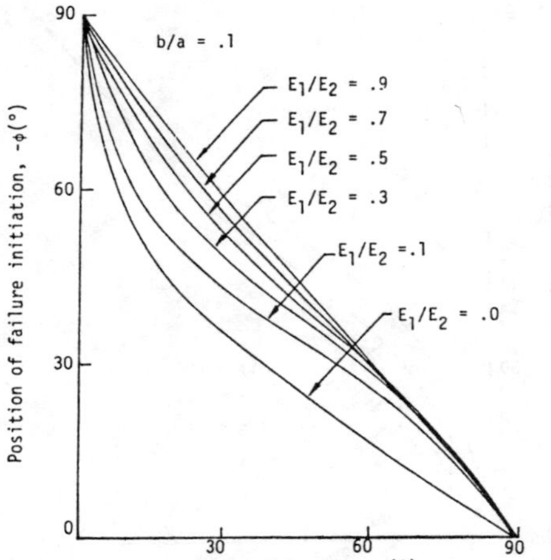

Fig. (8) - Variation of initial failure position with angle of loading: b/a = .1 and tensile load

Similar results are shown in Figure 9 for (b/a) = 0.01. The curves for (E_1/E_2) > 0.1 tends to intersect at $\gamma \simeq 55°$. The opposite trend is observed in Figure 10 when T is negative, i.e., the failure initiation ϕ which is now positive increases with increasing γ for all (E_1/E_2) ratios. This means that failure tends to initiate

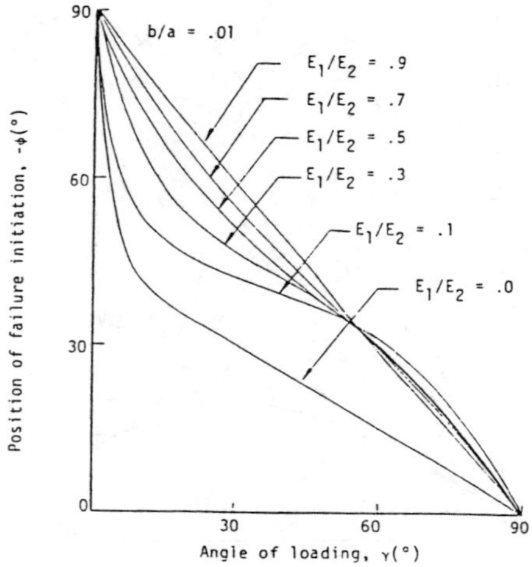

Fig. (9) - Variation of initial failure position with angle of loading: $b/a = .01$ and tensile load

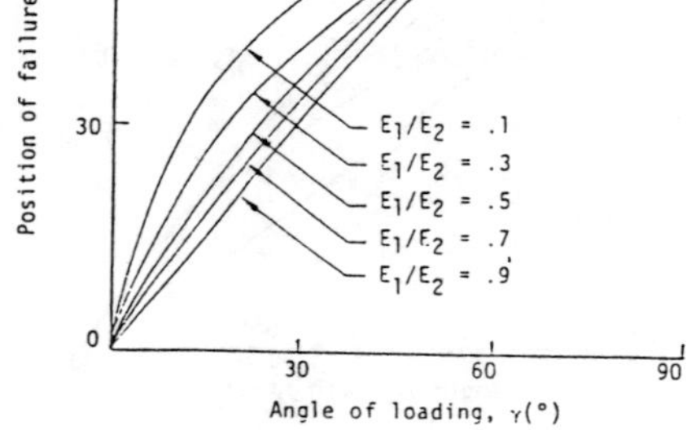

Fig. (10) - Variation of initial failure position with angle of loading: $b/a = .1$ and compressive load

in the direction of compression. The same has been observed by Kipp and Sih [12] for a cavity or $(E_1/E_2) = 0$.

FRACTURE TRAJECTORY

It is also of interest to find the fracture trajectory for elements that are outside of the interface layer and core region as shown in Figure 3. First of all, the locations of $(dW/dV)_{max}$ in the interface layer and $(dW/dV)_{min}$ for interior elements should agree closely for $r \simeq r_o$ and $\theta = \theta_o$. For points that are sufficiently far away and yet close enough to be influenced only by the initial undamaged configuration of the system, the $(dW/dV)_{min}$ condition can be used to determine the path of fracture*. A series of locations of (r_o, θ_o) will thus be calculated for the inclusion and medium having different (E_1/E_2) ratios.

Figures 11 through 13 illustrate the projected fracture paths for several angles of loading. All fracture paths are seen to initiate in an almost normal

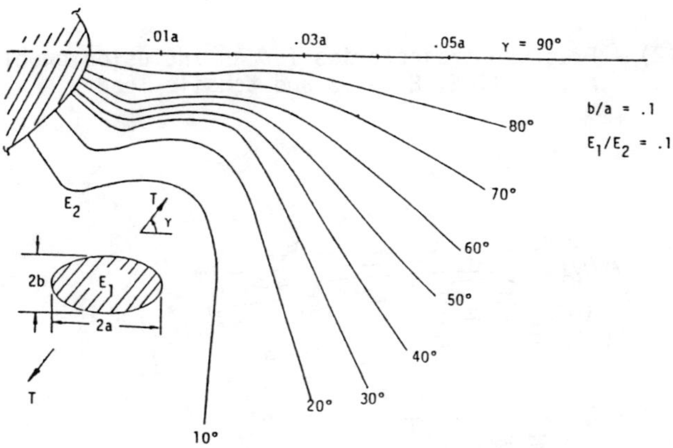

Fig. (11) - Fracture trajectories for an inclusion problem: $b/a = .1$, $E_1/E_2 = .1$ and tensile load

to the inclusion boundary. Depending on the ratio (E_1/E_2), the path then bends sharply towards the major axis of ellipse at first and then reverses its direction as the distance is increased. This effect is more pronounced for all small ratios of (E_1/E_2). For example, the fracture path for $(E_1/E_2) = 0.5$ bends at $(r_o/a) \simeq 0.046$ while the fracture path for $(E_1/E_2) = 0.1$ bends earlier at $(r/a) \simeq 0.0077$. In addition, the reversal of failure path direction also depends on the ratio (b/a).

For different stiffnesses of the inclusion, all the fracture paths tend to straight line when the loading angles are approximately between 70° and 90°. Com-

*For the case of a notch, Kipp and Sih [12] have shown that prediction based on the stress solution corresponding to the undamaged notch configuration is good up to a distance approximately equal to the length of the major axis of the elliptical notch.

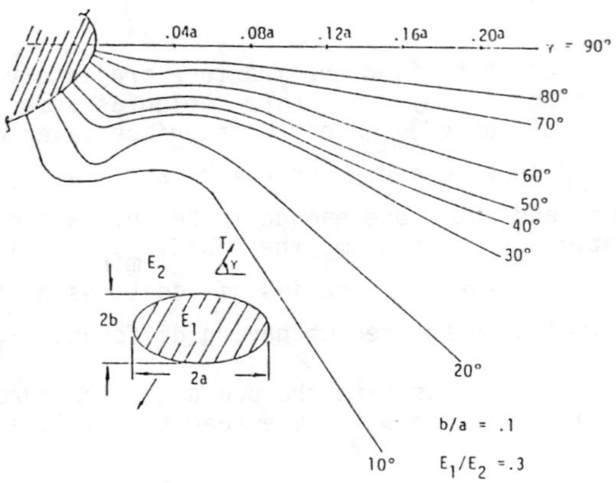

Fig. (12) - Fracture trajectories for an inclusion problem: $b/a = .1$, $E_1/E_2 = .3$ and tensile load

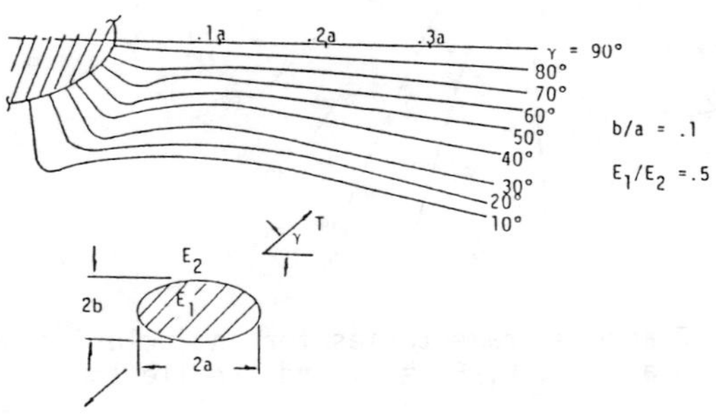

Fig. (13) - Fracture trajectories for an inclusion problem: $b/a = .1$, $E_1/E_2 = .5$ and tensile load

paring Figure 11 to 13 with Sih's results for the notch, the same trend has been observed for $E_1 < E_2$, i.e., the inclusion's modulus is lower than of the surrounding material.

CONCLUDING REMARKS

One of the important purposes of the application of the strain energy density theory was to discover the phenomenon of post-fracture behavior of an inclusion and to identify the effects due to the degree of material inhomogeneity.

A prerequisite of failure analysis is an accurate knowledge of the stress distribution within the system corresponding to a given damage state. If the damage changes with load, then the redistribution of stresses must be evaluated accordingly. The site of failure initiation may not, in general, coincide with the location of final termination. An interim period of subcritical material damage may exist such that damage follows certain trajectory. Therefore, by the results shown previously, it is not sufficient to just compute for the elevation of local stresses since the path of failure can be affected by the stress or energy state spread over a much larger region. Material inhomogeneity tends to complicate the solution even more since some of the stress and strain components acquire discontinuity across the interface at which the material properties undergo substantial changes.

REFERENCES

[1] Sih, G. C., "Plane extension of rigidly embedded line inclusions", Proc. 9th Midwestern Mech. Conf., pp. 61-79, 1965.

[2] Gdoutos, E. E., "Fracture phenomena in composites with rigid inclusions", Proc. 1st USA-Greece Sym. on Mixed Mode Crack Propagation, pp. 109-122, 1980.

[3] Bhargava, R. D. and Radhakrishna, H. C., "Two-dimensional elliptic inclusion", Proc. Camb. Phil. Soc. Math. Phys. Sci., Vol. 59, pp. 811-820, 1963.

[4] Eshelby, J. D., "The determination of the elastic field of an ellipsoidal inclusion and related problems", Proc. Roy. Soc. Ser. A, Vol. 241, pp. 376-396, 1957.

[5] Sendeckyj, G. P., "Elastic inclusion problems in plane elastostatics", Int. J. Solids Structure, Vol. 16, pp. 1535-1543, 1970.

[6] Donnell, L. H., "Stress concentration due to elliptical discontinuities in plates under edge forces", Theodore von Karman Anniversary Vol., p. 203, 1941.

[7] Hardiman, N. J., "Elliptic elastic inclusion in an infinite elastic plate", Q. J. Mech. Appl. Math., Vol. 7, pp. 226-230, 1952.

[8] Muskhelishvili, N. I., Some Basic Problems of the Mathematical Theory of Elasticity, Noordhoff, Groningen, 1953.

[9] Love, A. E. H., A Treatise on the Mathematical Theory of Elasticity, 4th ed., Cambridge, 1944.

[10] Sih, G. C., Methods of Analysis and Solutions of Crack Problems, Mech. of Fracture, Vol. 1, Noordhoff, 1973.

[11] Sih, G. C., "Stress analysis of notch problems", Mechanics of Fracture, Vol. 5, Noordhoff, 1978.

[12] Sih, G. C. and Kipp, M. E., "The strain energy density failure criterion applied to notched elastic solids", Int. J. Solids Structures, pp. 153-173, 1975.

[illegible] the loading [illegible] may exist [illegible] results shown [illegible] complicated [illegible] no substantial changes.

REFERENCES

[1] [illegible] Proc. 5th [illegible] Mech. [illegible]

[2] Soong, [illegible] Proc. 1st USA-Japan [illegible]

[3] [illegible] Proc. [illegible]

[4] [illegible]

[5] [illegible] Solids Structures [illegible]

[6] [illegible]

[7] [illegible] J. Mech. [illegible]

[8] [illegible]

[9] [illegible]

[10] [illegible]

[11] [illegible]

[12] [illegible] 1986.

ENERGY DENSITY REDISTRIBUTION CAUSED BY BROKEN FIBER

D. M. Chou

Institute of Fracture and Solid Mechanics, Lehigh University
Bethlehem, Pennsylvania 18015 USA

ABSTRACT

Fiber reinforced composites possess the dual properties of having high strength and damage tolerance. They can sustain a substantial local damage such as matrix cracking, fiber breaking and interface debonding without seriously affecting the structural integrity of the system. This work is concerned with the redistribution of volume energy density caused by a broken fiber. The analysis makes use of the incremental theory of plasticity and consists of a detail study of the energy states before and after the fiber breaks. A high rise of volume energy density prevails across the interface in the case of a broken fiber. This implies the likelihood of fiber/matrix debonding and/or matrix cracking. The local load transfer characteristics will, therefore, be greatly altered and become more representative of those in short-fiber composites. Such features are predicted by application of the strain energy density criterion that involves the determination of the maximum of the minimum strain energy density function. Numerical results are discussed and presented in graphical forms.

INTRODUCTION

The failure of fiber reinforced composites is, to say the least, not well understood. Cracks can easily be developed in both the fibers and matrix during the early stage of loading. This, in turn, will alter the load transfer characteristics. The way with which crack initiation and crack spreading alters composite behavior depends sensitively on load type and history. It is a cumulative damage process. Hence, no general conclusions can be made by testing composite specimens under specific loading conditions. In other words, the sequential order of the different failure modes is load history dependent. Although no attempt will be made to offer a general treatment of this problem, a possible method of approach will be offered, based on local and global stationary values of the strain energy density function [1,2] such that the effects of load, geometry and material on composite damage can be monitored by a single length parameter in terms of its overall integrity. This concept will be illustrated by investigating the state of affairs near a broken fiber embedded in a continuous matrix.

Previous works [3-6] on the fracture of fiber reinforced composites have been based mostly on the theory of elasticity where irreversible plastic deformation was not considered. Since the matrix material is usually much softer than that of the fiber, local yielding is eminent, particularly in regions near the interface and/or broken fibers. These effects will be reflected by the strain energy density contours and the numerical results to follow.

PROBLEM STATEMENT

Consider an axially symmetric fiber reinforced composite structure with length L = 30 cm and radius r_m = 6 cm as shown in Figure 1. The fiber is com-

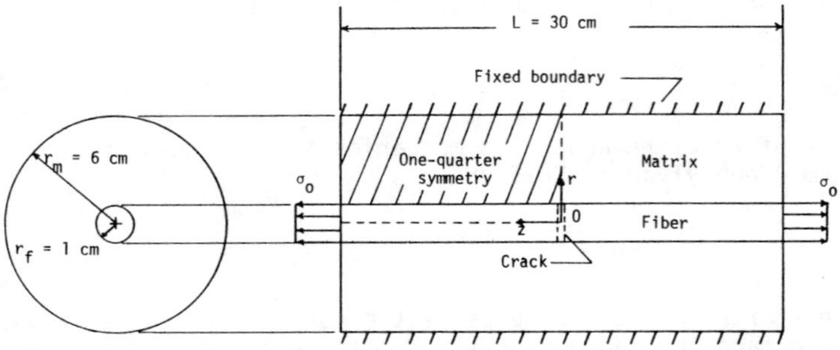

Fig. (1) - A stretched fiber embedded in matrix

pletely embedded in the matrix. It has a radius r_f = 1.0 cm and is extended by a stress with magnitude σ_o. Cylindrical polar coordinates (r,θ,z) are used with the origin centered in the composite. The circumferential surface is fixed such that it does not displace while continuity of stress and displacement is assumed to occur across the interface.

The uniaxial stress and strain curves for the matrix and fiber are shown in Figure 2. They are represented by the Ramberg-Osgood relation

$$\varepsilon = \begin{cases} \dfrac{\sigma}{E} & , \text{ for } \sigma \leq \sigma_y \\[2ex] \dfrac{\sigma}{E} + \dfrac{\alpha\sigma_y}{E}\left[\left(\dfrac{\sigma}{\sigma_y}\right)^{\beta} - 1\right], & \text{for } \sigma > \sigma_y \end{cases} \quad (1)$$

with α = 0.02 and β = 5.0 being the strain hardening coefficients for both the matrix and fiber which are also assumed to have the same Poisson's ratio ν = 0.3. The other material properties in equation (1) are

$$\begin{array}{ll} E_f = 207 \text{ GPa} & E_m = 20.7 \text{ GPa} \\ (\sigma_y)_f = 414 \text{ MPa} & (\sigma_y)_m = 100 \text{ MPa} \\ (dW/dV)_f = 19.3 \text{ MPa} & (dW/dV)_m = 22.2 \text{ MPa} \end{array} \quad (2)$$

where E is the Young's modulus and σ_y the yield strength. The subscripts f and m stand, respectively, for the fiber and matrix while $(dW/dV)_f$ and $(dW/dV)_m$ are the critical values at which failure is assumed to take place.

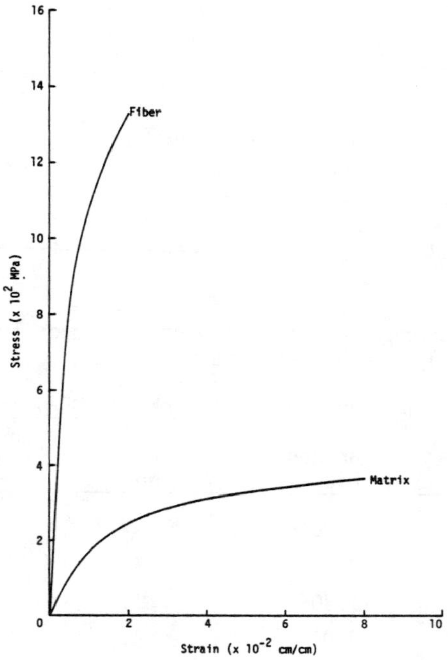

Fig. (2) - Stress and strain curves for fiber and matrix

A total of 28 isoparametric finite elements and 174 nodes are employed to model the composite geometry in Figure 1 with one-quarter symmetry. Special consideration is given to the elements in the immediate vicinity of the crack for the case of the broken fiber so that the desired numerical accuracy can be achieved. This is accomplished by embedding 64 Gaussian points in the element adjacent to the crack tip. Elastic-plastic stress analyses are performed for the situations of a continuous fiber and broken fiber. This is accomplished by utilizing the PAPST program [7] which incorporates the incremental theory of plasticity. Numerical results are obtained for σ_o = 2,500 MPa.

STRAIN ENERGY DENSITY CONTOURS

The strain energy density criterion [1,2] has been known to give good results for predicting the initiation of failure by fracture and/or yielding, a situation that is commonly found in fiber reinforced composites. Very briefly, the location of maximum of minimum dW/dV is identified with possible fracture initiation while the maximum of maximum dW/dV with failure by yielding. These failures are assumed to take place when the aforementioned stationary values of dW/dV become critical as given in equations (2) for the fiber and matrix.

Plotted in Figure 3 are global contours of the strain energy density function dW/dV. Their values for the case of continuous and broken fiber are given in Table 1. Note that their values are not significantly influenced by the broken fiber as they are sufficiently far removed from the local region A. The contours

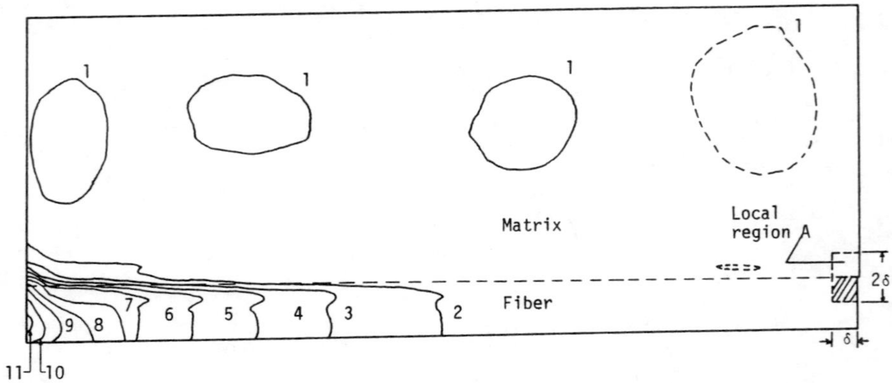

Fig. (3) - Strain energy density contours in composite: dotted contours occur only for broken fiber

Table 1 - Strain energy density dW/dV (MPa) in composite

Contour No.	Continuous Fiber	Broken Fiber	Contour No.	Continuous Fiber	Broken Fiber
1	0.0012	0.0016	6	3.40	3.40
2	0.68	0.67	7	4.80	4.70
3	1.40	1.30	8	6.10	6.10
4	2.00	2.00	9	7.50	7.40
5	2.70	2.70	10	9.50	10.00
			11	14.00	13.00

2 to 11 exhibit the effect of the applied stress σ_o. The two additional dotted contours occur only for the case of broken fiber.

Locally, however, the situation is different. Consider the region A of dimension $\delta \times 2\delta$ in Figure 3 that straddles across the interface that one-half is in the fiber and the other in the matrix. The dW/dV contours for the continuous fiber in Figure 4 and broken fiber in Figure 5 are now noticeably different. While the contours 1,2,...,10 in Figure 4 have different values as indicated in Table 2, their shapes or qualitative features are almost symmetric across the interface. Such a symmetry is no longer recognized for the broken fiber result in Figure 5. The rise in the intensity of dW/dV is apparent as the crack tip is approached. Contour no. 10 being closest to the crack tip attains the highest amplitude which is shown in Table 2.

INFLUENCE OF BROKEN FIBER

The influence of broken fiber can be best illustrated by considering the values of dW/dV averaged over a local region A with different sizes, say $\delta \times \delta$, in both the fiber and matrix. Three different values of δ = 0.10, 0.25 and 0.50 cm are depicted. Let $(dW/dV)_o$ and $(dW/dV)_a$ denote the average value of dW/dV for the continuous and broken fiber, respectively. Their values given in Tables 3 and 4 show that the influence of the broken fiber becomes more and more pronounced as the area over which their averages are taken tends to decrease. Almost an one

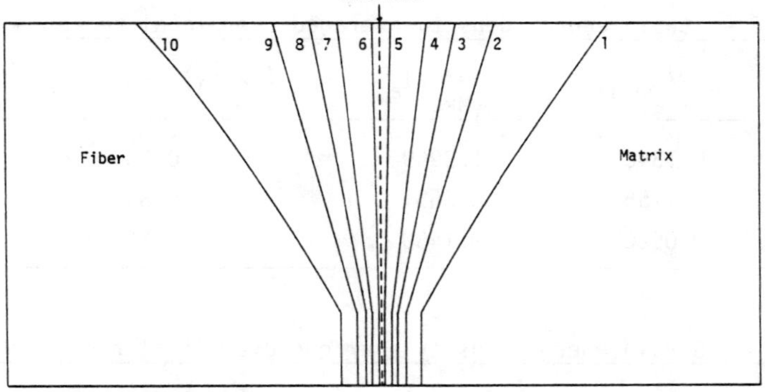

Fig. (4) - Enlarged strain energy density contours near z=0 for continuous fiber

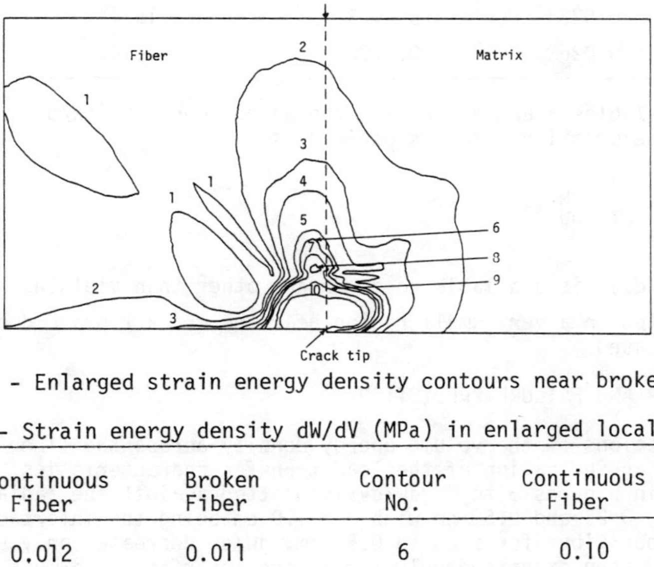

Fig. (5) - Enlarged strain energy density contours near broken fiber

Table 2 - Strain energy density dW/dV (MPa) in enlarged local region

Contour No.	Continuous Fiber	Broken Fiber	Contour No.	Continuous Fiber	Broken Fiber
1	0.012	0.011	6	0.10	0.91
2	0.020	0.16	7	0.13	1.10
3	0.036	0.31	8	0.15	1.20
4	0.052	0.46	9	0.16	2.10
5	0.085	0.76	10	0.17	3.00

order of magnitude increase in the ratio of $(dW/dV)_a/(dW/dV)_o$ is seen to occur in the fiber when δ is decreased from 0.50 cm to 0.10 cm. This effect is even more severe for the matrix according to the values given in Table 4.

Table 3 - Strain energy density averaged over local region A in fiber

δ (cm)	$(dW/dV)_o$ (MPa)	$(dW/dV)_a$ (MPa)	$(dW/dV)_a/(dW/dV)_o$	$(dW/dV)_p$ (MPa)
0.10	0.1346	1.3969	10.38	0.3005
0.25	0.1155	0.5651	4.89	0.0455
0.50	0.0920	0.1439	1.56	0.0102

Table 4 - Strain energy density averaged over local region A in matrix

δ (cm)	$(dW/dV)_o$ (MPa)	$(dW/dV)_a$ (MPa)	$(dW/dV)_a/(dW/dV)_o$	$(dW/dV)_p$ (MPa)
0.10	0.0358	1.1543	32.24	0.3336
0.25	0.0275	0.4163	15.14	0.0596
0.50	0.0262	0.1700	6.49	0.0134

Given in Tables 3 and 4 are also the energy density $(dW/dV)_p$ dissipated during plastic deformation which is defined as

$$\frac{dW}{dV} = \left(\frac{dW}{dV}\right)^* + \left(\frac{dW}{dV}\right)_p \tag{3}$$

The part $(dW/dV)^*$ is available to do damage other than yielding while $(dW/dV)_p$ is concentrated in a very small region near the crack tip and decreases quickly as δ is increased.

LOAD TRANSFER AND FAILURE PREDICTION

The variations of the volume energy density across the fiber and matrix interface give an indication of the load transfer characteristics. Referring to the results in Figures 6 to 8, dW/dV is plotted against the radial distance r for z = 0.10, 0.25 and 0.50 cm with r = 1.0 cm being the interface. In the case of a continuous fiber for z up to 0.50 cm, dW/dV decreases only slightly across the fiber and then drops suddenly across the interface to a much lower value in the matrix. It terminates at the composite boundary. When the fiber is broken, dW/dV near the crack at z = 0.1 cm in Figure 6 first decreases and then increases dramatically as the interface is approached. An equally pronounced drop in dW/dV occurs on the matrix side of the interface. This jump phenomenon dies off quickly as z is increased to 0.25 cm and 0.50 cm as shown in Figures 7 and 8. Exhibited in Figures 9 and 10 are the local elevation of dW/dV as a function of the variable z for r = 0.95 cm in the fiber next to the interface and r = 1.05 cm being in the matrix. The influence of the broken fiber diminishes as z is increased beyond z = 0.50 cm.

The sites of possible failure by fracture or cracking may be predicted from the strain energy density criterion which remains valid in the presence of plas-

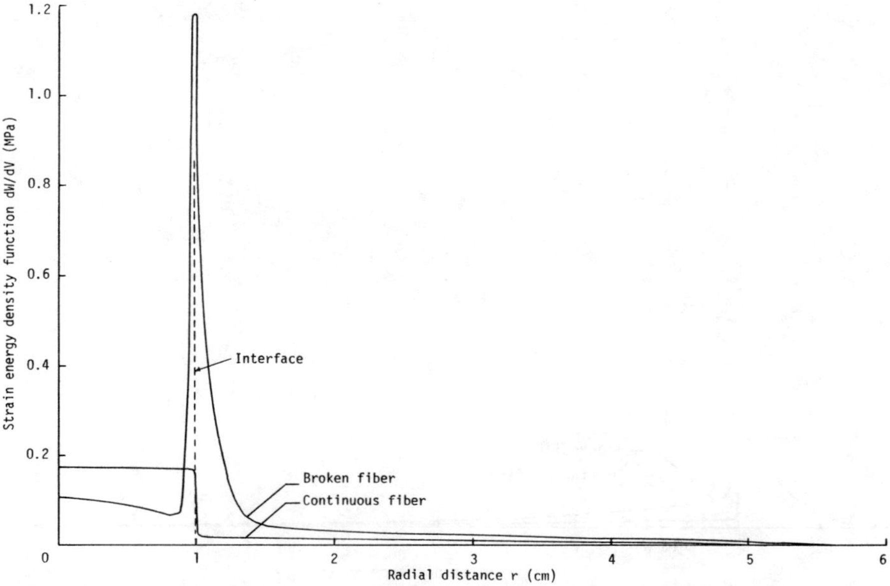

Fig. (6) - Comparison of strain energy density function at z = 0.1 cm for two cases

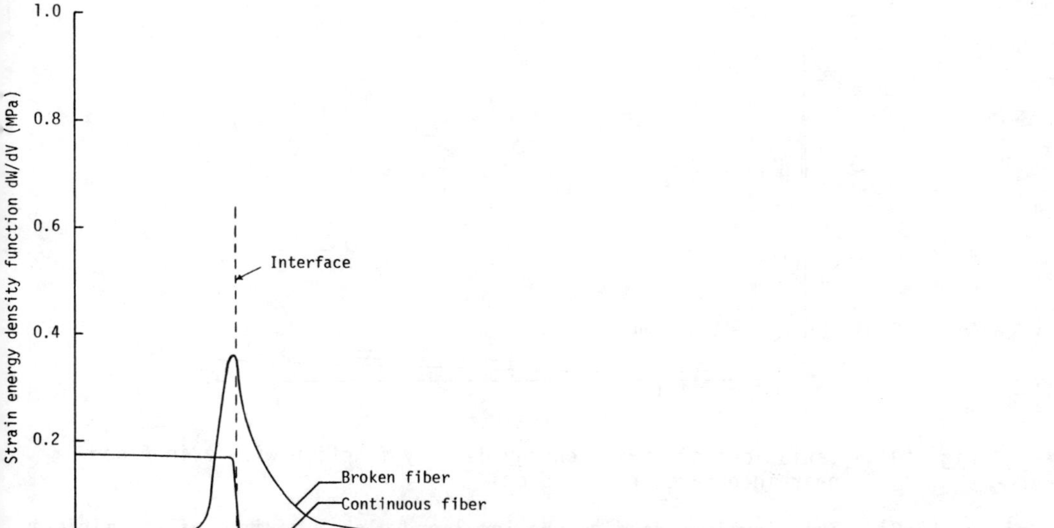

Fig. (7) - Comparison of strain energy density function at z = 0.25 cm for two cases

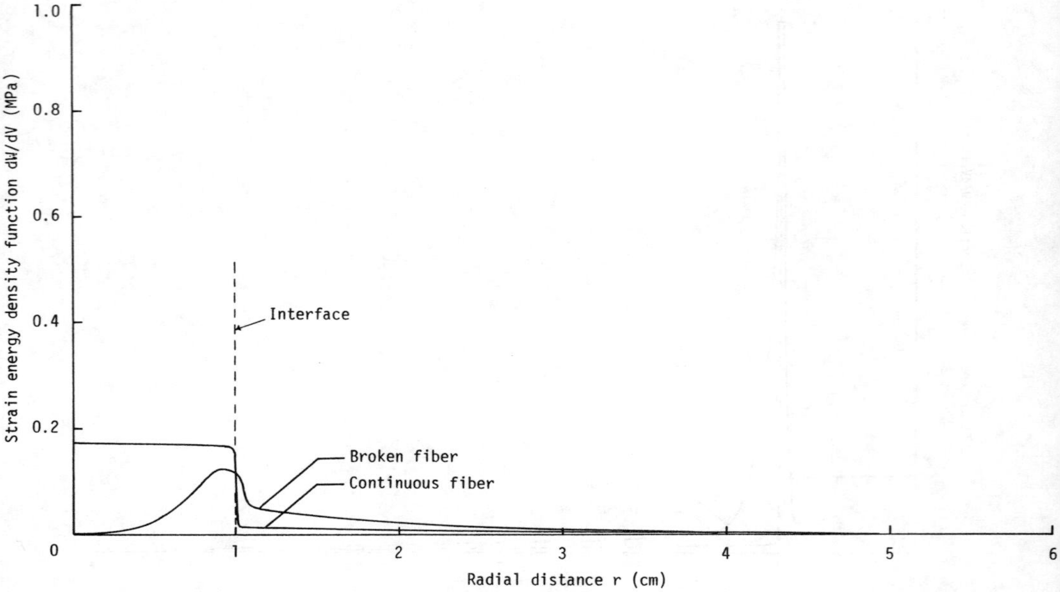

Fig. (8) - Comparison of strain energy density function at z = 0.5 cm for two cases

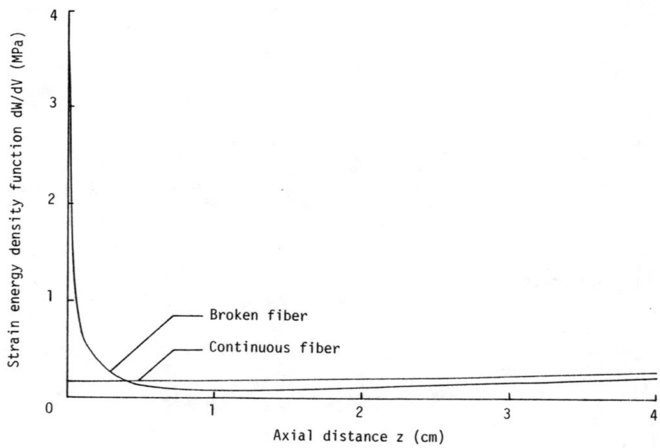

Fig. (9) - Variations of strain energy density function with z in fiber near interface (r = 0.95 cm)

ticity [8,9]. This involves finding the local and global maximum of the minimum strain energy density functions to be denoted, respectively, by $[(dW/dV)_{min}^{max}]_{\ell}$ and $[(dW/dV)_{min}^{max}]_{g}$. Their locations will be labelled as L and G. A small circle of radius r_o = 0.01 cm is chosen around the crack tip in Figure 11. It is outside

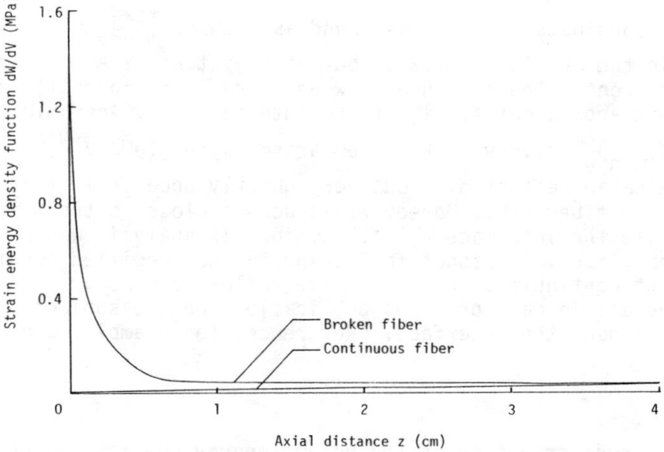

Fig. (10) - Variations of strain energy density function with z in matrix near interface (r = 1.05 cm)

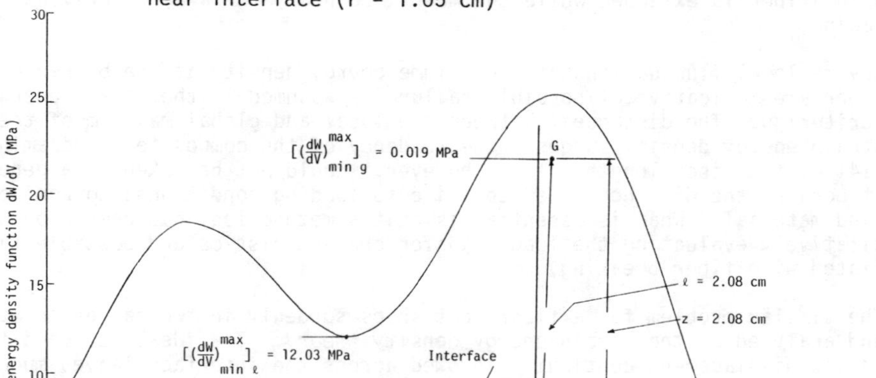

Fig. (11) - Variations of strain energy density function with θ around crack tip

of this region that dW/dV is computed as a function of the angle θ. The results are given in Figure 11. A maximum of $(dW/dV)_{min}$ is found at θ_o = 65.9° with $[(dW/dV)^{max}_{min}]_\ell$ = 12.03 MPa. Thus, the failure initiation site is predicted to occur in the matrix next to the crack tip. The global maximum of $(dW/dV)_{min}$ referred to the coordinates (r,θ,z) is found as $[(dW/dV)_{min}^{max}]_g = 0.019$ MPa. This again occurs in the matrix r = 1.2 cm but at a distance z = 2.08 cm further away from the crack front. The distance between L and G in Figure 11 is ℓ = 2.08 cm and suggest that approximately 14% of the interface is vulnerable to delamination. As the $[(dW/dV)_{min}^{max}]_g$ is very weak in comparison with $[(dW/dV)_{min}^{max}]_\ell$, it suggests that the delamination effect dies out very quickly once it initiated* from the broken end of the fiber z=0. Moreover, it occurs close to the matrix. This is basically because the interface r = 1.0 cm in this analysis was assumed to be perfect or ideal, i.e., it cannot fail owing to the specified condition of stress and displacement continuity. Only cohesive failure can occur. This has been discussed in detail in many previous publications on the subject of adhesive and cohesive failure near the interface. Reference, for example, can be made to the work in [10].

CONCLUSIONS

Significant redistribution of the volume energy density occurred near the region where an embedded fiber breaks in a composite. Analyzed is the situation where the fiber is extended while the matrix boundary far away is prevented from displacing.

Severe local fluctuations of the volume energy density at the broken end of the fiber are indicative of possible failure as assumed in the strain energy density criterion. The distance ℓ between the local and global maximum of the minimum strain energy density suggests the tendency of the composite to delaminate over 14% of the fiber length. This, however, should not be taken as a general result because the distance ℓ is sensitive to loading conditions, composite geometry and material. What is essential is that a methodology has been provided for quantitatively evaluating the load transfer characteristics and possible failure associated with fiber breaking.

The similar problem for a fiber that snaps suddenly in two can be solved easily and analyzed by the strain energy density theory. The idealized condition of stress and displacement continuity invoked across the interface leaves much to be desired. It can also affect the result quantitatively. This, however, is a topic beyond this discussion.

REFERENCES

[1] Sih, G. C., "Mechanics and physics of energy density theory", Theoretical and Applied Fracture Mechanics, Vol. 4, No. 3, pp. 157-173, 1985.

[2] Sih, G. C. and Chu, R. C., "Characterization of material inhomogeneity by stationary values of strain energy density", Theoretical and Applied Fracture Mechanics, Vol. 5, No. 2, 1986.

---

*Since dynamic effects associated with snapping of the fiber have been neglected, the estimate of 14% delamination is not unrealistic for the composite configuration considered even though $[(dW/dV)_{min}^{max}]_g$ may be weak.

[3] Sih, G. C., Hilton, P. D., Badaliance, R., Shenberger, P. S. and Villarreal, G., "Fracture mechanics for fibrous composites", American Society for Testing and Materials, STP 521, pp. 98-132, 1973.

[4] Sih, G. C. and Chen, E. P., "Fracture analysis of unidirectional composites", Journal of Composite Materials, Vol. 7, pp. 230-244, 1973.

[5] Sih, G. C., Chen, E. P. and Huang, S. L., "Fracture mechanics of plastic-fiber composites", Journal of Engineering Fracture Mechanics, Vol. 6, pp. 343-359, 1974.

[6] Sih, G. C., Chen, E. P., Huang, S. L. and McQuillan, E. J., "Material characterization on the fracture of filament reinforced composites", Journal of Composite Materials, Vol. 9, pp. 167-186, 1975.

[7] "Preliminary Documentation of PAPST Nonlinear Fracture and Stress Analysis by Finite Elements", David W. Taylor Naval Ship Research and Development Center, Bethesda, Maryland, 1981.

[8] Sih, G. C. and Madenci, E., "Fracture initiation under gross yielding: strain energy density criterion", Journal of Engineering Fracture Mechanics, 18, pp 667-677, 1983.

[9] Sih, G. C. and Madenci, E., "Crack growth resistance characterized by strain energy density function", Journal of Engineering Fracture Mechanics, 18, pp. 1159-1171, 1983.

[10] Sih, G. C., "Design of adhesive joints: stress and failure", Plastic Products Design Handbook, edited by Edward Miller, Marcel Dekker, Inc., pp. 344-350, 1983.

INFLUENCE OF STACKING SEQUENCE ON THE INSTABILITY OF LAYERED COMPOSITE

D. Y. Tzou

Institute of Fracture and Solid Mechanics, Lehigh University
Bethlehem, Pennsylvania 18015 USA

ABSTRACT

This work is concerned with the impact of the three layered system. The stacking sequence is altered to find the optimum system stability by application of the strain energy density theory. This involves the determination of the local and global stationary values of the strain energy density function. A characteristic length parameter, ℓ, is obtained such that it reflects the comparison of local geometry and material on the system instability. Results are presented graphically and discussed in detail.

INTRODUCTION

The prediction of life expectancy of structure components has been one of the problematic areas in designing modern structures. Due to the lack of a methodology for predicting different failure modes, the present-day approach is to switch from one criterion to another depending on the interaction of material type, loading history and geometric configurations. The end results are prone to inconsistencies and inaccurate predictions. When material inhomogeneity and/or anisotropy come into play as in the case of composites, failure mode assessment becomes even more dependent on the choice of criterion. A knowledge of the combined interaction of the various physical and loading parameters is necessary for evaluating the overall instability of the system. An unique concept has been advanced by Sih [1,2] who introduced a characteristic length parameter, say ℓ, for measuring the system instability. This parameter reflects the fluctuation of the energy density in space and time. Some simple examples can be found in [1,2].

The present study shall apply the ℓ-concept in conjunction with the strain energy density theory to evaluate the stability of a composite laminate. A three-layer system will be considered such that the stacking sequence can be varied to alter the dynamic load transfer characteristics. Both normal and shear impact are investigated. Determined specifically are the locations of the local and global stationary values of the strain energy density function, dW/dV. The distance between the maximum of the local and global minimum of dW/dV yields ℓ that is not only transient in character but changes location as the rate of load transfer and stacking sequence are altered. It tends to be load dominant under normal impact and stacking sequence dominant under shear impact. As it is expected, the layer system is more stable when sheared with ℓ being larger.

SPECIFICATION OF THE PROBLEM

A layered composite is made of three different materials as shown in Figure 1. The overall dimensions are 15.24 cm x 152.4 cm. Analyzed is the section ABCD

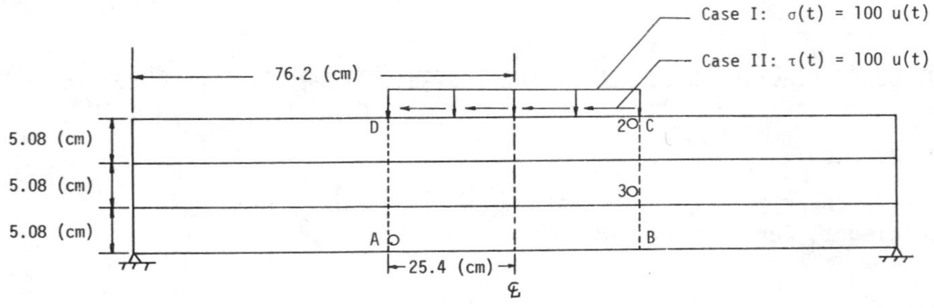

Fig. (1) - Geometrical dimension of a layered composite consisting of three different materials

with dimensions 15.24 cm x 50.8 cm. All materials are described by the Ramberg-Osgood relationship

$$\varepsilon = \begin{cases} \sigma/E & , \sigma \leq \sigma_{ys} \\ \{\sigma + \alpha[(\frac{\sigma}{\sigma_{ys}})^{\beta} - 1]\sigma_{ys}\}/E, & \sigma \geq \sigma_{ys} \end{cases} \tag{1}$$

displayed in Figure 2 by the curves 1, 2 and 3. E, σ_{ys}, α and β are Young's modulus, yielding stress, strain hardening coefficient and exponent, respectively. The loading conditions imposed on the top surface of the composite are normal and shear impact with u(t) being the unit step function as indicated in Figure 1. They are referred to as Case I and II, respectively.

Of interest are the magnitude and location $[(dW/dV)^{max}_{min}]_{\ell}$ and $[(dW/dV)^{max}_{min}]_{g}$. The former represents the local maximum of $(dW/dV)_{min}$ and the latter the global maximum of $(dW/dV)_{min}$. Their locations will be denoted by L and G with ℓ being the distance in between.

TRANSIENT STRESS ANALYSIS

The DEPA [3] program is applied to find the time-dependent elastic-plastic stress distribution in the layered medium. It is a finite element method FEM that employs twelve-node isoparametric elements analyzed in the Laplace transform domain. The advantage of this approach is that the numerical accuracy of the FEM

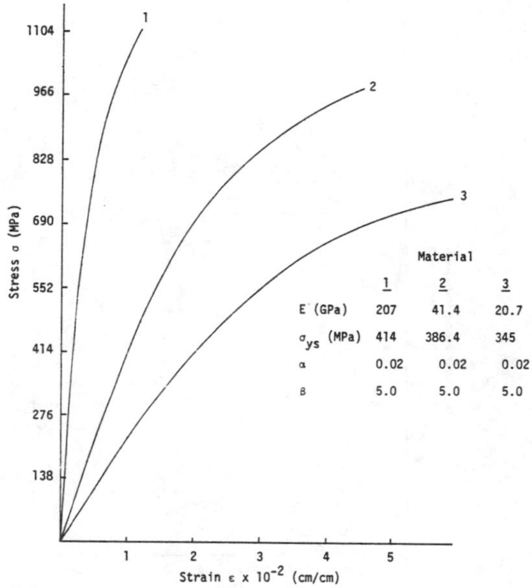

Fig. (2) - True stress and true strain curves for the materials in the layered composite

solution is not affected by the interaction between time increment Δt and grid sizes Δx and Δy. The grid pattern in the Laplace transform domain contains 15 elements and 100 nodes. The bandwidth for the active stiffness matrix is 24. Sixteen (16) Gaussian points of integration in each element are used to search for the local and global points where dW/dV possesses minima. J_2-flow theory in conjunction with incremental theory of plasticity is used. The transient response for all sequences of the layered composite occurs in time domain t = 0 (sec), 0.04 (sec) which corresponds to s = 0, 500 in the Laplace transform domain. Steady state solutions for all cases pertain to response after t = 0.04 sec.

NORMAL IMPACT

Figure 3 shows the temporal variation of strain energy density distribution along the central line of the composite with the stacking sequence* 3-2-1. Energy state in the composite is the highest at t = 0.01 sec and steady state is reached after t = 0.037 sec. Material difference in the layered composite can only be recognized after t = 0.02 sec which makes the strain energy density fluctuate. Global scan in the entire structure is then made to find the points L and G. As shown in Figure 4, small circles with radius 0.127 cm are attached to field points

*The leading number refers to the top layer.

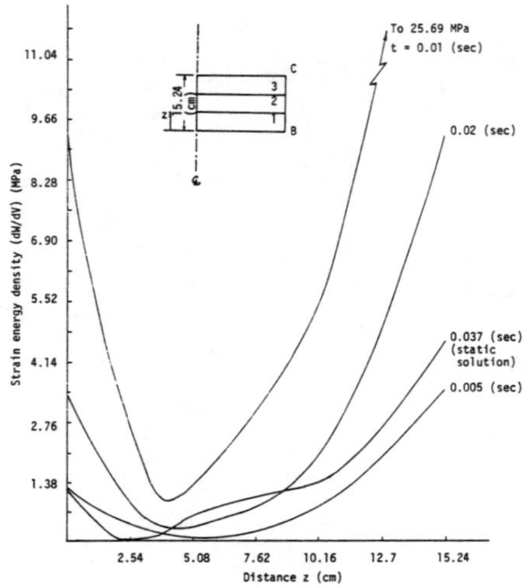

Fig. (3) - Time change of strain energy density distribution along the central line, stacking sequence 3-2-1

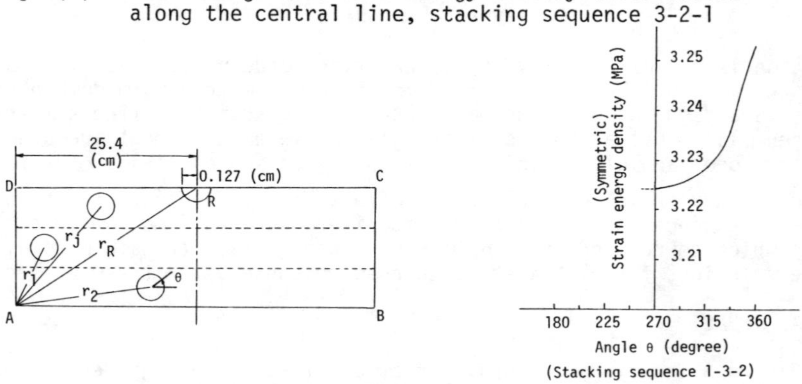

Fig. (4) - Global scan for $[(dW/dV)_{min}^{max}]_\ell$ and $[(dW/dV)_{min}^{max}]_g$ and strain energy density distribution along the kth circle

at which $[(dW/dV)_{min}^{max}]_\ell$ and $[(dW/dV)_{min}^{max}]_g$ are most likely to occur. Variation of strain energy density along the circumference of the circles can be obtained with the help of nodal values at Gaussian points in the neighborhood. Locations at which the energy density reaches maximum among the local and global minima can then be identified. Figure 5 shows the transient variation of relative distance ℓ between L and G in the time interval $t \in [0 \text{ sec}, 0.04 \text{ sec}]$. The temporal varia-

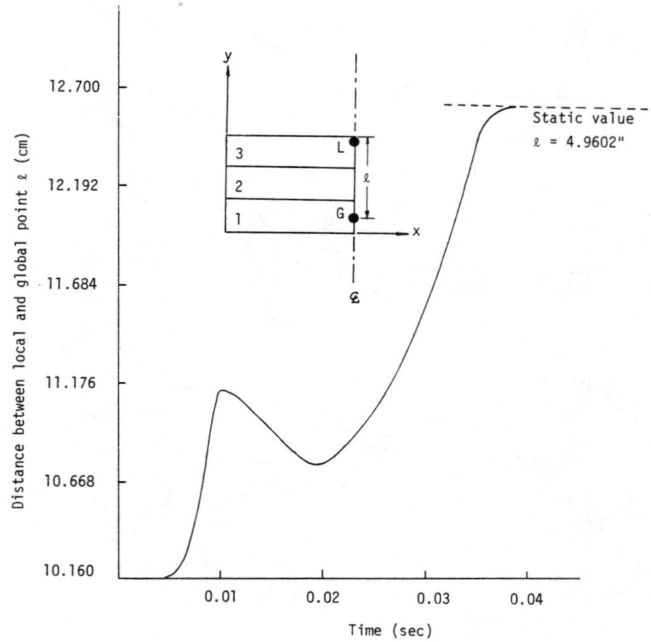

Fig. (5) - History of development of distance between $[(dW/dV)^{max}_{min}]_\ell$ and $[(dW/dV)^{max}_{min}]_g$ stacking sequence 3-2-1

tion of ℓ is not sensitive to the presence of stress waves. Small oscillation of ℓ can be seen for t ε [0.01 sec, 0.025 sec] but the total change of ℓ as referred to the steady state value at t ≃ 0.037 sec is about 20%. Should the dW/dV variation exceed the critical value $(dW/dV)_c$ at a point in the composite, the locations for both L and G must then be redetermined in terms of the new geometry to include the presence of cracks or other forms of damage. Figure 6 shows a typical energy density contour plot for the case of stacking sequence 3-2-1 after steady state is reached; i.e., t $\geq$ 0.04 sec. Figure 7 shows the relative distance ℓ between L and G for each case. The numerical values of ℓ and $[(dW/dV)^{max}_{min}]_\ell$ and $[(dW/dV)^{min}_{min}]_g$ for different stacking sequence of the layers are listed in Table 1. The case 2-3-1 displays the most unstable situation since not only ℓ has the largest value in this case but the local maximum of minimum strain energy density is the highest for normal impact. According to strain energy density criterion, the 2-3-1 combination is most likely to cause global instability once failure initiates from L as $[(dW/dV)^{max}_{min}]_\ell$ exceeds $(dW/dV)_c$. Similarly, the 3-2-1 combination also yields a relatively large ℓ that is nearly equal to the total thickness of the composite system. In both cases, G remained in layer 1. It is, therefore, desirable to place the stiffer layer, i.e., no. 1, closest to normal

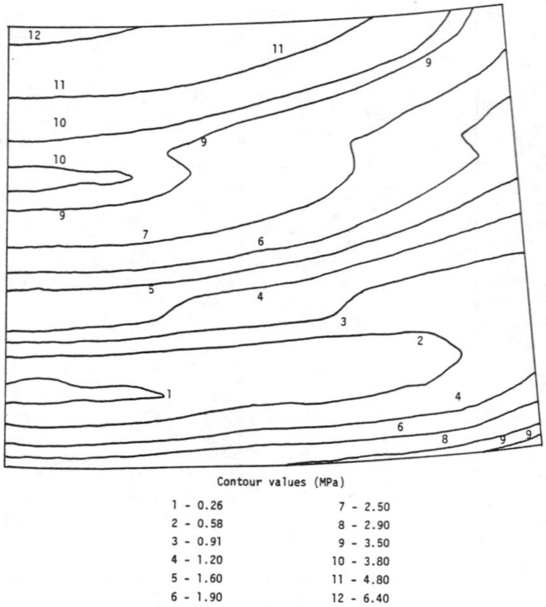

Fig. (6) - Strain energy density contour for the composite with stacking sequence 3-2-1 under normal impact

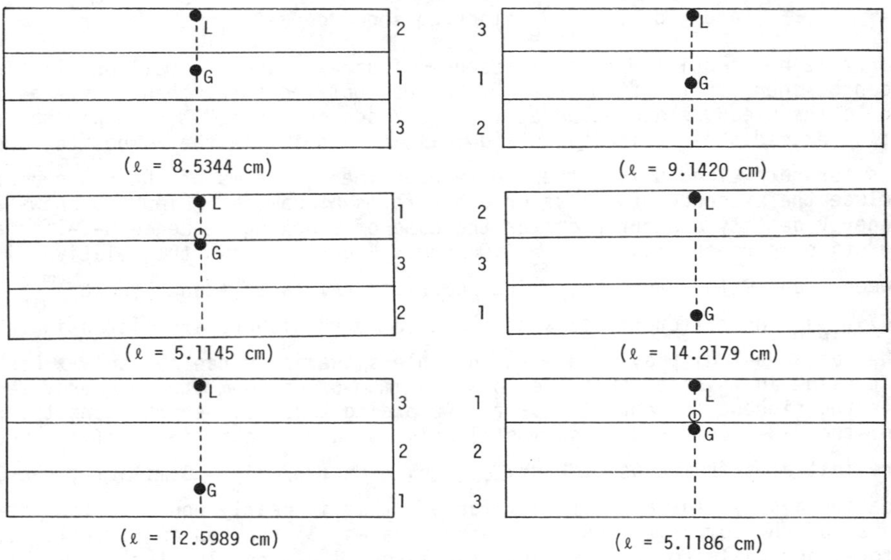

Fig. (7) - Relative distance ℓ for different stacking sequence after steady state normal impact

Table 1 - Energy levels of $[(dW/dV)_{min}^{max}]_{\ell}$ and $[(dW/dV)_{min}^{max}]_{g}$ for layered composite with different stacking sequence, normal impact

Stacking sequence	Distance ℓ (cm)	$[(dW/dV)_{min}^{max}]_{\ell}$ (MPa)	$[(dW/dV)_{min}^{max}]_{g}$ (MPa)
3-1-2	9.1420	3.9588	0.0850
2-1-3	8.5344	6.0895	0.2200
1-3-2	5.1145	0.3225	0.0186
2-3-1	14.2179	4.6230	0.0117
1-2-3	5.1186	4.2236	0.0345
3-2-1	12.5989	4.7066	0.1802

impact* in order to have ℓ kept as small as possible. Such are the combinations 1-3-2 and 1-2-3 shown in Figure 7. The precise values of ℓ, however, are suspect** because the locations of G are too close to the composite interface where the theory of plasticity may not yield reliable results. The mechanism of energy dissipation across composite interfaces requires a more in-depth investigation. The exchange of volume and surface energy as considered in [2,5,6] will shed more light on interface behavior.

SHEAR IMPACT

Figure 8 shows the oscillation of strain energy density at locations 1, 2 and 3 shown in Figure 1 for the layered composite subjected to shear impact load. At location no. 2, the energy density varied monotonically with time for $t \in [0 \text{ sec}, 5 \times 10^{-4} \text{ sec}]$ while those at points no. 1 and 3 oscillated initially and then converged to their static values at $t = 8.2 \times 10^{-7}$ sec and 1.45×10^{-6} sec, respectively. The oscillatory behavior of strain energy density for point 2 occurred at $t \in [3 \times 10^{-4} \text{ sec}, 2.5 \times 10^{-2} \text{ sec}]$ before it tends to the static value. This is also shown in Figure 8. Global search of points L and G is made as in the case of normal impact. Time change of distance ℓ is more complicated because of the lack of symmetry under shear. For a stacking sequence of 2-3-1, the variation of ℓ as a function of time is shown in Figure 9. A dramatic change of ℓ occurred at t near 10^{-6} sec. The point L moved diagonally while G remained still. This is indicative of the sudden change in the system instability. Figure 10 shows a typical contour pattern for strain energy density distribution in the layered composite under shear impact after steady state is developed ($t \geq 0.024$ sec). Ener-

* This observation was made by G. C. Sih [4] and pointed out to the author.

**The open circles for 1-3-2 and 1-2-3 may also be possible locations of G in Figure 7.

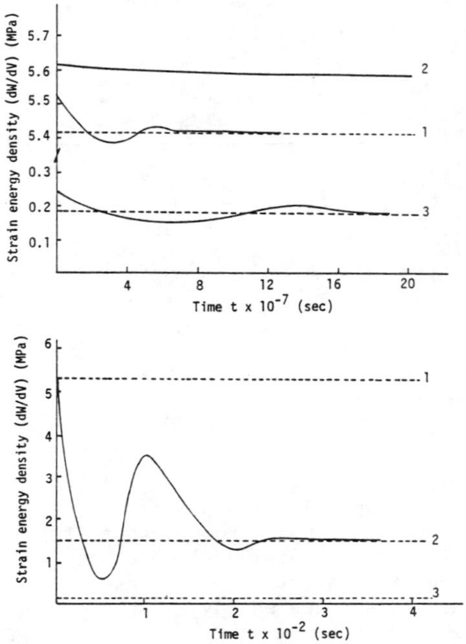

Fig. (8) - Temporal variation of strain energy density at typical locations in the composite subjected to shear impact

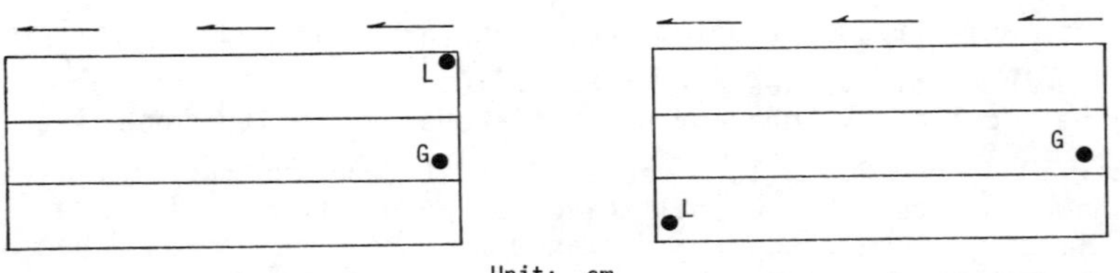

Unit: cm

$t < 10^{-6}$ (sec)	$t > 10^{-6}$ (sec)
G (49.30, 6.78)	G (49.30, 6.78)
L (50.06, 15.04)	L (1.04, 1.70)
ℓ = 8.29	ℓ = 48.54

Fig. (9) - Temporal variation of distance ℓ in the composite with stacking sequence 2-3-1 subjected to shear impact gy concentration can be observed at the interface between material no. 1 and the adjacent layers in all cases. This is shown in Figure 10 for the stacking se-

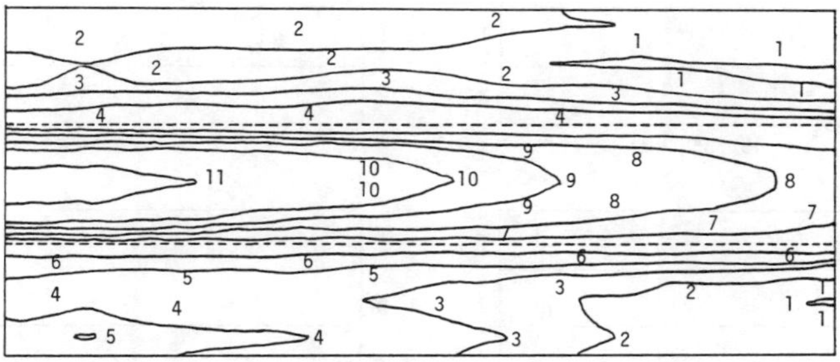

Contour values (MPa)

1 - 1.50	7 - 14.00
2 - 2.80	8 - 15.00
3 - 4.10	9 - 20.00
4 - 5.30	10 - 23.00
5 - 6.60	11 - 27.00
6 - 12.00	

Fig. (10) - Strain energy density contours in the composite with stacking sequence 3-1-2 under shear impact

quence 3-1-2 for which there are two interfaces between material no. 1 and other layers. Unlikely to the previous case, the locations of local and global points in the composite subjected to shear impact are more likely to be material-dominant. This is shown in Figure 11. Locations of L for all stacking sequences for shear impact are always located in the stiffer material, i.e., layer no. 1 while the locations of G are always located in the softer material, i.e., layer no. 3. Uncertainty again prevails [4] for the case 3-1-2 in Figure 11. The minimum values of dW/dV at the solid circle L and open circle are extremely close. Neither the numerical accuracy of the FEM or the approximation in the plasticity theory can change the outcome. Hence, no definite conclusion can be drawn in this case. Nevertheless, Table 2 provides the results for the general trend of ℓ as the stacking sequence is varied during shear impact. The most stable combination corresponds to 1-3-2 for which ℓ is the smallest and $[(dW/dV)^{max}_{min}]_{\ell}$ is the lowest. The central layer in this stacking sequence is the softest. This is beneficial as energy absorption requires a relatively soft backing layer after impact and applies to both normal and shear impact. The composite with this stacking sequence is thus expected to have the most stable response to external loads. A closer situation to this case is that of the sequence 1-2-3 for which the central layer can still absorb a certain amount of energy but not as efficient as the case 1-3-2.

CONCLUDING REMARKS

It can be concluded from this study that structural instability of layered composite is more likely to be load controlled under normal impact while material tends to dominate under shear impact. Obviously, the combined interaction of load, geometry and material can not be sorted out individually for systems that are en-

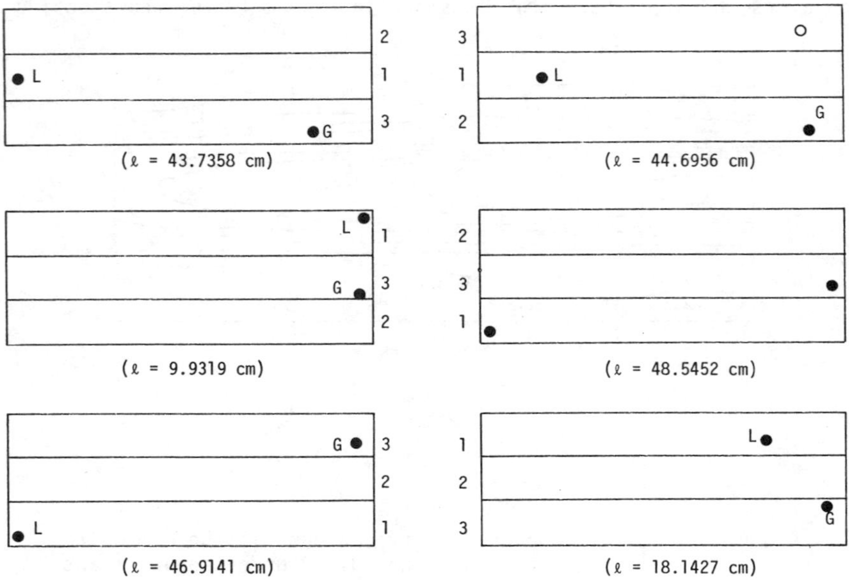

Fig. (11) - Locations of local and global points in the layered composite with different stacking sequence under impact shear

Table 2 - Energy levels of $[(dW/dV)_{min}^{max}]_\ell$ and $[(dW/dV)_{min}^{max}]_g$ for layered composite with different stacking sequence, shear impact

Stacking sequence	Distance ℓ (cm)	$[(dW/dV)_{min}^{max}]_\ell$ (MPa)	$[(dW/dV)_{min}^{max}]_g$ (MPa)
3-1-2	44.6956	21.3121	1.1225
2-1-3	43.7358	9.4815	0.2683
1-3-2	9.9319	0.2689	0.0088
2-3-1	48.5452	4.1976	0.2130
1-2-3	18.1427	0.4014	0.01124
3-2-1	46.9141	17.3142	1.2541

ergy dissipative. The ℓ-concept is powerful as it can measure the combined effect of all parameters for determining the overall structure instability. The stacking sequence of 1-3-2 is found to be the best for both normal and shear impact. The hardest layer should be next to impact followed by the softest layer [4]. The same method of analysis can be applied if cracks or defects are present before or after loading. This, of course, will alter the location and value of ℓ and hence the system instability. To be kept in mind is that more accurate assessments can be made by application of the strain energy density theory without invoking on the theory of plasticity.

REFERENCES

[1] Sih, G. C., "The strain energy density concept and criterion", J. of the Aeronautical Society of India, Vol. 37, No. 1, pp. 43-60, 1985.

[2] Sih, G. C., "Mechanics and physics of energy density and rate of change of volume with surface", J. Theoretical and Applied Fracture Mechanics, Vol. 4, No. 3, 1985.

[3] Sih, G. C. and Tzou, D. Y., "Dynamic Elastic Plastic Analysis (DEPA)", Institute of Fracture and Solid Mechanics Technical Report IFSM-86-140, 1986.

[4] Sih, G. C., Private communication.

[5] Sih, G. C. and Tzou, D. Y., "Nonhomogeneous energy dissipation ahead of a slowly growing crack: Part I - monotonic loading", Institute of Fracture and Solid Mechanics Technical Report IFSM-84-126, 1984.

[6] Sih, G. C. and Tzou, D. Y., "Cooling and heating of uniaxial tensile coupon", Institute of Fracture and Solid Mechanics Technical Report IFSM-85-137, 1985.

FRACTURE AND STRENGTHENING MECHANISMS IN DISCONTINUOUS SiC REINFORCED ALUMINUM COMPOSITES

T. G. Nieh and J. Wadsworth

Lockheed Palo Alto Research Laboratory, 0/93-10, B/204
Palo Alto, California 94304 USA

and

W. H. Kao

The Aerospace Corporation, Materials Laboratory
Los Angeles, California 90009 USA

ABSTRACT

Discontinuous SiC, particulate (p) or whisker (w), reinforced aluminum composites (SiC/Al) offer some unusual combinations of properties. The materials generally have excellent specific stiffness and strength comparable to the most advanced Al-Li alloys. Also, the materials exhibit reasonable creep resistance and good dimensional stability under high temperature environments. Despite all of the good features that the materials exhibit, they also have some deficiencies preventing them from wider applications. The two major ones are low tensile ductility and poor fracture toughness.

In this paper, an overview of the understanding of the metallurgical factors governing the strengthening and fracture behavior of the composites are given. Mechanical properties of both the particulate and whisker reinforced materials, as well as different aluminum matrices, e.g., 1XXX, 2XXX, 6XXX and 7XXX, are presented. Strengthening mechanisms in different aluminum alloys due to the presence of SiC are evaluated. Microstructural evidence, e.g., dislocation density, dispersion morphology and matrix substructure, is presented to help in the understanding of the strengthening mechanisms in these composites.

Metallurgical factors that may affect the fracture of SiC/Al are discussed. These include the SiC-Al interface, intermetallic inclusions, oxides, defective reinforcements, residual gas and dislocation structure. Some experiments have been conducted to evaluate the importance of each individual factor. Special efforts are made to isolate each factor during the experiments. The origins of fracture in the composites, based on the results of these experiments, are suggested.

INTRODUCTION

A. Fracture Mechanisms

Low fracture toughness has been one of the major barriers for immediate application of discontinuous SiC/Al composites (DSC). The toughness values as shown in Table 1 depend upon the aluminum matrix, orientation of fibers and heat

Table 1 - K_Q for SiC/Al composites

	K_Q, MPa $m^{1/2}$	Thickness, cm.
1100Al - 20% SiC_w	19.2	1.27
6061Al - 20% SiC_w	13.5	1.27
- 30% SiC_p	18.1	1.27
2124Al - 15% SiC_w	17.6	1.27
- 20% SiC_w	22.0/33.0	1.27/0.32

treatment [1,2]. The best value recorded is about 20 MPa $m^{1/2}$ which was obtained from 6061Al-20% SiC_w in the as-fabricated condition at which the tensile strength was only 359 MPa as compared to 582 MPa in the T6 condition. Fracture surfaces of regular composite samples, as shown in Figure 1, normally show significant lo-

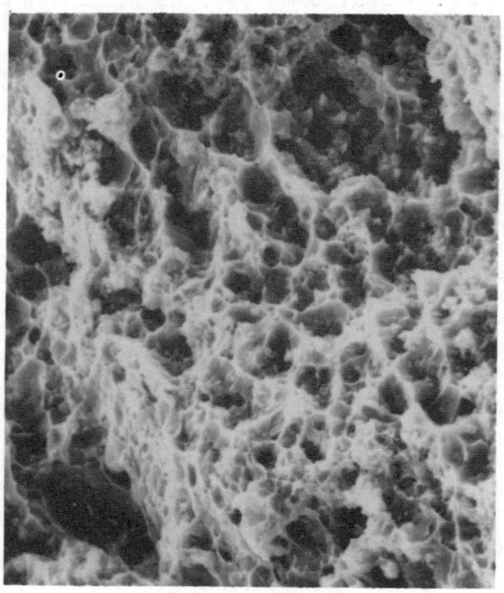

Fig. (1) - Fracture surface of a SiC/Al composite. The Al matrix exhibits significant local plasticity cal plasticity which seems to suggest the material can dissipate a large number of deformation energy [3]. Therefore, the low toughness value is difficult to understand. Fractographic analysis of SiC/Al composites is difficult due to the large amount of reinforcements in the matrix and, also, because of the many potential sites that may originate the crack. This difficulty is not only because of the problems in location of the fracture origin in the material, but also the complex nature of the fracture origin, e.g., the presence of intermetallics, whiskers, and whisker-matrix interfaces. So far, there is little or no understanding concerning the fracture mechanisms in this type of composite.

B. Strengthening Mechanisms

Although the elastic modulus of all SiC_w/Al (alloy) composites follows the rule-of-mixtures, there is no consistent relationship of the tensile yield or ultimate strength (YS and UTS) with the volume fraction of whisker reinforcement. As shown in Table 2, the YS of 6061-T6 is greatly increased by the addition of

Table 2 - Reinforced effect due to SiC

T6	E, GPa	σ_y, MPa	σ_t, MPa	e_f, %
1100	69.0	69.7	104.9	40
1100 - 20% SiC_p	89.0	115.2	210.0	14
1100 - 20% SiC_w	104.2	176.0	338.8	5
2124	69.7	398.1	538.2	13.5
2124 - 20% SiC_w	118.7	431.9	711.4	3.0
6061	69.0	262.2	293.3	8.5
6061 - 20% SiC_w	107.6	447.1	581.7	2.1
6061 - 20% SiC_p	102.1	396.8	483.0	3.5
7075	69.0	503.7	572.7	11
7075 - 20% SiC_w	109.0	448.5	631.4	3
7091 - 20% SiC_w	110.4	724.5	931.5	-

20 vol. % of SiC_w, whereas there is only a small improvement for 2124-T6 and 7075-T6. Even larger differences exist in the UTS, especially after artificial aging.

A peculiar response to the SiC_w reinforcement occurs with 7091 Al. Unpublished data for that composite, from Arco Chemicals (Silga) [4], is included in Table 2. The strength of the unreinforced alloy is essentially the same as that of 7091 but with 20 v/o SiC_w, the strength is increased by about 60 percent - an unprecedented, yet explainable, improvement. The 7091 aluminum alloy, which has almost the same chemical composition and mechanical strength as 7075 (except that the former contains about 0.5% excessive cobalt used to improve the stress corrosion cracking properties) can be reinforced by SiC reinforcement. Apparently, the Co addition in 7XXX alloy strongly affects the solute distribution so it balances out the "detrimental" effect that is due to the reinforcement.

Another example is the Mg+Cu containing alloys (2XXX series). Both 2024 and 2219 alloys can be reinforced by SiC, but their strength is less than expected theoretically [5]. Surprisingly, although the SiC/2024 composite responds partially to artificial aging, the one with a 2219 matrix does not at all. Similar results have been observed in 2024-aluminum containing various refractory-compound particulates, e.g., TiB_2, B_4C and $NiAl_3$ [6]. In contrast, amongst all the age-hardenable aluminum alloys, 6061 (Al-Mg-Si) shows the largest strength increase upon reinforcement [7]. So far, no explanation has been offered to interpret these results.

DISCUSSION

A. Fracture Mechanisms

As mentioned earlier, although SiC/Al composites exhibit many interesting properties, the materials also have some limitations, namely, low tensile ductilities and low fracture toughnesses. These limitations inhibit the use of SiC_w/Al in a wide range of structural applications. In order to improve these properties, it is necessary to understand the metallurgical factors that govern fracture. The fracture of SiC_w/Al composites is, in general, catastrophic, i.e., crack propagation or growth is in a relatively fast fashion. Therefore, it is reasonable to assume that the fracture of the composite is dominated by crack initiation. There are many potential crack initiation sites including: (1) the whisker-matrix interface, (2) oxides, (3) defective whiskers, (4) intermetallics, including impurities in SiC_w powder, (5) microstructural inhomogeneities, e.g., precipitate-free-zones, whisker-free-zones and heterogeneous dislocation distributions, and (6) residual gases, e.g., hydrogen, that are trapped during material processing. As many of these factors are interrelated, it is difficult, if not impossible, to separate each individual item and to evaluate it.

Direct observations of the fracture of SiC/Al, by conducting an in situ tensile test using High Voltage Transmission Electron Microscopy (HVTEM), for example, is impractical. This is because the diameter of the SiC whiskers is relatively large with respect to the finite penetration depth offered even by HVTEM. The effective volume for the electron beam is relatively small, which means an extensive TEM study is necessary to provide enough statistical information on the fracture mechanisms. In addition, because of the thinness of the TEM foil, the stress-strain state in the material, especially at the whisker-matrix inferface, may be significantly modified by the free surfaces. This certainly limits the usefulness of the results obtained from such a study.

There are some simple experiments that can be performed to provide useful information about the fracture mechanisms in SiC/Al. An example is in situ Auger fractographic study of the composite. In these experiments, pre-notched impact samples were first cooled in liquid nitrogen and then fractured in situ in an Auger Electron Spectroscope which was normally maintained at an extremely high vacuum (10^{-7} to 10^{-8} Pa). Analyses of fracture surfaces were conducted immediately following fracture. Figure 2 is a typical AES spectrum of the fracture surface of a 15% SiC_w/2124 Al impact sample immediately after fracture. The spectrum shows

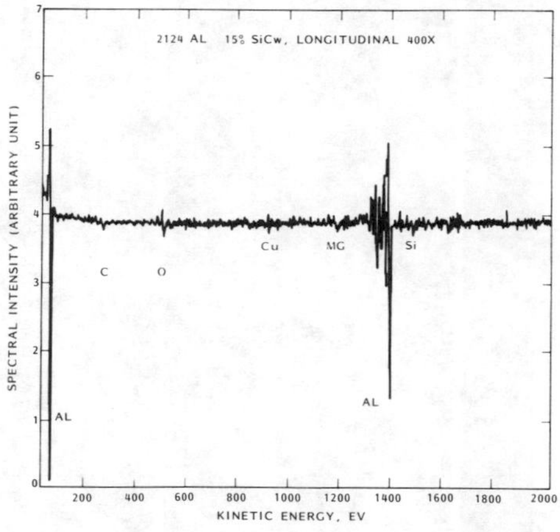

Fig. (2) - Auger spectrum from the fracture surface of 2124 Al-15% SiC_w composite

the presence of Al, Mg, Cu, C, O and a small amount of Si on the fracture surface. The amounts of Mg and Cu measured from the fracture surface are consistent with the values for their bulk concentrations in the 2124 Al alloy. Carbon and oxygen are two common absorbents found in many Auger analyses, even at extremely high vacuum. It is necessary to be careful, therefore, in interpreting their presence. Since the amount of SiC in the composite is relatively large (15% by weight), it is believed that the C signal is due to the whisker phase. However, the interpretation of the presence of oxygen is more subtle.

Al. Interface

One of the major findings from the general Auger spectra of the test sample is that the spectral intensity for Si, which comes mainly from SiC, is much too low to account for the proper bulk concentration of SiC. This observation indicates that there was little whisker pull-out. Also, sputter etching of the fracture surface causes the spectral intensity of Si to increase, indicating that the SiC that is present must be coated with Al from the matrix. Both results suggest that the bonding between SiC whiskers and 2124 Al matrix is quite good. This has also been proposed by Marcus [8] based on his study of SiC_w/6061 Al. This conclusion is further supported by scanning electron microscope examination of the fracture surface. Figure 3 is a fracture surface of 2124 Al/15% SiC_w. It is apparent that the amount of SiC whiskers exposed is lower than that expected from the bulk content of SiC in the composite. In addition, high magnification shows that most of the exposed whiskers are clearly coated with the Al matrix. Similar observations have been reported in SiC_w/6061 Al [8].

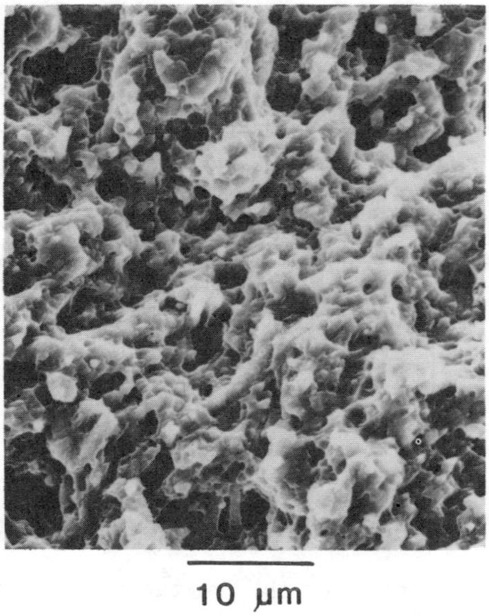

Fig. (3) - Typical fracture surface of a SiC/Al composite. The apparent amount of SiC exposed is lower than that expected from the bulk content of SiC in the composite

A2. Oxides

As mentioned earlier, Al is a strong oxide former and it tends to absorb oxygen from even an extremely high vacuum (10^{-7} Pa) environment. Consequently, oxygen analysis from the fracture surface must be conducted relatively quickly before the surface is contaminated. Also, it is necessary to determine whether the oxygen was absorbed from the environment or was actually in the material originally. The chemical state of the oxygen in the former is elemental and in the latter is an oxide. Since the Auger spectra for pure Al and Al_2O_3 are different, one can distinguish the chemical state of the oxygen on the fracture surface by measuring the shape of Al spectra, particularly in the high energy regime. Figure 4 shows a typical Al spectrum from the fracture surface. In order to make a direct comparison, the standard spectra [9] for Al and Al_2O_3 are also shown in the figure. The test results clearly show that the oxygen is in an elemental state and is not in an oxide form. The measured high-energy Al spectrum (from 1300 to 1400 ev) may not coincide exactly with that shown in the standard, which is often due to the off calibration of the instrument at the high-energy end. However, the most important feature is that the shape of the two spectra are almost identical. It should be noted that the tests do not exclude the existence of small amounts of Al_2O_3. They simply show that Al_2O_3 does not play an important role in the fracture of the composite. In particular, the Al_2O_3-Al interface does not act as a preferential

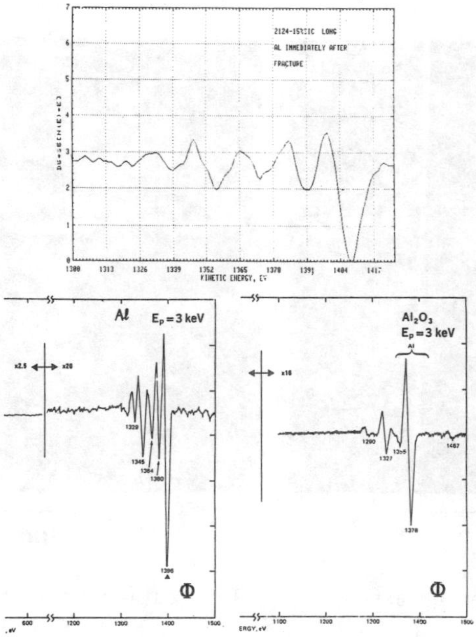

Fig. (4) - Al spectrum from the fracture surface of 2124 Al-15%SiC_w failed in high vacuum (10^{-7} Pa). In order to make a direct comparison, the standard spectra for Al and Al_2O_3 are also shown

path for crack propagation in the composite. Otherwise, one would have observed a highly distorted Al spectrum, namely, a spectrum from pure Al but strongly weighted by that from Al_2O_3, since the penetration depth of the Auger electrons is normally about two to three atomic layers.

A3. Defective Reinforcement

Defects in the whiskers are either present in the original whisker or are produced during consolidation of the composite. Figure 5 is an example showing the breakage of the whisker during material processing. In fact, many of the SiC whiskers have entirely lost their whisker identities and have virtually been converted into particulates. The damage in the whiskers not only reduces the load transfer capability of the whiskers, but also provides sites for crack initiation and propagation. Therefore, it is expected that both the mechanical strength and elongation to failure are affected by the whisker damage. The intrinsic defects in the whiskers cannot be observed in the photomicrographs but have been reported and discussed by many researchers using transmission electron microscopy [10,11]. In summary, the whisker, by its nature of growth, is highly defective. It has many microvoids inside the core region, microtwins and stacking faults along the length of whiskers, and often exhibits a serrated surface. All these characteristics can significantly degrade the mechanical properties of the composite by nu-

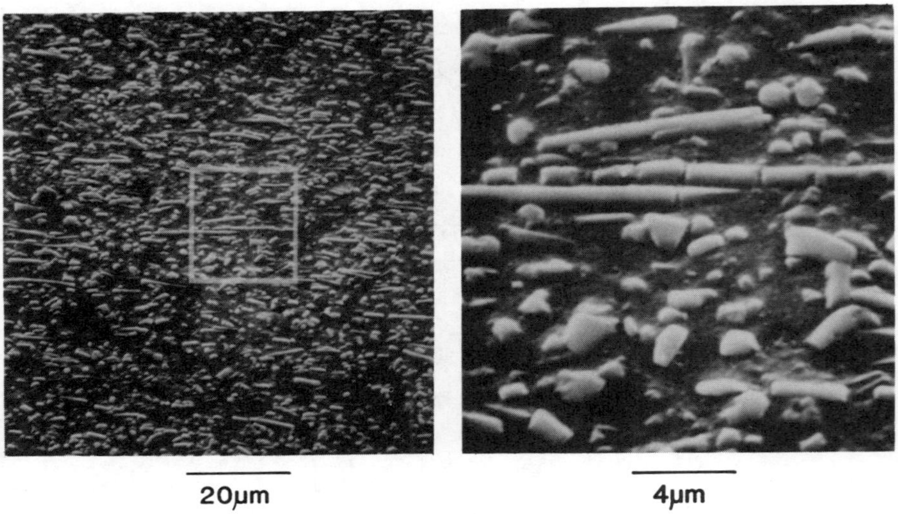

Fig. (5) - Whisker breakage during fabrication of a SiC/Al composite

cleating cracks either at the SiC surface or in the core. Since SiC is notch-sensitive, the stress that is required to break the whisker is much less than that predicted from the case of an ideal whisker. This low stress may further accelerate the fracturing processes and, consequently, result in low ductility and toughness of the composite. However, direct evidences of whisker breakage during plastic deformation is difficult, if not impossible, to obtain.

A4. Impurities

One of the important features in the microstructure of a SiC/Al composite is the high impurity content, as shown in Figure 6 [12]. The impurities are either from the original SiC powders or are products of the alloy matrix during consolidation. The former impurities are characterized as irregular particulates which normally show sharp edges and are of dimensions less than 5 μm, while the latter impurities are intermetallic inclusions which have a smoother morphology and occasionally show dimensions of roughly 10 μm. Also, in the optical micrograph, the intermetallics appear to be grey in contrast to the nearly black particles from the original SiC powders. The intermetallics are identified as Fe, Cr, Mn-rich particles produced during the composite fabrication. These intermetallics are potentially detrimental to toughness and tensile ductility of the composite.

Figure 7 is a typical fracture surface of the composite sheet material. It is noted that in addition to the large local plasticity in the matrix, which often has been observed, there is also a large number of brittle regions which are related to intermetallic inclusions. Measurements using EDAX from the fracture origin shows that the intermetallic inclusions contain Cr, Mn, Fe and Cu and have dimensions surprisingly, of over 50 μm. Intermetallic inclusions of that size have not been observed in the extruded sample. Apparently, this is due to the long-time, high-temperature exposure of the material during the rolling operation to produce

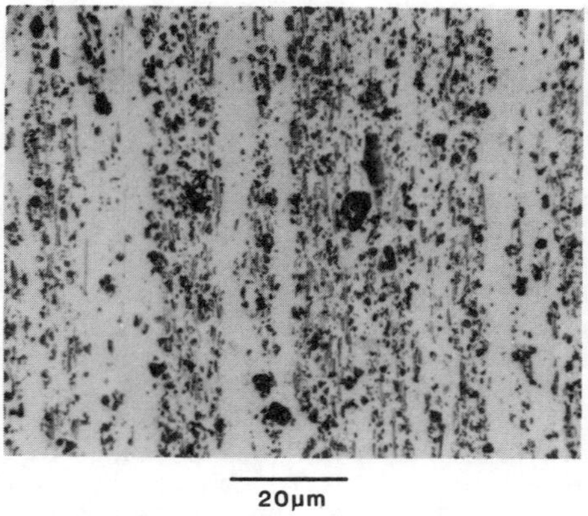

Fig. (6) - Microstructure of a SiC/Al composite. Large amount of impurities and intermetallics are present in the matrix. Whisker-free-zone is also one of the important features in the microstructure

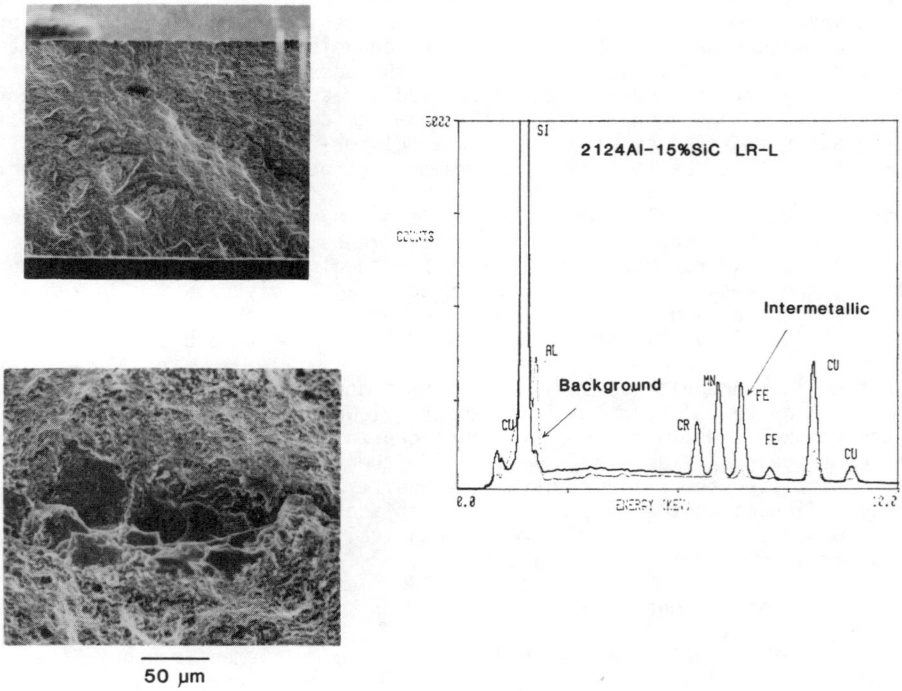

Fig. (7) - Fracture origin in rolled SiC/Al sheet. EDAX shows the intermetallics are rich in Cr, Fe, Mn and Cu

sheet. This observation demonstrates the importance of intermetallic inclusions to the fracture behavior of the SiC/Al composite and, in fact, it is the dominant factor controlling the fracture in the rolled material [12].

Another type of impurity which is also often observed on the fracture surface of the SiC/Al composite is an environmental contaminant. For example, it has been reported [13] that Ca-rich and Na-rich particles were observed on the fracture surfaces of 2124 Al-SiC. Neither Ca nor Na is expected to be present in the composite. They were apparently introduced during material fabrication. The effect of these contaminants on the fracture of the composites is not clear at present. However, the frequent exposure of these particles on the fracture surface of SiC/Al suggests that the bonding between them and the Al matrix is probably not strong.

It is difficult to evaluate the relative importance of intermetallics and impurities on the fracture behavior of SiC/Al composites. The exact conclusion, of course, depends on the size, distribution and nature of the intermetallics and impurities produced during the composite fabrication.

A5. Inhomogeneous Microstructure

One of the pronounced features in the microstructure of SiC/Al is the presence of SiC whisker-free zones (of 5 to 10 μm width) as shown in Figure 6. The effect of this inhomogeneous microstructure on the mechanical properties of the material is not yet clear. However, it may lead to inhomogeneous plastic flow when the material is under stress and, consequently, could result in premature tensile failure. In fact, it has been reported [12] that the ductility of SiC/Al composite can be greatly improved by eliminating the whisker-free zones by thermal rolling. Presumably, this is due to a more uniform microstructure in the rolled materials.

Other microstructural inhomogeneities, e.g., high dislocation density near the matrix-SiC interface have also been observed by some researchers [2,10]. Such inhomogeneities may contribute to local strengthening in the matrix. But more importantly, they result in heterogeneous precipitation distribution in the alloyed matrix during aging and this subsequently affects the mechanical properties of the composite [23]. This will be discussed in detail in Section B3.

Another inhomogeneity is originated from a nonuniform stress distribution when the composite is under loading. Recently, Nutt [24] has shown void formation near the whisker ends, where the stress concentration is very high. If Nutt's observation is generally applied, and it is further demonstrated to be a prevalent factor controlling the fracture of the composite, then his result will have a very significant impact on the development and application of DSC. This is because brittle nature of a DSC will be purely mechanical which cannot be overcome by metallurgical modifications in the material.

A6. Residual Hydrogen

It is generally believed that hydrogen is a major element responsible for poor weldments in PM alloys through the generation of microporosities during fusion welding. While the effect of hydrogen on the embrittlement behavior in Al alloys is still controversial [14], there is plenty of evidence [15,16] suggesting that the presence of hydrogen in Al may be detrimental to its ductility. This is the basis of studying the effect of vacuum degassing on the ductility of SiC/Al. The technique used to identify degassing species which evolved during heating is Thermal Analysis Mass Spectrometric (TA/MS) analysis. A brief description of the experimental apparatus has been reported elsewhere [17].

Figure 8 is a summary of the ion chromatrograms for the 2124 Al/SiC_w powder mixture after heating from a temperature of 25°C to 850°C. The degassed species

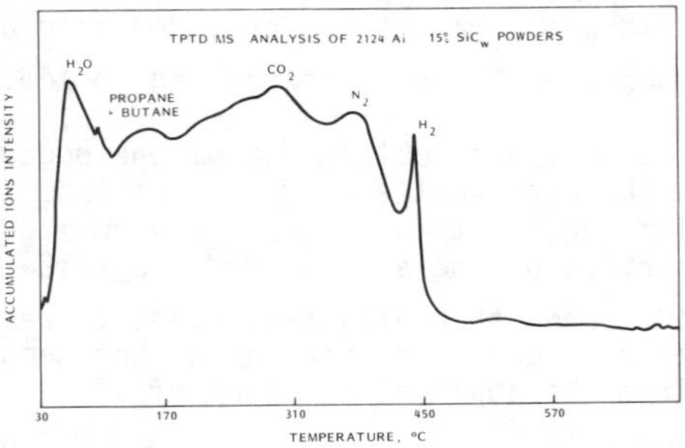

Fig. (8) - TA/MS analysis from 2124 Al-15%SiC whisker powder mixture. It shows 450°C in vacuum is sufficient to degas the powders

include H_2, H_2O, CO_2 and a series of organic decompositions. Emissions of H_2O occur at several stages and this is consistent with results from some of the other Al powders [18]. The 105°C peak may represent the emission of the physisorbed water, while the 270°C and 330°C peak emissions represent some specific oxide dehydration reactions. For example, the reaction $Al_2O_3 \cdot 3H_2O = Al_2O_3 \cdot H_2O + 2H_2O$ is known to occur readily at temperatures above 200°C. Also, the m/e = 44 spectrum is marked as CO_2, one of the common absorbents found in atmospheric environment. The spectrum may also be attributed to some organics, e.g., C_3H_8. Similarly, the m/e = 28 spectrum can be identified as N_2CO or C_2H_6. The various organic hydrocarbons appearing in the samples are believed to be due to the handling of the powder, as they are usually sorted in plastic bags prior to further processing. The data obtained from TA/MS analysis is very useful in terms of material fabrication, as it clearly demonstrates that a 450°C vacuum heat treatment is a feasible way to degas the powder mixture.

TA/MS analysis was also conducted with consolidated SiC_w/2124 Al. However, no gaseous species was evolved until the temperature almost reached the melting point of the Al alloy matrix. This was expected since the degassing mechanism for the powder is different from that for the consolidated material. For the former, it is a surface desorption phenomenon, which is primarily temperature dependent, but for the latter, it is a diffusional phenomenon, which is both temperature and time dependent. TA/MS analysis is not sufficient to provide data with the outgas schedule for the consolidated material. Additional information about the hydrogen diffusivity in Al, D_H is necessary. In spite of the large reported discrepancy on the activation energy for D_H [19-21], the actual values of D_H at 450-500°C range seem to be in good agreement from all research groups.

if one assumes the equation $X = (2D_H t)^{1/2}$, where X is the diffusion distance and t is the time, then an estimation of 20h at 500°C in vacuum would be sufficient to remove the excessive hydrogen from a 12.7 mm thick specimen. In fact, it has been shown [22] that vacuum (5×10^{-4} Pa) degassing of SiC_w/Al at 500°C for 24 h can reduce the hydrogen level by nearly half, namely, from 6-7 ppm for the as-received material to 3-4 ppm (by weight) for the degassed condition. A 2124 Al-15% SiC_w composite extrudate was, therefore, degassed at 500°C for 72 h in vacuum (10^{-4} Pa) and subsequently heat treated and tested.

However, the tensile test data obtained from the degassed samples are essentially the same within the data scatterband as that from the samples that were not degassed. These results suggest that either the vacuum degassing is insignificant to the mechanical properties of the as-received SiC_w/2124 Al extrudate, or the degassing schedule used is not thermally sufficient to remove the excessive residual hydrogen. However, vacuum degassing at higher temperature and longer time would be impractical from the application point of view.

B. Strengthening Mechanisms

B1. Dislocations

Some researchers believe that dislocation strengthening is the main mechanism responsible for the improvements in yield and tensile strengths in composites [25]. However, based on this argument, it is difficult to explain the difference in the strengthening observed in composites that have different alloy matrices. Since the mismatch in coefficient of thermal expansion between SiC and aluminum alloy matrix is about the same in all composites, the high dislocation densities caused by thermal mismatch between SiC and the matrix would be expected to be similar in composites with different aluminum alloy matrix. Consequently, it would be predicted that a similar amount of strength increase should be observed in all composites, regardless of their aluminum matrices, so long as the volume contents of SiC are the same. However, as shown in Table 2, this is obviously not the case. The results, therefore, suggest that dislocation strengthening cannot be dominant mechanism contributing to strengthening in composites.

B2. Dispersion

Dispersion strengthening has also been considered as a possible mechanism contributing to the strengthening in composites. However, this possibility was ruled out based on geometrical considerations [26]. This is because SiC whiskers, which are typically of 0.5 μm in diameter and 10 μm long, are relatively coarse to be considered as an effective dispersion phase. For example, even at a 20 vol% SiC reinforced level, geometric calculations based on a simple cubic array of dispersoid distribution shows the interparticle spacing is no less than 1 μm. Figure 9 is a representative microstructure of 6061 Al-20 vol% SiC which shows that the typical interparticle spacing is indeed of the order of micron. The Orowan bowing stress, with a value 1 μm interparticle spacing was found, in aluminum composites, is estimated to be only 10 MPa which is insignificant for strengthening of the matrix. In fact, it has been reported [10] that no threshold stresses could be found during creep of 6061 Al-20% SiC composites, demonstrating the ineffectiveness of the SiC dispersion upon strengthening.

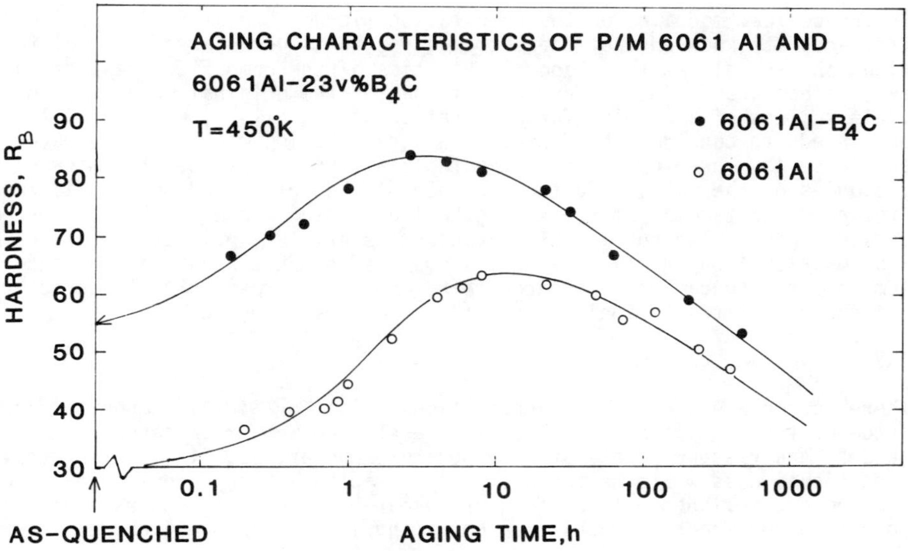

Fig. (9) - Direct comparison of the aging characteristics of 6061 Al with and without B_4C reinforcements. The aging response is much faster in the reinforced material

B3. Precipitation

It is important to note that a reinforced composite is metallurgically different from its unreinforced matrix. For example, the aging response in 6061 Al-20% B_4C is much faster than that in 6061 Al, as shown in Figure 9. The accelerated aging response can be explained if we considered that that the CTE ratio between a typical aluminum alloy and SiC is about 5:1. The large thermal mismatch between the matrices and reinforcement causes the matrices to yield and also generates a high density of dislocations [27] near the matrix-reinforcement interface during cooling from the processing or solutioning temperatures. This result is expected to be true for all Metal Matrix Composite systems. The consequence of high dislocation density in the matrix is actually two-fold. Since precipitation hardening is a diffusion-limited process, a high dislocation density, as well as the reinforcement-matrix interface, would greatly enhance the diffusion of the solute elements. Consequently, the aging rate of the composite matrix is faster than the unreinforced matrix. In fact, this has been demonstrated in SiC/Al [28] and B_4C/Al [29] composites. The second consequence of the high dislocation density, which is considered to be more significant, is the solute distribution in the composites. The strength of heat treatable aluminum alloys originates from uniformly distributed fine precipitates or GP zones in the lattice. Dislocations in the matrix, as well as at the SiC-Al interfaces, offer excellent sites for solute segregation and heterogeneous nucleation of precipitates. This leads to an inhomogeneous microstructure in the matrix. The heterogeneous precipitation in SiC/Al has been observed by Nutt and Carpenter [23]. Most MMC's show a very low elastic limit [30] which may be attributed to this microstructural inhomogeneity.

In theoretical modeling of the deformation micromechanism for composite materials, accurate knowledge of the matrix strength is necessary. The solute redistribution, as well as the change in kinetics of the composite, results in a matrix that has properties different from that of the original unreinforced matrix. It is less of a problem for non-heat treatable alloys, e.g., 1100 Al, since one only needs to consider the dislocation strengthening mechanism. However, it presents a serious problem for heat treatable alloys. Consequently, microstructural studies of the matrix, in the presence of the reinforced phases, is important to the understanding of the strengthening mechanism due to the reinforcements. Such a study would also supply information necessary for optimization of the matrix composition. Unfortunately, only a small effort has been made to study this problem so far. Fundamental research to relate the microstructural characteristics of the composites to their properties is urgently needed.

SUMMARY

There is ample evidence that applications exist for SiC/Al composite materials by virtue of their improved specific stiffness, good wear resistance, and, in some cases, their superior elevated-temperature properties. It is clear that many more additional uses would be found if the strengths and toughness were improved. This can be accomplished only by further research directed to understanding the strengthening and fracture mechanisms operating in those composites.

ACKNOWLEDGEMENT

This work was supported by the Lockheed Independent Research Program. The authors wish to thank Mr. D. J. Chellman for providing some unpublished data.

REFERENCES

[1] Min, B. K. and Morrow, J. D., "Aspects of basic mechanical characteristics of discontinuously reinforced metals", presented in Testing Technology of Metal Matrix Composites: 1st Symposium, held at Nashville, Tennessee, 18-20 November, 1985.

[2] Chellman, D. J., Unpublished data, Lockheed California Company, 1982.

[3] Divecha, A. P., Fishman, S. G. and Karmarkar, S. D., "Silicon carbide reinforced aluminum - a formable composite", J. Metals, Vol. 33, No. 9, 1981.

[4] Arco Chemicals, Silag Operation, Greer, S. C., 29651.

[5] Chellman, D. J. and Slaughter, H. C., "Elevated temperature properties of aluminum metal matrix composites", presented at 111th AIME Annual Meeting, Dallas, Texas, 1982.

[6] Nieh, T. G., Unpublished data.

[7] Hood, P. E., "Baseline properties and fabrication data for silicon carbide whisker reinforced aluminum alloy", Arco Metal/Silag Operation, NSWC Contract No. N60921-81-CA-328.

[8] Marcus, H. L., "Microscopic study of the influence of impurities on interface bonding", Annual Technical Report, ONR Contract No. N00014-83-K-0143, 1984.

[9] Davis, L. E., et al, "Handbook of Auger electron spectroscopy", 2nd edition, Physical Electronics Industries, Inc., pp. 43-47, 1976.

[10] Nieh, T. G., "Creep rupture of a silicon carbide reinforced aluminum composite", Met. Trans. Vol. 15A, p. 139, 1984.

[11] Nutt, S. R., J. Amer. Ceramic Soc., Vol. 67, No. 6, p. 428, 1984.

[12] Nieh, T. G., Rainen, R. A. and Chellman, D. J., "Microstructure and fracture in SiC whisker reinforced 2124 aluminum composite", W. C. Harrington, Jr., et al, ed., Proceedings 5th Int. Conf. on Composite Materials, AIME, p. 825, 1985.

[13] Nieh, T. G., Wadsworth, J. and Chellman, D. J., "In situ Auger fractographic study of a SiC whisker reinforced Al alloy", Scripta Metal., Vol. 19, p. 181, 1985.

[14] Bernstein, I. M. and Thompson, A. W., eds., "Hydrogen in metals", Proc. Int. Conf. on the Effect of Hydrogen on Materials Properties and Selection and Structural Design, ASM, pp. 249-276, 1973.

[15] Gest. R. J. and Troiano, A. R., "Corrosion", Vol. 30, No. 8, pp. 274-279, 1974.

[16] Hill, D. P. and Williams, D. N., "Relationship between hydrogen content and low ductility in Al-Li alloys", Final Report, Contract No. N00019-81-C0433, 1982.

[17] Moncur, J. G., Campa, A. B. and Pinoli, P. C., J. High Resolution Chromatography, Vol. 5, pp. 322-324, 1982.

[18] Mehrabian, R., "Rapid solidification processing principles and technologies, III", Proceedings, Conf. on Rapid Solidification Processing, NBS, p. 521, 1982.

[19] Hebenkamp, T., Schmidt, W. and Schlett, V., Acta Metal., Vol. 28, No. 12, p. 1715, 1980.

[20] Rapp, K. and Dovacs-Csenyi, E., Scripta Metal. 11, p. 921, 1977.

[21] Hashimoto, E. and Kino, T., J. Phys. F:Met. Phys. 13, p. 1157, 1983.

[22] Luhman, T. S. and Das, K. B., "Development of joint and joining techniques for metal-matrix composites", 4th Quarterly Prog. Rept., Contract DAAG46-82-C-0064, 1983.

[23] Nutt, S. R. and Carpenter, R. W., "Non-equilibrium phase distribution in an Al-SiC composite", Mat. Sci. Eng., 75, pp. 169-177, 1985.

[24] Nutt, S. R., "Interfaces and failure mechanism in Al-SiC composites", presented at 115th TMS-AIME Annual Meeting, March 2-6, 1986.

[25] Arsenault, R. J., "The strengthening of Al alloy 6061 by fiber and platelet silicon carbide", Mat. Sci. Eng., 64, pp. 171-181, 1984.

[26] Nardon, V. C. and Prewo, K. M., "On the strength of discontinuous silicon carbide reinforced Al composites", Scripta Met., Vol. 20, pp. 43-48, 1986.

[27] Arsenault, R. J. and Fisher, R. M., "Microstructure of fiber and particulate SiC in 6061 Al composites", Scripta Met., 17, p. 67, 1983.

[28] Coyne, E. J., Jr., "Heat treatment of a silicon carbide reinforced Al Matrix material".

[29] Nieh, T. G. and Karlak, R. F., "Aging characteristics of B_4C-reinforced 6061 Al", Scripta Metal., 18, p. 25, 1984.

[30] Nieh, T. G. and Chellman, D. J., "Modulus measurements in discontinuous reinforced Al composites", Scripta Metal., 18, p. 925, 1984.

ACOUSTIC EMISSION ANALYSIS OF COMPOSITE FAILURE MECHANISMS WITH SPECIAL REFERENCE TO THE MODELLING OF AMPLITUDE DISTRIBUTIONS USING A PERSONAL COMPUTER

Y. H. Wu, M. H. Yang, J. Y. Kuo and S. P. Chang

Industrial Technology Research Institute
Chutung, Hsinchu, Taiwan 31015, Republic of China

ABSTRACT

This paper describes the application of Acoustic Emission (AE) to the characterization and analysis of failure mechanisms in carbon fiber reinforced composite materials emphasizing the modelling of amplitude distributions using a personal computer. Representative laminates were tested in uniaxial tension and flexure. The failure processes were monitored and identified by AE.

The modelling technique, which was based on the statistics of extremes, is shown to provide a fairly good method of description for the fracture behavior in composite materials, although exact discrimination between failure modes needs further experimental investigation. The technique offers an unique combination of advantages in terms of facility and versatility of applications for the nondestructive testing of advanced composite materials.

INTRODUCTION

When a material is loaded to a certain stress level, high frequency stress waves are generated due to the abrupt irreversible changes of state in the micro and macro regions. Some of the most important AE-producing processes are: plastic deformations, formation and propagation of cracks, friction process in a crack, phase transformation, fiber breaking in composites, delamination and matrix cracking in composite materials [1]. The stress waves generated and the techniques for their measurement are called AE. Since the mid 1980's, applications of AE to materials research and characterization, nondestructive testing and structural integrity assessment have rapidly increased. While early research works were conducted on metals, recently AE technique has been widely used to study fracture process in composite members, because AE is the only NDT method capable of detecting failure in real time. For this reason, acoustic emission is ideally suited to monitoring the complicated failure process in composite materials, namely the progressive or simultaneous occurrence of matrix cracking, delamination and fiber breaking, these being arranged in the order of ascending energies [2,3]. Several authors [2,4,5] indicated that the slope of the cumulative amplitude distribution remained constant as long as a single type of emission source predominates. Therefore, values and variation of the slope may be used to discriminate between failure modes. Critical points of the amplitude spectrum may be identified by discontinuity in the slope.

The cumulative distribution F(V) is defined as the number of events whose peak amplitude exceeds V. The differential distribution function f(V) is proportional to the number of events whose peak amplitude falls within a narrow window centered about V [6,7]. Thus,

$$f(V) = \left|\frac{\Delta F(V)}{\Delta V}\right| \tag{1}$$

where ΔV is the width of the window.

For the development of mathematical models, it is convenient to define the following normalized functions:

$$\Phi(V) = F(V/V_t) \tag{2}$$

and

$$\psi(V) = f(V/V_t) \tag{3}$$

where V_t is a system constant.

Many models for amplitude distribution have been developed [7]. We propose to use the extreme values distribution function:

$$\Phi(V) = 1 - e^{-(V/V_t)^{-b}} \tag{4}$$

where V_t and b are constants. This function has a peak and a roll-off at low amplitudes and converges to power law with exponent-b in the high amplitude range.

Assume that Φ_1, Φ_2, Φ_3 represent the statistical extreme values functions modelling the cumulative amplitude distributions of the three basic failure modes: matrix cracking, delamination and fiber fracture, respectively, i.e.,

$$\Phi_n(V) = 1 - e^{-(V/V_n)^{-B_n}} \tag{5}$$

where n = 1,2,3 and V_n and B_n are constants. The value of $\Phi(V)$ is proportional to the number of events provided that the energy of the events is randomly distributed.

There are generally more than one failure mode during the composite deformation process. At any moment, the cumulative amplitude distribution of the entire process may be represented as a linear combination of individual distributions,

$$\Phi_t(V) = \Sigma N_k \Phi_k(V), \; k = 1,2,3 \tag{6}$$

for combination with the logarithmical electronics of the AE system, equation (6) may be written as:

$$\log\Phi_t(V) = \log N_k + \log\Phi_k(V) \tag{7}$$

and the log differential amplitude distribution is

$$\log\psi_t = \Sigma[\log N_k + \log B_k + \log(V/V_k)^{-B_k} + \log e^{-(V/V_k)^{-B_k}}] \tag{8}$$

EXPERIMENTAL

Three kinds of specimen were prepared for the experiment. The first was tensile and bending specimens which were cut from an 8 ply, 0 degree unidirectional laminate made of MRL7301/XA-S hot-melt prepregs and cured with a hot press. The second was tensile specimens which were fabricated from MRL7401/XA-S solvent-type prepregs and cured in an autoclave. The third was moulded MRL7401 neat resin tensile specimens. The tensile and bending specimens were in the form of 178 mm x 25 mm x 1.1 mm and 70 mm x 25 mm x 1.1 mm respectively. The end tabs were aluminum.

A 250 KN servohydraulic testing machine (SCHENCK Model PSB 250) was employed for carrying out the monotonically tensile or bending test at constant crosshead speed of 0.5 mm/min and 1 mm/min respectively. For AE measurement, two transducers with frequency responses of 30-70 kHz (Model D9241) and 150-900 kHz (Model S9223 M1), preamplifier with a 20-1000 kHz filter (Model 1801-20H) having a gain of 40dB and a minicomputer based AE system (DE1032D) having a total gain of 80dB and threshold of 32-40dB were used. The transducers were acoustically coupled to the broad face of the specimen using coupling agent.

RESULTS AND DISCUSSION

As shown in Table 1, test results were divided into five groups according to the failure modes observed. For the first two groups of tensile specimen, ma-

GROUP	SPECIMEN		σ_u	E_1	E_f	ϵ_f	TEST	FAILURE MODE	
			kg/mm²	GPA		%	METHOD	MAJOR	MINOR
1	COMP1	[0]8	168	–		–	TENSION	FIBER FRACTURE	MATRIX CRACKING + DEBONDING (OR DELAMINATION)
	COMP2	·	150	–		–			
	COMP4	·	159	105		1.4			
2	WUT0	·	153	111	130	1.3	TENSION	MATRIX CRACKING + DEBONDING (OR DELAMINATION)	FIBER FRACTURE
	WUT2	·	157	112	124	1.3			
	WUT5	·	169	104	117	1.5			
3	BEND1	·	141	110		–	BENDING	DELAMINATION + MATRIX CRACKING	
4	BEND2	·	–	145		–	BENDING	FIBER FRACTURE	MATRIX CRACKING
5	WUEP3		6.0	3.0	1.5	2.6	TENSION	RESIN CRACKING	
	WUEP4		4.9	3.0	2.0	2.0			
	WUEP5		6.3	3.0	1.5	2.8			

Table 1 - Summary of test results

trix cracked first following a concurrence of matrix crackings, debondings and/or fiber breakings which depended on the intrinsic property of the material. It was found that the bending specimens of GROUP 3 and GROUP 4 had failed by delamination and flexural failure respectively. GROUP 5 specimens of neat resin failed by resin cracking undoubtedly. Details are discussed as follows:

A. Fiber Fracture as the Major Failure Mode in Tensile Test

There were a series of fiber fractures before the onset of the ultimate failure in specimens of GROUP 1. With tensile specimen COMP4 as an example, Figure 1(a), the elastic modulus almost holds a constant value of E_1 before 2800 Kg and

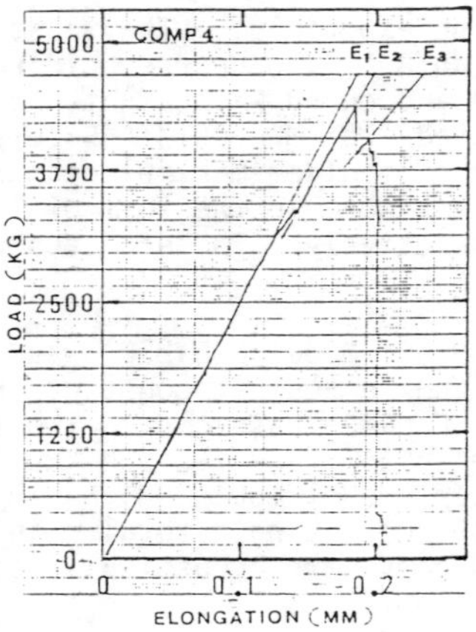

(a)

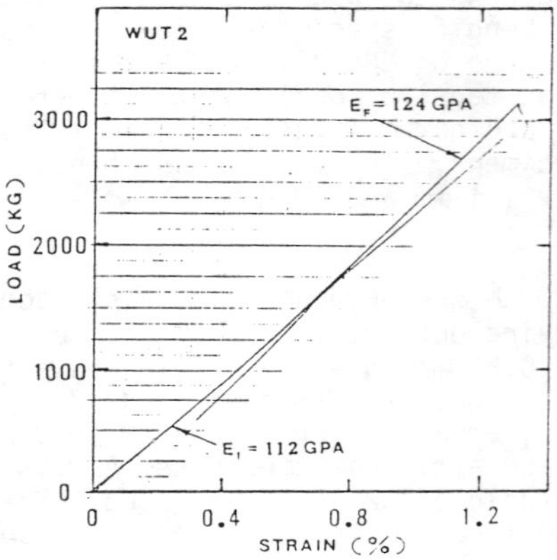

(b)

Fig. (1) - Characteristic load-deformation curves of specimen (a) COMP4 and (b) WUT2

as the load increased to ultimate failure, several severe load drops occurred between which two more distinctive moduli E_2 and E_3 were observed, where $E_1 > E_2 > E_3$. Referring to Figures 2(a) to 2(c), first AE appeared at about 940 Kg (21% Pmax), afterwards, the AE activity increased exponentially in the first stage and leveled off after a short period of high count rate in the second stage. Finally, a sharp increase of count rate occurred prior to the ultimate failure. In addition, the sound of fiber fractures had been heard clearly and continuously throughout the test. Those phenomena suggest that fiber fracture played an important role during the test. Examining the cumulative amplitude distributions in Figure 2(d) to 2(g), they all showed an inflection point at about 65dB, two linear slopes and a concave feature. The evidences mentioned above suggest that there were more than two different mechanisms and those events higher than 65dB might mainly result from fiber fractures.

B. Matrix Cracking and Debonding as the Major Failure Modes in Tensile Test

Specimens of GROUP 2 had exhibited different load-deformation characteristics from those of GROUP 1 as shown in Figure 1. As an example, the elastic modulus of specimen WUT2 increased continuously with loading since it deviated from the initial linear slope. According to the rule of mixture,

$$E_c = E_f V_f + E_m V_m \qquad (9)$$

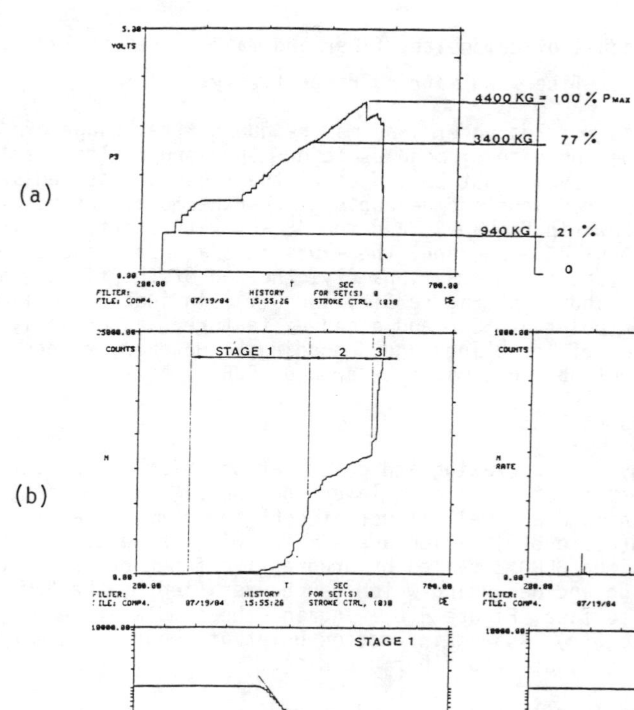

Fig. (2) - Plots of (a) load versus time, (b) cumulative counts versus time, (c) count rate versus time and (d)-(g) cumulative amplitude distributions of specimen COMP4

where E_c, E_f, E_m = Young's moduli of composite, fiber and matrix, respectively, and V_f, V_m = volume contents of fiber and matrix, respectively.

Variation of the composite modulus under load may be due to the change of fiber volume content. So, the increase of modulus before the abrupt ultimate fracture, which was mentioned above, might be caused by the fiber stress redistribution owing to the matrix crackings and debondings. Furthermore, let us examine the AE characteristics shown in Figure 3, and the AE activity started early from 119 Kg which was only about 4% Pmax, and, there was also a sudden increase of count rate around 1700 Kg (about 55% Pmax), namely, the end of the first stage, the cumulative amplitude distributions (Figure 3(e) and 3(f)) indicated two linear slopes with an inflection point at 50dB and a convex feature. Accordingly, it suggests the occurrences of matrix crackings and debondings. However, events of fiber fractures, which were not obvious, might be around 65dB.

C. Failure Modes of Bending Test

BEND 1 specimen failed by matrix crazing and delamination which can be verified by the failed specimen split at the middle layer and the SEM fractograph, Figure 4(a). This phenomenon is again well illustrated by the cumulative amplitude distributions, Figure 5(e) to 5(h), which are similar with those of specimen WUT2. On the contrary, specimen BEND 2 failed by progressive fiber breaking accompanied with matrix cracking and debonding which can be verified by the SEM fractograph of the fracture surface, Figure 4(b). Again, the cumulative amplitude distributions, Figure 6(e) to 6(h), showed inflection point at about 50dB, many high amplitude events being attributed to fiber breakings.

D. AE Characteristics of Tensile Testing of Neat Resin

Figure 7 shows the typical AE characteristics of the tensile testing of neat resin which is illustrated by specimen WUEP5. Three stages were noticed on the plot of events versus time and a sudden rise of count rate occurred between stages 1 and 2 with reference to the plot of count rate versus time. With reference to the cumulative amplitude distributions, they showed no obvious inflection point and a somewhat concave feature suggesting a more active low amplitude distribution.

E. Modelling the Amplitude Distribution of Failure Modes

As has been shown above, the cumulative amplitude distributions showed two inflection points at about 50dB and 65dB respectively, which were dependent on the failure mechanisms during deformation. While the high amplitude AE signals caused by fiber fractures were active, an inflection point appeared near 65dB and the curve was concave as shown in Figure 2. On the other hand, the cumulative amplitude distributions, Figure 3, became convex with an inflection point at 50dB when the low and medium amplitude signals (i.e., matrix cracking (crazing) and delamination or debonding) were the only AE sources. Accordingly, we propose to model the failure mechanisms in terms of statistical extreme values function (i.e., equation (5)), by comparing the log cumulative amplitude distribution, log differential amplitude distribution and linear differential amplitude distribution with relative test results of AE characteristics. Firstly, when the low and medium amplitude distributions were combined to form the total amplitude distribution, it was found that the slope of the curve decreased with the increasing dB value. Although the slope changes are not so clear as the test results, they showed a similar convex feature as shown in Figures 8(c) and 8(d). Meanwhile, if the high

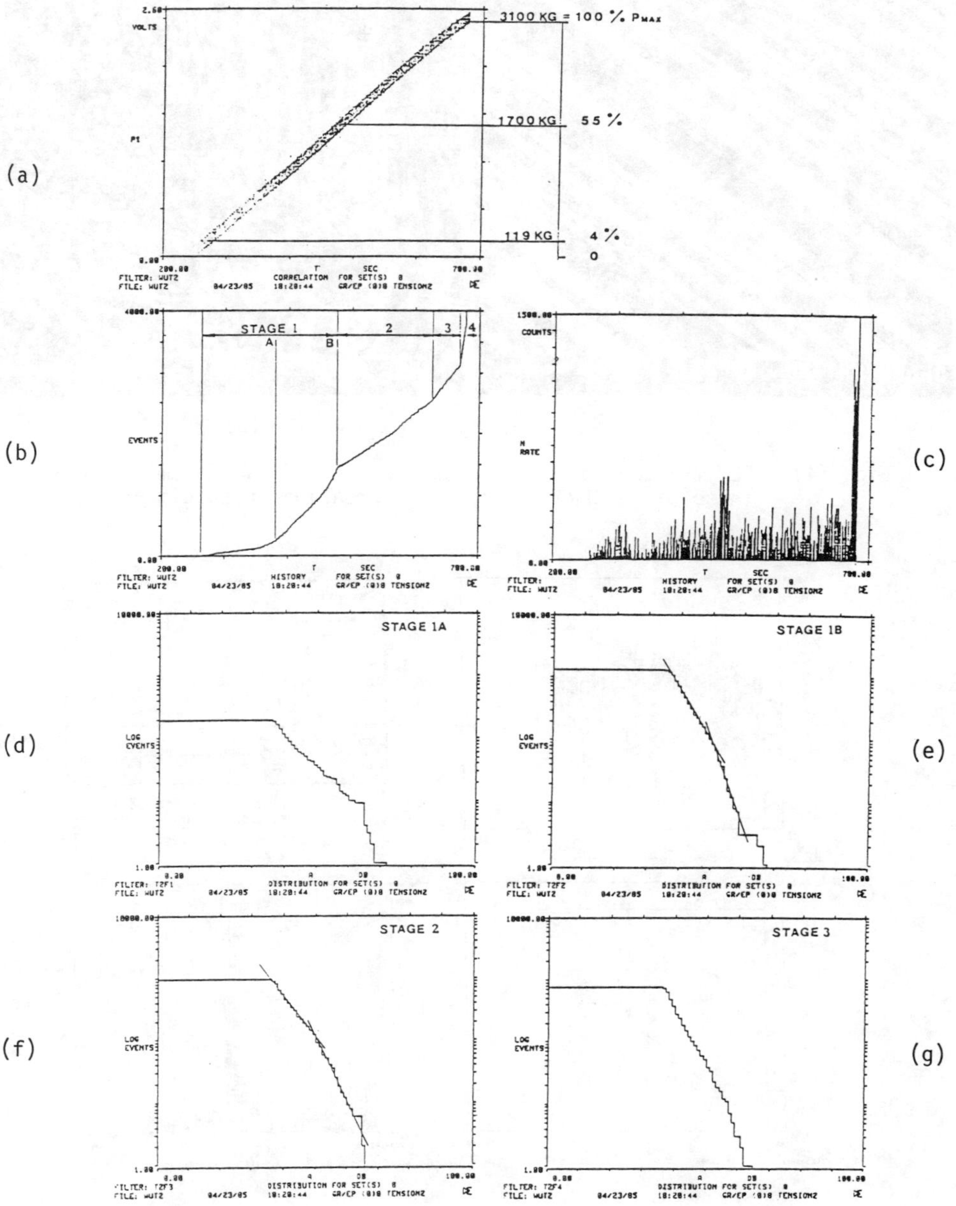

Fig. (3) - Plots of (a) load versus time, (b) cumulative events versus time, (c) count rate versus time and (d)-(g) cumulative amplitude distributions of specimen WUT2

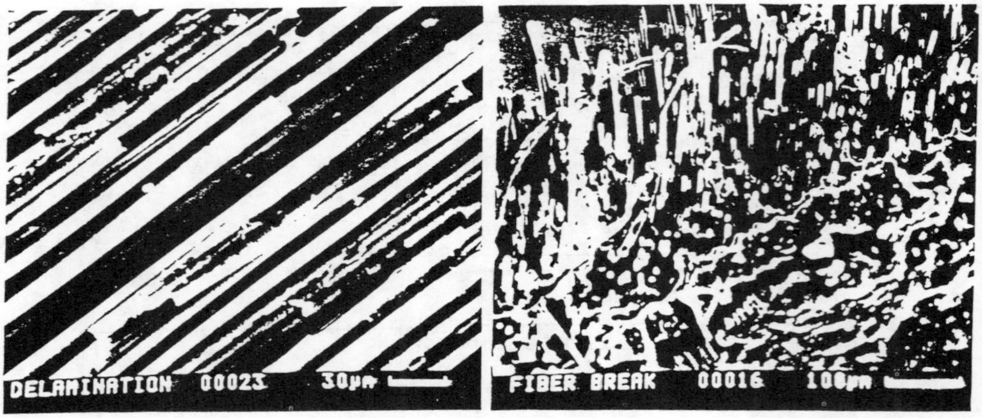

(a) (b)

Fig. (4) - SEM fractograph of fracture-surface of (a) delamination in specimen BEND 1, (b) fiber breaking in specimen BEND 2

(a) (b) (c) (d)

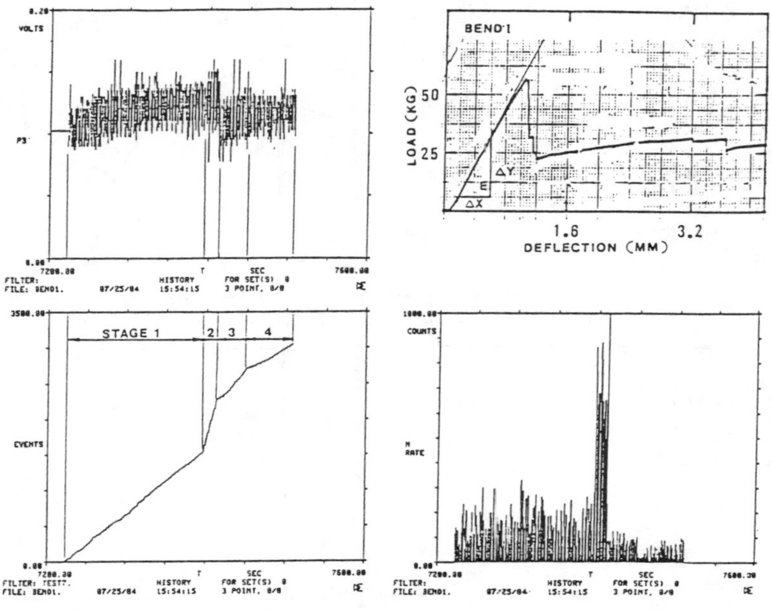

Fig. (5) - Plots of (a) load versus time, (b) load versus deflection, (c) cumulative events versus time, (d) count rate versus time

(e) (f) (g) (h)

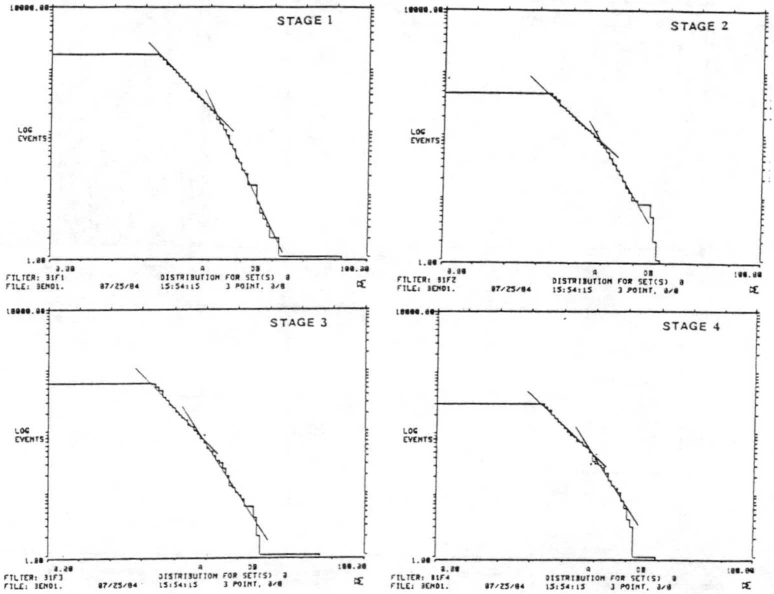

Fig. (5 continued) - Plots of (e)-(h) cumulative amplitude distributions of specimen BEND 1

(a) (b)

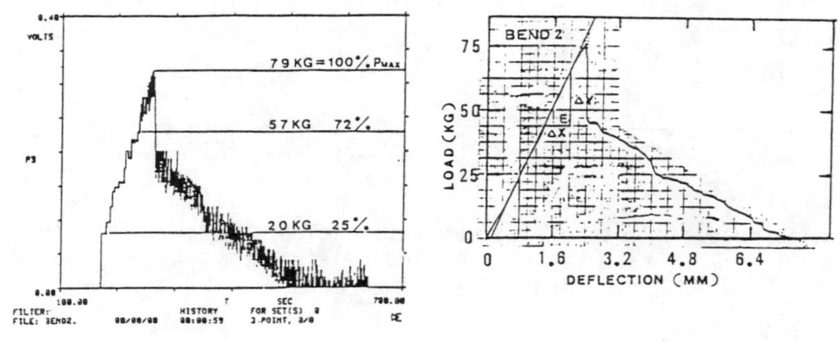

Fig. (6) - Plots of (a) load versus time, (b) load versus deflection

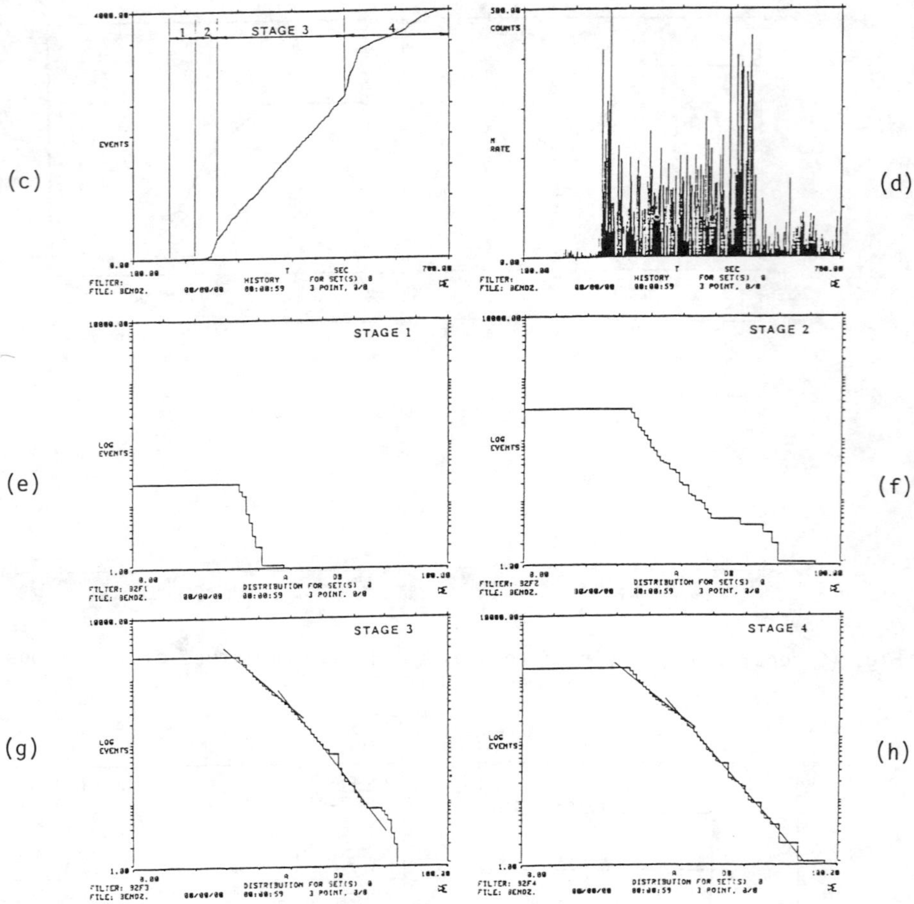

Fig. (6 continued) - Plots of (c) cumulative events versus time, (d) count rate versus time and (e)-(h) cumulative amplitude distributions of specimen BEND 2

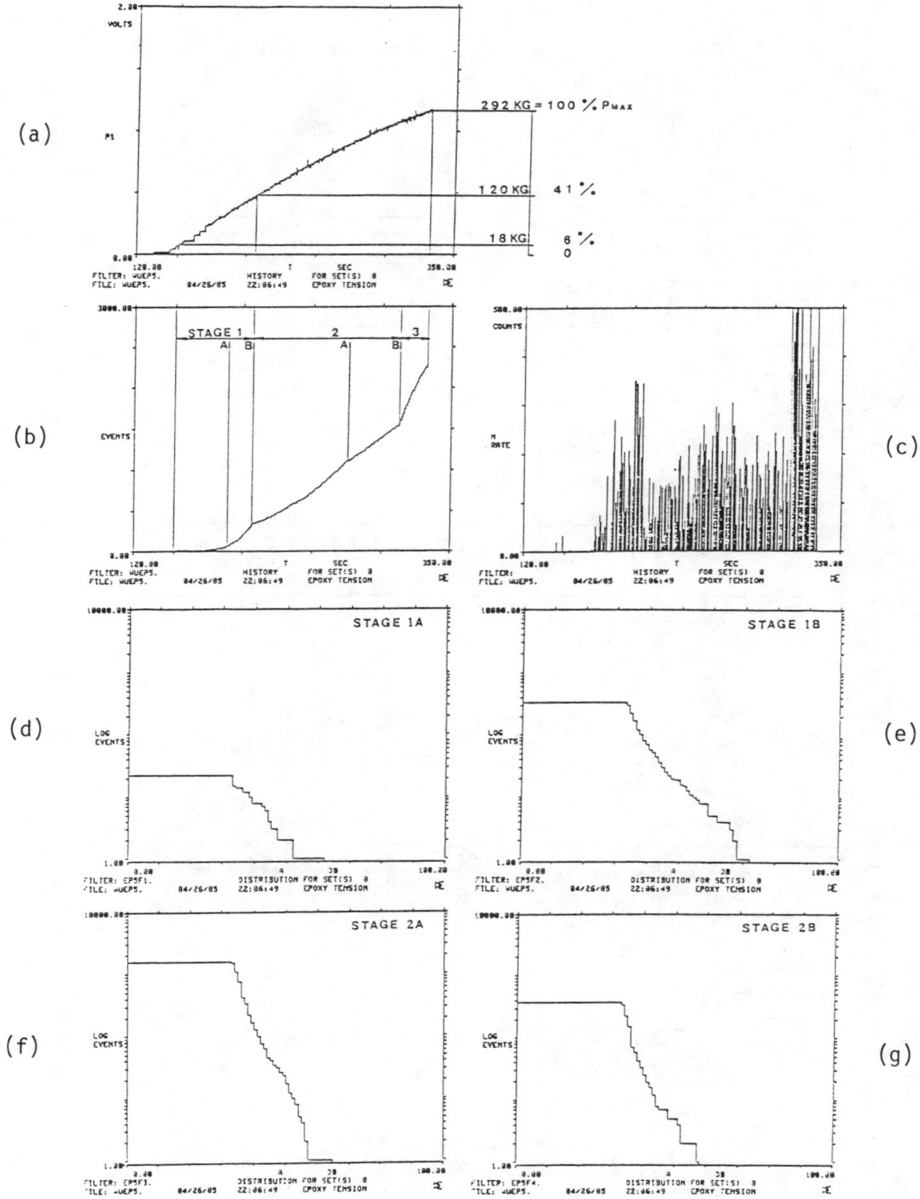

Fig. (7) - Plots of (a) load versus time, (b) events versus time, (c) count rate versus time and (d)-(g) cumulative amplitude distributions of specimen WUEP 5

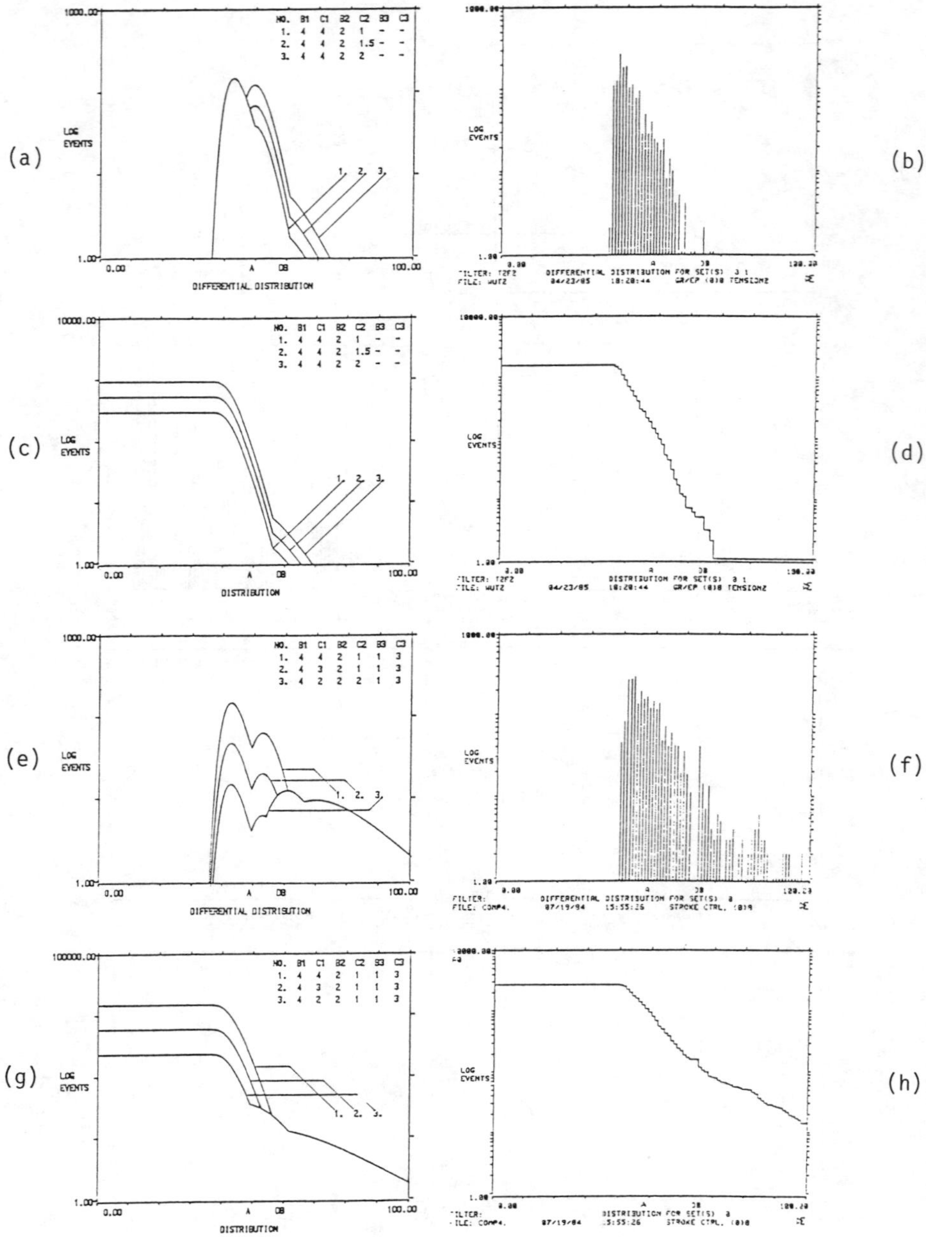

Fig. (8) - Amplitude distributions and modelling curves of specimen WUT2 (a)-(d) and specimen COMP4 (e)-(h)

amplitudes were added, the curves showed clearly a concave feature and an inflection point, Figure 8(g) and 8(h), which were similar with the test results. Moreover, the log differential amplitude distribution, Figure 8(a), 8(b), 8(e), 8(f), and the linear differential amplitude distributions match the typical experimental results very well, Figure 9.

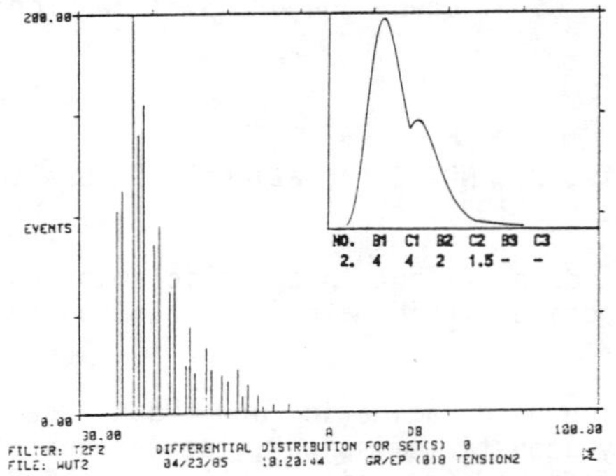

(a)

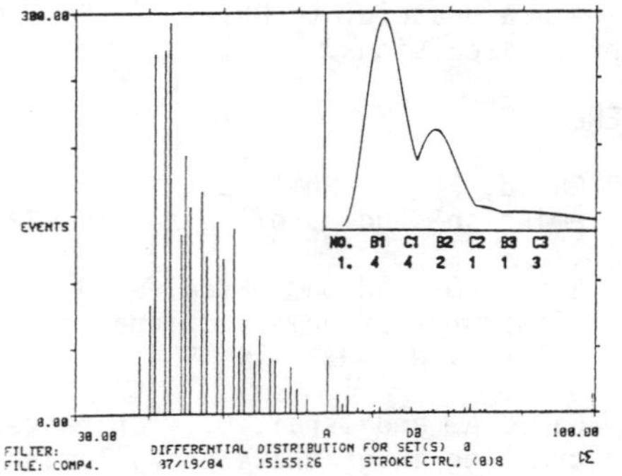

(b)

Fig. (9) - Plots of linear amplitude distributions and their relative modelling curves of specimen (a) WUT2 and specimen (b) COMP4

Finally, it was noticed that the time dependence of total events and total counts indicated a similar pattern showing three stages and a marked change of AE activity at the end of the first stage. This suggested a substantial damage occurred at that time. Nevertheless, the severity of the damage needs to be investigated further.

CONCLUSIONS

(1) The experimental amplitude distributions showed a practical usefulness of discriminating failure mechanisms in CFRP coupons. The presence of matrix cracking and debonding and/or fiber fracture were verified in terms of the change of linear slopes between which two inflection points appeared at about 50dB and 65dB respectively.

(2) The statistics of extremes model was shown to give good fits to the amplitude distributions. It is possible to give an insight into the deformation process in composites with such a modelling technique: verifying the failure mechanisms in terms of inflections and B_n parameters shown on the cumulative amplitude distribution and describing the severity of damage in terms of C_n parameters.

(3) The time dependence of total events or counts and count rate indicated a similar pattern showing three stages and a large value of count rate before the second stage, which suggested a more serious damage at that time. Accordingly, it is possible to build up a criteria for estimating the severity of damage, re-

sidual strength or integrity of composites in terms of the amplitude distributions, time dependence of events, counts and count rate accompanied with the statistics of extremes model.

ACKNOWLEDGEMENTS

We are grateful to Dr. A. A. Pollock, Dr. C. H. Cheng and Dr. S. T. Wu for helpful discussions.

REFERENCES

[1] Guild, F. J., Phillips, M. G. and Harris, B., NDT International, "Acoustic emission studies of damage in GRP", October 1980.

[2] Short, D. and Summerscales, J., "Amplitude distribution acoustic emission signatures of unidirectional fiber composite hybrid materials", Composites, Vol. 15, No. 3, July 1984.

[3] Garc, A. and Ishai, O., "Characterization of delamination and transverse cracking in graphite/epoxy laminates by acoustic emission", 1st International Symposium on Acoustic Emission from Reinforced Composites, SPI, July 19-21, 1983.

[4] Wadin, J. and Pollock, A. A., "Periodic proof testing of fibreglass booms through the use of non-kaiser emission responses", 9th Nat. Tech. Conf., SAMPE, Atlanta, Georgia, USA, pp. 519-560, 1977.

[5] Graham, L. J. and Elsley, R. K., "Characterization of acoustic emission signals and application to composite structures monitoring", AFML-TR-78-55, pp. 40-45, 1978.

[6] Pollock, A. A., "Acoustic emission amplitude", Non-Destructing Testing, October 1983.

[7] Pollock, A. A., "Acoustic emission amplitude distribution", International Advances in Nondestructive Testing, Vol. 7, pp. 215-239, 1981.

PREDICTION OF STIFFNESS DEGRADATION OF COMPOSITE LAMINATES UNDER FATIGUE LOADING*

J. N. Yang, A. Meskini and S. H. Yang

School of Engineering and Applied Science, The George Washington University
Washington, D.C. 20052 USA

ABSTRACT

It has been observed that the stiffness of composite laminates reduces monotonically under fatigue loadings. Hence, a phenomenological approach for estimating the fatigue damage in composite is the determination of the residual stiffness. Since the stiffness of composite laminates can be measured nondestructively in service, a time series analysis procedure is proposed herein for predicting the stiffness degradation of composite laminates. A second order autoregressive model is found to describe very well the behavior of the stiffness degradation of composite laminates. Available experimental data are used to verify the results of time series analysis. It is demonstrated that the experimental results correlate very well with the theoretical predictions.

INTRODUCTION

One of the important problems in the design and analysis of aircraft structures is the prediction of the fatigue behavior of composite laminates subjected to service loading spectra. From a maintenance standpoint, it is very important to determine nondestructively the extent of fatigue damage accumulation in composite materials after a certain period of service. For metallic materials, the fatigue damage can be characterized by the growth of a predominant crack and the crack can be detected using various available NDE techniques. For composite materials, however, the fatigue damage mechanisms are complicated, involving matrix cracking, fiber breakage, debonding, delamination, void growth, etc. To date, a satisfactory characterization for the fatigue damage in composite laminates has not been developed.

Intensive research for the fatigue of composite materials has been made recently, e.g., [1-9]. These research efforts are based on the residual strength degradation criterion. While such efforts are useful for the characterization of fatigue damage, the residual strength cannot be measured nondestructively. It has been observed that the stiffness of composite laminates reduces monotonically as the fatigue damage accumulates. Hence, a promising phenomenological approach

*This research was supported by Naval Air Systems Command under contract No. N00019-83-C-0346.

for characterizing the fatigue damage is the determination of the residual stiffness.

The time series analysis methodology is proposed herein to predict the stiffness degradation of composite laminates under fatigue loadings. Since the residual stiffness can be measured nondestructively in service, the method of time series analysis is a natural tool for the prediction purpose. Available experimental data for composite stiffness degradation are used to verify the proposed methodology. It is demonstrated that the experimental results correlate well with the time series analysis predictions.

TIME SERIES ANALYSIS APPROACH

A time series is a set of observations ordered in time (or in any other dimension). The observations of the stiffness, $\overline{G}(t)$, of a composite laminate is a typical time series. An important class of stochastic models for describing the time series is the Autoregressive Integrated Moving Average Model referred to as ARIMA model [10]. The mathematical model of ARIMA expresses the data at time t as a combination of past values of the series and past values of random errors. A special case of the ARIMA model which is useful in the representation of the composite stiffness data is the so-called autoregressive model. In this model, the current value of the stiffness is expressed as a finite, linear aggregate of previous values of the stiffness and a shock a_t.

Let the values of a stochastic stiffness process at equally spaced times t, t-Δt, t-2Δt,..., be denoted by $\overline{G}_t$, $\overline{G}_{t-1}$, $\overline{G}_{t-2}$,... The stiffness process $\{\overline{G}_t\}$ is a nonstationary time series which can be transformed into a stationary one by either differencing or subtracting the mean values from the original data as will be described later. Let G_t, G_{t-1}, G_{t-2},..., be the transformed stiffness corresponding to $\overline{G}_t$, $\overline{G}_{t-1}$, $\overline{G}_{t-2}$,..., such that the stochastic stiffness process $\{G_t\}$ is a stationary time series. Then

$$G_t = \phi_1 G_{t-1} + \phi_2 G_{t-2} + \ldots + \phi_p G_{t-p} + a_t \tag{1}$$

is called an autoregressive process of order p and is denoted by AR(p). In equation (1), ϕ_1, ϕ_2,..., ϕ_p are autoregressive parameters and $\{a_t\}$ is a white noise process, i.e., a_t, a_{t-1},..., a_1 are statistically independent normal random variables with zero mean and standard deviation σ_a. The degree of the autoregressive model, p, depends on the particular data set under consideration. The time series analysis procedures have been proposed by Box and Jenkins [10] which consist of the following three steps: (i) identification; (ii) estimation, and (iii) diagnostic checking. Each of these steps will be described briefly in the following paragraphs.

Identification of model: the basic tools in identifying an AR(p) model are ordinary and partial autocorrelation functions (or correlograms) [10]. The required number of terms, p, in the AR(p) model depends on the autocorrelation function. The autocorrelation function at jth lag, R_j (j = 1,2,...,3) for a time series with zero mean, is defined as

$$R_j = c_j/c_o \text{ for } j = 0,1,2,\ldots,m \tag{2}$$

in which

$$c_j = \frac{1}{N} \sum_{t=1}^{N-j} G_t G_{t+j} \text{ for } j = 0,1,2,\ldots,m \tag{3}$$

where N is the total number of data points in the time series, c_j is an estimate of the autocovariance and c_o is an estimate of the variance of the time series. Theoretically, the degree p should be chosen such that the autocorrelation beyond G_{t-p} is negligible. While the autocorrelation function of a time series may die down as the time lag p increases, it may also stretch out to infinity. Hence, the choice of an appropriate degree p is determined by the partial autocorrelation function. Suppose the time series is fitted by an AR(p) model, equation (1), with all the autoregressive parameters $(\phi_1,\phi_2,\ldots,\phi_p)$ being estimated. It is desirable to evaluate if the inclusion of an additional lag G_{t-p-1} (i.e., $\phi_{t-p-1} \times G_{t-p-1}$) in the model, namely AR(p+1) model, will represent the data more accurately. If the coefficient ϕ_{t-p-1} in the AR(p+1) model is larger, then G_{t-p-1} should be included in the model, i.e., AR(p+1) is better than AR(p); otherwise, it can be assumed that the AR(p) model is adequate for the time series. The highest order autoregressive coefficient, ϕ_p, of AR(p) model is referred to as the partial autocorrelation function at lag p. The partial autocorrelation function is a device for summarizing all information contained in the autocorrelation function of an AR(p) process in a small number of nonzero statistics. For an AR(p) model, only p such nonzero statistics are necessary rather than an infinite number of nonzero autocorrelations. The partial autocorrelation function may be estimated by fitting successively autoregressive processes AR(p) of orders p = 1,2,3... by least-squares procedures and picking out the estimates of the last coefficient fitted at each stage. The required number of terms, p, is determined such that the partial autocorrelation functions at a lag larger than p are negligibly small. The computations of both ordinary and partial autocorrelation functions from experimental data are described in [10].

Estimation of model parameters: after identifying a tentative model, the next step is to estimate the model parameters, ϕ_1, $\phi_2,\ldots,$ ϕ_p and σ_a from observed base-line data. Methods are available for the parameter estimation, such as the method of maximum likelihood or least-squares [10].

Diagnostic checking: diagnostic checking is used to determine if the fitted model is adequate in its relations to the data and to achieve model improvement. If any inadequacy is found, the interactive cycle of identification, estimation and diagnostic checking is repeated until a suitable model is found. One simple technique for diagnostic checking is to see if the residuals $(a_t,a_{t-1},\ldots,a_1)$ form a white noise process [10].

The purpose of time series analysis is to predict the stiffness G_{t+k} at times $t+k\Delta t$ for $k \geq 1$, using the available data up to time t, i.e., $G_1 = G(\Delta t)$, $G_2 = G(2\Delta t),\ldots,$ $G_t = G(t)$. Such a prediction is said to be made at origin t for lead time k. For illustrative purpose, consider an autoregressive model of degree 2, i.e., AR(2),

$$G_t = \phi_1 G_{t-1} + \phi_2 G_{t-2} + a_t \tag{4}$$

Then, an observed G_{t+k} can be expressed

$$G_{t+k} = \phi_1 G_{t+k-1} + \phi_2 G_{t+k-2} + a_{t+k} \text{ for } k = 1,2,\ldots \tag{5}$$

Let $\hat{G}_t(k)$ for $k = 1,2,\ldots$, be the minimum mean-square error forecast of G_{t+k}. Then, it can be shown [10] for the AR(2) model that

$$\hat{G}_t(k) = \phi_1 \hat{G}_t(k-1) + \phi_2 \hat{G}_t(k-2) \text{ for } k = 1,2,\ldots \tag{6}$$

in which $\hat{G}_t(0) = G_t$ and $\hat{G}_t(-1) = G_{t-1}$. Equation (6) will be used to forecast the stiffness G_{t+k} for $k = 1,2,\ldots$ using available observed data up to time t, i.e., $G_1, G_2, \ldots, G_t$.

THEORETICAL AND EXPERIMENTAL CORRELATIONS

Experimental data generated by General Dynamics, Fort Worth Division [11], are used herein for illustrative purpose. All tests were performed at 5 Hz in standard laboratory conditions of room temperature and ambient humidity. During each test, periodic stiffness measurements were taken with an extensometer. Test results for the stress ratio R = 0.1 at maximum stress levels of 45.0, 47.5, 50.0 and 52.5 ksi will be used. The material of the specimens in Hercules AS1/3502 and the specimen configuration with the stacking sequence is shown in Figure 1. In this configuration, ply-terminations specimens have two sets of double 0° plies

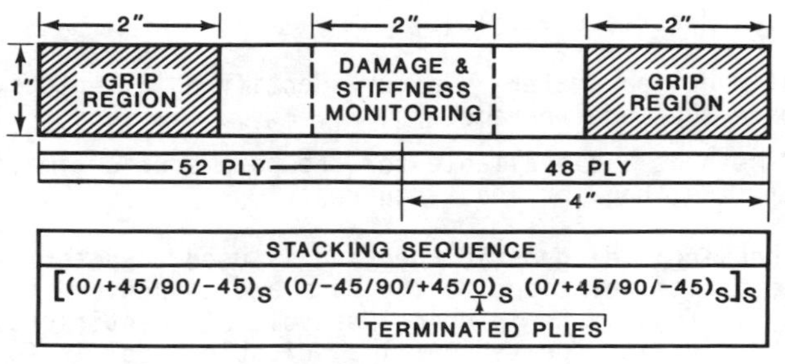

Fig. (1) - Specimen configuration and stacking sequence

terminated at the center of the specimen gage length. One side of these specimens is 52 plies and the other is 48 plies.

The stiffness data exhibit a sudden reduction ranging from 0 to 6% after one fatigue cycle, and then the degradation becomes more regular afterward. In order to obtain a homogeneous model, such an irregularity is removed by excluding the initial stiffness $\bar{G}(0)$, i.e., the stiffness data start from one fatigue cycle $\bar{G}(1)$. This approach appears to be reasonable because $\bar{G}(1)$ can be measured non-destructively. For the time series analysis, stiffness data should be measured at equally-spaced time intervals. In reality, however, this usually cannot be accomplished and interpolation procedures are needed to determine the missing data.

The autoregressive parameters $(\phi_1, \phi_2, \ldots, \phi_p)$ should be determined from some base-line test data. Experimental data for 5 specimens are available under each stress level. To demonstrate the capability of the time series analysis approach, test results of 4 specimens will be considered as base-line data from which the autoregressive parameters will be estimated. The established autoregressive model is then used to predict the stiffness degradation for the fifth specimen. Finally, the time series analysis predictions for the fifth specimen will be compared with the corresponding experimental data to demonstrate the correlation. Likewise, test results of different combinations of any 4 specimens will be considered as base-line data from which the stiffness degradation of the fifth specimen will be predicted and compared.

Let $\bar{G}(n)$ be the stiffness after n load cycles and t = logn be the number of load cycles in log scale. Then the residual stiffness G(t) expressed in term of t is normalized by the initial stiffness at n = 1 or t = 0, i.e., $G^*(t) = \bar{G}(t)/\bar{G}(0)$. The normalized stiffness $G^*(t) = G^*_t$ against the log cycle t constitute a nonstationary time series $\{G^*_t\}$ at discrete time t.

Let $\bar{N}$ be the average fatigue life of a total of five specimens and $T = \log\bar{N}$. The four base-line samples of stiffness degradation are considered to be periodic with a periodic of T, so that they can be connected together for the computation of autocorrelation and partial autocorrelation functions. The samples with a fatigue life longer than T were cut off at T and those with a shorter fatigue life were extrapolated up to T. Then, the nonstationary time series $\{G^*(t)\}$ for the base-line data is converted into a stationary time series $\{G(t)\}$ by substructing $G^*(t)$ from the population mean $\mu(t)$, i.e.,

$$G_t = G(t) = G^*(t) - \mu(t) \tag{7}$$

in which $\mu(t)$ is the mean value of four base-line samples at time t, i.e., $\mu_t = \mu(t) = [G^*_1(t) + G^*_2(t) + G^*_3(t) + G^*_4(t)]/4$ where $G^*_j(t)$ is the normalized stiffness at t for the jth sample (specimen). Thus, the stationary time series $\{G(t)\} = \{G_t\}$ represents the deviation of the normalized stiffness from the mean of the base-line data.

The autocorrelation functions using any combination of four base-line samples under different stress levels die down rapidly [12]. Moreover, all the partial autocorrelation functions show that only the first two values are significant [12]. It is further noticed that there is no periodic pattern in both autocorrelation and partial autocorrelation functions. This precludes the possibility of periodic (or seasonal) autoregressive model. As a result, the second order autoregressive model, AR(2), given by equation (4) is a reasonable representation for the stiffness degradation data.

The autoregressive parameters ϕ_1 and ϕ_2 as well as the standard deviation of σ_a of a_t are estimated using the method of least squares. The results of ϕ_1 and ϕ_2 are presented in Table 1 for different combinations of 4 specimens that

Table 1 - Estimated autoregressive parameters ϕ_1 and ϕ_2 for different stress levels and different combinations of the base-line data

Specimen No. Used as Base-Line Data	Stress Levels, ksi							
	45		47.5		50		52.5	
	ϕ_1	ϕ_2	ϕ_1	ϕ_2	ϕ_1	ϕ_2	ϕ_1	ϕ_2
1,2,3,4	1.432	-0.559	0.988	-0.033	0.864	0.109	1.274	-0.314
1,2,3,5	1.409	-0.533	1.259	-0.317	0.986	-0.068	1.253	-0.301
1,2,4,5	1.301	-0.391	0.981	-0.072	0.949	-0.011	1.213	-0.243
1,3,4,5	1.367	-0.503	0.981	-0.066	1.006	-0.021	1.356	-0.463
2,3,4,5	1.328	-0.499	1.074	-0.151	0.887	-0.030	1.196	-0.287

are considered as the base-line data. From the estimated autoregressive parameters, the residuals $(a_1, a_2, ...)$ and their autocorrelation function are computed. The results indicate that $\{a_t\}$ does represent a white noise process [12]. Thus, the selected model AR(2) given in equation (4) appears to be adequate for the time series $\{G_t\}$, representing the deviation of the normalized stiffness from the population mean $\mu(t)$.

The time series analysis methodology described above will be used to forecast the stiffness for each specimen up to the population mean life T. Then the forecast of G_t will be converted back to the normalized stiffness G_t^* by the addition of the mean function $\mu(t)$, see equation (7).

Consider the specimen group under a fatigue stress level of 45 ksi. The average fatigue life $\bar{N}$ is 546,000 cycles, i.e., $T = \log \bar{N} = 5.7$. Fifty one (51) data points are considered for the time series of each specimen, i.e., $\Delta t = T/50 = 0.114$. The time series for the stiffness degradation of specimen number 5 is plotted in Figure 2 as solid circles. For clarity of presentation, only 26 out of 51 data points are plotted. Assuming that the first eleven stiffness data points are measured in service, i.e., stiffness is known up to $t = 10\ \Delta t = 1.14$ or n = 14 cycles, and using the autoregressive parameters ϕ_1 and ϕ_2 shown in Table 1, a prediction of the stiffness degradation after 14 cycles up to 546,000 cycles is made. The results are presented in Figure 2 as open circles along with the experimental data shown by solid circles.

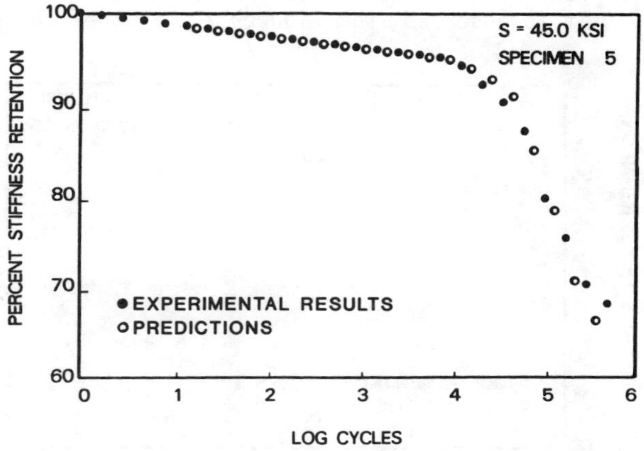

Fig. (2) - Prediction of stiffness retention for specimen no. 5 given observations up to 14 cycles under stress level 45.0 ksi

When the stiffness is known up to t = 3.42, i.e., n = 2,630 cycles, the predictions for the stiffness degradation from 3,630 cycles up to 546,000 cycles are depicted in Figure 3 as open circles. When the stiffness is known up to

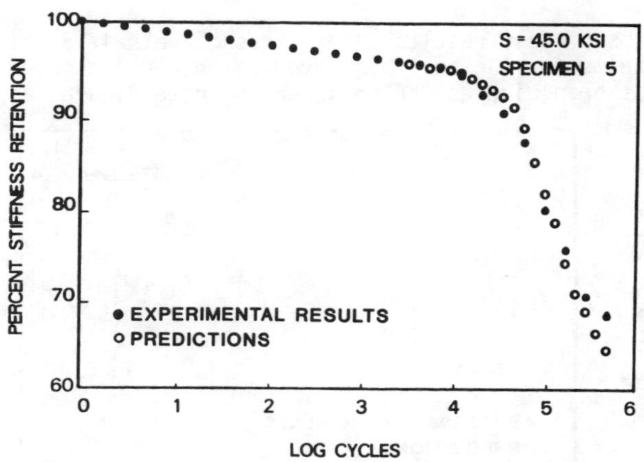

Fig. (3) - Prediction of stiffness retention for specimen no. 5 given observations up to 2,630 cycles under stress level 45.0 ksi

36,308 cycles, the predictions for stiffness degradation are displayed in Figure 4 as open circles. It is observed from Figures 2 to 4 that the predictions of

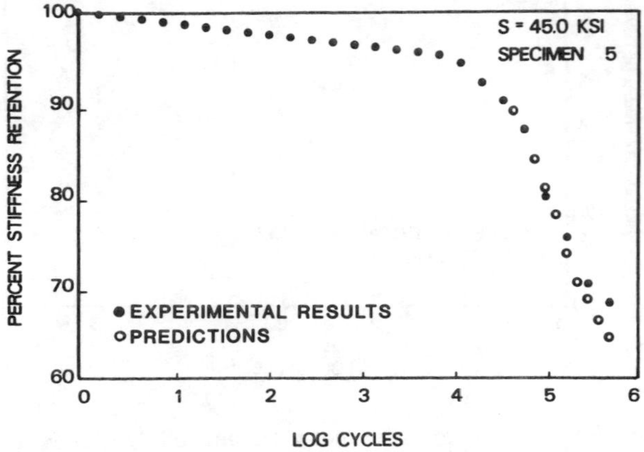

Fig. (4) - Prediction of stiffness retention for specimen no. 5 given observations up to 36,308 cycles under stress level 45.0 ksi

the stiffness degradation correlate well with the experimental results. In a similar manner, the predictions of the stiffness degradation for other specimens have been conducted and the results have a good correlation with the experimental data [12].

For other groups of test results under stress levels 47.5 ksi, 50 ksi and 52.5 ksi, some results for the stiffness prediction are presented in Figures 5 to 7, respectively, as open circles. Also shown in these figures as solid circles

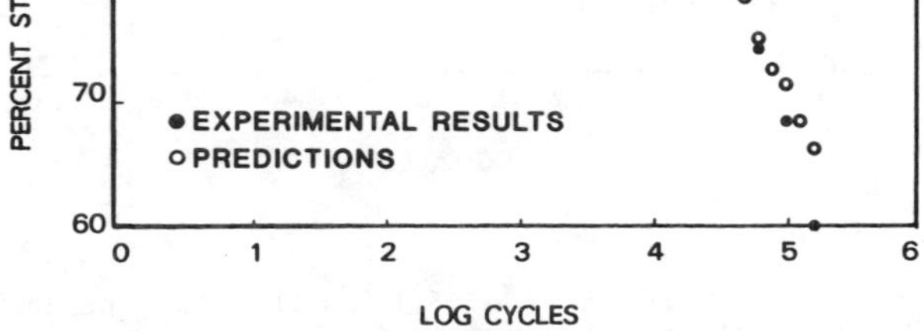

Fig. (5) - Prediction of stiffness retention for specimen no. 4 given observations up to 1,413 cycles under stress level 47.5 ksi

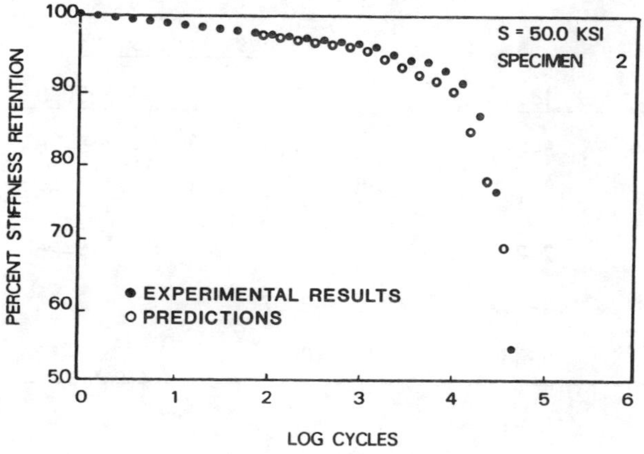

Fig. (6) - Prediction of stiffness retention for specimen no. 2 given observations up to 76 cycles under stress level 50.0 ksi

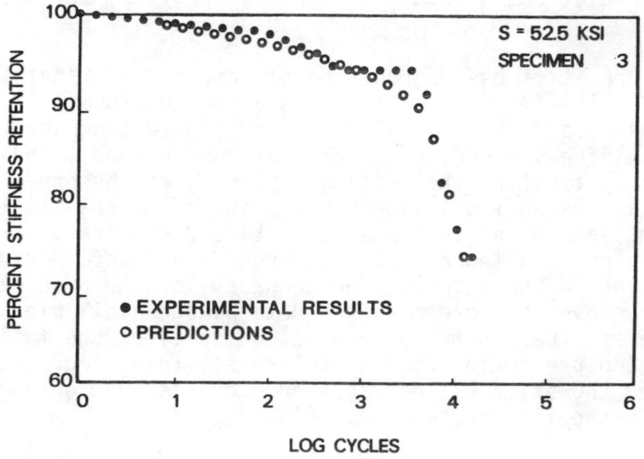

Fig. (7) - Prediction of stiffness retention for specimen no. 3 given observations up to 7 cycles under stress level 52.5 ksi

are the experimental data. Again, good correlations between the time series analysis predictions and the experimental data are observed.

Because of simplicity in presentation, the identification numbers for all test specimens used in this paper are different from those of [11]. The corresponding specimen identification numbers are shown in Table 2 for convenience of reference.

Table 2 - Specimen identification number used in [11]

Specimen No. Used	Stress Levels, ksi			
	45.0	47.5	50.0	52.5
1	5-22	4-23	5- 8	4- 9
2	11-12	6- 1	9-20	6- 5
3	7-28	13-19	11-26	13- 5
4	3-16	8- 7	7-14	8-23
5	9- 6	10-13	3- 2	10-27

CONCLUSION

The time series analysis methodology is proposed for application to fatigue of composite laminates. A second order autoregressive model is found to describe very well the behavior of the stiffness degradation data. This model is used to predict the future stiffness reduction for each composite specimen under cyclic loadings. The predictions have been shown to correlate very well with the experimental data.

In general, the stiffness degradation of composite laminates consists of three phases [1]. The first phase involves a sudden drop in stiffness in a few load cycles. The second phase involves a slow degradation and the last one involves a rapid stiffness reduction before fatigue failure. The first phase can be taken care of by treating the initial stiffness as the one after a sudden drop. The time series analysis is perfectly suited to the stiffness degradation prediction in the second and last phases. It is demonstrated that a knowledge of stiffness only up to a few cycles will provide very good prediction for the stiffness reduction in the future. The accuracy for the prediction of the stiffness reduction improves as more measured data become available. In general, a knowledge of the stiffness data up to about one half of the fatigue life will provide a very good prediction for the future stiffness degradation. Finally, investigations on the stiffness degradation mechanisms in composite laminates have been made in the literature, e.g., [13-18].

REFERENCES

[1] Salkind, M. J., "Fatigue of composites", in Composite Materials: Testing and Design, ASTM STP 497, pp. 143-169, 1972.

[2] Yang, J. N. and Du, S., "An exploratory study for fatigue of composites under spectrum loading", J. Composite Materials, Vol. 17, pp. 511-526, 1983.

[3] Yang, J. N. and Jones, D. L., "Fatigue of graphite/epoxy $[0/90+45/-45]_S$ laminates under dual stress levels", Composite Technology Review, Vol. 4, No. 3, pp. 63-70, 1982.

[4] Yang, J. N. and Sun, C. T., "Proof test and fatigue of composite laminates", J. of Composite Materials, Vol. 14, pp. 801-817, 1980.

[5] Yang, J. N. and Cole, R. T., "Fatigue of composite bolted joints under dual load levels", in Progress in Science and Engineering, Proc. 4th Int. Conf. on Composite Materials, ICCM-IV, Tokyo, pp. 333-340, October 24-26, 1983.

[6] Yang, J. N., "Fatigue and residual strength degradation for graphite/epoxy composites under tension-compression cyclic loading", J. of Composite Materials, Vol. 12, pp. 19-29, 1978.

[7] Yang, J. N. and Jones, D. L., "The effect of load sequence on statistical fatigue of composites", AIAA Journal, Vol. 18, No. 12, pp. 1525-1531, 1980.

[8] Hahn, T. H., "Fatigue behavior and life prediction of composite materials", Air Force Materials Lab., AFML-TR-78-43, 1978.

[9] Whitney, J. M., "Fatigue characterization of composite materials", in Fatigue of Fibrous Composite Materials, ASTM STP 723, pp. 133-151, 1981.

[10] Box, G. E. P. and Jenkins, G. M., Time Series Analysis: Forecasting and Control, Holden-Day, San Francisco, 1976.

[11] Ulman, D. A., et al, "Damage accumulation in composites", General Dynamics Fort Worth Division Technical Report FZM-7150, 1983.

[12] Meskini, A., "Prediction of stiffness degradation under fatigue loading", Sc.D. dissertation submitted to School of Engineering and Applied Science, The George Washington University, 1986.

[13] Highsmith, A. L. and Reifsnider, K. L., "Stiffness-reduction mechanisms in composite laminates", Damage in Composite Materials, ASTM STP 775, pp. 103-117, 1982.

[14] Withworth, H. A., "Characterization of the dynamic stiffness reduction of graphite/epoxy composite laminates under fatigue loading", D.Sc. dissertation, The George Washington University, 1983.

[15] Odom, E. M. and Adams, D. F., "Stiffness reductions during tensile fatigue testing of graphite/epoxy angle/ply laminates", Contract No. NAS1-16557, NASA-16557, Langley Research Center, 1982.

[16] O'Brien, T. K. and Reifsnider, K. L., "Fatigue damage: stiffness/strength comparison for composite materials", J. Testing and Evaluation, Vol. 5, 1977.

[17] Reifsnider, K. L., Stinchcomb, W. W. and O'Brien, T. K., "Frequency effects on a stiffness-based fatigue failure criterion in flawed composite specimens", Fatigue of Filamentary Composite Materials, ASTM STP 636, 1977.

[18] O'Brien, T. K., "Stiffness change as a nondestructive damage measurement", Mechanics of Nondestructive Testing, W. W. Stinchcomb, ed., pp. 101-121, 1980.

INNER MATERIALS DEGRADATION EFFECTS ON STIFFNESS AND DYNAMIC CRACK PROPAGATION IN COMPOSITES SUBJECTED TO CYCLIC LOADING

A. Kobayashi and N. Ohtani

Institute of Interdisciplinary Research, The University of Tokyo
Komaba-4-6-1, Meguro-ku, Tokyo, 153, Japan

ABSTRACT

Stiffness degradation due to cyclic loading, i.e., fatigue in the carbon fiber composites is investigated from the microscopic point of views and the dynamic crack propagation behaviour in the fatigue-degradated composites is compared to that of virgin, non-degradated ones.

It is found that debonding around the transverse-ply and the microcrack generation in the transverse-ply are influential to the stiffness degradation due to cyclic loading.

As for the dynamic crack propagation behaviour, almost no difference is observed between CFRP specimens with and without inner materials degradation due to fatigue, showing almost the same crack propagation velocities close to the elastic theoretical crack propagation velocity for both with and without fatigue degradation.

INTRODUCTION

In spite of much work done so far, there still exist many to be solved concerning the fatigue degradation domain, especially for composites.

In the present report, the materials degradation in terms of stiffness degradation in the carbon fiber composites (CFRP) is investigated from the microscopic point of views, in correlation with the inner materials degradation factors such as the microcrack and debonding with the aid of optical tools.

Next, the dynamic crack propagation profiles are investigated for both with and without inner fatigue degradated CFRP in the quasi-static mode I tension loading, to see how the inner degradation affects the crack propagation velocity profiles.

STIFFNESS DEGRADATION DUE TO INNER MATERIALS DEGRADATION INVESTIGATED THROUGH MICROSCOPIC APPROACH

A. Specimen and Loading

Six-plied, 8-satin woven carbon fiber epoxy composites with the carbon fiber content of 57% in volume were used. The fiber direction is orthogonally lengthwise and widthwise as shown in Figure 1.

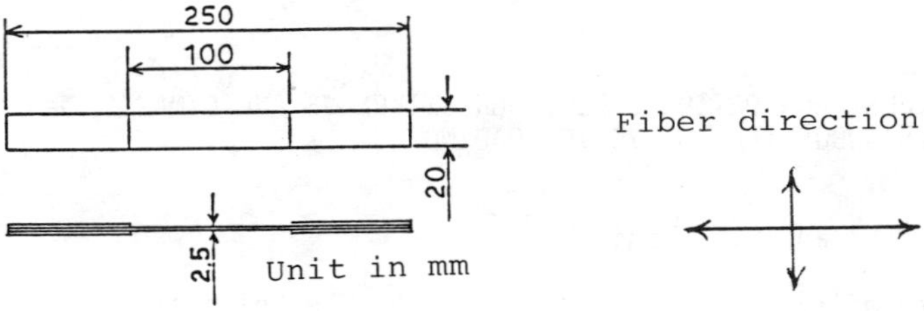

Fig. (1) - Specimen

The fatigue loading was applied by a Servopulser of Shimadzu's make, Japan, which is a hydraulic fatigue tester, of which the stress wave was of a sinusoidal one with a stress ratio of zero at 5 Hz. The distance between jaws for gripping was 100 mm.

The stress amplitude σ_a versus the number of cycle to failure N_f is shown in Figure 2.

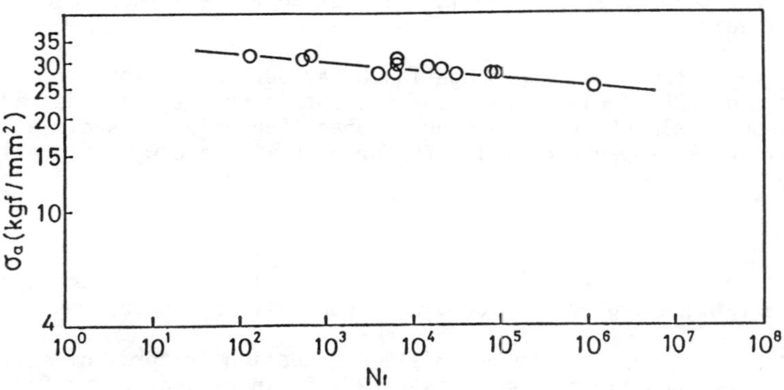

Fig. (2) - Stress amplitude versus number of cycle to failure

B. Stiffness Degradation Due to Cyclic Loading

Stiffness degradation obtained is shown in Figure 3. Note that a lower stress amplitude case of 25 Kgf/mm^2 shows more decrease in stiffness than in a 28 Kgf/mm^2 case. Probable reason for this would be due to heat accumulation inherent to longer fatigue life cycles for lower stress amplitude case. Kim [1] already obtained such tendency for CFRP.

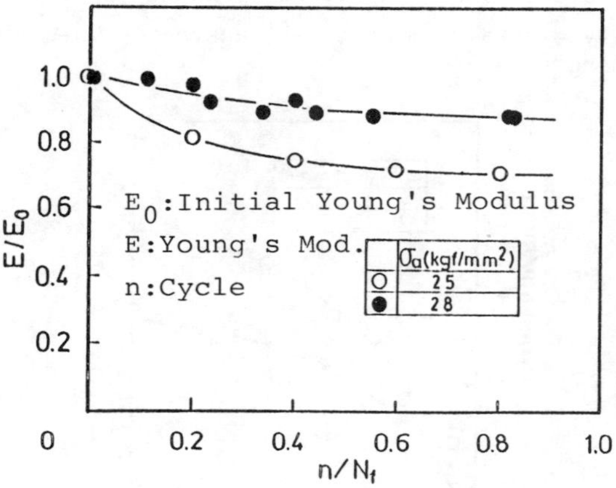

Fig. (3) - Degradation of stiffness versus cycle ratio

C. Microcrack Generation in Transverse-Ply

Microcracks generated in the transverse-ply, as shown in Figure 4, were measured at the middle part of a specimen alongside the whole width with 2 mm wide,

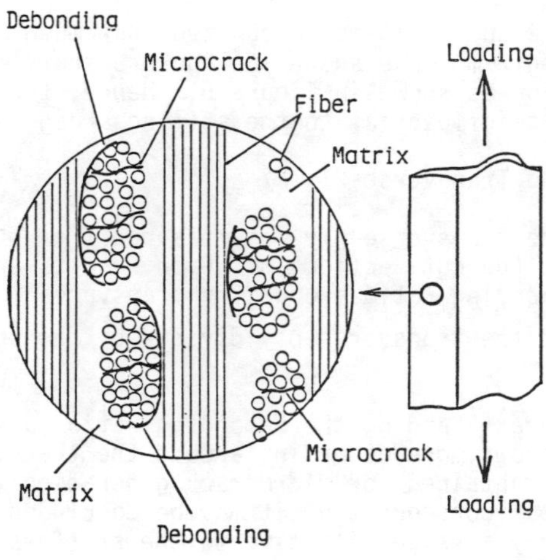

Fig. (4) - Cross-section of transverse-ply

i.e., 50 mm x 2 mm stripwise, by using an optical microscope with 50 to 400 times magnification. Then the average total summed microcrack length measured on both specimen surfaces was calculated on the photograph from the number of intersection of microcrack and a 1 mm x 1 mm mesh by employing the statistical approach [2].

Thus obtained result is shown in Figure 5, where $(L_A)_T$ is the transverse-ply microcrack density = total summed in - transverse - ply microcrack length divided by the scanning area.

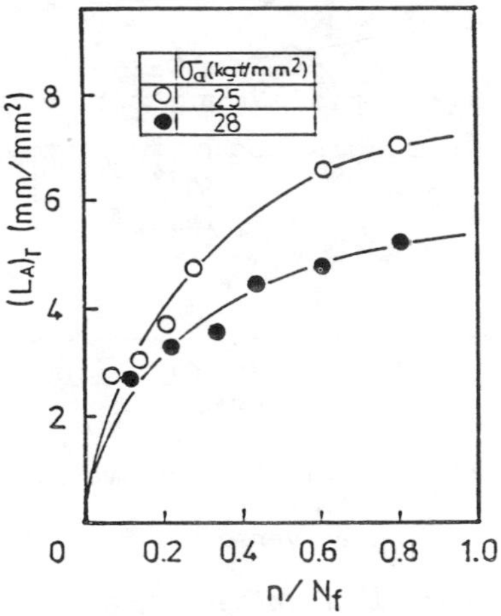

Fig. (5) - Transverse-ply microcrack density versus cycle ratio

In view of Figures 3 and 5, it is recognized that much more stiffness degradation observed in a 25 Kgf/mm^2 case shown in Figure 3 shows much more transverse-ply microcrack generation as shown in Figure 5. Hence, the microcrack generation in the transverse-ply is influential to the stiffness degradation.

D. Debonding Around Transverse-Ply

Debonding around the transverse-ply was measured from photomicrography at the end section and also at the cut-section, of which results are shown in Figure 6, where $(L_L)_{CT}$ is the debonding ratio around the transverse-ply = the summation of debonding length around the transverse-ply divided by the summation of transverse-ply boundary length.

Now, examining Figures 3 and 6, the debonding ratio curve having the greater stiffness degradation shows more debonding around the transverse-ply, which is quite the same tendency obtained for microcrack generation in the transverse-ply as just mentioned above. Consequently, it may be concluded that the debonding around the transverse-ply is also effective to the stiffness degradation.

DYNAMIC CRACK PROPAGATION VELOCITY PROFILES AFFECTED BY INNER MATERIALS DEGRADATION

A. Specimen and Loading

The specimen configuration is the same as previously described, except having a crack initiation notch which is prepared after fatigue loading, as shown in Figure 7.

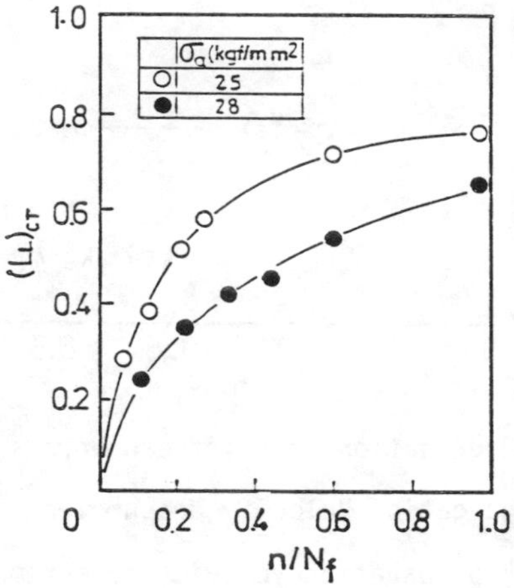

Fig. (6) - Debonding ratio around the transverse-ply versus cycle ratio

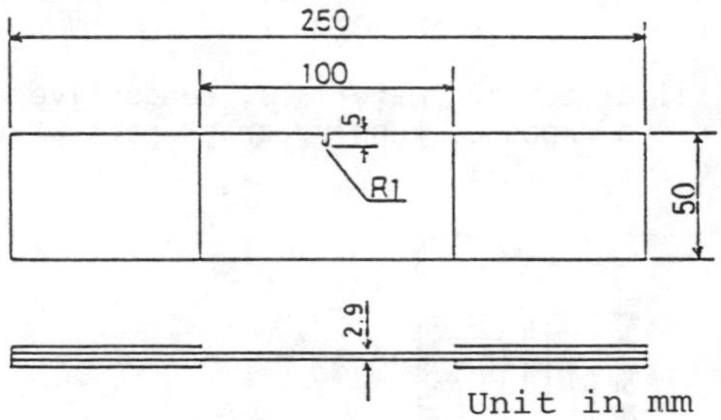

Fig. (7) - Specimen with a crack initiation notch

The fatigue loading producing the inner materials degradation was given by a hydraulic Servopulser of Shimadzu, Japan, as same as previously described.

The mode I tension loading to initiate the dynamic crack propagation was done by a tensile tester in the quasistatic manner at room temperature.

B. Stiffness Degradation Due to Cyclic Loading

The degradation of stiffness was obtained as shown in Figure 8. The degradation profile is somewhat different from the one shown in Figure 3 due to different specimen lots.

The virgin-state stiffness, say, the initial Young's modulus E_0 was 6400 Kgf/mm^2.

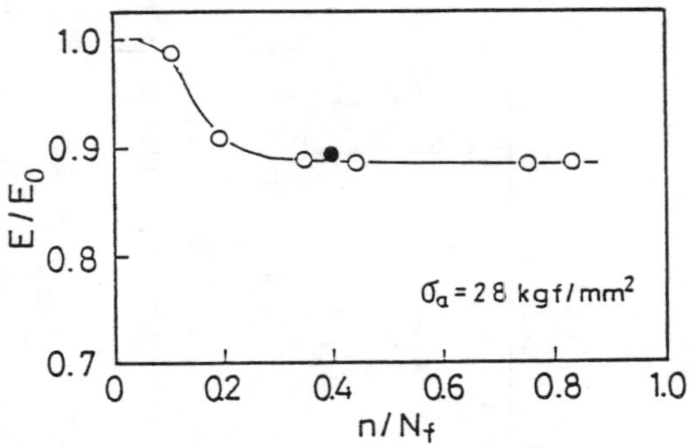

Fig. (8) - Degradation of stiffness versus cycle ratio

C. Dynamic Crack Propagation Velocity Measurement

For the dynamic crack propagation velocity measurement, both the velocity gages and the electric potential method were used. For the velocity gage application, thin epoxy coating was first applied on the CFRP specimen surface and then these velocity gages were applied on it, since the carbon is electric-conductive.

DuPont conductive silver coating materials, conductive wires in short, are placed perpendicular to the expected running crack path at the prescribed intervals as shown in Figure 9.

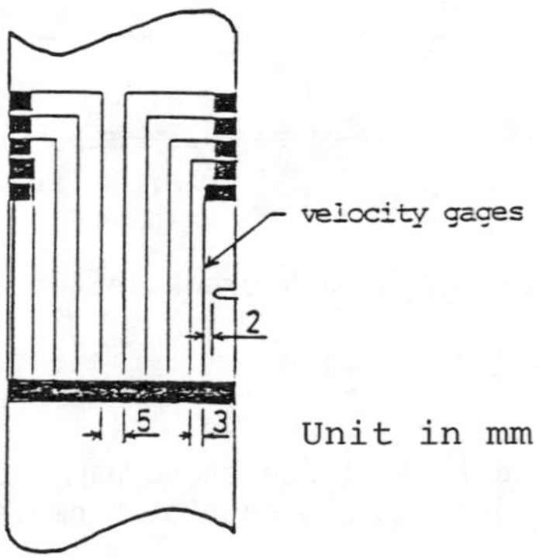

Fig. (9) - Velocity gage arrangement

In the course of crack advancement, these wires are broken by the running crack front, hence generating the electronic signal to be stored in a digital memory, from which the time required for a crack to run between two conductive wires can be obtained so that the average crack propagation velocity can be calculated by employing this time and the distance between two wires [3].

Another technique was also used. That is, the electric potential method shown in Figure 10. In Figure 10, the constant electric current is charged via grips.

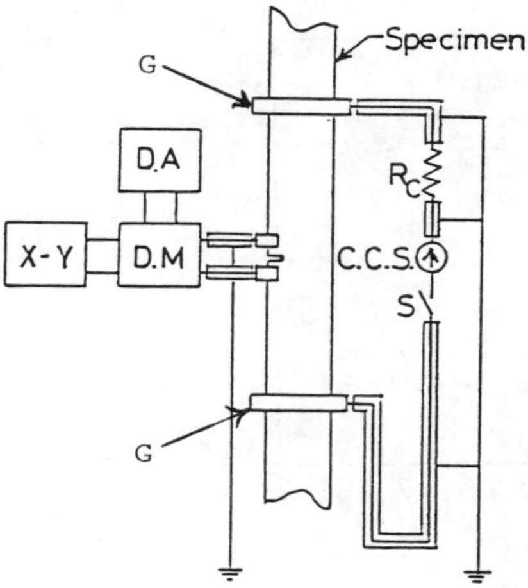

D.A. = Direct current amplifier
D.M. = Digital memory
X-Y = X-Y recorder
S = Switch
C.C.S. = Constant current source
R_c = Resistance
G = Grip

Fig. (10) - Electric potential method arrangement

As a crack advances, a change in electric current occurred can be detected and stored in a memory, from which the crack velocity can be calculated as mentioned previously, by using a calibration curve prepared in advance [4].

D. Dynamic Crack Propagation Velocity Profile

According to the conventional elastic crack propagation, the lower bound for crack velocity can be written as

$$\dot{C} = 0.38 \sqrt{\frac{E}{\rho}} \left(1 - \frac{C_0}{C}\right) \tag{1}$$

after Berry [5], where $\dot{C}$ = the crack velocity, ρ = density, C_0 = the initiation crack length and C = the arbitrary running crack length.

First, the crack propagation velocity profiles without any fatigue degradation is shown and compared with the theoretical curve by equation (1) in Figure 11.

It is observed that the experimental crack propagation velocity profiles denoted by two dotted lines showing the scatter ranges are close to the theoretical one of equation (1). Hence, the experimental curves could be regarded to be rather brittle.

Next, shown in Figure 12 is the crack propagation velocity profiles with and without inner fatigue degradation. The dotted lines show the experimental scatters for virgin-state, non-degradated specimens, while the solid lines for fatigue-degradated ones. In view of Figure 12, almost no inner fatigue degradation effects are observed.

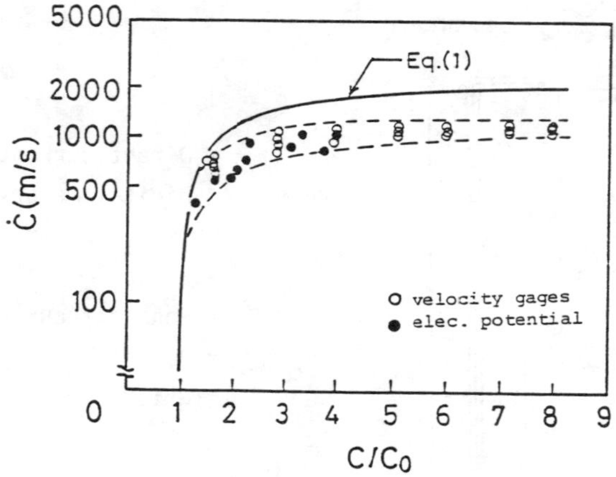

Fig. (11) - Dynamic crack propagation velocity profiles without fatigue degradation

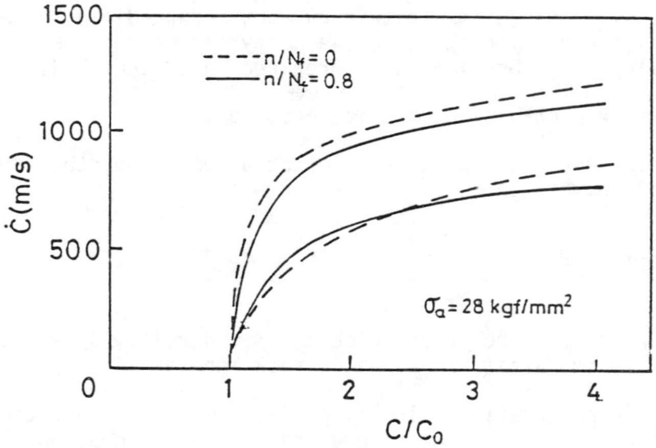

Fig. (12) - Dynamic crack propagation velocity profiles with and without fatigue degradation

CONCLUSIONS

Material degradation in terms of stiffness degradation is investigated under fatigue loading from microscopic approach. It is found that the microcrack generation in the transverse-ply and the debonding around the transverse-ply are influential to the stiffness degradation.

As for the fatigue degradation effects upon the dynamic crack propagation velocity profiles, almost no influence is observed and the obtained experimental profiles are rather close to the elastic theoretical ones, showing brittle-like crack propagation.

ACKNOWLEDGEMENTS

Grant-in-aid for Scientific Research supported by Ministry of Education is acknowledged. Toray Industries, Inc., is greatly appreciated for offering specimens.

REFERENCES

[1] Kim, R. Y., "Experimental assessment of static and fatigue damage of graphite/epoxy laminates", Proceedings, ICCM-III, Paris, France, 1980.

[2] Underwood, E. E., Quantitative Stereology, Addison-Wesley, Massachusetts, 1970.

[3] Kobayashi, A., Ohtani, N. and Sato, T., "Phenomenological aspects of viscoelastic crack propagation", J. Appl. Polymer Sci., 18, p. 1625, 1974.

[4] Kobayashi, A., Ohtani, N. and Munemura, M., "Phenomenological aspects of dynamic crack propagation in several aerospace materials", Proceedings, 12th Int. Symp. Space Tech. and Sci., Tokyo, Japan, p. 981, 1977.

[5] Berry J. P., "Some kinetic considerations of the Griffith criterion for fracture", Int. J. Mech. and Phys. Solids, 8, p. 194, 1960.

THE FATIGUE RESIDUAL STRENGTH IN CENTRALLY NOTCHED COMPOSITE MATERIALS

M.-H. R. Jen

Institute of Mechanical Engineering, National Sun Yat-Sen University
Kaohsiung, Taiwan 80424, Republic of China

ABSTRACT

This research is to predict the fatigue residual strength of centrally notched composite laminates with various layups. The low frequency fatigue loads include tension-tension tests and tension-compression tests at room moisture and room temperature with constant amplitude. The methods adopted here are by using a material modeling approach, fracture and fatigue mechanics and the finite element method (FEM).

For simplicity, after thousands of cycles, the geometry of a centrally circular hole of the deformed laminate was classified as (1) uniformly expanded hole into elliptical shape, (2) crack propagation at the hole edges transversely. The effect of geometry change during fatigue on residual strength was attributed to the elliptical hole, longitudinal splitting, matrix cracking, crack growth and local delamination. Due to the thin through-the-thickness notched laminate, two-dimensional plane stress FEM was used and interlaminar stresses were ignored.

The loss of stress concentration factor (SCF) is a reason for the increase of the residual strength. The SCF decreases while the elliptical hole becomes more slender. The residual strength and stiffness were determined by the material modeling with moduli reduction and damaged zone, and the numerical results were obtained by FEM. The analytical methods include classical lamination theory point stress criterion, ply discount method and netting analysis for type (1), fatigue and fracture mechanics, and modified Paris power law for type (2).

The correlation between the life and residual strength as predicted by this model numerically is found within acceptable errors in comparison with the experimental data.

INTRODUCTION

In recent years, composite materials have replaced metals in many applications because of their light weight, high strength and low cost. Composites also provide improved fatigue capability, good damage tolerance and the ability to be molded in complex aerodynamic configurations.

In many structural applications, notched composite material components cannot be avoided. However, the effects on the component itself and the whole structure caused by notches are very complex. The problem and the influence of a notched component can be found in the literature in the recent ten years. For simplicity, in this work, we are only concerned with the change of residual strength of a composite laminate with a central hole subjected to two types of fatigue loads, tension-tension and tension-compression (T-T and T-C), with constant amplitude at room temperature and low frequency.

A few investigators have reported the increase of residual tensile strength after loading for certain composite laminates. Most of these observations cannot be used as clearcut evidence, because of the experimental complexity involved. It is felt that comparing the residual mean strength with the mean top percent of the static strength leads to fairer conclusion [1]. Now, let us summarize a few of the experimental works that involve an increase of residual strength. Reifsnider et al [2] measured residual strength of boron/epoxy laminates with lapup of $[0/\pm45/0]_s$, and with a central hole. After cyclic loading, static residual strength was measured. Their results show that specimens with higher applied cycles had higher residual strength. Kulkarni et al [3] also observed an increase in residual strength after T-T fatigue in notched boron/epoxy laminates. Their results for notched coupons of $[0_2/\pm45]_s$ lapup with maximum fatigue stress equal to 80% of the static strength, and R = 0.1, show an increase of residual strength of 10% after 50000 cycles. The increase was 15% after 500000 cycles, and 16% after 1.5 million cycles.

Waddoups et al [4] observed an increase in residual strength in $[0/90]_s$ graphite/epoxy composite laminate with circular hole, subjected to fatigue loading. Chang et al [5] observed about 30% increase of residual strength in $[0/\pm45]_s$ graphite/epoxy coupons with a center slit under fatigue loading to 0.2 million cycles without failure. Their results were described by the nonlinear damage growth rate rule quite well. Chang et al [6] also found an increase in residual strength in $[0/\pm45/90]_s$ graphite/epoxy notched specimens.

The test data of Chou [7], Awerbuch and Hahn [8], indicate an increase in fatigue residual strength in unidirectional graphite/epoxy composites. There are also other test data of different composites and lapups which have an increase in residual strength in graphite/epoxy specimens with a central hole after fatigue loading, see [9] for T-T test and [10] for T-C test. It is also reported in [1] that graphite/epoxy (AS-3501-05) $[0_6]$ specimens without notches show an increase in residual (tension and compression) strength after T-T test. That behavior might be due to micro-damage of the coupon.

There are two common approaches to describe the notched strength phenomena; the first is by statistics using the strength-life equal rank assumption, and the second is due to use of fracture and fatigue mechanics. In this work, the latter approach is used in combination with the stress analysis theorem and FEM. Our goal is to work out an analytical method that closely correlates with experimental data.

The objective is to model the stress redistribution caused by damage development around a hole in a composite coupon, and to estimate the residual strength as a function of cycles of loading using those states and an appropriate failure theory. The present approach is the first reported attempt to account for both the relaxation of the stress concentration around the hole and the change in stiffness due to matrix cracking throughout the laminate [11]. A normal feature of the approach is the representation of the stress concentration change by an effective change in the hole from circular to elliptical.

ANALYTIS AND FORMULATION

Zweben [12] suggested three main approaches to analyze the problem. The first applies classical fracture mechanics (CFM) on a macroscopic level, treating composites as homogeneous, anisotropic materials. The second recognizes material heterogeneity and applies CFM to the problems of crack propagation in the matrix and fiber phases and interfaces separately. The third method, which might be called the material modeling approach, uses approximate models in order to represent the major effects of heterogeneity and to simplify the analysis. Actually, some factors are neglected such as frequency, temperature and moisture. Also, the interlaminar stresses which we do not consider in the process may be the controlling stress for residual tensile strength.

After the advantages and limitations of the three approaches are compared, the emphasis will be placed on material modeling for simple analysis. Waddoups et al [4] reported the fracture model by the application of CFM to laminated composites with a central hole. Also, we use a crack growth rate theorem and stress analysis theorem in the process to solve the problem. The stress analysis is based on treating the laminate as a homogeneous plate using a finite element technique. The strength analysis is based on the tensor polynomial failure criterion applied to each ply. The ultimate laminate failure strength is based on the last ply failure stress [13]. After the simulation of material modeling, we can find the residual tensile strength after the specific cycles of loading and see the change of residual tensile strength and a function of the increase of cycles.

After several thousand cycles, the geometry of a circular hole of the deformed laminate is categorized as (1) uniformly expanded hole into an elliptic shape; (2) symmetric crack propagation at the hole edge transversely; (3) irregularly deformed shape. However, the case of irregularly deformed shape is omitted here because it is very difficult to model. The case of transverse crack propagation is expressed in a compact form without the verification of numerical results, because some damage properties are still unknown.

Consider a composite laminate with a central hole shown in Figure 1. Our aim is to determine the residual strength of the notched laminate subjected to fatigue loading. For simplicity, we neglect the parameters that are not important, such as environmental effects. First, the aspect ratio of l and w should not be large in order to avoid the buckling by compressive loading. The environment of test is assumed to be room temperature and dry, with no temperature and moisture changes. The test frequency is constant at the normal low frequency. The stress amplitude is constant and the stress ratio is a fixed value. The geometry of the through-the-thickness notch will be circular hole or ellipse. First, we consider a graphite/epoxy composite laminate with a central circular hole deformed into elliptic hole when subjected to T-T test. There are several ways to find stress and strain fields for a notched plate with an elliptic hole due to uniaxial loading.

Once the stress concentration factor (SCF), K_t, obtained, we consider the failure criteria suggested by Whitney and Nuismer [14] and [15], such as the

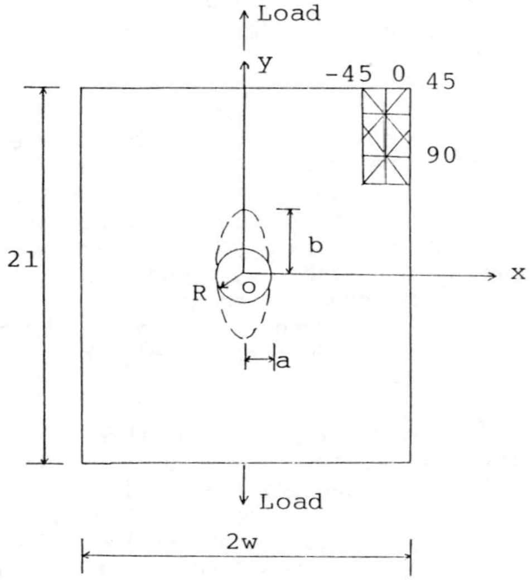

Fig. (1) - Laminate with a central circular hole and the deformed ellipse in x-y coordinates

point stress and average stress criteria. The first failure criterion referred to as the "point stress criterion"; it assumes failure to occur when σ_y at some distance, d_o, ahead of the hole first reaches the unnotched tensile strength of the material. The relation of notched and unnotched strength of the infinite width laminate is

$$\frac{\sigma_N}{\sigma_o} = \frac{2}{\{2+\xi_1^2+3\xi_1^4-(K_t-3)(5\xi_1^6-7\xi_1^8)\}} \tag{1}$$

where $\xi_1 = R/(R+d_o)$, d_o is characteristic distance and a material property; σ_o = unnotched strength, σ_N = notched strength.

Now we try to formulate the transverse crack propagation, see Figure 2, by using linear elastic fracture mechanics. The stress intensity factor (SIF) at the crack tip is

$$K_I = \sigma\sqrt{\pi a}f(a/r) \tag{2}$$

where a is characteristic dimension of the intense energy region.

Actually, Paris' law is not sufficient to express the defect initiation and growth in notched composite laminates when subjected to T-C tests. Delamination growth is a major concern to designers of composite structures and is usually planar and self similar. Using principles of fracture mechanics as the delamination grows along the interface, the growth rate (da/dN) is related to the strain energy release rate (G) by the power law [16,17].

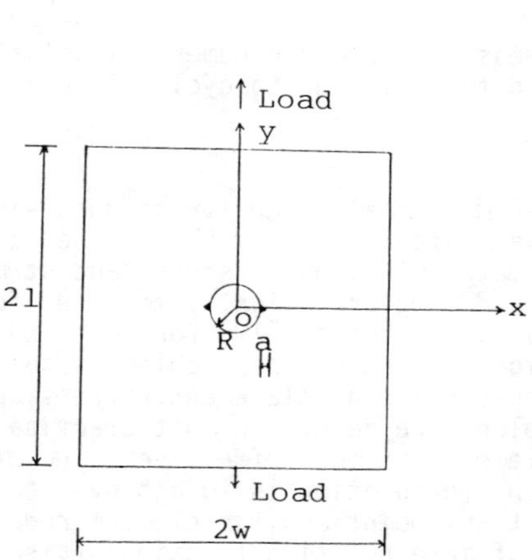

Fig. (2) - Fracture model (Mode I) for transverse crack growth

$$\frac{da}{dN} = A(G)^n \tag{3}$$

In general practice, G is the total strain energy release rate with components due to Mode I cracking. The growth rates of interfacial cracks and delaminations are particularly sensitive to the mode of cyclic loading. In the case of plane stress

$$G_I = K_I^2/E \tag{4}$$

Then,

$$\frac{da}{dN} = A\left(\frac{K_I^2}{E}\right) \tag{5}$$

Plugging equation (2) into equation (5), we get

$$\frac{da}{dN} = A\,\frac{[(\sigma\sqrt{\pi a}Y)^2]^n}{E^n} \tag{6}$$

where A and n are material properties, Y is f(a/r) and E is modulus changing with cycles. After rearrangement, we express it in the integral form

$$\int_{a_o}^{a_N} \frac{da}{a^n} = \int_0^N [A(\pi Y^2)^n \sigma^{2n}/E^n]dN \tag{7}$$

where σ is a function of cyclic loading and E is a function of (a,N). Then, we have

$$\sigma_c/\sigma_o = \sqrt{a_N/(r+a_N)} \tag{8}$$

However, it is not easy to obtain a numerical solution, because some micro-damage properties around the hole due to cyclic loading are not well understood.

NUMERICAL METHODS

In the case of two-dimensional plane stress quasi-isotropic, orthotropic and anisotropic composite laminates, we apply FEM to the laminate coupon with a center hole to get the approximate solutions of stress and strain fields. Two computer programs given in [18] and [19] are slightly modified for the composite materials. Here, we use the program described in [19] for time and space savings. In this research, a two-dimensional plane stress problem is solved by using the general formulation with the appropriate displacement, stress and strain variables, i.e., the stresses in the z-plane are zero. In most practical analysis, the use of isoparametric finite elements is more effective. The principal idea of the isoparametric finite element formulation is to achieve the relationship between the element displacements at any point and the element nodal point displacements directly through the use of interpolation functions (also called shape functions). This means that the transformation matrix is not evaluated; instead, the matrices corresponding to the required degree of freedom are obtained directly. Also, the continuum and structural isoparametric element formulations satisfy the convergence criteria (complete and compatible). For practical consideration in isoparametric element calculations, each element of the composite coupon has 8 nodes (8 nodes for plate, 9 nodes for shell) and 9 Gaussian points. The integration order is 3x3 and the degree of precision is 5. Each element has 8 nodes to eliminate the geometric error and has 9 Gaussian points to yield the closest results (the SCF is determined at element 1 Gaussian point 9).

From experimental data of notched composite specimens, we consider 1.5" x 8" specimens with a central hole with a diameter of 3/8". The ratio of hole diameter to width is 1:4, which is smaller than 0.33 [20]. In order to see the influences of length and width on the SCF, we reduce the length to half length and double the width to 3" with the length up to 10". One important point is that the aspect ratio should not be large enough to cause buckling of the specimen in compression. For symmetry, we only consider the first quarter of the coupons. For $[0/45_2/90]_s$ and $[0/45_4/90]_s$ layups, we consider the upper half of the coupons. To make meshes and nodes, the elements around the hole have to be as small as possible to meet the localized stress concentration.

Due to fatigue loads, the circular hole will be deformed into the elliptic hole with the assumption that the diameter of the semi-axis perpendicular to the loading is unchanged. The diameter parallel to loading is enlarged with the increase of the cycles. Although it is not yet possible to predict the relation of the enlarged diameter to cycles, we can obtain such data experimentally. We see that as the elliptic hole becomes slender, the SCF becomes smaller. That is, a stress concentration reduction which can contribute to the increase of residual strength.

From plate theory, laminate theory and the finite element analysis, we can obtain the stresses and strains at all the elements of every layer at the applied stress range from minimum to maximum at the first cycles. Similarly, we can obtain from X-ray radiographs of damage zones the different elliptic dimensions. The strength analysis is based on the tensor polynomial failure criterion and can be applied to each ply. Then, the initial laminate strength can be estimated by the first-ply failure theory on the conservative side [21]. But the ultimate laminate failure strength is based on the last-ply failure stress, that is used as base line information for comparison with the residual strength after N cyclic loads have been applied.

The diameter of the semi-axis parallel to loading direction enlarges with the increase of the number of applied cycles. Actually, the longitudinal semi-axis, b, is not a hole shape change. It is a superposition of effective notch determined by the damaged region around a hole. We consider tht total region of hole and effective notch as an elliptic hole. However, the mathematical relationship of the enlarged diameter to the number of applied cycles is difficult to predict. The determination of such a relationship requires extensive experimental data showing the deformed geometry by X-ray or C-scan at specific cycles of loads. The complexity of this problem is increased by the fact that the response depends on ply orientations, stacking sequence and geometrical factors.

Actually, the elastic moduli are also reduced during cyclic loading, in addition to the geometry effect changes. In order to better represent the real phenomena, we use discount theory to predict the last-ply failure of the laminate [22,23]. In many cases, laminates have considerable strength remaining after first ply failure and there arises the difficult and important problem of analytically determining subsequent failures. Initial failure of a layer may occur in the fiber or in the matrix mode. In the first case, the stiffness in the fiber direction E_1 is reduced, and in the second case, the elastic properties E_2 and G_{12} are reduced. These elastic properties are not reduced to zero, because the uncracked regions remain bonded to adjacent layers. A precise estimate of these stiffness reduction is difficult to make. The simple (but extreme) assumption is that E_1 reduces to zero in the fiber failure mode and that E_2 and G_{12} reduce to zero in the matrix mode. Since a laminate in most cases will not survive a fiber mode failure, the progressive failure cases of interest are initial and subsequent failures in the matrix mode. If it is assumed that, for failure of a lamina in the matrix mode, E_2 and G_{12} can be set equal to zero, eventually the assumptions of netting analysis result. For netting analysis, the ultimate load is defined by the state when E_2 and G_{12} vanish in all laminae.

We know that the internal stress redistribution is a driving force for subsequent subcritical element damage and critical element degradation. The basic nature of that stress redistribution can be established by considering the quasi-static strength of such a laminate since such cracks also form before fracture under quasi-static loading. When edge effects do not dominate the failure process, the ply discount method is a good engineering approach to estimate laminate strength. To apply the scheme, one must calculate ply stresses, use a failure criterion to predict first ply failure (matrix cracking), reduce the moduli in the broken plies (E_2, $G_{12} = 0$), recalculate ply stresses, test for second ply failure, etc., until last ply failure is predicted.

The literature indicates that there are several fracture models which have been used to determine the notched strength of composite laminates. After reviewing these methods, we first choose to consider the Waddoups, Eisenmann and Kaminski (WEK) fracture model [4]. The disadvantages of the WEK fracture model are (1) the SCF is not involved; (2) the change of hole geometry from circular to elliptic is not incorporated; (3) the finite width correction factor is not involved. Then we choose the Whitney and Nuismer (WN) fracture model to determine the notched strength.

RESULTS AND DISCUSSIONS

Now we attempt to use the WN fracture model to determine the residual tensile strength for a specific number of cycles of fatigue loading after comparing with several fracture models. At the beginning, we use the point stress criterion instead of average stress criterion for simplicity, although the latter generally has better accuracy than the former. From the test data for T300/5208 quasi-isotropic laminate [24], we have the residual tensile strength (S_n) at specific cycles of T-T fatigue loading and the corresponding geometry changes. The unnotched tensile strength for those specimens is 71 ksi. The point stress criterion is shown in equation (1). For different cracked situations in each b value, we assume that K_{tg} is equal to K_t when all 90, -45 and 45 degree plies have cracked, i.e., at b = 0.1875" K_{tg} = 1.98, at b = 0.5" K_{tg} = 1.73, at b = 0.8" K_{tg} = 1.67 and at b = 1.2" K_{tg} = 1.63, see column 2 in Table 1. With the K_{tg}

Table 1 - Notched strength for T300/5208 quasi-isotropic laminate with plies cracked

Plies Cracked	K_t	d_o (in)	N (Ksi)	K^*_{tg}	d^*_o (in)	S_n (Ksi)
N=10E0, b=0.1875"						
90	2.50	0.05	36.73	1.92	0.052	37
-45	2.34	0.05	36.50	1.92	0.052	
45	1.98	0.05	35.99	1.92	0.052	
0	1.19	0.05	34.91	1.92	0.052	
N=10E1, b=0.5"						
90	1.85	0.08	41.42	1.73	0.09	41
-45	1.82	0.08	41.35	1.73	0.09	
45	1.73	0.08	41.15	1.73	0.09	
0	1.33	0.08	40.29	1.73	0.09	
N=10E3, b=0.8"						
90	1.76	0.09	43.10	1.61	0.09375	44
-45	1.73	0.09	43.03	1.61	0.09375	
45	1.67	0.09	42.90	1.61	0.09375	
0	1.19	0.09	41.86	1.61	0.09375	
N=10E4, b=1.2"						
90	1.70	0.10	44.78	1.54	0.11	46
-45	1.68	0.10	44.74	1.54	0.11	
45	1.63	0.10	44.62	1.54	0.11	
0	1.21	0.10	43.73	1.54	0.11	

value known, we obtain d_o by intersecting the base normalized stress distribution curve which is the stress state for the circular hole (a=b=R) with no ply cracking. We use the base curve to represent the WN normalized point stress criterion, since it is suitable for the circular hole. The scheme to determine the characteristic length (d_o) at the unnotched laminate strength (failure) from the WN fracture model is shown in Figure 3(a), d_o determined at K_{tg} in normalized stress distribution is also shown in Figure 3(b).

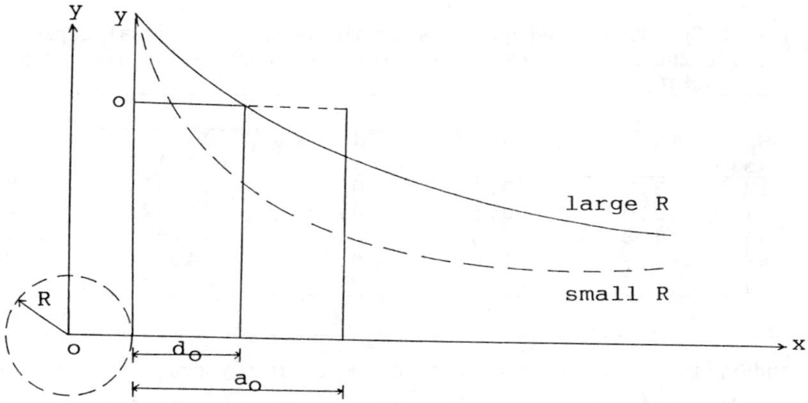

(a) Stress distribution at failure to determine d_o and a_o

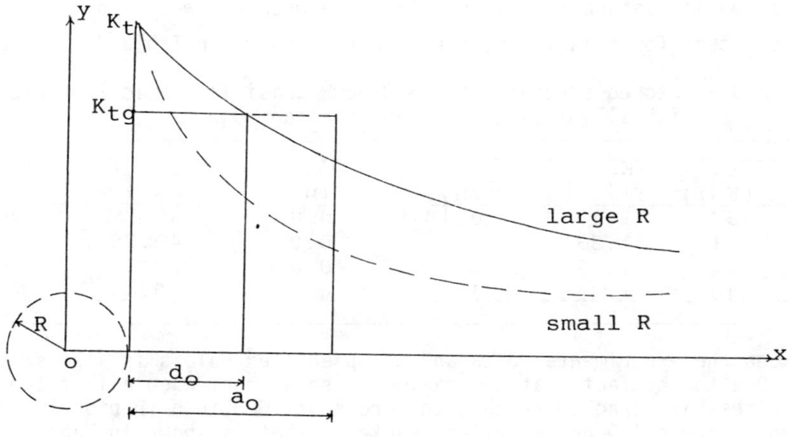

(b) Normalized stress at fialure to determine d_o and a_o

Fig. (3) - The point stress and average stress criteria in WN fracture model

K_t is the SCF at the hole edge and K_{tg} is the SCF (normalized by the un-notched strength) corresponding to the characteristic length. It is reasonable to have larger d_o when K_t is decreasing. Then we obtain d_o = 0.05, 0.08, 0.09 and 0.10" respectively, see column 3. That shows d_o has a slight change in the same layup. We conclude that d_o is not necessarily a material constant for the same layup. With d_o and K_t for each case, we can predict the notched tensile strength via equation (1), see column 4 in Table 1.

Then if we assume no ply cracking, we have the notched strength and the errors in comparison with experimental data listed in Table 2, where d_o is the same

Table 2 - T300/5208 notched quasi-isotropic laminate residual strength via the WN fracture model point stress criterion with no ply cracking

Cycles	S_n (Ksi)	K_t	b (in)	d_o (in)	σ_N (Ksi)	Error (%)
10E0	37	3.11	0.1875	0.05	37.31	0.84
10E1	41	2.07	0.5	0.08	41.92	2.24
10E3	44	1.87	0.8	0.09	43.35	1.48
10E4	46	1.77	1.2	0.10	44.94	2.30
2x10E5	31					

as Table 1 and K_t is the SCF at the hole edge without ply cracking (no moduli reduction). We also find that K^*_{tg} is close to K_{tg}, where K^*_{tg} is defined as the unnotched strength divided by the observed strength (S_n). Similarly, d^*_o determined by K^*_{tg} in the base curve is closed to d_o. That suggests that the empirical proof what we assumed is reasonable. To simplify the problem, we set d_o = 0.08" (average) as a constant for this layup; the result is shown in Table 3. We see the

Table 3 - Notched strength for T300/5208 quasi-isotropic laminate with all plies cracked and constant d_o

Cycles	S_n (Ksi)	K_t	b (in)	d_o (in)	σ_N (Ksi)	Error (%)
10E0	37	1.19	0.1875	0.08	39.99	8.08
10E1	41	1.33	0.5	0.08	40.29	1.73
10E3	44	1.19	0.8	0.08	39.99	9.11
10E4	46	1.21	1.2	0.08	40.04	12.96

error between the experimental data and the predicted values is also small and acceptable. Due to the fact that the maximum displacement when all of the (90, -45, 45 and 0) plies have cracked reaches the order of 10E5 inch (rupture), the laminate failed before all 0 degree plies cracked. That is shown in Table 1, K_{tg} at d_o is assumed to equal K_t for the situation of all 90, -45 and 45 degree plies cracked. So K_t at that stage is the lower limit. Obviously, K_t and d_o obtained without ply cracking defines the upper limit.

There are two sets of fatigue data from [25] which show the compliances, residual strengths and geometry changes of graphite/epoxy T300/934 laminates with a central circular hole for both $[0/\pm45/90]_{2s}$ layups. Following the same procedure, the K_{tg} equals to K_t when all the -45 and 45 plies have cracked for each b value with the $[0/\pm45/0]_{2s}$ layup, and when all of the 90, -45 and 45 plies have cracked for the $[0/\pm45/90]_{2s}$ layup. The d_o was also determined from K_{tg} in the base curve (a = b = 0.06, no ply cracking). When we have determined K_t and d_o, we can compute the notched strength. The predicted notched strengths and errors in comparison with empirical data are shown in Table 4 and Table 5. We see that

Table 4 - Notched strength for T300/934 $[0/\pm45/0]_{2s}$ laminate with a central circular hole due to T-T fatigue loading

Cycles	S_n (MPa)	a (in)	b (in)	K_t	d_o (in)	σ_N (MPa)	Error (%)
10E0	505	0.06	0.06	2.53	0.02	472.17	6.50
10E3	525	0.06	0.11	2.01	0.03	531.05	1.15
10E4	560	0.06	0.12	1.99	0.03	530.50	5.27
10E5	613	0.06	0.16	1.88	0.04	590.61	3.65
10E6	682	0.06	0.48	1.71	0.05	638.57	6.37
2x10E6	707	0.076					
5x10E6	Failure						

Table 5 - Notched strength for T300/934 $[0/\pm45/90]_{2s}$ laminate with a central circular hole due to T-T fatigue loading

Cycles	S_n (MPa)	a (in)	b (in)	K_t	d_o (in)	σ_N (MPa)	Error (%)
10E0	370	0.06	0.06	3.11	0.03	364.38	1.52
10E3	374	0.06	0.10	2.51	0.03	352.76	5.68
10E4	415	0.06	0.11	2.27	0.04	388.24	6.45
10E5	433	0.06	0.20	1.87	0.04	382.00	11.78
10E6	401	0.06	0.48	1.77	0.04	380.48	5.12
2x10E6	Failure						

the WN fracture model can be used to predict the notched strength. The point stress criterion is a simple way to find the notched strength with increasing d_o while K_t is decreasing. In Tables 4 and 5, we see that compliance increases with increasing cycles, i.e., the modulus decreases with increasing cycles. That will result in a stress concentration loss and the characteristic length will become larger with increasing cycles. Additionally, if we plot the major effective notch axis (b) versus log cycles for these layups, we do not find a clearly linear relation. In our analysis, we assume that when all plies which crack reach a saturated pattern, then the stress concentration factor is the lower bound, i.e., the most damaged case. As for the observed data in Table 1, we do not know exactly the number of cycles for which the 90, -45 and 45 plies become saturated. However, when all plies are not cracked and the laminate is subjected to fatigue loading, that case is the upper bound on the stress concentration factor. As a matter of fact, it is generally not possible to have plies not cracked under T-T fatigue tests. S the predicted residual strength should be within the upper and lower bounds. Finally, the single ply (0,45,-45,90) notched strengths are computed.

SUMMARY AND CONCLUSIONS

This research concerns the use of a stress redistribution material modelling method to predict the notched strength of damaged laminates, including theoretical, numerical and experimental analyses. We considered the case of a circular hole deformed into an effective elliptic shape. This model is developed in detail and shows satisfactory agreement with experimental data. This approach can be summarized as: geometry changes, moduli reduction, the WN fracture model, last ply failure theory, the ply strength is predicted. Using this method, we have found that the predicted notched strength for a laminate and single ply is within acceptable error in comparison with experimental data.

Due to the scarcity of observed data and the lack of statistics for available data, the present model has been verified for only three data sets in T-T tests. In T-C tests, edge delamination dominates, i.e., tensile strain to failure controls. Additionally, we need to modify a_c as the critical delamination length, ε_c, where a_c is the critical crack length obtained from Paris law.

However, the present work suggests that even a very approximate account of stress redistribution associated with damage accumulation caused by fatigue loading of centrally notched composite laminate coupons can be used to obtain engineering estimates of residual strength.

REFERENCES

[1] Chou, P. C., et al, "Statistical analysis of strength and life of composite materials", AFWAL-TR-80-4049, 1980.

[2] Reifsnider, K. L., et al, "Frequency effects on a stiffness-based fatigue criterion in flawed composite specimens", ASTM STP 636, pp. 171-184, 1977.

[3] Kulkarni, S. V. et al, "Fatigue of notched fiber composite laminates, part II: analytical and experimental evaluation", NASA-CR-145039, Materials Science Corp.

[4] Waddoups, M. E., et al, "Macroscopic fracture mechanics of advanced composite materials", J. of Comp. Matls., Vol. 5, pp. 446-454, 1971.

[5] Chang, F. H., et al, "A study of fatigue damage in composites by nondestructive testing technique", ASTM STP 636, pp. 57-72, 1977.

[6] Chang, F. H., et al, "Real-time characterization of damage growth in graphite/epoxy laminates", J. of Comp. Matls., Vol. 10, pp. 182-192, 1976.

[7] Chou, P. C. and Croman, R., "Residual strength in fatigue based on the strength-life equal rank assumption", J. of Comp. Matls., Vol. 12, pp. 177-194, 1978.

[8] Awerbuch, J. and Hahn, H. T., "Fatigue and proof-testing of unidirectional graphite/epoxy composites", ASTM STP 636, pp. 248-266, 1977.

[9] Askins, D. R., "Development of engineering data on advanced composite materials", AFWAL-TR-81-4172, 1982.

[10] Lauraitis, K. N., et al, "Advanced residual strength degradation rate modeling for advanced composite materials, Vol. II-Task II and III", AFWAL-TR-79-3095, 1981.

[11] Jen, M.-H. R., "The residual strength determination due to fatigue loading by fracture mechanics in notched composite materials", Ph.D. Dissertation, VPI&SU, 1985.

[12] Zweben, C., "Fracture mechanics and composite materials: a critical analysis", ASTM STP 521, pp. 65-97, 1973.

[13] Soni, S. R., "Failure analysis of composite laminates with a fastener hole", ASTM STP 749, pp. 145-164, 1981.

[14] Nuismer, R. J. and Whitney, J. M., "Uniaxial failure of composite laminates containing stress concentrations", ASTM STP 593, pp. 117-142, 1975.

[15] Whitney, J. M. and Nuismer, R. J., "Stress fracture criterion for laminated composites containing stress concentration", J. of Comp. Matls., Vol. 8, pp. 253-265, 1974.

[16] Wilkins, J. D., et al, "Characterizing delamination growth in graphite/epoxy", ASTM STP 775, 1982.

[17] O'Brien, T. K., "Characterization of delamination onset and growth in a composite laminate", ASTM STP 775, pp. 140-168, 1982.

[18] Reddy, J. N., An Introduction to the Finite Element Method, McGraw-Hill, 1984.

[19] Bathe, K. J., Finite Element Procedures in Engineering Analysis, Prentice-Hall, 1982.

[20] Hong, C. S. and Crews, J. H., Jr., "Stress-concentration factors for finite orthotropic laminates with a circular hole and uniaxial loading", NASA TP 1469, pp. 1-26, 1979.

[21] Tsai, S. W. and Hahn, H. T., "Introduction to Composite Materials, Technomic Publishing Co., Inc., 1980.

[22] Rosen, B. W., "Failure of fiber composite laminates", Proc. of the IUTAM Symposium, p. 105, 1982.

[23] Reifsnider, K. L., et al, "Damage mechanics and NDE of composite laminates", Proc. of the IUTAM Symposium, p. 399, 1982.

[24] Stinchcomb, W. W., et al, "Fatigue damage in notched composite laminates under tension-tension cyclic loads", NASA NAG-1-232, Langley Research Center, 1982.

[25] Ramani, S. V. and Williams, D. P., "Notched and unnotched fatigue behavior of angle-ply graphite/epoxy composites", ASTM STP 636, pp. 27-46, 1977.

SECTION IV

COMPOSITE LAMINATES

HIGH SPEED ASPERITY EXCITATION OVER A LAYERED MEDIUM*

F. D. Ju and J.-C. Liu

Mechanical Engineering Department, University of New Mexico
Albuquerque, New Mexico 87131 USA

ABSTRACT

The paper considers the thermo-mechanical effect of an asperity traversing at a high speed over a semi-infinite medium with a thin hard surface layer. The analyses emphasize the effect of the friction force, leading to the initiation of thermo-mechanical cracking, or "heat-checking" in the substrate, the surface layer or their interface. For hard coating layers, the initiation of crack will occur at a depth where the maximum temperature gradient occurs, which defines the thermal layer. It is found that, for a thick coating layer, the crack will occur inside the surface layer where the stress state is little affected by the substrate; the estimate of fracture initiation is the same as that of a single material of the coating. Yet, for modified surfaces of thin coating layer, of which the thickness is of the same order of the thermal layer, the interaction between the surface layer and the substrate, through their differences in mechanical and thermal properties, greatly influences the combined stress state near the interface, and thus the crack initiation in its neighborhood.

INTRODUCTION

The present paper deals with a general class of problems for thermo-mechanical cracking that is caused by a moving asperity traversing at a high speed over a semi-infinite medium with a thin surface coating layer. Such a thin surface layer is commonly designed by surface modification to improve the wear property of the substrate such as composite, which is generally of poor resistance to friction wear. The hard coating, while improving wear of sliding surfaces, may cause localized "hot spots" at a high relative sliding speed, from thermoelastic instability [1-3], leading eventually to cracking [4-6], a phenomenon called "heat-checking". The localized hot spots manifest solid contacts of asperities. The actual contact area of a single asperity is only a small fraction of the nominal contact area of the sliding surface [7]. Burton [8] considered 10^{-4} as a possible areal ratio (the ratio between the actual contact area and the nominal area at the design interface). A small design pressure, distributed over a large design contact area, may thus result in some extremely high contact pressure under the asperity, which traverses at high speed of the order 10 to 100 mps over the wear

*The work was performed under ONR Grant No. N00014-84-K-0252.

surface. As a consequence, even at reasonable coefficient of Coulomb friction, the friction force and the rate of frictional work can be very high at the asperity contact. The frictional excitation of the asperity thus imparts a mechanical load as well as a thermal load, through frictional heating, onto the surface boundary of the coating layer. The high speed traversing heat source behaves like a thermal shock. It is the combined thermo-mechanical stress state that leads to the initiation of cracks in the surface layer, the substrate or their interface. The stress states depend on the asperity characteristics, the coating layer thickness and the material properties of both the coating layers and the substrate. The paper will expound their effects on the stress states and on the locations of the maximum stresses.

MATHEMATICAL MODEL

For hard coating surface layers, the surface plastic damage, as indicated by experiments [9,10], is subgranular ($\sim 5\mu$). Fracture usually initiates as cohesive failure in the subsurface region, where the material remains elastic. The effect of the surface plasticity may be adequately represented for the present purpose by being included in the coefficient of Coulomb friction. A two-dimensional representation of the high speed asperity excitation is shown in Figure 1. The asperity

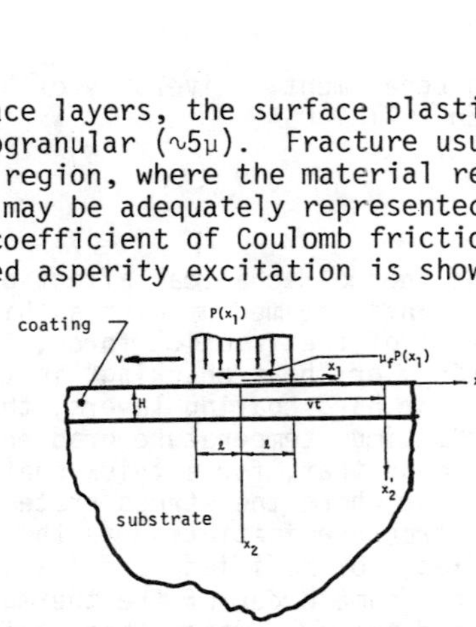

Fig. (1) - Two-dimensional asperity

is characterized by the contact pressure $P(x_1)$, its size L, the coefficient of friction μ_f, and its traversing speed V, all of which are invariant to time. The layer thickness, H, is uniform. For the present purpose, it is postulated that there is no flaw in the material. With respect to the material coordinates $(x_1'-x_2')$, the governing equations are the thermoelastic Navier's equation and the Fourier equation

$$(\lambda_\beta+\mu_\beta)\partial_{ij}u_j^\beta + \mu_\beta\partial_{jj}u_i^\beta = \rho_\beta\ddot{u}_i^\beta + (3\lambda_\beta+2\mu_\beta)\alpha_\beta\partial_i T^\beta \quad (1)$$

$$\kappa_\beta\partial_{jj}T^\beta = \dot{T}^\beta \quad (2)$$

where (λ,μ) are the Lame's elastic coefficients, u_i is the displacement field, T is the temperature field, α is the coefficient of thermal expansion, κ is the thermal diffusivity, i and j index the coordinates and β indexes I and II for the surface layer and the substrate respectively. The indicial summation convention and the Schouten's partial derivative notation, $\partial_i = \partial/\partial x_i$, are used. Equations (1) and (2) require explicit time-dependent solutions. The analytical complexity may be alleviated with the use of the convective coordinates (x_1-x_2), which are fixed to the asperity, that $x_i' = x_i + Vt\delta_{i1}$. Equations (1) and (2) thus become

$$\partial_{ij}u_j^\beta + (1-2\nu_\beta)\partial_{jj}u_i^\beta = M_\beta^2(1-2\nu_\beta)\partial_{11}u_i^\beta + 2(1+\nu_\beta)\alpha_\beta\partial_i T^\beta \tag{3}$$

$$\kappa_\beta\partial_{jj}T^\beta = V\partial_1 T^\beta \tag{4}$$

where ν is the Poisson's ratio and $M_\beta = [V^2\rho/\mu]^{1/2}$ is the Mach number of shear in the β region. Time is no longer an explicit variable. The boundary conditions are: on the surface, $x_2 = 0$

$$\sigma_{2i}^I = -R_i \tag{5}$$

$$\kappa_I\partial_2 T^I = -VR_1 \tag{6}$$

where κ_I is the thermal conductivity in the surface layer, $R_1 = \mu_f P(x_1)$ and $R_2 = P(x_1)$ in the contact region and zero elsewhere. The convective heat loss at free surface, being of small order, is neglected without loss of generality. Regular conditions hold at infinity $x_i x_i = \infty$

$$\sigma_{ij}^\beta = 0,\ T^\beta = 0 \tag{7}$$

The mechanical and the thermal fields must also satisfy the continuity conditions at the layer/substrate interface, $x_2 = H$

$$u_i^I = u_i^{II},\ \sigma_{2j}^I = \sigma_{2j}^{II} \tag{8}$$

$$T^I = T^{II},\ \kappa_I\partial_2 T^I = \kappa_{II}\partial_2 T^{II} \tag{9}$$

It is noticed that the governing differential, equation (3), and the boundary condition, equation (5), are both non-homogeneous. By separating the fields into two parts, such that

$$\{\sigma_{ij}^\beta, u_i^\beta\} = \{\sigma_{ij}^\beta, u_i^\beta\}_T + \{\sigma_{ij}^\beta, u_i^\beta\}_M$$

The first set shall satisfy the non-homogeneous differential equation with the homogeneous boundary conditions; while the second set shall satisfy the homogeneous differential equation but with the non-homogeneous boundary conditions. Essentially the solutions are delineated into one resulting from temperature field and another from mechanical traction. They are respectively the thermal stress field and the mechanical stress field. For high speed asperity excitation over a hard surface, the thermal stress field dominates the critical stress state. Ju and Huang [11,12] used an asperity traversing speed of 10 → 15 mps to illustrate that, at the critical depth, the thermal stress accounts for 85% of the maximum thermo-mechanical principal stress. The emphasis of the present paper will, therefore, be on the thermal effect of the asperity on the layered medium. The solutions are obtained in the Fourier transformed space. Numerical solutions are obtained for individual cases by numerical integration of the inverse transform.

THERMO-MECHANICAL PARAMETERS

The analysis of the thermo-mechanical field in a layered medium involves the matching of the mechanical and thermal properties of both the surface coating layer and the substrate. The asperity characteristics play also an important role. Those individual influences can be protracted from works on the parametric analysis of high speed asperity excitation over a single material by Ju and Huang [11-13]. The parameters as inferred from the mathematical model are tabulated as follows:

Table 1 - Asperity parameters

V	- Asperity traversing speed
$P(x_1)$	- Asperity pressure
L	- Asperity contact area width
b	- Contact area length

Table 2 - Material parameters

(λ,μ) or (E,ν)	- Mechanical constitutive coefficients
ρ	- Mass density
μ_f	- Coefficient of Coulomb friction
α	- Coefficient of thermal expansion
κ	- Thermal diffusivity
k	- Thermal conductivity

A. Asperity Parameters

The traversing speed of the asperity (V) influences both the convective terms in equations (3) and (4) and the thermal flux input, equation (6). The speed, therefore, affects not only the magnitude of temperature rise but also the depth, from the wear surface, of the point where the temperature gradient is maximum. For the case of a single asperity, the depth of maximum temperature gradient (ζ_m) can be obtained [4] from

$$\int_0^\infty \theta^{-2}(\zeta_m^2-\theta)f(\xi,\eta,\theta)\exp(-\zeta_m^2/\theta)d\theta = 0 \tag{10}$$

where (ξ,η) are the surface coordinates modified by the asperity size with ξ in the direction of asperity traverse, ζ_m is the depth coordinate modified by the asperity size, and

$$f(\xi,\eta,\theta) = \frac{1}{4\sqrt{\pi\theta}}\left[\mathrm{erf}\left(\frac{\eta+t}{\sqrt{2\theta}}\right) - \mathrm{erf}\left(\frac{\eta-t}{\sqrt{2\theta}}\right)\right]\left[\mathrm{erf}\left(\frac{\xi'+1}{\sqrt{2\theta}}\right) - \mathrm{erf}\left(\frac{\xi'-1}{\sqrt{2\theta}}\right)\right] \tag{11}$$

where $\xi' = \xi - Pe\theta/2$, $Pe = VL/\kappa$ is the Peclet number and $t = b/L$ is the aspect ratio of the asperity contact area. It is apparent that both the aspect ratio and the Peclet number, thus the speed V, affect the depth ζ_m, which defines the thickness of the surface thermal layer. For the problem of a single wear medium, it is at the thermal layer depth (ζ_m) that the maximum principal stress occurs. At an aspect ratio of 10 and a Peclet number of 1000, the depth of thermal layer ζ_m is approximately 0.1. In a two dimension formulation, for which the aspect ratio t is infinite, the thermal layer ζ_m is approximately 0.16 at the same Peclet number. The depth of thermal layer is all the more important for the asperity excitation over layered media. Ju and Chen [15] demonstrated that the worst stress state results, for varying coating layer thickness, when the thermal layer thickness is in the neighborhood of the coating layer/substrate interface.

The assumption of uniform distribution of asperity pressure and two-dimensional analysis will result in stress much lower than that of nonuniform distribution of pressure and in three-dimensional analysis. It was shown [12-14] that by assuming an uniform distribution of the asperity load over the contact area, the predicted stress level could be 40% less than that of a nonuniform distribution of an equal total load. While the shape of the contact area does not affect the stress level, the aspect ratio does. Increasing the aspect ratio lowers the thermal stress level. In the limit of an infinite aspect ratio (a plane strain model), the corresponding maximum thermal stress is less than 20% of that at the aspect ratio of one. However, because of mathematical simplicity, the assumption of uniform distribution of asperity pressure and two-dimensional analysis may be justified to be used in study of interaction among various parameters without actually predicting the stress level for fracture initiation.

B. Material Parameters

As the emphasis of the paper is on the friction force and the rate of frictional work, simple Coulomb law is used, $F_f = \mu_f F_n$, for analytical convenience. The rate of frictional work heats the surface layer of the wear medium to raise the surface temperature, which in turn, increases the Coulomb coefficient μ_f with resulting further increase in friction force [8]. The present analysis avoids the iteration by presuming a relatively high constant coefficient to anticipate a steady state high temperature field. The mass density (ρ) affects the stress field, equation (3), through the Mach number (M_β) and the thermal field through the thermal diffusivity ($\kappa = k/\rho c$). Since the Mach number is generally of very small order in this class of problems, the effect of the mass density is principally combined with the specific heat (c) to define the thermal capacity (ρc).

The mechanical constitutive coefficients and the coefficient of thermal expansion affects the thermal stress field through the material compliances and the thermal dilatation. Stiffer material and higher thermal expansion are known to contribute toward higher thermal stress. The effects of thermal parameters ($k, \kappa, \rho c$ - conductivity, diffusivity and capacity) can be best illustrated with the maximum principal thermal stress at the depth of thermal layer, Figure 2, in which Curve 1 is plotted based on the properties of a known wear material, Stellite III. Curves 2 and 3 illustrate the decrease and increase of the principal thermal stresses corresponding to doubling and halving thermal diffusivity due to changes in conductivity while keeping the capacity invariant. However, when

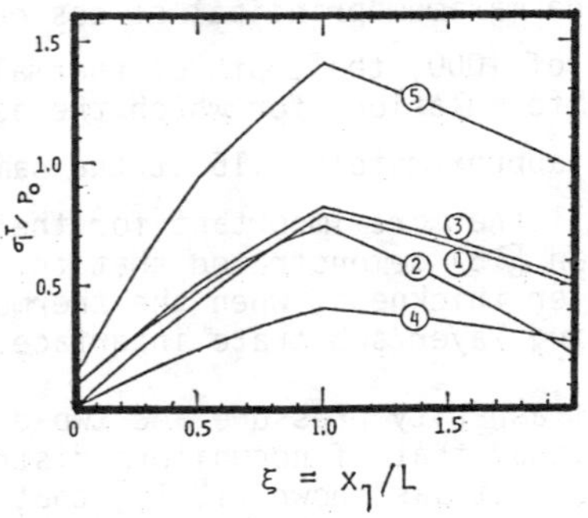

1. $\kappa_1 = \kappa$, $k_1 = k$, $(\rho c)_1 = \rho c$
2. $\kappa_2 = 2\kappa$, $k_2 = 2k$, $(\rho c)_2 = \rho c$
3. $\kappa_3 = \kappa/2$, $k_3 = k/2$, $(\rho c)_3 = \rho c$
4. $\kappa_4 = \kappa/2$, $k_4 = k$, $(\rho c)_4 = 2\rho c$
5. $\kappa_5 = \kappa$, $k_5 = k/2$, $(\rho c)_5 = \rho c/2$

Fig. (2) - Effect of thermal parameter variations of the principal thermal stress

halving of the thermal diffusivity is caused by a doubling in the thermal capacity rather than halving the thermal conductivity, the principal thermal stress reduces almost in half as shown in Curve 4. Finally, in Curve 5, the thermal diffusivity is kept the same as in Curve 1, but both the thermal conductivity and the thermal capacity are halved. The ensuing increase in thermal stress is most noticeable. The illustration demonstrates that the thermal diffusivity is indeed a derivative thermal parameter. It is the individual value of the thermal capacity, the thermal conductivity or their combination that determines the thermal field, thus the stress field that leads to thermo-mechanical cracking of the wear material.

The mechanical and thermal impedance matching between the surface layer and the substrate therefore will consider the differences in mechanical constitutive coefficient, coefficient of thermal expansion, thermal conductivities and thermal diffusivities.

COATING LAYER THICKNESS EFFECT

The estimate of thermo-mechanical cracking from high speed asperity excitation in the layered medium is more complicated than that of a single wear material. The latter, as shown by Ju and Huang [11-13], predicts the fracture initiation in hard wear material to be cohesive failure at the depth of thermal layer, where the maximum temperature gradient occurs. For layered media, the initiation may occur inside the surface layer or in the substrate through cohesive failure or at the coating/substrate interface through shear delamination. Ju and Chen [15] had shown that for thick coating layer, for which thickness is of the order of the asperity size, the critical stress will be inside the coating at the thermal layer. The present paper thus will address the stress states in the surface layer as well as in the substrate at various coating thickness, taking into consideration of the effects of mechanical and thermal impedance mismatchings.

A. Effects of Differences in Mechanical Impedance

The effects of both the elastic coefficients and the coefficient of thermal expansion are mechanical. The solutions in thermal stresses found them indeed as a group, $\alpha E/(1-\nu)$, which defines the mechanical impedance for both the surface layer and the substrate. Figure 3 illustrates the maximum principal thermal

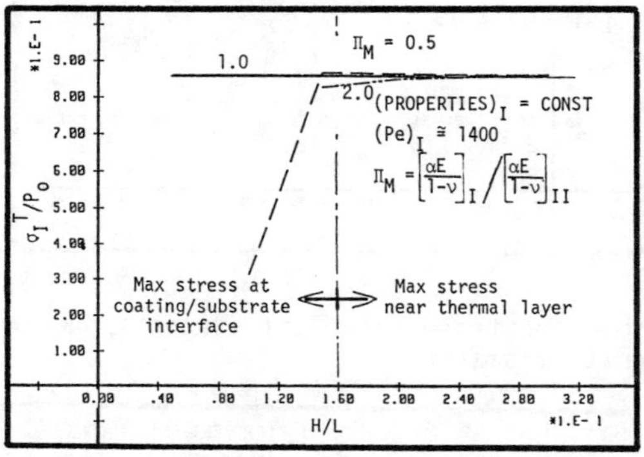

Fig. (3) - Maximum surface layer principal stresses due to mechanical property mismatch

stresses in the surface layer, σ_I^T, as a function of the coating thickness. The dimensionless parameter Π_M defines the quotient of the mechanical impedances of the coating layer and the substrate, respectively. $\Pi_M = 0.5$, 1.0 and 2.0 indicate, respectively, a harder substrate, same material and a softer substrate. For coating thickness less than the thermal layer, the maximum thermal stress in the coating layer occurs in the neighborhood of the coating/substrate interface. With increase of the coating thickness, more than the thermal layer, the maximum thermal stress is in the neighborhood of the thermal layer with diminishing effect of the substrate. It is noticed that, after two thermal layer thickness, the result is no different from that of a single material. Figure 4 shows the corresponding maximum principal thermal stresses in the substrate, σ_{II}^T, as a function of the coating thickness. Since the thermal layer for thin coating is inside the substrate, the maximum thermal stress will occur therein. The maximum substrate stress for coating thicker than the thermal layer will be at the coating/substrate interface. The prediction of fracture initiation must depend on the cohesive strength of individual materials of the coating and the substrate.

B. Effect of Differences in Thermal Properties

Figure 5 shows the maximum principal thermal stresses in the surface layer, σ_I^T, as functions of the coating thickness, with the conductivity ratio, $\Pi_k = k_I/k_{II}$, as the parameter. Similar to the case of mechanical mismatch, the maximum stresses in thick coatings are at the thermal layer and approaches that of the single wear material. When the substrate is less conductive, $\Pi_k = 10.0$, the ther-

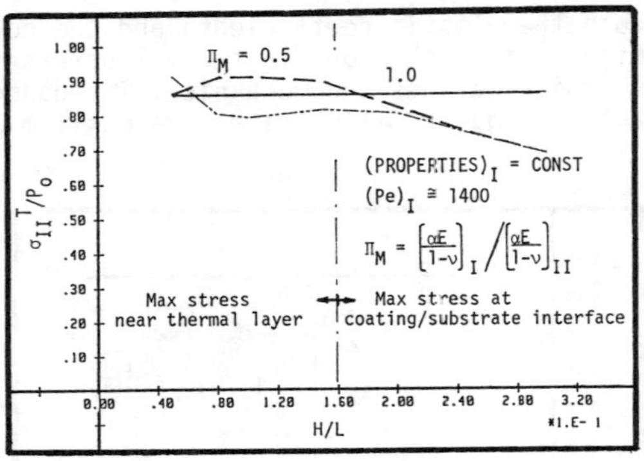

Fig. (4) - Maximum substrate principal stresses due to mechanical property mismatch

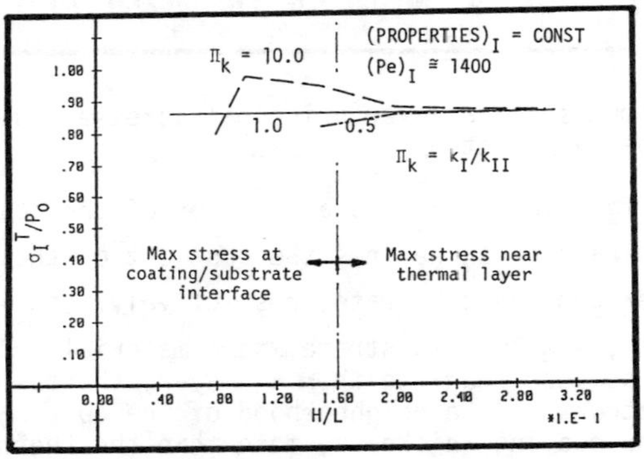

Fig. (5) - Maximum surface layer principal stresses due to mismatch in thermal conductivity

mal stress is higher in the surface layer, especially on the coating/substrate interface for thin coatings. When the substrate is more conductive, $\Pi_k = 0.5$, more heat is readily transferred to the substrate. The thermal stress is correspondingly reduced. The maximum principal thermal stresses in the substrate has a similar trend, as shown in Figure 6.

The specific heat (c) of the wear material does not vary significantly. However, the density difference between the coating and some of the light metallic composites can be large. Figure 7 illustrates a combined curve for maximum principal thermal stresses, indicating that the maximum stress for thin coating is inside the substrate but in the surface layer for thick coatings. In both cases, the stresses are evaluated at the thermal layer.

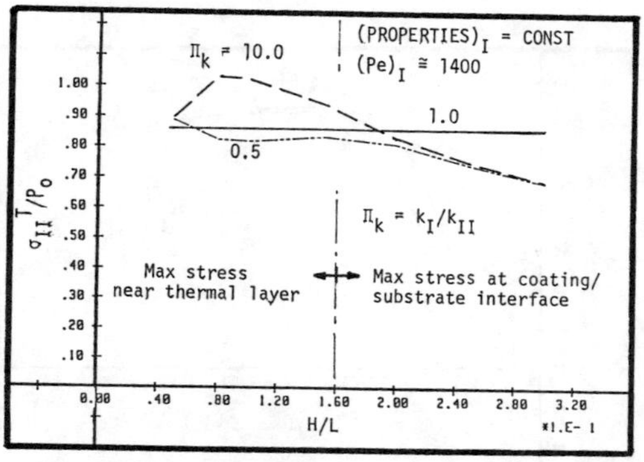

Fig. (6) - Maximum substrate principal stresses due to mismatch in thermal conductivity

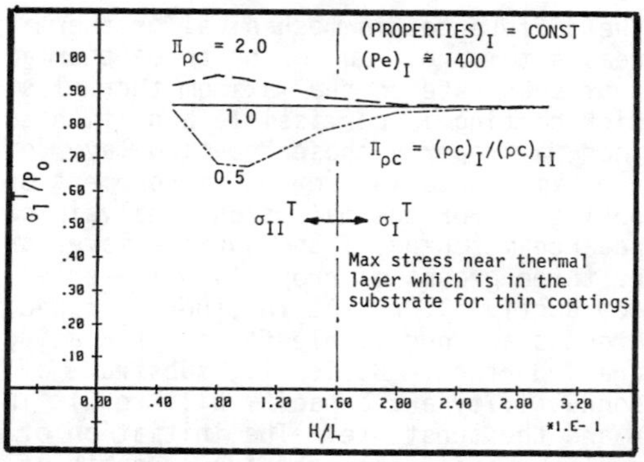

Fig. (7) - Maximum principal stresses due to thermal capacity mismatch

SHEARING STRESS AT THE COATING/SUBSTRATE INTERFACE

Because of the asymmetrical distribution of the temperature gradient, the direction of the maximum temperature gradient is as anticipated, not perpendicular to the wear surface. Although the slanting angle is less than 8°, the resulting value of shearing stress at the coating/substrate interface is not to be ignored. In Figure 8, the curve for $\Pi_k = 1.0$ can be considered as the shearing stress at different depths from the wear surface for a single wear material. The worst case shows the maximum shearing stress at almost 30% of the maximum principal stress. The value is significant in designing the coating/substrate bonding strength to avoid shear delamination. The difference in thermal conductivities of the coating and the substrate is shown to affect the interface shearing stress for thin coatings only. If the coating/substrate bonding of a thin coating is a primary consideration, the substrate may purposely, by design, be less conductive.

CONCLUSIONS

The paper demonstrates the significant effect of coating thickness. The highest stress occurs when the coating thickness is in the neighborhood of the thermal

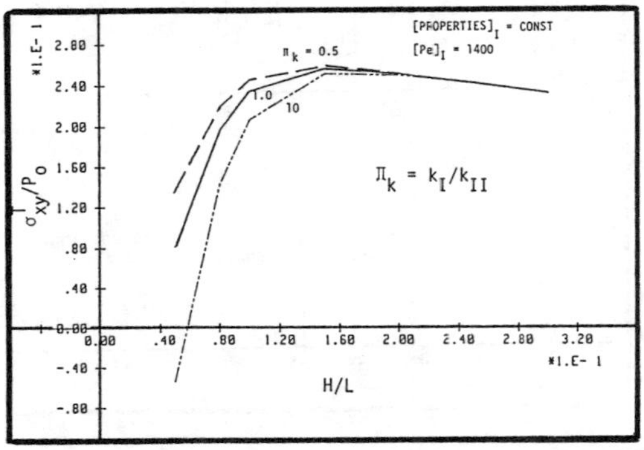

Fig. (8) - Interface shear due to mismatch in thermal conductivity

layer, regardless of the differences in mechanical or thermal properties. As the coating thickness increases to more than two or three thermal layer thickness, there is little effect of substrate on the maximum thermal stress in the coating layer. Hence, if a thick coating is permissible and if the asperity size can be statistically ascertained, a properly chosen coating layer of half the thermal layer thickness can be estimated to insure the improvement on wear resistance for materials such as composites. For thermal stress analysis of a layered media, in order to study the impedance matching of the surface layer and the substrate, it is adequate to consider three groups of property differences, the mechanical, $\alpha E/(1-\nu)$, the thermal conductivity, k, and the thermal capacity, ρc. While the effects of their differences are negligible for thick coatings, the effects on thin surface coating are important. Basically, substrate of low mechanical property, higher thermal conductivity and capacity will result in lower stresses in both the surface layer and the substrate. The initiation of cracks will occur wherever the stress level surpasses the cohesive strength of the respective material. The critical locations in the surface layer or in the substrate is either the surface layer/substrate interface or the thermal layer, depending on the coating thickness relative to the thermal layer thickness.

The shearing stress at the surface layer/substrate interface is by no means trivial, depending again on the surface coating thickness. The interface shear reaches a maximum when the coating thickness is in the neighborhood of the thermal layer. The knowledge is important to the design of bonding of the surface coating. Any shear crack, resulted from the shear delamination at the interface, could cause further extension at repeated asperity excitations. The analysis of the flawed layered media is currently under investigation.

REFERENCES

[1] Archard, J. F., "The temperature of rubbing surfaces", Wear, Vol. 2, 6, pp. 438-455, 1959.

[2] Burton, R. A., "Thermomechanical effects in sliding wear", Workshop on Frictionally Induced Thermal Deformation and Wear, Annapolis, Maryland, June 1979.

[3] Burton, R. A., Nerlikor, V. and Kilaparti, S. R., "Thermoelastic instability in a seal-like configuration", Wear, Vol. 24, pp. 177-188, 1973.

[4] Dow, T. A., "Thermoelastic effects in brakes", Wear, Vol. 59, pp. 213-221, 1980.

[5] Neumann, F. K. and Spies, F., "Grinding cracks", pp. 228-233, "Crack due to frictional heating", Source Book in Failure Analysis, ASM, pp. 337-340, 1974.

[6] Kennedy, F. E. and Karpe, S. L., "Thermocracking of a mechanical face seal", Wear, Vol. 79, pp. 21-36, 1982.

[7] Greenwood, J. A. and Williamson, J. B., "Contact of nominally flat surfaces", Proc. Roy. Soc. London, Ser. A., Vol. 295, p. 300, 1966.

[8] Burton, R. A., "Friction and wear", Chap. 2, Tribology, ed. A. Z. Szeri, McGraw-Hill, pp. 17-38, 1980.

[9] Blau, P. J., "The role of metallurgical structure in the integrity of sliding solid contacts", Solid Contact and Lubrication, ASME AMD, Vol. 39, pp. 135-191, 1980.

[10] Ruff, A. W. and Blau, P. J., "Studies of microscopic aspects of wear processes in metals", National Bureau of Std. NBSIR 80-2058, June 1980.

[11] Ju, F. D. and Huang, J. H., "Thermomechanical cracking due to moving friction loads", UNM Report No. ME-125(84)ONR-233-2, May 1984.

[12] Huang, J. H. and Ju, F. D., "Thermomechanical cracking due to moving friction load - part II. A three dimensional model of a single asperity", Wear, Vol. 102, pp. 87-97, 1985.

[13] Ju, F. D. and Huang, J. H., "The asperity and material parameters in thermomechanical cracking due to moving friction load", Proc. Mech. Failures Prevention Group, April 1985.

[14] Ju, F. D., "The asperity and material parameters in thermomechanical cracking due to moving friction load", UNM Report No. ME-136(86)ONR-233-3, February 1986.

[15] Ju, F. D. and Chen, T. Y., "Thermomechanical cracking in layered media from moving friction load", Jour. Tribology, Vol. 106, pp. 513-518, 1984.

ANALYSIS OF ASYMMETRIC LAMINATES

H. Chin*, J. K. Chen and C. T. Sun

School of Aeronautics and Astronautics, Purdue University
West Lafayette, Indiana 47907 USA

ABSTRACT

Linear small deflection laminated plate theory, large deflection theory, and a theory obtained from modifying large deflection plate theory were used to investigate the behavior of asymmetric cross-plied laminates subjected to in-plane loads. Comparison of theoretical solutions and experimental results was presented. It was found that classical linear theory for thin laminates was completely inadequate in analyzing asymmetric laminates even for small loads.

INTRODUCTION

Asymmetric composite laminates are seldom used in structures mainly because of the difficulty in controlling their configurations after curing. In reality, however, asymmetric laminates may result from delamination and surface damage in a symmetric laminate. This necessitates analysis of asymmetric laminates.

In this paper, asymmetric cross-plied laminates subjected to in-plane loadings are considered. Specifically, linear small deflection laminated plate theory [1] was evaluated. If the extension-bending coupling of the laminate is strong, then linear theory is totally inadequate even for very small in-plane loads. In this case, the effect of geometrical nonlinearity must be included. A finite element plate program based upon the Mindlin plate theory and Von Karman's large deflection theory was used to analyze asymmetric laminates under in-plane loads. Although nonlinear analysis can provide more accurate results, it is computationally expensive. In view of this, a linear theory which was derived from Von Karman's large deflection plate theory was used for analysis. The result was excellent.

Experiments on asymmetric laminates were also conducted at a higher temperature at which asymmetric laminates became flat. Comparison of the theoretical solutions and experimental data is presented.

SIMPLIFIED LARGE DEFLECTION THEORY

Consider an asymmetric cross-plied laminate subjected to an uniform in-plane load N_x° along the two straight edges $x = \pm a$. The governing equations are thus in-

---
*On leave from Aeronautical Research Laboratory, Taichung, Taiwan, R.O.C.

dependent of y-axis. Equilibrium equations based on the Von Karman large deflection theory are given by [2]

$$N_{x,x} = 0 \tag{1}$$

$$M_{x,xx} + N_x w_{,xx} = 0 \tag{2}$$

where N_x and M_x are the in-plane stress resultants and w is transverse deflection of the laminate.

From equation (1), we conclude that

$$N_x = \text{constant} = N_x^\circ \tag{3}$$

and, consequently, equation (2) becomes

$$M_{x,xx} + N_x^\circ w_{,xx} = 0 \tag{4}$$

In the large deflection plate theory, we have [2]

$$N_x = A_{11}u_{,x} - B_{11}w_{,xx} + \frac{1}{2} A_{11}w_{,x}^2 \tag{5}$$

$$M_x = B_{11}u_{,x} - D_{11}w_{,xx} + \frac{1}{2} B_{11}w_{,x}^2 \tag{6}$$

If the slope of the transverse deflection, $w_{,x}$ is assumed to be small, then the nonlinear term in equations (5) and (6) can be neglected. Thus, we arrive at the classical definitions:

$$N_x = A_{11}u_{,x} - B_{11}w_{,xx} \tag{7}$$

$$M_x = B_{11}u_{,x} - D_{11}w_{,xx} \tag{8}$$

Substitution of equation (8) into equation (4) leads to

$$D_{11}w_{,xxxx} - B_{11}u_{,xxx} = N_x^\circ w_{,xx} \tag{9}$$

Although equation (9) has the same form as that for incremental deformation in a laminated plate under initial stress [1], the displacements u and w should not be confused with incremental displacements that are to be added to the initial deformed state. As is clear from our derivation, u and w are displacements measured from the original stress-free state.

Also note that if equation (5) or (7) is substituted into equation (9), then a nonlinear equation results. Dropping the nonlinear terms at this stage would lead to classical linear theory. Hence, although equation (9) appears as a linear differential equation, it contains the effect of nonlinear terms.

Using equations (7) and (3), we obtain

$$A_{11}u_{,x} = N_x^\circ + B_{11}w_{,xx} \tag{10}$$

Substitution of equation (10) into equation (9) yields

$$w_{,xxxx} = \frac{N_x^\circ}{(D_{11}-B_{11}^2/A_{11})} w_{,xx} \tag{11}$$

The term on the right-hand-side of equation (11) indicates the effect of in-plane load N_x° on transverse deflection. It is interesting to note that a larger value of B_{11} (strong extension - bending coupling) has an apparent effect of a large value of N_x°. Thus, the in-plane load effect on bending can be enhanced by increasing the value of the coupling coefficient.

The general solution for equation (11) is

$$w(x) = C_1 \cosh kx + C_2 \sinh kx + C_3 x + C_4 \tag{12}$$

where

$$k = [N_x^\circ/(D_{11}-B_{11}^2/A_{11})]^{1/2} \tag{13}$$

If the origin of the coordinate axis is chosen to locate at the mid-span, then $C_2 = C_3 = 0$ because of symmetry. The two constants C_1 and C_4 are determined by the boundary conditions

$$w = 0,\ M_x = 0 \text{ at } x = \pm a \tag{14}$$

We obtain

$$C_1 = \frac{B_{11}}{A_{11}\cosh ka} \qquad C_4 = -\frac{B_{11}}{A_{11}} \tag{15}$$

The in-plane displacement can be obtained from equation (10) as

$$u(x) = \left(\frac{B_{11}^2 k}{A_{11}^2 \cosh ka}\right) \sinh kx + \frac{N_x^\circ}{A_{11}} x \tag{16}$$

Maximum deflection occurs at the mid-span, $x = 0$:

$$w_{max} = C_1 + C_4 = -\frac{B_{11}}{A_{11}} \left(1 - \frac{1}{\cosh ka}\right) \tag{17}$$

It is interesting to note that

$$w_{max} \to -B_{11}/A_{11} \text{ as } N_x^\circ \to \infty \tag{18}$$

Thus, there exists a bound for deflection as N_x° increases.

The longitudinal strain at the mid-plane is given by

$$\varepsilon_x^\circ = u_{,x} = \frac{B_{11}^2 k^2}{A_{11}^2 \cosh ka} \cosh kx + \frac{N_x^\circ}{A_{11}} \tag{19}$$

From equation (19), we note that

$$\varepsilon_x^\circ = \frac{N_x^\circ}{A_{11}} \text{ as } N_x^\circ \to \infty \qquad (20)$$

Thus, the extension-coupling effect diminishes as N_x° increases. The same conclusion for the longitudinal strain ε_x over the thickness can be reached. Hence, for large in-plane loads, strains can be obtained from the governing equations by setting $B_{11} = 0$.

CLASSICAL LINEAR THEORY

The difference between the above formulation and laminated plate theory lies in the extra term on the right-hand-side of equation (9) in the former. The displacement solutions according to classical theory can be obtained from equation (9) by setting $N_x^\circ = 0$ and equation (10). The results are

$$w(x) = [B_{11}/2A_{11}(D_{11}-B_{11}^2/A_{11})]N_x^\circ x^2 - [B_{11}a^2/2A_{11}(D_{11}-B_{11}^2/A_{11})]N_x^\circ \qquad (21)$$

$$u(x) = [D_{11}/A_{11}(D_{11}-B_{11}^2/A_{11})]N_x^\circ x \qquad (22)$$

Consider the deflection at $x = 0$,

$$w_{max} = -\frac{B_{11}a^2N_x^\circ}{2A_{11}(D_{11}-B_{11}^2/A_{11})} \qquad (23)$$

Thus, $|w_{max}| \to \infty$ as $N_x^\circ \to \infty$. A great discrepancy is noted by comparing equations (17) and (23).

The longitudinal strain at the mid-plane of the laminate is obtained from equation (22). We have

$$\varepsilon_x^\circ = [D_{11}/A_{11}(D_{11}-B_{11}^2/A_{11})]N_x^\circ \qquad (24)$$

which is quite different from that given by equation (19). In particular, for large values of N_x°, the strain given by equation (24) is still affected by the coupling coefficient B_{11} and tends to be larger than the one in equation (20). However, when $B_{11} = 0$, both values coincide.

ANALYSIS USING LARGE DEFLECTION THEORY

More accurate solutions are obtained by using the complete large deflection plate theory. In this study, a nine-node isoparametric finite element, developed in [3], based on the Mindlin plate theory and Von Karman large deflection assumptions, was employed for large deflection analyses. Although composite plates are weak in transverse shear resistance, transverse shear deformation may still be insignificant for very thin plates. Therefore, the reduced integration scheme was employed to improve the performance of the element. The order of the Gauss quadratures used to evaluate the stiffness matrices and the internal force vector depends upon whether the item under consideration is a function of membrane, bending, or transverse shear behavior. In this study, the 3x3 Gaussian rule was used to compute the items for in-plane and bending deformations while the reduced 2x2 rule was employed for transverse shear deformation.

Laminate plates examined in this study measure 9" x 1.5" $[0_4/90_4]$. These are subjected to uniform in-plane load N_x° at the two edges. The edges along the y-direction are assumed to be simply-supported but movable in the x-direction. Due to symmetry, only a quadrant of the plate is used for analysis. Twenty-four elements are used. This mesh is shown to yield converged solutions.

Since there exists a coupling effect of bending and in-plane deformation for asymmetric laminates, the solution procedure is the same as the conventional approach for solving large deflection problems. The Newton-Raphson method was adopted in the numerical analysis here. The iteration procedure was carried out until a convergence tolerance of 0.1% was met.

EXPERIMENTAL PROCEDURE

The $[0_4/90_4]$ cross-plied laminate fabricated from Fiberite-1076E graphite/epoxy prepreg was chosen for our experiment. A number of 9" x 1.5" specimens were cut from the panel. To simulate the simply supported boundary condition on the edges, a set of self-aligning ball bearings with outside diameter .75 inch was pressed into a hole of the same diameter at both end tabs. Due to the self-alignment of the bearing, the out-of-plane rotation was not constrained. The specimen was then installed into the test fixture with a .25 inch diameter steel pin.

Because of residual curing stresses, asymmetric laminates curved up in room temperature. However, the analysis was performed based on the assumption that the laminate was initially flat. To make a meaningful comparison, the experiment was conducted in an oven at an elevated temperature (320°F) at which the curing stresses would be neutralized and the specimen showed the configuration of a flat plate.

High temperature resistance strain gages were used to record strains on the upper and lower surfaces. Strain readings were taken with an automatic strain indicator and recorder. A dial-gage was used to record the total elongation of the specimen through the movement of the fixture.

RESULTS AND DISCUSSION

In the numerical analysis, two cross-plied laminates, $[0_4/90_4]$ and $[0_2/90_2/0_2/90_2]$, were considered. The length of both laminates was chosen to be L = 9 inches. The elastic properties of a graphite/epoxy lamina of thickness 0.0055 inch are given by

$E_1 = 20 \times 10^6$ psi $\qquad E_2 = 1.4 \times 10^6$ psi

$\nu_{12} = 0.30 \qquad G_{12} = G_{13} = G_{23} = 0.7 \times 10^6$ psi

The plate stiffnesses for the $[0_4/90_4]$ laminate are

A_{11} = 431400 lb/in B_{11} = -3737 lb-in/in D_{11} = 57 lb-in

and for the $[0_2/90_2/0_2/90_2]$ laminate, we have

A_{11} = 431400 lb/in B_{11} = -1869 lb-in/in D_{11} = 57 lb-in

Figures 1 and 2, respectively, show the out-of-plane deflections w(x) of the two laminates subjected to in-plane load N_x° = 100 lb/in. It is obvious that the solution of classical linear theory drastically deviates from that of nonlinear theory even at very low loads. The simplified nonlinear theory, however, provides very good agreement with the nonlinear theory.

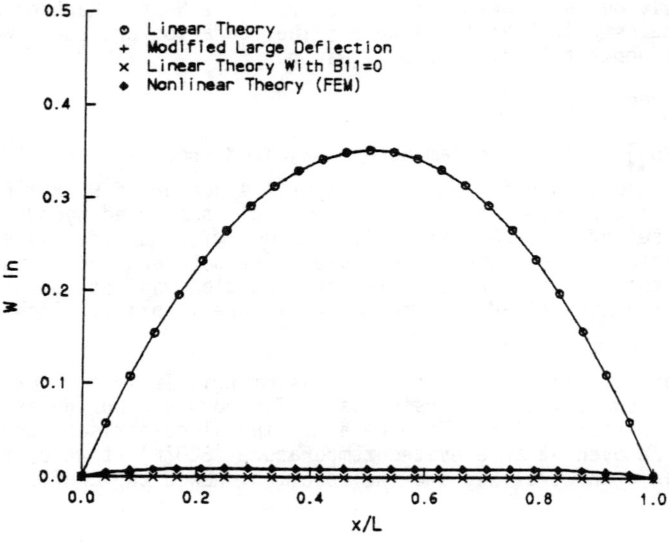

Fig. (1) - Out-of-plane deflection w of 9" x 1.5" $[0_4/90_4]$ laminate subjected to uniform axial load N_x° = 100 lb/in

Substituting the plate stiffness into equation (17), the midspan deflections w_{max} of $[0_4/90_4]$ and $[0_2/90_2/0_2/90_2]$ laminates are bound to be 0.00866 inch and 0.00432 inch, respectively. Note that as $N_x^\circ \to \infty$, the value of w_{max} for $[0_4/90_4]$ and $[0_2/90_2/0_2/90_2]$ laminates according to the nonlinear analysis approach 0.008663 inch and 0.004333 inch, respectively. In contrast, the linear analysis, predicts that w_{max} will go to infinity for both cases. Since the value of the coupling term B_{11} of the $[0_2/90_2/0_2/90_2]$ laminate is less than that of the $[0_4/90_4]$ laminate, the latter produces much more severe transverse deflection. It is interesting to note that the solutions predicted using the simplified large deflection theory agree very well with those obtained by using the large deflection plate finite element. Another solution which was obtained by setting the coupling term B_{11} equal to zero yields zero deflection as expected.

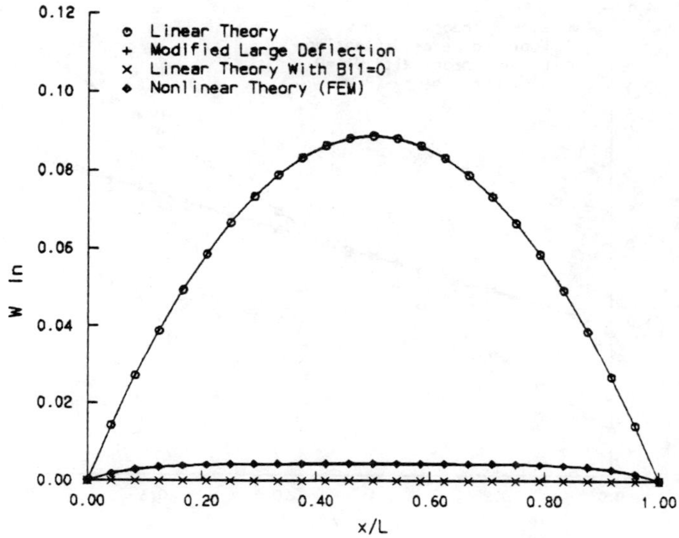

Fig. (2) - Out-of-plane deflection w of 9" x 1.5" $[0_2/90_2/0_2/90_2]$ laminate subjected to uniform axial load N_x° = 100 lb/in

In Figure 3, axial displacements of a 9" x 1.5" $[0_4/90_4]$ laminate predicted by different methods are presented. It is noted that at x = 0, the slopes of u (i.e., $u,_x = \varepsilon_x^\circ$) from the solutions of the linear theory, modified large deflection theory, and nonlinear finite element method are 0.000536 inch, 0.0002318 inch, and 0.0002327 inch, respectively. At x = 4.45 inch, these strains are 0.000536 inch, 0.000503 inch, and 0.000507 inch, respectively. Note that the linear theory produced larger deformation than the nonlinear theory.

According to equations (17), (18), (21) and (22), out-of-plane deflection w_{max} and axial strain ε_x° of 9" x 1.5" $[0_4/90_4]$ and $[0_2/90_2/0_2/90_2]$ laminates as function of in-plane load are shown in Figures 4 and 5. It is noted that the value of w_{max} converges to an identical value with a small load. The axial strain became linear proportional to the small load.

Normal stresses are shown in Figures 6 and 7 for the two laminates, respectively. It is interesting to note that the classical linear solution and that with B_{11} = 0 form the upper bound and lower bound, respectively, of the nonlinear solution. The solution according to the simplified large deflection theory is again excellent. The solution obtained using the classical linear theory is nowhere near the nonlinear solution. However, the uncoupled linear theory (B_{11} = 0) yields good solutions in the region away from the two ends.

Figure 8 shows the comparison of experimental and analytical results for axial displacement as a function of the applied load N_x°. As seen from the figure, there

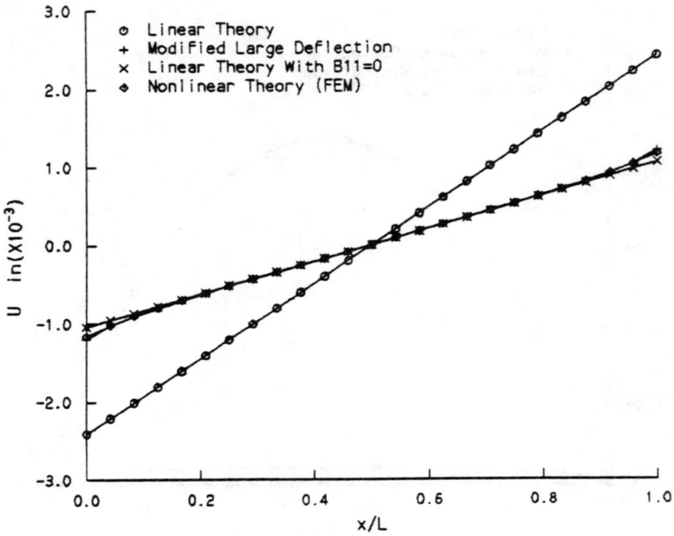

Fig. (3) - Axial displacement u_x of 9" x 1.5" x $[0_4/90_4]$ laminate subjected to uniform axial load N_x° = 100 lb/in

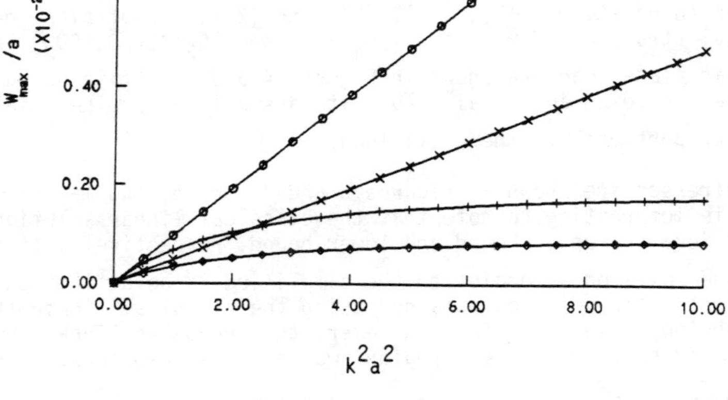

Fig. (4) - Maximum out-of-plane deflection w_{max} versus in-plane load N_x°

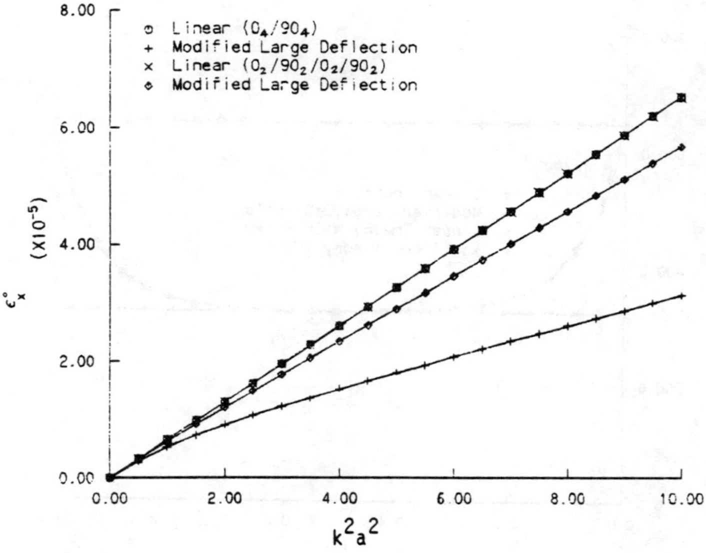

Fig. (5) - Axial strain ε_x° versus in-plane load N_x°

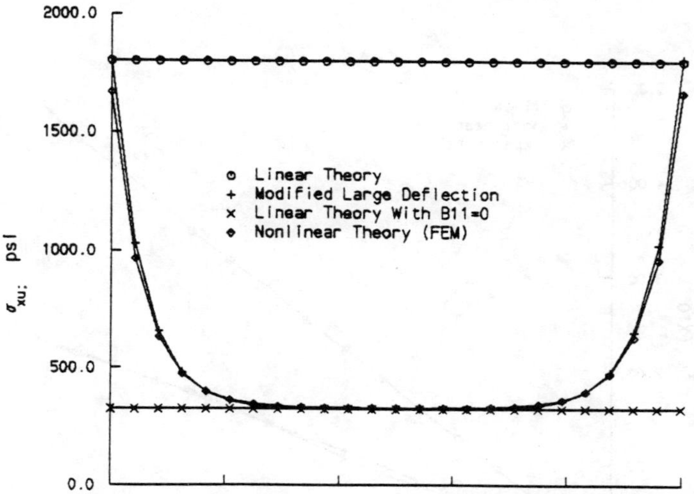

Fig. (6) - Axial stresses at the upper surface along the length of $[0_4/90_4]$ specimen subjected to N_x° = 100 lb/in

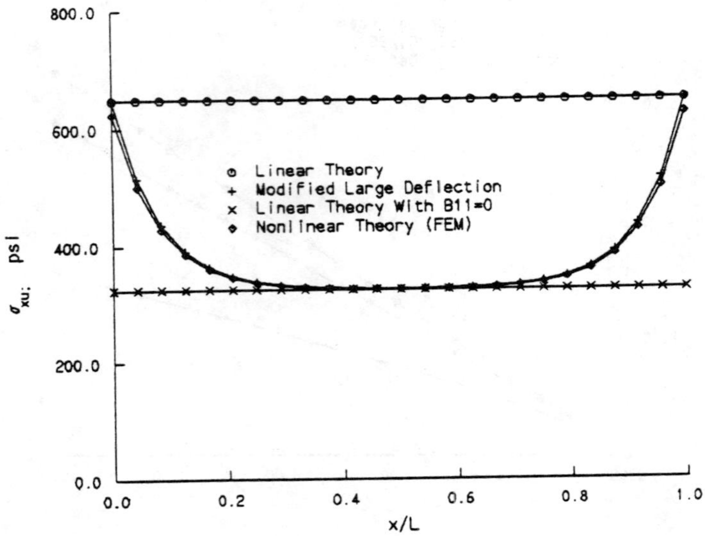

Fig. (7) - Axial stresses at the upper surface along the length of $[0_2/90_2/0_2/90_2]$ specimen subjected to N_X° = 100 lb/in

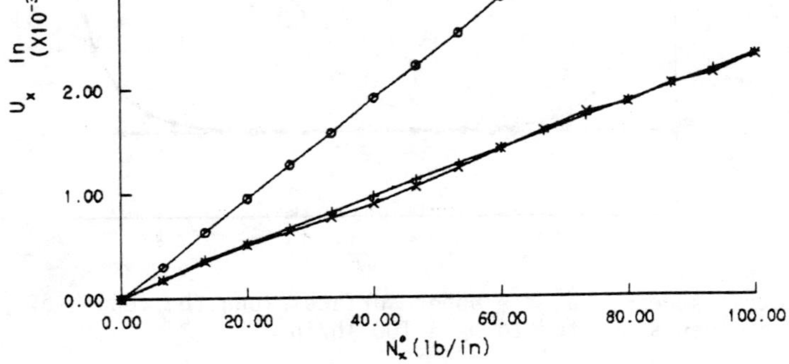

Fig. (8) - Elongation of specimen as function of applied load N_X°

is good agreement between the experimental data and those obtained by the non-linear finite element method. In contrast, large errors are evident in the linear theory.

The comparisons of surface strains ε_x is shown in Figure 9. The strains on the upper and lower surfaces were measured at the center of the specimen. The

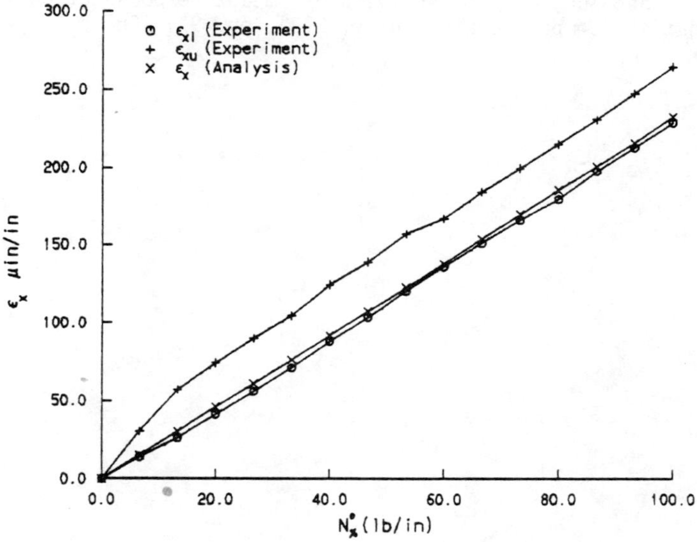

Fig. (9) - Upper and lower surface strain as function of applied load N_x°

analytical results were predicted by using large deflection finite elements. Since the strains on both surfaces are very close in numerical computation, only one curve was presented. Although not shown, the strains predicted by simplified deflection theory agreed with the finite element solution very well. It is seen that the experimental data and analytical solutions correlate very well for one surface. Difference is noted at the upper surface (90° layer). This discrepancy could be caused by the initial surface imperfection due to thermal effects.

CONCLUSION

In this paper, it has been shown that classical linear laminated plate theory is completely inadequate for analysis of asymmetric laminates even for small in-plane loads. A simplified large deflection theory which included the in-plane stress effect on bending was employed for such cases and was shown to be accurate as compared with solution using large deflection plate finite element method.

REFERENCES

[1] Ashton, J. E. and Whitney, J. M., Theory of Laminated Plates, TECHNOMIC Publishing Co., 1970.

[2] Chia, C. Y., Nonlinear Analysis of Plates, McGraw-Hill, 1980.

[3] Chen, J. K. and Sun, C. T., "Analysis of impact response of buckled composite laminates", Composite Structures, 3(2), pp. 97, 1985.

THE FAILURE CHARACTERISTICS OF A GR/EP LAMINATE

D.-G. Hwang

Chung Shan Institute of Science and Technology
P.O. Box 1-26-2, Lung-Tan, Taiwan 32500, Republic of China

and

H. T. Hahn

Washington University
St. Louis, Missouri 63130 USA

ABSTRACT

Failure behavior of a $[0_2/90/\pm45]_s$ AS/3501-5A Gr/Ep laminate has been theoretically and fractographically investigated. The failure characteristics, including first ply-failure (FPF), crack multiplication, delamination growth, and 0° fiber breaks, have been identified throughout static and fatigue tests of the laminate using microscopic and replication methods. The modulus change has also been correlated to the crack density and delamination growth in fatigue.

INTRODUCTION

When a multi-directional composite laminate is subjected to either static or cyclic load, cracks develop in the weaker plies prior to ultimate failure. These cracks multiply progressively throughout the specimen volume in contrast to the formation and growth of a single dominant crack observed in homogeneous materials. The ply failure can also be accompanied by delamination between adjacent plies. Both ply failure and delamination are mostly dominated by matrix cracking and interfacial debonding. However, a few fiber breaks are also observed. The ply failure and delamination result in a reduction of strength and stiffness. The exact combination of these two types of damage varies widely depending on the material properties of the laminate, its lamination, including stacking sequence and layer thickness, the type of loading, and environment [1-9].

The ply cracks develop parallel to the fibers when the tensile stress normal to the fibers exceeds a threshold level which is the transverse tension strength. The corresponding laminate stress applied is called the first ply-failure (FPF) stress. Thus, below the FPF stress no ply fails. Naturally, the FPF stress depends on the type of laminate, its constituent ply properties, and residual stresses present within the laminate, including curing and swelling stresses [3, 10-14].

For most Gr/Ep composites the difference between cure temperature and room temperature is in the range of 150°C. When this temperature difference is coupled with the large thermal expansion mismatch between the fiber and matrix, the resulting residual stresses can be large enough to cause premature matrix cracking [13,14]. In this study, the replication method is employed together with photomicrography to monitor the FPF and transverse crack multiplication during static tests.

Delamination may develop from the ply cracks or from the interlaminar stresses developed at the stress-free edges. The interlaminar stresses are highly localized, and their signs and magnitudes depend on the stacking sequence and ply thicknesses [4,15-17]. Furthermore, delamination may grow more easily under cyclic loading. Delamination renders the stresses in a laminate redistributed, and influences the residual strength, fatigue life, and residual stiffness [15,18-22]. Therefore, the growth of delamination should be taken into account, especially for fatigue investigation.

Stiffness change is directly related to the damage [4,19,21,22] and may be generated very early in the load history for multi-directional laminates, rendering the laminates incapable of meeting the originally specified structural requirements. O'Brien [21] and Kim [4] have shown that stiffness reduction depends on the fatigue stress level and may amount to 10% to 20% of the original stiffness in quasi-isotropic Gr/Ep laminates.

This paper is in part intended to improve the understanding of static and fatigue failure processes and mechanisms for the Gr/Ep laminate through microscopic examinations. Specifically, the crack density and delamination are monitored along the edges of the specimen by replication method and are correlated to the stiffness loss.

EXPERIMENTAL PROCEDURE

The material system used for this study is $[0_2/90/\pm45]_s$ AS/3501-5A graphite/epoxy. The standard cure cycle with the exception of post cure was employed to fabricate 610 mm x 610 mm (24" x 24") panels with an average thickness of 1.334 mm (0.0525") by the University of Dayton Research Institute. The average fiber content is 72% by volume.

All specimens were 12.7 mm wide and 152 mm long with 76 mm of gage length. G-10 woven glass/epoxy 1.6 mm thick was used for end tabs. The adhesive was Epon 828 Resin combined with Versamid-140. The panels were cut into test specimens by using an 8" diameter, 0.090" x 1/8" thick, 60/80 grit impregnated watersaw blade. All specimens were kept in the laboratory after preparation. No preconditioning was applied to the specimens before testing.

All mechanical tests were performed on an MTS servo-controlled closed-loop testing machine. Cyclic tests were run between fixed upper and lower load limits at the stress ratio R of 0.1 and the frequency f of 5 Hz, and a sinusoidal wave form was employed. The loading rate in static was 100 N/S. Axial strain was monitored with an one inch clip-gage extensometer. A few specimens had a strain gage of type EA-06-125 TM-120 mounted to obtain typical stress-strain relation in static tension.

The test matrix employed in the present study is shown in Table 1, where $\overline{X}$ is the average static strength and S is the maximum fatigue stress. Fatigue tests at $S/\overline{X}$ = 0.60, 0.70, and 0.80 were carried out after proof tests to the respective fatigue stresses only. Some specimens had one of the edges polished to allow microscopic examinations of crack formation and delamination at various stages of static and fatigue tests. Simultaneously, replication method was employed for these edge-polished specimens to monitor the crack multiplication in each off-axis ply and the development of delamination. Modulus was measured intermittently of two specimens at each fatigue stress level.

Table 1 - Test matrix

Test Series	Objective	Number of specimens
Static	X	20
Static*	Crack Characteristics	2
Fatigue*	Crack Characteristics, Delamination and Modulus Change	2(S/X=0.60) 2(S/X=0.70) 2(S/X=0.80)

* Specimens were edge-polished and replication method and microscopic examinations were employed.

RESULTS AND DISCUSSION

Static Failure Process

The static strength data for present laminate are listed in Table 2 and a typical stress-strain relation in axial and lateral directions for specimen with a two-dimensional strain gage is shown in Figure 1. Note that the axial strain is quite linear up to failure while the lateral strain shows a slight nonlinearity.

Table 2 - Mechanical properties of present laminate

No. of Samples	20
Ultimate Stress	
Average X, MPa	788.11
C.O.V.,%[a]	5.21
Ultimate Strain	
Average m/m	1.13%
C.O.V., %	5.43
Modulus	
Average E, GPa	69.49
C.O.V.,%	5.29

Note: [a]C.O.V.: Coefficient of Variation

Photomicrographs of crack pattern at different stress levels are shown in Figure 2. The failure sequence of the present laminate in static tension was found to be essentially the same as that of quasi-isotropic laminate [1,3,4,7]. That

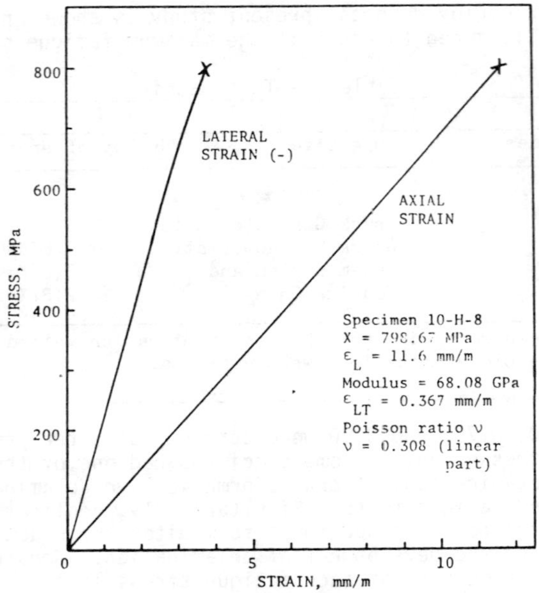

Fig. (1) - Stress-strain relation in a static tension

is, as the load is increased, transverse cracks appear first in the 90° plies, next in the +45° plies, and finally in the -45° plies, which are in the middle of the laminate.

In Figure 2(b), the first crack is seen to have occurred in the 90° ply when the applied stress reached $0.40\overline{X}$ (315 MPa, $0.45\overline{X}$ of the specimen A-b-4). The number of cracks increases rapidly as the stress is increased from $0.40\overline{X}$ to $0.50\overline{X}$ and further to $0.60\overline{X}$. Above $0.60\overline{X}$, the number of cracks in the 90° plies increases slowly and much more so especially after $0.90\overline{X}$. The first +45° ply crack occurred at $0.50\overline{X}$, followed by more cracks with increasing stress. Most of these cracks appeared at the tips of cracks in the 90° plies and extended instantly to the interface between +45° and -45° plies at an angle approximately 60° to the interface. Very few cracks occurred in the -45° plies before final failure, and only 2 were found in specimen A-b-4, and none in specimen A-b-6, within 3" gage length before final failure. No delamination occurred before final failure. After final failure, delamination occurred just around the tip of failure site.

An interesting discovery is the 0° fiber breakage. Figures 2(d), (e) and (f) show 0° fiber breaks around $0.70\overline{X}$ (552 MPa) and more at $0.80\overline{X}$ on the polished surfaces. Figure 3 taken from a replicate, also shows fiber breaks.

Figure 4 (specimen A-b-4) is the replication result. By examining these large-scale pictures, the failure sequence can be identified.

The crack density in each ply was counted from replicates, and is plotted as a function of applied stress in Figure 5, where the solid curves represent the mean crack densities from two specimens, A-b-4 and A-b-6. Figure 6 shows the

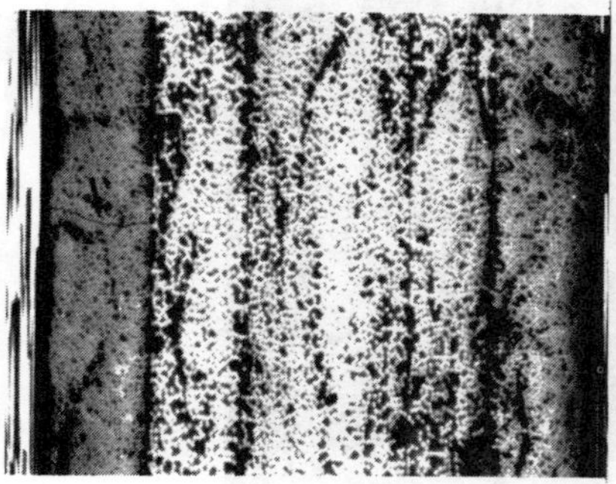

[A] $\sigma/\bar{X}$ = 0.40

[B] $\sigma/\bar{X}$ = 0.50

[C] $\sigma/\bar{X}$ = 0.60

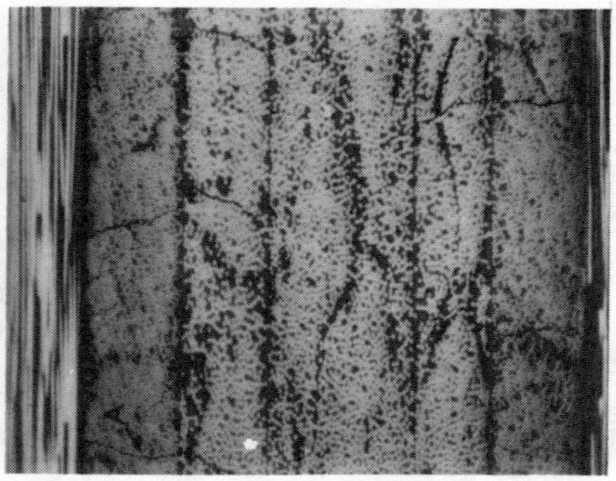

[D] $\sigma/\bar{X}$ = 0.70

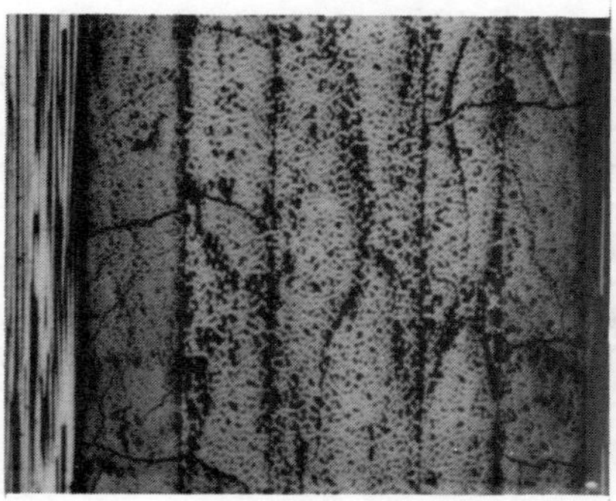

100 X

[E] $\sigma/\bar{X}$ = 0.80

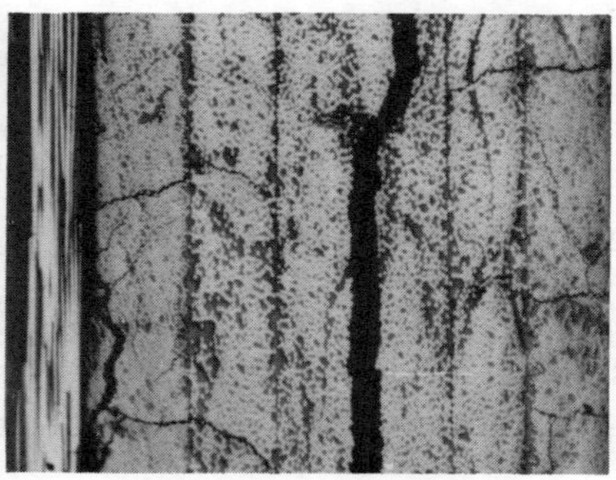

[F] After fracture, X = 0.89 $\bar{X}$ [701.4 MPa]

Fig. (2) - Static failure sequence for specimen A-b-4

$\sigma/\bar{X} = 0.80$ 200X

Fig. (3) - 0° fiber breaks in static test

corresponding average crack spacing versus the applied stress. The tendency of crack density for each off-axis ply is the same as the results of [4], and also fit the theoretical predictions in [23].

From Figure 6, we determine the average crack spacings at failure to be about 0.299 mm for 90° ply cracks and 0.529 mm for +45° ply cracks. The data are summarized together with the data from [1,4] in Table 3. There are more cracks per unit length in the 90° plies of the laminate that has a lower ratio of 90° plies to 0° plies [1]. The $[0_2/90/\pm45]_s$ laminate has 25% more cracks in the 90° plies than the $[0/90/\pm45]_s$ laminate. This tendency also holds true for +45° plies.

FIRST PLY - FAILURE (FPF) PREDICTION

The first step in the damage process is the cracking of the weakest ply. The so-called first ply-failure (FPF) is defined as the first occurrence of a crack within a laminate subjected to, usually, a tensile loading [3]. The initiation of this cracking process occurs at a load which can be predicted with reasonable accuracy if thermal residual stresses and moisture-related stresses are properly included in a failure criterion such as the maximum stress or maximum strain criterion, or the polynomial criterion [6] at the lamina level.

In this study, we use the laminated plate theory, including curing and hygrothermal residual stresses, and the maximum stress failure criterion to predict the FPF stress and compare it with the experimental data. The corresponding laminate stress resultant N_X^{FPF} can be calculated by following equations [3,25]:

$$\sigma_i^R = Q_{ij}A_{jk}^{-1}N_k^N - Q_{ij}(\alpha_j\Delta T + \beta_j C) \qquad (1)$$

$$N_X^{FPF} = \frac{A_{xx}A_{yy}-A_{xy}^2}{Q_{22}A_{yy}-Q_{12}A_{xy}}\,(X_T-\sigma_2^R) \qquad (2)$$

(a) Before test

20X

(b) $\sigma/\overline{X} = 0.40$

(c) $\sigma/\overline{X} = 0.50$

(d) $\sigma/\overline{X} = 0.60$

(e) $\sigma/\overline{X} = 0.70$

(f) $\sigma/\overline{X} = 0.80$

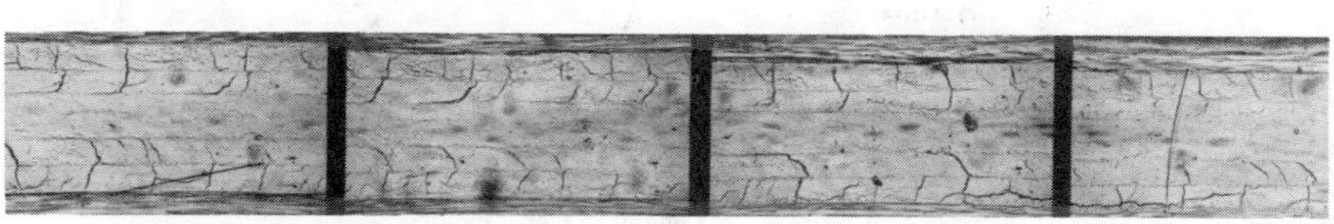

Fig. (4) - Replication inspection of crack multiplication in static test specimen: A-b-4; maximum strength: $0.89\overline{X}$ (701.4 MPa)

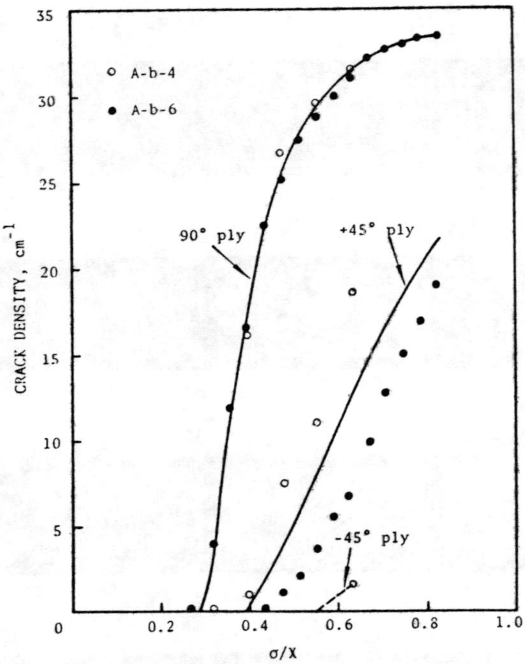

Fig. (5) - Crack density increasing with applied stress

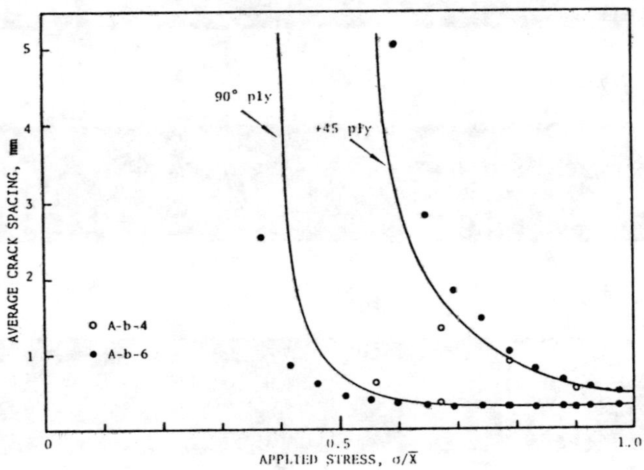

Fig. (6) - The correlation between average crack spacing and applied stress for each angle plies

Table 3 - Minimum crack spacing for each off-axis ply

	Minimum average crack spacing, mm		
	90° ply	+45° ply	-45° ply
Present study	0.299	0.529	6.36[c]
Reference 1[a]	0.393	0.737	1.64
Reference 4[b]	0.413	1.114	almost no crack[c] in -45° plies

[a] T300/5208 Gr/Ep $[0/90/\pm45]_s$

[b] AS/3501 Gr/Ep $[0/90/\pm45]_s$ laminate and data calculated from Figure 6 of Reference 4

[c] Only one specimen had cracks in -45° plies

where:

σ_i^R = residual stresses

Q_{ij} = reduced stiffnesses at the final temperature of interest

A_{ij} = laminate stiffness

N_k^N = non-mechanical stress resultants due to hygrothermal strains

α_j = thermal expansion coefficients

ΔT = temperature change; positive for temperature increase and negative for decrease

β_j = moisture swelling coefficients

C = moisture concentration, always positive

X_T = transverse strength of lamina

Note that the subscript 1 is associated with the fiber direction of 90° plies and subscript x with the loading direction. Therefore, the FPF stress is

$$\sigma_{FPF} = N_x^{FPF}/h \tag{3}$$

where h is the thickness of laminate.

The typical properties of AS/3501 Gr/Ep used in the calculation are listed below:

E_{11} = 138 GPa; E_{12} = 7.1 GPa; E_{22} = 8.96 GPa; X_L = 1447 MPa; ν_{12} = 0.30; X_T = 51.7 MPa; Q_{11} = 138.8 GPa; Q_{12} = 2.704 GPa; Q_{22} = 9.013 GPa; Q_{66} = 7.1 GPa.

Thermal expansion coefficients are

$$\alpha_x = -0.3,\ \alpha_y = \alpha_z = 28.1\ (\mu m/m)/°K$$

Swelling coefficients are

$$\beta_x = 0,\ \beta_y = \beta_z = 0.44\ m/m$$

The stress-free curing temperature T_o and the final temperature are 177°C and 22°C, respectively, and the moisture content in the laboratory environment is c = 0.56% which was measured by drying the specimens in a vacuum oven at 70°C, 600 hours. The final results are

$$\begin{aligned} \sigma_2^R &= 15.17\ \text{MPa} \\ \sigma_{FPF} &= 316.48\ \text{MPa} \end{aligned} \tag{4}$$

The foregoing analytical prediction is in excellent agreement with the experimental data. The experimental FPF stress is 315.42 MPa ($0.4\overline{X}$) for specimen A-b-4 and 275.84 ∿ 315.24 MPa (0.35X ∿ 0.40X) for specimen A-b-6. There is only 0.39% deviation from the prediction for specimen A-b-4. Thus, the FPF stress can be calculated by the maximum stress failure criterion as long as the curing stress and moisture content are properly determined and incorporated into the analysis.

THEORETICAL PREDICTION OF AVERAGE CRACK SPACING IN STATIC LOADING

It was postulated in most papers that the number of transverse cracks in each ply reached a limiting value after which no further cracks occurred in spite of increased stress or fatigue cycles. Reifsnider et al [1] referred to this limiting state as the characteristic damage state (CDS) and concluded that the CDS is a well-defined material property independent of load history. The question now arises whether or not this CDS will be reached before final failure of the present laminate.

The first transverse crack develops when the transverse stress in 90° plies reaches the transverse ply strength X_T, as discussed in the preceding section. After the FPF occurs, the stress will be redistributed with the maximum value lower than X_T. With an increase in the applied stress, the maximum transverse stress will again reach X_T, and more transverse cracks will occur. In this section, the Vasil'ev model described in [26] is modified to investigate the crack multiplication.

Consider a long $[0/90]_s$ laminate subjected to load P. The stresses in the outer longitudinal layers are denoted by the subscripts 1 and 3 and those in the middle transverse layer by the subscript 2, Figure 7.

The following three assumptions are held: (1) the longitudinal strain and the transverse strain are the same for all layers; (2) the deformation in the y

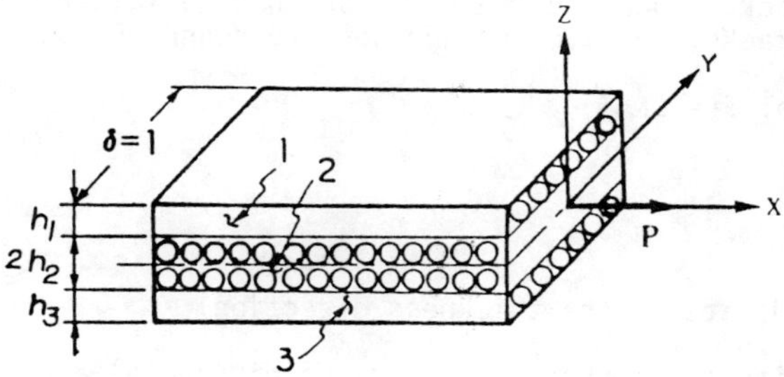

Fig. (7) - Schematic diagram of a symmetric laminate

direction is rather small because of the smallness of the Poisson's ratio and can be assumed to be zero; and (3) the bond between the layers is not disturbed after the cracks occur. We may assume that at a certain load P_1 at which $\sigma_{x2} = X_T$ a crack appears in a certain section of 90° ply and that this crack is located at the coordinate origin x = 0. Here, σ_{x2} is the transverse stress in 90° ply.

By using the equilibrium equation with the assistance of the strain compatibility equation and the least work theorem, we can finally obtain the following variational equation [26]:

$$\sigma^{iv} - 2a^2\sigma'' + b^4\sigma = 0 \tag{5}$$

where

$$\sigma = \sigma_2^o - \sigma_{x2} \tag{6}$$

Note that σ_2^o is the transverse stress in 90° ply without cracks, which can be calculated by the laminated plate theory, and that the prime denotes the derivative with respect to x, and

$$a^2 = \frac{E_2\left[\frac{h_1+h_2}{G_1} - \frac{\nu_2}{E_2}(2h_2+3h_1) + \frac{\nu_1 h_1}{E_1}\right]}{3h_2^3-10h_2^2(h_1+h_2)+15h_2(h_1+h_2)^2+3h_1^3} \tag{7}$$

$$b^4 = \frac{60E_2\left(\frac{1}{h_2E_2} + \frac{1}{h_1E_1}\right)}{3h_2^3-10h_2^2(h_1+h_2)+15h_2(h_1+h_2)^2+3h_1^3} \tag{8}$$

Note that E_1, E_2 are longitudinal and transverse Young's moduli, G_1 is shear modulus, ν_1, ν_2 are longitudinal and transverse Poisson's ratio, respectively, for unidirectional composite.

When the applied stress is increased, a new crack will develop a distance 2K away from this crack. Locating the new coordinate origin x = 0 at the midpoint between the two cracks, we have the appropriate boundary condition as

$$\sigma_{x2} = \tau_2 = 0 \text{ at } x = \pm K \tag{9}$$

where

$$\tau_2 = \sigma' z \tag{10}$$

and z is the coordinate in the thickness direction.

Solving equation (5) with the boundary condition (9) and (10) simultaneously, we have

$$\sigma_{x2} = \sigma_o^2 \left[1 - \frac{\alpha \cosh k_1 X \cos k_2 X}{\beta} - \frac{\sinh k_1 X \sinh k_2 X}{\beta}\right] \tag{11}$$

where

$$\alpha = \frac{k_2 \sinh k_1 K \cos k_2 K + k_1 \cosh k_1 K \sin k_2 k}{k_2 \cosh k_1 K \sin k_2 K - k_1 \sinh k_1 K \cos k_2 K} \tag{12}$$

$$\beta = \cosh k_1 K \cos k_2 K + \sinh k_1 K \sin k_2 K \tag{13}$$

and

$$k_1,\ k_2 = \sqrt{1/2(b^2 \pm a^2)} \tag{14}$$

The maximum value of σ_{x2} occurs at x = 0. Therefore, the new crack will occur there when

$$\sigma_{x2} = X_T = \sigma_2^o\ [1 - \alpha/\beta] \tag{15}$$

This is the general form of stress state from which the average crack spacing K can be obtained with the assistance of equations (2) and (4).

For the present laminate, the results are shown in Figure 8 where the stacking sequences $[0_2/90]_s$ and $[0_2/90]_{2s}$ are used to simulate the present laminate.

From Figure 8, we see the trend of analytical prediction is similar to the experimental result. The average crack spacing is reduced rapidly right after the first ply failure stress is exceeded, and then more gradually afterwards. Before the applied stress reaches 400 MPa, the experimental data are higher than the theoretical predictions, and when the applied stress is greater than 400 MPa, the experimental data are between the two predictions. Note that 400 MPa is the stress level at which the +45° ply cracks occurred.

Because we use $[0_2/90]_s$ to simulate $[0_2/90/\pm 45]_s$ and this is done by the ±45 plies taking no load, and represents the upper bound of the average crack spacing.

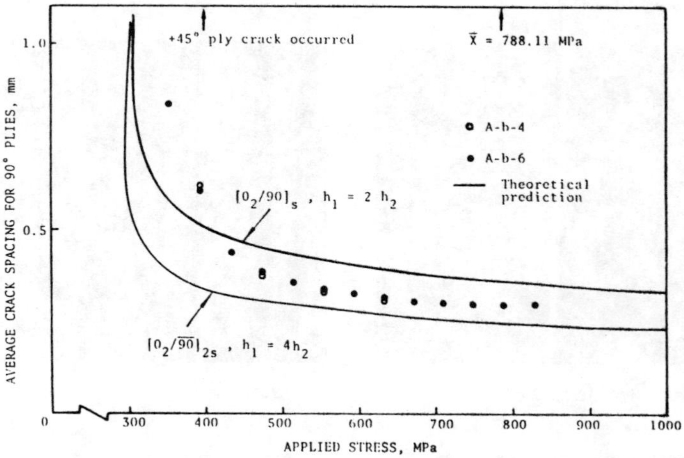

Fig. (8) - The correlation between average crack spacing and applied stress for 90° ply

In contrast to this, the use of $[0_2/90]_{2s}$ to simulate $[0_2/90/\pm45]_s$ represents the lower bound of average crack spacing because we use the 0° plies to substitute for the ±45 plies. Figure 8 shows this point obviously.

The present results indicate that the crack spacing reaches a limiting value where the applied stress is about 1,000 MPa. But the strength of the present laminate is only 870 MPa; therefore, the limiting state will not be reached for the present laminate.

We can use this result to explain in Figure 8 why the number of 90° ply cracks in specimen A-b-4 appears to be increasing whereas the number of 90° ply cracks in specimen A-b-6 appears to be near an equilibrium state. Note that specimen A-b-4 is relatively weaker ($0.89\bar{X}$) and specimen A-b-6 is relatively stronger ($1.08\bar{X}$).

Thus, in the present laminate, no CDS occurs. The same result was also obtained for quasi-isotropic Gr/Ep laminate in [4] and [27]. It is obvious that a conflict exists between the present study and [1] in both experimental results and interpretation, and that further investigation merits attention.

FATIGUE FAILURE PROCESS

Ply failures in fatigue are similar to those in static tension. However, fatigue cycling causes additional cracks, especially in -45° plies. Figure 9 (a-f) shows the sequence of crack occurrence and delamination at different cycles for the maximum fatigue stress of $0.70\bar{X}$. Some of these cracks were linked to those in the adjacent plies, and then axial cracks between plies occurred; this is the initiation of delamination. The typical cracking pattern just described is photomicrographically shown in Figure 9 (b and c).

The axial cracks propagate in cycling, and then connect to each other and form local delamination. Delamination is most frequent either between 90° and

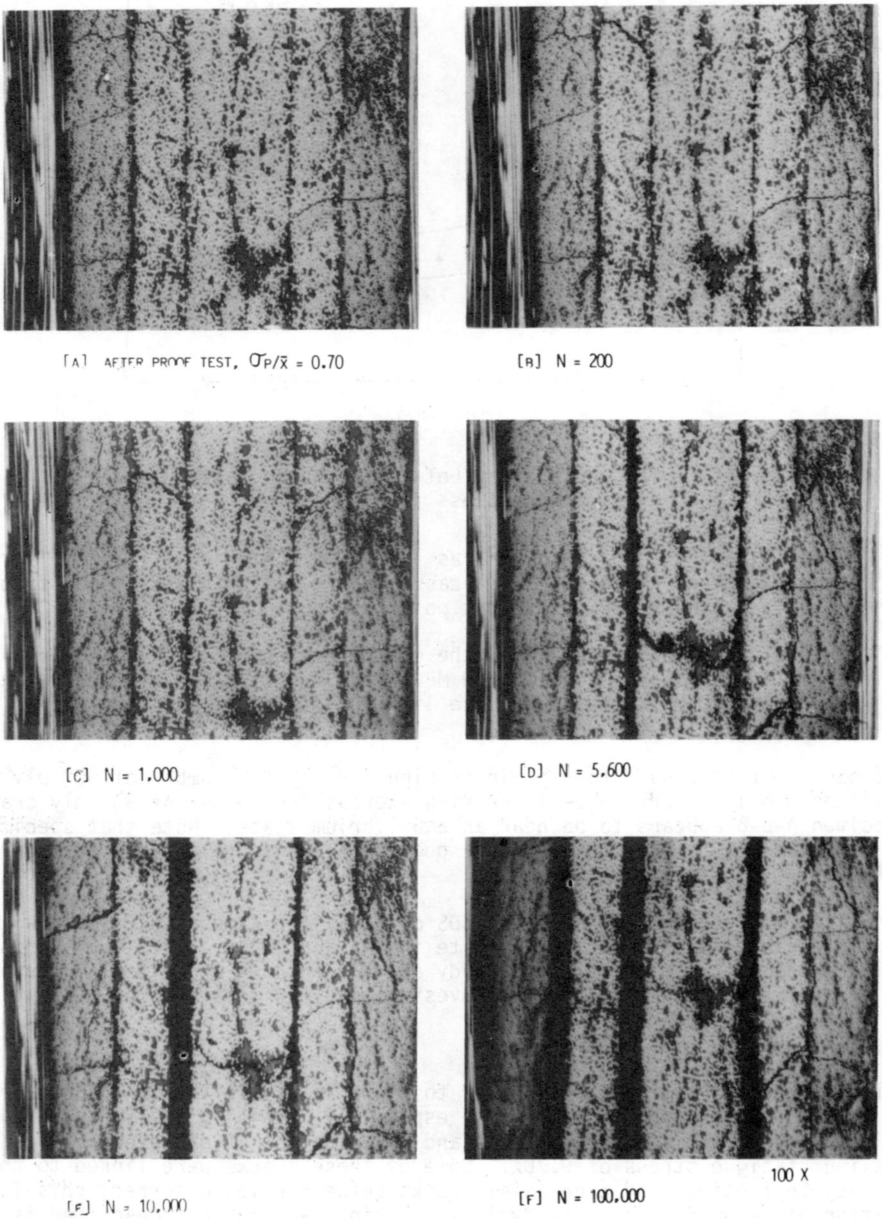

Fig. (9) - Fatigue failure sequence at $S/\overline{X}$ = 0.70, $\sigma_p/\overline{X}$ = 0.70, fatigue life N_F = 705,438 cycles

+45° plies or between +45° and -45° plies. Delamination is not found between the two -45° plies, probably because of no interlaminar shear stress. Interlaminar shear stress exists in other interfaces, especially +45°/-45° interface.

Local delaminations propagate and widen, and a fully developed delamination takes over in a fairly early stage of fatigue life. Figures 10 and 11 are each a series of photographs from replicates which show the growth process of cracks and delamination.

Figure 12 shows the 0° fiber breaks on the polished surface at 50,000 cycles at $0.70\overline{X}$. These figures clearly show 0° fiber breaks, a piece of 0° ply torn out, and axial cracks in 0° plies. A few 0° fiber breaks occurred after the first cycle, and more breaks did not occur until 50,000 cycles was reached, when the 0° fiber damage phenomena occurred. The fibers inside the specimen were examined by burning the specimen and removing the epoxy. Even after 10^6 fatigue cycles, very few fiber breaks exist in 0° plies, Figure 13. The 0° fiber breaks are found only at the edges of the specimen, Figure 13. Therefore, the 0° fiber breaks observed at the polished surfaces are believed to have resulted from the damage which was produced during specimen preparation or the polishing process.

It is believed that the final fatigue failure is due to abrupt fracture of 0° fibers. When the laminate residual strength is lower than the applied stress, the laminate then fails.

CRACK CHARACTERISTICS IN FATIGUE

In this section, crack characteristics including crack density, average crack spacing, saturated number of cracks and 90° ply thickness effect, are investigated for different fatigue stress levels.

The crack density and average crack spacing in each off-axis ply for different fatigue stress levels are plotted as functions of fatigue life in Figures 14 and 15, respectively. In all cases, the crack density reaches a steady state in fairly early stage of fatigue life, and the onset of delamination seems to precede the stabilization of crack multiplication. Table 4 shows the correlation between the number of cycles when noticeable delamination occurs, N_{DL}, and the percentage of crack density at N_{DL} over the final saturation crack density. It is found that the noticeable delamination occurred about 2.5% life regardless of the fatigue and proof stress levels. The crack density in each off-axis ply almost reached saturation. Note that the effect of proof test on both residual strength and residual life has been found negligible by Hahn and Hwang, and additional ply failures caused by proof test are not critical to the final failure [28].

Ryder et al [29] have reported that when significant visible delamination occurred, the remaining cycles to failure could be either small or large. However, the variation range of N_{DL}/N_F in the present study is very small compared to the data in [29] and it seems to have a consistent result in the present study.

Table 5 shows the correlation between saturation crack density (and equilibrium crack spacing and total crack number per centimeter) and fatigue stress level including the proof test results. The coefficient of variation (C.O.V.) for 90° and +45° plies is under 7% and is 15% for -45° plies, while the total crack density data show the C.O.V. better than the data of each off-axis ply. Figure 16

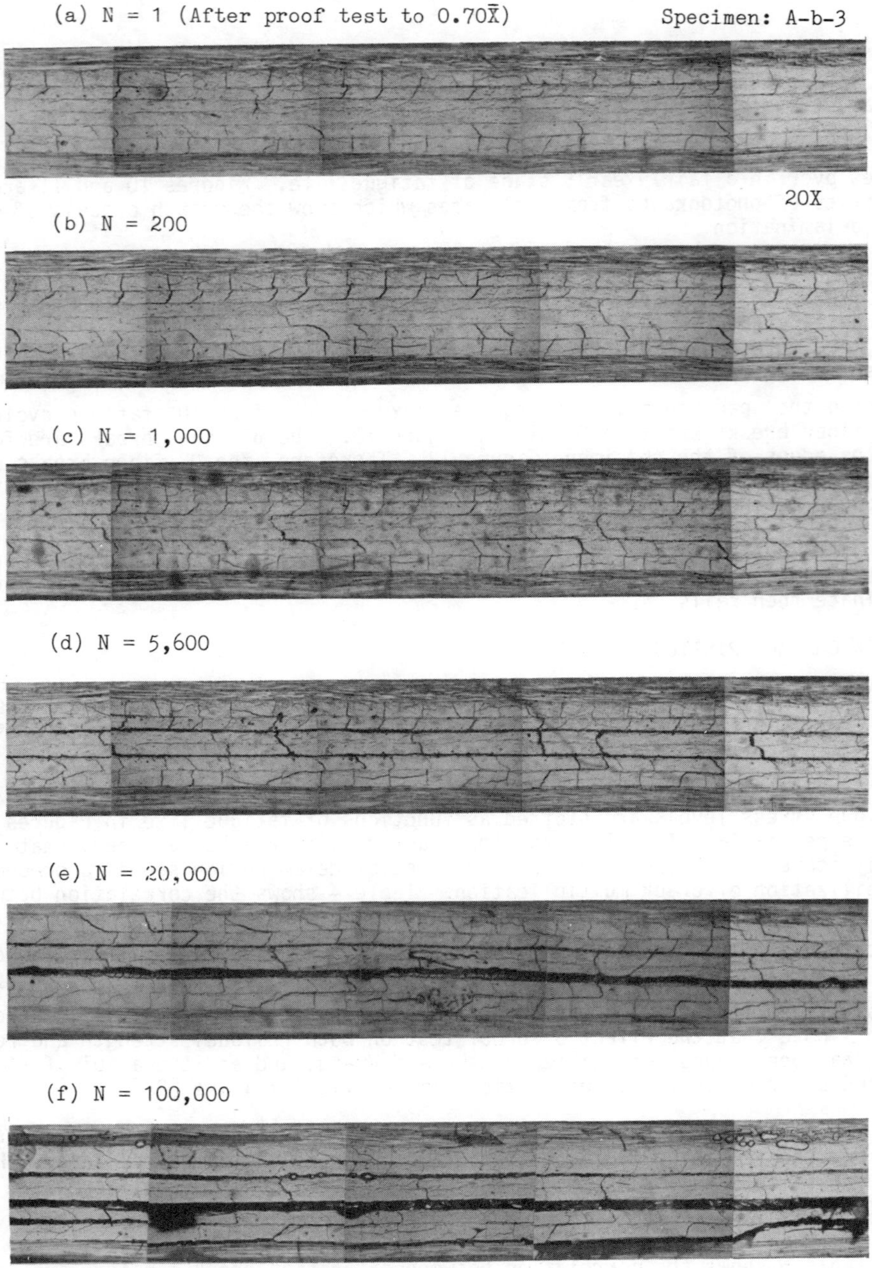

Specimen failed at 705,238 cycles

Fig. (10) - Replication inspection of crack multiplication and delamination propagation in fatigue ($S/\overline{X} = 0.70$)

(a) N = 1 (After proof test to $0.80\bar{X}$) Specimen: A-b-2

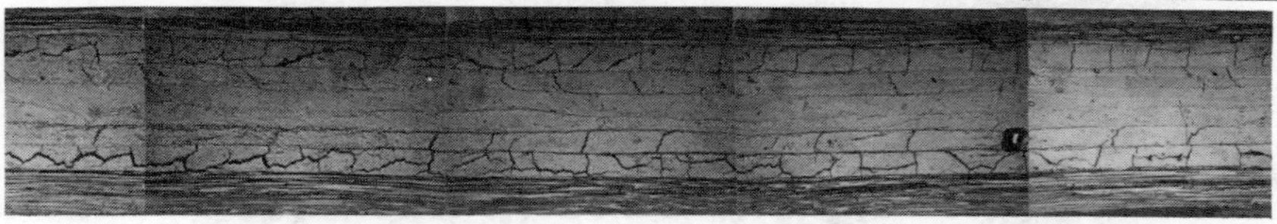

20X

(b) N = 100

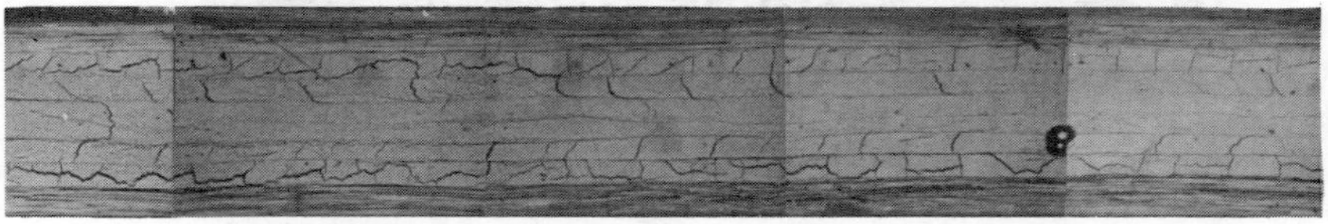

(c) N = 500

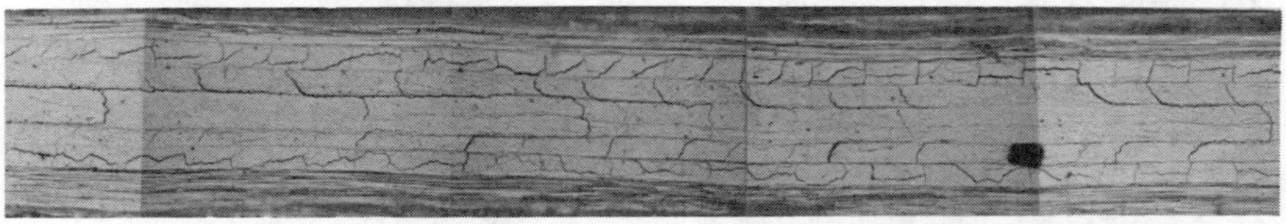

(d) N = 1,000

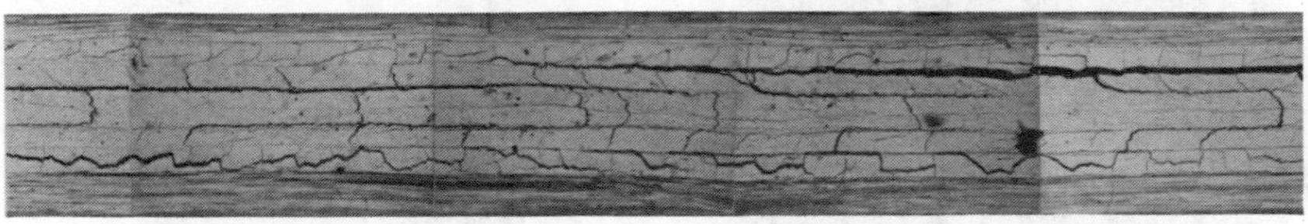

(e) N = 2,000

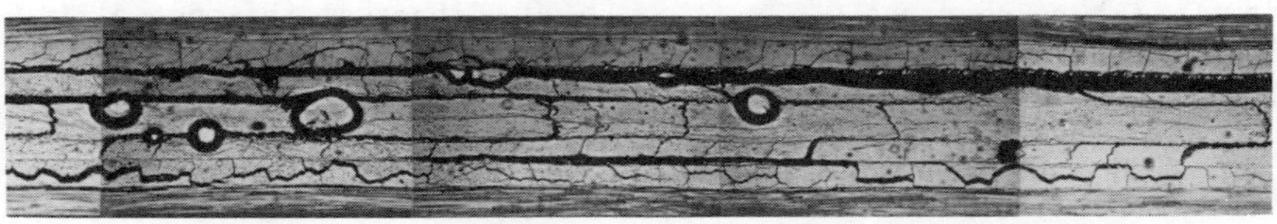

Specimen failed at 83,201 cycles

Fig. (11) - Replication inspection of crack multiplication and delamination propagation in fatigue ($S/\bar{X}$ = 0.80)

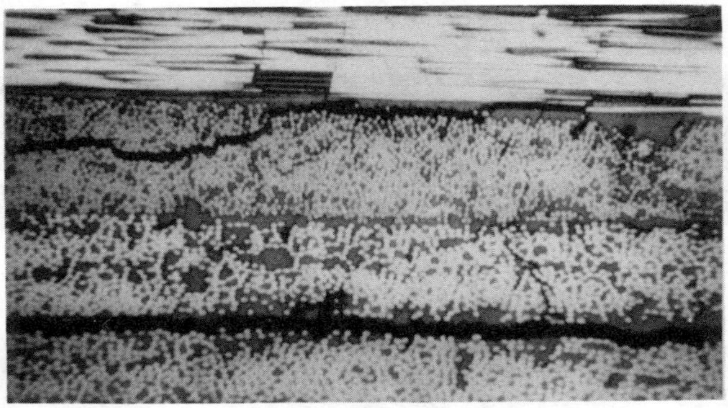

N = 50,000 cycles Specimen: A-b-3 100X

Fig. (12) - 0° fiber breaks

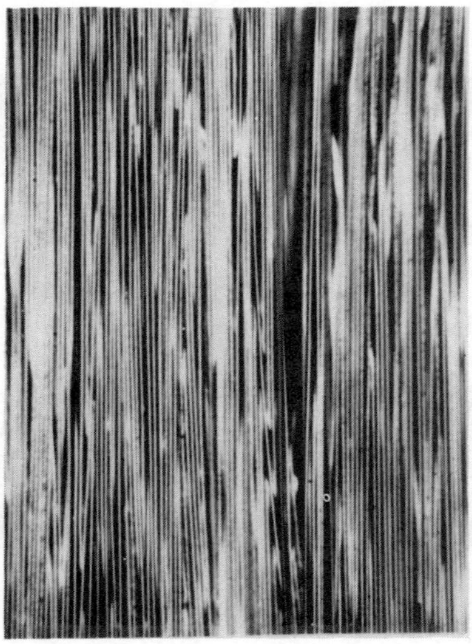

(a) 0° fibers near the 0°-90° interface and at the middle of the specimen

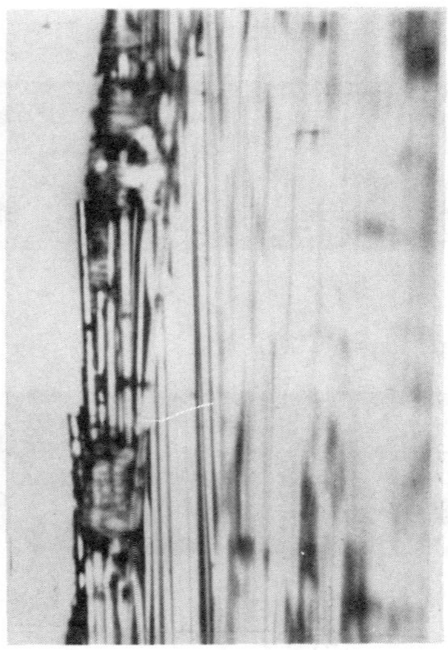

(b) 0° fibers at the edge of the specimen

Fig. (13) - 0° fiber examination by burning the surviving specimen

shows the correlation between saturation crack density in two plies of the same angle and fatigue stress level. Because of the low C.O.V., we can use a horizontal line to present the saturation crack density regardless of the fatigue stress. The data in [1] and [4] are also plotted in this figure. Reference [1] reports that the total number of cracks decreases when the fatigue stress is greater than $0.67\overline{X}$, and that below this stress the total number of cracks is in-

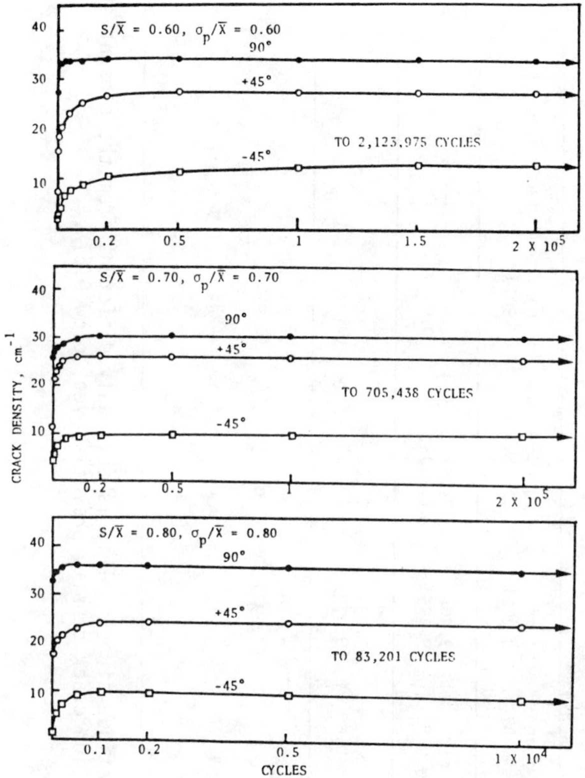

Fig. (14) - Crack density increasing with fatigue cycles

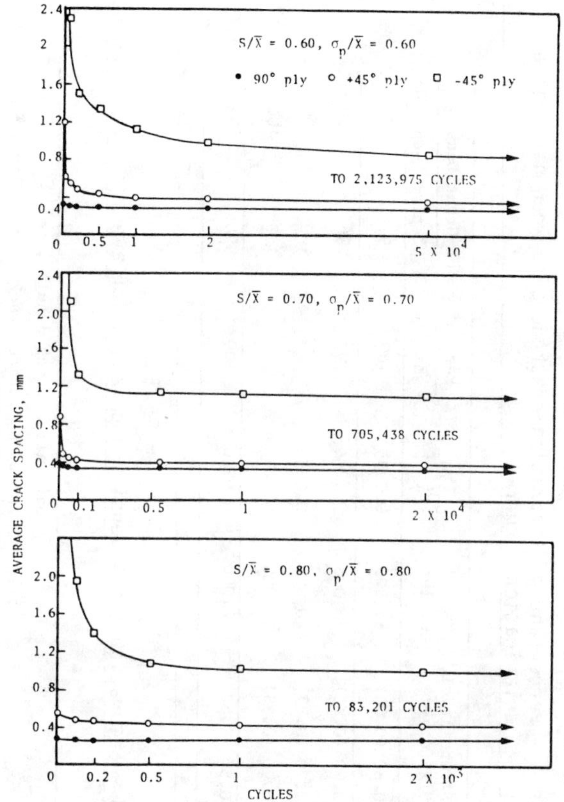

Fig. (15) - Average crack spacing decreasing with fatigue cycles

Table 4 - Percentage of saturation crack density when noticeable delamination occurred

Stress level, S/X (σ_p/X)	N_{DL} Cycles	N_{DL}/N_F	Percentage of Crack Density		
			90° ply	+45° ply	-45° ply
0.60 (0.60)	50,000	0.0235	100	97	82
0.70 (0.70)	20,000	0.0284	99	96	100
0.70 (0.87)	40,000	0.0250	100	100	100
0.70 (0.95)	30,000	0.0286	100	100	100
0.80 (0.80)	2,000	0.0240	100	98	92

Table 5 - Saturation crack density and saturation average crack spacing

Stress level	Saturation crack density, cm^{-1} (Saturation average crack spacing, mm)			
	90° ply	+45° ply	-45° ply	Total crack no. per cm
S/X = 0.60 σ_p/X = 0.60	34.25 (0.292)	27.95 (0.358)	13.39 (0.747)	151.2
S/X = 0.70 σ_p/X = 0.70	30.71 (0.326)	26.77 (0.374)	8.66 (1.155)	132.3
S/X = 0.70 σ_p/X = 0.87	34.25 (0.292)	22.80 (0.438)	11.03 (0.907)	136.2
S/X = 0.70 σ_p/X = 0.95	33.90 (0.295)	24.81 (0.403)	11.81 (0.847)	140.9
S/X = 0.80 σ_p/X = 0.80	35.83 (0.279)	24.80 (0.403)	10.24 (0.977)	141.7
Average	3.379 (0.297)	25.32 (0.395)	11.03 (0.927)	140.5
C. O. V., %	4.97	6.96	14.29	4.52

dependent of fatigue stress. This is obviously different from the present study where a saturation crack density prevails even below $0.67\overline{X}$.

In Figure 14, the number of cracks reaches a steady state quite early and no further cracks occur until failure. Reifsnider et al [1] refers to this limiting state as a CDS, as mentioned before. No existence of CDS was found in static failure but CDS does exist in fatigue loading. We had already shown in the preceding discussion that the equilibrium state occurred around 0.025 N_F for the present laminate. Thus CDS could be a fatigue property for multi-directional laminate.

Reifsnider [1] found the number of cracks in 90° plies was a function of the total thickness of contiguous 90° plies. Kim [4] and Wang et al [15,23] confirmed this observation. Table 6 summarizes between saturation crack spacing in 90° plies and percentage of contiguous 90° plies. The smooth curves are obtained from the available reports. That more cracks occurred in fatigue tests can also be seen from Figure 17.

MODULUS CHANGE IN FATIGUE

The data for modulus change at stresses 0.60, 0.70, and 0.80$\overline{X}$ in fatigue are plotted against cycles and cycles ratio in Figures 18 and 19, respectively. It

Table 6 - Effect of 90° ply thickness on cracks

(A) Static test

Stacking sequence	Percentage of 90° plies	Saturation crack spacing of 90° ply, mm	Reference
$[0_2/90/\pm45]_s$	20	0.318	Present study
$[0/90/\pm45]_s$	25	0.393	[4]
$[0/90/\pm45]_s$	25	0.413	[4]
$[\pm25/90_2]_s$	50	1.887	[15]
$[\pm25/90_3]_s$	60	2.50	[15]

(B) Fatigue test

Stacking sequence	Percentage of 90° plies	Saturation crack spacing of 90° ply, mm	Reference
$[0_2/90/\pm45]_s$	20	0.296	Present study
$[0/90/\pm45]_s$	25	0.33	[1]
$[0/90/\pm45]_s$	40	0.53	[1]
$[0/90]_s$	50	0.81	[1]

Note: The data are all for Gr/Ep laminate.

is seen from Figure 18, that the longer the life, the greater the modulus change and the later the occurrence of rapid modulus change. No abrupt modulus change was observed just before the final failure.

For low stress levels, the modulus change occurred in the fairly early stage, then remained constant until final failure. For high stress fatigue, e.g., $S/\overline{X}$ = 0.80, if the specimen is weak, e.g., specimens A-b-12 (N_F = 267 cycles), A-b-13 (N_F = 887 cycles) and A-b-14 (N_F = 8,075 cycles), there is no time for delamina-

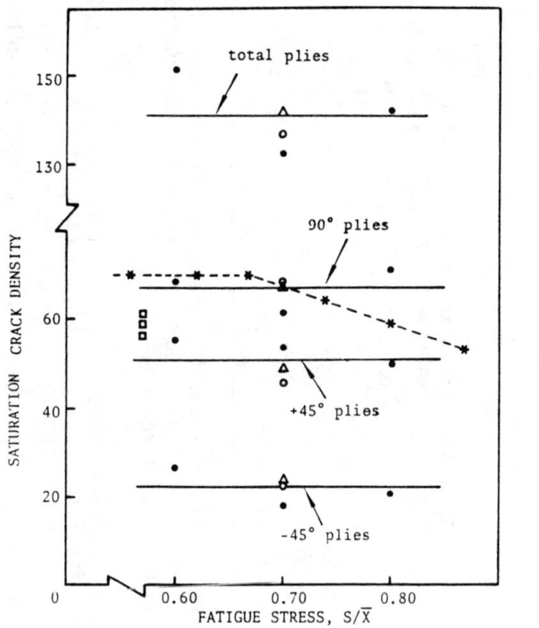

• $\sigma_p = S$ ○ $\sigma_p/\overline{X}=0.87$ △ $\sigma_p/\overline{X} = 0.95$ for present study

* Reference [4], total number of cracks data for $[0/90/\pm45]_s$ T300/5208 laminate

□ Reference [1], total number of cracks data for $[0/90/\pm45]_s$ AS/3501 laminate

Fig. (16) - Correlation between saturation crack density and fatigue stress

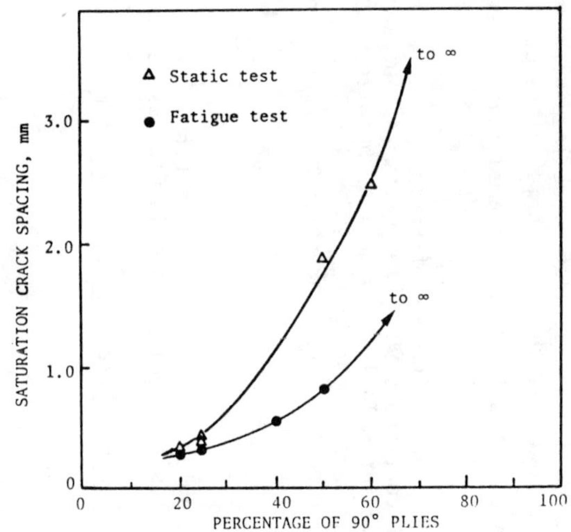

Fig. (17) - Correlation between equilibrium crack spacing of 90° plies and percentage of 90° plies

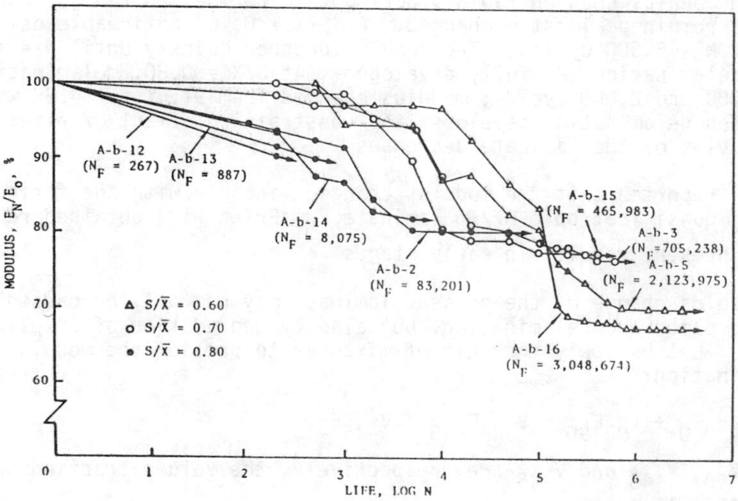

Fig. (18) - Correlation between modulus and fatigue life

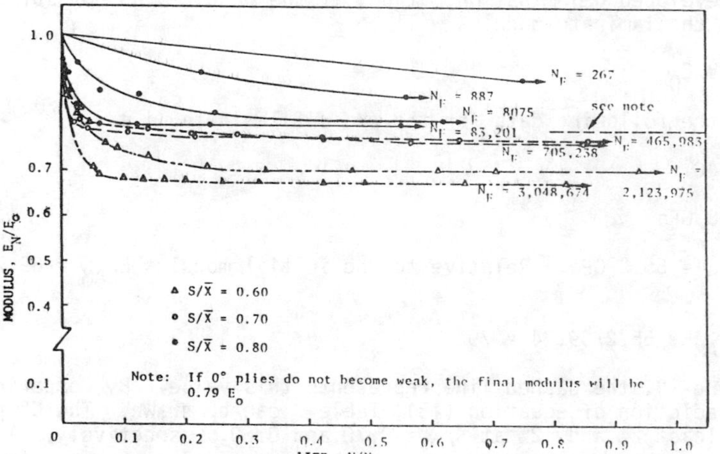

Fig. (19) - Correlation between modulus and cycle ratio

tion to develop fully; therefore, the modulus appears to keep decreasing. But for a strong specimen, e.g., specimen A-b-2 (N_F = 83,201 cycles), the modulus curve looks more like that at low stresses. The lower the fatigue stress, the later the occurrence of 0° fiber breakage observed on the polished surface.

Usually, the modulus change depends considerably on delamination. For example, at $S/\overline{X}$ = 0.60, delamination grew significantly between 10,000 and 150,000 cycles, and modulus changed rapidly during this period. After 150,000 cycles, the modulus remained almost unchanged. At $S/\overline{X} = 0.70$, noticeable delamination occurred at N = 5,600 cycles. The modulus dropped quickly until N = 100,000 cycles when delamination was fully developed. At $S/\overline{X} = 0.80$, delamination occurred between 1,000 and 2,000 cycles; modulus changed from 91.6% to 84.3% within the period. When delamination develops, the constraining effect of plies is reduced, and the modulus of the laminate decreases.

Kim [4] reported that the modulus changed linearly with the fatigue life for $[0/90/\pm45]_s$ quasi-isotropic Gr/Ep laminate. O'Brien [19] obtained results similar to the present results in early stages.

The modulus change of the present laminate may not only be caused by failure of off-axis plies and delamination, but also by degradation of 0° plies caused by fatigue. Let us apply the rule of mixtures to predict the modulus E_{LAM} after full delamination:

$$E_{LAM} = V_0E_0 + V_{90}E_{90} + V_{+45}E_{+45} + V_{-45}E_{-45} \tag{16}$$

where V_0, V_{90}, V_{+45} and V_{-45} are, respectively, the volume fractions of 0°, 90°, +45° and -45° plies.

Suppose that 90° and ±45° plies cannot take any load because of ply failures and fully developed delamination. Then, if the 0° plies do not suffer any modulus change, the laminate modulus will be

$$E'_{LAM} = V_0E_0 \tag{17}$$

Now, using the following data for the present laminate, i.e.,

$$V_0 = 0.40$$

$$E_0 = 138 \text{ GPa}$$

We find $E'_{LAM} = 55.2$ GPa. Relative to the initial modulus $E_{LAM} = 69.74$ GPa, we have

$$E'_{LAM}/E_{LAM} = 55.2/69.74 = 79\% \tag{18}$$

In Figure 19, the dashed line represents this value. By comparing the test data and prediction of equation (18), Table 7 can be drawn. The 0° ply modulus changed at least 2% and 12% at $S/\overline{X} = 0.70$ and 0.60, respectively. In fact, the 0° ply modulus change should be greater than these values because the 90° and ±45° plies can still take portions of applied load.

Delamination of composite laminates has been studied by several investigators [1,8,15,19,20,23]. In the present study, a simple model presented by O'Brien [19] is employed to investigate the correlation between modulus change and delamination growth in fatigue. Delamination growth has been idealized as shown in Figure 20, and the modulus, E, of a partially delaminated specimen can be correlated to the delamination ratio by the following equation.

$$\frac{E-E_{LAM}}{E^*-E_{LAM}} = \frac{a}{b} \tag{19}$$

Table 7 - Modulus change of 0° ply

Fatigue stress $S/\bar{X}$	Specimen No.	Final modulus E'_{LAM}/E_{LAM}	E'_{LAM}/E_{LAM} predicted by Eq. [18]	0° ply modulus change
0.60	A-b-5	70.2%	79.09%	12.1%
	A-b-16	67.6%		14.5%
0.70	A-b-3	76.5%	79.09%	3.27%
	A-b-15	77.5%		2.01%
0.80	A-b-2	80.5%	79.09%	--
	A-b-14	82.0%		--
	A-b-12	89.9%		--
	A-b-13	90.1%		--

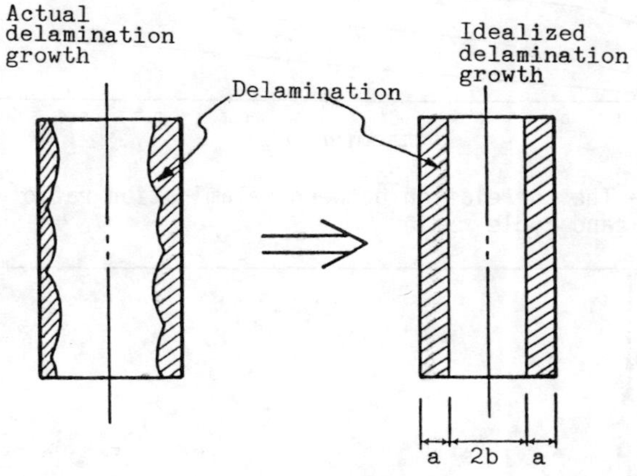

Fig. (20) - Schematic diagram for idealized delamination growth

where a/b is defined as the delamination ratio and E* is the modulus of a laminate completely delaminated along one or more interfaces.

In the present study, the values of E_{LAM}, E* and E have been obtained experimentally. Thus, a correlation between the delamination ratio a/b and fatigue life can be obtained from equation (19) with knowledge of E, E* and E_{LAM} determined by experiments.

The correlations between delamination ratio a/b, modulus change E/E_{LAM} and average crack spacing versus fatigue life are shown in Figures 21 and 22. In Figure 21, the delamination ratio shows the same trend as the modulus change. The delaminations will grow through the entire width around 25% of life regardless of the fatigue stress. This is approximately the life at which the modulus remains unchanged.

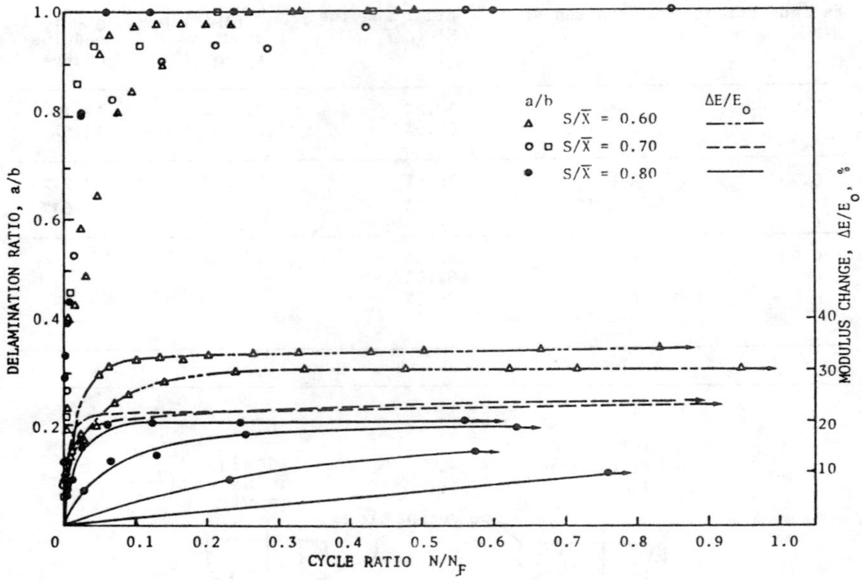

Fig. (21) - The correlation between delamination ratio, modulus change and cycle ratio

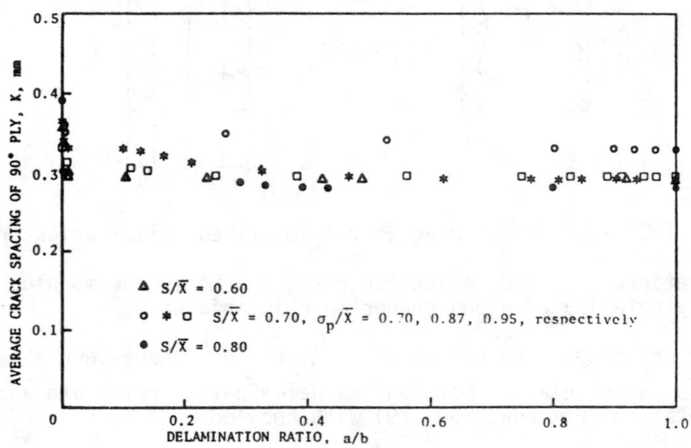

Fig. (22) - The correlation between average crack spacing and delamination ratio

Figure 22 shows a correlation between the average crack spacing in the 90° plies and the delamination ratio. Saturation crack spacing is reached almost immediately at a/b = 0. This means that the 90° ply cracks reach steady state before delamination occurs. This point has been demonstrated in previous section. The same results also can be obtained for the +45° and -45° ply cracks.

CONCLUSIONS

The failure behavior of a $[0_2/90/\pm45]_s$ AS/3501-5A Gr/Ep laminate has been theoretically and fractographically investigated. The failure characteristics, including first-ply failure (FPF), crack multiplication, delamination growth, and 0° fiber breaks, have been identified throughout static and fatigue tests of the laminate using microscopic and replication methods.

Theoretical prediction of FPF stress has been successfully executed by using laminated plate theory including hygrothermal behavior. Average crack spacing also has been properly predicted by modifying the model of Vasil'ev et al. The modulus change with the increasing fatigue cycles also has been obtained and quantitatively related to delamination.

Major findings can be summarized as follows:

(1) The moisture content in the laboratory environment is about 0.56% for the present laminate and will result in about 15 MPa residual stress in the transverse direction of 90° plies. The FPF stress was found to be around 315 MPa which is in excellent agreement with the predicted value (316 MPa).

(2) The failure processes in the present laminate are similar to those in quasi-isotropic laminate. The constituent plies fail in the following order: 90° plies, +45° plies, and -45° plies. In static tension, very few cracks were found in -45° plies and delamination is negligible. However, the final fatigue failure is frequently preceded by extensive delamination which exists between plies with different orientations. Almost no 0° fiber breaks were found away from the polished surface. Therefore, the final failure seems to be the result of localized fracture of the 0° plies.

(3) The replication method can be used to monitor the crack multiplication and delamination growth. In static tests, the number of cracks increases rapidly as the stress is increased from $0.40\overline{X}$ to $0.50\overline{X}$ and further to $0.60\overline{X}$. Above $0.60\overline{X}$, the number of cracks in the 90° plies increases slowly and much more so especially after $0.90\overline{X}$. No existence of characteristic damage state (CDS) for crack multiplication was found in static tension because the strength is not high enough to allow saturation of cracks. But in fatigue, the CDS does exist at a fairly early stage (2.5% life) and crack density is independent of the fatigue stress levels.

(4) The prediction of average crack spacing by a modified model of Vasil'ev et al agrees well with the experimental data in static tests.

(5) The saturation crack density is independent of fatigue stress. The number of cracks in 90° plies is a function of the percentage of contiguous 90° plies. The higher the percentage, the smaller the saturation crack density.

(6) The noticeable delamination occurs around 25% of life regardless of the fatigue stress level. When noticeable delamination occurs, the crack density in off-axis ply almost reaches the CDS, and the modulus changes considerably.

(7) The higher the fatigue stress, the smaller the modulus change before failure. The modulus reduction depends on fatigue life; they are about 20%, 22% and 30-33% for the fatigue stress levels of 0.80, 0.70 and $0.60\overline{X}$, respectively.

(8) Delamination grows through the entire width of specimen before about 25% of life is spent, regardless of fatigue stress level. Thereafter, no obvious modulus change can be found.

ACKNOWLEDGEMENTS

This work was supported by the Materials Research Laboratory, Washington University in St. Louis, Missouri. The authors wish to thank colleagues for their assistance.

REFERENCES

[1] Reifsnider, K. L., Henneke, E. G., II and Stinchcomb, W. W., Technical Report AFML-TR-76-81, Air Force Materials Laboratory, April 1976; Part II, June 1977; Part III, June 1978; Part IV, June 1979.

[2] Broutman, L. J. and Sahu, S., Proc. 24th SPI Conference, 11-D, 1969.

[3] Kim, R. Y. and Hahn, H. T., Journal of Composite Materials, Vol. 13, pp. 2-16, 1979.

[4] Kim, R. Y., Advances in Materials, Proc. 3rd Int. Conf. on Composite Materials, Pergamon, pp. 1015-1028, 1980.

[5] Hahn, H. T. and Tsai, S. W., Journal of Composite Materials, Vol. 8, pp. 288-305, 1974.

[6] Tsai, S. W. and Hahn, H. T., Inelastic Behavior of Composite Materials, C. T. Herakovich, ed., AMD-Vol. 13, pp. 73-96, ASME, 1975.

[7] Hahn, H. T., Technical Report AFML-TR-78-43, Air Force Materials Laboratory, 1978.

[8] Wang, A. S. D., Advances in Materials, Proc. 3rd Int. Conf. on Composite Materials, Pergamon, pp. 170-185, 1980.

[9] Sendeckyj, G. P. and Stalnaker, H. D., Composite Materials: Testing and Design, ASTM STP 617, p. 39, 1977.

[10] Hahn, H. T. and Kim, R. Y., Advanced Composite Materials - Environmental Effects, ASTM STP 658, pp. 98-120, 1978.

[11] Doner, D. R. and Novak, R. C., 24th Annual Technical Conference, The Society of the Plastics Industry, No. 2-D, 1969.

[12] Crossman, F. W., Mauri, R. E. and Warren, W. J., NASA Contractor Report 3189, 1979.

[13] Hahn, H. T. and Pagano, N. J., Journal of Composite Materials, Vol. 9, 1975.

[14] Pagano, N. J. and Hahn, H. T., Composite Materials: Testing and Design, ASTM STP 617, J. G. Davis, ed., p. 317, 1977.

[15] Crossman, F. W., Warren, W. J., Wang, A. S. D. and Law, G. E., Jr., Journal of Composite Materials, Supplement, Vol. 14, p. 88, 1980.

[16] Pipes, R. B. and Pagano, N. J., Journal of Composite Materials, Vol. 4, p. 538, 1970.

[17] Wang, A. S. D. and Crossman, F. W., Journal of Composite Materials, Vol. 11, p. 92, 1977.

[18] Reifsnider, K. L., Henneke, E. G. and Stinchcomb, W. W., Composite Materials: Testing and Design, ASTM STP 617, p. 93, 1977.

[19] O'Brien, T. K., Technical Report NASA TM-81940, 1981.

[20] Wang, A. S. D., Drexel University Report, 1978.

[21] O'Brien, T. K., Ph.D. Dissertation, Virginia Polytechnic Institute and State University, Virginia, 1978.

[22] O'Brien, T. K. and Reifsnider, K. L., Journal of Composite Materials, Vol. 15, p. 55, 1981.

[23] Wang, A. S. D. and Crossman, F. W., Journal of Composite Materials, Supplement, Vol. 14, p. 71, 1980.

[24] Nguyen, N. Q. and Kardos, J. L., Technical Report AFML-TR-79-4035, Air Force Materials Laboratory, 1979.

[25] Tsai, S. W. and Hahn, H. T., Introduction to Composite Materials, Technomic, Inc., 1980.

[26] Vasil'ev, V. V., Dudchenko, A. A. and Elpat'evskii, A. N., Translated from Mekhanika Polimerov, No. 1, pp. 144-147, 1970.

[27] Mills, J. S., Herakovich, C. T. and David, J. G., Jr., Technical Report, VPI-E-79-35, 1979.

[28] Hahn, H. T. and Hwang, D. G., Composite Materials: Testing and Design, Sixth Conference, ASTM STP 787, pp. 247-273, 1982.

[29] Ryder, J. T. and Walker, E. K., Technical Report AFML-TR-76-241, Air Force Materials Laboratory, 1976.

THE IMPROVEMENT OF NOTCHED STRENGTH PREDICTION FOR LAMINATED COMPOSITES

T.-H. G. Hu

AeroStructures, Inc.
Arlington, Virginia 22202 USA

and

D. G. Goetschel

Corning Glass Works
Corning, New York 14831 USA

ABSTRACT

Two modifications of the Average Stress Criterion [1] are proposed to improve the accuracy of notched strength prediction for composites. The first one is to insert a smooth continuous weighting function into the equation for the calculation of average stress. The other is to use an area integral instead of a line integral to calculate the weighted average stress.

The experimental data from uniaxial loading condition are used to assess the validity and accuracy of these modifications. A large-scale testing program (850 specimens) are conducted to generate the experimental data. These specimens are made of Fiberite HY-E 3048 AIK unidirection prepreg tape consisting of Union Carbide T-300 graphite fibers and Fiberite's low temperature 948A1 epoxy resin. Five replicates for each configuration seem to be an optimum compromise between expense and accuracy (90% confidence level and 10% error). Notch type, notch size and lay-up are parameters for this investigation.

The theoretical stress distribution of notched specimen is obtained from a finite element analysis. The most suitable of the proposed modifications can be determined by best fitting to all experimental data.

Overall, the Average Stress Criterion gives reasonably good strength prediction for notched composite laminates. But the best correlation with experimental data is obtained by performing a weighted average of the stress distribution. Identical results are obtained from failure criteria applied at ply level and at laminate level for specimen under uniaxial stress (i.e., the laminates with symmetric type of notch subject to uniaxial loading).

POSITION OF PROBLEM

Notched strength is one of the subjects on which extensive research has been conducted for composites. Different assumptions, concepts, and criteria were introduced to describe the characteristics of notched strength: Point Stress Criterion and Average Stress Criterion by Whitney and Nuismer [1], Intense Energy Region Assumption by Waddoups, Eisenmann and Kamaniker [2], Inherent Flow Concept by Cruse [3], Modified Point Stress Criterion by Karlak [4], another Modified Point Stress Criterion by Pipes, Wetherhold and Gillespie [5], Micromechanic Criterion by Mar and Lin [6], Progressive Failure Criterion by Lo, Wu and Konish [7], Strain Failure Criterion by Poe [8], etc. The detail review has been done by Awerbush and Madhukur [9]. From these works, some conclusions can be drawn:

1. Stress concentration factor is not the only dominant factor to prescribe the notched strength of composite structures.

2. Good agreement between all models and testing data can be achieved provided that the parameters are properly determined.

3. The strength does drop sharply with the introduction of a small notch in almost all laminates. The Average Stress Criterion model fits the data better in the small notch size range.

4. The Characteristic Dimensions of the Point Stress Criterion and its modifications as well as the Average Stress Criterion are not pure material constants. However, some successful applications have been reported.

5. Only symmetric notches (with respect to loading axis) are used for experimental investigation. Center hole and center crack are the two most common.

6. The statistical reliability of the test data has either not been reported or not been conducted.

7. The application of LEFM (Linear Elastic Fracture Mechanics) in composites is inconclusive, even though some successful correlations with testing data from fracture model is reported.

The task of current report is to construct a data basis with a certain level of confidence to assess two modifications on the Average Stress Criterion with assumption of constant Characteristic Dimension. Also, the LEFM will not be used, in other words, cracks will be modified as a slot or slit with a finite radius at the end.

PROPOSED MODELS

First, the Point Stress Criterion (PSC) and the Average Stress Criterion (ASC) are introduced. (See Figure 1). Point Stress Criterion assumes that failure occurs when the stress over some distance, d_o (Characteristic Dimension), away from the discontinuity is equal to or greater than the unnotched laminate strength, i.e.,

$$\sigma_o = \sigma|_{x=(R+d_o)}$$

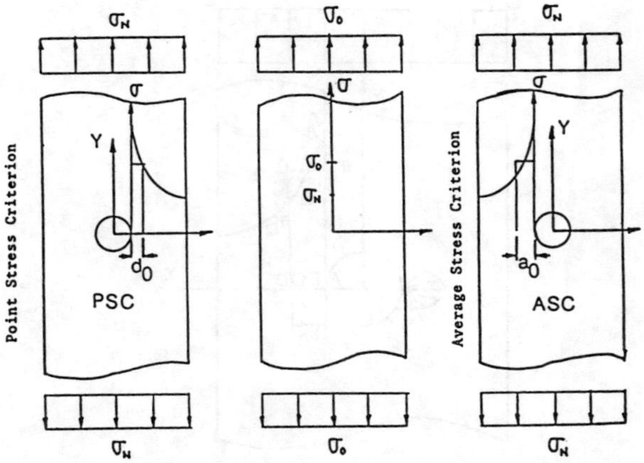

Fig. (1) - The point stress criterion and the average stress criterion

Average Stress Criterion assumes that failure occurs when the average stress over some distances, a_o (Characteristic Dimension), equals the unnotched strength, i.e.,

$$\sigma_o = \frac{\int_R^{R+a_o} \sigma dx}{a_o}$$

The Characteristic Dimensions are also assumed to be material constants for both criteria.

In the following proposed modifications of Average Stress Criterion, composite material is assumed to be homogeneous but anisotropic. The average laminate stresses are used.

Modification PA1. A smooth, continuous weighting function is inserted into the equation for the calculation of average stress, i.e.,

$$\sigma_a = \frac{\int_R^{R+a_o} \sigma f(x)dx}{\int_R^{R+a_o} f(x)dx}$$

The supportive argument is as follows: (see Figure 2).

Consider four points A, B, C, D in the highly stressed area. Points A, B, and C are located on the path of the line integral, and point D is infinitely close to point C, but it does not contribute to the integral. If we carefully examine the situation, we can find that the stress difference between point C and D is small, yet the stresses at points A, B and C contribute equally to the calculation of weighted stress at all. This is not reasonable. Inserting a smooth, continuous weighting function into the equation for the calculation of

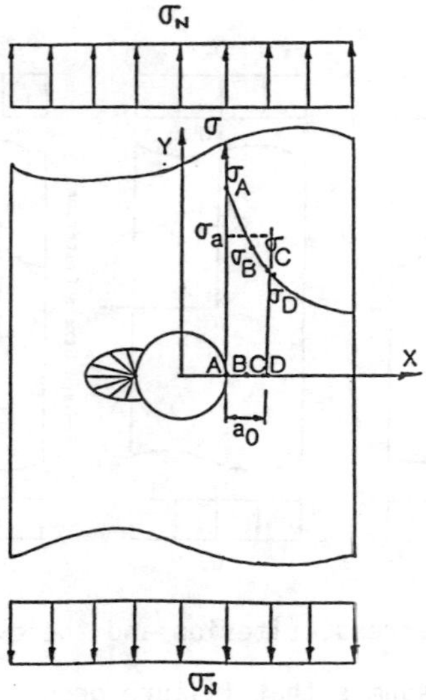

Fig. (2) - Supportive plot for modifications

average stress is expected to improve the accuracy of the predictions.

Modification PA2. An area integral instead of the line integral is used to calculate the weighted average stress, i.e.,

$$\sigma_a = \frac{\int_A \sigma(x,y)g(x,y)dA}{\int_A g(x,y)dA}$$

The Average Stress Criterion assumes that stress distribution is as important as the stress concentration itself. The decay direction is not necessarily limited to one direction, see Figure 2. Area integral might yield better results and also eliminate the possible difficulty in determining the integral path for notched laminates under multi-axial stresses.

The weighting function and corresponding characteristic dimension are assumed to be material constants, independent of laminate construction, notch type and notch size. With certain types of weighting functions, these two modifications can be reduced to be the Point Stress Criterion, the Average Stress Criterion and Conventional Stress Concentration Factor Failure Criterion.

Another possible physical explanation for these modifications is as follows: It is well known that composites can redistribute stress concentrations [10] and in practical situations, this exact stress distribution is unknown. However, the approximate stress distribution corresponding to a linear elastic material (i.e., ignoring the stress redistribute) can be readily calculated using Finite Element Method. The weighting function can be seen as a way of correcting this approximate stress distribution so that the product of the approximate stress distribution and the weighting function would be closer to the actual stress distribution.

TEST PROGRAM AND TEST RESULT

Experimental data with high level of confidence is the best tool to assess the validity and accuracy of the various proposed criteria. A large scale testing program has been designed to provide testing data with high confidence level for composites.

The scatter in the experimental data could be reduced by two means:

1. Use of 5 testing replicates for each configuration. With assumption of a 2-parameter Weibull model, the relationship between the number of specimens and the confidence level in term of the shape parameter can be obtained [11]. Five replicates yield a 95 confidence level and 5% error with a shape parameter of 24 or a 90% confidence level and 10% error with a shape parameter of 10. Five replicates of each specification seems to be the optimum compromise between accuracy and expenses. The shape parameter is about 12 [12] from the data analysis for the current testing program.

2. Reducing possible error sources. All the specimens in this testing program are manufactured from the same material batch, cured, machined, and notched using the same procedure to avoid possible manufacturing and testing error.

The specimens were made of Fiberite HY-E 3048 AIK unidirection prepreg tape consisting of Union Carbide T-300 Graphite fibers and Fiberite's low temperation 948 Al epoxy resin.

Following the recommended procedure of fabrication and curing, a 12" by 12" plate of a particular laminate configuration is made. Then, strips with dimensions of 12" by 1.000 + 0.010" are obtained by using a carbide band saw, and a power sander with abrasive papers of different grades. There were a total of 13 different laminate constructions.

Various notch types of circular hole, slant slot, slant slit, double edge, semi-hole, double edge slit and shoulder (see Figure 3), are machined in with proper procedure and tools (diamond core drill, 4-flute carbide end mill with diameter of 0.065" and jewelry blades with thickness of 0.016").

End tabs cut from glass fiber laminate were bonded on each end of the specimens with super glue. To reduce the possible stress concentrations from the grips, an EPON 828 resin was put on the transition portion where the tabs taper down to the specimen.

All specimens were tested on the INSTRON Testing Machine, Model 1333. These specimens were subjected to uniaxial extension by clamping both ends in the friction fixture. Stroke control mode was chosen due to safety considerations. The extension speed was about 1 mm/min (0.00066 in/sec).

The response of each specimen was recorded in the form of a load-extension plot. The detail testing configuration and mean strength of the specimens are shown in Table 1 to Table 7. Material properties of the lamina is shown in the Appendix. Detail failure process and data on each test data are recorded [12].

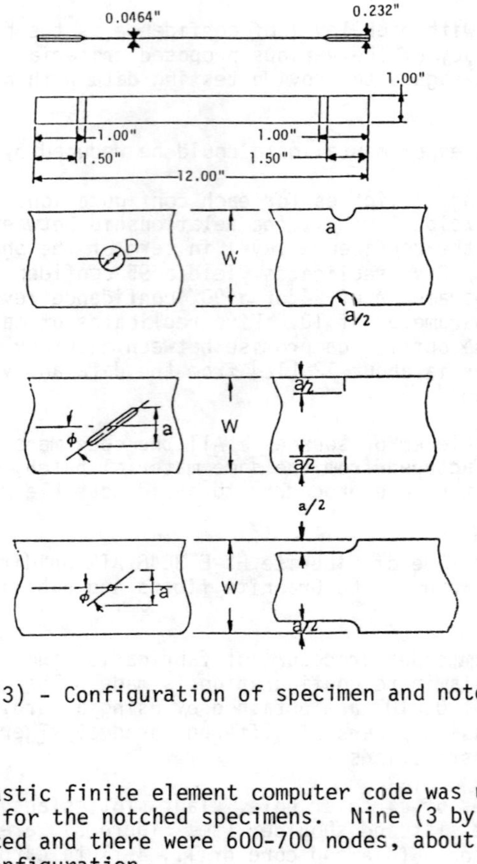

Fig. (3) - Configuration of specimen and notch type

COMPUTER SIMULATION

A 2-D linear elastic finite element computer code was used to obtain the stress distribution for the notched specimens. Nine (3 by 3) noded Lagrangian elements were selected and there were 600-700 nodes, about 1300 DOF (degree of freedom) for each configuration.

The Lagrangian polynomials are used to calculate the weighted average stress for the line integral, and Gaussian integration is used for the area integral. The weighting functions for both modifications and the shape of the characteristic area of modification PA2 are shown in Figure 4 to Figure 6. It is noted that the options on choice of weighting functions is infinite. The testing data from circular hole cases will be used to determine the suitable weighting function and the corresponding characteristic dimension. Then the data from other notch types will be used to justify the proposed modifications. The error index (RASE) is constructed as the basis of these assessments.

$$RASE = \left[\frac{\sum_{i=1}^{N} ER_i^2}{N}\right]^{1/2}$$

where N is the total data points, and

Table 1 - Mean failure stress (ksi) for specimens with a circular hole

Lay-up \ Dia. (in.)	0.000	0.065	0.125	0.250	0.375	0.500
$(90/\pm45/0)_s$	85.14	67.63	61.17	48.42	37.79	30.01
$(\pm45/0/90)_s$	70.30	54.67	48.91	41.52	35.22	27.03
$(0/\pm45/90)_s$	78.29	59.48	54.46	43.22	37.06	29.92
$(0/\pm45)_s$	100.11	74.61	67.76	49.60	41.92	32.57
$(0_2/\pm45)_s$	142.87	110.29	97.79	89.44	73.06	56.76
$(0_4/\pm45)_s$	178.21	145.02	139.05	116.84	91.47	70.41
$(0/\pm15)_s$	139.95	126.13	103.58	91.08	72.19	56.33
$(0/\pm30)_s$	115.47	88.66	79.33	56.04	47.25	37.85
$(0/\pm75)_s$	80.99	70.94	63.05	50.71	39.97	33.51
$(0/\pm60)_s$	94.23	75.35	62.46	52.83	41.80	32.95
$(0/90)_{2s}$	126.3	98.52	91.33	74.51	61.99	49.20
$(\pm45)_{2s}$	19.66	19.31	----	14.80	----	9.69
$(90/\pm30)_s$	52.08	52.71	----	48.71	----	25.61
$(0/\pm45/90)_s$	77.52	66.66	55.81	44.26	----	30.09

Table 2 - Mean failure stress (ksi) for specimens of $(0/\pm45/90)_s$ with a slant slot

Angle \ Size	a=0.0625"	a=0.250"	a=0.500"	a=0.750"
$\phi=30°$	59.90	44.91	24.64	10.22
$\phi=45°$		44.13	28.98	12.35
$\phi=60°$		46.70	31.56	14.28
$\phi=90°$		46.36	32.02	13.49

Table 3 - Mean failure stress (ksi) for specimens of $(0_2/\pm45)_s$ with a slant slot

Angle \ Size	a=0.0625"	a=0.250"	a=0.500"	a=0.750"
$\phi=30°$	110.29	90.07	44.89	21.90
$\phi=45°$		75.92	44.43	24.83
$\phi=60°$		77.71	54.43	25.51
$\phi=90°$		74.95	45.13	23.52

Table 4 - Mean failure stress (ksi) for specimens of $(0/\pm45/90)_s$ with a slant slot

Angle \ Size	a=0.250"	a=0.500"	a=0.750"
ϕ=30°	43.46	27.03	14.63
ϕ=45°	44.02	28.83	13.59
ϕ=60°	47.29	27.91	16.33
ϕ=90°	45.35	31.95	14.80

Table 5 - Mean failure stress (ksi) for specimens of $(0_2/\pm45)_s$ with a slant slot

Angle \ Size	a=0.250"	a=0.500"	a=0.750"
ϕ=30°	82.05	50.58	31.18
ϕ=45°	81.27	50.09	20.35
ϕ=60°	77.11	51.69	22.31
ϕ=90°	73.51	47.41	23.21

Table 6 - Mean failure stress (ksi) for specimens of $(0/\pm45/90)_s$ with an edge notch

Notch \ Size	a=0.250"	a=0.500"	a=0.750"
Double Edge Semi-Hole	54.20	36.31	20.78
Double Edge Fine Slot	48.59	36.99	24.47
Shoulder	50.51	36.65	18.75

Table 7 - Mean failure stress (ksi) for specimens of $(0_2/\pm45)_s$ with a base notch

Notch \ Size	a=0.250"	a=0.500"	a=0.750"
Double Edge Semi-Hole	81.67	60.50	34.29
Double Edge Fine Slot	84.66	62.50	34.97
Shoulder	83.91	53.97	-----

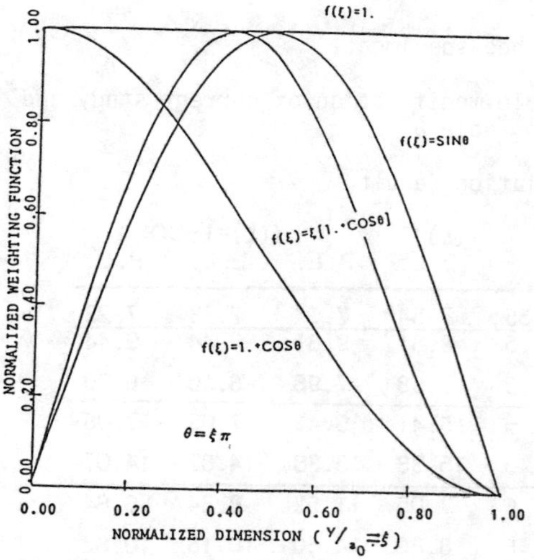

Fig. (4) - Weighting functions for proposed modification PA1

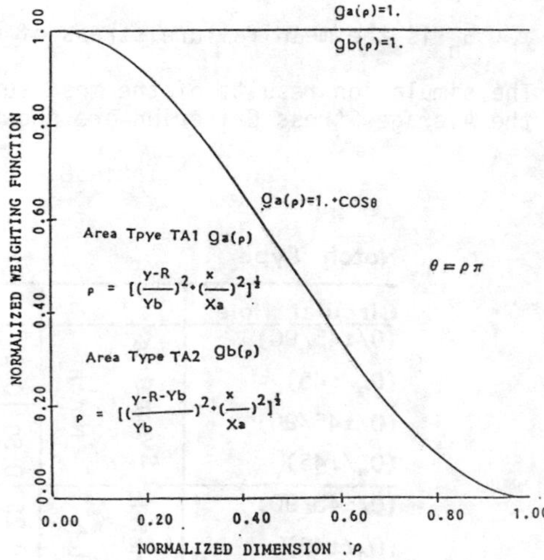

Fig. (5) - Weighting functions for proposed modification PA2

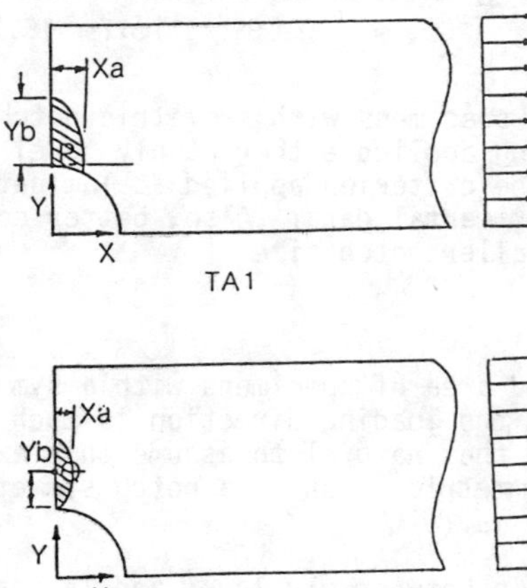

Fig. (6) - Schematic characteristic area for proposed modification PA2

$$ER = [(FSR)_{exp} - (FSR)_{pred.}]/(FSR)_{exp} * 100$$

$$FSR = \frac{\sigma_o}{\sigma_n}$$

where

σ_o is the mean failure stress of control specimen

σ_n is the mean failure stress of notched specimen.

The simulation results of the most suitable modification of current study and the Average Stress Criterion are shown in Table 8.

Table 8 - Simulation result

Notch Type					$f(\xi)=1$		$f(\xi)=1+COS\theta$	
					L.L.	P.L.	L.L.	P.L.
Circular Hole				55	7.54	7.54	7.23	7.23
$(0/\pm45/90)_s$	Symmetric	Notch	0.25"	5	9.81	9.81	9.44	9.44
$(0_2/\pm45)_s$	Symmetric	Notch	0.25"	5	7.98	7.98	6.50	6.50
$(0/\pm45/90)_s$	Symmetric	Notch	0.50"	5	5.41	5.41	7.08	7.08
$(0_2/\pm45)_s$	Symmetric	Notch	0.50"	5	15.38	15.38	14.07	14.07
$(0/\pm45/90)_s$	Non-symmetric	Notch	0.25"	6	7.05	11.52	8.34	6.64
$(0_2/\pm45)_s$	Non-symmetric	Notch	0.25"	6	8.86	14.60	10.15	10.82
$(0/\pm45/90)_s$	Non-symmetric	Notch	0.50"	6	22.28	30.48	15.40	23.64
$(0_2/\pm45)_s$	Non-symmetric	Notch	0.50"	6	27.28	46.90	20.67	41.80
OVERALL				99	11.17	16.24	9.95	14.06

Identical results for specimens with symmetric notches are obtained with the proposed modification being applied either at ply level (P.L.) or at laminate level (L.L.). Overall, the criterion applied at laminate level seems to have better correlation with experimental data. Also, better correlation with testing data is obtained for a smaller notch size.

UNIAXIAL STRESS PROBLEM

In the highly stressed area of specimens with a symmetric notch under uniaxial loading, the stress along the loading direction is much higher than the other stress components. It is thus natural to assume this axial stress to dominate the failure behavior. The symmetric notch is a notch symmetric with respect to a line parallel to the loading axis.

The stress relationship between ply level and laminate level can be constructed. Through the matrix operation [13], the fiber stress for each individual ply can be expressed as follows:

For notched specimen:

$$\sigma_{1i} = [(Q_{11}\cos^2\theta_i + Q_{12}\sin^2\theta_i)C_{xx} + (Q_{11}\sin^2\theta_i + Q_{12}\cos^2\theta_i)C_{xy}]F(x,y)\sigma_n$$

For unnotched specimen:

$$\sigma_{1i} = [(Q_{11}\cos^2\theta_i + Q_{12}\sin^2\theta_i)C_{xx} + (Q_{11}\sin^2\theta_i + Q_{12}\cos^2\theta_i)C_{xy}]\sigma_o$$

where

x-y is the global rectangular coordinate system;

1-2 is the local rectangular coordinate system defined in each ply as aligned with the fiber direction;

$\sigma_n F(x,y)$ is the stress distribution for notched laminate.

Proposed criterion at laminate level. The weighted average stress can be obtained. For unnotched specimen,

$$\sigma_a = \sigma_o$$

For notched specimen,

$$\sigma_a = \frac{\int \sigma_n(o,y)f(y)dx}{\int f(y)dy}$$

The failure stress ratio, then, is

$$\frac{\sigma_o}{\sigma_n} = \frac{\int F(o,y)f(y)dy}{\int f(y)dy}$$

Proposed criterion at ply level. Since a 0 ply is the Principal Load-Bearing Lamina, the criterion is applied at this layer. The Principal Load-Bearing Lamina is identified as the layer with the largest fiber stress in composite laminate. The weighted average stress for unnotched and notched specimens can be obtained.

For unnotched specimen,

$$\sigma_a = (Q_{11}C_{xx} + Q_{12}C_{xy})\sigma_o$$

For notched specimen,

$$\sigma_a = \frac{\int F(o,y)f(y)dy}{\int f(y)dy}[(Q_{11}C_{xx} + Q_{12}C_{xy})\sigma_n]$$

The failure stress ratio is

$$\frac{\sigma_o}{\sigma_n} = \frac{\int F(o,y)f(y)dy}{\int f(y)dy}$$

Identical strength prediction can be obtained by applying Weighted Average Criterion either at laminate level or at ply level if the notch is symmetric.

CONCLUSION

The Weighted Average Stress Criterion (with weighting function of $1+\cos(\theta)$) applied at laminate level yields the best correlation with the current experimental data. However, the Average Stress Criterion yields a quite reasonable prediction.

APPENDIX

- Lamina Material Property

  E1 = 19.420 Msi

  E2 = 1.228 Msi

  E6 = 0.725 Msi

  ν = 0.294

  t = 0.0058 in

- Lamina Strength Data

  X = 194.1 Ksi

  Y = 3.408 Ksi

  S = 8.624 Ksi

REFERENCES

[1] Whitney, J. M. and Nuismer, R. J., "Stress fracture criteria for laminated composites containing stress concentrations", J. Comp. Matl., Vol. 9, p. 253, 1974.

[2] Waddoups, M. E., Eisenmann, J. R. and Kaminski, B. E., "Macroscopic fracture mechanics of advanced composites materials", J. Comp. Matl., Vol. 5, p. 446, 1971.

[3] Cruse, T. A., "Tensile strength of notched composites", J. Comp. Matl., Vol. 7, p. 340, 1972.

[4] Karlak, R. F., "Hole effects in a related series of symmetrical laminates", Proc. of Failure Modes in Composite 111, ASM, Chicago, 1977.

[5] Pipes, R. B., Wetherhold, R. C. and Gillespie, J. W., "Notched strength of composite materials", J. Comp. Matl., Vol. 13, 1979.

[6] Mar, J. W. and Lin, K. Y., "Fracture mechanics correlation for tensile failure of filamentary composites with holes", J. Aircraft, Vol. 14, No. 7.

[7] Lo, K. H., Wu, E. M. and Konishi, Y., "Failure strength of notched composite laminates", J. Comp. Matl., Vol. 17, 1983.

[8] Poe, C. C., Jr., "Fracture toughness of fibrous composite materials", NASA Technical Paper 2370, 1984.

[9] Awerbuch, J. and Madkukar, M. S., "Notched strength of composite laminates: predictions and experiments - a review", J. of Reinforced Plastics and Composites, Vol. 4, No. 1, 1985.

[10] Highsmith, A., Reifonider, K. L. and Stinchcomb, W. W., "Stress redistribution in composite laminate", Composite Technology Review.

[11] Park, W. J., "On estimation of sample size for testing composite materials", J. Comp. Matl., Vol. 13, p. 219, 1979.

[12] Hu, T. H., "Improving the strength prediction for advanced composite material with stress concentration", Ph.D. Thesis, August 1985.

[13] Jones, R. B., "Mechanics of composite materials", Scripta Book Company.

FREE VIBRATIONS OF ELLIPTICAL LAMINATED PLATES

C.-C. Lin and S.-C. Lin

Institute of Applied Mathematics, National Chung-Hsing University
Taichung, Taiwan, Republic of China

ABSTRACT

This paper is concerned with an analytical study on free vibrations of elliptical laminated plates with simply supported or clamped edges. Antisymmetric cross-ply and angle-ply laminates, as well as symmetric angle-ply laminates are investigated. Hamilton's principle and Rayleigh-Ritz method are employed to obtain the solutions for natural modes and natural frequencies. Numerical results for fundamental frequencies of elliptical laminated plates are obtained for various aspect ratio a/b, material property ratio E_L/E_T, and ply orientation θ. The coupling effects between bending and stretching are also discussed.

INTRODUCTION

The fiber-reinforced composite materials have been increasingly used in aerocraft and other engineering structures in recent years. However, composite materials have many characteristics that are different from more conventional engineering materials. Therefore, research works on mechanical behavior of composite structures became considerably important and attracted many investigators both academically and experimentally.

In the known literature, most vibration analyses on composite laminates are concerned with rectangular boundary but much less about elliptical boundary. However, investigated works with valuable results on vibrations of homogeneous isotropic elliptical plates have been presented by many researchers. For examples, using Mathieu functions, Shiboaka [1] found the exact solution on the transverse vibration of an elliptical plate with clamped edge. Employing Rayleigh-Ritz method, Leissa [2] obtained an approximation on vibration of a simply-supported elliptical plate. Mazumdar [3] determined the approximate solutions on transverse vibrations of simple and clamped elliptical plates by the method of constant deflection lines. Sato [4] following the procedure of [1] thus derived the exact solution for vibration of simple elliptical plate. Furthermore, Bert [5] and Tomar and Gupta [6] investigated on vibrations of orthotropic elliptic clamped plates with non-uniform thickness. Dharmarajan and Chou [7] obtained approximate solutions for vibrations of elliptical laminated plates with clamped or simple edge. To the authors' knowledge, there is no known solution concerned with vibration of "anisotropic" elliptical laminated plate given in the literature.

In this paper, an analytical study concerned with free vibrations of elliptical laminated plates is presented. Antisymmetric cross-ply and angle-ply laminates, as well as symmetric angle-ply laminates with clamped and simple edges are investigated. Hamilton's principle and Rayleigh-Ritz method are employed to obtain the solutions for natural modes and natural frequencies. Numerical results for fundamental frequencies of elliptical laminates are obtained for various aspect ratio a/b, material property ratio E_L/E_T and ply-orientation θ. The coupling effects between bending and stretching are also discussed.

FORMULATION

Each laminated plate considered in this study consists of even number of identical orthotropic composite laminae stacking antisymmetrically or symmetrically about the middle surface of the plate. The Cartesian coordinate system (x,y,z) is used for the formulation which the xy-plane coincides with the middle plane of the elliptic plate as shown in Figure 1. The displacement components in the x, y and z directions at middle plane are denoted by u^o, v^o and w, respectively.

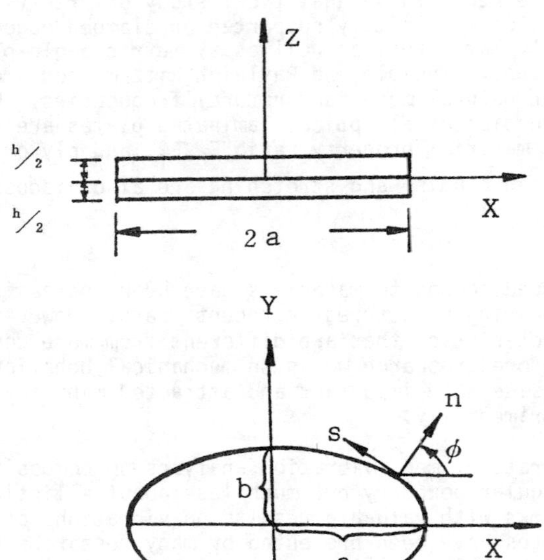

Fig. (1) - Geometry and coordinates of elliptical laminated plates

According to the classical lamination plate theory [8], the displacements at any point in the laminate are given by

$$u = u^o(x,y,t) - z\,\frac{\partial w}{\partial y}$$
$$v = v^o(x,y,t) - z\,\frac{\partial w}{\partial y} \tag{1}$$
$$w = w(x,y,t)$$

From equation (1), we have the strain-displacement relations

$$\begin{Bmatrix} \varepsilon_x \\ \varepsilon_y \\ \gamma_{xy} \end{Bmatrix} = \begin{Bmatrix} u^o_{,x} \\ v^o_{,y} \\ u^o_{,y} + v^o_{,x} \end{Bmatrix} + z \begin{Bmatrix} -w_{,xx} \\ -w_{,yy} \\ -2w_{,xy} \end{Bmatrix} \tag{2}$$

where comma in the subscripts denotes partial differentiation with respect to its followed variables.

The constitutive equation referred to the material principal axes (L,T) of a lamina is

$$\begin{Bmatrix} \sigma_L \\ \sigma_T \\ \tau_{LT} \end{Bmatrix} = \begin{bmatrix} Q_{11} & Q_{12} & 0 \\ Q_{12} & Q_{22} & 0 \\ 0 & 0 & Q_{66} \end{bmatrix} \begin{Bmatrix} \varepsilon_L \\ \varepsilon_T \\ \gamma_{LT} \end{Bmatrix} \tag{3}$$

where Q_{ij} are the stiffness coefficients. The constitutive equation of any (the kth) layer in the laminate referred to the plate geometry axes (x,y) is given by

$$\begin{Bmatrix} \sigma_x \\ \sigma_y \\ \tau_{xy} \end{Bmatrix}^{(k)} = \begin{bmatrix} \bar{Q}_{11} & \bar{Q}_{12} & \bar{Q}_{16} \\ \bar{Q}_{12} & \bar{Q}_{22} & \bar{Q}_{26} \\ \bar{Q}_{16} & \bar{Q}_{26} & \bar{Q}_{66} \end{bmatrix}^{(k)} \begin{Bmatrix} \varepsilon_x \\ \varepsilon_y \\ \gamma_{xy} \end{Bmatrix} \tag{4}$$

where

$$[\bar{Q}]^{(k)} = [T]^{-1}[Q][T^T]^{-1}$$

$$[T] = \begin{bmatrix} c^2 & s^2 & 2cs \\ s^2 & c^2 & -2cs \\ -cs & cs & c^2-s^2 \end{bmatrix}, \quad c = \cos\theta_k, \; s = \sin\theta_k \tag{5}$$

where θ_k is the angle between the fiber direction L of the kth layer and the x-axis.

The total energy of a laminated plate is defined as [8]

$$\pi = U - T \tag{6}$$

where U is the strain energy of bending given by

$$U = \frac{1}{2} \iiint_V (\sigma_x \varepsilon_x + \sigma_y \varepsilon_y + \tau_{xy} \gamma_{xy}) dV \tag{7}$$

here the CPT assumptions have been employed and T is the kinetic energy given by

$$T = \frac{1}{2} \iint_A \rho h \left(\frac{\partial w}{\partial t}\right)^2 dA \tag{8}$$

here ρ is the mass density of the plate and in-plane inertias and rotary inertias have been neglected.

The substitutions of equations (2) and (4) into (7) lead to [8]

$$\begin{aligned} U = \frac{1}{2} \iint_A \{ & A_{11} u^{o2}_{,x} + 2A_{12} u^o_{,x} v^o_{,y} + A_{22} v^{o2}_{,y} + 2A_{16} u^o_{,x}(u^o_{,y} + v^o_{,x}) \\ & + 2A_{26} v^o_{,y}(u^o_{,y} + v^o_{,x}) + A_{66}(u^o_{,y} + v^o_{,x})^2 - B_{11} u^o_{,x} w_{,xx} \\ & - 2B_{12}(v^o_{,y} w_{,xx} + u^o_{,x} w_{,yy}) - B_{22} v^o_{,y} w_{,yy} - 2B_{16}(u^o_{,y} w_{,xx} \\ & + v^o_{,x} w_{,xx} + 2u^o_{,x} w_{,xy}) - 2B_{26}(u^o_{,y} w_{,yy} + v^o_{,x} w_{,yy} + 2v^o_{,y} w_{,xy}) \\ & - 4B_{66}(u^o_{,y} + v^o_{,x}) w_{,xy} + D_{11} w^2_{,xx} + 2D_{12} w_{,xx} w_{,yy} + D_{22} w^2_{,yy} \\ & + 4D_{16} w_{,xx} w_{,xy} + 4D_{26} w_{,yy} w_{,xy} + 4D_{66} w^2_{,xy} \} dA \end{aligned} \tag{9}$$

where the extensional stiffnesses A_{ij}, the coupling stiffnesses B_{ij}, and the bending stiffnesses D_{ij} of laminate are defined, respectively, by

$$\begin{aligned} A_{ij} &= \sum_{k=1} Q^{(k)}_{ij}(h_k - h_{k-1}) \\ B_{ij} &= \frac{1}{2} \sum_{k=1} \overline{Q}^{(k)}_{ij}(h^2_k - h^2_{k-1}) \\ D_{ij} &= \frac{1}{3} \sum_{k=1} \overline{Q}^{(k)}_{ij}(h^3_k - h^3_{k-1}) \end{aligned} \tag{10}$$

here, h_k and h_{k-1} represent the z coordinates of the upper and lower surfaces of the kth layer.

For free vibration, the displacements are harmonic in time as the following form

$$\begin{aligned} u^o(t) &= u^o(x,y)\sin\omega t \\ v^o(t) &= v^o(x,y)\sin\omega t \\ w^o(t) &= w(x,y)\sin\omega t \end{aligned} \tag{11}$$

Thus, equation (8) becomes

$$T = \frac{1}{2} \iint_A \omega^2 \rho w^2 dA \tag{12}$$

According to Hamilton's principle, the equilibrium condition for the laminate vibration is

$$\pi = U-T = \text{stationary value}$$

or (13)

$$\delta\pi = \delta U - \delta T = 0$$

The Rayleigh-Ritz method in conjunction with the energy principle is employed in this study. Now the functional π is a function of u^o, v^o and w, and, according to the Ritz method, our problem is to find the minimum of certain functions by assuming the displacements in the form of a finite series:

$$u^o = \sum_{i=1}^{m} u_i \Phi_i(x,y)$$

$$v^o = \sum_{i=1}^{n} v_i \Theta_i(x,y) \tag{14}$$

$$w = \sum_{i=1}^{p} w_i \Psi_i(x,y)$$

where the terms u_i, v_i, w_i are undetermined coefficients and the functions $\Phi_i(x,y)$, $\Theta_i(x,y)$, $\Psi_i(x,y)$ are chosen such that they at least satisfy the geometric boundary conditions. With equation (14), the functional π is a function of u_i, v_i and w_i and thus the condition (13) leads to the following m+n+p equations:

$$\frac{\partial\pi(u_i,v_i,w_i)}{\partial u_j} = 0, \; j = 1,2,\ldots,m$$

$$\frac{\partial\pi(u_i,v_i,w_i)}{\partial v_j} = 0, \; j = 1,2,\ldots,n \tag{15}$$

$$\frac{\partial\pi(u_i,v_i,w_i)}{\partial w_j} = 0, \; j = 1,2,\ldots,p$$

Equation (15) represents a system of linear simultaneous equations in terms of the m+n+p coefficients u_i, v_i and w_i, it can be represented by the matrix equation

$$[K]\begin{Bmatrix}\{u_i\}\\ \{v_j\}\\ \{w_k\}\end{Bmatrix} - \omega^2\left[\begin{array}{c:c}[0] & [0]\\ (m+n)x(m+n) & (m+n)xp\\ \hdashline [0] & [M]\\ px(m+n) & pxp\end{array}\right]\begin{Bmatrix}\{u_i\}\\ \{v_j\}\\ \{w_k\}\end{Bmatrix} = 0 \tag{16}$$

where [K] resulted from partial differentiations of U with respect to u_i, v_j, w_k and is a (m+n+p)x(mxn+p) symmetric matrix due to U, which is a homogeneous quadratic function of u_i, v_j, w_k; [M] is obtained from the partial differentiations of T with respect to w_k.

For the convenience to find the eigenvalue problem (16), we may reduce the dimension of the matrix equation (16) by the following procedure. First, rewrite equation (16) in the form

$$\left[\begin{array}{c:c}[K_1] + [0] & [K_2] + [0]\\ (m+n)x(m+n) & (m+n)xp\\ \hdashline [K_2]^T + [0] & [K_3] - \quad [M]\\ px(m+n) & pxp\end{array}\right]\left\{\begin{array}{c}\{u_i\}\\ \{v_j\}\\ (m+n)x1\\ \hdashline \{w_k\}\\ px1\end{array}\right\} = 0 \tag{17}$$

which can be spread into two matrix equations

$$[K_1]\begin{Bmatrix}\{u_i\}\\ \{v_j\}\end{Bmatrix} + [K_2]\{w_k\} = 0 \tag{18}$$

$$[K_2]^T\begin{Bmatrix}\{u_i\}\\ \{v_j\}\end{Bmatrix} + [K_3^*]\{w_k\} = 0 \tag{19}$$

where

$$[K_3^*] = [K_3] - \omega^2[M]$$

From equation (18), we obtain

$$\begin{Bmatrix}\{u_i\}\\ \{v_j\}\end{Bmatrix} = -[K_1]^{-1}[K_2]\{w_k\} \tag{20}$$

The substitution of equation (20) into (19) yields

$$[\Lambda]\{w_k\} = 0 \tag{21}$$

where

$$[\Lambda] = [K'] - \omega^2[M]$$
$$[K'] = [K_3] - [K_2]^T[K_1]^{-1}[K_2] \tag{22}$$

$[\Lambda]$ is a pxp matrix, for a nontrivial solution of $\{w_k\}$, the determinant of $[\Lambda]$ must be zero, i.e.,

$$\det[\Lambda] = 0 \tag{23}$$

The natural frequencies ω are determined from the characteristic equation (23), and the natural modes thus can be found from equations (21), (22) and (14).

SOLUTIONS OF ELLIPTICAL LAMINATES

The boundary of the elliptical laminated plate, see Figure 1, considered in this study can be described by the standard elliptical equation

$$\frac{x^2}{a^2} + \frac{y^2}{b^2} = 1 \tag{24}$$

Two types of boundary conditions (clamped and simply supported) are considered

$$\text{clamped B.C.'s:} \quad w = \frac{\partial w}{\partial n} = u_n^o = u_s^o = 0 \tag{25}$$

$$\text{simple B.C.'s:} \quad w = M_n = u_n^o = u_s^o = 0 \tag{26}$$

where u_n^o and u_s^o denote the normal and tangential displacements at any boundary point. Let ϕ be the angle between the n-direction and the x-axis, see Figure 1, at any boundary point, then

$$\frac{\partial w}{\partial n} = \frac{\partial w}{\partial x}\frac{\partial x}{\partial n} + \frac{\partial w}{\partial y}\frac{\partial y}{\partial n} = \frac{\partial w}{\partial x}\cos\phi + \frac{\partial w}{\partial y}\sin\phi$$
$$u_n^o = u^o\cos\phi + v^o\sin\phi \tag{27}$$
$$u_s^o = -u^o\sin\phi + v^o\cos\phi$$

According to Ritz method, now we choose the displacement functions in the clamped elliptical laminate to be of the form [9]

$$u^o = u_1 x\psi$$
$$v^o = v_1 y\psi \tag{28}$$
$$w = w_1\psi^2 + w_2\psi^3$$

where

$$\psi = 1 - \frac{x^2}{a^2} - \frac{y^2}{b^2}$$

Using equation (28), the corresponding stiffness matrix [K'] and mass matrix [M] can be obtained

$$[M] = 2\pi ab\rho h \begin{bmatrix} \frac{1}{10} & \frac{1}{12} \\ \frac{1}{12} & \frac{1}{14} \end{bmatrix} \tag{29}$$

for symmetric angle-ply laminate, $B_{ij} \equiv 0$,

$$[K'] = [K_3] = \frac{2\pi\alpha_1}{ab} \begin{bmatrix} 4 & 3 \\ 3 & 18/5 \end{bmatrix} \tag{30a}$$

where

$$\alpha_1 = D_{11}(\frac{b}{a})^2 + D_{22}(\frac{a}{b})^2 + \frac{2}{3}(D_{12}+2D_{66}) + \frac{8}{3\pi}(D_{16}\frac{b}{a} + D_{26}\frac{a}{b})$$

for antisymmetric cross-ply laminate (0°/90° alternations),

$$[K'] = \frac{2\pi}{ab} \begin{bmatrix} 4\alpha_2 - \beta_2 & , 3\alpha_2 - \frac{3}{4}\beta_2 \\ 3\alpha_2 - \frac{3}{4}\beta_2, & \frac{18}{5}\alpha_2 - \frac{9}{16}\beta_2 \end{bmatrix}$$

$$\alpha_2 = D_{11}(\frac{b^2}{a^2} + \frac{a^2}{b^2}) + \frac{2}{3}(D_{12}+2D_{66}) \tag{30b}$$

$$\beta_2 = \frac{B_{11}^2}{2d}[A_{11}(\frac{a^2}{b^2} + \frac{b^2}{a^2}) + \frac{A_{66}}{3}(\frac{b^4}{a^4} + \frac{a^4}{b^4}) + \frac{2}{3}(A_{12}+A_{66})]$$

$$d + (A_{11}^2 - \frac{A_{12}^2}{9}) + \frac{A_{66}}{3}[A_{11}(\frac{a^2}{b^2} + \frac{b^2}{a^2}) - \frac{2}{3}A_{12}]$$

for antisymmetric angle-ply laminate (θ/-θ alternations),

$$[K'] = \frac{2}{ab} \begin{bmatrix} 4\alpha_3 - \beta_3, & 3\alpha_3 - \beta_3' \\ 3\alpha_3 - \beta_3', & \frac{18}{5}\alpha_3 - \beta_3^* \end{bmatrix} \tag{30c}$$

where

$$\alpha_3 = D_{11}(\frac{b}{a})^2 + D_{22}(\frac{a}{b})^2 + \frac{2}{3}(D_{12}+2D_{66})$$

$$\beta_3 = \frac{2}{\pi^2 d'}(4e_1e_4^2 - \frac{8}{3}e_3e_4e_5 + 4e_2e_5^2)$$

$$\beta_3' = \frac{2}{\pi^2 d'}(e_1e_4e_6 - \frac{1}{12}e_3e_4e_7 - \frac{1}{12}e_3e_5e_6 + e_2e_5e_7)$$

$$\beta_3^* = \frac{2}{\pi^2 d'} \left(\frac{1}{4} e_1 e_6^2 - \frac{1}{12} e_3 e_6 e_7 + \frac{1}{4} e_2 e_7^2\right)$$

$$d' = \left(A_{11}A_{22} - \frac{A_{12}^2}{9}\right) + \frac{A_{66}}{3}\left(A_{22}\frac{a^2}{b^2} + A_{11}\frac{b^2}{a^2} - \frac{2}{3}A_{12}\right)$$

$$e_1 = A_{22} + \frac{A_{66}}{3}\left(\frac{b}{a}\right)^2$$

$$e_2 = A_{11} + \frac{A_{66}}{3}\left(\frac{a}{b}\right)^2$$

$$e_3 = A_{12} + A_{66}$$

$$e_4 = B_{16} + \frac{B_{26}}{3}\left(\frac{a}{b}\right)^2$$

$$e_5 = B_{26} + \frac{B_{16}}{3}\left(\frac{b}{a}\right)^2$$

$$e_6 = B_{16} + B_{26}\left(\frac{a}{b}\right)^2$$

$$e_7 = B_{26} + B_{16}\left(\frac{b}{a}\right)^2$$

For the simple elliptical laminate, we choose the displacement functions to be of the form

$$\begin{aligned} u^o &= u_1 x\psi \\ v^o &= v_1 y\psi \qquad , \quad \psi = 1 - \frac{x^2}{a^2} - \frac{y^2}{b^2} \\ w &= w_1\psi + w_2\psi^2 \end{aligned} \tag{31}$$

The corresponding stiffness matrix [K'] and mass matrix [M] thus are

$$[M] = 2\pi ab\rho h \begin{bmatrix} \frac{1}{6} & \frac{1}{8} \\ \frac{1}{8} & \frac{1}{10} \end{bmatrix} \tag{32}$$

for symmetric angle-ply laminate,

$$[K'] = [K_3] = \frac{2\pi}{ab} \begin{bmatrix} 2(\alpha_4 + 2D_{12}), & \beta_4 \\ \beta_4 & , 4\left(\alpha_4 + \alpha_4' - \frac{\beta_4}{3}\right) \end{bmatrix} \tag{33a}$$

for antisymmetric cross-ply laminate,

$$[K'] = \frac{2\pi}{ab}\begin{bmatrix} 2(\alpha_5 + 2D_{12}), & 0 \\ 0 & , 4(\alpha_5 + \alpha_4') - \beta_2 \end{bmatrix} \tag{33b}$$

where

$$\alpha_5 = D_{11}(\frac{b^2}{a^2} + \frac{a^2}{b^2})$$

for antisymmetric angle-ply laminate,

$$[K'] = \frac{2\pi}{ab}\begin{bmatrix} 2(\alpha_4 + 2D_{12}) - \frac{\overset{*}{\beta_3}}{4}, & \beta_5 \\ \beta_5 & , 4(\alpha_4 + \alpha_4') - \beta_3 \end{bmatrix} \tag{33c}$$

where

$$\beta_5 = \frac{2}{\pi^2 d'} (2e_1e_4e_6 - \frac{2}{3} e_3e_4e_7 - \frac{2}{3} e_3e_5e_6 + 2e_2e_5e_7)$$

With those known mass matrices [M] and stiffness matrices [K'] in equations (29), (30), (32) and (33), the natural frequencies ω can be determined from equation (23) by computer.

NUMERICAL RESULTS AND DISCUSSIONS

In order to verify the validity of the displacement functions chosen in this study, comparisons between the present results and the exact solutions obtained by Shiboaka [1] and Sato [4] are shown in Figure 2 for fundamental natural fre-

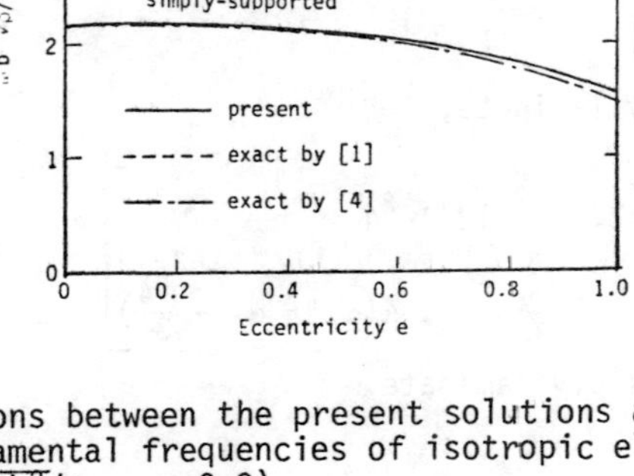

Fig. (2) - Comparisons between the present solutions and exact solutions for fundamental frequencies of isotropic elliptical plates. ($e = \sqrt{a^2-b^2}/a$, $\nu = 0.3$)

quencies of homogeneous isotropic elliptical plates versus aspect ratio. Comparisons between the present results and those of Dharmarajan and Chou [7], who applied the method of constant deflection lines, are listed in Table 1 for fundamental frequencies of orthotropic boron-aluminum composite elliptical plates

Table 1 - Fundamental frequencies of orthotropic boron-aluminum composite elliptical plates $\omega a^2\sqrt{\rho/D_{11}}$

	A. 25% Boron ($\frac{D_{22}}{D_{11}} = 0.65$, $\frac{D_{12}}{D_{11}} = 0.18$, $\frac{D_{66}}{D_{11}} = 0.25$)				B. 50% Boron ($\frac{D_{22}}{D_{11}} = 0.60$, $\frac{D_{12}}{D_{11}} = 0.15$, $\frac{D_{66}}{D_{11}} = 0.26$)			
	(i) Clamped Edge		(ii) Simple Edge		(i) Clamped Edge		(ii) Simple Edge	
a/b	present	[7]	present	[7]	present	[7]	present	[7]
1.0	9.094	8.711	4.352	4.092	8.948	8.564	4.248	3.990
1.1	9.913	9.488	4.742	4.474	9.729	9.302	4.616	4.329
1.2	10.857	10.396	5.195	4.910	10.632	10.165	5.045	4.770
1.3	11.926	11.433	5.710	5.410	11.655	11.156	5.535	5.231
1.5	13.949	13.890	6.922	6.605	14.058	13.508	6.692	6.342
2.0	22.734	22.125	10.961	10.511	22.052	21.403	10.574	10.016
3.0	47.457	46.801	23.025	22.376	45.915	45.240	22.214	21.435
5.0	127.687	126.990	62.200	60.719	123.429	122.710	60.062	58.142

with various aspect ratios. All the results show that they are in excellent agreement and the present results appearing a little higher agree with that Rayleigh-Ritz method provides an upper bound for the eigenvalues.

The essentials of the present study are concentrated on those of antisymmetric cross-ply and angle-ply, as well as symmetric angle-ply laminated anisotropic elliptical plates. All the laminae in each laminated plate are identical in thickness and material properties given as below:

$$\frac{E_L}{E_T} = 1\text{-}40, \quad \frac{G_{LT}}{E_T} = 0.5, \quad E_T = 1.0 \times 10^6 \text{ psi}, \quad \nu_{LT} = 0.25$$

Figures 3 and 4 illustrate the variations of fundamental frequencies of antisymmetric cross-ply laminated elliptical plates versus the aspect ratio a/b and

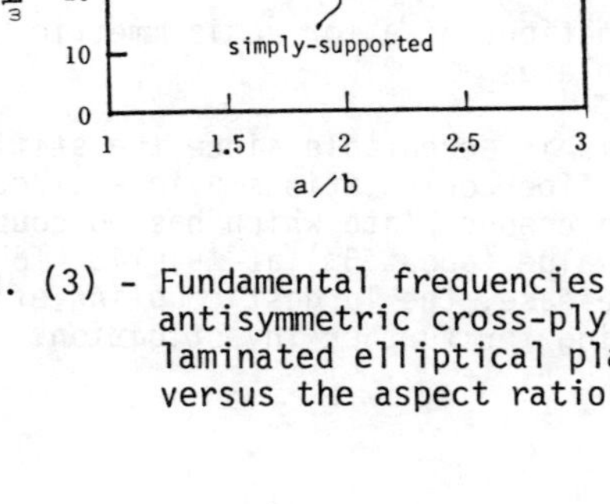

Fig. (3) - Fundamental frequencies of antisymmetric cross-ply laminated elliptical plates versus the aspect ratio

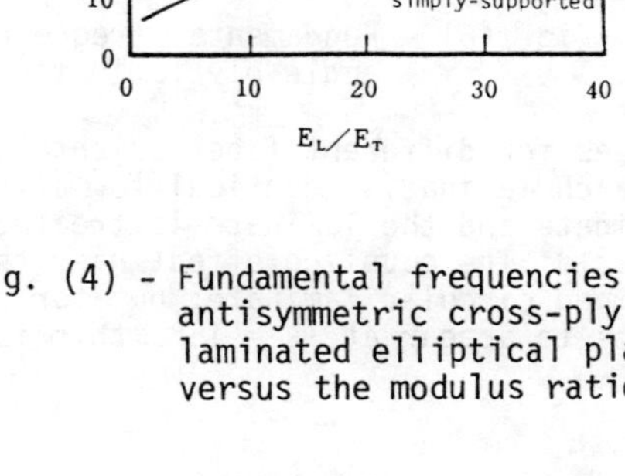

Fig. (4) - Fundamental frequencies of antisymmetric cross-ply laminated elliptical plates versus the modulus ratio

the modulus ratio E_L/E_T. Obviously, the frequencies increase as the ratio E_L/E_T increases but decrease with increasing aspect ratio. Fundamental frequencies as functions of the ply-orientation θ of antisymmetric angle-ply laminated elliptical plates are shown in Figure 5; results show that the value of frequency in-

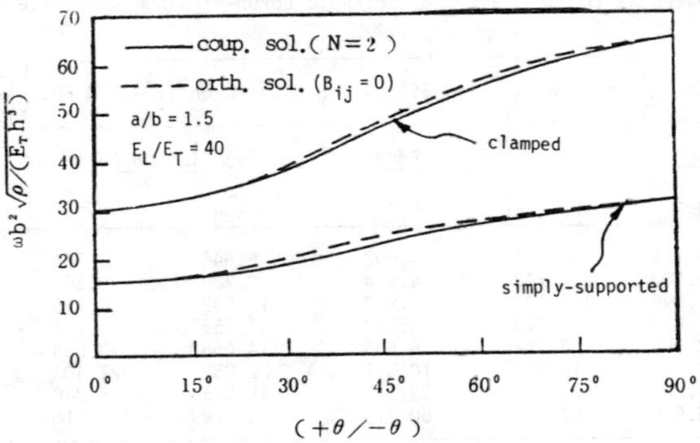

Fig. (5) - Fundamental frequencies as functions of θ for antisymmetric angle-ply laminated elliptical plates

creases as the fiber direction approaches the minor axis y (i.e., θ → 90°), this is reasonable. It can be seen that the coupling effect between extension and bending are not significant for these cases.

Figure 6 presents the results for antisymmetric angle-ply laminated square plates. Results indicate that the orthotropic solutions ($B_{ij} \equiv 0$) have constant

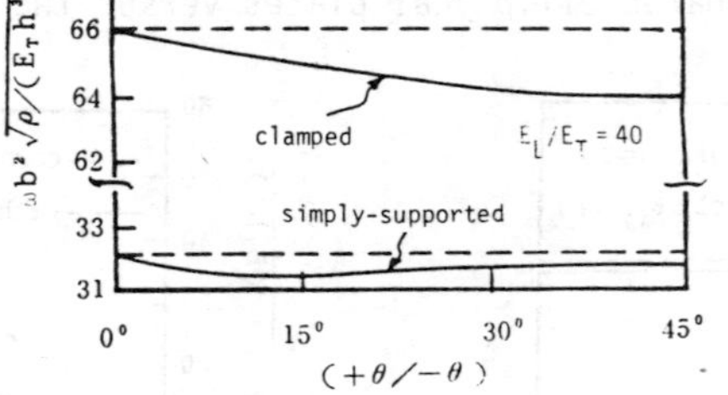

Fig. (6) - Fundamental frequencies as functions of θ for antisymmetric angle-ply laminated circular plates

values for different fiber orientations θ, this is acceptable since the stiffness of each lamina is identical for all different fiber orientations θ in a circular laminate and the laminate is treated as a homogeneous plate which has no coupling effect. The coupling effect has its largest value (about 3%) at ±θ = 45° for clamped circular laminate; but for simple edge case, the largest coupling effect seems to appear at ±θ = 15°, this is interesting for further investigation.

For symmetric angle-ply laminated circular plates, results are calculated with displacement field in equations (28) or (31) for the case of all layers with the same fiber direction (all θ) and for the case of alternating plies, i.e., $(\pm\theta)_s$, and are presented in Figures 7 and 8 in which orthotropic solutions of

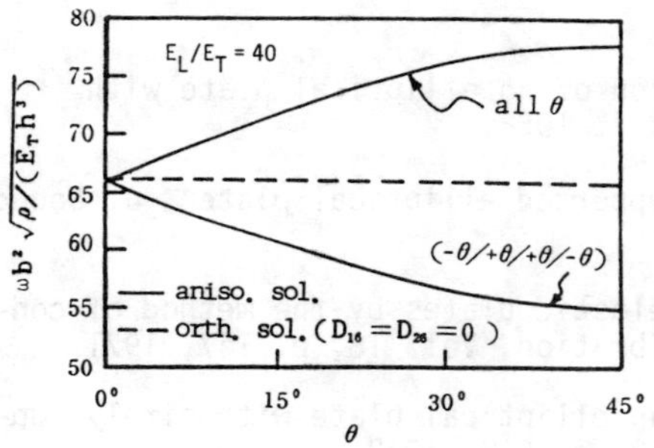

Fig. (7) - Fundamental frequencies of symmetric angle-ply laminated circular clamped plate

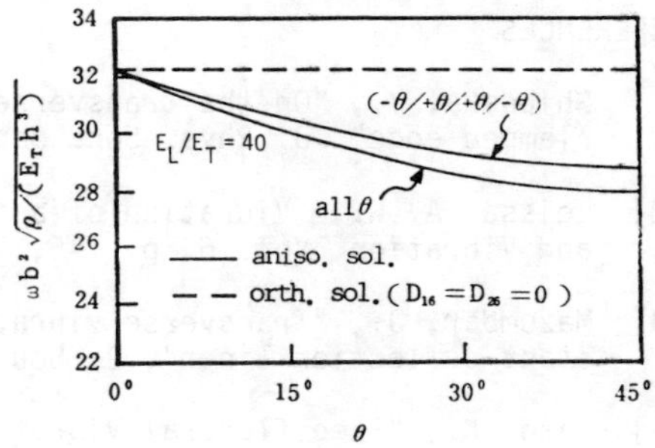

Fig. (8) - Fundamental frequencies of symmetric angle-ply laminated circular simple plate

neglecting the D_{16} and D_{26} terms are included. There reveal some unreasonable results: 1) the anisotropic solution for the case of all θ is higher than the orthotropic solution and the solution for the case of $(\pm\theta)_s$ in the clamped circular laminate, 2) the orthotropic solution is constant value for different orientations θ but the anisotropic solution for the case of all θ is not (actually the circular laminate of all θ is identical for different orientations of θ). Therefore, the assumed displacement field in equations (28) or (31) is not suitable for the symmetric angle-ply laminated circular plates.

CONCLUDING REMARKS

From our investigation, we have the conclusion as follows:

1) The solution model presented in this study is effective and gives acceptable results except for the cases of symmetric angle-ply laminated circular plates.

2) The effect of boundary conditions on the natural frequency of laminated elliptical plate is considerably significant, fundamental frequency of clamped elliptical plate is about two times of the simple elliptical plate.

3) The coupling between bending and extension for elliptical laminated plate is not significant.

4) In order to verify the accuracy of those results obtained by the present solution model for the anisotropic laminates being examined and to get a reasonable solution for the symmetric angle-ply laminated circular plate,

further investigation by choosing more terms in the series of displacement functions (for more accurate vibration mode) by the Ritz method or by using other numerical techniques (e.g., the finite element method) are suggested.

EFERENCES

[1] Shiboaka, Y., "On the transverse vibration of an elliptical plate with clamped edge", J. Phys. Soc. Japan, p. 797, 1956.

[2] Leissa, A. W., "Vibration of a simply-supported elliptical plate", J. Sound and Vibration, Vol. 6, p. 145, 1967.

[3] Mazumdar, J., "Transverse vibration of elastic plates by the method of constant deflection lines", J. Sound and Vibration, Vol. 18, p. 147, 1971.

[4] Sato, K., "Free flexural vibrations of an elliptical plate with simply supported edge", J. Acous. Soc. Am., Vol. 52, p. 919, 1972.

[5] Bert, C. W., "Fundamental frequencies of orthotropic plates with various platforms and edge conditions", Shock and Vibration Bulletin, p. 89, 1977.

[6] Tomar, J. S. and Gupta, A. K., "Vibrations of an orthotropic elliptic plate of non-uniform thickness and temperature", Vol. 96, p. 29, 1984.

[7] Dharmarajan, S. and Chou, F. H., "Vibration of composite plates of arbitrary shape by the method of constant deflection lines", Shock and Vibration Bulletin, p. 99, 1977.

[8] Ashton, J. E. and Whitney, J. M., "Theory of Laminated Plates, Technomic Publishing Co., Inc., 1970.

[9] Kicher, T. P., "The analysis of unbalanced cross-plied elliptical plate under uniform pressure", J. Composite Materials, Vol. 13, p. 424, 1969.

THE FLEXURAL ANALYSIS OF LAMINATED COMPOSITES USING A TRIANGULAR PLATE BENDING ELEMENT

A.-D. Han, F.-Y. Shieh

ARL/CSIST
TAichung, Taiwan, Republic of China

and

C.-C. Lin

National Chung-Hsing University
Taichung, Taiwan, Republic of China

ABSTRACT

In this paper, a high-order triangular plate bending element which was developed by Cowper et al is modified to study the problems of bending of symmetric composite laminates subject to uniformly distributed load. The element consists of six deflection parameters, namely the transverse displacement, its first and second partial derivatives as the generalized degree of freedom per node, yielding a total of 18 d.o.f. for the element. The center (maximum) displacements, maximum stresses, as well as moments at plate center and middle point of free edge are obtained for symmetric rectangular laminated plates. The effects of elastic modulus ratio (longitudinal/transverse) and ply orientation of laminae, plate geometry aspect ratio and various boundary conditions are investigated. Comparisons are made with other available solutions and results show that they are in excellent agreement.

INTRODUCTION

Because of their high performance advantages in strength-to-weight ratio and utility of optimum design through the variation of fibre orientations, stacking pattern, and choice of fiber and matrix materials, it has been stimulated the increasingly use of composites (especially fibre-reinforced composites) in aerocraft and other industries. The fundamental development in the mechanics of composite materials has been accomplished by many scholars, e.g., [1-3]. However, practical solutions have been given in the literatures still limited to simple geometry and boundary conditions. For composite laminates with more general geometrical configuration and boundary conditions, the finite element analysis can be adopted as a powerful technique. Rectangular element is the most popular and easiest to deal with but it gives poor approximation for those problems with irregular geometrical boundary, in this case, the triangular element is suggested.

In this paper, a high-order triangular plate bending element developed by Cowper et al [4] is modified to study the bending problems of symmetric composite laminates subject to uniformly distributed load. Results are compared with alternative analytical and numerical solutions obtained by other investigators.

BASIC EQUATIONS

It is assumed that each lamina in the composite plate is homogeneous orthotropic materials and is in a state of plane stress. Then, the stress-strain relations to the material axes (L,T) are given by

$$\begin{Bmatrix} \sigma_L \\ \sigma_T \\ \tau_{LT} \end{Bmatrix} = \begin{bmatrix} C_{11} & C_{12} & 0 \\ C_{12} & C_{22} & 0 \\ 0 & 0 & C_{66} \end{bmatrix} \begin{Bmatrix} \varepsilon_L \\ \varepsilon_T \\ \gamma_{LT} \end{Bmatrix} \tag{1}$$

In general, the material principal axes (L,T) of each lamina are not coincident with the geometry axes (x,y) of the plate. Let the principal direction L (filament direction) of the kth-layer in the laminate make an angle θ_f with the x-axis. Then, after coordinate transformation, we have the constitutive equation referred to the geometry coordinates as the form

$$\begin{Bmatrix} \sigma_x \\ \sigma_y \\ \tau_{xy} \end{Bmatrix} = [Q] \begin{Bmatrix} \varepsilon_x \\ \varepsilon_y \\ \gamma_{xy} \end{Bmatrix} \tag{2}$$

where

$$[Q] = [T]^{-1}[C][T]^T$$

$$[T] = \begin{bmatrix} m^2 & n^2 & 2mn \\ n^2 & m^2 & -2mn \\ -mn & mn & m^2-n^2 \end{bmatrix}, \; m = \cos\theta_f, \; n = \sin\theta_f \tag{3}$$

in which Q_{ij} (i,j = 1,2,6) are the so-called "reduced stiffnesses".

Consider a composite plate consisting of n orthotropic layers. The plate is thin and is limited to small deformation so that the classical lamination plate theory (CPT) is applied. Then, by the definitions

$$[N_x, N_y, N_{xy}] = \int_o^h [\sigma_x, \sigma_y, \tau_{xy}] dz \tag{4a}$$

$$[M_x, M_y, M_{xy}] = \int_o^h [\sigma_x, \sigma_y, \tau_{xy}] z dz \tag{4b}$$

the in-plane stress resultants {N} and bending moments {M} of the laminate are given by

$$\left\{\frac{N}{M}\right\} = \left[\frac{A}{B}\frac{B}{D}\right]\left\{\frac{\varepsilon^\circ}{\kappa^\circ}\right\} \tag{5}$$

where

$$[A_{ij}, B_{ij}, D_{ij}] = \int_{-h/2}^{h/2} Q_{ij}[1, z, z^2]dz \tag{6}$$

and the middle surface strains $\{\varepsilon^\circ\}$ and curvatures $\{\kappa^\circ\}$ are related to the middle surface displacements u°, v°, w as follows:

$$\begin{Bmatrix} \varepsilon_x^\circ \\ \varepsilon_y^\circ \\ \gamma_{xy}^\circ \end{Bmatrix} = \begin{Bmatrix} \dfrac{\partial u^\circ}{\partial x} \\ \dfrac{\partial v^\circ}{\partial y} \\ \dfrac{\partial u^\circ}{\partial y} + \dfrac{\partial v^\circ}{\partial x} \end{Bmatrix}, \begin{Bmatrix} \kappa_x^\circ \\ \kappa_y^\circ \\ \kappa_{xy}^\circ \end{Bmatrix} = \begin{Bmatrix} -\dfrac{\partial^2 w}{x^2} \\ -\dfrac{\partial^2 w}{y^2} \\ -2\dfrac{\partial^2 w}{\partial x \partial y} \end{Bmatrix} \tag{7}$$

The presence of B_{ij} implies coupling between bending and extension of a laminate. For a laminate which is symmetric in both geometric and material properties about the middle surface, the coupling vanishes (i.e., $B_{ij} \equiv 0$).

For the convenience of preliminary design purpose, we may treat this middle-plane symmetric laminate as an anisotropic homogeneous plate in which the equivalent stress-strain relationship is

$$\{\sigma\}_{xy} = [G]\{\varepsilon\}_{xy} \tag{8}$$

where G_{ij} are the equivalent stiffness coefficients of the laminated plate and are given by

$$G_{ij} = \frac{12}{h^3} D_{ij} \tag{9}$$

FINITE ELEMENT ANALYSIS

A. Displacement Function

The geometry and coordinate system for a typical triangular finite element are shown in Figure 1, in which (x,y,z) is the global coordinate system and (ξ,η,ζ) is the local coordinate system. The x-y and ξ-η planes coincide with the middle plane of the element. There are six degree-of-freedom ($w, w_\xi, w_\eta, w_{\xi\xi}, w_{\xi\eta}, w_{\eta\eta}$) at each of the three vertices, hence it has eighteen degree-of-freedom for a triangular element. According to [4], we assume the displacement function $w(\xi,\eta)$ within a triangular element to be a polynomial of degree five as the following form

$$w(\xi,\eta) = \sum_{p=0}^{5}\sum_{i=0}^{p} a_{pi}\, \xi^{p-i}\eta^{i} \equiv \sum_{k=1}^{20} a_k \xi^{m_k}\eta^{n_k}, \quad (a_{51} = 0) \tag{10}$$

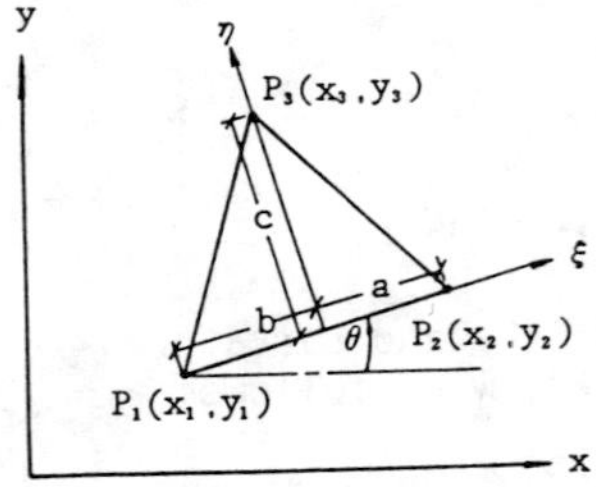

Fig. (1) - Geometry and coordinates of a triangular finite element

The vanishing of the term $\xi^4\eta$ in equation (10) ensures that the normal slope w_n along the boundary η=0 to be a cubic polynomial in ξ. The same constraint is also imposed to the other two boundaries ($\overline{p_1p_3}$ and $\overline{p_2p_3}$), from which the number of independent coefficients (a_{pi}) is thus reduced to eighteen. The generalized displacements at each vertice (node) can be expressed in terms of local coordinate parameters a, b and c, see Figure 1, with coefficients containing a_{pi}'s. Introducing column matrix {w*} and {A} such as

$$\{w^*\}^T = [w_1, w_{1\xi}, \ldots, w_{1\eta\eta}; w_2, w_{2\xi}, \ldots, w_3, \ldots]_{1x18} \tag{11a}$$

$$\{A\}^T = [a_{00}, a_{10}, a_{11}, \ldots, a_{55}]_{1x20} \equiv [a_1, a_2, \ldots, a_{20}] \tag{11b}$$

where the subscripts 1, 2 and 3 in $\{w^*\}^T$ denote the node (vertice) numbers, we may express the relation between {w*} and A by the following matrix equation

$$\{\frac{w^*}{0}\} = [T_1]\{A\} \tag{12}$$

where $[T_1]$ is a 20x20 regular matrix whose elements are polynomials of a, b and c [4]. The two constraints along the boundaries $\overline{p_1p_3}$ and $\overline{p_2p_3}$ have been added into equation (12) - the last two rows.

B. Stiffness and Consistent Load Matrices

The stiffness matrix of an element is obtained by calculating the strain energy from the relation

$$U_e = \frac{1}{2}\iiint_{vol.} \{\varepsilon\}^T[G^*]\{\varepsilon\}dvol = \frac{1}{2}\iint_A \frac{h^3}{12}\{\kappa^*\}^T[G^*]\{\kappa^*\}dA \tag{13}$$

where [G*] is the equivalent stress-strain relation for anisotropic materials defined in local (ξ,η) coordinate system. The relationship between [G*] and [G] is

$$[G^*] = [T][G][T]^T \tag{14}$$

where the transformation matrix [T] has the form as shown in equation (4) with θ_f replaced by θ - the angle between the ξ- and x-axes. The vector $\{\kappa^*\}$ in equation (13) is the middle-surface curvature referred to local coordinate system such that

$$\{\kappa^*\}^T = - [w_{\xi\zeta} w_{\eta\eta} 2w_{\xi\zeta}] \tag{15}$$

by substituting equations (10) and (15) into equation (13) and taking integrations, the strain energy may be expressed as a quadratic form in the a_{pi}'s as the form

$$U_e = \frac{1}{2} \{A\}^T[k]\{A\} \tag{16}$$

The elements of matrix [k] are given below:

$$\begin{aligned} k_{ij} = \frac{h^3}{12} & G^*_{11} m_i m_j (m_i-1)(m_j-1) \cdot F(m_{ij}-4, n_{ij}) \\ & + G^*_{22} n_i \cdot n_j (n_i-1)(n_j-1) \cdot F(m_{ij}, n_{ij}-4) \\ & + [G^*_{12} m_i n_j (m_i-1)(n_j-1) + G^*_{12} m_j n_i (M_j-1)(N_i-1) \\ & + 4G^*_{66} m_i m_j n_i n_j] \cdot F(m_{ij}-2, n_{ij}-2) \qquad (i,j = 1,\ldots,20) \\ & + 2G^*_{16} \cdot [m_i n_i m_j (m_j-1) + m_j n_j m_i (m_i-1)] \cdot F(m_{ij}-3, n_{ij}-1) \\ & + 2G^*_{26} \cdot [m_i n_i n_j (n_j-1) + m_j n_j (n_i)(n_i-1)] \cdot F(m_{ij}-1, n_{ij}-3) \end{aligned} \tag{17}$$

where

$$\begin{aligned} & F(m,n) = c^{n+1}[a^{m+1} - (-b)^{m+1}]m!n!/(m+n+2)! \\ & m_{ij} = m_i + m_J, \quad n_{ij} = n_i + n_J \end{aligned} \tag{18}$$

in which m_i and n_i are the powers of ξ and η, respectively, in equation (10).

By means of equations (12) and (16), the strain energy may be expressed in terms of $\{w^*\}$

$$U_e = \frac{1}{2} \{w^*\}^T[k_1]\{w^*\} \tag{19}$$

where

$$[k_1] = [T_2]^T[k][T_2] \tag{20}$$

and the 20x18 matrix $[T_2]$ consists of the first eighteen columns of $[T_1]^{-1}$. It remains to transform $[k_1]$ to global coordinate system. Let $\{w\}$ be the generalized displacement vector to global coordinate system, then we may express the relation between $\{w^*\}$ and $\{w\}$ by the following matrix equation

$$\{w^*\} = [R]\{w\} \tag{21}$$

here the 18x18 matrix [R] is the transformation matrix which can be easily derived. From equations (19)-(21), the strain energy for the element in the global coordinate system is

$$U_e = \frac{1}{2} \{w\}^T[K]\{w\} \tag{22}$$

where

$$[K] = [R]^T[k_1][R] \tag{23}$$

Similarly, the consistent load matrix is established by calculating the virtual work done by the applied load. If the intensity of the applied load is q_o, then the work done for a single element is

$$V_e = \iint_A q_o\, w(\xi,\eta)d_\xi d_\eta = \{A\}^T\{p\} = \{w\}^T\{P\} \tag{24}$$

where

$$p_i = \iint_A q_o\, \xi^{m_i}\eta^{n_i}d\xi d\eta, \qquad (i = 1,2,\ldots,20)$$

$$\{P\} = [R]^T[T_2]^T\{p\} \tag{25}$$

here {P} is the consistent load matrix for a single element in the global coordinate system.

The total potential energy of the plate is

$$\pi = \sum_{\text{elements}} (U_e - V_e) \tag{26}$$

Using the variational principle, $\delta\pi = 0$, we obtain

$$[K]\{w\} = \{P\} \tag{27}$$

From equation (27) and the boundary and compatibility conditions, we can solve the overall static problems of the laminated plates.

NUMERICAL EXAMPLES AND DISCUSSIONS

All of the numerical examples considered here are symmetric rectangular laminated plates subject to uniformly distributed load. The symmetric properties enable us to do finite element model analysis only requiring for one quadrant of the plate. Four types of orthotropic materials are selected in the analysis:

1. $E_L/E_T = 3.0$, $G_{LT}/E_T = 0.5$, $\nu_{LT} = 0.25$ ($E_T = 2.6 \times 10^6$ psi) (Glass/Epoxy)

2. $E_L/E_T = 31.2$, $G_{LT}/E_T = 0.9$, $\nu_{LT} = 0.31$ ($E_T = 1.02 \times 10^6$ psi) (Graphite/Epoxy)

3. $E_L/E_T = 25.0$, $G_{LT}/E_T = 0.5$, $\nu_{LT} = 0.25$

4. $E_L/E_T = 40.0$, $G_{LT}/E_T = 0.5$, $\nu_{LT} = 0.25$

($E_T = 1.0 \times 10^6$ psi)

The center deflection and moments are nondimensionalized as follows:

$$\bar{w}_c = E_T h^3 \times 10^3/q_o a^4 \cdot w(0,0),\ \bar{M}_{xc} = M_x(0,0)/q_o a^2,\ \bar{M}_{yc} = M_y(0,0)/q_o a^2$$

Table 1 lists the center displacements and moments for the two-layer (0°/0°) square laminated plates of materials 1 and 2, with various mesh sizes in comparison with those obtained by using the mixed rectangular linear/quadratic elements by [5] and the exact solution of [6]. The center displacements and stresses for a three-layer (0°/90°/0°) square laminated plate of material 3 are shown in Table 2 in comparison with the results in [6-8], results in [7] are obtained by using

Table 1 - Center deflections and moments of (0°/0°) laminated simple square plate. (a=b=10", h=0.1")

		present				[5]				[6]
	mesh →	2x2	3x3	4x4	5x5	4x4 L	8x8 L	2x2 Q	4x4 Q	exact sol.
material 1	$\bar{w}_c$	29.5153	29.5153	29.5153	29.5153	29.5670	29.5288	31.6232	29.5113	29.5154
	$\bar{M}_{xc}$	0.07613	0.07626	0.07628	0.07628	0.07737	0.07654	0.07687	0.07627	0.07628
	$\bar{M}_{yc}$	0.02739	0.02751	0.02754	0.02755	0.02785	0.02763	0.02785	0.02756	0.02756
2	$\bar{w}_c$	5.0336	5.0336	5.0336	5.0336	5.0444	5.0357	5.0466	5.0324	5.0335
	$\bar{M}_{xc}$	0.12462	0.12637	0.12654	0.12658	0.12845	0.12704	0.13094	0.12261	0.12659
	$\bar{M}_{yc}$	0.00311	0.00323	0.00325	0.00326	0.00314	0.00324	0.00337	0.00327	0.00327

Table 2 - Center deflections, moments and maximum stresses of a (0°/90°/0°) laminated simple square plate. (a=b=10", h=0.1"; material 3)

	present			[7]		[8]		[6]
mesh →	1x1	3x3	5x5	4x4 L	8x8 L	FSDT	HSDT	CPT exact
$\bar{w}_c$	6.6407	6.6601	6.6601	6.7171	6.7018	6.6970	6.7050	6.6601
$\bar{M}_{xc}$	0.12748	0.12970	0.12984	-	-	-	-	0.12985
$\bar{M}_{yc}$	0.00641	0.00839	0.00845	-	-	-	-	0.00847
$\bar{\sigma}_x$	0.7929	0.8067	0.8075	0.7724	0.7986	-	-	0.8072
$\bar{\sigma}_y$	0.1376	0.1984	0.1910	0.1977	0.1939	-	-	0.1927
$\bar{\tau}_{xy}$	0.04545	0.04328	0.04287	0.03650	0.04275	-	-	0.04275

$\bar{\sigma}_x = \mu\sigma_x(0,0,\pm h/2)$, $\bar{\sigma}_y = \mu\sigma_y(0,0,\pm h/2)$, $\bar{\tau}_{xy} = \mu\tau_{xy}(a/2,b/2,\pm h/6)$, $\mu = h^2/q_o^2$

a mixed shear flexible linear rectangular element with 8 d.o.f. and [8] are obtained by using the first-order shear deformation theory (FSDT) and the higher-order shear deformation theory (HSDT). A comparison between the results obtained by the present method and Reddy [9] who used a 2x2 mesh of linear/Hermite cubic element displacement model is presented in Table 3 for a quasi-isotropic $(0°/\pm45°/90°)_s$ laminated plate of material 4. All of the results show that both the ac-

Table 3 - Center deflections and moments of a quasi-isotropic $(0°/\pm45°/90°)_s$ laminated simple square plate of material 4

	present			[9]
mesh →	1x1	3x3	5x5	2x2 H
$\overline{w}_c$	3.7323	3.7581	3.7605	3.7690
$\overline{M}_{xc}$	0.11688	0.12478	0.12958	-
$\overline{M}_{yc}$	0.00326	0.00384	0.00408	-

curacy and convergence of the present method are excellent and are better than the rectangular element models. Results in Table 2 also show that the discrepancy between the CPT solution and HSDT solution is less than 0.7% for a symmetric laminated thin square plate (h/a = 0.01); thus, for this type of thin laminated plate, the use of CPT analysis is sufficient.

The effects of material anisotropy (E_L/E_T ratio) and plate aspect ratio (a/b) on the center (maximum) deflection and bending moment for a three-layer (0°/90°/0°) cross-ply plate of material 3 are illustrated in Figures 2 and 3. Results show that the material anisotropy and plate aspect ratio have strong effects on the deflection but are moderate on the moments. The deflection increases rapidly with the aspect ratio a/b varies from 1 to 3 and reaches its maximum value at approximately a/b = 5, then decreases slightly and approaches a constant value. The center bending moment, M_{xc}, has its maximum value at the aspect ratio a/b = 2 and M_{yc} has its maximum value at a/b = 5. Results in Figure 2 show that the center deflection decreases as the material anisotropy gets stronger.

For the alternating $\pm\theta_f$ angle-ply laminate which consists of infinite (many) number of orthotropic laminae, it can be treated as a homogeneous orthotropic plate. The equivalent elastic coefficients G_{ij} of the boron/epoxy and glass/epoxy $\pm\theta_f$ angle-ply laminated plates used here are directly adopted from [10] and listed in Table 4. The effects of fiber orientation, boundary conditions and plate aspect ratio on the deflections and moments are shown in Tables 5 and 6 for these $\pm\theta_f$ angle-ply laminated rectangular plates, results obtained by Hussain and Srinivas [10] are also included for comparison. The agreements between the present solutions and [10] are excellent. For square laminated plates, the effect of fiber orientation on the deflection is not significant for the case with all edges simply-supported and is greatly significant for those with S-C-S-C and S-F-S-F boundary conditions, for rectangular laminated plates, this effect is considerably large for all types of boundary conditions. In general, this effect is strongest

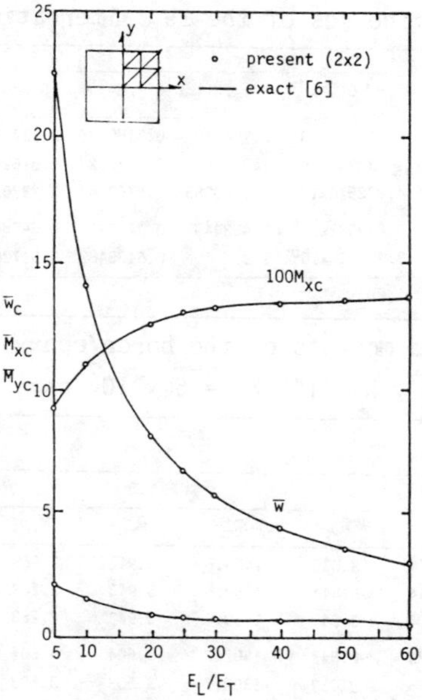

Fig. (2) - Effect of material anisotropy on the center deflection and moments of a three-layer $(0°/90°/0°)_s$ laminated simple plate. (a=b=10", h=0.1", material 3)

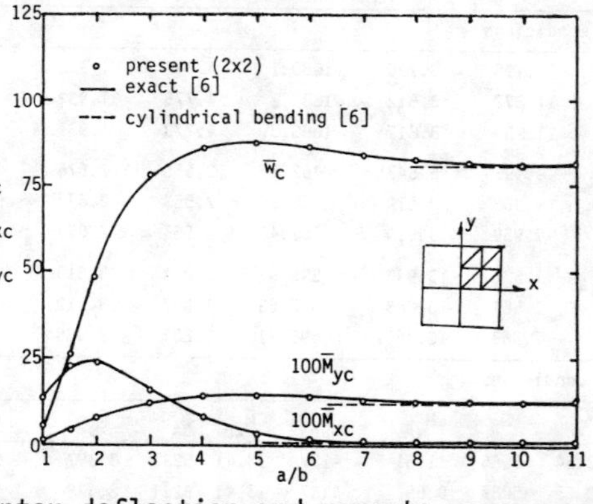

Fig. (3) - Center deflection and moments versus aspect ratio for a three-layer $(0°/90°/0°)_s$ laminated simple plate. (b=10", h=0.1", material 3)

for the S-F-S-F case. Results also show that the deflection w_A at the middle point of free edge is larger than the center deflection w_c for all types of S-F-S-F rectangular plates, this means that the plate is not a cylindrical bending form but is slightly a saddle form.

Table 4 - Equivalent stiffnesses of the $\pm\theta_f$ alternating angle-ply laminates

material	$\pm\theta_f$	G_{11}	G_{22}/G_{11}	G_{12}/G_{11}	G_{66}/G_{11}
Boron/ Epoxy	30°	2.487037×10^7	0.1943106	0.3180775	0.3375236
	45°	1.238941×10^7	1.0	0.8372123	0.8762567
	60°	4.832576×10^6	5.1463985	1.6369543	1.7370310
Glass/ Epoxy	30°	5.269582×10^6	0.4524497	0.2788481	0.3478621
	45°	3.499728×10^6	1.0	0.5133482	0.6172642

Table 5 - Deflections and moments of the boron/epoxy rectangular laminated plates. (b=10", h=0.1", $E_f = 6 \times 10^7$ psi)

(a) S-S-S-S boundary conditions

	a/b	0.3			1.0			3.0		
$\pm\theta_f$	mesh	$\bar{w}_c$	$\bar{M}_{xc}$	$\bar{M}_{yc}$	$\bar{w}_c$	$\bar{M}_{xc}$	$\bar{M}_{yc}$	$\bar{w}_c$	$\bar{M}_{xc}$	$\bar{M}_{yc}$
	3x3	3593.0	11.974	3.836	1459.6	5.942	2.268	132.78	1.571	0.862
30°	5x5	3593.2	11.994	3.844	1459.6	5.945	2.269	132.78	1.569	0.860
	[10]	3593.2	12.004	3.847	1459.6	5.943	2.268	132.78	1.569	0.860
	3x3	6065.8	10.383	8.842	1302.7	3.604	3.604	70.11	0.936	1.093
45°	5x5	6065.8	10.375	8.839	1302.7	3.603	3.603	70.11	0.936	1.092
	[10]	6065.7	10.373	8.836	1302.6	3.603	3.603	70.12	0.937	1.093
	3x3	11994.	8.442	15.052	1459.6	2.268	5.942	43.35	0.420	1.305
60°	5x5	11994.	8.424	15.031	1459.6	2.269	5.945	43.35	0.42	1.307
	[10]	11994.	8.425	15.030	1459.6	2.268	5.943	43.33	0.421	1.307

(b): S-C-S-C boundary conditions

	3x3	3546.6	11.798	3.796	1080.1	4.762	1.931	43.33	0.741	0.434
30°	5x5	3550.6	11.872	3.812	1080.2	4.775	1.932	43.32	0.734	0.429
	[10]	3552.2	11.887	3.817	1080.3	4.773	1.931	43.33	0.734	0.429
	3x3	5759.4	9.995	8.547	762.36	2.555	2.676	18.18	0.385	0.458
45°	5x5	5770.8	9.958	8.519	762.36	2.553	2.673	18.18	0.381	0.453
	[10]	5772.7	9.959	8.519	762.46	2.555	2.673	18.16	0.382	0.453
	3x3	10440.	7.626	13.943	595.62	1.233	3.510	9.30	0.149	0.468
60°	5x5	10449.	7.564	13.863	595.63	1.237	3.512	9.29	0.147	0.462
	[10]	10450.	7.563	13.861	595.71	1.237	3.509	9.31	0.147	0.462

(c): S-F-S-F boundary conditions

		$\bar{w}_c$	$\bar{M}_{xc}$	$\bar{w}_A$	$\bar{M}_{xA}$	$\bar{w}_c$	$\bar{M}_{xc}$	$\bar{w}_A$	$\bar{M}_{xA}$	$\bar{w}_c$	$\bar{M}_{xc}$	$\bar{w}_A$	$\bar{M}_{xA}$
	3x3	3791	12.54	5056	11.81	4137	13.41	5231	8.397	5477	13.69	6174	9.849
30°	5x5	3791	12.56	5087	9.592	4137	13.41	5231	8.338	5477	13.69	6174	9.854
	[10]	3785	12.53	4927	7.837	4093	13.28	5146	8.185	5403	13.52	6088	9.719
	3x3	7821	12.87	10642	7.131	9584	14.56	11853	5.900	16107	14.29	1744	8.687
45°	5x5	7824	12.86	10657	5.823	9584	14.56	11853	5.879	16107	14.30	1744	8.691
	[10]	7785	12.80	10420	5.152	9458	14.39	11674	5.783	15892	14.13	1721	8.576
	3x3	20384	13.05	26428	8.449	25647	13.81	29967	9.287	35866	12.96	3722	11.56
60°	5x5	20386	13.03	26428	8.226	25647	13.82	29967	9.285	35906	12.97	3722	11.55
	[10]	20225	12.94	25925	8.012	25293	13.65	29530	9.148	35400	12.81	3673	11.40

$\bar{w}_c = \frac{E_f}{q_o a^2 \times 10^2} w(0,0)$, $\bar{M}_{xc} = \frac{100}{q_o a^2} M_x(0,0)$, $\bar{M}_{yc} = \frac{100}{q_o a^2} M_y(0,0)$, $\bar{w}_A = \frac{E_f}{q_o a^2 \times 10^2} w(0, b/2)$, $\bar{M}_{xA} = \frac{100}{q_o a^2} M(0, b/2)$.

Table 6 - Deflections and moments of the glass/epoxy laminated square plates. (a=b=10", h=0.1", $E_f = 1.06 \times 10^6$ psi)

$\pm\theta_f$	mesh	B.C.'s - S-C-S-C			S-S-S-S			S-F-S-F			
		$\bar{w}_c$	$\bar{M}_{xc}$	$\bar{M}_{yc}$	$\bar{w}_c$	$\bar{M}_{xc}$	$\bar{M}_{yc}$	$\bar{w}_c$	$\bar{M}_{xc}$	$\bar{w}_A$	$\bar{M}_{xA}$
30°	3x3	697.63	3.6316	2.5019	1147.5	5.5088	3.0705	3210.7	12.512	3781.8	12.461
	5x5	697.65	3.6406	2.4973	1147.5	5.5097	3.0720	3210.7	12.515	3781.3	12.456
	[10]	697.68	3.642	2.497	1147.5	5.510	3.072	3177.0	12.397	3719.8	12.251
45°	3x3	572.14	2.4279	2.9010	1069.1	3.9593	3.9593	5066.8	12.802	5977.8	11.640
	5x5	572.15	2.4336	2.8960	1069.1	3.9594	3.9594	5066.8	12.806	5977.8	11.635
	[10]	572.2	2.434	2.896	1069.1	3.959	3.959	5005.1	12.666	5885.6	11.453

*The dimensionless parameters $\bar{w}_c,\ldots,\bar{M}_{xA}$ are defined in Table 5.

CONCLUDING REMARKS

A higher-order triangular plate bending element developed by Cowper et al [4] has been presented with some modifications for solving the bending problems of symmetric composite laminates subject to uniformly distributed load. Several examples with various material properties, ply orientations, plate aspect ratio, and boundary conditions have been investigated. According to the results obtained in this study, we have some conclusions as follows:

1. Because of requiring the satisfactions of compatibility conditions for the displacements and normal slopes on the boundaries of all triangular elements, the present method gives excellent accuracy and convergent speed for the solutions.

2. The present method provides a direct calculation for the bending moments at each node from the given values of curvatures (w_{xx}, w_{xy}, w_{yy}) at each node.

3. The deflections, moments and stresses are considerably affected by the lamination arrangement and the material anisotropy for various plate aspect ratio and boundary conditions.

4. For the alternating $\pm\theta_f$ angle-ply laminates with S-F-S-F boundary conditions, the deflections at the middle points of free edges are larger than the center deflection, that is, the plate deforms into a slightly saddle form. However, the bending moments M_{XA} at the middle points of free edges are smaller than moment M_{xc} at plate center which is in contradiction to the homogeneous isotropic plate.

5. The application of the present method can be extended to the vibration and buckling problems which are currently undertaken by the authors.

REFERENCES

[1] Tsai, S. W., Introduction to Composite Materials, Technomic Publishing Co., Inc., 1980.

[2] Whitney, J. M. and Leissa, A. W., "Analysis of heterogeneous anisotropic plates", Trans. ASME, Journal of Applied Mechanics, Vol. 36, No. 2, p. 261, 1969.

[3] Jones, R. M., Mechanics of Composite Materials, Scripta Book Co., Washington, D.C., 1975.

[4] Cowper, G. R., Kosko, G. M. and Olson, M. D., "A high precision triangular plate-bending element", Aeronautical Report LR-514, National Research Council of Canada, December 1968.

[5] Tsay, C. T. and Reddy, J. N., "Bending, stability and free vibration of thin orthotropic plates by simplified mixed finite elements", J. Sound and Vibration, 59(2), p. 307, 1978.

[6] Ashton, J. E. and Whitney, J. M., Theory of Laminated Plates, Technomic Publishing Co., Inc., 1970.

[7] Putcha, N. S. and Reddy, J. N., "A mixed shear flexible finite element for the analysis of laminated plates", Computer Methods in Applied Mechanics and Engineering, 44, p. 213, 1984.

[8] Reddy, J. N., "A simple higher-order theory for laminated composite plates", Trans. ASME, Journal of Applied Mechanics, Vol. 51, p. 745, 1984.

[9] Reddy, J. N., "On mixed and displacement finite element models of a refined shear deformation theory for laminated anisotropic plate", 4th International Conference on Applied Numerical Modeling, Tainan, Taiwan, ROC, p. 64, 1984.

[10] Hussain, S. A. and Srinivas, S., "Flexure of rectangular composite plates", Fibre Science and Technology, 8, p. 59, 1975.

LAMINATED COMPOSITES OF AMORPHOUS ALLOYS FOR EMI SHIELDING

D.-R. Huang, P.-C. Yao and S.-E. Hsu

Chung Shan Institute of Science and Technology
P.O. Box 1-26-1, Lungtan, Taiwan 325, Republic of China

ABSTRACT

Recently, ferromagnetic amorphous alloys have been used as new electromagnetic interference (EMI) shielding materials, especially for magnetic effect, because of their good magnetic, electric and mechanical properties. Also, the primary advantage is that the magnetic and electric properties are not evidently degraded under the bending, flexing and mechanical shock. We have made some EMI shields by combining the amorphous ribbons via composite technology. The cylindrical and plate shields were laminated and molded by bonding six-layer amorphous ribbons $Fe_{40}Ni_{38}M_{04}B_{17}Si$ with epoxy resin. We have tested the mechanical properties: tensile strength 27 ksi, tensile modulus 3×10^7 psi and fracture strain 80 ∿ 85%. The magnetic shielding effectiveness of cylindrical and plate shields at field strength 0.1 Oe ∿ 3 Oe over broad frequency ranges 60 Hz ∿ 1 kHz were 20 dB ∿ 40 dB. The Ac 60 Hz magnetic shielding effectiveness between theoretical calculation and experimental data for a six-layer shield are similar. An excellent electric field shielding effectiveness above 60 dB was tested over frequencies 10 kHz ∿ 1 GHz.

INTRODUCTION

Amorphous alloy is a new material made by rapid quenching technique with quenching rate about 10^6°C/sec. There are many applications and developments for amorphous ribbons due to their unusual properties. Ferromagnetic amorphous ribbons can contain their good magnetic and electric properties even after the bending and flexing. We can fabricate a rigid shape of EMI shield by ferromagnetic amorphous ribbons via composite technology [1,2]. The cylindrical and plate shields in our experiments were made by laminating and bonding the amorphous ribbons with epoxy resin. The purpose of this paper is to compare the results between experimental data and theoretical calculations of the composite shields. To design a good magnetic shield, the data of theoretical calculations is a valuable evaluation for the electronic engineering designers.

EXPERIMENT

Ferromagnetic amorphous ribbons of the composition $(FeNi)_{78}M_{04}B_{17}Si$ were made by melt-spinning technique [3]. The ribbons used in this experiment were about

30 μm in thickness and 20 mm in width. The magnetization of amorphous ribbons was measured as a function of the applied magnetic field ($0 \le H \le 10$ kOe) in vibrating sample magnetometer (VSM, EG&G, 155-III) [4]. The $\overline{DC} \sim \overline{AC}$ 10 kHz low field soft magnetic properties such as permeability (μ), coercive force (Hc), remanence (Br),... etc., were measured by B-H meter (Walker MH-20) [5]. The electric resistivity ρ (μΩ-cm) was tested by four points method.

Cylindrical shields were fabricated by convoluting six-layer amorphous ribbons on a 20 cm long pyrex tube with outer diameter 3.5 cm. The intermedium between ribbon layers were bonded by epoxy resin (or reinforced by adding glass fabric). Two type composite shields (parallel ribbons and cross ribbons) were made as shown in Figure 1. The thicknesses of composite shields were about 2 mm and 3.5 mm (with glass fabric). Plate shields were fabricated by laminating six-

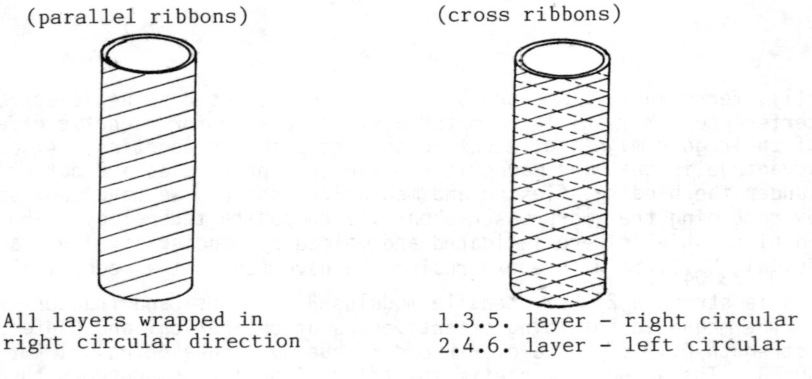

Fig. (1) - Cylindrical shield (6-layer ribbons)

layer amorphous ribbons with epoxy resin (or adding glass fabric). The ribbons were overlapped slightly. Also, two plate composite shields were made as Figure 2. The size of shield was a square plate with dimension 30 cm x 30 cm x 2 mm or 30 cm x 30 cm x 3.5 mm (with glass fabric).

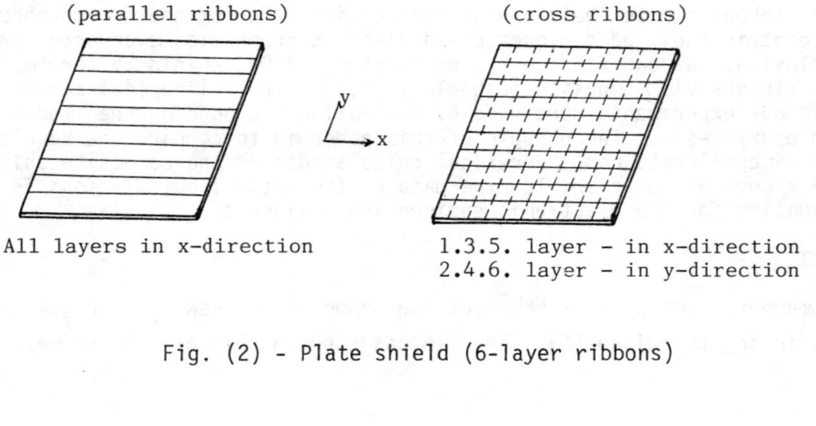

Fig. (2) - Plate shield (6-layer ribbons)

The mechanical properties of laminated composites were determined by linear stress-strain curves with longitudinal tensile specimens, Figure 3.

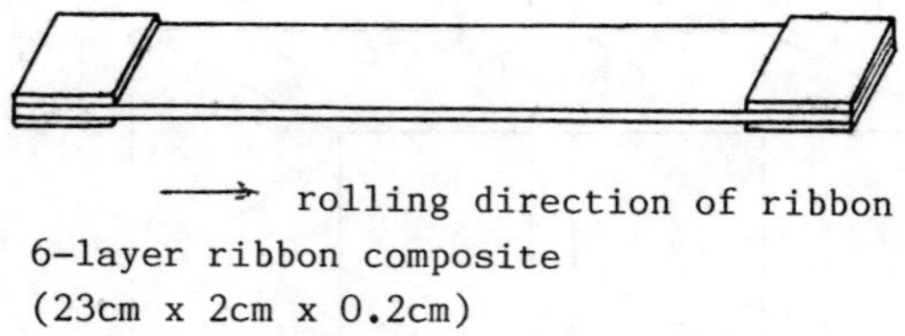

Fig. (3) - Longitudinal tensile specimen

The magnetic shielding effectivenesses (60 Hz ∿ 1 kHz) were tested by following the procedures described in ASTM standard A698-74 with slight modification. A pair of Helmholtz coils with 30 cm in diameter was used for testing cylindrical composite shields [5,6], Figure 4. The magnetic shielding effectivenesses (60 Hz ∿ 1 kHz) of plate composite shields were measured by small loop-to-loop method

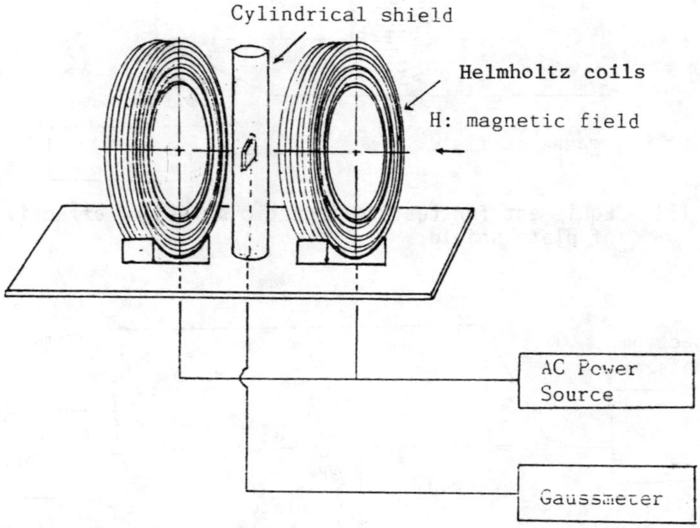

Fig. (4) - Equipment for testing magnetic shielding effectiveness of cylindrical shield by following the procedure of ASTM A698-74

with modification [7,8], Figure 5. Electric shielding effectiveness was obtained by an insertion loss technique based on the MIL-STD-285 over an effective frequency range from 10 kHz to 1 GHz in shield rooms. A variable frequency electric field was generated by an antenna from one shielded room and transmitted through a hole (with or without plate shield) into an adjacent shielded room where a pick up antenna received the field signal to a spectrum analyzer [6], Figure 6.

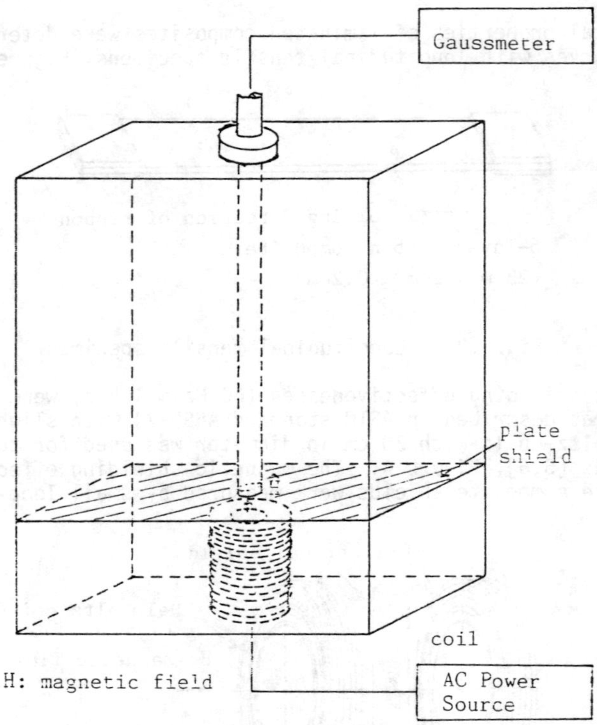

Fig. (5) - Equipment for testing magnetic shielding effectiveness of plate shield

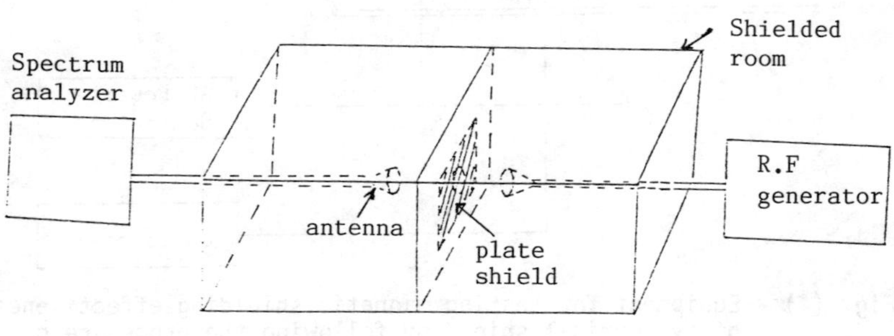

Fig. (6) - Measurement of EM-shielding effectiveness over broad frequency range (10 kHz $\sim$ 1 GHz)

DEFINITION AND THEORY

EMI shielding effect can be divided into three parts. We may define the shielding effectiveness as follows:

Magnetic shielding effectiveness:

$$S_H = 20 \log_{10} \frac{H_o}{H_i} \text{ (dB)}$$

where H_o: the magnetic field strength unshielded

H_i: the magnetic field strength measured with shield

Electric shielding effectiveness:

$$S_E = 20 \log_{10} \frac{E_o}{E_i} \text{ (dB)}$$

where E_o: the electric field intensity detected by antenna without a shield

E_i: the electric field intensity detected by antenna with a shield

EM - wave shielding effectiveness

$$S_P = \log_{10} \frac{P_o}{P_i} \text{ (dB)}$$

where P_o: the plane wave energy intensity detected without a shield

P_i: the plane wave energy intensity detected with a shield

For a multi-layer cylindrical shield, the total magnetic shielding effectiveness at DC or low frequency magnetic field (H_o) can be derived from Maxwell's equation [9] as

$$S_H = 1 + \sum_{i=1}^{n} [S_i + S_1 \cdot S_2 (1 - \frac{A_1}{A_2}) \ldots S_i (1 - \frac{A_i - 1}{A_i})]$$

where

$$S_i = \frac{1}{2} \frac{\mu_i t_i}{R_i}$$

For a 6-layer amorphous ribbon cylindrical shield, if each layer has the same value of thickness and permeability, we can get the magnetic shielding effectiveness as

$$S_H = 1 + \frac{\mu t}{2} (\frac{1}{R_1} + \ldots + \frac{1}{R_6}) + (\frac{\mu t}{2})^6 (1 - \frac{A_1}{A_2}) \ldots (1 - \frac{A_5}{A_6}) = 1 + \frac{\mu t}{2} \times$$

$$\times (\frac{1}{R_1} + \ldots + \frac{1}{R_6}) + (\frac{\mu t}{2})^6 (1 - \frac{R_1}{R_2}) \ldots (1 - \frac{R_5}{R_6}) / R_1 \ldots R_6$$

where

where

$$A_1 = 2\pi R_1 \ell \ \ldots \ A_6 = 2\pi R_6 \ell, \text{ and } \frac{A_1}{A_2} = \frac{R_1}{R_2}, \ \ldots \ \frac{A_5}{A_6} = \frac{R_5}{R_6}$$

$$R_1 = 35 \text{ mm}, \ R_2 = R_1 + \Delta R + t, \ \ldots \ R_6 = R_5 + \Delta R + t$$

The above values (A,R,r) of composite cylindrical shield with six-layer amorphous ribbons can be obtained as shown in Figure 7. For theoretical calculation, we

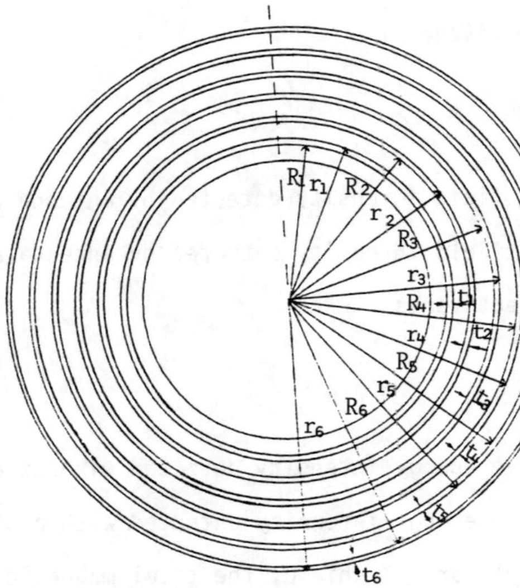

thickness of ribbon
$t = (t_1 + \cdots \ t_6)/6$
thickness of resin
$\Delta R = (\Delta R_1 + \cdots \ \Delta R_6)/6$
where $\Delta R_1 = r_1 - R_1, \cdots \Delta R_6 = r_6 - R_6$

Fig. (7) - Composite cylindrical shield with 6-layer ribbons

can measure the thickness of intermedium (ΔR) and thickness of amorphous ribbon (t) by using the SEM photograph as Figure 8. The mechanical properties of laminated composites can be estimated by the rule of "mixture calculation" [10,11].

$$Sc = SrVr + SmVm$$

$$Ec = ErVr + EmVm$$

where S is tensile strength; E is Young's modulus; c is composite; r is ribbon and m is mixture.

RESULTS AND DISCUSSION

The magnetic and electric properties of ferromagnetic amorphous ribbon with different constituents vary evidently. For Fe-Ni base amorphous ribbons, we have

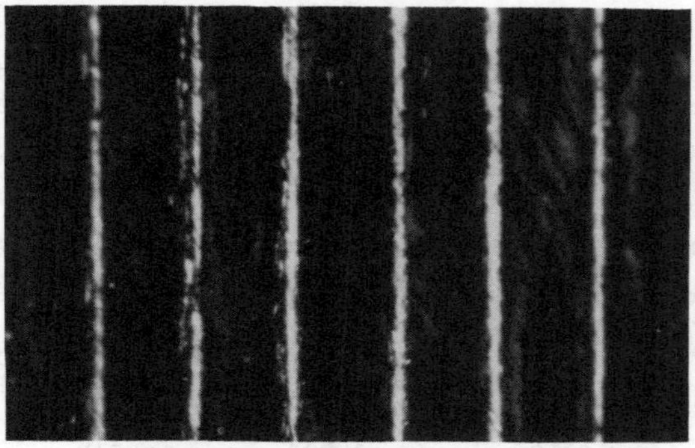

| ΔR_1 || ΔR_2 || ΔR_3 || ΔR_4 || ΔR_5 || ΔR_6 || ΔR_7 |

t_1 t_2 t_3 t_4 t_5 t_6

thickness of resin

$$\Delta R = \frac{(\Delta R_1+\Delta R_2+\Delta R_3+\Delta R_4+\Delta R_5+\Delta R_6+\Delta R_7)}{7} \simeq 0.25\text{mm}$$

thickness of ribbon

$$t = (t_1+ t_2+ t_3+ t_4+ t_5+ t_6)/6 = 30\mu\text{m} \simeq 0.03\text{mm}$$

thickness of shield

$$T_1 = 6t + 7\Delta R \simeq 2.0\text{mm (without fabric)}$$

Fig. (8) - Photograph of 6-layer ribbon shield by SEM

measured the magnetic and electric properties as Table 1. We find that Fe-rich amorphous ribbon $Fe_{78}M_{04}B_{17}Si$ possess high saturation magnetization σ_s, low

Table 1 - Composition dependence of magnetic and electrical properties for $(FeNi)_{78}M_{04}B_{17}Si$ amorphous ribbons

Constituent Properties	$Fe_{78}M_{04}B_{17}Si$	$Fe_{40}Ni_{38}M_{04}B_{17}Si$	$Fe_{18}Ni_{60}M_{04}B_{17}Si$
Saturation magnetization σ_s	163 emu/g	96 emu/g	45 emu/g
Max. permeability μ_{max} (as-cast)	18,000	21,000	17,000
Initial permeability μ_i (as-cast)	2,000	6,000	3,500
Coercivity Hc (as-cast)	0.1 Oe	0.05 Oe	0.07 Oe
Resistivity ρ	140 $\mu\Omega$-cm	158 $\mu\Omega$-cm	166 $\mu\Omega$-cm
Curie temp. Tc	∿ 370°C	∿ 350°C	∿ 330°C

electric resistivity and higher curie temperature Tc. While medium Ni amorphous ribbon possess the large permeabilities (μ_i, μ_{max}) and lower coercivity Hc, we may use $Fe_{40}Ni_{38}M_{04}B_{17}Si$ as good magnetic shielding material, especially for low fields.

Some mechanical properties of laminated composite made by $Fe_{40}Ni_{38}M_{04}B_{17}Si$ amorphous ribbon bonded with resin are listed in Table 2. The data were obtained

Table 2 - Mechanical properties of laminated composite ($Fe_{40}Ni_{38}M_{04}B_{17}Si$ ribbons - epoxy resin)

(Longitudinal)	Laminated Composite	Ribbon	Resin	Mixture Calculation
Tensile Strength	27 ksi	245 ksi	0.19 ksi	26.1 ksi
Young's Modulus	3×10^7 psi			
Poisson's Ratio	0.34			
Fracture Strain	80 ∿ 85%			
Volume Fraction	100%	10.5%	89.5%	

by average values of 10 tested specimens. The tensile strength obtained by mixture calculation is also listed in the table. The error value of tensile strength between experiment and mixture calculation is about 4%. We may estimate the mechanical properties of composite shield by mixture calculation.

From the theory of electromagnetic shielding, the magnetic shielding efficiency of magnetic material is dominately determined by the material's permeability (lower than 10 kHz). Since the permeability of a magnetic material is not a constant value, but a function of applied field, we would obtain the shielding effectiveness as a function of applied field H.

Figure 9 shows the AC 60 Hz magnetic shielding effectiveness as a function of applied magnetic field (0.1 Oe ∿ 10 Oe). We find that the maximum shielding effectiveness of the four curves are near the applied field 2∿3 Oe, which is the field to produce the maximum permeability of as-cast $Fe_{40}Ni_{38}M_{04}B_{17}Si$ amorphous ribbons. Also, we find that the shields with glass fabric possess higher magnetic shielding effectiveness which may be induced by the larger intermedium distance ΔR between ribbons. The result is the same as the formula of magnetic shielding theory

$$S_H \sim S_1 \cdot S_2 (1 - \frac{A_1}{A_2}) \ldots S_6 (1 - \frac{A_5}{A_6}) = S_1 \cdot S_2 (1 - \frac{R_1}{R_2}) \ldots S_6 (1 - \frac{R_5}{R_6})$$

the larger ΔR will produce the larger S_H. The magnetic shielding effectiveness of multi-layer cylindrical shield (without glass fabric) can be calculated by the theoretical formula as shown in Figure 10. It indicates that the curve of theoretical calculation is very close to the experimental data.

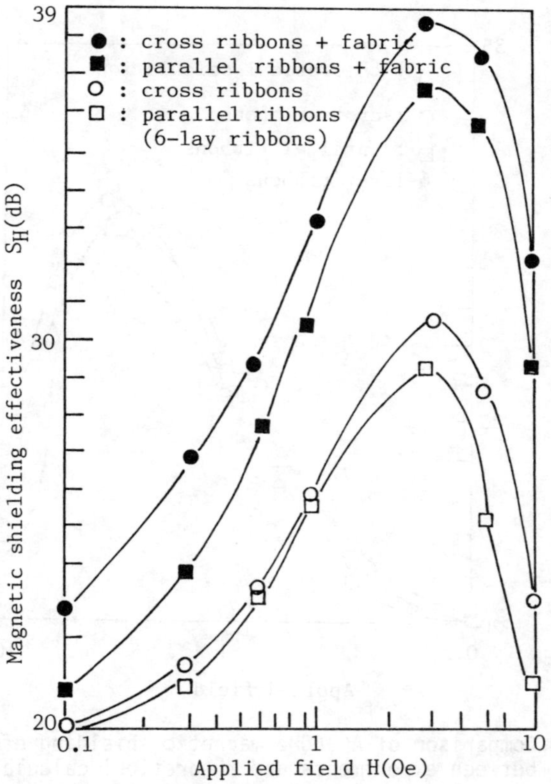

Fig. (9) - AC 60Hz magnetic shielding effectiveness as a function of applied field for cylindrical shields

For AC fields, the magnetic shielding phenomenon is a combination of shunting effect and skin effect. At lower frequency, the applied magnetic field is dominately shunted through the low reluctance shield (as in DC field). While at higher frequency, the skin effect also occurred by an induction of the eddy current in the shield which contracts the applied field. Figures 11 and 12 show the magnetic shielding effectivenesses of cylindrical and plate shields under different frequencies (60 Hz ∿ 10 KHz). For a given magnetic field H = 1 Oe, the magnetic shielding effectiveness of cylindrical shield is only a slight increase at higher frequency, but the magnetic shielding effectivenesses of plate shields are increased evidently. The reason is that the skin effect induced on the surface of shield which is perpendicular to the applied field, the plate shields possess larger plane area to produce the skin effect.

Because of the good electric conductivity of amorphous ribbons, the electric shielding effectiveness of plate shields are higher than 60 dB over broad frequency ranges (10 KHz ∿ 10 GHz) for a given field strength, Figure 13.

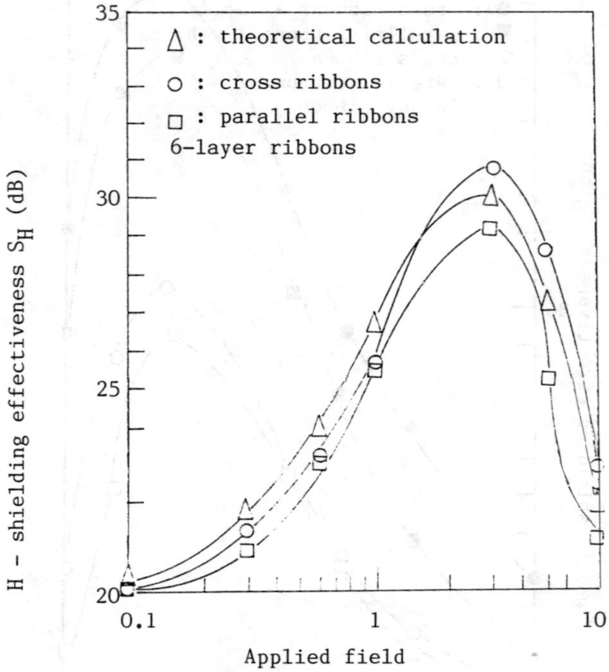

Fig. (10) - Comparison of AC 60Hz magnetic shielding effectiveness between experiments and theoretical calculation

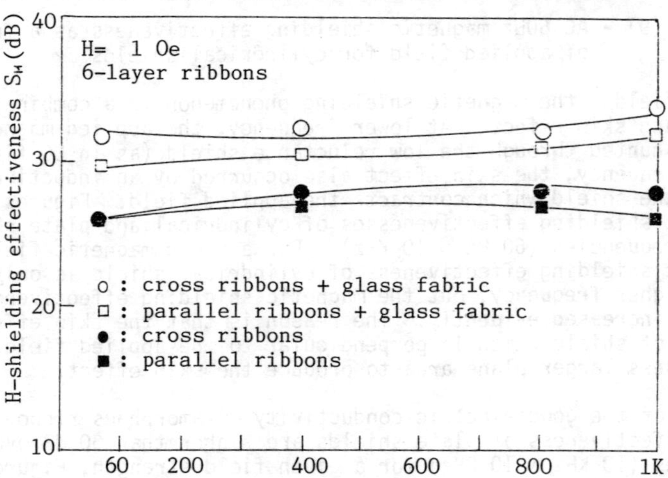

Fig. (11) - Magnetic shielding effectiveness of cylindrical shields under different frequencies (60Hz ∿ 1KHz)

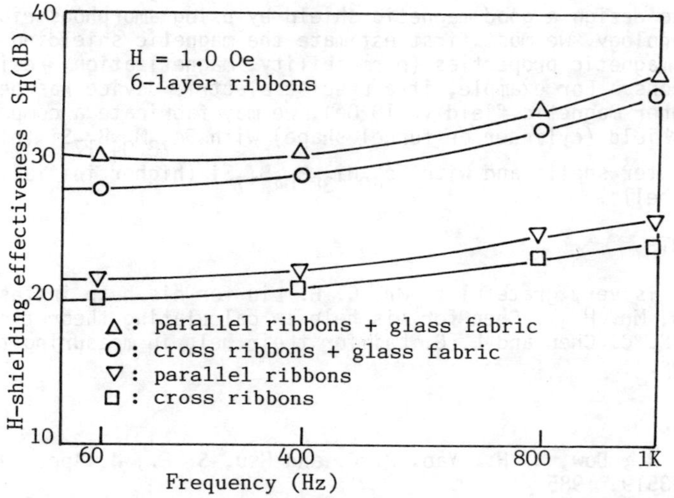

Fig. (12) - Magnetic shielding effectivenesses of plate shields under different frequencies (60Hz ∼ 1 KHz)

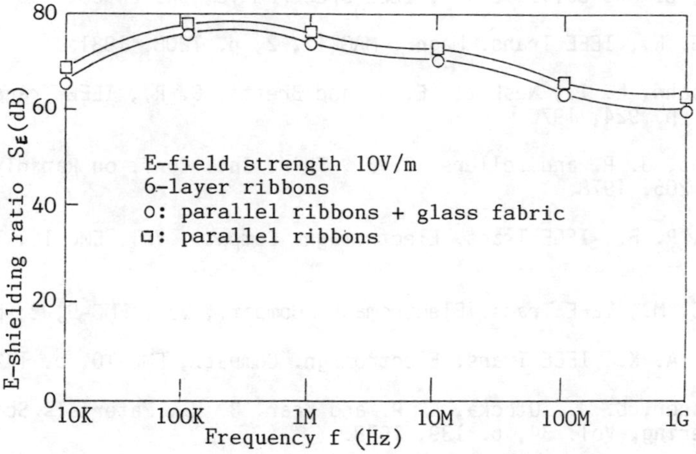

Fig. (13) - Electric shielding effectiveness of plate shields under broad frequency range (10KHz ∼ 1GHz)

CONCLUSION

An excellent EMI shielding is essential to many electronic devices such as cathode ray tube, RF-filter, switching power supply, etc. With good magnetic, electric and mechanical properties, ferromagnetic amorphous ribbons offer electric engineers a new option to design complicated shape EMI shield, especially for magnetic shield.

In order to design a good magnetic shield by using amorphous ribbons via composite technology, we must first estimate the magnetic shielding efficiency by the electromagnetic properties (permeability, magnetization, resistivity) of amorphous ribbons. For example, if a precise electric device has been interfered by a higher magnetic field (> 10 Oe), we may fabricate a composite multilayer ribbon shield (cylinder or funnel shape) with $Fe_{78}M_{04}B_{17}Si$ (high magnetization) in the outer shells and with $Fe_{40}Ni_{38}M_{04}B_{17}Si$ (higher initial permeability) in the inner shells.

ACKNOWLEDGEMENTS

The author is very grateful to Mr. C. H. Liu for his help in testing mechanical properties, Mr. H. Y. Chen for his help in calculating theoretical data by computer, Mr. S. C. Chen and L. G. Lai for their help in measuring the shielding effectivenesses.

REFERENCES

[1] Huang, D. R., Dow, W. H., Yao, P. C. and Hsu, S. E., J. Appl. Phys., 57(1), pp. 3517-3519, 1985.

[2] Huang, D. R., Dow, W. H., Yao, P. C. and Hsu, S. E., Symposium on MRS 1985 Fall Meeting, Session 5P20, 1985.

[3] Raskin, D. and Davis, L. A., IEEE Spectrum 18, pp. 28-33, 1981.

[4] Kanl, S. N., IEEE Trans. Magn., MAG-17, 2, p. 1208, 1981.

[5] Mendelsohn, L. I., Nesbitt, E. A. and Bretts, G. R., IEEE Trans. Magn., MAG-12, p. 924, 1976.

[6] Dismukos, J. P. and Sellers, G. J., Third Int. Conf. on Rapidly Quenched Metal, 205, 1978.

[7] Schulz, R. B., IEEE Trans. Electromagn. Compat., Vol. EMC-10, 1, p. 95, 1968.

[8] Ryan, C. M., IEEE Trans. Electromagn. Compat., Vol. EMC-9, 2, p. 83, 1967.

[9] Thomas, A. K., IEEE Trans. Electromagn. Compat., EMC-10, p. 142, 1968.

[10] Goldwasser, D. J., Otocka, E. P. and Kear, B. H., Materials Science and Engineering, Vol. 34, p. 139, 1978.

[11] Goldwasser, D. J. and Kear, B. H., Mater. Sci. Eng., Vol. 23, p. 237, 1975.

STRESS AND STRENGTH ANALYSIS OF FINITE COMPOSITE LAMINATES WITH UNLOADED FASTENER HOLES

C.-C. Ko

ARL/CSIST
Taichung, Taiwan, Republic of China

C.-C. Lin

National Chung-Hsing University
Taichung, Taiwan, Republic of China

and

C.-L. Ong

MRDC/CSIST
Taichung, Taiwan, Republic of China

ABSTRACT

An analytical study concerned with the stress distributions around an unloaded fastener circular or elliptical hole in a finite composite laminate and its strength is proposed. The Airy stress function expressed in terms of full Laurent's series and a complex variable technique in conjunction with the "boundary collocation method" are employed in this study. Once the stress is determined for each point in the laminate, the on-axis stress of each single layer can be evaluated. Thus, the laminate strength can be predicted by using the concept of characteristic curve and the Yamada-Sun failure criterion. Numerical solutions for various plate and hole sizes, layups, material properties and loading conditions are examined. Results are compared with some given experimental data. Some conclusions are drawn.

INTRODUCTION

Because of their favorable performance advantages, the fiber-reinforced composite materials have been increasingly used in industry in recent years. However, composite materials have many characteristics that are different from more conventional engineering materials. It is known that the maximum value of stress concentrations around an open fastener circular hole in an infinite isotropic plate subject to uniform tensile load is three times of the applied stress and the value will be higher with an elliptical hole. Due to the material anisotropy, the stress concentrations for composites can be much more severe. Moreover, because the general lack of ductility (assumed to be linearly elastic up to failure), composites are very sensitive to stress concentrations and are easy to failure when this high stress reaches the proportional limit. Therefore, the study on the stress concentrations and strength prediction is an important course.

In the 1960's, Lekhnitskii [1] and Savin [2] used a complex variable method developed by Muskhelishvili to investigate the stress concentration around a hole in an infinite anisotropic plate. Later, Garbo and Ogonowski [3,4] predicted the strength of composite laminates with unloaded fastener holes using a closed-form analytic approach. However, to the authors' knowledge, no known exact solutions are available for stress concentrations of finite anisotropic plates with holes but some numerical solutions have been accomplished. Ogonowski [5] employed the complex variable technique in conjunction with boundary collocation procedure to determine the stress concentrations about a circular hole in a finite laminated plate. Gerhardt [6] used a hybrid finite element approach to investigate the same problem.

In this paper, we present an analytical study on the stress distributions around unloaded fastener circular or elliptical holes in finite composite laminates and the strength predictions. Similar to [5], the stress function expressed in terms of full Laurent's series and a complex variable technique in conjunction with the "boundary collocation method" are employed in the present study. The laminate strength is predicted by using the concept of "characteristic curve" [7] and the Yamada-Sun failure criterion [8]. Numerical solutions for various plate and hole sizes, layups, material properties and loading conditions are obtained and compared with some other available results.

METHOD OF STRESS ANALYSIS

The geometry and coordinates of a finite composite laminate are shown in Figure 1. To determine the stress and strength for this laminate, the method of

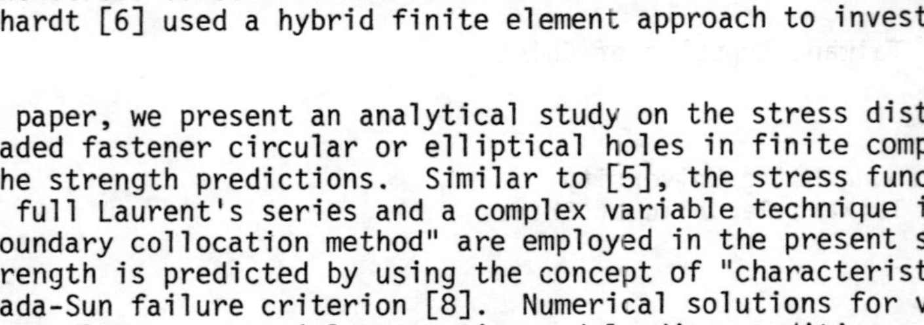

(a) Laminate configuration. (b) Loading conditions.

Fig. (1) - Generalized problem of interest

analysis is based on (1) classical lamination plate theory, (2) laminate effective moduli, (3) anisotropic theory of elasticity, (4) least square boundary collocation approximation, (5) characteristic radius, (6) Yamada-Sun failure criterion.

Consider an orthotropic lamina with principal axes L and T, the constitutive equation for plane state of stress referred to principal axes is given by [9]

$$\begin{Bmatrix} \sigma_L \\ \sigma_T \\ \tau_{LT} \end{Bmatrix} = \begin{bmatrix} Q_{11} & Q_{12} & 0 \\ Q_{12} & Q_{22} & 0 \\ 0 & 0 & Q_{66} \end{bmatrix} \begin{Bmatrix} \varepsilon_L \\ \varepsilon_T \\ \varepsilon_{LT} \end{Bmatrix} \qquad (1)$$

The fiber direction L makes an angle θ_f with the geometric axis x. After coordinate transformations, we can obtain the constitutive equation of this lamina referred to the geometric axes (x,y) as

$$\begin{Bmatrix} \sigma_x \\ \sigma_y \\ \tau_{xy} \end{Bmatrix} = \begin{bmatrix} \overline{Q}_{11} & \overline{Q}_{12} & \overline{Q}_{16} \\ \overline{Q}_{12} & \overline{Q}_{22} & \overline{Q}_{26} \\ \overline{Q}_{16} & \overline{Q}_{26} & \overline{Q}_{66} \end{bmatrix} \begin{Bmatrix} \varepsilon_x \\ \varepsilon_y \\ \gamma_{xy} \end{Bmatrix} \qquad (2)$$

where $[\overline{Q}] = [T]^{-1}[Q][T]^{-T}$ and $[T]$ is the coordinate transformation matrix [9].

The laminated plate consisting of N orthotropic laminae is considered to be symmetric and subjected to inplane loads. Thus, the relationship between the inplane stress resultants and the strains (constant through the thickness) is

$$\begin{Bmatrix} N_x \\ N_y \\ N_{xy} \end{Bmatrix} = \begin{bmatrix} A_{11} & A_{12} & A_{16} \\ A_{12} & A_{22} & A_{26} \\ A_{16} & A_{26} & A_{66} \end{bmatrix} \begin{Bmatrix} \varepsilon_x \\ \varepsilon_y \\ \gamma_{xy} \end{Bmatrix} \qquad (3)$$

where

$$A_{ij} = \sum_{k=1}^{N} \overline{Q}_{ij}^{(k)} h_k$$

in which h_k is the thickness of kth layer. For the purpose of convenience in practical analysis, we may consider the composite laminate as a homogeneous anisotropic plate with equivalent stiffnesses G_{ij} defined as

$$\begin{Bmatrix} \sigma_x \\ \sigma_y \\ \tau_{xy} \end{Bmatrix} = \begin{bmatrix} G_{11} & G_{12} & G_{16} \\ G_{12} & G_{22} & G_{26} \\ G_{16} & G_{26} & G_{66} \end{bmatrix} \begin{Bmatrix} \varepsilon_x \\ \varepsilon_y \\ \gamma_{xy} \end{Bmatrix} \qquad (4)$$

where $G_{ij} = A_{ij}/h$ and h is the plate thickness.

Now the composite laminates to be examined in this study are symmetric and orthotropic, i.e., the laminate with ply-orientations in the type $(0^\circ_i/90^\circ_j/\pm\theta_k)_s$. Then according to Lekhnitskii's theory of two-dimensional anisotropic elasticity [10], the stress function F(x,y), satisfying the equilibrium and compatibility conditions, is the solution of the biharmonic equation

$$S_{22}\frac{\partial^4 F}{\partial x^4} + (2S_{12}+S_{66})\frac{\partial^4 F}{\partial x^2 \partial y^2} - S_{11}\frac{\partial^4 F}{\partial y^4} = 0 \tag{5}$$

where S_{ij} are laminate compliance coefficients. The characteristic equation of equation (5) is

$$S_{11}\mu^4 + (2S_{12}+S_{66})\mu^2 + S_{22} = 0 \tag{6}$$

There is no real root in equation (6). Hence, it has two pairs of conjugate complex roots which are denoted by μ_1, $\bar{\mu}_1$ and μ_2, $\bar{\mu}_2$ (it is easily seen $\mu_1 = \mu_2 = i$ for isotropic material). Since the stress function must be real, hence we may express F(x,y) as the form

$$F = 2R_e[F_1(z_1) + F_2(z_2)] \tag{7}$$

where

$$z_1 = x + \mu_i y,\ z_2 = x + \mu_2 y \tag{8}$$

Introducing the functions

$$\phi_1(z_1) = \frac{\partial F}{\partial z_1},\ \phi_2(z_2) = \frac{\partial F}{\partial z_2} \tag{9}$$

the general expressions for the stress and displacement components are obtained

$$\begin{aligned}
\sigma_x &= 2\mathrm{Re}[\mu_1^2\phi_1'(z_1) + \mu_2^2\phi_2'(z_2)]\\
\sigma_y &= 2\mathrm{Re}[\phi_1'(z_1) + \phi_2'(z_2)]\\
\tau_{xy} &= -2\mathrm{Re}[\mu_1\phi_1'(z_1) + \mu_2\phi_2'(z_2)]\\
u &= 2\mathrm{Re}[p_1\phi_1(z_1) + p_2\phi_2(z_2)]\\
v &= 2\mathrm{Re}[q_1\phi_1(z_1) + q_2\phi_2(z_2)]
\end{aligned} \tag{10}$$

where

$$\begin{aligned}
p_1 &= S_{11}\mu_1^2 + S_{12}\\
p_2 &= S_{11}\mu_2^2 + S_{12}\\
q_1 &= S_{12}\mu_1 + S_{22}/\mu_1\\
q_2 &= S_{12}\mu_2 + S_{22}/\mu_2
\end{aligned}$$

The functions $\phi_1(z_1)$ and $\phi_2(z_2)$ are to be determined by satisfying the boundary conditions. The laminate considered here is a type of doubly-connected region which outer boundary is rectangular and inner boundary is circular or elliptic. We obtain the solutions $\phi_k(z_k)$ by mapping conformally the region outside the elliptic hole onto the exterior of the unit circle $|\xi|\geq|$ under the mapping function [10]

$$\xi_k = [z_k + \sqrt{z_k^2-(a^2+\mu_k^2 b^2)}]/(a-i\mu_k b), \ (k = 1,2) \tag{11}$$

where a and b are the lengths of the principal semi-axes of the ellipse and $i = \sqrt{-1}$. The seeking stress function may be expressed as

$$\phi_k'(z_k) = \psi_k(\xi_k) = \sum_{m=1}^{\infty} H_{km}\xi_k^{-m} \tag{12}$$

After integration

$$\phi_k(z_k) = A_k \ell n \xi_k + \sum_{m=1}^{\infty} A_{km}\xi_k^{-m} \tag{13}$$

On the other hand, the stress function associated with the interior of the rectangular boundary may be represented by a Taylor series

$$\phi_k(z_k) = \sum_{m=1}^{\infty} B_{km}z_k^m \tag{14}$$

Thus the stress function in the doubly-connected region can be obtained by the superposition of equations (13) and (14)

$$\phi_k(z_k) = A_k \ell n \xi_k + \sum_{m=1}^{\infty} (A_{km}\xi_k^{-m} + B_{km}z_k^m), \ (k = 1,2) \tag{15}$$

The constants A_k are to be determined from the internal boundary conditions and the conditions of single-valuedness on displacements, the coefficients A_{km} and B_{km} are determined from the internal and external boundary conditions.

For a finite rectangular plate with elliptic or circular hole, the external and internal boundaries cannot be described exactly by a coordinate system. In order to determine the stress function in the doubly-connected region, we shall take the procedures:

(1) Truncate the Laurent series (15) to a finite series containing the positive and negative power terms.

(2) Select certain suitable points on the external and internal boundaries and impose the stresses on these boundary points to satisfy the boundary conditions by the "boundary collocation method".

For more accurate results, we shall use the least-squares boundary collocation technique, in which the boundary points are selected such that the number of equations (the imposed conditions) is greater than the number of unknown coefficients in the truncated Laurent series. These unknown coefficients are then solved from the over-determined system of simultaneous linear equations by minimizing the average squared error of the conditions on boundary points. Once the stress function $\phi_k(z_k)$ are obtained, the stresses and displacements at any point in the laminate can be determined.

PREDICTION OF FAILURE

When the stress distribution around the perimeter of the hole is given, the failure strength and location of failure initiation can be predicted by comparing with a failure criterion. To account for composite laminate inelastic or nonlinear behavior at the hole boundary, the "characteristic dimension" concept by Whitney and Nuismer [7] has been adopted in cooperation with the Yamada-Sun [8] failure criterion in this paper. Laminate failure is predicted by comparing elastic stress distributions with the Yamada-Sun failure criterion

$$\left(\frac{\sigma_1}{X}\right)^2 + \left(\frac{\tau_{12}}{S_c}\right)^2 = e^2, \begin{cases} e>1 & \text{failure} \\ e=1 & \text{critical} \\ e<1 & \text{safe} \end{cases} \qquad (16)$$

on a ply-by-ply basis at a characteristic dimension away from the hole boundary ($r = R_c$). The parameter X in equation (16) is the axial strength of a single layer and S_c is the rail-shear strength of a symmetric cross-ply laminate. Laminate failure is assumed to occur when first ply fiber failure is predicted by equation (16).

In this paper, strength predictions were made by the following procedures: (1) laminate stiffnesses are calculated using classical lamination plate theory; (2) calculate laminate stresses relative to load factor by means of the method described in the previous section; (3) determine relative laminae stresses in principal material axes; (4) using Yamada-Sun criterion (16) on a ply-by-ply basis applied at characteristic curve to get the value e which is proportional to the applied load p; (5) calculate load for a first-ply failure by the formula

$$P_f = \frac{p}{e}$$

thus P_f is the predicted laminate strength.

NUMERICAL RESULTS AND DISCUSSIONS

In order to check the accuracy of the numerical results obtained by the present method, we first present the classical example of infinite isotropic plate with circular hole. The tangential stress σ_θ distributions around the hole obtained by the present method in comparison with exact solutions are shown in Figure 2; it can be seen that they are in excellent agreement.

Figure 3 illustrates the stress distributions σ_θ around the circular hole in a quasi-isotropic finite laminated square plate subject to various inplane loadings. It has been known that for an infinite isotropic plate under pure shear τ, the stress distribution σ_θ about the circular hole in the plate is given by $-4\tau\sin 2\theta$. It can be seen clearly from Figure 3 that the stress σ_θ has its highest stress concentration of value about 5τ at $\theta = 45°$ and the distribution is almost according to the curve of $-5\tau\sin 2\theta$ for the case of finite plate subject to pure shear τ. Actually, the behavior of stress concentrations in the quasi-isotropic laminate considered here is the same as the homogeneous isotropic plate because they have the similar material properties. The stress distributions σ_θ is constant for the case when the plate is subjected to biaxial tensile uniform loads σ and its value

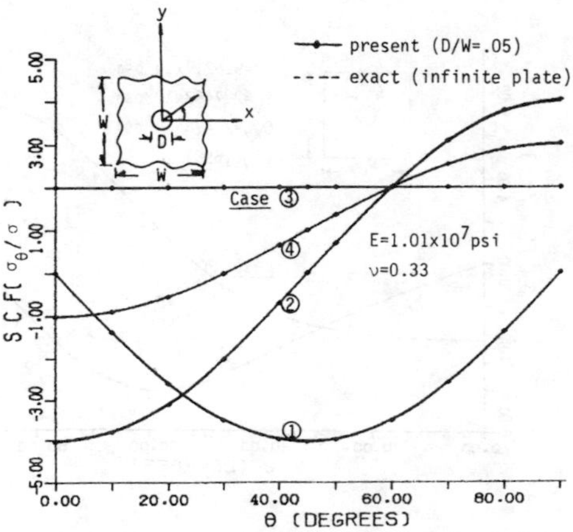

Fig. (2) - Comparisons between the present solutions and exact solutions of infinite plate analysis for the stress concentrations around the circular hole in isotropic plates

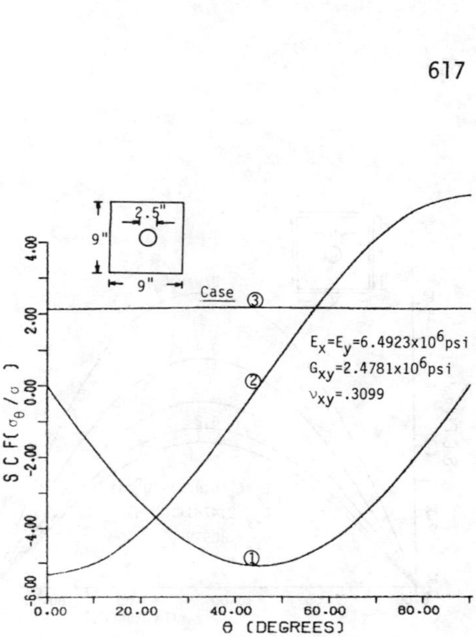

Fig. (3) - Stress distributions σ_θ around the circular hole in a quasi-isotropic $(0°/\pm45°/90°)_s$ laminated plate

(about 2.2σ) is lower than the maximum stress concentrations (3σ) in an infinite isotropic plate subject to uniaxial tensile load. For the case when the plate subject to biaxial tensile-compressive uniform loads, the stress distribution σ_θ has the same behavior as the plate subject to pure shear with a phase angle of 45°, this is similar to that of infinite isotropic plate.

Figure 4 illustrates the variation of stress concentrations around the circular hole in a finite quasi-isotropic square plate under pure shear with various hole sizes (the hole diameter D to plate width W ratios). Clearly, the stress concentration rises as the ratio D/W increases and its distribution about the hole may be represented by the curve $-\sigma_{max}\sin2\theta$. When the ratio D/W = 20, the highest stress concentration is about four times of the applied shear load τ and the plate can be treated as an infinite plate.

The stress distributions σ_θ along the circular boundary in a symmetric $(0_2°/\pm45°)_s$ laminated finite plate subject to pure shear or biaxial stresses are shown in Figure 5. The highest stress concentration occur at the points θ = ±90° with magnitude about seven times of the applied tensile-compressive stresses and at the points θ = 60°, 240° with magnitude about six times of the applied shear. It can be seen that the behavior of stress concentration around the hole is different from the quasi-isotropic or isotropic plate, and that the maximum value of stress concentration is higher for the orthotropic laminated plate due to the anisotropy of the laminate.

Stress concentrations in finite symmetric $(0_2°/\pm45°)_s$ laminated plates with unloaded fastener elliptical holes are also examined and results for the cases under

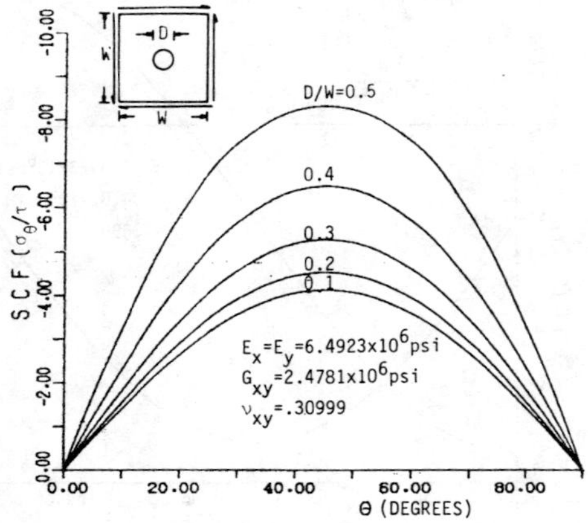

Fig. (4) - Effect of hole size on the stress concentrations in a finite quasi-isotropic $(0°/\pm45°/90°)_s$ plate under pure shear

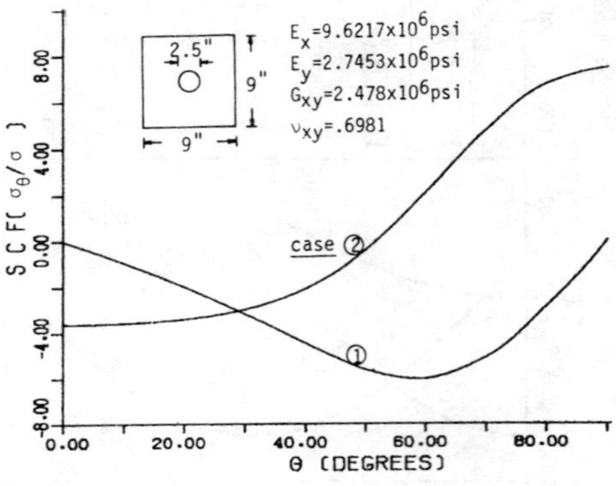

Fig. (5) - Stress concentrations around the circular hole in a $(0^\circ_2/\pm45°)_s$ laminated plate

pure shear and biaxial stresses are given in Figure 6. The maximum stress concentration of the plate subject to biaxial stresses is located at (or near) the elliptical-boundary points on the major axis (θ = 0°, 180°). For the case when plate subject to pure shear, the location of maximum stress concentration is closer to the major axis for elliptic hole than for circular hole and approaches to the major axis as the aspect ratio a/b (major to minor) of hole size increases. Figure 7 illustrates the stress distribution σ_θ around a narrow elliptic hole (with

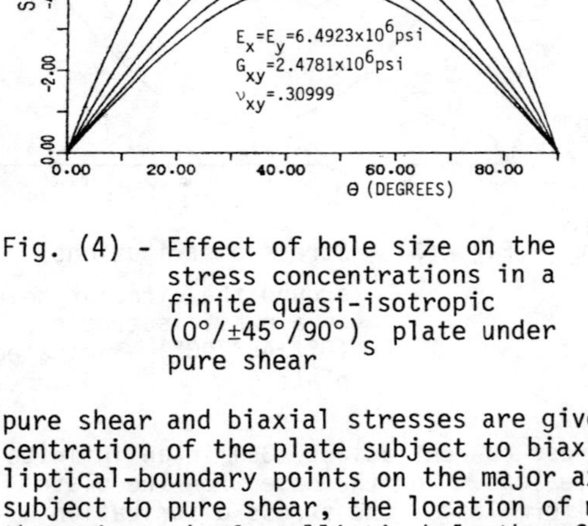

Fig. (6) - Stress concentrations around the elliptical hole in a $(0^\circ_2/\pm45°)_s$ laminated plate

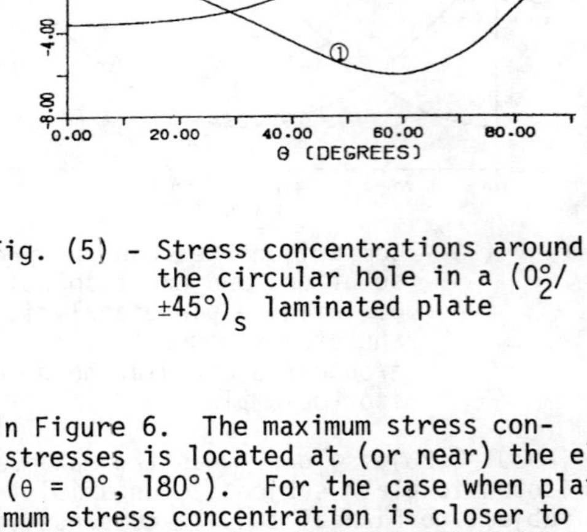

Fig. (7) - Stress concentrations around a narrow elliptical hole in a $(0^\circ_2/\pm45°)_s$ laminated plate

ratio 2a/ab = 2.0/0.2 = 10) in a finite symmetric $(0^\circ_2/\pm45^\circ)_s$ laminated plate. Obviously, there is a very high maximum stress concentration at (or near) the tip of narrow hole; therefore, this narrow elliptic hole can be considered as a crack.

Comparisons between the present results and those of Garbo and Ogonowski [3,4] for strength predictions of Graphite-Epoxy AS/3501-6 rectangular laminate are shown in Figure 8 with different layups $(0^\circ_\alpha/\pm45^\circ_\beta/90^\circ_\gamma)_s$, here α, β and γ denote the per-

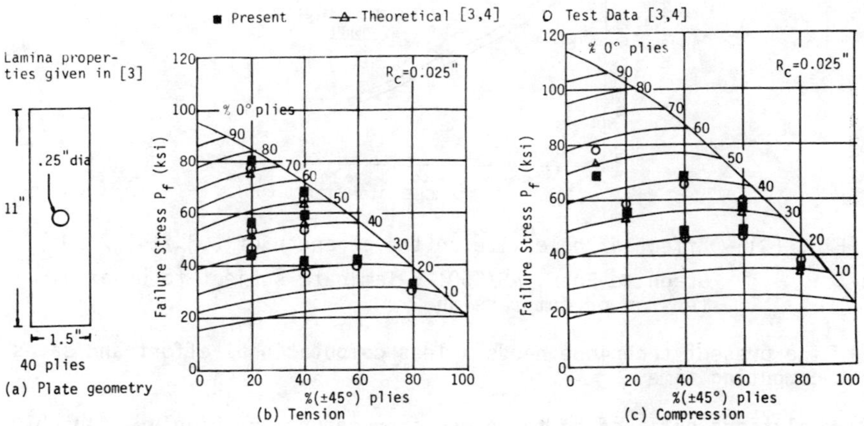

Fig. (8) - Failure stress predictions using characteristic dimension (R_c) compared to test data for Graphite-Epoxy AS/3501-6 $(0^\circ_\alpha/\pm45^\circ_\beta/90^\circ_\gamma)_s$ rectangular laminate

centages of the indicated plies, results show that they are in good agreement. Correlation of predicted effect of hole size on $(0^\circ_{70}/\pm45^\circ_{20}/90^\circ_{10})_s$ laminate strength is illustrated in Figure 9. All of the results reflect that the value of strength and location of failure initiation depend on layup and hole size, a plate with larger hole has the less strength. Results obtained by the present procedure (stress concentration based on finite laminate analysis and strength prediction using the Yamada-Sun failure criterion) show that failure initiations occur on the hole boundary and at 90° to the applied load direction for all cases except the $(0^\circ_{70}/\pm45^\circ_{20}/90^\circ_{10})_s$ laminate which initiates to failure at 62°. These are same as the predicted origins of failure by [3] (stress concentration based on infinite plate analysis and strength prediction using the Tsai-Hill failure criterion) except the $(0^\circ_{70}/\pm45^\circ_{20}/90^\circ_{10})_s$ laminate which was at 65°.

CONCLUSIONS

An analytical procedure to predict stress distributions around an open fastener hole and plate strength of a finite composite laminate subject to various inplane loadings has been presented and verified with known solutions. From the results obtained, we have the following conclusions:

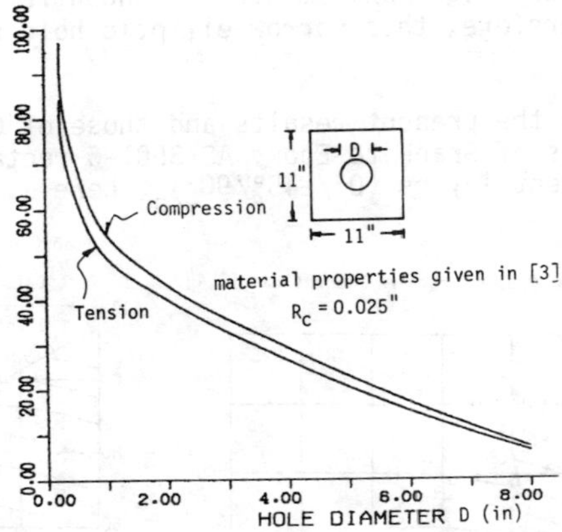

Fig. (9) - Effect of hole size on the strength of $(0^\circ_{70}/\pm45^\circ_{20}/90^\circ)_s$ Graphite-Epoxy AS/3501-6 laminate subject to uniaxial tension or compression

(1) The present technique needs a less computational effort and saves the computing time.

(2) Solutions obtained by the present procedure are highly acceptable.

(3) The maximum stress concentration rises as the hole size (i.e., D/W ratio) increases and plate with larger hole has the less strength.

(4) The variation of stress distribution along the elliptical boundary is more severe than the circular hole and it may produce a very high maximum stress concentration at (or near) the tip of narrow hole; therefore, a very narrow elliptic hole can be considered as a crack.

(5) Due to the anisotropy, the maximum stress concentration in an orthotropic laminated plate is higher than an isotropic plate.

(6) The value of strength and location of failure initiation of a composite laminate depend on its layup and hole size.

ACKNOWLEDGEMENTS

This paper is a part of the research work supported by a research grant from the National Science Council, Republic of China.

REFERENCES

[1] Lekhnitskii, S. G., Anisotropic Plates, Gordon and Breach, 1968.

[2] Savin, G. N., "Stress concentration around holes", NASA-TT-F-607, 1968.

[3] Garbo, S. P. and Ogonowski, J. M., "Strength prediction of composite laminate with unloaded fastener holes", AIAA Journal, Vol. 18, p. 585, 1980.

[4] Garbo, S. P., "Compression strength of laminates with unloaded fastener holes", AIAA/ASME/ASCE/AHS 21st SDM Conference, p. 291, 1980.

[5] Ogonowski, J. M., "Analytical study of finite geometry plates with stress concentrations", AIAA/ASME/ASCE/AHS 21st SDM Conference, p. 694, 1980.

[6] Gerhardt, "A hybrid/finite element approach for stress analysis of notched anisotropic materials", ASME Journal of Applied Mechanics, Vol. 51, p. 804, 1984.

[7] Whitney, J. M. and Nuismer, R. J., "Stress fracture criteria for laminated composites containing stress concentrations", J. of Composite Materials, Vol. 8, p. 253, 1974.

[8] Yamada, S. E. and Sun, C. T., "Analysis of laminate strength and its distribution", J. Composite Materials, Vol. 12, p. 275, 1978.

[9] Jones, R. M., Mechanics of Composite Materials, Scripta Book Co., 1975.

[10] Lekhnitskii, S. G., Theory of Elasticity of an Anisotropic Elastic Body, Holden-Day, 1963.

FREE VIBRATION AND BUCKLING ANALYSIS OF THICK INITIAL STRESS LAMINATE

T.-Y. Ye and C.-L. Wu

Industrial Technology Research Institute
Hsinchu, Taiwan, Republic of China

ABSTRACT

The finite strip method is applied to analyze the vibration and buckling problems of the thick fiber-reinforced composite initial stress laminates. The effects of the transverse shear deformation and rotary inertia are considered in this study.

In solving the problems, the harmonic terms of the finite strip's shape function will couple in the stiffness matrix, so we use two terms of harmonic function. The values from the above two harmonic functions coincide with that calculated by the three-dimensional finite element method.

It is concluded that the buckling and vibration coefficients are significantly affected by the laminating stacking sequence and geometrical parameters (thickness, length, width). We also show that those have more significant effects when the boundary constraints are more rigid.

INTRODUCTION

The finite strip method, which is a semi-analytical, semi-numerical process, has important advantage for a range of plate problems. For isotropic materials, the free vibration and stability problems of thin plate using the finite strip method has been considered. Benson and Hinton [2], using a thick plate approach, have included the effects of transverse shear deformation and rotary inertia to give a thick finite strip solution. This paper extends the free vibration and stability problems of isotropic materials to orthotropic materials.

Most of the free vibration and stability problems of the fiber reinforced composite have been using the thin plate laminate theory. However, when the thickness of composite plate increases, the transverse shear deformation increases too. The thin plate laminate theory cannot give accurate solutions. Present study is to employ the finite strip method to investigate the free vibration and stability problems of thick composite plate. For isotropic materials, each harmonic term of simply supported strip will uncouple and each harmonic term's stiffness matrix can be solved independently. But, in fibre reinforced composite material, each harmonic term of simply supported strip will couple and each harmonic term's stiffness matrix cannot be solved independently. For fibre reinforced composite material, the final solution is obtained by using the total stiffness matrix. The present analysis uses two harmonic terms to solve the free vibration and stability problems.

The results of finite strip method using two harmonic terms coincide very well with the results of finite element method which was based on two-dimensional thick plate theories and three-dimensional elasticity. The effects of thickness, length, stacking sequence, elastic constants are also studied.

THE ELASTIC THICK PLATE UNDER INITIAL STRESS

The basic assumptions regarding the form of the Mindlin plate displacements are that

$$\begin{aligned} \bar{u} &= u(x,y) + z\psi_x(x,y) \\ \bar{v} &= v(x,y) + z\psi_y(x,y) \\ \bar{w} &= w(x,y) \end{aligned} \tag{1}$$

where $\bar{u}$, $\bar{v}$ and $\bar{w}$ are final displacements, u, v and w are the displane (z=0) displacements, and ψ_y are rotations of the normal lines of the undeformed plate in in the x-z and y-z planes, respectively. These normal lines are not necessarily perpendicular to the midplane after deformation and consequently, transverse shear deformation is permitted.

For Mindlin plate, the relevant Green's strain vector is given as

$$\varepsilon = \begin{bmatrix} \varepsilon_x \\ \varepsilon_y \\ \gamma_{xy} \\ \gamma_{xz} \\ \gamma_{yz} \end{bmatrix} = \begin{bmatrix} \frac{\partial\bar{u}}{\partial x} \\ \frac{\partial\bar{v}}{\partial y} \\ \frac{\partial\bar{u}}{\partial y} + \frac{\partial\bar{v}}{\partial x} \\ \frac{\partial\bar{w}}{\partial x} + \frac{\partial\bar{u}}{\partial z} \\ \frac{\partial\bar{w}}{\partial y} + \frac{\partial\bar{v}}{\partial z} \end{bmatrix} + \begin{bmatrix} \frac{1}{2}(\frac{\partial\bar{u}}{\partial x})^2 + \frac{1}{2}(\frac{\partial\bar{v}}{\partial x})^2 + \frac{1}{2}(\frac{\partial\bar{w}}{\partial x})^2 \\ \frac{1}{2}(\frac{\partial\bar{u}}{\partial y})^2 + \frac{1}{2}(\frac{\partial\bar{v}}{\partial y})^2 + \frac{1}{2}(\frac{\partial\bar{w}}{\partial y})^2 \\ \frac{\partial\bar{u}}{\partial x}\cdot\frac{\partial\bar{u}}{\partial y} + \frac{\partial\bar{v}}{\partial x}\cdot\frac{\partial\bar{v}}{\partial y} + \frac{\partial\bar{w}}{\partial x}\cdot\frac{\partial\bar{w}}{\partial y} \\ \frac{\partial\bar{u}}{\partial x}\cdot\frac{\partial\bar{u}}{\partial z} + \frac{\partial\bar{v}}{\partial x}\cdot\frac{\partial\bar{v}}{\partial z} + \frac{\partial\bar{w}}{\partial x}\cdot\frac{\partial\bar{w}}{\partial z} \\ \frac{\partial\bar{u}}{\partial y}\cdot\frac{\partial\bar{u}}{\partial z} + \frac{\partial\bar{v}}{\partial y}\cdot\frac{\partial\bar{v}}{\partial z} + \frac{\partial\bar{w}}{\partial y}\cdot\frac{\partial\bar{w}}{\partial z} \end{bmatrix}$$

or

$$(\varepsilon) = (e) + (\alpha) \tag{2}$$

The final state of stress can be given as

$$\begin{bmatrix} \bar{\sigma}_{xx} \\ \bar{\sigma}_{yy} \\ \bar{\sigma}_{xy} \\ \bar{\sigma}_{xz} \\ \bar{\sigma}_{yz} \end{bmatrix} = \begin{bmatrix} \sigma_{xx} \\ \sigma_{yy} \\ \sigma_{xy} \\ \sigma_{xz} \\ \sigma_{yz} \end{bmatrix} + \begin{bmatrix} \sigma_x^o \\ \sigma_y^o \\ \sigma_{xy}^o \\ 0 \\ 0 \end{bmatrix}$$

or

$$(\bar{\alpha}) = (\alpha) + (\alpha^o) \tag{3}$$

where (σ) are the incremental stresses and (σ^o) are the initial stresses. It is assumed that

$$\bar{\alpha}_{zz} = \bar{\alpha}_{zz} = 0$$

For orthotropic materials, the stress-strain relationship of filament direction can be written as

$$(\sigma) = (d)(\varepsilon)$$

where

$$(d) = \begin{bmatrix} d_{11} & d_{12} & 0 & 0 & 0 \\ d_{12} & d_{22} & 0 & 0 & 0 \\ 0 & 0 & d_{33} & 0 & 0 \\ 0 & 0 & 0_k & 2d_{44} & 0 \\ 0 & 0 & 0 & 0_k & 2d_{55} \end{bmatrix} \quad d_{11} = \frac{E_1}{1-\nu_{12}\nu_{21}} \quad d_{12} = \frac{\nu_{12}E_2}{1-\nu_{12}\nu_{21}}$$

$$d_{22} = \frac{E_2}{1-\nu_{12}\nu_{21}} \tag{4}$$

$$d_{33} = G_{12} \quad d_{44} = G_{13} \quad d_{55} = G_{23}$$

If the angle between loading direction and filament direction is θ, the stress-strain relation can be expressed as

$$(\sigma) = (D)(\varepsilon) \tag{5}$$

$$\begin{bmatrix} \sigma_x \\ \sigma_y \\ \sigma_{xy} \\ \sigma_{xz} \\ \sigma_{yz} \end{bmatrix} = \begin{bmatrix} d'_{11} & d'_{12} & d'_{13} & 0 & 0 \\ d'_{21} & d'_{22} & d'_{23} & 0 & 0 \\ d'_{31} & d'_{32} & d'_{33} & 0 & 0 \\ 0 & 0 & 0 & \kappa^2 d'_{44} & \kappa^2 d'_{45} \\ 0 & 0 & 0 & \kappa^2 d'_{45} & \kappa^2 d'_{55} \end{bmatrix} \begin{bmatrix} \varepsilon_x \\ \varepsilon_y \\ \varepsilon_{xy} \\ \varepsilon_{xz} \\ \varepsilon_{yz} \end{bmatrix}$$

where

$$d'_{11} = d_{11}\cos^4\theta + d_{22}\sin^4\theta + (2d_{12}+4d_{33})\sin^2\theta\cos^2\theta$$

$$d'_{12} = d_{12} + (d_{11}+d_{22}-2d_{12}-4d_{33})\sin^2\theta\cos^2\theta$$

$$d'_{22} = d_{11}\sin^4\theta + d_{22}\cos^4\theta + (2d_{12}+4d_{33})\sin^2\theta\cos^2\theta$$

$$d'_{13} = (d_{11}-d_{12}-2d_{33})\sin\theta\cos^3\theta + (d_{12}-d_{22}+2d_{33})\sin^3\theta\cos\theta$$

$$d'_{23} = (d_{11}-d_{12}-2d_{33})\sin^3\theta\cos\theta + (d_{12}-d_{22}+2d_{33})\sin\theta\cos^3\theta$$

$$d'_{33} = (d_{11}+d_{22}-2d_{12}-2d_{33})\sin^2\theta\cos^2\theta + d_{33}(\sin^4\theta+\cos^4\theta)$$

$$d'_{44} = d_{44}\cos^2\theta + d_{55}\sin^2\theta$$

$$d'_{45} = \sin\theta\cos\theta(d_{55}-d_{44})$$

$$d'_{55} = d_{55}\cos^2\theta + d_{44}\sin^2\theta$$

The strain energy density can be expressed as

$$W = (\sigma^o)^T(\varepsilon) + \frac{1}{2}(\sigma)^T(\varepsilon) \tag{6}$$

By substituting equations (2) to (5) into equation (6) and neglecting higher order terms in displacement gradients, we obtain the following expression

$$w = (\sigma^o)^T(e) + (\sigma) + \frac{1}{2}(e)^T(D)(e) \tag{7}$$

The incremental potential energy is written as

$$\begin{aligned}\Delta W &= w - (\sigma^o)^T(e)\\ &= \frac{1}{2}(e)^T(D)(e) + (\sigma^o)^T(\alpha)\\ &= \frac{1}{2}(e)^T(D)(e) + \sigma_x^o\left(\frac{1}{2}\left(\frac{\partial\bar{u}}{\partial x}\right)^2 + \frac{1}{2}\left(\frac{\partial\bar{v}}{\partial x}\right)^2 + \frac{1}{2}\left(\frac{\partial\bar{w}}{\partial x}\right)^2\right) + \sigma_y^o\left(\frac{1}{2}\left(\frac{\partial\bar{u}}{\partial y}\right)^2\right.\\ &\quad \left.+ \frac{1}{2}\left(\frac{\partial\bar{v}}{\partial y}\right)^2 + \frac{1}{2}\left(\frac{\partial\bar{w}}{\partial y}\right)^2\right) + \sigma_{xy}^o\left(\frac{\partial\bar{u}}{\partial x}\cdot\frac{\partial\bar{u}}{\partial y} + \frac{\partial\bar{v}}{\partial x}\cdot\frac{\partial\bar{v}}{\partial y} + \frac{\partial\bar{w}}{\partial x}\cdot\frac{\partial\bar{w}}{\partial y}\right)\end{aligned} \tag{8}$$

By substituting equation (1) into equation (8) and integrating over the thickness of the plate, the incremental potential energy per unit area can then be written as

$$\Delta U = \int_{-h/2}^{h/2} W dz = \frac{1}{2}(\bar{e})^T(\bar{D})(\bar{e}) + \frac{1}{2}\left(\frac{\partial u}{\partial x}, \frac{\partial u}{\partial y}\right)(N)\begin{Bmatrix}\frac{\partial u}{\partial x}\\ \frac{\partial u}{\partial y}\end{Bmatrix} + \frac{1}{2}\left(\frac{\partial u}{\partial x}, \frac{\partial u}{\partial y}\right)(M)\begin{Bmatrix}\frac{\partial \psi_x}{\partial x}\\ \frac{\partial \psi_x}{\partial y}\end{Bmatrix}$$

$$+ \frac{1}{2}\left(\frac{\partial \psi_x}{\partial x}, \frac{\partial \psi_x}{\partial y}\right)(M)\begin{Bmatrix}\frac{\partial u}{\partial x}\\ \frac{\partial u}{\partial y}\end{Bmatrix} + \frac{1}{2}\left(\frac{\partial \psi_x}{\partial x}, \frac{\partial \psi_x}{\partial y}\right)(M^*)\begin{Bmatrix}\frac{\partial \psi_x}{\partial x}\\ \frac{\partial \psi_x}{\partial y}\end{Bmatrix} + \frac{1}{2}\left(\frac{\partial v}{\partial x}, \frac{\partial v}{\partial y}\right)(N)\begin{Bmatrix}\frac{\partial v}{\partial x}\\ \frac{\partial v}{\partial y}\end{Bmatrix}$$

$$+ \frac{1}{2}\left(\frac{\partial v}{\partial x}, \frac{\partial v}{\partial y}\right)(M)\begin{Bmatrix}\frac{\partial \psi_y}{\partial x}\\ \frac{\partial \psi_y}{\partial y}\end{Bmatrix} + \frac{1}{2}\left(\frac{\partial \psi_y}{\partial x}, \frac{\partial \psi_y}{\partial y}\right)(M)\begin{Bmatrix}\frac{\partial v}{\partial x}\\ \frac{\partial v}{\partial y}\end{Bmatrix}$$

$$+ \frac{1}{2}\left(\frac{\partial \psi_y}{\partial x}, \frac{\partial \psi_y}{\partial y}\right)(M^*)\begin{Bmatrix}\frac{\partial \psi_y}{\partial x}\\ \frac{\partial \psi_y}{\partial y}\end{Bmatrix} + \frac{1}{2}\left(\frac{\partial w}{\partial x}, \frac{\partial w}{\partial y}\right)(N)\begin{Bmatrix}\frac{\partial w}{\partial x}\\ \frac{\partial w}{\partial y}\end{Bmatrix} \qquad (9)$$

$$(\bar{D}) = \int_{-h/2}^{h/2} (D)dz \begin{bmatrix} D_{11} & D_{12} & D_{13} & 0 & 0 & 0 & 0 & 0 \\ D_{21} & D_{22} & D_{23} & 0 & 0 & 0 & 0 & 0 \\ D_{31} & D_{32} & D_{33} & 0 & 0 & 0 & 0 & 0 \\ 0 & 0 & 0 & D^*_{11} & D^*_{12} & D^*_{13} & 0 & 0 \\ 0 & 0 & 0 & D^*_{21} & D^*_{22} & D^*_{23} & 0 & 0 \\ 0 & 0 & 0 & D^*_{31} & D^*_{32} & D^*_{33} & 0 & 0 \\ 0 & 0 & 0 & 0 & 0 & 0 & \kappa^2 D_{44} & \kappa^2 D_{45} \\ 0 & 0 & 0 & 0 & 0 & 0 & \kappa^2 D_{54} & \kappa^2 D_{55} \end{bmatrix}$$

$$D_{ij} = \int_{-h/2}^{h/2} d'_{ij}dz, \quad (\sigma^o) = \begin{bmatrix} \sigma_x^o & \sigma_{xy}^o \\ \sigma_{xy}^o & \sigma_y^o \end{bmatrix} \quad (N) = \int_{-h/2}^{h/2} (\sigma^o)dz, \quad (M) = \int_{-h/2}^{h/2} (\sigma^o)zdz$$

$$D_{ij}^* = \int_{-h/2}^{h/2} d'_{ij}z^2dz, \quad (M^*) = \int_{-h/2}^{h/2} (\sigma^o)z^2dz$$

$$(\bar{e})^T = \left(\frac{\partial u}{\partial x}, \frac{\partial v}{\partial y}, \frac{\partial u}{\partial y}, \frac{\partial v}{\partial x}, \frac{\partial \psi_x}{\partial x}, \frac{\partial \psi_y}{\partial y}, \frac{\partial \psi_x}{\partial y}, \frac{\partial \psi_y}{\partial x}, \theta_x, \theta_y\right)$$

$$\theta_x = \psi_x + \frac{\partial w}{\partial x}, \quad \theta_y = \psi_y + \frac{\partial w}{\partial y} \tag{10}$$

THE THICK FINITE STRIP METHOD

The thick finite strip geometry is illustrated in Figure 1. The two y = constant edges are simply supported and the following displacements are chosen to satisfy the boundary conditions of y = constant edges.

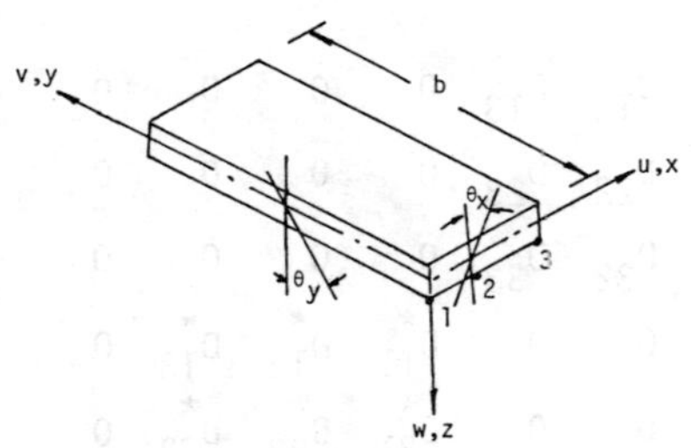

Fig. (1) - The finite strip of thick plates

$$u = \sum_{m=1}^{n} \sum_{i=1}^{3} N_i u_i^m \sin \frac{m\pi y}{b}$$

$$v = \sum_{m=1}^{n} \sum_{i=1}^{3} N_i v_i^m \cos \frac{m\pi y}{b}$$

$$w = \sum_{m=1}^{n} \sum_{i=1}^{3} N_i w_i^m \sin \frac{m\pi y}{b}$$

$$\psi_x = \sum_{m=1}^{n} \sum_{i=1}^{3} N_i \psi_{xi}^{m} \sin \frac{m\pi y}{b}$$

$$\psi_y = \sum_{m=1}^{n} \sum_{i=1}^{3} N_i \psi_{xi} \cos \frac{m\pi y}{b} \qquad (11)$$

where m is the harmonic considered, and shape functions N_i are

$$N_1 = -\frac{\xi}{2}(1-\xi),\ N_2 = 1-\xi^2,\ N_3 = \frac{\xi}{2}(1+\xi) \qquad (12)$$

The generalized displacement parameter vector for node i is written as

$$(\delta_i^m) = (u_i^m,\ v_i^m,\ w_i^m,\ \psi_{xi}^m,\ \psi_{yi}^m)^T \qquad (13)$$

Using the standard finite strip procedures, the buckling problem becomes the following eigenvalue problem

$$([K] - \lambda[K_s])\{\delta\} = 0 \qquad (14)$$

where λ is now the buckling coefficient. $[K_s]$ is geometric stiffness matrix.

The vibration problem becomes

$$([K] + [K_s] - w^2[M_s])\{\delta\} = 0 \qquad (15)$$

where w is frequence coefficient. $[M_s]$ is mass matrix.

NUMERICAL RESULTS

To evaluate the accuracy of the finite strip method, the results compared with the published results. A number of cases of orthotropic property, orientation, aspect ratio, thickness ratio, stacking sequence and boundary conditions were studied. In all cases, the plates were subjected to an uniform in-plane loading as shown in Figure 2.

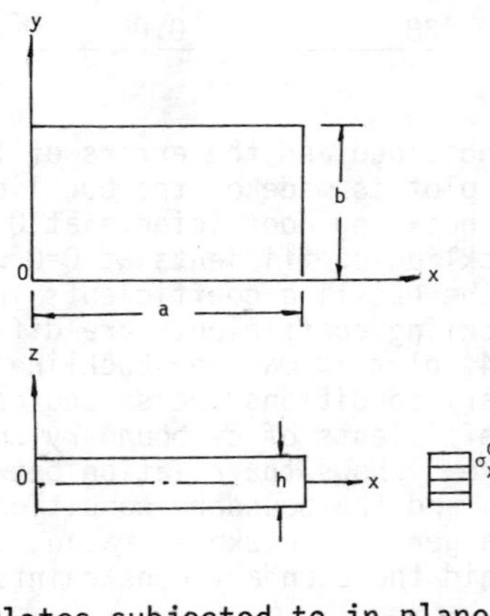

Fig. (2) - Plates subjected to in-plane stresses

The geometrical and material constants are:

a	= the length of plate
b	= the width of plate
h	= the thickness of plate
ss	= the boundary conditions are all simply supported
cs	= the boundary conditions are clamped at x=0, a and simply supported at y=0, b
E_1	= the modulus in the filament direction
E_2	= the modulus in the transverse direction
G_{12}, G_{23}	= the shear modulus
ν_{12}, ν_{23}	= Poisson's ratio

The parameters of following results are defined as a/b = 1, h/b = 0.1, E_1/E_2 = 10, G_{12}/E_2 = 0.6, G_{23}/E_2 = -.5, $\nu_{12} = \nu_{23}$ = 0.25, stacking sequence $(0/90/45/-45)_s$, except the parameter is a variable parameter.

BUCKLING RESULTS

Table 1 compares the buckling coefficients obtained using the three dimensional theory (3) with buckling coefficients obtained using the finite strip method. A

Table 1 - The accuracy of buckling coefficient (κ) obtained by finite strip method with B.C. ss, (0/90/±45) $\kappa = - N_x b^2/E_2 h^2$

E_1/E_2	10	20	30	40
Finite Strip	9.9775	15.925	21.059	25.535
3-D Finite Element	10.0417	15.9153	20.9614	25.5436
Error (%)	0.438	0.06	0.466	0.75

reasonable correction is obtained and the errors of finite strip method are less than 0.75%. In Figure 3, plot is made of the buckling coefficients versus fibre orientation. The maximum buckling coefficients at Q = ±45 when boundary condition is ss, and the maximum buckling coefficients at Q=0 when boundary condition is cs. Table 2 and Table 3 show the buckling coefficients in different initial stress and stacking sequence, the buckling coefficients are different when stacking sequence is different. In Figure 4, plot is made of buckling coefficients versus aspect ratio (a/b) and the boundary conditions are ss and cs respectively. It can be seen that the buckling coefficients of cs boundary condition are larger than ss boundary condition. Figure 5 shows the relation between the buckling coefficients and thickness ratio (h/b), and the boundary conditions are ss and cs respectively. It can be seen that the larger the thickness ratio, the smaller the buckling coefficients, and the more rigid the boundary constraints, the greater the effect.

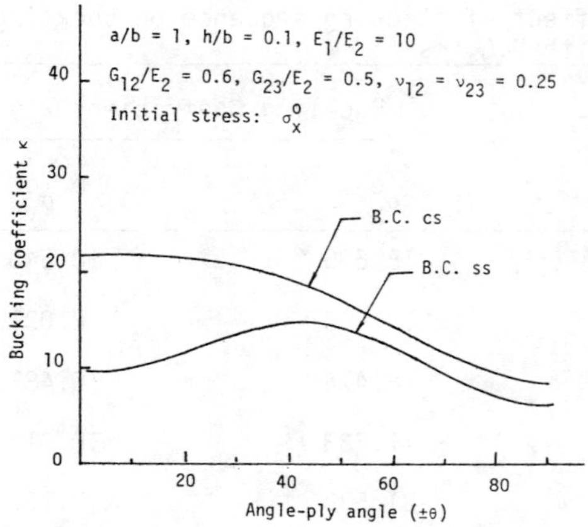

Fig. (3) - Variation of buckling coefficients with angle-ply angle (±θ)

Table 2 - Effect of stacking sequence on buckling coefficients with B.C. ss

Stacking Sequence	Buckling Coefficient $\kappa = -(N_x/E_2)(BL^2/T^3)$		
	σ_x^o	$\sigma_x^o\ \sigma_y^o$	σ_{xy}^o
$[0/90/45/-45]_s$	10.599	5.2794	8.2027
$[90/45/-45/0]_s$	11.321	5.8839	9.1446
$[4S/-45/0/90]_s$	14.077	7.0397	11.121
$[-45/0/90/45]_s$	11.846	5.9273	9.3677
$[0/0/0/0]_s$	9.6508	4.8249	7.4389
$[90/90/90/90]_s$	6.4567	4.8126	7.4278
$[45/-45/45/-45]_s$	14.802	7.4019	11.539

Figures 6 and 7 show the relation between the buckling coefficients and materials constants ratios E_1/E_2 and G_{12}/E_2 respectively. It can be seen that buckling coefficients increase when material constant ratio increases.

Table 3 - Effect of stacking sequence on buckling coefficients with B.C. cs

Stacking Sequence	Buckling Coefficient $\kappa = -(N_x/E_2)(BL^2/T^3)$		
	σ_x^o	$\sigma_x^o\ \sigma_y^o$	σ_{xy}^o
$[0/90/45/-45]_s$	18.695	10.540	15.476
$[90/45/-45/0]_s$	13.663	8.031	12.099
$[45/-45/0/90]_s$	18.476	10.485	16.125
$[-45/0/90/45]_s$	17.733	10.012	15.456
$[0/0/0/0]_s$	21.538	7.776	17.421
$[90/90/90/90]_s$	8.715	5.714	8.481
$[45/-45/45/-45]_s$	18.137	10.357	16.099

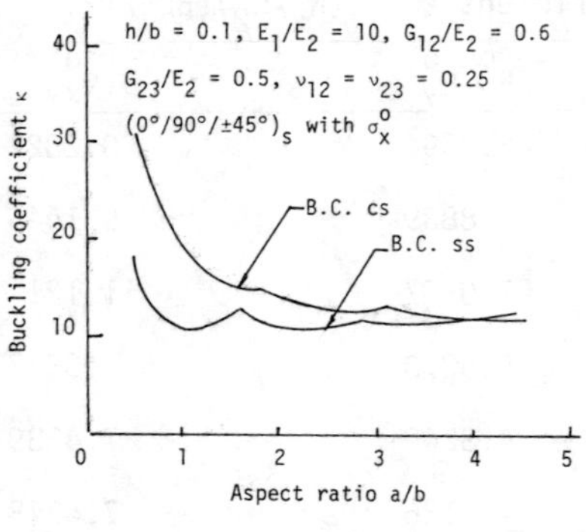

Fig. (4) - Variation of buckling coefficients with aspect ratio (a/b)

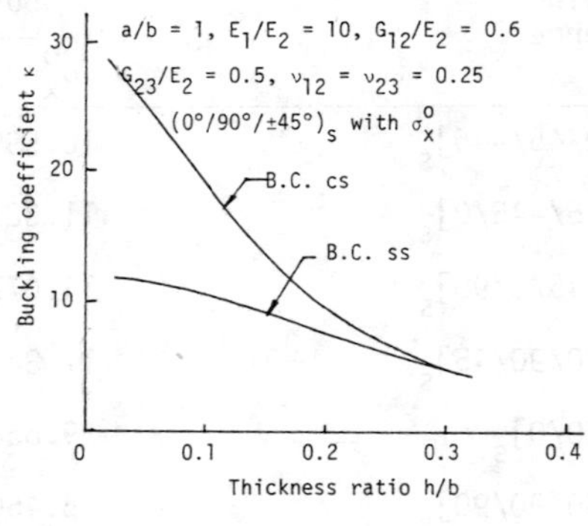

Fig. (5) - Variation of buckling coefficients with thickness ratio (h/b)

VIBRATION RESULTS

In Figure 8, the frequence coefficients are plotted against the fibre orientation for different B.C.'s. Both Figure 8 and Figure 4 give the same results. In the case of square plate, the maximum frequence Ω occurs at Q=0 when the B.C. is cs. Table 4 and Table 5 show the Ω in different initial stress and stacking sequence, the Ω are different when stacking sequence is different. Figure 9 shows the relation between Ω and aspect ratio (a/b), the B.C.'s are ss and cs respectively. It can be seen that Ω decreases when aspect ratio increases. Figure 10 and Figure 11 show the relation between Ω and E_1/E_2 and G_{12}/E_2 respectively.

It can be seen that Ω increases when material constant ratio increases.

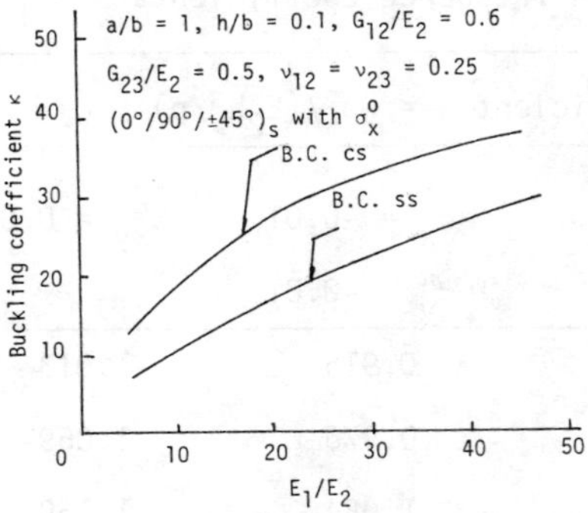

Fig. (6) - Variation of buckling coefficients with E_1/E_2

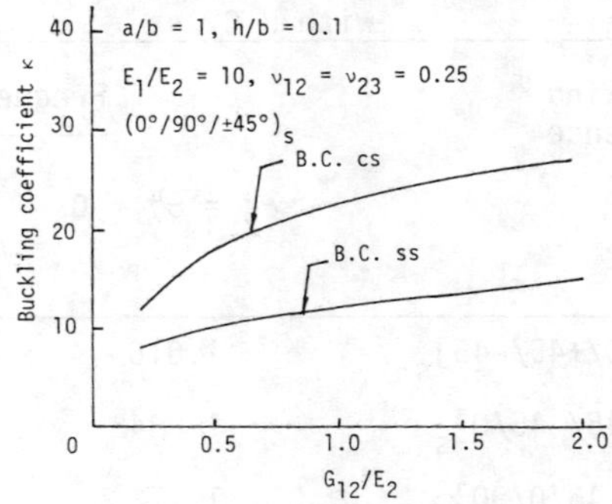

Fig. (7) - Variation of buckling coefficients with G_{12}/E_2

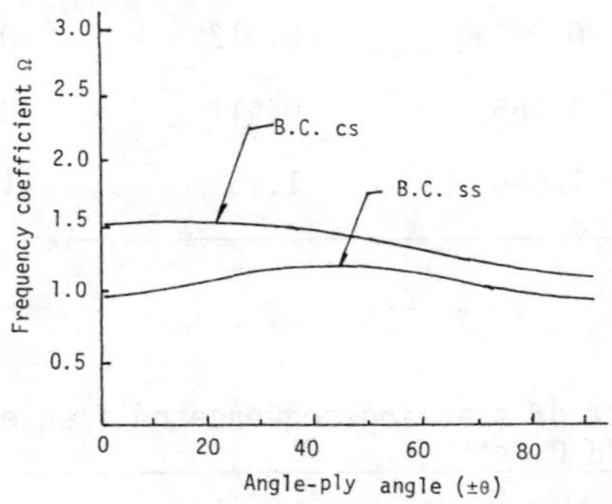

Fig. (8) - Variation of frequence coefficients with angle-ply (±θ). (a/b = 1, h/b = 0.1, E_1/E_2 = 10, G_{12}/E_2 = 0.6, G_{23}/E_2 = 0.51)

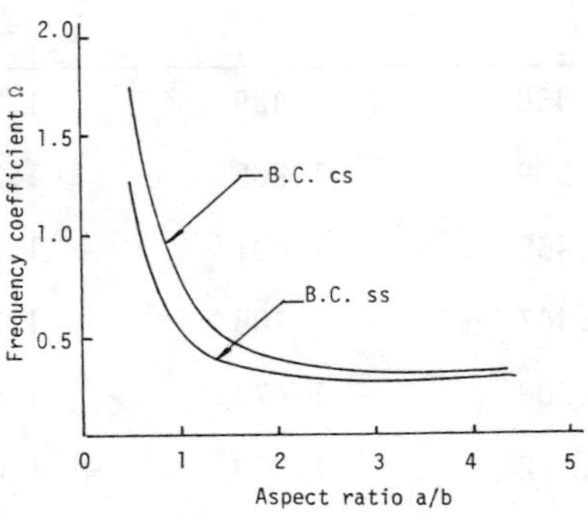

Fig. (9) - Variation of frequence coefficients with aspect ratio (a/b). (h/b = 0.1, E_1/E_2 = 10, G_{12}/E_2 = 0.6, G_{23}/E_2 = 0.5 $[0/90/\pm45]_s$)

Table 4 - Effect of stacking sequence on frequence coefficients with B.C. ss

Stacking Sequence	Frequence Coefficient $\Omega = wb^2/(E_2h^3/p)$			
	$\sigma_x^4 = \sigma_y^4 = 0$	$\sigma_x^4 = -0.01$	$\sigma_x^4 = -0.01$ $\sigma_y^4 = -0.01$	$\sigma_{xy}^4 = 0.01$
$[0/90/+45/-45]_s$	1.016	0.966	0.915	1.013
$[90/45/-45/0]_s$	1.084	1.025	0.976	1.069
$[45/-45/0/90]_s$	1.172	1.129	1.084	1.168
$[-45/0/90/+45]_s$	1.084	1.036	0.987	1.091
$[0/0/0/0]_s$	0.965	0.912	0.856	0.963
$[90/90/90/90]_s$	0.965	0.911	0.855	0.962
$[45/-45/45/-45]_s$	1.000	1.158	1.115	1.198

Table 5 - Effect of stacking sequence on frequence coefficients with B.C. cs

Stacking Sequence	Frequence Coefficient $\Omega = wb^2/(E_2h^3/p)$			
	$\sigma_x^4 = \sigma_y^4 = 0$	$\sigma_x^4 = -0.01$	$\sigma_x^4 = 0.01$ $\sigma_y^4 = 0.01$	$\sigma_{xy}^4 = 0.01$
$[0/90/45/-45]_s$	1.466	1.429	1.394	1.464
$[90/0/-45/0]_s$	1.309	1.265	1.226	1.306
$[45/-45/0/90]_s$	1.468	1.431	1.396	1.464
$[-45/0/90/45]_s$	1.447	1.408	1.372	1.453
$[0/0/0/0]_s$	1.508	1.473	1.439	1.508
$[90/90/90/90]_s$	1.112	1.070	1.023	1.123
$[45/-45/45/-45]_s$	1.460	1.422	1.387	1.457

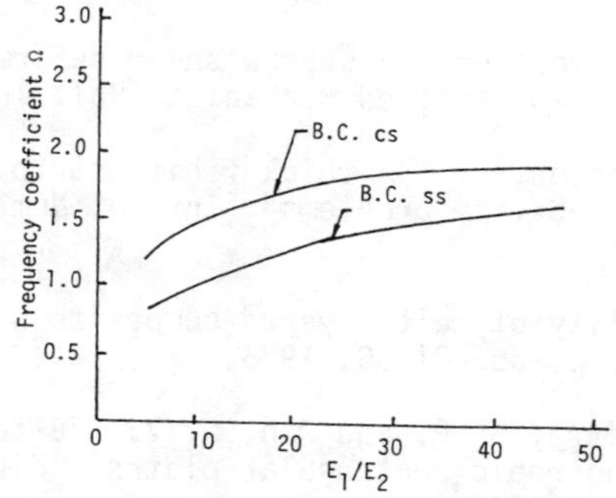

Fig. (10) - Variation of frequence coefficients with E_1/E_2. ($a/b = 1$, $h/b = 0.1$, $G_{12}/E_2 = 0.6$, $G_{23}/E_2 = 0.5$ $[0/90/\pm 45]_s$)

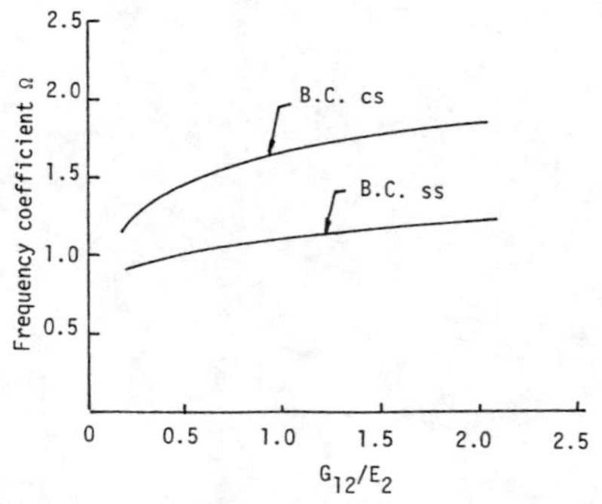

Fig. (11) - Variation of frequence coefficients with G_{12}/E_2. ($a/b = 1$, $h/b = 0.1$, $E_1/E_2 = 10$ $[0/90/\pm 45]_s$)

CONCLUSIONS

A thick finite strip method has been used to solve the buckling and free vibration problems of the fibre reinforced composite materials. The initially stressed Mindlin plates are subject to x, y and xy stresses.

The following results are obtained:

(1) The buckling and frequence coefficients decrease when the aspect ratio increases.

(2) The more rigid boundary condition enlarges the effect of aspect ratio, thickness ratio and material constants.

(3) The maximum values of the buckling and frequence coefficients are at ±45 when B.C. is ss and 0 when B.C. is cs.

REFERENCES

[1] Reissner, E., "The effect of transverse shear deformation on the bending of elastic plates", J. of Applied Mechanics, Vol. 18, pp. 31-38, 1951.

[2] Benson, P. R. and Hinton, E., "A thick finite strip solution for static, free vibration and stability problems", Int. J. Num. Meth. Engng., Vol. 10, pp. 665-678, 1976.

[3] Noor, A. K., "Stability of multilayered composite plates", Fibre Science and Technology, Vol. 8, pp. 81-89, 1975.

[4] Chelladura, T., Shastry, B. P. and Rao, G. V., "Effect of fibre orientation on stability of orthotropic rectangular plates", Fibre Science and Technology, Vol. 20, pp. 121-134, 1984.

SECTION V

STRESS/STRENGTH AND INSTABILITY ANALYSIS

ANALYSIS OF STRESS SINGULARITIES OF THE CRACK TIP IN THE TWO-DIMENSIONAL ORTHOTROPIC ELASTIC BODY

Y. Fujitani

Hiroshima University
Shitami Saijo-cho Higashihiroshima-city 724, Japan

ABSTRACT

This study deals with the analysis of the stress singularities of the crack tip in the two dimensional orthotropic elastic body. Two types of orthotropic materials are considered in this paper. One is a polar anisotropic material; another is a rectilinear anisotropic material. The singular solution of the crack tip in the case of the polar anisotropic material is derived by analytical method. On the other hand, since it is very difficult to obtain analytically the singular solution in the case of the rectilinear anisotropic material, we shall try to obtain this solution by the finite element method which has been proposed by the author. The intensity of the stress singularity and its stress and displacement modes are shown together with some figures and these characteristics of the stress singularities are discussed.

INTRODUCTION

Singular solutions play an important role in the evaluation of the strength of the materials in the linear fracture mechanics. Recently, the various advanced composite materials have been developed and they have been used practically as the structural members. It is necessary to know the stress singularity of the crack tip in these composite materials.

The stress singularity of the two dimensional crack with various boundary conditions and general opening angles was first analyzed by M. L. Williams [1,2]. Y. Fujitani [3], Y. Yamada and H. Okumura [4] proposed the method of the finite element analysis of the two dimensional stress singularity which method have been originally proposed by Z. P. Bazant and L. Estensorro [5] in the analysis of the three dimensional surface crack problem.

In this paper, the stress singularity of the crack tip in a polar anisotropic material and a rectilinear anisotropic material are analyzed. The former one is analyzed by M. W. William's method and the latter one is analyzed by finite element method. The outline of which will be described in the following sections.

ANALYSIS OF STRESS SINGULARITIES OF CRACK TIP IN A POLAR ANISOTROPIC MATERIAL

A. General Solution of the Stress and Displacement Components at the Crack Tip

The origin of the two dimensional polar coordinates (r,θ) is taken at the crack tip as shown in Figure 1. The stress-strain relations of a polar anisotropic material in case of plane stress are given as follows (see Figure 2):

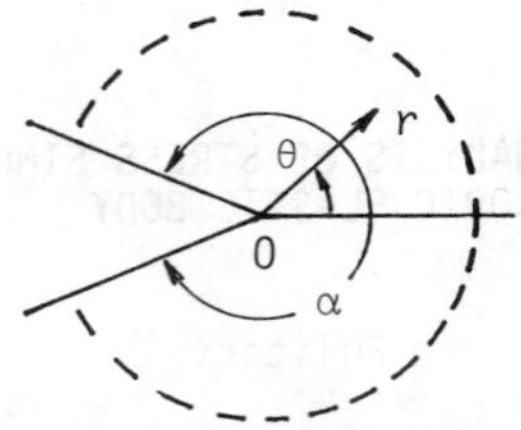

Fig. (1) - r-θ coordinates

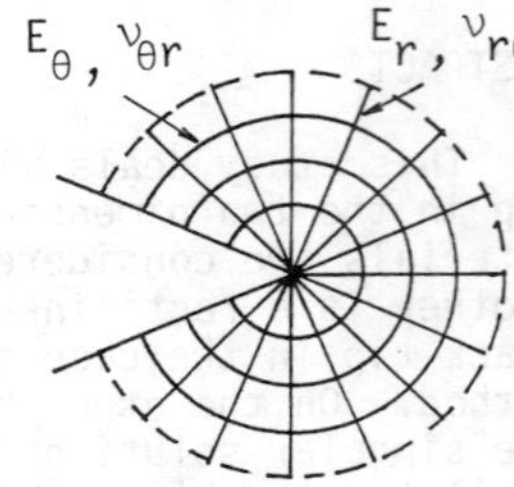

Fig. (2) - Polar anisotropy

$$\begin{Bmatrix}\sigma_r\\ \sigma_\theta\\ \tau_{r\theta}\end{Bmatrix} = \frac{1}{1-\nu_{r\theta}\nu_{\theta r}}\begin{bmatrix}E_r & E_\theta\nu_{\theta r} & 0\\ E_r\nu_{r\theta} & E_\theta & 0\\ 0 & 0 & G_{r\theta}(1-\nu_{r\theta}\nu_{\theta r})\end{bmatrix}\begin{Bmatrix}\varepsilon_r\\ \varepsilon_\theta\\ \gamma_{r\theta}\end{Bmatrix} \quad (1)$$

In which, $E_r\nu_{r\theta} = E_\theta\nu_{\theta r}$. $E_r, \nu_{r\theta}, E_\theta, \nu_{\theta r}$ imply the Young's modulus, Poisson's ratios of the radial direction and the circumferential direction respectively. In this analysis, the shear modulus $G_{r\theta}$ is assumed as following equation:

$$G_{r\theta} = E_rE_\theta/(E_r+E_\theta+2E_r\nu_{r\theta}) \quad (2)$$

To investigate the mechanical behaviour in the crack tip, we assume the displacement components $u_r(r,\theta), u_\theta(r,\theta)$ as the following form of separation of variables with parameter λ [1].

$$u_r(r,\theta) = r^\lambda u_r^*(\theta), \quad u_\theta(r,\theta) = r^\lambda u_\theta^*(\theta) \quad (3)$$

Substituting equation (3) into the strain-displacement relations and equation (1), and substituting this stress components into the equation of equilibrium, the following equation are derived finally:

$$\frac{d^4u_r^*}{d\theta^4} + B\frac{d^2u_r^*}{d\theta^2} + C\,u_r^* = 0, \quad \frac{d^4u_\theta^*}{d\theta^4} + B\frac{d^2u_\theta^*}{d\theta^2} + C\,u_\theta^* = 0 \quad (4)$$

in which, $B = \lambda^2(m+1)/m+2$, $C = (\lambda^2-m)(\lambda^2-1)/m$ (5)

and $m = E_\theta/E_r$ implies the degree of the orthotropy of the given materials. The root of the characteristic equation of the differential equation (4) is expressed as follows:

$$D^2 = (-B\pm\sqrt{B^2-4C})/2 \quad (6)$$

The following formulation shall be classified by the 3 cases of $C > 0$, $C = 0$ and $C < 0$.

(1) C > 0

In this case, equation (6) has 4 imaginary roots ($D_1 i, -D_1 i, D_2 i, -D_2 i$). Therefore, the general solution of $u_r^*(\theta)$, $U_\theta^*(\theta)$ can be expressed as follows:

$$\left.\begin{aligned} u_r^*(\theta) &= a_1\sin D_1\theta + a_2\cos D_1\theta + a_3\sin D_2\theta + a_4\cos D_2\theta \\ u_\theta^*(\theta) &= b_1\sin D_1\theta + b_2\cos D_1\theta + b_3\sin D_2\theta + b_4\cos D_2\theta \end{aligned}\right\} \quad (7)$$

in which, $b_1 = -p_1 a_2, \quad b_2 = p_1 a_1, \quad b_3 = -p_2 a_4, \quad b_4 = p_2 a_3$ (8)

$$P_i = \frac{(\lambda^2 - m)(1+m+2\nu_{r\theta}) - (m - \nu_{r\theta}^2) D_i^2}{[\lambda(m+\nu_{r\theta})(\nu_{r\theta}+1) - \{m^2 + 2(\nu_{r\theta}+1)m - \nu_{r\theta}^2\}] D_i} \quad (i=1,2) \quad (9)$$

The stress components are expressed as follows:

$$\left.\begin{aligned} \sigma_r(r,\theta) &= E_\theta r^{\lambda-1}\{a_1R_1\sin D_1\theta + a_2R_1\cos D_1\theta + a_3R_2\sin D_2\theta + a_4R_2\cos D_2\theta\} \\ \sigma_\theta(r,\theta) &= E_\theta r^{\lambda-1}\{a_1S_1\sin D_1\theta + a_2S_1\cos D_1\theta + a_3S_2\sin D_2\theta + a_4S_2\cos D_2\theta\} \\ \tau_{r\theta}(r,\theta) &= E_\theta r^{\lambda-1}\{a_1T_1\sin D_1\theta + a_2T_1\cos D_1\theta + a_3T_2\sin D_2\theta + a_4T_2\cos D_2\theta\} \end{aligned}\right\} \quad (10)$$

in which,

$$R_i = \frac{\lambda + \nu_{r\theta} - \nu_{r\theta} p_i D_i}{m - \nu_{r\theta}^2}, \quad S_i = \frac{\lambda\nu_{r\theta} + m - m p_i D_i}{m - \nu_{r\theta}^2}, \quad T_i = \frac{D_i + (\lambda-1) p_i}{1+m+2\nu_{r\theta}} \quad (11)$$

When $0 < \lambda < 1$, the stress become infinite at the crack tip, therefore, λ imply the degree of the stress singularity.

(2) C = 0

In this case, equation (6) has 2 imaginary roots ($\pm\sqrt{\lambda^2+3}\, i$) and 0 root. The general solution of the displacement components are expressed as follows:

$$\left.\begin{aligned} u_r^*(\theta) &= a_1\sin\sqrt{\lambda^2+3}\,\theta + a_2\cos\sqrt{\lambda^2+3}\,\theta + a_3\theta + a_4 \\ u_\theta^*(\theta) &= b_1\sin\sqrt{\lambda^2+3}\,\theta + b_2\cos\sqrt{\lambda^2+3}\,\theta + b_3\theta + b_4 \end{aligned}\right\} \quad (12)$$

(3) C < 0

In this case, equation (6) has 2 imaginary roots ($D_1 i, -D_1 i$) and 2 real roots ($D_3, -D_3$). The general solution of the displacement components are expressed as follows:

$$\left.\begin{aligned} u_r^*(\theta) &= a_1\sin D_1\theta + a_2\cos D_1\theta + a_3\sinh D_3\theta + a_4\cosh D_3\theta \\ u_\theta^*(\theta) &= b_1\sin D_1\theta + b_2\cos D_1\theta + b_3\sinh D_3\theta + b_4\cosh D_3\theta \end{aligned}\right\} \quad (13)$$

The general solution of the stress components in the cases above (2),(3) are obtained similarly.

B. Derivation of the Characteristic Equation with respect to the Stress Singularity

The boundary condition at the radial edge in the stress singular analysis of the crack tip with opening angle $\pm\alpha/2$ is given as follows:

$$\left.\begin{aligned} \sigma_\theta(r,\alpha/2) &= \tau_{r\theta}(r,\alpha/2) = 0 \\ \sigma_\theta(r,-\alpha/2) &= \tau_{r\theta}(r,-\alpha/2) = 0 \end{aligned}\right\} \quad (14)$$

Substituting equation (10) etc. into equation (14), the following characteristic equations governing the intensity of the singularity λ can be derived:

(1) $C > 0$

$$(z_1\cos\frac{D_1}{2}\alpha\sin\frac{D_2}{2}\alpha-\sin\frac{D_1}{2}\alpha\cos\frac{D_2}{2}\alpha)(z_1\sin\frac{D_1}{2}\alpha\sin\frac{D_2}{2}\alpha-\cos\frac{D_1}{2}\alpha\sin\frac{D_2}{2}\alpha) = 0 \qquad (15)$$

(2) $C = 0$

$$\sin\frac{\sqrt{\lambda^2+3}}{2}\alpha\left\{\sin\frac{\sqrt{\lambda^2+3}}{2}\alpha-\frac{\sqrt{\lambda^2+3}}{2}\alpha\cos\frac{\sqrt{\lambda^2+3}}{2}\alpha\right\} = 0 \qquad (16)$$

(3) $C < 0$

$$(z_2\cos\frac{D_1}{2}\alpha\sinh\frac{D_3}{2}\alpha+\sin\frac{D_1}{2}\alpha\cosh\frac{D_3}{2}\alpha)(z_2\sin\frac{D_1}{2}\alpha\cosh\frac{D_3}{2}\alpha-\cos\frac{D_1}{2}\alpha\sinh\frac{D_3}{2}\alpha) = 0 \qquad (17)$$

where,

$$z_i = \sqrt{(-1)^{i-1}\times\frac{(m+1)\lambda^2+2m-\lambda\sqrt{(m-1)^2\lambda^2+8m(m+1)}}{(m+1)\lambda^2+2m+\lambda\sqrt{(m-1)^2\lambda^2+8m(m+1)}}} \quad (i=1,2) \qquad (18)$$

C. Results and Considerations

Figure 3 shows the relationship between λ and m at $\alpha = 360°$ and $\alpha = 320°$. The straight line is the solution of MODE I (symmetric, opening mode) and the broken line the MODE II (antisymmetric, shearing mode). The dot-dash-line show the line of $\lambda^2 = m$ corresponding to the case $C = 0$. From these figures, it is observed that, when $m > 1$, there are only one weaker singular solution than the well-known $1/\sqrt{r}$ singularity in the case of isotropic material ($m = 1$), but, when $m < 1$, there are stronger singular solutions. It can be considered as this reason that, when $m > 1$ (i.e. $E_\theta > E_r$), the stress singularity at the crack tip can be softened for the sake of the larger circumferential rigidity than the radial one, but, when $m < 1$ ($E_\theta < E_r$), its effect go by contraries and some singular modes can exist for the weaker circumferential rigidity.

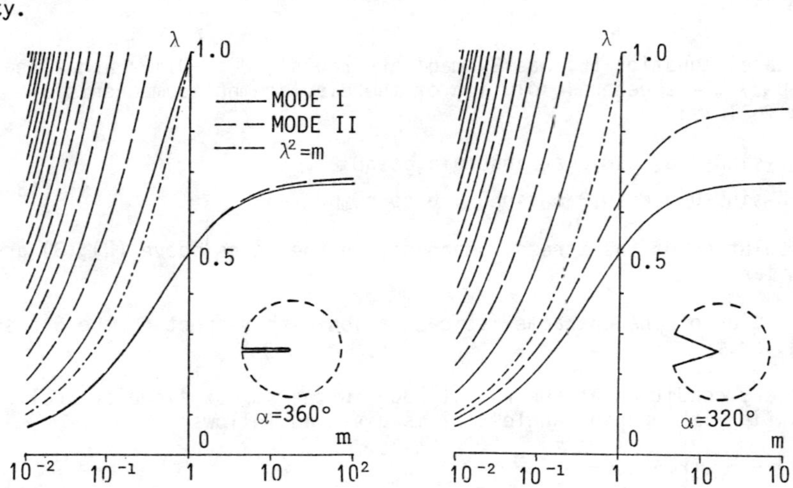

Fig. (3) - Stress singularity (polar anisotropic material)

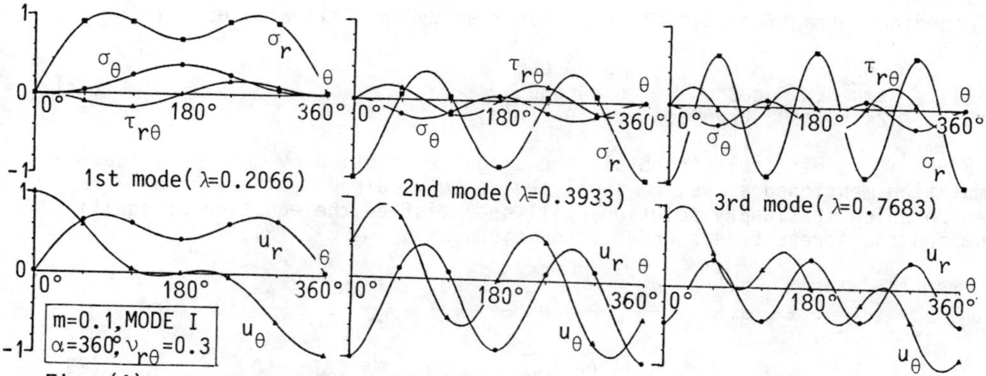

Fig. (4) - Stress and displacement modes (polar anisotropic material)

Figure 4 shows the first, second and third mode of the stress and displacement components at $m = 0.1$, $\alpha = 360°$ corresponding to $\lambda = 0.2066$, $\lambda = 0.3933$ and $\lambda = 0.7683$ respectively in MODE I.

ANALYSIS OF STRESS SINGULARITIES OF CRACK TIP IN A RECTILINEAR ANISOTROPIC MATERIAL

A. Virtual Work Equation governing the Stress Distribution near the Crack Tip

The stress-strain relations of a rectilinear anisotropic material in case of plane stress are given as follows (see Figure 5):

$$\begin{Bmatrix}\sigma_x\\ \sigma_y\\ \tau_{xy}\end{Bmatrix} = \frac{1}{1-\nu_{xy}\nu_{yx}}\begin{bmatrix}E_x & E_y\nu_{yx} & 0\\ E_x\nu_{xy} & E_y & 0\\ 0 & 0 & G_{xy}(1-\nu_{xy}\nu_{yx})\end{bmatrix}\begin{Bmatrix}\varepsilon_x\\ \varepsilon_y\\ \gamma_{xy}\end{Bmatrix} \tag{19}$$

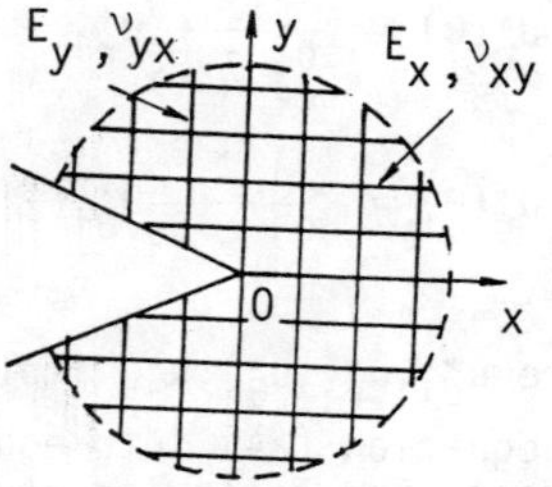

Fig. (5) - Rectilinear anisotropy

In which, $E_x\nu_{xy} = E_y\nu_{yx}$. $E_x, \nu_{xy}, E_y, \nu_{yx}$ are the Young's modulus, Poisson's ratios of the x and y direction respectively in Cartesian coordinates.
In this analysis, the shear modulus G_{xy} is assumed as following equation:

$$G_{xy} = E_xE_y/(E_x+E_y+2E_x\nu_{xy}) \tag{20}$$

By transforming this stress-strain relations of the x-y coordinates system into the one of r-θ coordinate system, the following relations are obtained:

$$\begin{Bmatrix}\sigma_r\\ \sigma_\theta\\ \tau_{r\theta}\end{Bmatrix} = [T]^T[D][T]\begin{Bmatrix}\varepsilon_r\\ \varepsilon_\theta\\ \gamma_{r\theta}\end{Bmatrix} \tag{21}$$

in which, [D] denotes the stress-strain matrix in equation (19), and [T] denotes the coordinate transformation matrix expressed by the following equation:

$$[T] = \begin{bmatrix} \cos^2\theta & \sin^2\theta & -\sin\theta\cos\theta \\ \sin^2\theta & \cos^2\theta & \sin\theta\cos\theta \\ 2\sin\theta\cos\theta & -2\sin\theta\cos\theta & \cos^2\theta-\sin^2\theta \end{bmatrix} \quad (22)$$

Since it is difficult to obtain the singular solution by the analytical formulation mentioned above, we shall try to solve directly the virtual work equation which stationary solutions will be satisfied the equation of equilibrium and the stress free boundary conditions at $\theta = \pm\alpha/2$ [3]:

$$\int_0^\infty\int_{-\alpha/2}^{\alpha/2}\{-(\frac{\partial\sigma_r}{\partial r} + \frac{1}{r}\frac{\partial\tau_{r\theta}}{\partial\theta} + \frac{\sigma_r - \sigma_\theta}{r})\delta u_r - (\frac{1}{r}\frac{\partial\sigma_\theta}{\partial\theta} + \frac{\partial\tau_{r\theta}}{\partial r} + \frac{2\tau_{r\theta}}{r})\delta u_\theta\}rdrd\theta + \int_0^\infty(\tau_{r\theta}\delta u_r + \sigma_\theta\delta u_\theta)_{\theta=\pm\alpha/2}dr = 0 \quad (23)$$

Assuming $u_r(r,\theta), u_\theta(r,\theta)$ by equation (3) and using the strain-displacement relations and the stress-strain relations (21), all terms in the virtual work equation (23) can be expressed by $u^*_r(\theta), u^*_\theta(\theta)$ which are defined by the integration of θ direction.

B. Finite Element Analysis of the Stress Singularities

We shall divide the domain at the crack tip into several finite elements radially as shown in Figure 6 and we assume the distribution of the displacement $u^*_r(\theta), u^*_\theta(\theta)$ in an element as following form:

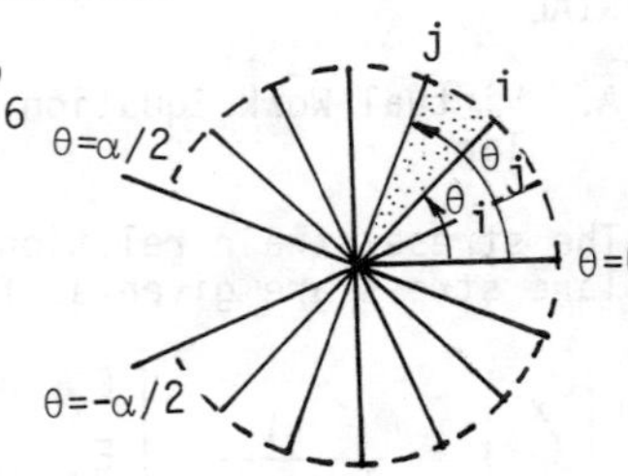

Fig. (6) - Finite element

$$\left.\begin{aligned} u^*_r(\theta) &= \frac{\theta_j - \theta}{\theta_j - \theta_i}u^*_{ri} + \frac{\theta - \theta_i}{\theta_j - \theta_i}u^*_{rj} \\ u^*_\theta(\theta) &= \frac{\theta_j - \theta}{\theta_j - \theta_i}u^*_{\theta i} + \frac{\theta - \theta_i}{\theta_j - \theta_i}u^*_{\theta j} \end{aligned}\right\} \quad (24)$$

where $u^*_{ri}, u^*_{rj}, u^*_{\theta i}, u^*_{\theta j}$ imply the displacement at nodal line i and j. Substituting equation (24) into equation (23) and integrating this equation about each element, the following equation is obtained:

$$\Sigma(\{\delta u^{*e}\}^T[k^e]\{u^{*e}\}) = 0 \quad (25)$$

in which Σ imply summation of all element matrices around the crack tip, $[k^e]$ is the element stiffness matrix and $\{u^e\}$ is the element nodal displacement vector. The more detail explanation can be found in the reference [3].

Now, the value of the determinant of the global stiffness matrix assembled by the all elements must be zero so as to have a nontrivial solution, therefore the characteristic equation governing the stress singularity is expressed as follows:

$$[K(\lambda,\alpha,m)] = 0 \quad (26)$$

We can obtain the eigen value λ in this determinant by a primitive searching technique of zero value of the equation.

C. Convergency of the Finite Element Solution

The Gaussian numerical integration was adopted in order to derive the element stiffness matrix from the virtual work equation (23). Figure 7 shows the relationship between λ and the number of elements (N) corresponding to the given number of the Gaussian integration points (NG) in case of $\alpha = 360°$, $m = 10$, MODE I. This convergency shows that a good solution can be obtained if $N > 30$ and $NG > 3$.

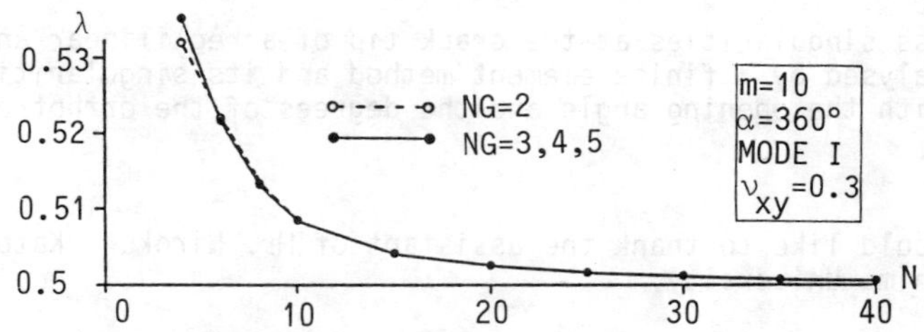

Fig. (7) - Convergency of the solution λ by finite element method

D. Results and Considerations

Figure 8 shows the relationship between λ and m at $\alpha = 270°, 300°, 360°$ in MODE I,II. The number of elements and the number of Gaussian integration points used were 40 and 4 respectively. From this figure, it is observed that, when $\alpha = 360°$, the stress intensity of the rectilinear anisotropic material for all the value of m in both case of MODE I,II equal to the one of the isotropic material. This result can be also confirmed by the existing condition of J

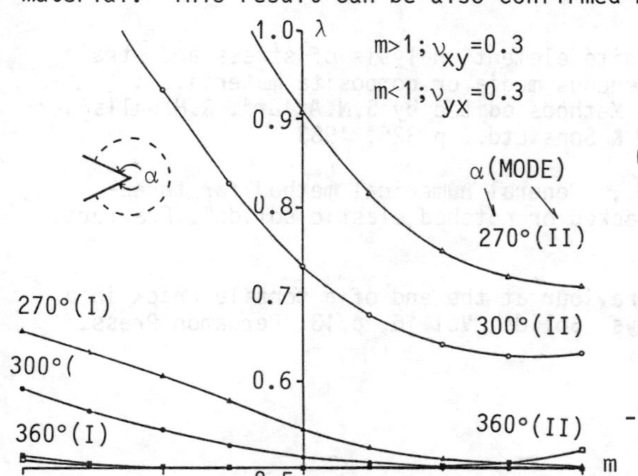

Fig. (8) - Stress singularity

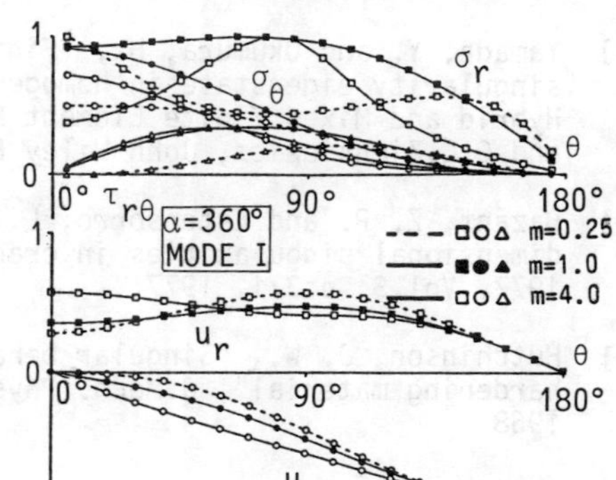

Fig. (9) -
Stress and displacement modes
(rectilinear anisotropy)

integral value in the case which the material property surrounding the crack tip do not change under the crack propagation [6]. And it is observed that, when $\alpha < 360°$, the stress singularities are small than $1/\sqrt{r}$. Figure 9 shows the stress and displacement mode corresponding to m = 0.25, 1.0, 4.0 at α = 360°, in MODE I.

CONCLUSIONS

(1) The characteristic equation with respect to the stress singularity at the crack tip of a polar anisotropic material were obtained analytically in an explicit form of the opening angles and the degrees of the orthotropy.

(2) The stress singularities at the crack tip of a rectilinear anisotropic material were analysed by a finite element method and its singularities were also discussed with the opening angle and the degrees of the orthotropy.

ACKNOWLEDGEMENTS

The author would like to thank the assistant of Mr. Hirokuni Kato, graduate student of Hiroshima University.

REFERENCES

[1] Williams, M. L., "Stress singularities from various conditions in angular corners of plates in extension", J.Appl. Mech.,19(4), p.526, 1952

[2] Williams, M. L., "Surface stress singularities resulting from various boundary conditions in angular corners of plates under bending", Proc. of the first U.S. National Congress of Applied Mechanics, ASME, p.325, 1951

[3] Fujitani. Y., "Finite element analysis of the singular solution in crack problem", Numerical Methods in Fracture Mechanics edited by A.R.Luxmoore and D.R.J.Owen, P.398, 1978

[4] Yamada, Y. and Okumura, H., "Finite element analysis of stress and strain singularity eigenstate in homogeneous media or composite materials", Hybrid and Mixed Finite Element Methods edited by S.N.Atluri, R.H.Gallagher and O.C.Zienkiewicz, John Wiley & Sons Ltd., p.325, 1983

[5] Bazant, Z. P. and Estensorro, L., "General numerical method for three-dimensional singularities in cracked or notched elastic solids", Fracture 1977, Vol.3, p.371, 1977

[6] Hutchinson, J. W., "Singular behaviour at the end of a tensile crack in a hardening material", J.Mech. Phys. Solids, Vol.16, p.13, Pergamon Press, 1968

PREDICTION OF TENSILE BEHAVIOR OF STRAIN SOFTENED COMPOSITES BY FLEXURAL TEST METHODS

T.-J. Chuang

Ceramics Division, National Bureau of Standards
Gaithersburg, Maryland 20899 USA

and

Y.-W. Mai

Department of Mechanical Engineering, University of Sydney
Sydney, N.S.W. 2006, Australia

ABSTRACT

A simple scheme of predicting tensile properties of a composite exhibiting strain-softening behavior from flexural tests is developed based on establishment of a relationship between bending and tensile properties. It is shown that strain-softening materials give higher bending strengths than tensile strengths, a result consistent with experimental observation. Also, given the load-displacement diagram obtained from a flexural test, it is possible to predict the entire tensile response. Bending data of polymer concrete are used to demonstrate the principles.

INTRODUCTION

Many structural composites of engineering significance like polymer concrete, fibre cement and other non-metal aggregates display what is called "strain-softening" behavior. Briefly, this means that in an uniaxial tension test, the stress-strain relationship is linear until the maximum tensile stress or strength (σ_m) is reached. Further straining hereafter results in a decaying stress which is the post-peak characteristic of the composite with a localized strain-softening zone. Strictly speaking, the strain in this localized zone is undefined due to a loss of a continuum state wherein a nascent crack opens up with untorn ligaments bridging the interface. One way to define this behavior is to give the stress (σ) versus crack width opening displacement (δ) relationship [1,2]. Alternatively, instead of a discrete crack model, the strain across the softening zone may be regarded as uniform in a macroscopic sense and as representative of the zone opening if the microstructural details are neglected. In this way, a stress-strain relationship can still be defined [3,4]. The opening displacement then is related to the post-peak strain by the fracture process zone width.

In fracture mechanics of such a material, knowledge of the stress-strain/crack opening relation is most important as this enables us to compute the specific work of fracture in the (localized) softening zone, as well as the complete crack growth resistance (R) curve [5]. Considerable efforts have, therefore, been made in the past to design proper tensile experiments in order to obtain the complete uniaxial stress-strain curve. Particular attention has been given to specimen grip design, closed-loop strain control and testing machine stiffness [1,2,6]. In view of these experimental complications in uniaxial testing, it is desirable to explore whether such information can be extracted from simple bending test data, for bending tests are more stable and easier to perform. In the present paper we derive the equations relating bending to tensile response and illustrate the working principles of predicting the tensile constitutive law from these equations. An example is given to demonstrate the procedure from a bending experiment performed on polymer concrete.

ANALYSIS

The flexural behavior (curvature, deflection and stress distribution, etc.) of a strain-softening composite in an uniform rectangular beam is analyzed here within the framework of simple beam theory. The basic assumptions and restrictions will first be given. The analysis pertains to a four-point loading configuration because of its popularity in laboratory experimentation and its generality (three-point bending, for instance, can be regarded as a limiting case where the minor span $2\ell = 0$). In this case, the bending moment in the outer spans varies linearly from zero at the outer supports to a maximum value M at the loading lines. This maximum moment is sustained in the inner-span uniformly and has the value $M = P(L-\ell)/2$ where P is the total applied load and 2L is the major span.

A. Basic Assumptions and Limitations

In order to perform the bending analysis, the constitutive laws for uniaxial tension and compression of the beam element must be specified. A literature review of the strain-softening behavior reveals that in general the tensile stress-strain relationships follow a power-law equation in the softening regime. Figure 1(a) gives a schematic diagram of such a plot on which the current analysis is based. When $\varepsilon \leq \varepsilon^*$, where ε^* is the maximum elastic strain at which the softening behavior begins, the elastic behavior dominates with the stresses varying linearly from zero to σ_m. σ_m is the tensile strength or the maximum tensile stress that the material can sustain. Upon further straining such that $\varepsilon^* < \varepsilon < \varepsilon_o$, ε_o being the strain at complete separation, the material enters the strain-softening region where the tensile stress (σ) is related to ε in the following way:

$$\sigma = \sigma_m \left[1 - \left(\frac{\varepsilon-\varepsilon^*}{\varepsilon_o-\varepsilon^*}\right)^n\right] \qquad (1)$$

where n is the softening exponent. The parameters σ_m, ε^*, ε_o and n are considered to be material constants.

The compression behavior, on the other hand, will be assumed linearly elastic throughout, with the same Young's modulus as in tension (i.e., $E = \sigma_m/\varepsilon^*$). This is justified in view of the fact that the compressive strength is at least an or-

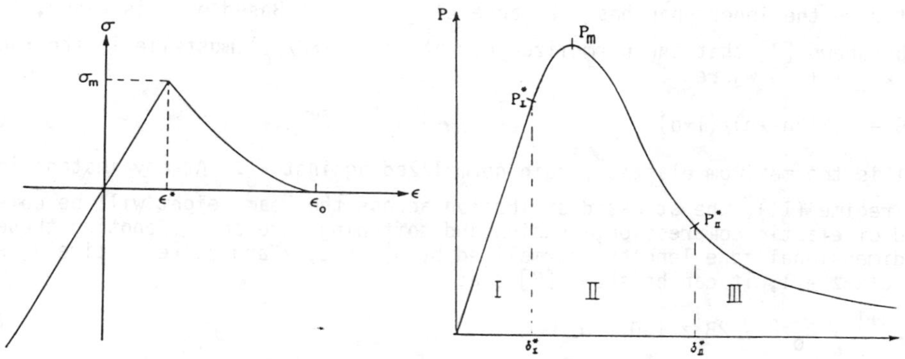

Fig. (1) - (a) Normal stress-strain curve showing strain-softening behavior; (b) Load-deflection curve for bending test

der of magnitude higher than the tensile strength for a typical strain-softened composite. Accordingly, the compression side of a beam under bending is always in the elastic regime.

In order to formulate the load-point deflection in a general manner, a width of 2w of the softening zone is assumed to develop in the inner span whenever the strain of the outer-fibre at the tension side reaches ε^*. In the present analysis, this width is treated as a constant; its value being dictated by such variables as notch width, grain size, etc. The zone width can affect the stability condition of the beam under constant cross-head speed conditions [7].

Other assumptions employed in the simple beam theory are also retained in the current investigation. These include: (1) Bernoulli's hypothesis which dictates that planar sections remain plane during bending; (2) small displacement in comparison to beam height and (3) contributions of shear deformation are not significant. Once the softening zone of 2w has developed, only monotonic loading is allowed thereby avoiding the complication of residual stress generation.

B. Bending Analysis

Because of the complexity of the constitutive laws, see Figure 1(a), the bending behavior of the composite beam can be divided into three stages: (I) elastic regime; (II) strain-softening zone growth regime and (III) crack growth regime. Detailed analysis is given in [7]; the results are summarized here.

In the elastic regime (I), the maximum strains of the beam are less than ε^* and the beam is completely elastic and reversible. Therefore, the conventional beam theory in strength of materials applies. At the end of (I), the applied load P_I^* and load-point deflection δ_I^*, Figure 1(b), are related to σ_m and ε^* by $P_I^* = \sigma_m BH^2/3(L-\ell)$ and $\delta_I^* = 2\varepsilon^*(L-\ell)(L+2\ell)/3H$ where B and H are the dimensions of the beam cross-section. The total elastic energy stored in the beam is $U = P_I^*\delta_I^*/2 = (1/9)\varepsilon^*(L+2\ell)(\sigma_m BH)$. A portion of this amount, $(\varepsilon^*\ell)(\sigma_m BH)/3$, is stored in the inner span.

In the strain-softening zone growth regime (II), the load-point deflection, δ is in excess of δ^*_I and the outer fiber strain of the strain-softening material located in the inner span has the range $\varepsilon^* \leq \varepsilon \leq \varepsilon_o$. Based on this range, it can be shown [7] that the normalized curvature κ (=KH/ε_o) must lie in the range $2\lambda \leq \kappa \leq 1 + \sqrt{G}$ where

$$G = \lambda(\lambda+2n-\lambda n)/(1+n) \tag{2}$$

and λ is the maximum elastic strain normalized against ε_o. At any instant in this regime (II), the stress distribution across the beam height will be composed of elastic compression, tension and softening stresses. Denoting these non-dimensional zone lengths (normalized by H) as c, x and z, respectively, such that c+x+z = 1, it can be shown [7] that

$$z^{n+1} + B_o z^2 - 2B_o z + D = 0 \tag{3a}$$

where

$$B_o = x^{n-1}(1+n)(1-\lambda)^n/2\lambda^n \tag{3b}$$

and

$$D = (1-2x)B_o \tag{3c}$$

based on the requirement of horizontal force balance. Closed form solutions for z in terms of κ are available only when n is an integer and less than 4. An approximate solution $z = 1 - \sqrt{(2x)}$, for small λ, is also available but underestimates the softening zone size by about 9 percent at the end of regime (II).

The normalized moment ($m = M/((1/6)\sigma_m BH^2)$) versus curvature relationship, $m = f(\kappa)$ has the following form:

$$m = 2C^3/x + 2x^2 + 6xz + 3z^2 - 6\left(\frac{\lambda}{1-\lambda}\right)\frac{z^{n+1}}{x^n}\left(\frac{x}{n+1} + \frac{z}{n+2}\right) \tag{4a}$$

and

$$\kappa = \lambda/x \tag{4b}$$

based on the requirement of $M = \int \sigma y dy$ where y is the distance from the neutral axis. At the end of regime II, when $k = 1 + \sqrt{G}$, we have

$$m^*_{II} = F/(1+\sqrt{G})^2 \tag{5a}$$

where

$$F = 3G + 2G\sqrt{G}/\lambda - \lambda^2 + 3n(1-\lambda)^2/(n+2) \tag{5b}$$

In the crack growth regime (III), the maximum strain exceeds ε_o and a crack will initiate from the tensile edge of the beam. The term "crack" is defined here as separation of "fibre" rather than discontinuity in the continuum sense. Hence, any possible stress elevation near the "crack" tip is not considered. This means that we consider the material as notch-insensitive. Under these conditions, we find that the crack length a is related to κ by $a = 1 - (1+\sqrt{G})/\kappa$ and the moment-curvature relation takes the simple form

$$m = F/\kappa^2 \tag{6}$$

where F has been defined in equation (5b). This simple expression is remarkable in that the carrying moment decays asymptotically as the inverse square of the increasing curvature. It should be noted that the correctness of these analytical expressions can be confirmed by checking the boundary point between regime (II) and (III) at which $\kappa = 1 + \sqrt{G}$. For example, substituting $\kappa = 1 + \sqrt{G}$ in equation (6), we obtain an expression which agrees with that given by equation (5a).

The load-point deflection δ in regimes (II) and (III) has contributions from both elastic and softening materials. It can be shown from a simple geometric sketch and from the method of strength of materials, that normalized load-point deflection is given by a summation of these two contributions:

$$\delta = (L-\ell)\tan[\sin^{-1}(w\kappa\varepsilon_0/H)] + (2/3)\lambda\varepsilon_0 mJ \tag{7a}$$

where J is a beam geometric factor:

$$J = [(L-w)^3 + 2(\ell-w)^3 - 3(L-w)(\ell-w)^2]/(L-\ell)H^2 \tag{7b}$$

C. Theoretical Results

We have presented in Section B the necessary equations for describing the behavior of a beam under four-point bending for all three characteristic regimes. We shall now show the theoretical results using representative values of $n = 0.21$ and $\lambda = 0.0173$. First, we can proceed to evaluate the moment-curvature law for the softening material by using moment equations of (4a) and (4b). Expressions for regimes (I) and (III) are analytical, but for regime (II), m must be computed numerically; there is no closed form solution for z from equation (3a). The numerical results are presented in Figure 2 as a three-stage curve plotted in m-κ space. The terminal points of regime (II), X and Y indicated in the figure, agree with the end point and the initiation point of (I) and (III), respectively, in bth magnitude and slope, indicating the correctness of the numerical scheme for regime (II).

For engineering applications, it is often required to determine the strength (i.e., the maximum load-bearing capacity) of a given material. The peak load P_m for this material with $n = 0.21$ and $\lambda = 0.0173$ as shown in Figure 2 is about 20% higher than the maximum elastic load. It should be noted that because $m = M/(1/6) \times \sigma_m BH^2$, the normalized peak load (P_m) is identical to the peak moment (m_m). To examine the bending strength for a wide variety of materials, the curves m_m as a function of n are plotted in Figure 3 for three different values of λ. It shows that increasing n leads to higher bending strength. Reducing λ for a given n yields similar results. This can be explained physically in terms of the enhancement of the energy absorption capabilities for increasing n or reducing λ. Figure 3 also shows that for strain-softening materials with $n > 0.1$, the ratio of bending to tensile strength is always greater than 1. This result is in agreement with our experience with this kind of composite. For "near" non-softening

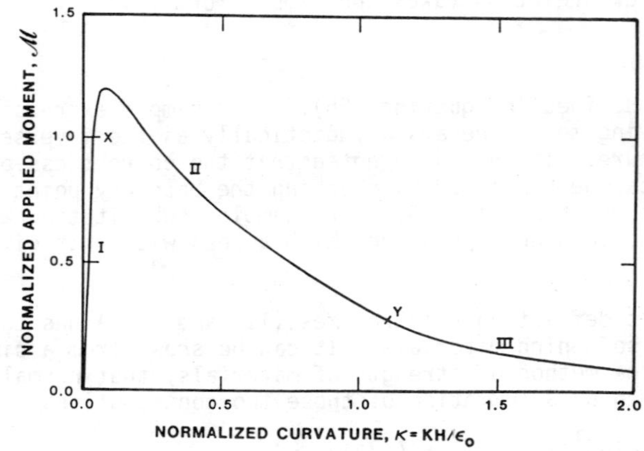

Fig. (2) - Normalized plot of moment (m) versus curvature (κ) in bending for n = 0.21, λ = 0.0173

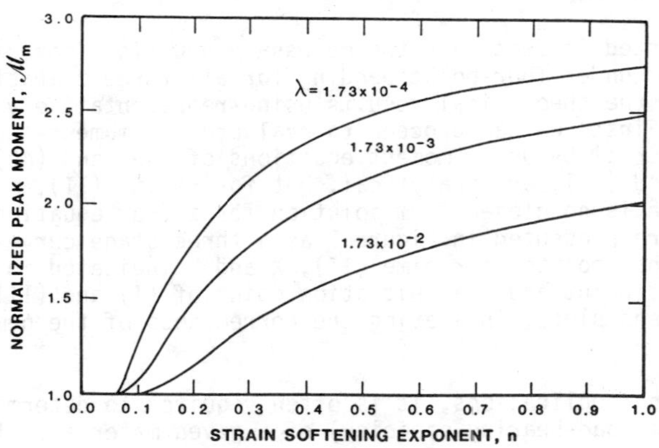

Fig. (3) - Variation of bending strength with strain-softening exponent for three values of λ

brittle materials where $n < 0.1$, however, $m_m = 1.0$ so that both tensile and bend strength are equal.

Secondly, by treating the curvature as an independent variable, the load-deflection curve obtained from a four-point bending test can be predicted for a specified material, softening zone width and specimen geometry. Assuming non-dimensional $L = 4$, $\ell = 1$, $\varepsilon_o = 0.0145$ and $w = 0.56$, Figure 4 gives the theoretically predicted curve for P versus δ. The points X and Y on the curve divide the

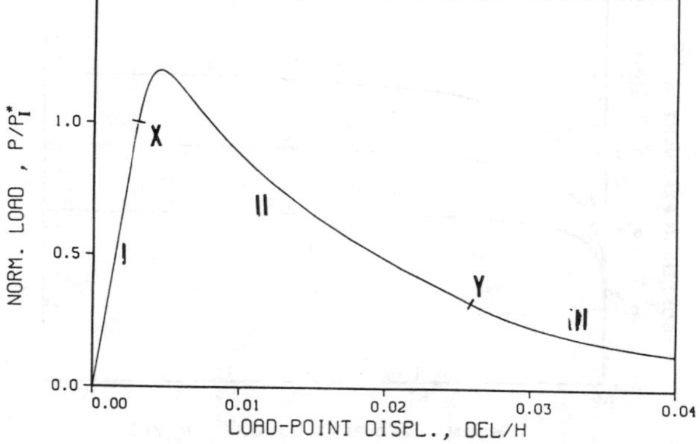

Fig. (4) - Load-deflection diagram predicted for the specific material and specimen geometry

flexural behavior into the three characteristic regimes. It is interesting to note that the deviation from linearity in the P versus δ curve at point X can be attributed to the development of the strain-softening zone. This is consistent with the conventional understanding of microstructure in which localized microcracking introduces non-linearity.

D. The Inversion Problem

Thus far, we are able to predict the load-deflection curve for a four-point bend beam when the material parameters (n,λ,w) and beam geometry are given. We now pursue the inversion problem, namely, given a flexural load-displacement diagram, can we determine the material's strain-softening behavior in tension.

From the P-δ curve obtained in the laboratory, we can determine the following values, P_I^* at X, P_{II}^* at Y and P_m, the peak load. This means that both m_m and m_{II}^* are known. Now from Figure 3, we have the following functional relation in numerical form

$$m_m = f(\lambda,n)$$

and

$$m_{II}^* = F/(1+\sqrt{G})^2$$

from equation (6). These two equations form a system of non-linear algebraic equations for the unknowns λ and n. The task is to solve them in terms of m_m and and m_{II}^* which are prescribed by the test data. Figures 5 and 6 give the solutions for n and λ for three representative values of P_{II}^*/P_I^*. From Figure 5, it is noted that n varies between 0.2 and 0.4 in practical cases. Further, n is quite insensitive to the normalized peak load. On the other hand, the solutions for λ vary

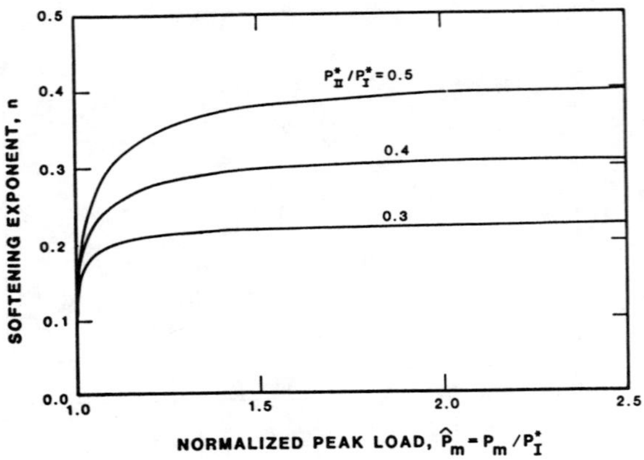

Fig. (5) - Solution of n as a function of normalized peak load and P_{II}^*

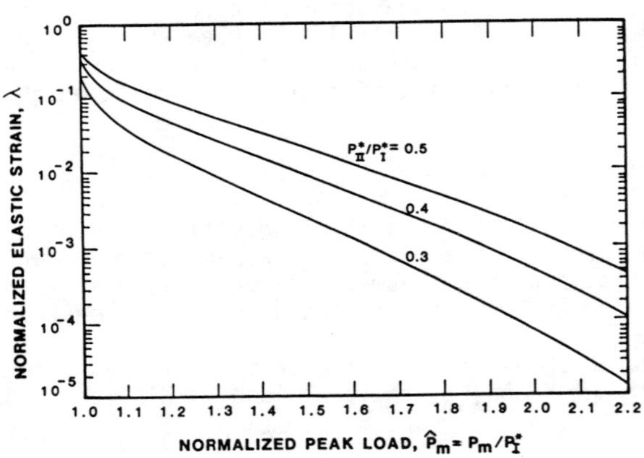

Fig. (6) - Solution of λ as a function of peak load and P_{II}^*

widely from 0.6 down to 10^{-5} when normalized peak loads change from 1.0 to 2.2. Once n and λ are determined, the complete tensile σ-ε curve including σ_m, ε* and ε_o can be obtained from the information gathered at the point X. However, to convert the tensile σ-ε relation to the σ-δ curve, we need to know the zone size w_t in tension. By assuming that there is a relationship between w_t and w, say w $= \alpha w_t$ where α is a material constant, we can determine w_t by extracting w from the deflection data δ_{II}^* at Y (or at peak load). This gives the following linear relationship:

$$w = H\delta_{II}^*/(L-\ell)(1+\sqrt{G})\varepsilon_o - (2/3)\lambda F(L^3+2\ell^3-3L\ell^2)/(1+\sqrt{G})^3(L-\ell)^2 \quad (8)$$

This expression is derived from equations (7) when $w \ll \ell$ and $\varepsilon_o \ll 1$.

E. Example on Polymer Concrete

In this section, we shall use the theory developed so far to predict the tension behavior of a strain softening composite, namely polymer concrete, from the bending curves presented in [8]. Table 1 summarizes the data obtained from the flexural load-deflection curve given in [8].

Table 1 - Data for polymer concrete [8]

P^*_I (N)	P_m (N)	m_m	P^*_{II}/P^*_I	H (mm)	δ^*_{II}/H	L (mm)	ℓ (mm)
97	116	1.2	0.3	7.5	0.027	60	15

We shall now show how to deduce the tensile σ-δ curve from the data on flexural P-δ curve listed in Table 1. The first step is to find values of n and λ respectively from Figures 5 and 6. The normalized values of P_m and P^*_{II} against P^*_I are found to be 1.2 and 0.3 from the table. Thus, from Figures 5 and 6, we obtain n = 0.21 and λ = 0.0173. The next step is to calculate σ_m and ε^* from the point X. This gives σ_m = 19.4 MPa and ε^* = 0.00025. This tensile strength is in very good agreement with the indirect splitting tensile strength of 18 + 2.4 MPa measured in [9]. The complete σ-ε curve in tension is now available.

To convert the σ-ε curve to the σ-δ curve, we must acquire the softening zone size w_t developed in simple tension. Assuming $w_t = (1/4)w$ where w is calculable from equation (8) when δ^*_{II} is measured. Thus, we obtain w_t = 1.05 mm from w = 4.20 mm. The crack opening displacement, δ, is simply related to the strain within the process zone, ε, by $\delta = \varepsilon w_t$ and the complete tensile σ-δ curve is thus uniquely determined. Table 2 gives the solutions of the relevant parameters obtained from the present theory using the input data from Table 1. Figure 7 gives the resulting σ-δ curve.

Table 2 - Predicted parameters for tension behavior

n	λ	σ_m (MPa)	ε^*	ε_o	w (mm)	w_t (mm)
0.21	0.0173	19.4	0.00025	0.0145	4.20	1.05

CONCLUSION

The tensile behavior of a composite exhibiting strain-softening character can be predicted from the load-deflection diagram obtained in a bending experiment. We are currently performing an uniaxial tension test program in an attempt to verify these theoretical predictions.

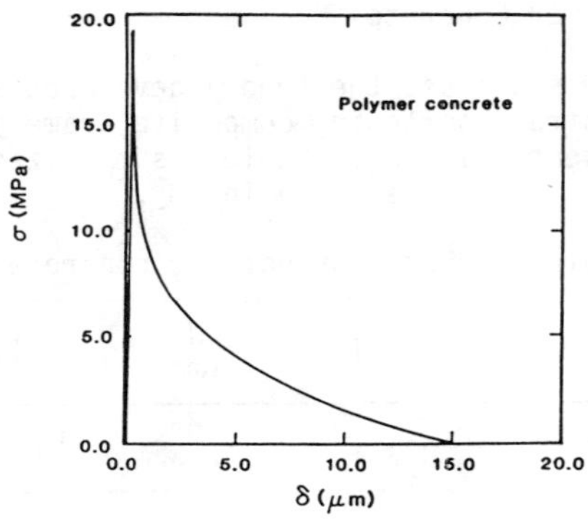

Fig. (7) - Predicted stress versus crack face opening relationship for polymer concrete

ACKNOWLEDGEMENT

The work was supported by the U.S. Department of Energy AR&TD Fossil Energy Materials Program under Interagency Agreement DE-AI05-800R20679 with NBS. The authors are grateful to P. Swanson for reviewing the manuscript.

REFERENCES

[1] Peterson, P. E., "Crack growth and development of fracture zones in plain concrete and similar materials", Report TVBM 1006, Lund Inst. Tech., Lund December 1981.

[2] Krech, W. W., "The energy balance theory and fracture energy measurements for uniaxial tension", in Proc. 3rd Int. Soc. Rock Mech. Congr., Vol. 2A, Denver, Colorado, pp. 167-173, 1974.

[3] Dougill, J. W., "On stable progressively fracturing solids", J. Appl. Math. Phys. (ZAMP), Vol. 27, pp. 423-437, 1976.

[4] Bazant, Z. P. and Oh, B. H., "Crack band theory for fracture of concrete", Materials and Structures, Vol. 16, No. 93, pp. 155-178, 1983.

[5] Shah, S. P., "Application of fracture mechanics to cementitious composites", Martinus Nijhoff Publishers, 1985.

[6] Gopalaratnam, V. S. and Shah, S. P., "Softening response of plain concrete in direct tension", ACI J. 82, p. 310, 1985.

[7] Chuang, T.-J. and Mai, Y.-W., "The relationship between tensile and bending properties of a strain-softening material", submitted to J. Appl. Mech., 1986.

[8] Krause, R. F. and Fuller, E. R., in Chevron-Notched Specimens: Testing and Stress Analysis, ASTM STP 855, p. 309, 1984.

[9] Krause, R. F., "Testing geothermal-well cements: strength measurements following exposures to simulated geothermal fluids", National Bureau of Standards Internal Report No. 80-2099-4, July 1980.

DYNAMIC VISCO-ELASTIC PROPERTIES OF GLASS FIBRE REINFORCED POLYPROPYLENE

L. W. Chen and P. H. Lin

Department of Chemical Engineering, National Taiwan University
Taipei, Taiwan, Republic of China

ABSTRACT

A series of experiments were taken to determine the physical properties of FRPP, especially the dynamic viscoelastic properties, from which a correlation among properties with varying interfacial strength was made. Results showed that high interfacial strength reduces damping above Tg while enhance it below Tg. Elastic moduli increased with interfacial force as expected, but loss moduli was high with strong interfacial bonding. By further observation on the interfacial morphology, transcrystalline was found on surface of glass fibre coupled with silane coupling agent. This explains the high loss moduli behavior of materials when interfacial slippage exists.

INTRODUCTION

Fibre glass in polypropylene matrix enhances the tensile strength and moduli while accompanying a reduction of impact strength [1]. Modification of interfacial bonding to a certain level, which generally is done by coupling agent, is necessary to improve the impact strength [2]. Lots of studies were made on the reinforced efficiency of silane coupling agent which showed the high increase in the interfacial bonding force [3,4]. Different concentration of silane results in different interfacial force, this is the basic parameter of our study.

The study of dynamic viscoelastic properties helps to understand the internal energy dissipation as heat of polymer under a periodic stress [5]. Glass fibre may affect the mechanism such as to reduce damping with high interfacial bonding [6,7]; increase the loss moduli when interface slippage occurs [8], or even after Tg of the matrix to a higher (lower) temperature [6]. Loss tangent (damping) is a function of the morphology of matrix, bonding force between fibre-matrix interface and fibre properties [9]. Lots of the study on dynamic viscoelastic properties of polymer reinforced by glass fibre were for thermosetting system for the ease of preparation. Few has been studied on PP-glass fibre system. This study also aimed to measure the dynamic viscoelastic properties on a controlled condition by varying interfacial force with silane treated, and glass fibre contended, from which some correlation to physical properties is found. To substantiate these phenomena, some subsidiary test of morphology and density were made.

EXPERIMENT

Chopped strand E-glass fibre was heat-burned in an oven at 600°C for 45 minutes to remove the original sizing, then dipped in silane (KBM 603) coupling agent solution of concentration from 0.1 to 2 wt% for 10 mins, dryed at 80°C for 10 hours. PP powder-fibre mixture of 15-25 wt% was extruded through Brobender extruder, and pelletized for compress mold. All test specimens were compress molded through a fastidious devoiding process at 220°C, the total molding time is 45 min. Void content was determined by a comparison of the calculated and measured density of materials. Tensile strength and moduli test were made according to ASTM D368 Type IV, which impact test by JIS K7110 Izod type with notch. Interfacial force test, set up is shown as Figure 1, the glass rod with diameter of 0.5-0.6 mm was dipped into extruded PP ribbon and cooled to be set on Instron for measurement.

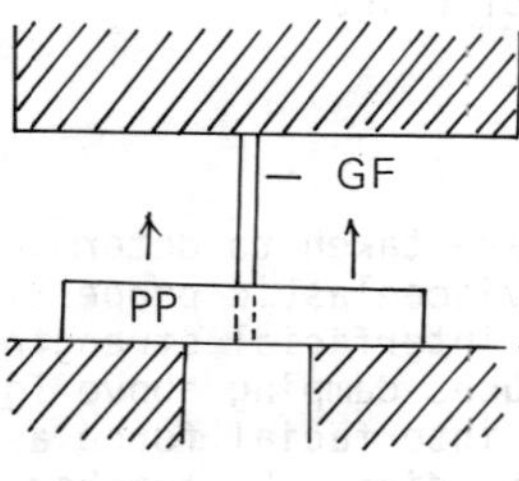

Fig. (1) - Schematic diagram for PP-glass fiber interfacial strength measurement

Samples of dimensions 5 cm x 0.1 cm x 0.05 cm were prepared for dynamic viscoelastic test in Reovibron DDV-II at frequency of 3.5 Hz and 110 Hz, temperature range from -75°C to 130°C. For the observation of morphology, glass fibre both coupled and uncoupled with silane compound was mixed with PP powder and compress between two stainless steel plate at 150 Kpa for 20 mins, and cooled to form a film of thickness 20 μm for observation at polarized microscope.

RESULTS AND DISCUSSIONS

Voids contained that affects some physical properties is unavoidable in our case. A highest of 5 vol% of voids content was measured, Figure 2, which is not higher than injection molding product [10]. It was found that (Figure 2) coupling agent reduced the void content which in other words, silane compound promoted the wetting between fibre-matrix interface. The interfacial strength test, shown in Figure 3, indicated the actual increase in bonding force by coupling agent treatment. Two main contributions to the interfacial forces, i.e., the shrinkage of matrix on fibre and the bonding between matrix and fibre were considered to affect the interfacial force. Therefore, the increase in interfacial force in the experiment is obviously contributed by bonding force of interface since the shrinkage were all the same for the materials tested.

Tensile strength and moduli, Figure 4, is as expected, increased by increasing of glass fibre content and coupling agent. It agrees with that the strong interfacial force is a determinative factor as well as stress transfer from matrix to glass fibre. Figure 5 showed that glass fibre reduced the Izod impact strength of PP, but a 2% of silane treatment regained the strength to the level of the matrix

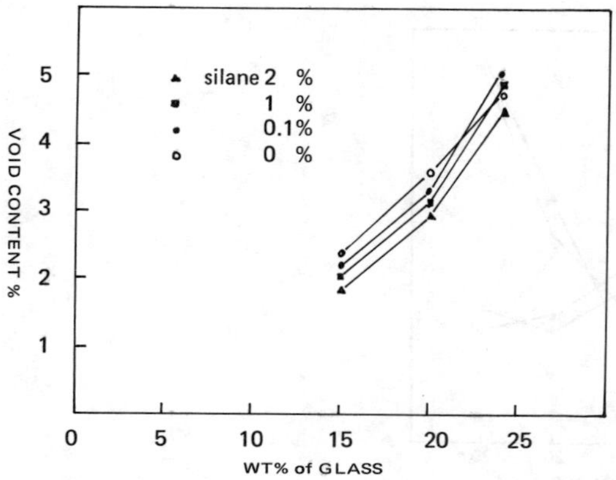

Fig. (2) - Void content of PP reinforced by glass fiber treated with different concentration of silane coupling agent

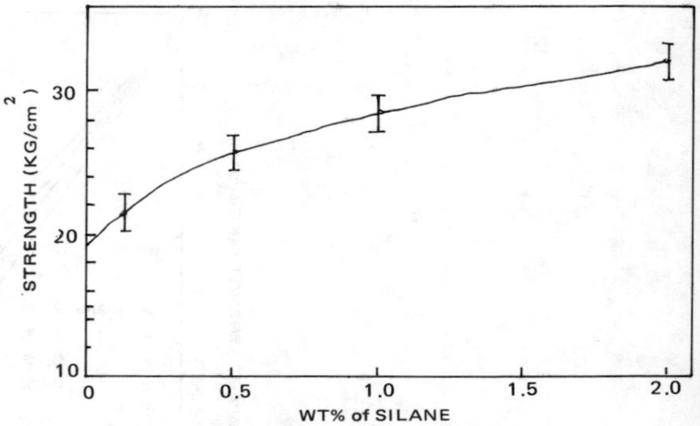

Fig. (3) - Interfacial strength of PP and glass fiber treated with different concentration of silane coupling agent

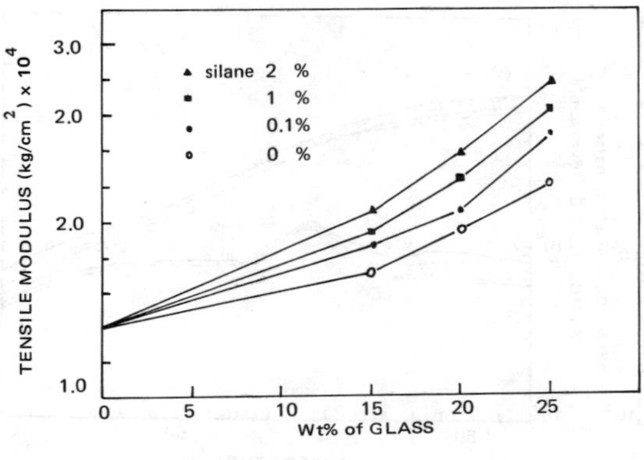

Fig. (4a) - Tensile modulus of PP reinforced by glass fiber treated with different concentration of silane coupling agent

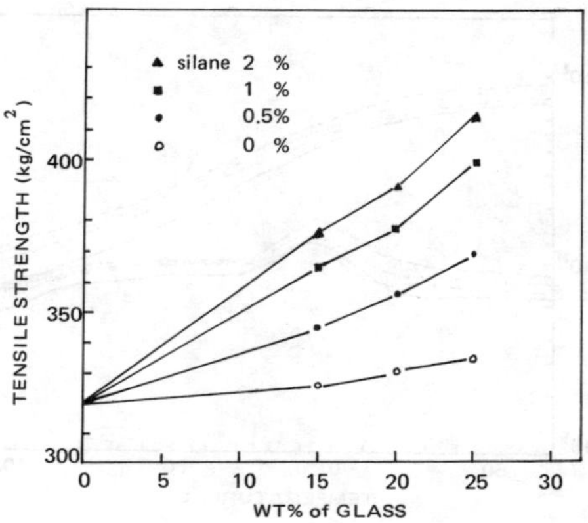

Fig. (4b) - Tensile strength of PP reinforced by glass fiber treated with different concentration of silane

at 25% glass content. An observation by SEM at impact fracture surface of glass fibre coupled with silane showed that all glass fibre was covered with a layer of PP when pulled out. It suggests that a new fracture energy dissipation mechanism along fibre is controlling besides the fibre fracture, fibre pull-out and matrix fracture energy. Some are consistent to Ramsteiner's study [1].

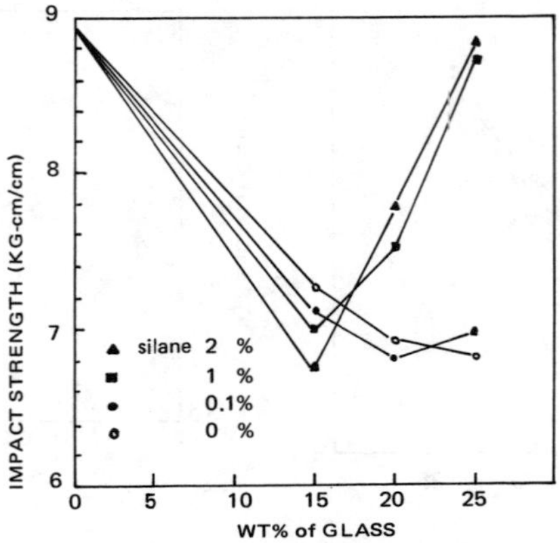

Fig. (5) - Impact strength of PP reinforced by glass fiber treated with different concentration of silane coupling agent

Smooth curves are drawn through all dynamic viscoelastic data because of the aggregation of data points taken (2°C each), comparison of storage moduli, Figure 6, with elastic moduli, both have the same trend for all composition of glass fibre content. Since both describe the behavior of elastic parts of polymer, which is not affected through the existence of rigid glass fibre, a strong interface helps for the transfer of stress to fibre that contribute to the increase of storage moduli.

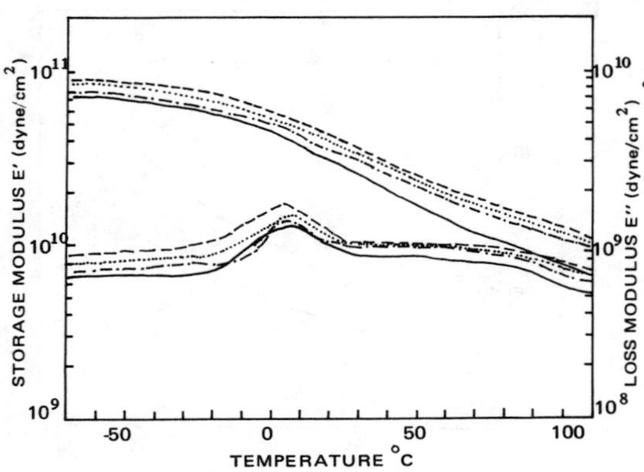

Fig. (6a) - Storage and loss moduli of pp reinforced by 25% glass fibre treated with 2% (---), 1% (...), 0.1% (-.-), 0% (—) silane coupling agent

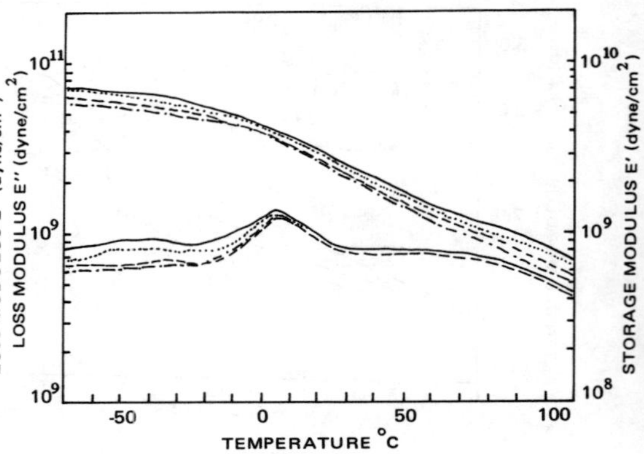

Fig. (6b) - Storage and loss moduli of PP reinforced by 15% glass fibre treated with 2% (—), 1% (...), 0.1% (---), 0% (-.-) silane coupling agent

For loss moduli, Figure 6, contrary to the study of Gibson who suggests that any mechanism which would increase the stiffness would also tend to reduce the loss moduli and vice versa [11], a weak interfacial bonding (uncoupled) does not increase E" in energy dissipation by interfacial slippage in comparison to a strong (coupled) one. The comparison of impact data with loss moduli shows a similar trend also at 25% glass fibre content. It is suggested that high E" would lead to a high impact strength. But because of the difference between the energy dissipation mechanism of dynamic test which is the friction heat loss by polymer molecules and polymer-fibre interface, from the impact fracture which is the debonding fracture of matrix and matrix fibre interface energy loss. It is suggested that some correlation between the impact fracture and dynamic energy loss needs further study.

Figure 7 is the loss tangent (damping) of the materials which shows a decrease of damping for higher interfacial force above Tg and an increase below Tg. The

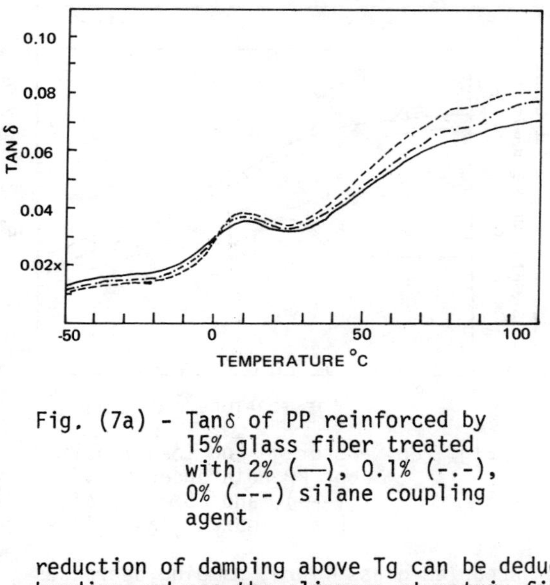

Fig. (7a) - Tanδ of PP reinforced by 15% glass fiber treated with 2% (—), 0.1% (-.-), 0% (---) silane coupling agent

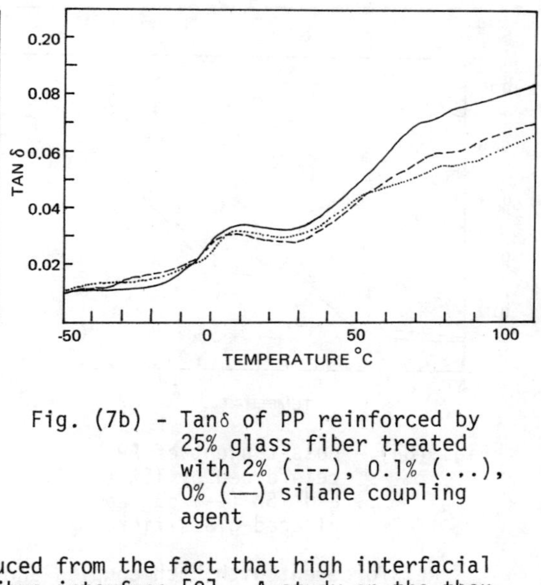

Fig. (7b) - Tanδ of PP reinforced by 25% glass fiber treated with 2% (---), 0.1% (...), 0% (—) silane coupling agent

reduction of damping above Tg can be deduced from the fact that high interfacial bonding reduces the slippage at matrix-fibre interface [9]. A study on the thermal expansion coefficient, α_L, with exactly the same material of our study [12], shows a much lower α_L for glass fibre treated with 2% silane, Table 1. We can expect an increase in dimensional stability for high interfacial bonding which can be verified from damping data. It is reasonable for the fact that rigid glass fibre with strong bonding to PP matrix would limit the molecular motion when temperature changed. The effect of glass fibre on dynamic viscoelastic properties, Figures 8 and 9, shows that the increase of uncoupled glass fibre content results in an obvious increase storage moduli but a slightly almost indistinguishable increase of loss moduli. Another case when coupled glass fibre is used, both storage and loss moduli were increased from 15% to 25% of glass fibre content.

It is expected for storage moduli to increase, while there seems to be a new mechanism of energy dissipated by silane treatment to increase the loss moduli with low possibility of interfacial slippage.

Table 1 - Thermal expansion coefficient and specific volume of PP composite

	Specific Volume (cm^3/g)	$\alpha_L \times 10^{-5}$ ($\frac{1}{°C}$)	$\beta_\ell \times 10^5$ ($\frac{1}{°C}$)	
PP	1.136	12.3	85.36	Treated
10 GF	1.0615	11.70	83.3	Treated
15 GF	1.031	11.51	82.8	Treated
30 GF	0.9523	10.70	81.5	Treated
15 CNGF	1.0471	11.58	84.2	Not Treated

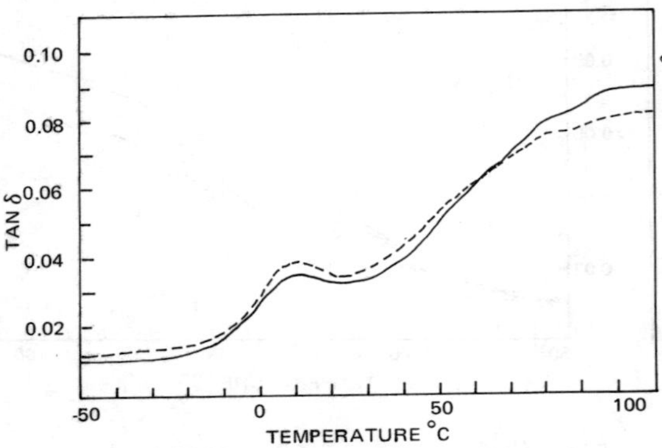

Fig. (8a) - Loss tangent of PP reinforced by 15% (---), and 25% (—) of heat-cleaned glass fibre

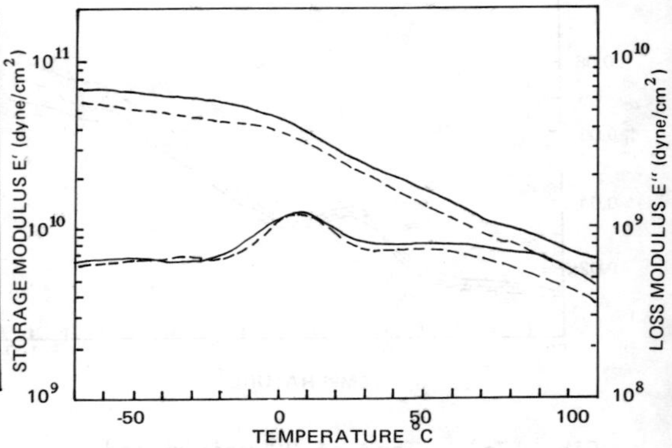

Fig. (8b) - Storage and loss moduli of PP reinforced by 15% (---), and 25% (—) of heat-cleaned glass fibre

Reinforcement with both coupled and uncoupled glass fibre, damping decreases as fibre content increased. This is consistent to Nielson's study [6] due to the increase of E' than E" while tanδ = E"/E'.

Study on morphology at glass-matrix interface for the interpretation of new energy mechanism was made by observation on polarized microscope. Figure 10 shows an interesting result that transcrystalline of PP was found at surface of glass fibre when coupled by 2 wt% of silane, where only spherlite morphology was found on uncoupled glass fibre. According to Kwei's study on the dynamic viscoelastic properties of transcrystalline of PP that the loss moduli of transcrystalline is 3 to 4 times higher than that of spherlite [13]. This may explain the phenomena that loss moduli of composite materials with coupled glass fibre is high though no slippage occurs.

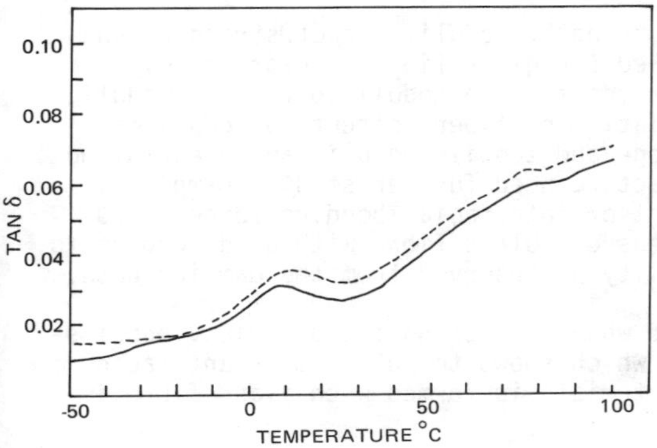

Fig. (9a) - Loss tangent of PP reinforced by 15% (---), and 25% (—) of glass fibre treated with 2% silane coupling agent

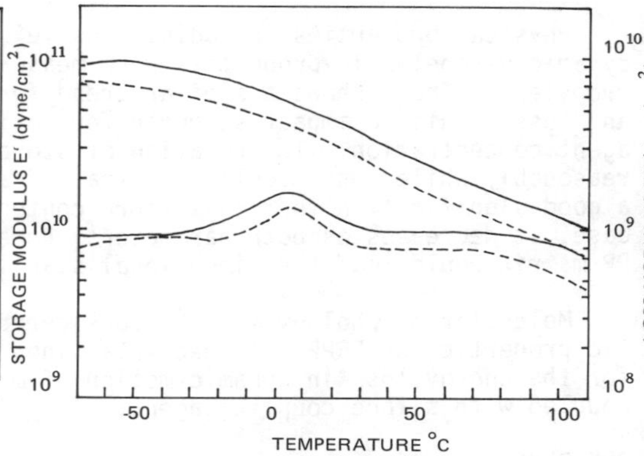

Fig. (9b) - Storage and loss moduli of PP reinforced by 15% (---), and 25% (—) glass fibre treated with 2% of silane

Fig. (10) - Polarized microscopy of PP-glass fibre system; (a), (b) show no transcrystalline generated at heat-cleaned fibre surface, (c) and (d) show the transcrystalline region generated at 2% silane treated glass surface

CONCLUSIONS

Physical properties including tensile strength, moduli, impact strength, and dynamic viscoelastic properties were measured for glass fibre reinforced polypropylene. There shows a similar trend for the storage moduli to tensile moduli, and loss moduli to impact strength for variation of fiber content, or coupling agent concentration. The relation of storage and tensile moduli can be explained reasonably while loss moduli and impact fracture need further study. Damping is a good signification for glass fibre content or interfacial bonding force in our case, it decreases as both parameters increased. Glass fibre with good bonding to PP matrix would lead the dimensional stability as observed from the damping data.

Molecular morphology must be considered when discussing the dynamic viscoelastic properties of FRPP. Transcrystalline, which shows to be an important factor for the energy loss in dynamic motion of material, is formed when glass fibre is coupled with silane coupling agent.

REFERENCES

[1] Ramsteiner, F. and Theysoln, R., "Tensile and impact strength of unidirectional short-fiber reinforced thermoplastic", Composites, p. 111, 1979.

[2] Lee, L. H., "Strength-composition relation of random short glass fiber-thermo plastics composites", Polym. Eng. Sci., 9 (3), p. 213, 1969.

[3] Sterman, S. and Marsden, J. G., "Filler-silane interaction", Mod. Plast., p. 260, 1963.

[4] Sterman, S. and Marsden, J. G., "Silane coupling agent as integral blends in resin-filler system", Mod. Plast., p. 205, 1963.

[5] Ferry, J. D., Viscoelastic Properties of Polymer, John Wiley, Inc., 3rd ed., 1980.

[6] Lewis, T. D. and Nielson, L. E., "Dynamic mechanical properties of particulate-filled composite", J. Appl. Polym. Sci., 14, p. 1449, 1970.

[7] Gibson, R. F., "Complex moduli of chopped and continuous fiber composites", J. Composite Mater., 14, p. 155, 1980.

[8] Shank, K. and Lakkad, S. C., "Damping in fiber reinforced plastics", J. Mater. Sci., 1, p. 53, 1980.

[9] Nelson, D. J. and Hancock, J. W., "Interfacial slip and damping in fibre reinforced composites", J. Mater. Sci., 13, p. 2429, 1978.

[10] Brace, F. B., "Mechanical properties of discontinuous fiber reinforced thermoplastics II", Polym. Eng. Sci., 1516, p. 428, 1975.

[11] Gibson, R. F., "Dynamic mechanical behavior of fiber-reinforced composite, measurement and analysis", J. Comp. Mater., 10, p. 325, 1976

[12] Sung, C. M., Master Thesis, National Taiwan University, 1984.

[13] Kwei, T. K. et al, "Dynamic mechanical properties of the transcrystalline region in two polymer", J. Appl. Phys., 38 (6), p. 2512, 1967.

GENERIC BUCKLING OF SPECIALLY ORTHOTROPIC CYLINDRICAL SHELLS

I.-H. Yang and W.-S. Kuo

Department of Mechanical Engineering, Chinese Military Academy
Feng-Shan, Taiwan, Republic of China

ABSTRACT

This paper presents the generic buckling results for specially orthotropic cylindrical shells under the loading conditions of axial compression, lateral pressure and the combination of the two. Affine transformations are used in analyzing the problem. With the help of the global constants in orthotropic affine space, full spectrum of buckling results are valid for any specially orthotropic materials. The solution trends indicate that the critical loads decrease with decreasing the generalized rigidity ratio of the material for moderate aspect ratios of cylindrical shells. Study is also performed on the buckling of short cylindrical shells under axial compression, and asymptotical results show that the buckling behavior of an orthotropic, simply-supported, short shell approaches that of a specially orthotropic, simply-supported plate.

INTRODUCTION

Orthotropic cylindrical shells are one of the most common structures in the aerospace industry. This is due to the advantages of mechanical properties of composite material. Therefore, aside from stress analysis, a better understanding of their buckling behavior is very essential. A number of papers have dealt with the problem [1-3]. However, most of the studies were done on relatively few cases of numerical results. The main reason for this is the number of parameters, whose relative importance and bounds are unknown, in the equations making a complete solution impossible. As a result, nothing is said about the general trends of the buckling behavior for all orthotropic materials. For instance, how do the buckling loads depend on the individual elastic constant? Without this information, it is difficult to design and optimize a structure. Therefore, the necessity of complete solutions of buckling for all orthotropic cylindrical shells is evident.

In this paper, an unified approach is developed to obtain the generic solutions of the problem. With the help of affine transformation [4-7], a generalized form of equations with only three generalized parameters whose characteristic and bounds are known can be formulated. When the transformation is done, the advantages of the method are apparent. First, only a small amount of numerical work is needed for all composite materials. Second, sufficient information is provided for a parametric study of the problem.

The analysis considers an orthotropic circular cylindrical shell, as shown in Figure 1. The shell is simply-supported at the edges and subjected to the loading conditions of axial compression, lateral pressure and the combination of

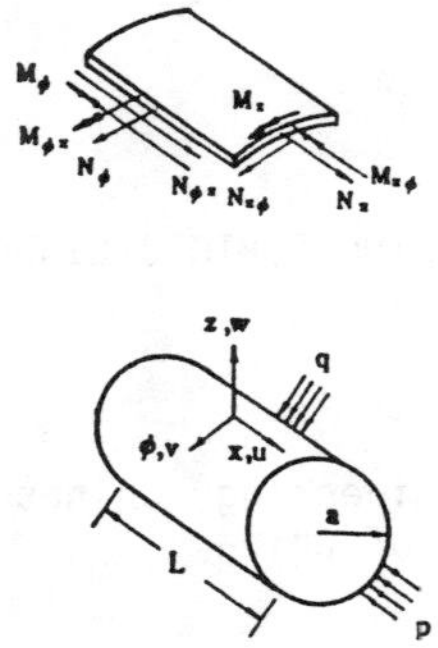

Fig. (1) - Shell element and cylindrical shell

the two. Study is also performed on the buckling problem of short shells (defined here by $L/a \le 0.1$) under axial compression. A comparison is made for buckling behavior of orthotropic short shells and of orthotropic plates.

ORTHOTROPIC SHELL THEORY

As shown in Figure 1, we consider a thin, circular cylindrical shell subjected to axial compression and lateral pressure. According to shell theory, the displacements of the shell can be expressed in terms of mid-surface displacements as

$$u_a = u(x,\phi) - \frac{z}{a} w'(x,\phi)$$

$$v_a = v(x,\phi) \cdot (1 + \frac{z}{a}) - \frac{z}{a} w^{\cdot}(x,\phi) \qquad (1)$$

$$w_a = w(x,\phi)$$

where u, v, w are displacements of the mid-surface of the shell in the directions of longitudinal, circumferential and radial axes x, ϕ and z, respectively. a is the radius of the mid-surface and z is the transverse distance measured from mid-surface.

The strain-displacement relationships in cylindrical coordinate are

$$\varepsilon_x = \frac{u'}{a} - \frac{z}{a^2} w''$$

$$\varepsilon_\phi = \frac{v^{\cdot}}{a} - \frac{z}{a(a+z)} w^{\cdot\cdot} + \frac{w}{a+z} \qquad (2)$$

$$\gamma_{x\phi} = \frac{u^{\cdot}}{a+z} + \frac{a+z}{a^2} v' - \frac{w'^{\cdot}}{a} (\frac{z}{a} + \frac{z}{a+z})$$

where prime and dot are dimensionless differential operators defined as

$$(\ )' \equiv a\frac{\partial(\ )}{\partial x} \qquad (\ )^{\cdot} \equiv \frac{\partial(\ )}{\partial\phi}$$

In this paper, the cylindrical shells are comprised of specially orthotropic materials. The stress-strain relations can be written in the form

$$\begin{aligned}\sigma_x &= Q_{11}\varepsilon_x + Q_{12}\varepsilon_\phi \\ \sigma_\phi &= Q_{12}\varepsilon_x + Q_{22}\varepsilon_\phi \\ \tau_{x\phi} &= Q_{66}\gamma_{x\phi}\end{aligned} \tag{3}$$

According to shell theory, the resultants of the shell are obtained by integrating over the thickness, i.e.,

$$\begin{aligned}[N_x, N_{x\phi}, M_x, M_{x\phi}] &= \int_{-h/2}^{h/2} [\sigma_x,\ \tau_{x\phi},\ -z\sigma_x,\ -z\tau_{x\phi}] \cdot (1 + \tfrac{z}{a})dz \\ [N_\phi, N_{\phi x}, M_\phi, M_{\phi x}] &= \int_{-h/2}^{h/2} [\sigma_x,\ \tau_{\phi x},\ -z\sigma_\phi,\ -z\tau_{\phi x}]dz\end{aligned} \tag{4}$$

GOVERNING EQUATIONS

In considering the equilibrium of a shell element, all primary deformation-dependent resultants shall be taken into account. Therefore, a set of very accurate Flügge-type equations [8] are used

$$aN_x' + aN_{\phi x}^{\cdot} - pu'' - qa(\ddot{u} - w') = 0 \tag{5a}$$

$$aN_\phi^{\cdot} + aN_{x\phi}' - M_\phi^{\cdot} - M_{x\phi}' - pv'' - qa(\ddot{v} + w^{\cdot}) = 0 \tag{5b}$$

$$M_\phi^{\cdot\cdot} + M_{x\phi}'^{\cdot} + M_{\phi x}'^{\cdot} + M_x'' + aN_\phi + pw'' + qa(u' - v^{\cdot} + \ddot{w}) = 0 \tag{5c}$$

where axial compression p are force per unit length and lateral pressure q are force per unit area.

All forces and moments can be expressed in terms of mid-surface displacements of the shell by substituting equations (1), (2) and (3) into equations (4). Consequently, equations (5) can be rewritten in the form

$$(A_{11}-p)u'' + (A_{66} + \frac{D_{66}}{a^2} - qa)\ddot{u} + (A_{12}+A_{66})v'^{\cdot}$$

$$+ (A_{12}+qa)w' - \frac{D_{11}}{a^2} w''' + \frac{D_{66}}{a^2} w'^{\cdot\cdot} = 0 \tag{6a}$$

$$(A_{12}+A_{66})u'^{\cdot} + (A_{66} + 3\,\frac{D_{66}}{a^2} - p)v'' + (A_{22}-qa)\ddot{v}$$

$$- (\frac{D_{12}}{a^2} + 3\,\frac{D_{66}}{a^2})w''^{\cdot} + (A_{22}-qa)\dot{w} = 0 \tag{6b}$$

$$(A_{12}+qa)u' + \frac{D_{66}}{a^2} u'^{\cdot\cdot} - \frac{D_{11}}{a^2} u''' - (\frac{D_{12}}{a^2} + 3\,\frac{D_{66}}{a^2})v''^{\cdot}$$

$$+ (A_{22}-qa)\dot{v} + \frac{D_{22}}{a^2} w^{\cdot\cdot\cdot\cdot} + 2(\frac{D_{12}}{a^2} + 2\,\frac{D_{66}}{a^2})w''^{\cdot\cdot}$$

$$+ (2\,\frac{D_{22}}{a^2} + qa)w^{\cdot\cdot} + (A_{22} + \frac{D_{22}}{a^2})w + \frac{D_{11}}{a^2} w'''' + pw' = 0 \tag{6c}$$

where the stiffness are as follows

$$(A_{ij},D_{ij}) = \int_{-h/2}^{h/2} Q_{ij}(1,z^2)dz$$

AFFINE TRANSFORMATIONS

The buckling equations for orthotropic cylindrical shells are given in equations (6). However, there are so many parameters (A_{ij} and D_{ij}) in the equations that this probably explains why most of the works on the problem have been restricted to few materials.

In order to fulfill the transformation such that the solution for entire spectrum of composite materials are possible, following nondimensional terms are introduced

$$D^* = \frac{A_{12}+2A_{66}}{\sqrt{A_{11}A_{22}}} \qquad \alpha \quad \sqrt{\frac{A_{22}}{A_{11}}}$$

$$\varepsilon = \frac{A_{12}}{A_{12}+2A_{66}} \tag{7}$$

$$p_o = \frac{p}{A_{11}A_{22}} \qquad q_o = \frac{qa}{\sqrt{A_{11}A_{22}}}$$

The key parameters of this analysis are generic terms D^*, ε and α, called generalized rigidity ratio, generalized Poisson's ratio and principal rigidity ratio, respectively. The limits for all composite materials have been established as $0 \leq D^* \leq 1$, $0 \leq \varepsilon \leq 0.65$ and $0 < \alpha \leq 1$ (for isotropic materials, $D^* = 1$, $\varepsilon = \nu$ and $\alpha = 1$).

By employing (7), equations (6) can be re-casted as

$$(\frac{1}{\alpha} - p_o)u'' + [\frac{D^*}{2}(1-\varepsilon) - q_o]\ddot{u} + \frac{D^*}{2}(1+\varepsilon)v'^{\cdot} + (\varepsilon D^* + q_o)w'$$

$$+ k[\frac{D^*}{2}(1-\varepsilon)\ddot{u} - \frac{1}{\alpha}w''' + \frac{D^*}{2}(1-\varepsilon)w'^{\cdot\cdot}] = 0 \qquad (8a)$$

$$\frac{D^*}{2}(1+\varepsilon)u'^{\cdot} + [\frac{D^*}{2}(1-\varepsilon) - p_o]v'' + (\alpha - q_o)\ddot{v} + (\alpha - q_o)w^{\cdot}$$

$$+ k(3\frac{D^*}{2}(1-\varepsilon)v'' - \frac{D^*}{2}(1+\varepsilon)w''^{\cdot}] = 0 \qquad (8b)$$

$$(\varepsilon D^* + q_o)u' + (\alpha - q_o)\dot{v} + \alpha w + q_o w^{\cdot\cdot} + k[\frac{D^*}{2}(1-\varepsilon)u'^{\cdot\cdot}$$

$$- \frac{1}{\alpha}u''' - \frac{D^*}{2}(1+\varepsilon)v''^{\cdot} + \alpha w^{\cdot\cdot\cdot\cdot} + 2D^* w''^{\cdot\cdot} + \frac{1}{\alpha}w''''$$

$$+ 2\alpha w^{\cdot\cdot} + \alpha w] + p_o w'' = 0 \qquad (8c)$$

where

$$k = \frac{h^2}{12a^2}$$

BUCKLING OF THE SHELL

A model solution to the homogeneous equations (8) that satisfies the simply-supported boundary conditions is

$$u = U_o \cos m\phi \cos\frac{\lambda x}{a}$$

$$v = V_o \sin m\phi \sin\frac{\lambda x}{a} \qquad \lambda = \frac{n\pi a}{L} \qquad (9)$$

$$w = W_o \cos m\phi \sin\frac{\lambda x}{a}$$

Substitution of equations (9) into the differential equations (8) yields the following eigenvalue problem for buckling load

$$a_{ij} \cdot \delta_j = 0$$

where the displacement vector $\{\delta\} = [U_o, V_o, W_o]_r^T$ and a_{ij} are symmetric coefficients listed in Appendix A.

Critical loads of axial compression p_o or lateral pressure q_o are obtained by allowing the determinant of the coefficient matrix a_{ij} to vanish.

RESULTS AND DISCUSSION

Although the lowest buckling load of simply-supported or clamped beams are believed to occur at the lowest mode of buckling; however, this statement is not necessarily true for shells. Additionally, unlike the behavior of beam, the buckling loads of the shells are insensitive to the length of the structure. This is because, in the shell structures, a slight change in curvature may lead to a substantial change in the structure's behavior. Therefore, a detailed examination of shell structures for a full range of buckling modes is necessary. In this paper, all buckling loads are determined by choosing the lowest critical load for m = 0, 1,2,...,10. The "festoon-shaped" buckling curves are plotted with respect to the aspect ratio λ, as shown in Figures 2, 3 and 4 for buckling of the shell under axial

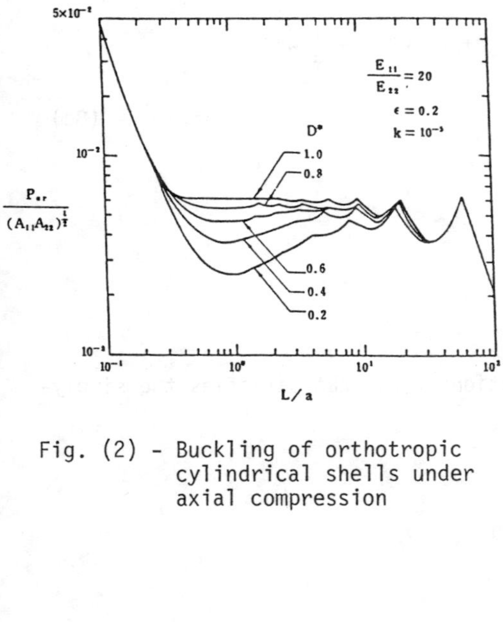

Fig. (2) - Buckling of orthotropic cylindrical shells under axial compression

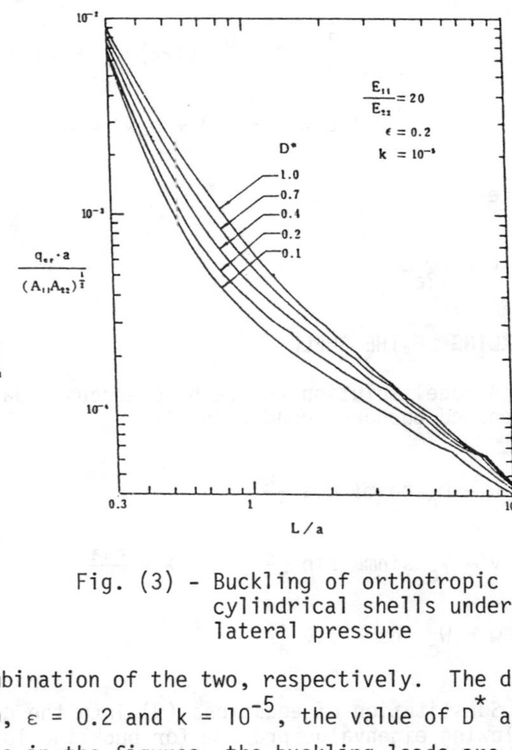

Fig. (3) - Buckling of orthotropic cylindrical shells under lateral pressure

compression, lateral pressure and combination of the two, respectively. The details of the figures are $E_{11}/E_{22} = 20$, $\varepsilon = 0.2$ and $k = 10^{-5}$, the value of D^* are taken from zero to one. As we can see in the figures, the buckling loads are decreased with decreasing the generalized rigidity ratio D^* of the material for moderate aspect ratios of the cylindrical shells. The detailed trends of buckling loads with respect to D^* are shown in Figure 5 for the axial compression case.

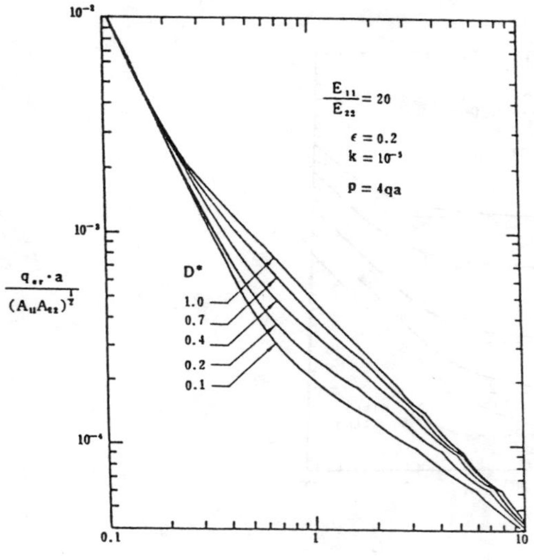

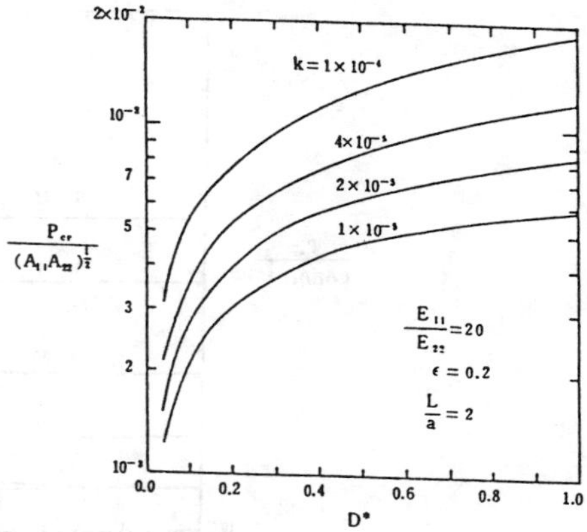

Fig. (5) - Critical axial loads of orthotropic cylindrical shells versus D^* with k as a parameter

Fig. (4) - Buckling of orthotropic cylindrical shells under combined axial compression and lateral pressure

Shells with extremely small aspect ratios having the properties of long-stripped plates and whose buckling loads are independent of D^* will be discussed.

In analyzing the buckling of short cylindrical shells, a variable, β, is introduced according to Kirchhoff hypothesis

$$\beta = \frac{h}{L}$$

and then (10)

$$k = \frac{L^2}{12a^2} \beta^2$$

Substituting equations (10) into (8), by prescribing $\beta = 1/15$ to $1/50$, the critical loads of axial compression for short cylindrical shell with β as a parameter are shown in Figure 6. In examining the results of Figure 6, one may conclude that the buckling behavior of a simply-supported short shell approach that of a simply-supported plate, whose thickness and length correspond to that of the shell (the width of plate is taken as circumference of the shell). The full derivation of buckling of long-stripped plates are given in Appendix B.

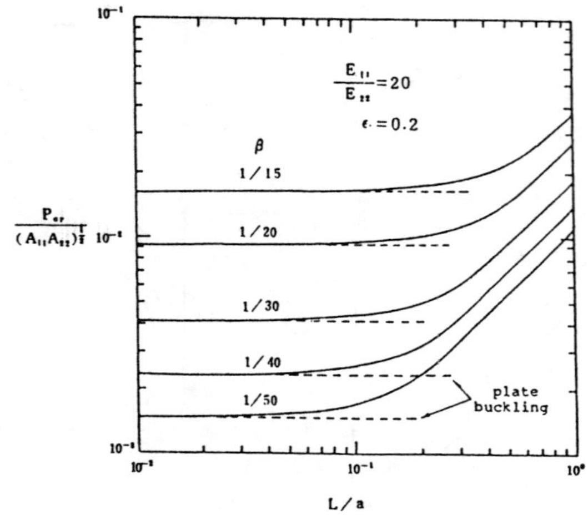

Fig. (6) - Critical axial loads of orthotropic cylindrical shells and plates with β as a parameter

Although there are three parameters, i.e., E_{11}/E_{22}, ε and k, whose values are prescribed in Figures 2 to 4, through numerical comparisons done by the authors, the facts of shell behavior are: a) the generalized Poisson's ratio ε has hardly any effect on buckling loads (similar results are concluded by Jones and Hennemann [2]; they found less than 1% of relative buckling load under lateral pressure increases as ν_{12} is increased from 0.1 to 1.9), and b) the buckling loads of thin cylindrical shells are approximately proportional to $\sqrt{k}$ in the axial compression case, and proportional to k in the lateral pressure case. Therefore, the independent parameters of the problem are the generalized rigidity ratio, D^*, and principal rigidity ratio, α, only. By varying D^* from 0 to 1, we just need a few figures, with different value of E_{11}/E_{22}, to represent the buckling loads of any orthotropic cylindrical shell.

CONCLUDING REMARKS

The advantages of the affine transformations have been proven in this paper by the analysis of the generic buckling behavior of a simply-supported orthotropic cylindrical shell under axial compression, lateral pressure and combined loading with satisfactory results. The rigidity of an orthotropic shell/plate can be considered equivalent to that of an isotropic shell/plate by a factor of generalized rigidity ratio. However, the importance of the bounds of D^* cannot be overlooked. The studies of the bounds of D^* by using the micromechanics of composite materials have been done by the authors [9]. It is concluded that D^* varies between 0 and 1 for all single layer laminates. The results should be very valuable in analyzing the structural behavior of composite structures.

APPENDIX A: COEFFICIENTS a_{ij} OF BUCKLING EQUATIONS

$$a_{11} = \frac{\lambda^2}{a^2} + \frac{D^*}{2}(1-\varepsilon)m^2(1+k) - p_o\lambda^2 - q_o m^2$$

$$a_{12} = -\frac{D^*}{2}(1+\varepsilon)\lambda m$$

$$a_{13} = -\varepsilon D^*\lambda - \frac{k}{\alpha}\lambda^3 + k\frac{D^*}{2}(1-\varepsilon)\lambda m^2 - q_o\lambda$$

$$a_{22} = \frac{D^*}{2}(1-\varepsilon)\lambda^2(1+3k) + \alpha m^2 - p_o\lambda^2 - q_o m^2$$

$$a_{23} = \alpha m + k\frac{D^*}{2}(3-\varepsilon)\lambda^2 m - q_o m$$

$$a_{33} = \alpha + k[\alpha m^4 + 2D^*\lambda^2 m^2 + \frac{\lambda^4}{\alpha} - 2\alpha m^2 + \alpha] - p_o\lambda^2 - q_o m^2$$

APPENDIX B: BUCKLING OF LONG-STRIPPED PLATES

Referring to [5], the closed form solution for buckling load of a simply-supported plate is given by

$$\frac{P_{cr}}{\sqrt{D_{11}D_{22}}} = 2D^*\frac{\pi^2}{b^2} + \frac{1}{m^2}\sqrt{\frac{D_{22}}{D_{11}}}\frac{a^2}{b^4}\pi^2 + m^2\sqrt{\frac{D_{11}}{D_{22}}}\frac{\pi^2}{a^2}$$

where a, b are plate dimensions in x,y directions, respectively.

By considering a single-layered plate with a small aspect ratio $a/b \to 0$, the plate bending stiffness D_{ij} can be expressed as

$$D_{ij} = \frac{h^2}{12}\cdot A_{ij}$$

Substituting D_{ij} into the previous equation, we get the buckling load of long-stripped plate in the form

$$\frac{P_{cr}}{\sqrt{A_{11}A_{22}}} = m^2\left(\frac{A_{11}}{A_{22}}\right)^2\frac{\pi^2}{12}\beta^2$$

where $\beta = h/a$. The lowest buckling load can be obtained by taking $m = 1$ in the equation.

ACKNOWLEDGEMENTS

The authors wish to express their sincere appreciation to Dr. E. J. Brunelle for his encouragement and for kindly giving me his papers.

REFERENCES

[1] Jones, R. M. and Morgan, H. S., "Buckling and vibration of cross ply laminated circular cylindrical shells", AIAA Journal, Vol. 13, pp. 664-671, 1975.

[2] Jones, R. M. and Hennemann, J. C. F., "Effect of prebuckling deformations on buckling of laminated composite circular cylindrical shells", AIAA Journal, Vol. 18, pp. 110-115, 1980.

[3] Whitney, J. M. and Sun, C. T., "Buckling of composite cylindrical characterization specimens", Journal of Composite Materials, Vol. 9, pp. 138-148, 1975.

[4] Brunelle, E. J., "The fundamental constants of orthotropic affine plate/slab equations", AIAA Journal, Vol. 23, pp. 1957-1961, 1985.

[5] Brunelle, E. J. and Oyibo, G. A., "Generic buckling curves for specially orthotropic rectangular plates", AIAA Journal, Vol. 21, pp. 1150-1156, 1983.

[6] Brunelle, E. J., "Eigenvalue similarity rules for symmetric cross-ply laminated plates", AIAA Journal, Vol. 24, pp. 151-154, 1986.

[7] Oyibo, G. A. and Brunelle, E. J., "Vibrations of circular orthotropic plates in affine space", AIAA Journal, Vol. 23, pp. 296-300, 1985.

[8] Flügge, W., Stresses in Shell, Springer-Verlag, 2nd edition, 1973.

[9] Yang, I. H. and Kuo, W. S., "The global constants in orthotropic affine space", unpublished work.

HIGH PERFORMANCE EPOXY MATRIX: THE PROPERTIES OF MODIFIED EPOXIES FOR GRAPHITE COMPOSITES

G. Lee and W.-Y. Chen

Chung Shan Institute of Science and Technology
P.O. Box 1-26-2, Lung-Tan, Taiwan, Republic of China

ABSTRACT

The 4, 4' - diamino diphenyl methane (TGDDM) epoxy is the currently commercial epoxy resin in aerospace. As a potential flexibilizer for graphite composites [1], amineterminated butadiene - nitrile (ATBN) rubber is used for TGDDM epoxy resin to improve toughness properties. In the ATBN modified formulation, tanδ, the dynamic properties, is moderately downward shift in temperature, while ATBN content increases. The morphology of fracture surface revealed that this system has better ductile fracture with an increase of ATBN content.

INTRODUCTION

Recently, there has been great interest in the development of rubber-modified matrix for composite. A more ductile matrix is desirable, because it is believed to have potential in improving composite performance. Some high molecular weight polymers with relatively flexible aliphatic structures, which are known as flexibilizers, are added to aromatic epoxy formulations to prolong elongation and reduce inherent brittleness of the cured resin.

The 4, 4' - diamino diphenyl methane (TGDDM) epoxy, one of the most important uses of epoxy resins as matrix materials, is modified with liquid rubber to improve its inherent brittleness [2,3]. The liquid rubber used is a butadiene/acrylonitrile copolymer, containing approximately 16% acrylonitrile. In comparison to the commonly used CTBN (carboxy - terminated - butadiene - nitrile), the significant advantage of using ATBN is that it can only react with epoxy function group in the amine cured epoxy system, while CTBN can react with both amine and epoxy function groups. Meanwhile, the number of active hydrogens in an ATBN molecular which can react with epoxy resin is twice as many as that of active carboxyl groups in a CTBN molecular. During the progress of curing, some of the rubber can coagulate to form a second phase of spherical particles in epoxy matrix that can promote toughening, while the other rubber that does not form a second phase can enhance ductility [5].

EXPERIMENTAL

A. Materials

The resin system contains TGDDM epoxy (Ciba-Geigy MY 720) and a novolac epoxy cured with diaminodiphenyl sulfone (DDS). The liquid rubber used is ATBN (B. F. Goodrich, 1300 x 16).

B. Impact Strength Properties

Izod impact test was used according to ASTM D256 (method C and E).

C. Electron Micrographs

Transmission electron micrographs (TEM) were obtained to make a scruting in the presence of rubber domains. Each specimen was stained with 1% Osmium tetroxide solution posterior to being cut with Ultrathin microtome. Scanning electron micrographs (SEM) were examined. The specimen was fractured immediately, when removed from liquid nitrogen. Then the fracture surface was coated with a thin layer of gold. The fracture surface of impact tested specimens were also performed.

D. Thermal Analysis

Thermomechanical analysis (TMA) was employed to determine glass transition temperatures. Thermal gravimetric analysis (TGA) was carried out to examine weight loss at a heating rate of 20°C/min, and heat distortion temperature (HDT) was also measured in accordance with ASTM D648 method.

E. Dynamic Mechanical Properties

The dynamic mechanical properties of the casting epoxy resin, 12 mm in width and 3 mm in thickness, were determined by using the Rheometrics Dynamic Spectrometer (RDS-7700). These experiments were performed in dry nitrogen atmosphere at a heating rate of 5°C/min, each step with one-minute soaking time, in the temperature range -130 to 270°C with 0.15% strain and at a rate of 1 rad/S.

F. The Mechanical Properties of Epoxy/ST-3000 Composite

Mechanical property data of the Epoxy/ST-3000 composite were generated on unidirectional test coupons. The tensile strength and modulus were determined by using the ASTM D2344 method.

RESULTS AND DISCUSSIONS

The rubber used is a ATBN with an amine equivalent weight of 900, whose structure is

$$H_2N(CH_2\text{-}CH{=}CHCH_2)_x\ [(CH_2\text{-}\underset{C\equiv N}{CH})_y]_z\ NH_2$$

These amine end groups, similar to those of amine hardener, can easily react with TGDDM to form a network structure.

The Izod impact energy is plotted as a function of ATBN content in Figure 1. Both the notched and the unnotched methods were carried out to make a scrutiny

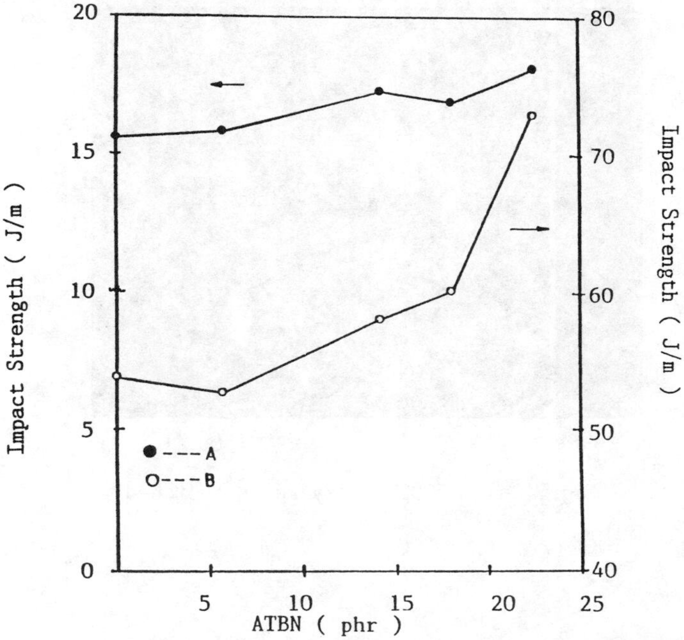

Fig. (1) - Izod impact strength of epoxy resin with ASTM D256.
(A) method C (notched); (B) method E (unnotched)

into the fracture initiation and propagation. The notch produces a stress concentration which promotes a brittle, rather than a ductile fracture. The impact strength increases slowly with an increase of ATBN content in the notched method. The unnotched impact is obtained by reversing the position in the vise of a notched specimen to determine the energy required to initiate and propagate a fracture. The value obtained is conspicuously increasing with the increase of ATBN content. There is a significant improvement in impact strength at the ATBN content of 14 phr and 23 phr.

Figure 2 shows the scanning electron photomicrographs of the fracture surface of impact tested specimens. Compared with the notched impact, the unnotched impact causes a more coarse surface, Figures 2a and 2b. The fracture surface of impact tested specimens which does not contain ATBN rubber is more brittle than that does, Figures 2b and 2c. There are many special circular fracture surfaces, less than 20 μm, in the rubber modified system, Figure 3. But they are absent in the unmodified system, Figure 3a. The number of circles on the fracture surface increases with the ATBN content.

Figure 4 shows transmission electron photomicrographs of Osmium dyed sections of the ATBN-modified epoxy resins. In these systems, Osmium tetraoxide only reacts with the carbon double bonds. Since the carbon double bonds are present in the butadiene portion of the rubber molecules and absent in the epoxy, the rubber

Fig. (2a) - TGDDM epoxy resin (notched)

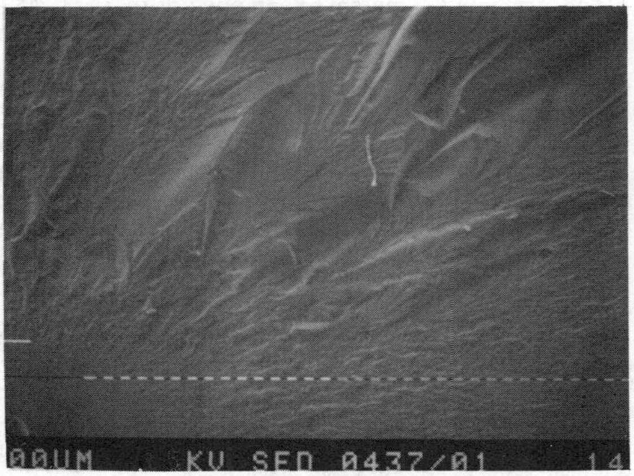

Fig. (2b) - TGDDM epoxy resin (unnotched)

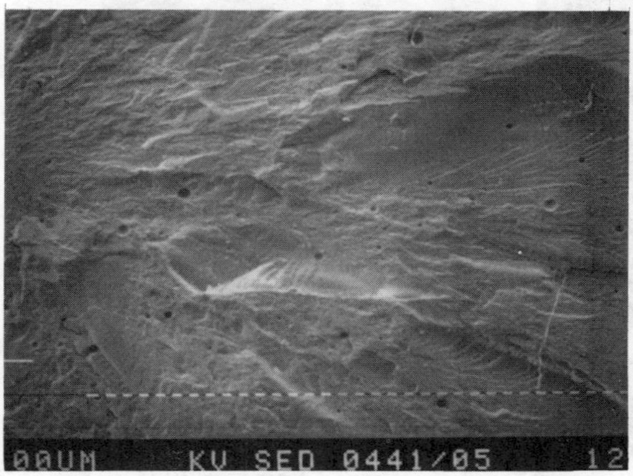

60 μ

Fig. (2c) - TGDDM + 23 phr ATBN (unnotched)

Fig. (2) - Scanning electro photographs showing fracture surface of impact test

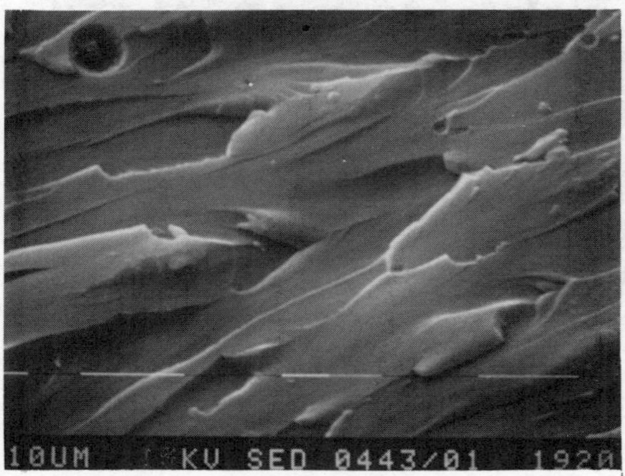

10 μ

Fig. (3a) - TGDDM epoxy resin

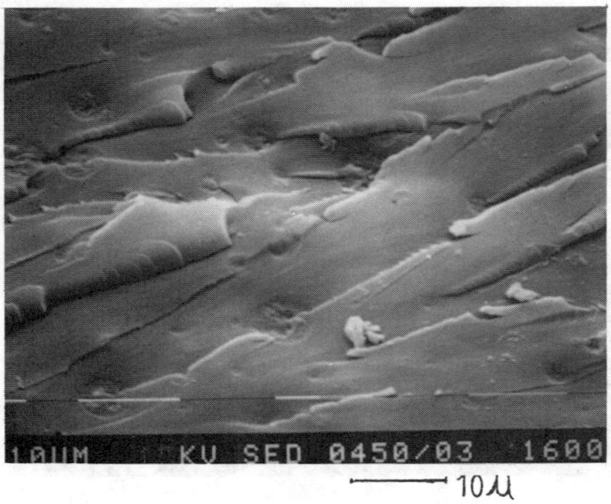

Fig. (3b) - TGDDM + 14 phr ATBN

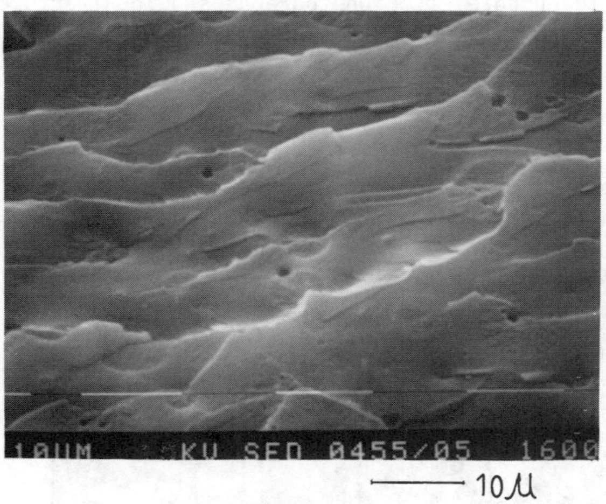

Fig. (3c) - TGDDM + 23 phr ATBN

Fig. (3) - Scanning electro photomicrographs showing fracture surface of unnotched impact test

molecules are selectively stained. The stained rubber domains appear as darkened areas in the transmission electron micrographs, while the epoxy phase appears white. The transmission electron photomicrographs were analyzed with Image Analyzer to measure the average domain size, volume fraction and number of domains/area. Table 1 lists the analytic data. The average diameter of the rubber particles (0.5 to 3 μm in diameter) is about 1.5 - 2.0 μm.

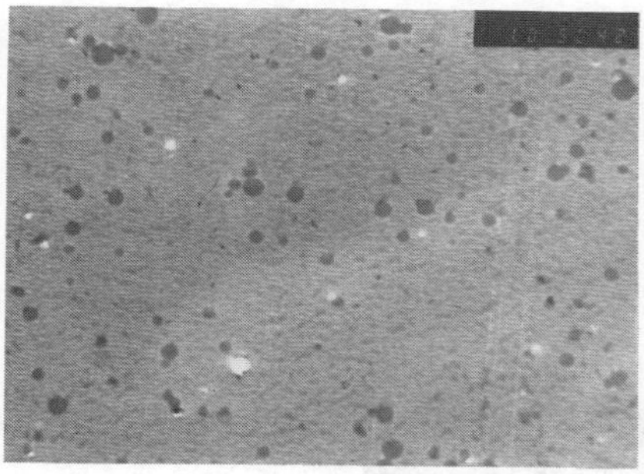

10μ

Fig. (4a) - TGDDM + 2.2 phr ATBN

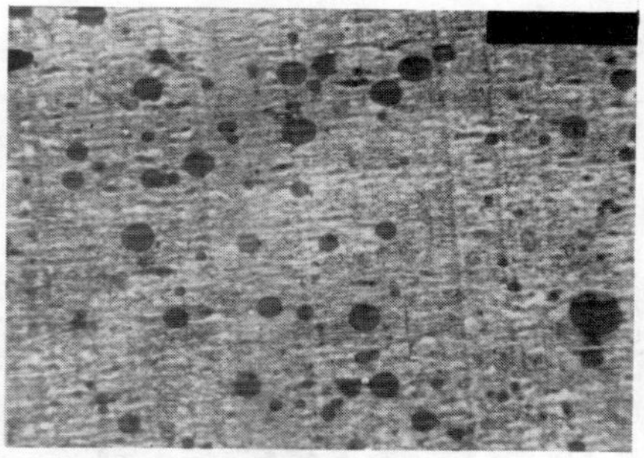

10μ

Fig. (4b) - TGDDM + 6.8 phr ATBN

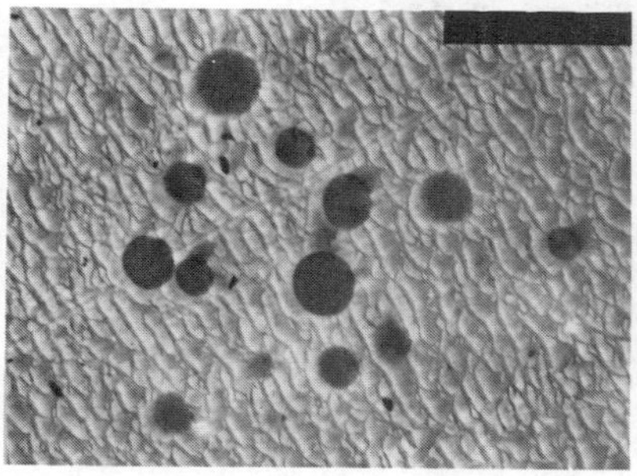

5μ

Fig. (4c) - TGDDM + 2.2 phr ATBN

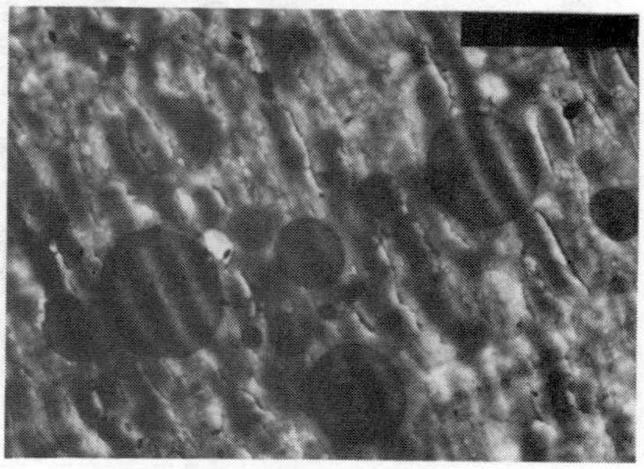

5μ

Fig. (4d) - TGDDM + 6.8 phr ATBN

Fig. (4) - Transmission electron photomicrographs showing Osimium dyed epoxy resins

The presence of rubber domains was found to have the influence on the fracture mechanisms of the epoxy resin system. Figure 5 shows the scanning electron photomicrographs of these epoxy resins fractured immediately out of liquid nitrogen. There are many differences in the fracture surface between unmodified and modified epoxy resins. The topography of the unmodified fracture surface is streamlike, while that of the modified fracture surface appears many special circular fracture surfaces. Meanwhile, a large number of the convex and concave hemispherical surfaces were found in these circular surfaces. The size of the hemispherical surfaces appearing in the rubber-modified system is nearly the same as that of the rubber particle measured from the transmission electron photographs.

In addition to promoting the fracture toughness, the effects of the participation of ATBN rubber on the other properties will be investigated.

Table 1 - Morphological data from image analyzer

No.	ATBN (phr)	D	ϕ_2	N/100 μ^2
1	2,2	1,5	3,7	2
2	3,6	1,9	7,31	2
3	6,8	1,7	8,25	4

D: average domain size (μm); ϕ_2: area fraction of rubber particles; N/100μ^2: number of domains on 100μ^2 of test plane.

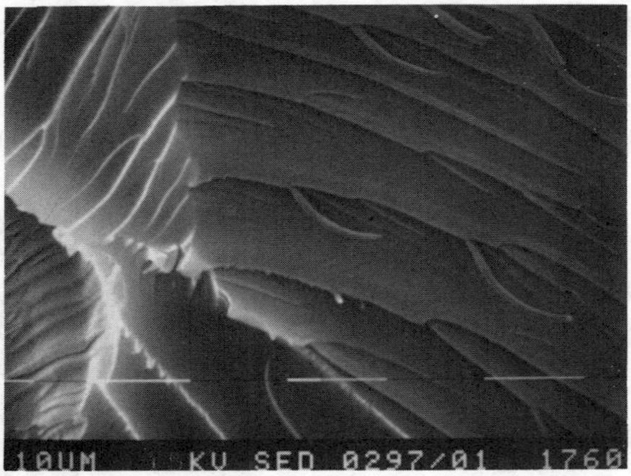

Fig. (5a) - TGDDM epoxy resin

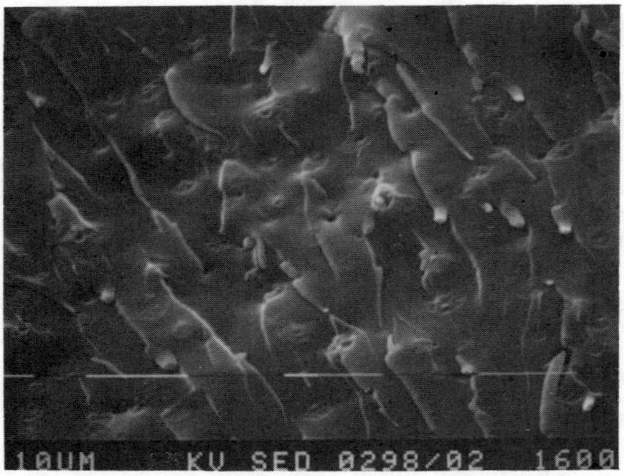

Fig. (5b) - TGDDM + 6 phr ATBN

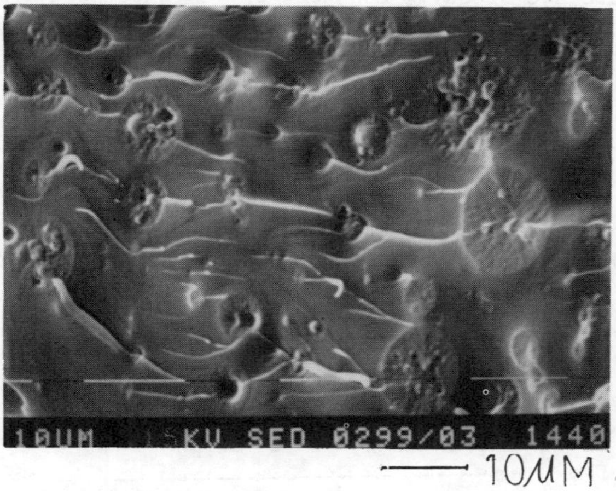

Fig. (5c) - TGDDM + 14 phr ATBN

Fig. (5) - Scanning electro photomicrographs of epoxy resins fractured immediately out of liquid nitrogen

Figure 6 shows the results of thermal gravimetric analysis. The range of ATBN content was from 0 to 28 phr. The weight loss of these resin systems becomes violent at 350°C. It goes down rapidly with the increase of ATBN content when the temperature is higher than 400°C.

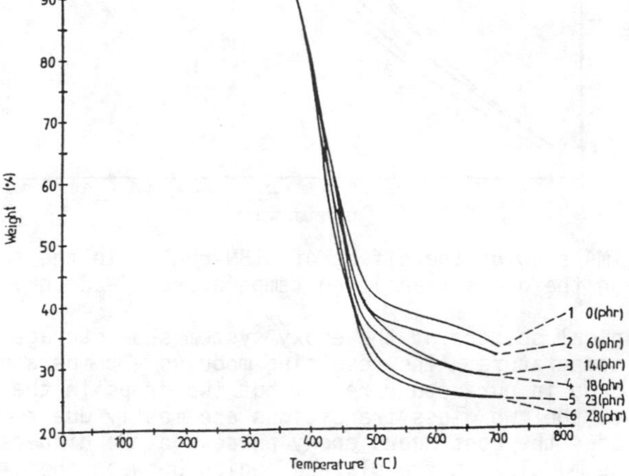

Fig. (6) - Thermal gravimetric analysis of epoxy resins

Figures 7 and 8 indicate that the heat distortion temperature and the glass transition temperature are steadily downward shift in temperature with ATBN content.

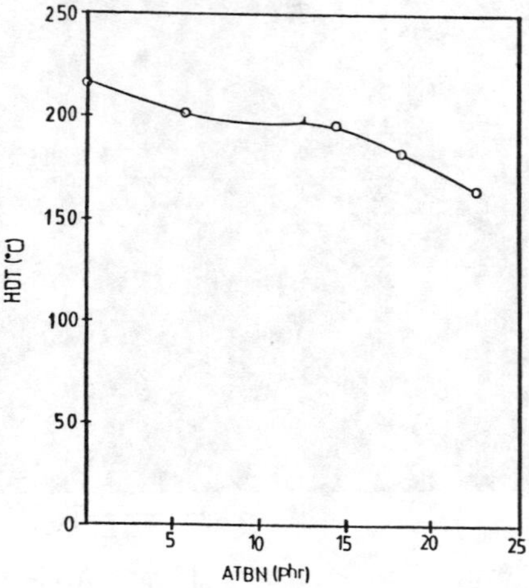

Fig. (7) - Heat distortion temperature of epoxy resins

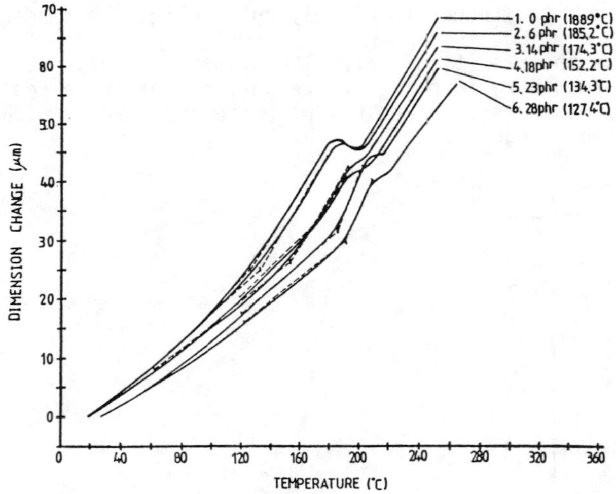

Fig. (8) - TMA showing the effect of ATBN rubber in the epoxy resins on the glass transition temperature (0-28 phr)

Dynamic mechanical spectra of the epoxy system show storage modulus and tanδ as a function of temperature. The resulting modulus - temperature curve, Figure 9, has one steep drop in pure epoxy resin, but two drops in the ATBN modified epoxy resin. Since the two glass transitions are mostly due to the two-phase system which includes the continuous epoxy phase and the dispersed rubber domains. The decrease of the modulus in the plateau region between the two glass transitions depends upon the increase of the rubber ratio. The phenomenon of the glass

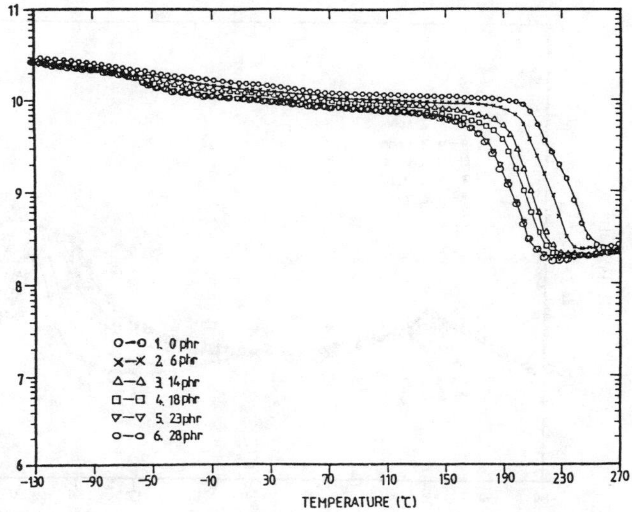

Fig. (9) - The modulus-temperature curve showing the effect of ATBN (0-28 phr) rubber in the epoxy resins on the dynamic mechanical properties

transition about - 60°C becomes clear with the increase of ATBN content. The glass transition temperature which appears around 200°C decreases rapidly as the ATBN content increases.

The tanδ - temperature spectrum for the epoxy system exhibits two viscoelastic transitions, α and β peaks, Figure 10. The β peak was attributed to the crankshaft rotation of the glycidyl amine linkage after reaction of the epoxide ring and the amine [6,7]. As the ATBN emerges in epoxy resin, the β peak is observed to split into two peaks which implies that new amine - epoxide reactions are taking place, i.e., the amine in ATBN reacts with epoxide resin. The area of the splitting peak increases with the ATBN content.

The α peak was identified as the glass transition of the system. It expands broadly and is moderately downward shift in temperature when the ATBN content increases. The α transition chiefly depends on the ratio of the rubber in epoxy resin. The two-phase system has a greater damping compliance than the pure epoxy system.

Table 2 presents mechanical properties of ATBN modified epoxy/ST-3000 composite. The ATBN rubber added to this epoxy system can extend elongation and enhance tensile strength.

CONCLUSIONS

The participation of ATBN rubber in the TGDDM epoxy system generates the dispersed rubber phase and the continuous epoxy phase. Special circle fracture surface made up with many convex and concave hemispherical surfaces promotes impact and toughness.

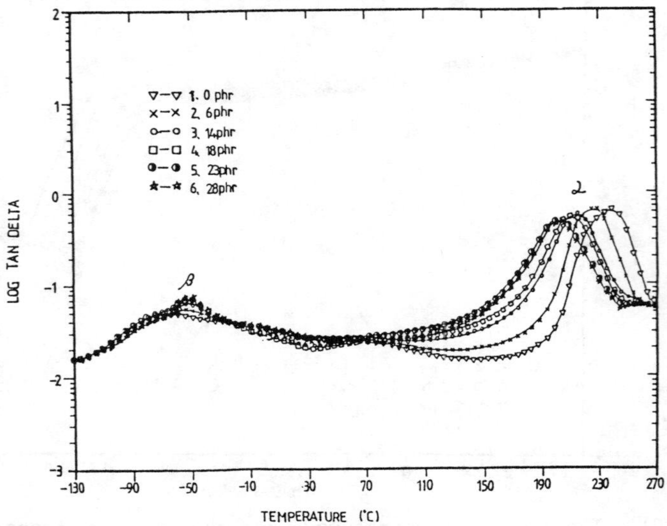

Fig. (10) - The tanδ-temperature curve showing the effect of ATBN rubber (0-28 phr) in the epoxy resins on the dynamic mechanical properties

Table 2 - Mechanical properties of ATBN modified epoxy/ST-3000 composite

Test	R.T. Test	C.O.V.	Fiber Volume
Tensile Strength (ksi)	234	3.7%	54%
Tensile Modulus (msi)	16.2	2.5%	54%
Tensile Strain (ε%)	1.53	3.8%	54%
Short Beam Shear Strength (ksi)	8.53	8.0%	65%

Besides the fracture toughness, the rubber additive steadily lowers the glass transition temperature and heat distortion temperature, and has a very little effect on weight loss below 350°C.

The dynamic modulus - temperature spectrum indicates one steep drop in pure epoxy resin, but two drops in the ATBN modified epoxy resin. Since the ATBN modified epoxy resin is a two-phase system. The β peak, in tanδ - temperature curve, splits into two peaks as the ATBN presents in the epoxy resin. This implies that new amine - epoxide reactions are taking place. The two-phase system has a greater damping compliance than the pure epoxy system.

Mechanical properties of the graphite composites for this new system revealed that ATBN rubber is a potential flexibilizer for TGDDM epoxy resin.

ACKNOWLEDGEMENTS

The authors express their appreciation to Mr. S. F. Jang, S. L. Liu, P. H. Sung and Ms. H. T. Liu for their technical help and to the members of the Materials R&D Center and the Fourth Department, Chung Shan Institute of Science and Technology, for their instrumentation support.

REFERENCES

[1] Cranmer, J. H., SAMPE Quarterly, p. 1, 1985.

[2] Lee, B. L., Lizak, C. M., Riew, C. K. and Moulton, R. J., Proceedings of 12th National SAMPE Conference, 12, p. 265, 1980.

[3] Nir, Z., Gilwee, W. J., Kourtides, D. A. and Parker, J. A., Polymer Composites, 6(2), p. 72, 1985.

[4] Ting, R. Y. and Moulton, J., Proceedings of the 12th National SAMPE Conference, 12, p. 265, 1980.

[5] Manzione, L. T., Gillham, J. K. and Mcpherson, C. A., Technical Report AD-A090122.

[6] Ibrahim, A. M. and Seferis, J. C., Polymer Composites, 6, p. 47, 1985.

[7] Chu, H. S. and Seferis, J. C., Polymer Composites, 5, p. 124, 1984.

SECTION VI

JOINT AND STRUCTURE DESIGN

FREE AND FORCED NONLINEAR DYNAMICS OF COMPOSITE SHELL STRUCTURES

Y. C. Wu, T. Y. Yang

School of Aeronautics and Astronautics, Purdue University
West Lafayette, Indiana 47907 USA

and

S. Saigal

Department of Mechanical Engineering, Worcester Polytechnic Institute
Worcester, Massachusetts 01609 USA

ABSTRACT

Natural frequencies and forced responses of laminated composite shells of general form are investigated using a high-order curved shell finite element. The effect of geometric nonlinearity is included in the determination of the forced responses which are obtained using Newmark's numerical integration scheme. A class of shell problems which are general with respect to geometry; boundary conditions; number, stacking sequence, and angle of orientation of the lamina; and load function can be solved using the present formulations. Numerical results and their comparisons with existing solutions are presented for cylindrical shells, shells of positive and negative Gaussian curvature, spherical caps, and shells of translation with different boundary conditions.

INTRODUCTION

Research in the area of finite element analysis of shell structures has experienced many advances in the past two decades. Yang and coworkers [1-3] have provided brief reviews of these developments in their papers on linear and nonlinear, static and dynamic analysis of shell structures. Due to the increased popularity of laminated composite plate and shell panels in aerospace and other technical applications, some efforts have been directed towards inclusion of such materials in the existing formulations for isotropic materials. A brief account of the literature on nonlinear static analysis of composite shells was given by Saigal et al [4].

In the area of dynamic analysis of composite shells, most of the research has been directed towards determining the free vibration response. A good overview of previous work has been given by Bert and Kumar [5]. Recently, Sheinman and Greif [6] presented an exhaustive study of natural frequencies of cylindrical, spherical, and conical shells made of laminated composite materials.

The transient response analysis of composite shells has received very little attention. Sheinman and Greif [6] determined the linear response of a cylindrical shell under a triangular pulse loading using conical frustum elements. Recently, Reddy and Chandrashekhara [7] presented dynamic responses of cylindrical shells and shells of translation of composite construction. A shear-flexible doubly curved finite element was employed in [7] and the effects of geometric nonlinearity were included. To the authors' best knowledge, no other works on transient analysis of laminated composite shells are available in the literature.

The present paper presents an unified approach to the dynamic analysis of laminated composite shells. The free vibration responses of a cylindrical shell, shells with positive and negative Gaussian curvature, and spherical shell with different composite constructions and boundary conditions are first studied. The nonlinear transient responses of a cylindrical shell, a cylindrical panel, and a shell of translation are studied next. A numerical study of the effect of number and orientation of lamina is also presented. A 48 degree-of-freedom (d.o.f.) shell finite element based on the Love-Kirchhoff theory is used for the present study. A large radius-to-thickness ratio is assumed for which shear deformation and rotatory inertia may be neglected [8]. A good correlation with existing alternative solutions for the examples studied is seen.

THEORY

A 48 d.o.f. Love-Kirchhoff shell finite element [1-3] based on classical lamination theory was used for nonlinear static analysis of complex structures [4]. This element has now been extended for free and forced nonlinear responses of composite laminated shells. Details on theoretical formulations, inclusion of composite material properties, and solution procedures are available in [1-4]. Some of the salient features of this element are summarized below.

The thin shell finite element is quadrilateral in shape and has four nodal points. The nodal displacement vector is given as

$$\{q\}_i = [u u_\xi u_\eta u_{\xi\eta} v v_\xi v_\eta v_{\xi\eta} w w_\xi w_\eta w_{\xi\eta}]_i^T$$

where u, v and w are the Cartesian displacement components and ξ and η are the curvilinear coordinate directions. Each displacement component is interpolated using bicubic Hermitian polynomials. Since the element strain-displacement relations are formulated in the Cartesian system of coordinates, the rigid body modes are accurately accounted for [9]. Also, high-order approximation of each displacement component avoids the problem of membrane-locking [3] which is of concern in the formulation of curved elements. The geometry of the middle-surface of the element is defined using variable-order polynomials [9]. This allows the modeling of a wide variety of shells with quite general shape and with arbitrary distribution of curvatures.

For the free vibration responses, the shell is assumed to vibrate in a normal mode with a natural circular frequency ω. The displacement components u, v and w are then proportional to a simple harmonically varying function of ωt, where t is the time. Using these assumptions and applying the principle of stationarity to the Lagrangian yields the equation of motion as

$$[K]\{q\} - \omega^2[M]\{q\} = 0 \tag{1}$$

where [K] and [M] are the stiffness and mass matrices, respectively, and {q} is the assembled vector of nodal displacements. The natural frequencies ω are obtained by solving the eigenvalue problem of equation (1).

For the nonlinear transient analysis, the geometric nonlinear effects are included using an incremental formulation based on the Lagrangian system of description of motion. The resulting nonlinear equations are solved by using a numerical integration procedure based on Newmark's generalized operator. Equilibrium is assured for each step through Newton-Raphson iterations till the convergence criteria [3] is satisfied within a specified tolerance. A tolerance of 0.1% was selected for all examples in the present paper.

NUMERICAL RESULTS

A general computer program was written for the procedure outlined above. All computations were made using a CYBER 205 supercomputer at Purdue University. For all the results obtained, consistent mass matrices were formulated. The free vibration results are presented first and the geometrically nonlinear transient responses are presented next for a variety of shell examples. Comparisons with existing alternative solutions are made when available.

FREE VIBRATION RESPONSE

Clamped-Clamped Thin Cylindrical Isotropic Steel Shell. The formulations for mass and stiffness matrices were first checked by determining the natural frequencies of a thin cylindrical shell with radius R = 3 in, R/h = 300, L/R = 4, E = 30 x 10^6 psi, ν = 0.3, ρ = 0.733 x 10^{-3} slugs/in^3, where L, h are the length and thickness of the cylinder, respectively and E, ν and ρ are the modulus of elasticity, Poisson's ratio and mass density of the isotropic material, respectively.

A 5x3 (longitudinal x circumferential) mesh was used to model a strip containing half of the circumferential wave. Table 1 shows the comparison of natural frequencies obtained using a sixteen d.o.f. ring finite element [10]. A good agreement of results is seen.

Two-Layered Cross Ply Boron-Epoxy Cylinder. A boron-epoxy, two-layer, cross ply laminated cylinder, a shell of interest in airframe applications was studied. The geometric and material data are radius = 2.481 in., length = 31.5 in., thickness = 0.01 in., E_1 = 30 x 10^6 psi, E_2 = 2.7 x 10^6 psi, ν_1 = 0.28, G_1 = 0.75 x 10^6 psi, ρ = 192 x 10^{-6} lb-sec^2/in^4.

The natural frequencies obtained for the axial half wave number m=1 were plotted in Figure 1 for the cases of both ends clamped and both ends simply-supported. Comparisons with the results of Abhat and Wilcox [11] based on Kirchhoff theory and Donnell's assumption and with the results of Bert et al [12] based on Vlasov's theory were also shown in Figure 1 and a good comparison was seen.

Multilayered Graphite Epoxy Cylindrical Shell. Cylindrical shells of radius R = 64 in and length L = 40 in simply supported at the ends were studied. The shell thickness was assumed to be made of 4 plies of graphite epoxy with the prop-

Table 1 - Natural frequencies (in Hertz) of clamped-clamped thin cylindrical steel shell (R/h = 300, L/R = 4, R = 3 in) for axial half wave number m=1

No. of Circumferential Waves n	Natural Frequency (Hertz) [10]	Present
3	1167	1168.6
4	781	774.2
5	600	587.7
6	558	545.8
7	615	607.2
8	736	733.3
9	899	900.0
10	1094	1096.7
11	1314	1318.9
12	1559	1564.6
13	1827	1833.0
14	-	2123.8

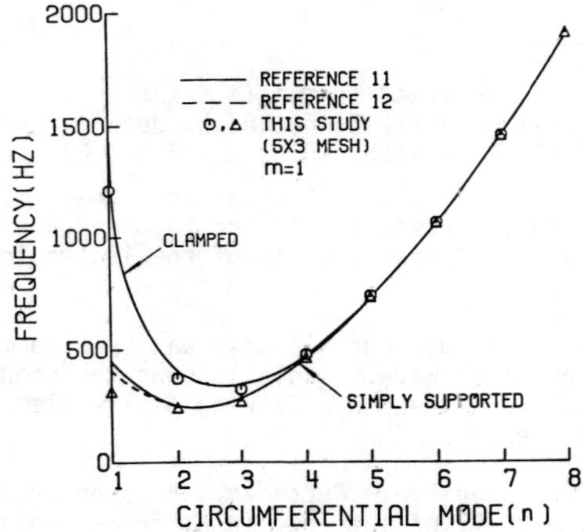

Fig. (1) - Free vibrations of two-layered, cross ply, boron-epoxy cylinder (m = axial half wave number)

erties $E_1 = 30 \times 10^6$ psi, $E_2 = 0.75 \times 10^6$ psi, $G_{12} = 0.75 \times 10^6$ psi, $\nu_{12} = 0.25$, $\rho = 0.146 \times 10^{-3}$ lb-sec^2/in^4 and t = 0.011 in for each ply. Four different arrangements of these plies were considered. This problem was previously studied by McElman and Knoell [13] using an analytical formulation.

A 5x3 mesh was used to model a strip containing half of the circumferential wave. The natural frequencies corresponding to the axial half wave number m=1 were shown in Figures 2 and 3. Results obtained in [13] are also shown in these figures for comparison and a good agreement is seen.

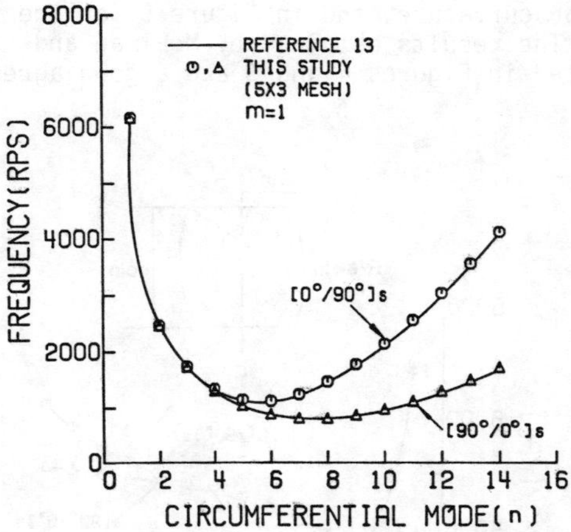

Fig. (2) - Natural frequencies for four ply graphite epoxy cylindrical shell with symmetrical ply construction

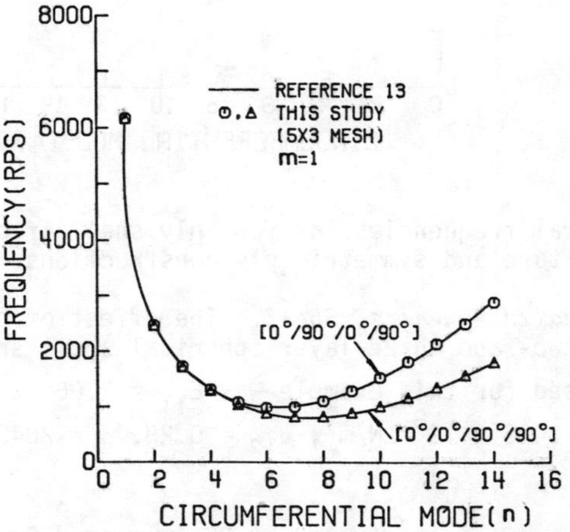

Fig. (3) - Natural frequencies for four ply graphite epoxy cylindrical shell

Shells with Positive and Negative Gaussian Curvature. The free vibration responses of shells of revolution with positive and negative Gaussian curvature made of graphite epoxy were studied next. The ply properties are the same as in the previous example. The geometry of the shells with positive and negative Gaussian curvature was shown in Figures 4 and 5, respectively. The shells were assumed to be simply supported at the ends.

A 5x3 mesh was used in each case to model a strip containing half of the circumferential wave. The frequency curves were shown in Figure 4 for the shell with positive Gaussian curvature, and in Figure 5 for the shell with negative Gaussian curvature. The results obtained by McElman and Knoell [13] for both cases were also plotted in Figures 4 and 5 and a good agreement was seen.

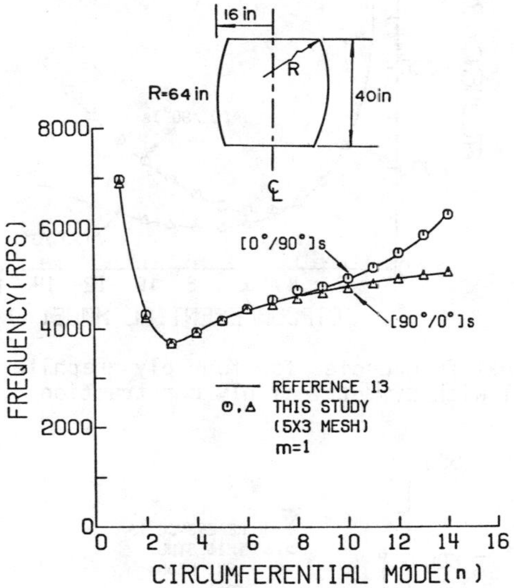

Fig. (4) - Natural frequencies for four ply shell of positive Gaussian curvature and symmetric ply constructions

Boron-Epoxy Laminated Spherical Shell. The effect of orientation angle on the frequencies for two- and three-layer spherical shell shown in Figure 6 was studied. The data used for this example was $E_{11} = 2.069 \times 10^8$ kN/m², $E_{22} = 0.1862 \times 10^8$ kN/m², $G_{12} = 0.0448 \times 10^8$ kN/m², $\nu_{12} = 0.28$, $\rho = 2048$ kg-mass/m³ and total thickness = 0.05 cm.

This problem was previously studied by Sheinman and Greif [6] using conical frustum elements. The present results were obtained using six elements to model a 10° segment of the shell. The results were plotted in Figure 6 along with those given in [6]. Good agreement is seen.

Forced Vibration Response. The nonlinear transient responses of a variety of shell examples and their comparisons with existing results are presented. The responses were obtained using Newmark's numerical integration scheme for which the parameters α and β were taken to be 0.5 and 0.25, respectively.

Clamped Laminated Cylindrical Shells Under Internal Pressure. The dynamic responses of laminated cylindrical shells clamped at both ends and under step

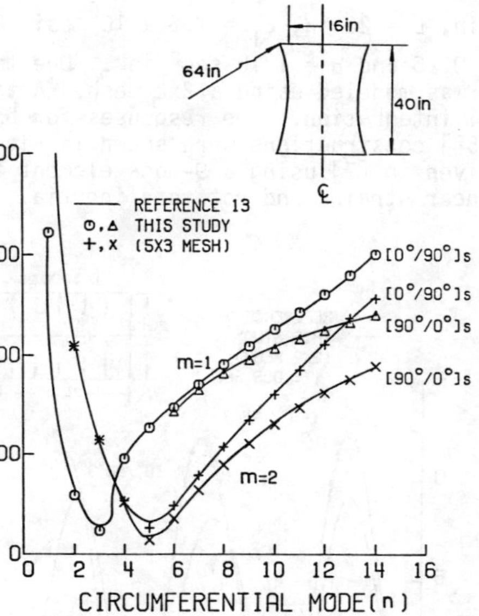

Fig. (5) - Natural frequencies for four-ply shell of negative Gaussian curvature and symmetric ply constructions

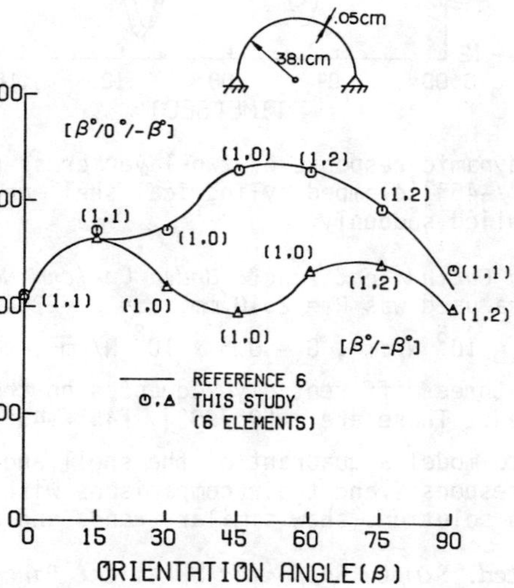

Fig. (6) - Effect of ply orientation angle on frequencies for two- and three-layer spherical shell. Numbers in parentheses indicate the mode wave numbers (ℓ,n); ℓ = meridional wave number, n = circumferential wave number

load of infinite duration were studied. The geometric and material data used was R = 20 in, h = 1 in, L = 20 in, $E_1 = 7.5 \times 10^6$ psi, $E_2 = 2 \times 10^6$ psi, G = 1.25×10^6 psi, $\nu = 0.25$ and $\rho = 1$ lb-sec^2/in^4. Due to double symmetry, a quadrant of the shell was modeled using a 2x2 mesh. A time step $\Delta t = 0.005$ sec was used for numerical integration. The responses for both cross-ply [0°/90°] and angle-ply [45°/-45°] constructions were shown in Figure 7. The results were compared with those given in [7] using a 9-node element that includes the effects of transverse shear strains and rotatory inertia. A good correlation of results is seen.

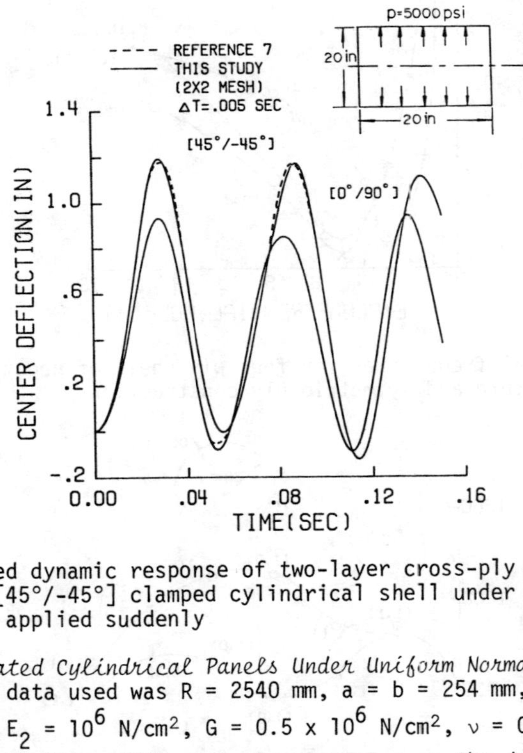

Fig. (7) - Forced dynamic response of two-layer cross-ply [0°/90°] and angle ply [45°/-45°] clamped cylindrical shell under internal step pressure applied suddenly

Clamped Laminated Cylindrical Panels Under Uniform Normal Pressure. For this example, the data used was R = 2540 mm, a = b = 254 mm, h = 2.54 mm, $E_1 = 25 \times 10^6$ N/cm^2, $E_2 = 10^6$ N/cm^2, G = 0.5×10^6 N/cm^2, $\nu = 0.25$ and $\rho = 1$ N-sec^2/in^4. The effects of three different ply sequences on the dynamic responses of the panel were studied. These are: [0°/90°], [45°/-45°] and $[0°/-45°/90°/45°]_{sym}$. A 2x2 mesh was used to model a quadrant of the shell and a time step $\Delta t = 0.05$ sec was used. The time responses and their comparisons with results of [7] are shown in Figure 8. The two solutions show similar trends and are in good agreement.

Two-Layer Laminated, Simply Supported Spherical Panels Under External Pressure. The dynamic responses of a two-layer angle-ply [-45°/45°] laminated, simply supported, spherical panel under step external pressures of different magnitude were next studied. The geometric and material data used was R = 1000 cm, a = b = 50 cm, h = 1 cm, $E_1 = 25 \times 10^6$ N/cm^2, $E_2 = 10^6$ N/cm^2, G = 0.5×10^6 N/cm^2, ν

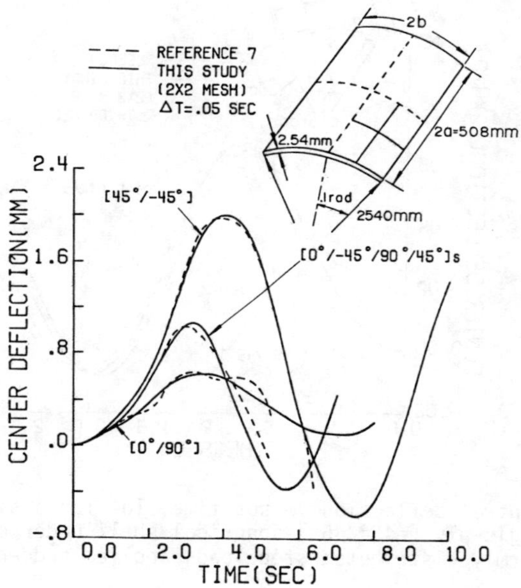

Fig. (8) - Transient response of clamped cylindrical panel under suddenly applied step internal pressure with [0°/90°], [45°/-45°] and $[0°/-45°/90°/45°]_{sym}$ ply arrangements

= 0.25 and ρ = 1 N-sec^2/cm^4. This problem was previously solved in [7] where the type of simple supports used was not given. In the present study, the simply supported edges are restricted to move in the direction of edge line, but are free to move in the direction perpendicular to the edge line.

A 2x2 mesh was used to model a quadrant of the shell and a time step Δt = 0.05 sec was used. The nonlinear dynamic responses were shown in Figure 9 along with their comparisons with results from [7]. A good agreement is seen.

CONCLUSIONS

A general formulation of a 48 d.o.f. Love-Kirchhoff shell element is applied to the free and forced nonlinear dynamic responses of laminated anisotropic composite shells. Numerical results are obtained for laminated shells with a wide variety of geometry, composite construction and boundary conditions. Good agreement of the solutions obtained with existing results establishes the validity of the present formulations for composite shell dynamics. The present shell element has computational and formulational simplicity compared to three-dimensional or degenerated shell elements. It provides an efficient tool for parametric studies for more efficient dynamic response analysis.

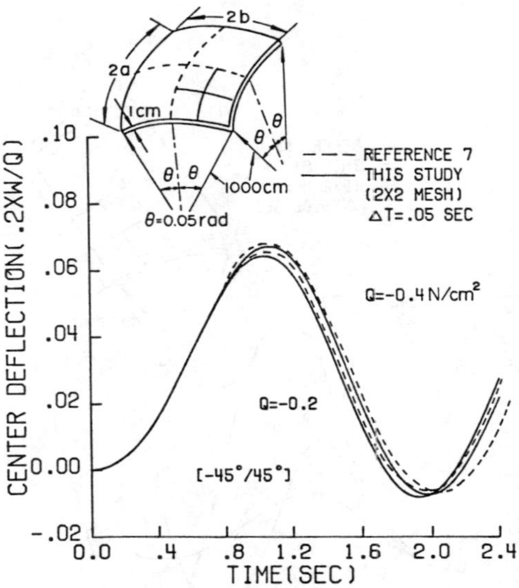

Fig. (9) - Central deflection versus time plot for a simply-supported angle-ply [-45°/45°] spherical shell under external uniformly distributed step loads applied suddenly

REFERENCES

[1] Yang, T. Y. and Kapania, R. K., "Shell elements for cooling tower analysis", Journal of Engineering Mechanics, ASCE, 109, p. 1270, 1983.

[2] Yang, T. Y. and Saigal, S., "A curved quadrilateral element for static analysis of shells with geometric and material nonlinearities", International Journal for Numerical Methods in Engineering, 21, p. 617, 1985.

[3] Saigal, S. and Yang, T. Y., "Nonlinear dynamic analysis with a 45 d.o.f. curved thin shell element", International Journal for Numerical Methods in Engineering, 21, p. 1115, 1985.

[4] Saigal, S., Kapania, R. K. and Yang, T. Y., "Geometrically Nonlinear Finite Element Analysis of Imperfect Laminated Shells", Journal of Composite Materials, 20, p. 197, 1986.

[5] Bert, C. W. and Kumar, M., "Vibration of cylindrical shells of bimodulus composite materials", Journal of Sound and Vibration, 81, p. 107, 1982.

[6] Sheinman, I. and Greif, S., "Dynamic analysis of laminated shells of revolution", Journal of Composite Materials, 18, p. 200, 1984.

[7] Reddy, J. N. and Chandrashekhara, K., "Geometrically nonlinear transient analysis of laminated doubly curved shells", International Journal of Nonlinear Mechanics, 20, p. 79, 1985.

[8] Whitney, J. and Sun, C. T., "A refined theory for laminated anisotropic, cylindrical shells", Journal of Applied Mechanics, ASME, 41, p. 471, 1974.

[9] Moore, C. J., Yang, T. Y. and Anderson, D. C., "A new 48 d.o.f. quadrilateral shell element with variable-order polynomials and rational B-spline geometries with rigid body modes", International Journal for Numerical Methods in Engineering, 20, p. 2121, 1984.

[10] Shivakumar, K. N. and Krisna Murthy, A. V. K., "A high precision ring element for vibrations of laminated shells", Journal of Sound and Vibration, 58, p. 311, 1978.

[11] Abhat, O. B. and Wilcox, M. W., "Dynamic response of laminated composite circular shells with freely supported or clamped edges", in Developments in Mechanics, 7, Proceedings of the 13th Midwest Mechanics Conference, p. 579, 1973.

[12] Bert, C. W., Baker, J. L. and Eagle, D. M., "Free vibrations of multilayer anisotropic cylindrical shells", Journal of Composite Materials, 3, p. 480, 1969.

[13] McElman, J. A. and Knoell, A. C., "Vibration and buckling analysis of composite plates and shells", Journal of Composite Materials, 5, p. 529, 1971.

RESPONSE OF ORTHOTROPIC CYLINDRICAL SHELLS SUBJECTED TO IMPACT LOADING

C. C. Chao, T. P. Tung and S. C. Lee

National Tsing Hua University
Hsinchu, Taiwan, Republic of China

ABSTRACT

This paper presents the analytical solution for dynamic response of orthotropic cylindrical shells subjected to impact loading with simple support boundary conditions at both ends. Effects of transverse shear strain and rotatory inertia are considered in the formulation of five equations of motion according to the improved theory. Free vibration is first treated for the normal mode expansion to be employed in the forced motion studies for the cylindrical shell. Special forms of loading are suddenly applied at the outer surface of the cylinder which may be a unit step pressure concentrated at the top of the mid-span, distributed along a line segment in the axial or circumferential direction or over a partial area or a unit impulse. Also investigated is the case of moving load across the span of the cylinder at various speeds. Numerical results are obtained through theoretical prediction.

INTRODUCTION

Recently, the impact problem has drawn attention from many authors in the fields of research such as penetration mechanics, armor resistance, car collision and aircraft crash. The basic interest is the dynamic response of structures acted upon by impact loading. Early studies in the subject area include transient and forced motion of isotropic linearly elastic plates and cylindrical shells with time-dependent boundary conditions and acted upon by time dependent surface tractions. In dealing with these dynamic boundary-value problems, existing theories ranging from the membrane theory, the classic shell theory, the improved shell theory to the three dimensional exact solution of elasticity, all depends on the degree of realism desired for representation of the problems concerned.

Mindlin demonstrated the influence of rotatory inertia and shear deformation on the flexural motion of isotropic elastic plates [1]. Similar studies were reported on the wave motion of thick cylindrical shells by Mirsky and Herrmann [2]. A comparison of shell theory with elastic theory was made by Weingarten and Reismann on a finite, moderately thick isotropic elastic cylindrical shell subjected to external step pressure over partial area [3] and its forced motions investigated by Huang in consideration of longitudinal tensile step forces at both ends [4]. As to the subject area of composites, many research papers have been published on the dynamics of composite and sandwich plates [5-7]. Recently, the present author conducted investigations concerning dynamic response of an orthotropic cylindrical shell under harmonic loading and the dynamic buckling of orthotropic spherical shells under uniform step pressure [8,9].

In the present paper, we are concerned with the problem of impact response of orthotropic cylindrical shells subjected to special forms of loading suddenly applied at the outer surface of the cylinder which may be a unit step pressure concentrated at the top of the mid-span, distributed along a line segment in the axial or circumferential direction, or over a small area, or a concentrated unit impulse. Also included in this study is the case of moving load across the span at the various speeds. The cylindrical shell, initially at rest, is simply supported at both ends. Considering the effects of transverse shear deformation and rotatory inertia, a set of five partial differential governing equations is set up in terms of mid-surface displacements and cross-section rotations. Analytic solutions are obtained by normal mode expansion and numerical results presented for the various loading cases.

GOVERNING EQUATIONS

Consider a linearly elastic, orthotropic circular cylindrical shell of length L, moderate wall thickness h and radius a at the mid-surface with the principal material axes (x_1,x_2,x_3) being aligned in the longitudinal, circumferential and radial (x,ϕ,z) directions respectively. Subjected to the time dependent surface tractions $(p_x,p_\phi,\ldots)$ at the outer surface, the forced motion of the cylinder can be represented by u,v,w the corresponding displacements of the mid-surface and ψ_x, ψ_ϕ, rotations of the cross-sections perpendicular to the respective axes. Note that $\psi_\phi = \psi_\phi$ (actual) - v/a is the plate equivalent for the cylindrical shell for the convenience of analysis [8]. For an arbitrary point away from the mid-surface, the displacements can be written as

$$\begin{aligned} u_a(x,\phi,z,t) &= u(x,\phi,t) - z\psi_x(x,\phi,t) \\ v_a(x,\phi,z,t) &= v(x,\phi,t) - z\psi_\phi(x,\phi,t) \\ w_a(x,\phi,z,t) &= w(x,\phi,t) \end{aligned} \tag{1}$$

According to the improved shell theory, we write the following equations of motion in terms of stress resultants

$$\begin{aligned} N_x' + N_{\phi x}^{\cdot} + ap_x &= \hat{\rho}au_{,tt} - \hat{I}\Psi_{x,tt} \\ N_\phi^{\cdot} + N_{x\phi}' - Q_\phi + ap_\phi &= \hat{\rho}av_{,tt} - \hat{I}\psi_{\phi,tt} \\ Q_\phi^{\cdot} + Q_x' + N_\phi + ap_r &= -\hat{\rho}aw_{,tt} \\ M_x' + M_{\phi x}^{\cdot} - aQ_x + m_x &= -\hat{I}(u_{,tt} - a\psi_{x,tt}) \\ M_\phi^{\cdot} + M_{x\phi}' - aQ_\phi + m_\phi &= -\hat{I}(v_{,tt} - a\psi_{\phi,tt}) \end{aligned} \tag{2}$$

where

$$(N_x, N_{x\phi}, M_x, M_{x\phi}, Q_x) = \int_{-h/2}^{h/2} (\sigma_x, \tau_{x\phi}, -\sigma_x z, -\tau_{x\phi} z, -\tau_{xz})(1 + \frac{z}{a})dz$$

$$(N_\phi, N_{\phi x}, M_\phi, M_{\phi x}, Q_\phi) = \int_{-h/2}^{h/2} (\sigma_\phi, \tau_{\phi x}, -\sigma_\phi z, -\tau_{\phi x} z, -\tau_{\phi z})dz \tag{3}$$

$$(\hat{\rho}, \hat{I}) = \sum_{k=1}^{N} \rho_k \int_{z_{k-1}}^{z_k} (1, z^2)dz$$

$$(\ )' = a \frac{\partial(\ )}{\partial x}, \ (\ )^{\cdot} = \frac{\partial(\ )}{\partial \phi}$$

The equations of motion can be rewritten in terms of displacements by using the stress-strain relations of the orthotropic material $\{\sigma\} = [Q]\{\varepsilon\}$, the strain-displacement relations of small strain. For simplicity, we introduce

$$(\bar{L}, \bar{h}) = (L, h)/a, \ (x, z) = (\bar{x}, \bar{z})/a$$

$$(\bar{u}, \bar{v}, \bar{w}) = (u, v, w)/a, \ (\bar{\psi}_x, \bar{\psi}_\phi) = (\psi_x, \psi_\phi)$$

$$(\bar{A}_{ij}, \bar{D}_{ij}, \bar{F}_{ij}) = \frac{1}{A} \sum_{k=1}^{N} Q_{ij}^{(k)} \int_{\bar{z}_{k-1}}^{\bar{z}_k} [1, \bar{z}^2, (1+\bar{z})^{-1}]d\bar{z} \tag{4}$$

$$(\bar{p}_x, \bar{p}_\phi, \bar{p}_r) = \frac{a}{A}(p_x, p_\phi, p_r), \ (\bar{m}_x, \bar{m}_\phi) = \frac{1}{aA}(m_x, m_\phi)$$

$$(\bar{\rho}, \bar{I}) = \frac{a^3}{A} \sum_{k=1}^{N} \rho_k \int_{\bar{z}_{k-1}}^{\bar{z}_k} (1, \bar{z}^2)d\bar{z}$$

where

$$A = \sum_{k=1}^{N} Q_{11}^{(k)}(z_k - z_{k-1})$$

Substituting these in equation (2) and dropping the overbars, one obtains:

A. Equations of Motion

$$A_{11}u'' + F_{66}u^{\cdot\cdot} + (A_{12}+A_{66})v'^{\cdot} + A_{12}w' - D_{11}\psi_x'' + (F_{66}-A_{66})\psi_x^{\cdot\cdot}$$
$$+ p_x = \rho u_{,tt} - I\psi_{x,tt}$$

$$(A_{12}+A_{66})u'^{\cdot} + A_{66}v'' + F_{22}v^{\cdot\cdot} - F_{44}v + (F_{22}+F_{44})w^{\cdot} - D_{66}\psi_\phi''$$
$$+ (F_{22}-A_{22})\psi_\phi^{\cdot\cdot} - F_{44}\psi_\phi + p_\phi = \rho v_{,tt} - I\psi_{\phi,tt}$$

$$A_{12}u' + (F_{22}+F_{44})v^{\cdot} - A_{55}w'' - F_{44}w^{\cdot\cdot} + F_{22}w + A_{55}\psi_x'$$
$$+ (F_{22}+F_{44}-A_{22})\psi_\phi^{\cdot} + p_r = -\rho w,_{tt}$$
$$-D_{11}u'' + (F_{66}-A_{66})u^{\cdot\cdot} + A_{55}w' + D_{11}\psi_x'' + (F_{66}-A_{66})\psi_x'' - A_{55}\psi_x$$
$$+ (D_{12}+D_{66})\psi_\phi'^{\cdot} + m_x = -I(u,_{tt}-\psi_{x,tt})$$
$$-D_{66}v'' + (F_{22}-A_{22})v^{\cdot\cdot} - F_{44}v + (F_{22}+F_{44}-A_{22})w^{\cdot} + (D_{12}+D_{66})\psi_x'^{\cdot}$$
$$+ D_{66}\psi_\phi'' + (F_{22}-A_{22})\psi_\phi^{\cdot\cdot} - F_{44}\psi_\phi + m_\phi = -I(v,_{tt}-\psi_{\phi,tt}) \quad (5)$$

These are the equations of an orthotropic cylindrical shell in which the mid-surface is a unit circle and its dimensions, stiffnesses, loading, mass and inertia distributions are in proper proportion with the given case. Once the problem of unit circle cylindrical shell is solved, the response of actual case can be found accordingly.

B. Boundary Conditions

The cylinder is simply supported at both ends. For all ϕ and t,

$$v = w = \psi_\phi = N_x = M_x = 0 \text{ at } x = 0, L \quad (6)$$

C. Initial Conditions

The cylinder is initially at rest. For all x and ϕ,

$$\left.\begin{aligned} &u = v = w = \psi_x = \psi_\phi = 0 \\ &u,_t = v,_t = w,_t = \psi_{x,t} = \psi_{\phi,t} = 0 \end{aligned}\right\} \text{ at } t = 0 \quad (7)$$

FREE VIBRATION ANALYSIS

In preparation for the normal mode expansion for the forced motion, we first treat the problem of free vibration by setting $p_x = p_\phi = p_r = m_x = m_\phi = 0$ in equations (5).

A. Natural Frequencies

To satsify the boundary conditions as specified in equation (6), we assume

$$u(x,\phi,t) + U_{mn}\cos m\phi \cos\lambda x e^{i\omega t}$$
$$v(x,\phi,t) = V_{mn}\sin m\phi \sin\lambda x e^{i\omega t}$$
$$w(x,\phi,t) = W_{mn}\cos m\phi \sin\lambda x e^{i\omega t} \qquad \lambda = n\pi/L \quad (8)$$
$$\psi_x(x,\phi,t) = \Psi_{xmn}\cos m\phi \cos\lambda x e^{i\omega t}$$
$$\psi_\phi(x,\phi,t) = \Psi_{\phi mn}\sin m\phi \sin\lambda x e^{i\omega t}$$

Substituting in equation (5) with zero loading terms, we have an eigenvalue problem of the form

$$([a] - \omega^2[b])\{c\} = \{0\} \tag{9}$$

where

$$\{c\} = (U_{mn} V_{mn} W_{mn} \Psi_{xmn} \Psi_{\phi mn})^T \tag{10}$$

For $m > 1$, $[a] =$

$$\begin{bmatrix} -A_{11}\lambda^2 - F_{66}m^2 & (A_{12}+A_{66})m\lambda & A_{12}\lambda & D_{11}\lambda^2+(A_{66}-F_{66})m^2 & 0 \\ & -A_{66}\lambda^2 - F_{22}m^2 - F_{44} & -(F_{22}+F_{44})m & 0 & (A_{22}-F_{22})m^2 + D_{66}\lambda^2 - F_{44} \\ & & A_{55}\lambda^2 + F_{44}m^2 + F_{22} & -A_{55}\lambda & (F_{22}+F_{44}-A_{22})m \\ & \text{Symmetric} & & -D_{11}\lambda^2 - A_{55} + (A_{66}-F_{66})m^2 & (D_{12}+D_{66})m\lambda \\ & & & & -D_{66}\lambda^2 - F_{44} + (A_{22}-F_{22})m^2 \end{bmatrix}$$

$[b] =$

$$\begin{bmatrix} -\rho & 0 & 0 & I & 0 \\ & -\rho & 0 & 0 & I \\ & & \rho & 0 & 0 \\ & \text{Symmetric} & & -I & 0 \\ & & & & -I \end{bmatrix}$$

For $m = 0$, the second and fifth rows and columns are deleted in both [a] and [b]. All terms containing m no longer remain in the resulting 3 x 3 matrices. The natural frequencies can be obtained from the eigenvalues for the various modes.

B. Normal Modes, Orthogonality

The eigenvector along with the associated spatial harmonic functions constitutes the normal modes, namely,

$$u_{mn} = U_{mn}\cos m\phi\cos\lambda x \qquad v_{mn} = V_{mn}\sin m\phi\sin\lambda x$$
$$w_{mn} = W_{mn}\cos m\phi\sin\lambda x \qquad \psi_{xmn} = \Psi_{xmn}\cos m\phi\cos\lambda x \tag{11}$$
$$\psi_{\phi mn} = \Psi_{\phi mn}\sin m\phi\sin\lambda x \qquad \lambda = \pi n/L$$

By virtue of Hamilton's principle and the existence of strain energy function of the composites, one obtains the orthogonality relation for the normal modes

$$\begin{aligned}\int_x\int_\phi[u_{mi}(\rho u_{nj} - I\psi_{xnj}) + v_{mi}(\rho v_{nj} - I\psi_{\phi nj}) + \rho w_{mi}w_{nj} \\ - I\psi_{xmi}(u_{nj} - \psi_{xnj}) - I\psi_{\phi mi}(v_{nj} - \psi_{\phi nj})]ad\phi dx \\ = \begin{cases} 0 \text{ for } m \neq n,\ i \neq j \\ 1 \text{ for } m = n,\ i = j \text{ after normalization}\end{cases}\end{aligned} \tag{12}$$

DYNAMIC RESPONSE

A. Normal Mode Expansion

The orthotropic cylindrical shell is subjected to time varying surface loads p_x, p_ϕ, p_r, m_x and m_ϕ. By using the normal mode expansion, the solution of the dynamic response can be expressed in the form

$$\begin{aligned} u(x,\phi,t) &= \sum_{m=0}^{\infty}\sum_{n=1}^{\infty} u_{mn}(x,\phi)q_{mn}(t) \\ v(x,\phi,t) &= \sum_{m=0}^{\infty}\sum_{n=1}^{\infty} v_{mn}(x,\phi)q_{mn}(t) \\ w(x,\phi,t) &= \sum_{m=0}^{\infty}\sum_{n=1}^{\infty} w_{mn}(x,\phi)q_{mn}(t) \\ \psi_x(x,\phi,t) &= \sum_{m=0}^{\infty}\sum_{n=1}^{\infty} \psi_{xmn}(x,\phi)q_{mn}(t) \\ \psi_\phi(x,\phi,t) &= \sum_{m=0}^{\infty}\sum_{n=1}^{\infty} \psi_{\phi mn}(x,\phi)q_{mn}(t) \end{aligned} \tag{13}$$

Note that n = 0 corresponds to longitudinal motion only, which is irrelevant to the present problem of study. Substituting equations (13) into equations (5), one obtains

$$\sum_{m=0}^{\infty}\sum_{n=1}^{\infty} (q_{mn,tt} + \omega^2 q_{mn})(\rho u_{mn} - I\psi_{xmn}) = -p_x$$
$$\sum_{m=0}^{\infty}\sum_{n=1}^{\infty} (q_{mn,tt} + \omega_{mn}^2 q_{mn})(\rho v_{mn} - I\psi_{mn}) = -p_\phi$$
$$\sum_{m=0}^{\infty}\sum_{n=1}^{\infty} (q_{mn,tt} + \omega_{mn}^2 q_{mn})(\rho w_{mn}) = p_r$$

$$\sum_{m=0}^{\infty}\sum_{n=1}^{\infty} I(q_{mn,tt} + \omega_{mn}^2 q_{mn})(-u_{mn} + \psi_{xmn}) = -m_x$$

$$\sum_{m=0}^{\infty}\sum_{n=1}^{\infty} I(q_{mn,tt} + \omega_{mn}^2 q_{mn})(-v_{mn} + \psi_{\phi mn}) = -m_\phi \tag{14}$$

Multiplying each of these equations by the various normal modes respectively, and integrating over x and ϕ, we have

$$q_{mn,tt}(t) + \omega_{mn}^2 q_{mn}(t) = F_{mn}(t) \tag{15}$$

where

$$F_{mn}(t) = \iint_{x\phi}(-p_x u_{mn} - p_\phi v_{mn} + p_r w_{mn} + m_x \psi_{xmn} + m_\phi \psi_{\phi mn}) a d\phi dx$$

By variation of parameters, we find the general solution of the dynamic problem as

$$q_{mn}(t) = q_{mn}(0)\cos\omega_{mn}t + (1/\omega_{mn})q_{mn,t}(0)\sin\omega_{mn}t$$
$$+ (1/\omega_{mn})\int_0^t F_{mn}(\tau)\sin\omega_{mn}(t-\tau)d\tau \tag{16}$$

In combination with the various normal modes, we can calculate the dynamic response with good accuracy by choosing a proper number of terms for m = 0,1,...M and n = 1,...N.

B. Dynamic Loading and Stresses

In the present study, the orthotropic cylinder, initially at rest, is subjected to radially inward directed loading p_r suddenly applied at the outer surface with all the rest $p_x = p_\phi = m_x = m_\phi = 0$. Zero initial conditions denote $q_{mn}(0) = 0$, $q_{mn,t}(0) = 0$. There are six specific forms of unit impact loading.

Concentrated Load (x = L/2, ϕ = 0)

$$p_r = -(x - L/2)\delta(\phi)H(t)$$
$$q_{mn}(t) = -(1/\omega_{mn})^2 W_{mn}\sin(n\pi/2)(1 - \cos\omega_{mn}t) \quad m = 0,1,...M;\ n = 1,...N \tag{17}$$

where $\delta(x - L/2)$, $\delta(\phi)$ are delta functions and H(t) is the Heaviside's step function.

Line Load Along a Longitudinal Segment ($2L/5 \leq x \leq 3L/5$; $\phi = 0$)

$$p_r = -[H(x - 2L/5) - H(x - 3L/5)]\delta(\phi)H(t)$$

$$q_{mn}(t) = (1/\omega_{mn})^2(L/n\pi)W_{mn}[\cos(3n\pi/5) - \cos(2n\pi/5)](1 - \cos\omega_{mn}t) \tag{18}$$

$$m = 0,1,...M;\ n = 1,...N$$

Line Load Along a Circumferential Segment $(x = L/2;\ -2° \le \phi \le 2°)$

$$p_r = -[H(\phi + 2°) - H(\phi - 2°)]\delta(x - L/2)H(t)$$

$$q_{mn}(t) = -(1/\omega_{mn})^2(2/m)W_{mn}\sin(n\pi/2) \cdot \sin(m\pi/90)(1 - \cos\omega_{mn}t),\ m \ge 1,\ n \ge 1$$

$$q_{on}(t) = -(1/\omega_{on})^2(\pi/45)W_{on}\sin(n\pi/2)(1 - \cos\omega_{on}t),\ n \ge 1 \tag{19}$$

Load Over a Small Area $(2L/5 \le x \le 3L/5;\ -2° \le \phi \le 2°)$

$$p_r = -[H(\phi + 2°) - H(\phi - 2°)][H(x - 2L/5) - H(x - 3L/5)]H(t)$$

$$q_{mn}(t) = -(1/\omega_{mn})^2(2L/mn\pi)W_{mn}\sin(m\pi/90)[\cos(3n\pi/5) - \cos(2n\pi/5)] \times (1 - \cos\omega_{mn}t),\ m \ge 1,\ n \ge 1 \tag{20}$$

$$q_{on}(t) = -(1/\omega_{on})^2(L/45n)W_{on}[\cos(3n\pi/5) - \cos(2n\pi/5)](1 - \cos\omega_{on}t)$$

Impulsive Load $(x = L/2,\ \phi = 0°)$

$$p_r = -\delta(x - L/2)\delta(\phi)\delta(t)$$

$$q_{mn}(t) = -(1/\omega_{mn})W_{mn}\sin(n\pi/2)\sin\omega_{mn}t \quad m \ge 0,\ n \ge 1 \tag{21}$$

Moving Load Across the Span

$$p_r = -\delta(x - Vt)\delta(\phi)$$

$$q_{mn}(t) = -(1/2\omega_{mn})W_{mn}[(\sin n'\pi t - \sin\omega t)/(n'\pi + \omega_{mn}) - (\sin n'\pi t + \sin\omega t)/(n'\pi - \omega_{mn})] \tag{22}$$

$$n' = nV/L,\ m = 0,1,\ldots M,\ n = 1,\ldots N$$

DYNAMIC STRESS LEVEL

Once the dynamic response is known as a result of the various time dependent coefficients $q_{mn}(t)$ and normal modes, the stress level in the cylindrical shell can be determined from the strain components such as

$$\varepsilon_x = -\sum_{m=0}^{M}\sum_{n=1}^{N}\{\lambda(U_{mn} - z\Psi_{xmn})\cos m\phi \sin\lambda x\}q_{mn}(t)$$

$$\varepsilon_\phi = \frac{1}{a+z}\{\sum_{m=0}^{M}\sum_{n=1}^{N}[m(V_{mn} - z\Psi_{\phi mn}) + W_{mn}]\cos m\phi \sin\lambda x\}q_{mn}(t) \tag{23}$$

RESULTS AND DISCUSSION

The present theory of forced motion for orthotropic, moderately thick-walled cylindrical shells proves to be satisfactory when reduced to the isotropic case in comparison of results with those of [3], even without knowing the details of the material properties they used. Geometry and material data of the present study are as follows.

Mid-surface radius a = 10 cm

Length to radius ratio L/a = 5, Thickness to radius ratio h/a = 0.1

E_1 = 120 Gpa, E_1/E_2 = 25, $G_{12} = G_{13} = 0.5\ E_2$

$G_{23} = 0.2\ E_2$, ν_{12} = 0.25, $\rho = 2.547 \times 10^{-6}$ N - sec²/cm⁴

Subjected to the various cases of impact loading, dynamic responses of the orthotropic cylinder are presented as time history of the two most important events, namely, the radial deflection and longitudinal stress level in the non-dimensional form of w/h and σ_x/E_{11} versus $\tau = C_{13}t/L$ in Figures 1, 2 or Vt/L in

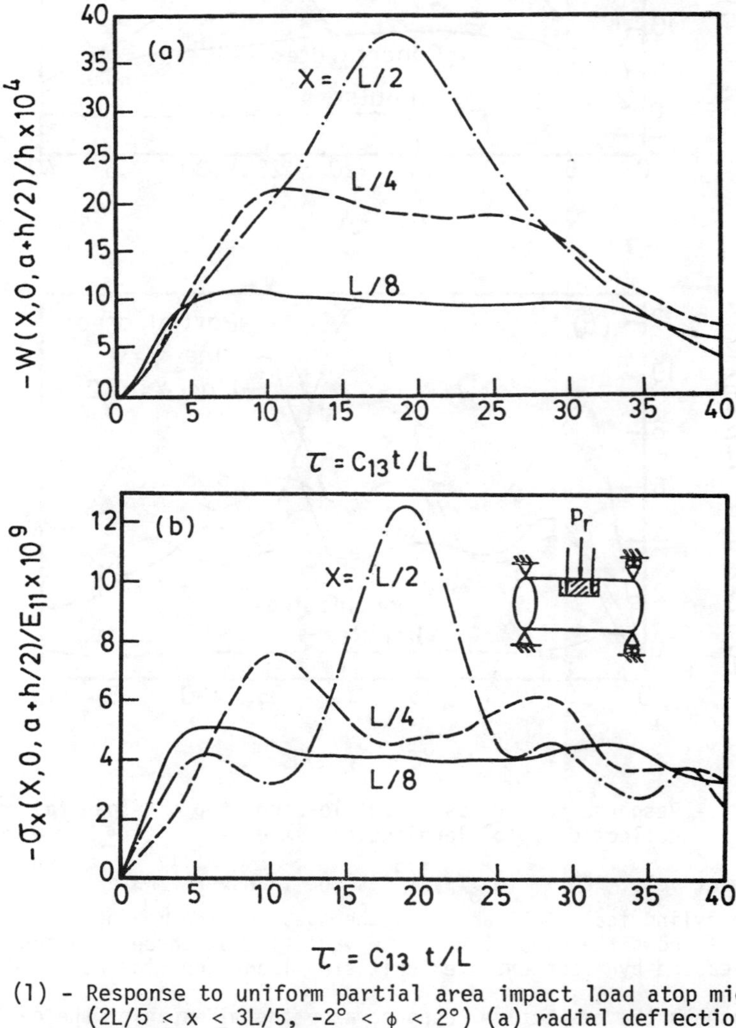

Fig. (1) - Response to uniform partial area impact load atop mid-span ($2L/5 \leq x \leq 3L/5$, $-2° \leq \phi \leq 2°$) (a) radial deflections, (b) longitudinal stresses

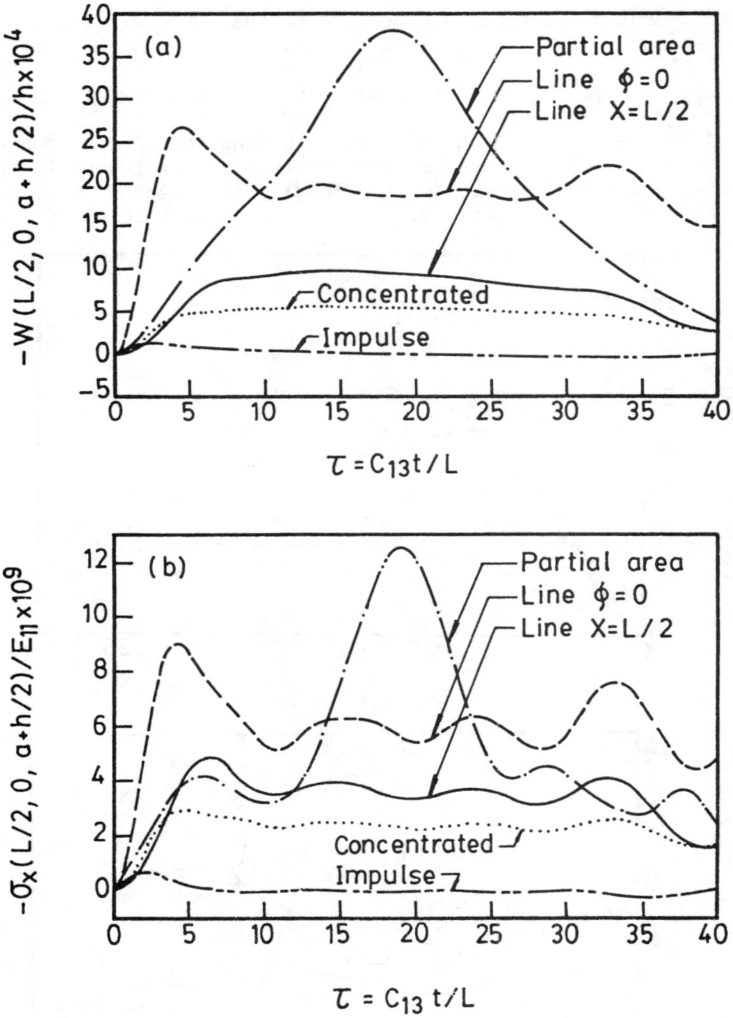

Fig. (2) - Response to various impact loading atop mid-span (a) radial deflections, (b) longitudinal stresses

Figure 3 respectively. $C_{13} = (G_{13}/\rho)^{1/2}$ is the transverse wave velocity along length of the cylindrical shell and V is the speed of the moving load across the span. Figure 1 shows the dynamic response variation at three locations x = L/8, L/4 and L/2, caused by a concentrated unit step load atop mid-span.

Figure 2 shows the respective response atop the mid-span of the cylinder acted upon there by an unit impulse, and a point concentrated, two line segments, and

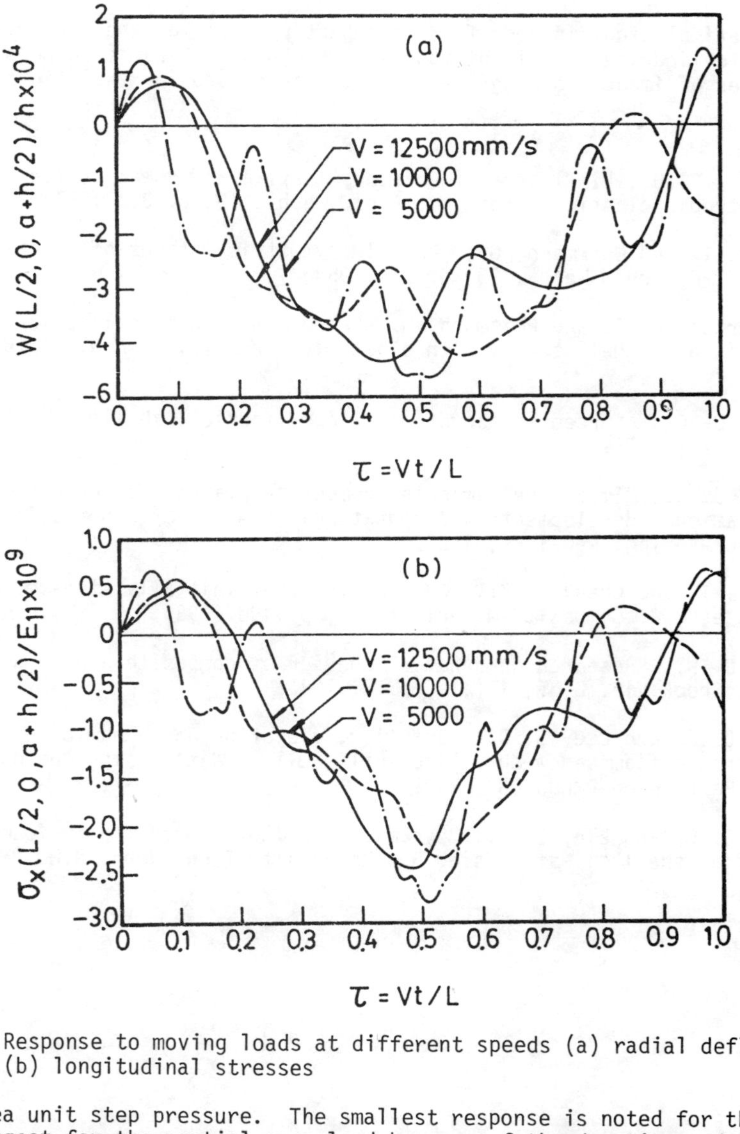

Fig. (3) - Response to moving loads at different speeds (a) radial deflections, (b) longitudinal stresses

partial area unit step pressure. The smallest response is noted for the impulse and the largest for the partial area load because of the duration and relative magnitudes. Figure 3 shows the deflection and stress level atop mid-span when a moving load travels across the span at three different speeds. The fastest load has the smallest response. In all cases, the stress fluctuation follows the deflection variation except for in a much sharper manner.

CONCLUSION

A theoretical study is done for the impact response of the linearly elastic, orthotropic cylindrical shells of moderate thickness. Dynamic responses of the various cases of impact loading are predicted in closed-form solutions.

REFERENCES

[1] Mindlin, R. D., "Influence of rotatory inertia and shear on flexural motion of isotropic elastic plates", J. Appl. Mech., 18, p. 31, 1951.

[2] Mirsky, I. and Herrmann, G., "Axially symmetric motions of thick cylindrical shells", J. Appl. Mech., 25, p. 97, 1958.

[3] Weingarten, L. I. and Reismann, H., "Forced motion of cylindrical shells, a comparison of shell theory with elasticity theory", AIAA J., 11(6), p. 769, 1973.

[4] Huang, C. C., "Forced motion of elastic cylindrical shells", J. Appl. Mech., 42, p. 321, 1975.

[5] Reddy, J. N., "Impact on laminated composite plates: a review of recent computational developments", Computational Aspects of Penetration Mechanics, Springer-Verlag, New York, 1982.

[6] Sun, C. T. and Chen, J. K., "On the impact of initially stressed composite laminates", J. Composite Materials, 19, p. 490, 1985.

[7] Bert, C. W., "Research on dynamic behavior of composite and sandwich plates - IV", Shock Vib. Dig., 17(11), p. 3, 1985.

[8] Chao, C. C. and Lee, S. C., "Buckling, vibration and dynamic behavior of orthotropic foam sandwich cylindrical shells", Natl. Tsing Hua University Tech. Rept. NTHU-PME85-002, 1985.

[9] Chao, C. C. and Lin, I. S., "Dynamic buckling of orthotropic foam sandwich spherical shells", Natl. Tsing Hua University Tech. Rept. NTHU-PME86-001, 1986.

EVALUATION ON AT-3 COMPOSITE SPEEDBRAKE BY STATIC TEST AND ANALYSIS

R. C. Chu, L. J. Hsu and Y. J. Chiou

Aeronautic Research Laboratory, AIDC/CSIST
Taiwan, Republic of China

ABSTRACT

The composite speedbrake of the AT-3 aircraft has been designed and manufactured using graphite epoxy. A concentrated effort has been made to analyze this composite structure by using a FEM with equivalent orthotropic material properties. The FEM results are verified by static test under simulated aerodynamic loadings.

Successful completion of the static test has qualified the structural design and it proves the consistency between test data and analysis results.

INTRODUCTION

There are more and more composite laminate structures being applied on the flight vehicle. The classical laminate theory has been widely used for the analysis of the composite laminate structures and it has proven to be a valid tool based on many laboratory tests.

Due to the complication of a structure compared to a test specimen, the application of the classical laminate theory needs to be very carefully applied. Anyway, by the use of linear elastic finite element analysis, the vulnerable points of a composite laminate structure can be predicted and the first possible failure location on the structure can also be predicted.

This paper will evaluate the finite element moment results of the AT-3 jet trainer composite speedbrake through static test results. The AT-3 speedbrake consists of skin, rib, beams, torque tube and hinges. Only the rib and the beam are made of symmetrically layered graphite epoxy composite. When the speedbrake is being pushed to the upper end point, the speedbrake takes its maximum aerodynamic load. This load is 5115 lbs. The aerodynamic load will be simulated during the static test and the failure points of the speedbrake shall be checked out.

In order to satisfy the anisotropic property of the carbon fibered composite, the plate element and the equivalent stiffness concept are applied in the finite element analysis. The speedbrake structural model contains 498 nodal points and 539 elements. The results of the finite element analysis shall be validated through the static test. Realistic loading for the static test article is accomplished using a whiffle tree arrangement.

DESIGN [2]

The design criteria imposed on the speedbrake structure is based on two principles: strength and stiffness. This criteria covers the laminate stacking sequence, ply drop off and nonbuckling structure up to ultimate load. Design allowables were taken from coupon tests. Approximately 400 specimens have been tested to obtain design allowables. In general, the stiffness of the composite speedbrake was designed approximately equivalent to the metal stiffness. The speedbrake (as shown in Figure 1) is approximately 672 square inches in area with a width 28 inches and a length 32 inches. The speedbrake pivots about two hinge

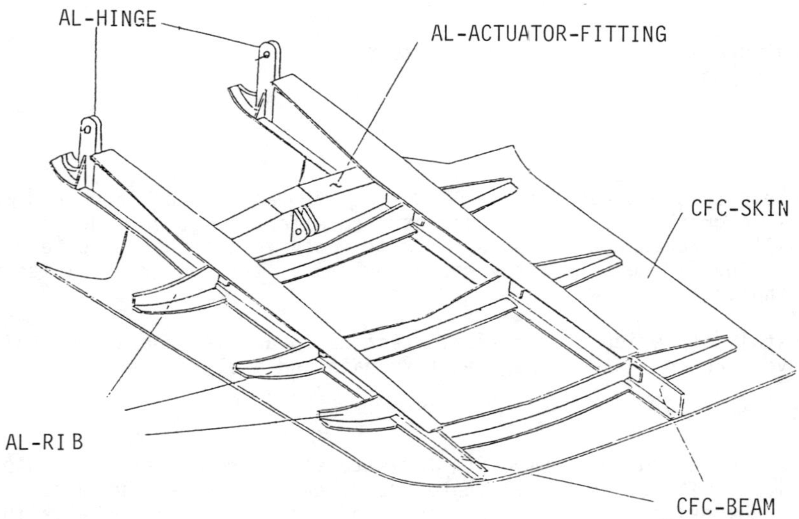

Fig. (1) - Speedbrake configuration

arms that extend out from the fuselage and is actuated by one hydraulic cylinder. The torque tube connects to the actuator and hinge arm, I beam to construct the main frame of the speedbrake structure. All aero-loads are taken by the skin and transferred on to the I beams and the torque tube. The torque tube is supported by three hinge points.

The principle composite component of the speedbrake consists of two I beams and one solid laminate skin. The entire assembly is mechanically fastened using conventional aircraft fasteners. In order to simplify the manufacturing process, all of the composite laminates are a symmetrically balanced [0°/±45°/90°] family system. The stacking sequence and thickness are shown in Figure 2.

ANALYSIS [1]

The internal loads, stress and strain for the composite speedbrake are obtained from the results of the finite element analysis. The finite element model used in this analysis includes triangular, quadrilateral plate elements and bar elements as displayed in Figure 3. Both the graphite/epoxy skin and I beam are modeled by plate elements.

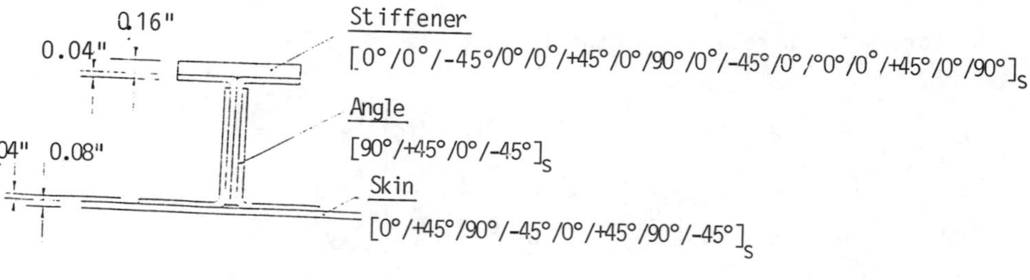

Fig. (2) - Stacking sequence of laminate

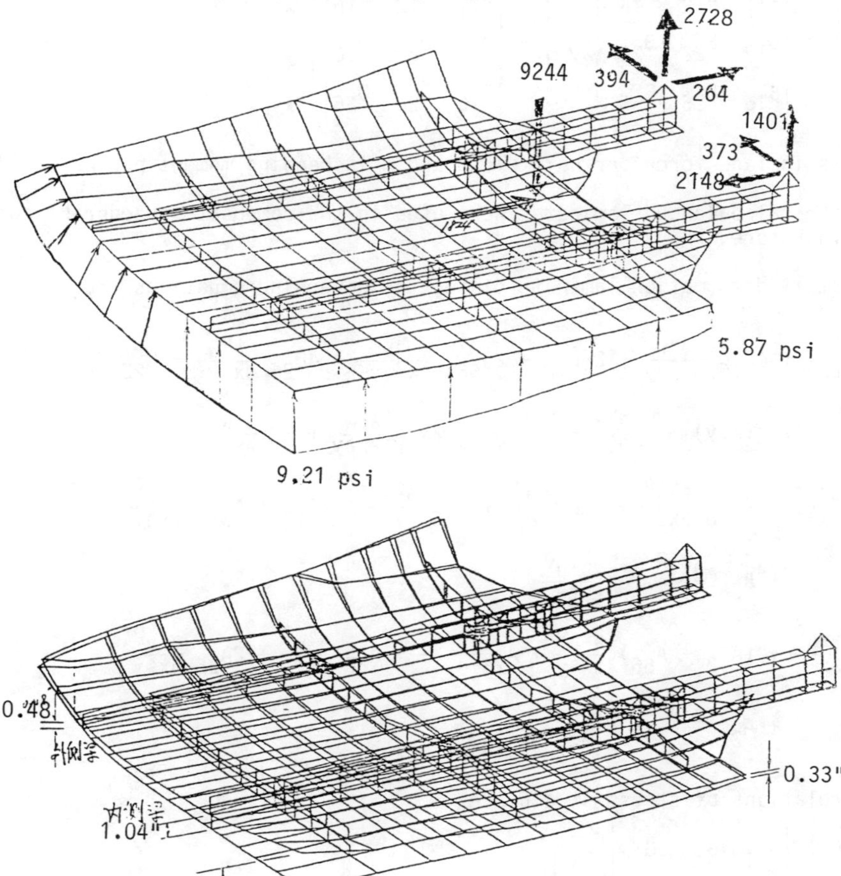

Fig. (3) - Load, reactions and displacement of FEM analysis

A simple method of introducing the material properties of composite structure into the FEM is used here. This method is based on the equivalent stiffness concept. By this concept, the stiffness coefficients G_{ij} are derived.

The force-strain relation of the laminate is as follows:

$$\begin{Bmatrix} N_x \\ N_y \\ N_{xy} \end{Bmatrix} = \begin{bmatrix} A_{11} & A_{12} & A_{16} \\ A_{12} & A_{22} & A_{26} \\ A_{16} & A_{26} & A_{66} \end{bmatrix} \begin{Bmatrix} \varepsilon_x^o \\ \varepsilon_y^o \\ \gamma_{xy}^o \end{Bmatrix} + \begin{bmatrix} B_{11} & B_{12} & B_{16} \\ B_{12} & B_{22} & B_{26} \\ B_{16} & B_{26} & B_{66} \end{bmatrix} \begin{Bmatrix} K_x \\ K_y \\ K_{xy} \end{Bmatrix}$$

$$\begin{Bmatrix} M_x \\ M_y \\ M_{xy} \end{Bmatrix} = \begin{bmatrix} B_{11} & B_{12} & B_{16} \\ B_{12} & B_{22} & B_{26} \\ B_{16} & B_{26} & B_{66} \end{bmatrix} \begin{Bmatrix} \varepsilon_x^o \\ \varepsilon_y^o \\ \gamma_{xy}^o \end{Bmatrix} + \begin{bmatrix} D_{11} & D_{12} & D_{16} \\ D_{12} & D_{22} & D_{26} \\ D_{16} & D_{26} & D_{66} \end{bmatrix} \begin{Bmatrix} K_x \\ K_y \\ K_{xy} \end{Bmatrix}$$

where N is inplane force per unit width and M is bending moment per unit width.

Symmetrical balanced laminates were used in the composite speedbrake components. Therefore, $[B_{ij}] = 0$.

For small deformation laminate plate, the governing equations are:

$$D_{11} \frac{\partial^4 w}{\partial x^4} + 4D_{16} \frac{\partial^4 w}{\partial x^3 \partial y} + 2(D_{12} + 2D_{66}) \frac{\partial^4 w}{\partial x^2 \partial y^2} + 4D_{26} \frac{\partial^4 w}{\partial x \partial y^3} + D_{22} \frac{\partial^4 w}{\partial y^4}$$

$$= q(x,y) - \rho \frac{\partial^2 w}{\partial t^2} + N_x \frac{\partial^2 w}{\partial x^2} + 2N_{xy} \frac{\partial^2 w}{\partial x \partial y} + N_y \frac{\partial^2 w}{\partial y^2}$$

$$A_{11} \frac{\partial^2 u^o}{\partial x^2} + 2A_{16} \frac{\partial^2 u^o}{\partial x \partial y} + A_{66} \frac{\partial^2 u^o}{\partial y^2} + A_{16} \frac{\partial^2 v^o}{\partial y^2} + (A_{12} + A_{66}) \frac{\partial^2 v^o}{\partial x \partial y}$$

$$+ A_{26} \frac{\partial^2 v^o}{\partial y^2} = \rho \frac{\partial^2 u^o}{\partial t^2}$$

$$A_{16} \frac{\partial^2 u^o}{\partial x^2} + (A_{12} + A_{66}) \frac{\partial^2 u^o}{\partial x \partial y} + A_{26} \frac{\partial^2 u^o}{\partial y^2} + A_{66} \frac{\partial^2 v^o}{\partial x^2} + 2A_{26} \frac{\partial^2 v^o}{\partial x \partial y}$$

$$+ A_{22} \frac{\partial^2 v^o}{\partial y^2} = \rho \frac{\partial^2 v^o}{\partial t^2}$$

The relations of stress-strain are

$$\begin{Bmatrix} \sigma_x \\ \sigma_y \\ \tau_{xy} \end{Bmatrix} = \begin{vmatrix} G_{11} & G_{12} & G_{16} \\ G_{12} & G_{22} & G_{26} \\ G_{16} & G_{26} & G_{66} \end{vmatrix} \begin{Bmatrix} \varepsilon_x \\ \varepsilon_y \\ \gamma_{xy} \end{Bmatrix}$$

There are two kinds of reduced stiffness coefficients. G_{ij} can be used in the composite element analysis as either

$$G_{ij} = \frac{1}{h} A_{ij} \text{ or } G_{ij} = \frac{12}{h^3} D_{ij}$$

The principle of choosing G_{ij} depends on what kind of internal force the element experiences. For the composite speedbrake skin, using

$$G_{ij} = \frac{12}{h^3} D_{ij}$$

for the flanges of the I beam, $G_{ij} = \frac{1}{h} A_{ij}$ would be better, because the skin and flanges of the beam are primarily loaded by moments and inplane forces separately.

TEST

The qualification plan for the speedbrake includes full scale static test and quasi-aero loading. The static test was tested in February of 1985 and failed at 170% of design ultimate load (DUL).

Prior to initiation of static testing, a strain and deformation survey was conducted. Increments of 10% DLL (Design Limit Load) were applied up to 30% DLL for the design condition. Some of the load rigging arrangement had to be modified to get better agreement with boundary conditions.

Test Setup. The speedbrake was fitted with an array of reaction pads bonded to the OML (Outer Mould Line) and installed in the test fixture as shown in Figure 4. With the speedbrake in the maximum open position, loading was applied through the actuator and reacted at the hinge arms and the reaction pads attached to the whippletree system.

Test Result. Failure of the structure occurred in the torque tube attachment at 170% DUL. Apparently, the failure mode was a mixture of net tension and cleavage of bolted joint.

The external load and strain relations of test data at critical points of skin and flanges of the beam has +6% $\sim$ +12% deviation with analysis results, Figure 5. After DUL, nonlinear geometrical deformation were shown in external load and displacement relation diagrams.

CONCLUDING REMARKS

At present, the composite speedbrake design and fabrication of the static test is complete. The flight speedbrake is being manufactured. Delivery to ARL/AIDC will meet the product schedule. The program has demonstrated a good static test of the design and has verified the weight savings. Actual weight of the static speedbrake shows a weight savings of 8 lbs. (36%) over the current metal speedbrake.

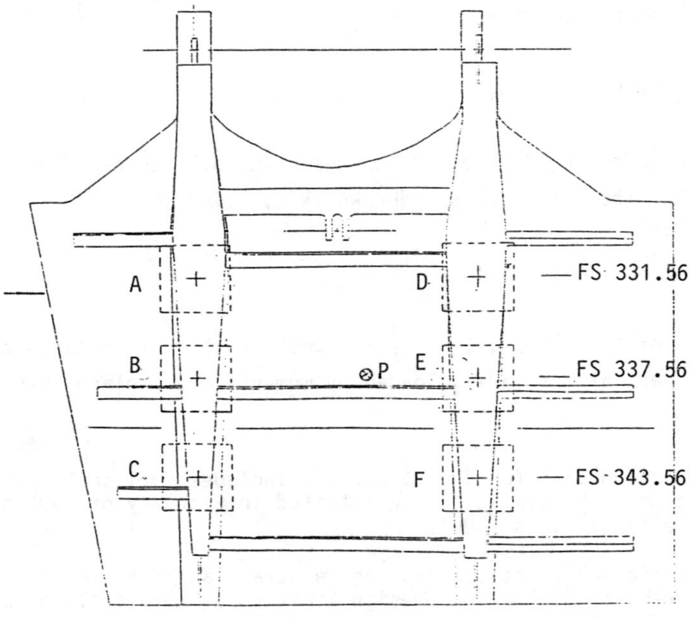

SPEED BRAKE STATIC TEST INSTALLATION

Fig. (4) - Pad location and installation of static test

Successful completion of the static test has qualified the design and has provided the AT-3 aircraft with a composite speedbrake that is lighter than its metal counterpart.

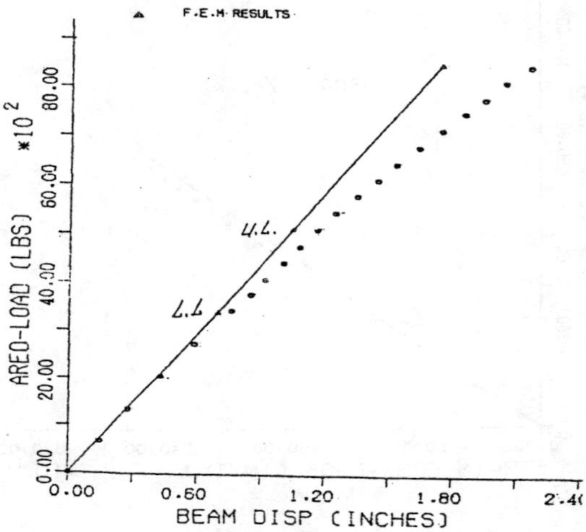

Displacement of inner beam

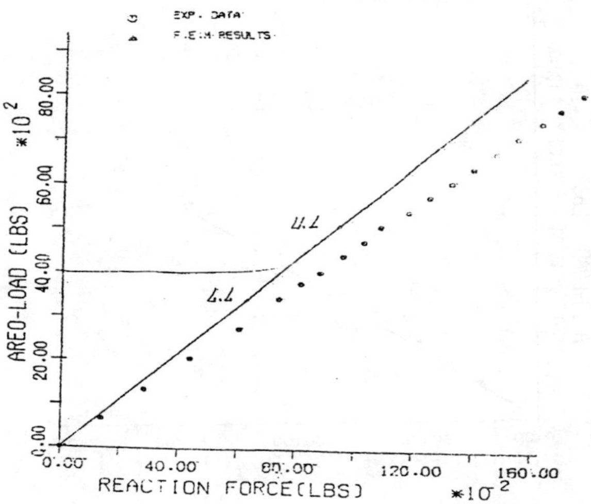

Reaction of dummy actuator

Fig. (5) - Comparison between analysis and test data

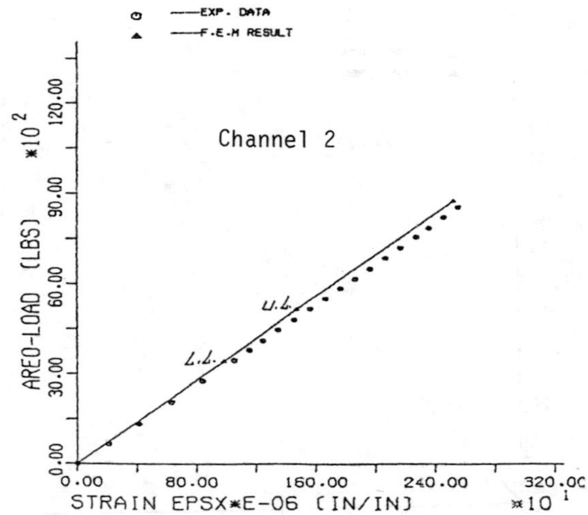

Strain of skin

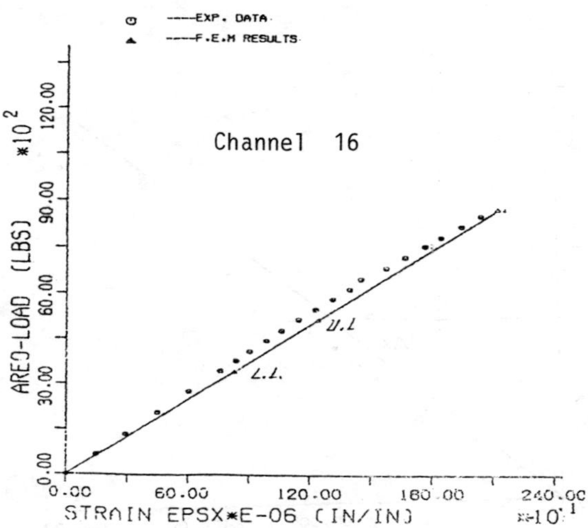

Strain at stiffener

Fig. (5 continued) - Comparison between analysis and test data

REFERENCES

[1] Hsu, L. J., "CFRP speedbrake stress analysis", Technical Report T3S1-018, August 1984.

[2] "Advanced composite design guide", Vol. II, Analysis, AFML, 1971.

A SIMPLE NUMERICAL METHOD FOR ANALYZING ELASTO-PLASTIC STRESS DISTRIBUTION AND PROGRESSIVE FAILURE IN ADHESIVE LAYER OF BONDED JOINTS

S. Amijima and T. Fujii

Faculty of Engineering, Doshisha University
Kyoto, Japan

and

T. Tanaka

Kobe Steel Industries
Kobe, Japan

ABSTRACT

A simple numerical method to analyze elasto-plastic stress distribution and progressive failure in adhesive layer of bonded joints is given. The method is based on the finite element procedure and one dimensional approximation for adherends. The numerical calculation on a single lap joint subjected to four point bending load was conducted and its result was compared with the experimental result. Good agreements except the final failure load were obtained between both results. A difference of elasto-plastic behaviors between single and double lap joints was also given. This method has some advantages that it is not necessary to consider the stress singularity in the adhesive layer and short computing time derived from its simplicity and a small number of total degrees of freedom.

INTRODUCTION

Some attempts to analyze elasto-plastic stress distribution in two dimensional space for adhesive bonded joints using nonlinear finite element programs (for example, NONSAP) are made [1,2]. The von Mises yield criterion and others are used. The joint model must be divided into extremely small elements in order to obtain high accuracy near the discontinuities at the lap ends since the stress singularities occur at such points under fully elastic condition. Therefore, we cannot avoid some uncertainties when distinguishing whether the element yields or not. We also have similar hesitation in the case of failure analysis of adhesive layer using the stress failure criteria because the stresses vary even in a small element. In practice, some assumptions are made in order to judge the yield and failure conditions of the element (Adams et al evaluate the adhesive stresses at the integrating point nearest the adherend in the element [3]), a great amount of computing time and much labor due to a large number of elements makes the cost unacceptable especially for the progressive failure analysis of the joint.

This paper shows a simple numerical method to analyze the elasto-plastic stress distribution and progressive failure in the adhesive layer of bonded joints. If we adopt an assumption of linear interpolation for displacements in the adhesive layer similar to the Goland-Reissner's treatment, not only the stress singularity but also long computing time can be avoided. The present method is based on the finite element procedure and one dimensional approximation for adherends with the assumption above mentioned.

FINITE ELEMENT FORMULATION

A. Assumption

Figure 1 shows stresses and stress resultants acting on an infinitesimal element of the adhesive bonded structure. The following assumptions are made for

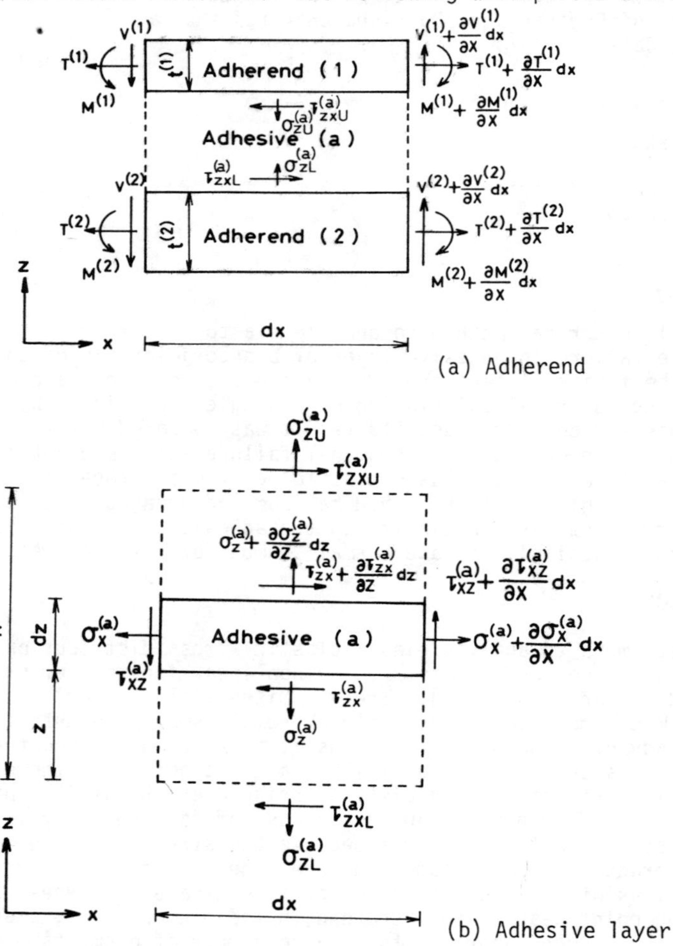

Fig. (1) - Adhesive bonded structural element

formulating the finite element equation in order to simplify the problem.

(1) Each adherend remains within an elastic range, and both deformations and strains of the adherend are small.

(2) Cross-section of the adherend is still plane and perpendicular to the material axis of it after deformed. Namely, the adherend is treated as a one dimensional beam or plate.

(3) Displacements in x direction (parallel to the longitudinal axis of the joint) and z direction (perpendicular to it) of the adhesive layer are uniquely determined by surface deformations of two adherends.

(4) Two dimensional stress state is held in the adhesive layer.

Upper suffixes (1), (2) and (a) in the figure designate the adherend (1), (2) and the adhesive layer, respectively.

B. Displacements in Adhesive Layer

Infinitesimal increments of displacement in the adhesive layer ($du^{(a)} = du^{(a)}(x,z)$ in x direction and $dw^{(a)} = dw^{(a)}(x,z)$ in z direction) are determined by displacements of upper and lower adherends as shown in Figure 2.

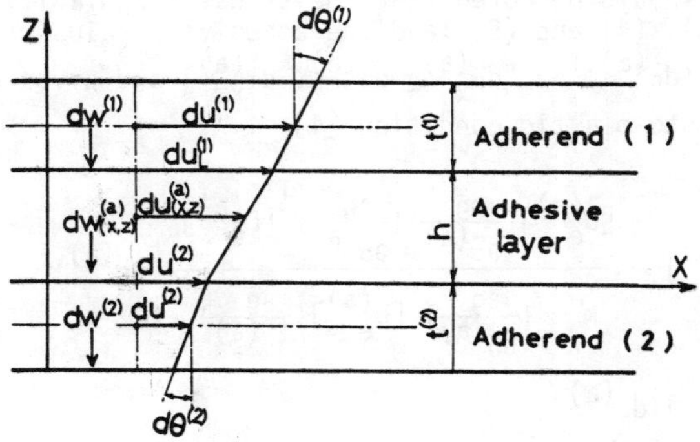

Fig. (2) - Displacements in adhesive layer

$$du^{(a)} = du_U^{(2)} + (du_L^{(1)} - du_U^{(2)}) \frac{z}{h} = \frac{z}{h} du^{(1)} + (1 - \frac{z}{h}) du^{(2)}$$

$$- \frac{1}{2} t^{(1)} \frac{z}{h} d\theta^{(1)} + \frac{1}{2} t^{(2)} (1 - \frac{z}{h}) d\theta^{(1)} \quad (1)$$

$$dw^{(a)} = dw^{(2)} + (dw^{(1)} - dw^{(2)}) \frac{z}{h} \quad (2)$$

where z is measured from the boundary between the adhesive layer and the adherend (2).

C. Stress State in Adhesive Layer

We obtain strain increments $\{d\varepsilon^{(a)}\}^T = [d\varepsilon_x^{(a)} d\varepsilon_z^{(a)} d\gamma_{xz}^{(a)}]$ in the adhesive layer from equations (1) and (2) as

$$d\varepsilon_x^{(a)} = \frac{\partial}{\partial x}(du^{(a)}) = \frac{z}{h}\frac{d}{dx}(du^{(1)}) + (1 - \frac{z}{h})\frac{d}{dx}(d\theta^{(1)})$$

$$- \frac{1}{2}t^{(1)}\frac{z}{h}\frac{d}{dx}(d\theta^{(1)}) + \frac{1}{2}t^{(2)}(1 - \frac{z}{h})\frac{d}{dx}(d\theta^{(1)}) \quad (3)$$

$$d\varepsilon_z^{(a)} = \frac{\partial}{\partial z}(dw^{(a)}) = \frac{1}{h}(dw^{(1)} - dw^{(2)}) \quad (4)$$

$$d\gamma_{xz}^{(a)} = \frac{\partial}{\partial z}(du^{(a)}) + \frac{\partial}{\partial x}(dw^{(a)}) = \frac{1}{h}(du^{(1)} - du^{(2)}) - \frac{1}{2h}(t^{(1)}d\theta^{(1)}$$

$$+ t^{(2)}d\theta^{(2)}) + \frac{z}{h}\frac{d}{dx}(dw^{(1)}) + (1 - \frac{z}{h})\frac{d}{dx}(dw^{(2)}) \quad (5)$$

In this case, $d\varepsilon_x^{(a)}$ and $d\gamma_{xz}^{(a)}$ vary not only along x axis but also through the thickness of the adhesive layer since $du^{(a)}$ includes a variable x as shown in equation (1). It should be noted that the stress singularities do not occur because equations (3), (4) and (5) lead the adhesive strains to have finite values. Stress increments $\{d\sigma^{(a)}\}^T = [d\sigma_x^{(a)}, d\sigma_z^{(a)}, d\tau_{xz}^{(a)}]$ are given by the following equation under elasto-plastic condition [4]

$$\{d\sigma^{(a)}\} = [D_e^{(a)}] - \frac{[D_e^{(a)}]\{\frac{\partial g}{\partial \sigma^{(a)}}\}\{\frac{\partial g}{\partial \sigma^{(a)}}\}^T[D_e^{(a)}]}{k + \{\frac{\partial g}{\partial \sigma^{(a)}}\}^T[D_e^{(a)}]\{\frac{\partial g}{\partial \sigma^{(a)}}\}}\{d\varepsilon^{(a)}\}$$

$$= [D_{ep}^{(a)}]\{d\varepsilon^{(a)}\} \quad (6)$$

where $[D_e^{(a)}]$ is the elastic stress-strain matrix. In general, the adhesive thickness is much less than the width of the joint, so the plane strain condition should be considered. "g" is the plastic potential and "k" is a parameter associated with the isotropic or kinetic strain hardening. We consider here the von Mises yield condition as the plastic potential g. The matrix $[D_{ep}^{(a)}]$ represents the instantaneous elastic plastic stress-strain matrix. Initial stresses caused by shrinkage of the adhesive also can be included in the constitutive equation if desired.

D. Finite Element Equilibrium Equation

An adhesive bonded part of the joint is divided into finite elements as shown in Figure 3. Nodal points with displacement increments (axial displacement, de-

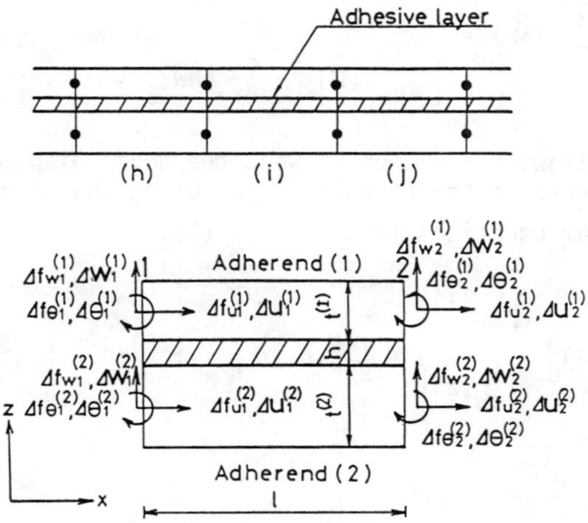

Fig. (3) - Finite element division for adhesive bonded structure

flection and rotation) and incremental forces (axial force, shear force and moment) are placed on the neutral axis of each adherend, and four nodes are assigned to each element. A conventional beam type element [5] is used for the adherends of the element. Finitesimal increments $\{\Delta u\}^T = [\Delta u^{(1)}, \Delta u^{(2)}]$ of axial displacement on the material axes of the adherends in the element are interpolated linearly by the shape function [Nu], [5].

$$\{\Delta u\} = \begin{bmatrix} [Nu][0] \\ [0]\ [Nu] \end{bmatrix} \begin{Bmatrix} \Delta u_1^{(1)} \\ \Delta u_2^{(1)} \\ \Delta u_1^{(2)} \\ \Delta u_2^{(2)} \end{Bmatrix} = [Nu^*]\{\Delta qu\} \tag{7}$$

Similarly, finitesimal increments of deflection $\{\Delta w\}$ are also interpolated by the cubic shape function [Nw].

$$\{\Delta w\} = \begin{bmatrix} [Nw][0] \\ [0]\ [Nw] \end{bmatrix} \{\Delta qw\} = [Nw^*]\{\Delta qw\}$$

$$\{\Delta q\}^T = [\Delta u_1^{(1)}, \Delta u_2^{(1)}, \Delta w_1^{(1)}, \Delta\theta_1^{(1)}, \Delta w_2^{(1)}, \Delta\theta_2^{(1)}$$
$$\Delta u_1^{(2)}, \Delta u_2^{(2)}, \Delta w_1^{(2)}, \Delta\theta_1^{(2)}, \Delta w_2^{(2)}, \Delta\theta_2^{(2)}] \tag{8}$$

The rotation increments $\{\Delta\theta\}$ is given by differentiating the $\{\Delta w\}$ with x.

$$\{\Delta\theta\} = \begin{Bmatrix} \Delta\theta^{(1)} \\ \Delta\theta^{(2)} \end{Bmatrix} = \frac{d}{dx}\{\Delta W\} = \begin{bmatrix} -\frac{d}{dx}[Nw] & [0] \\ [0] & -\frac{d}{dx}[Nw] \end{bmatrix}\{\Delta qw\} = [N\theta^*]\{\Delta qw\} \quad (9)$$

We also obtain the following relation between the nodal displacement increments and the strain increments in the adhesive layer using the strain-nodal displacement matrix $[B^{(a)}]$ from equations (3), (4) and (5).

$$\Delta\varepsilon_x^{(a)} = [\tfrac{z}{h}\ (1-\tfrac{z}{h})]\frac{d}{dx}\begin{Bmatrix} \Delta u^{(1)} \\ \Delta u^{(2)} \end{Bmatrix} + \frac{1}{2}[-t^{(1)}\tfrac{z}{h},\ -t^{(2)}(1-\tfrac{z}{h})]\frac{d}{dx}\begin{Bmatrix} \Delta\theta^{(1)} \\ \Delta\theta^{(2)} \end{Bmatrix}$$

$$= [B_x^{(a)}]\{\Delta q\} \quad (10)$$

$$\Delta\varepsilon_z^{(a)} = \frac{1}{h}[1 -1]\begin{Bmatrix} \Delta w^{(1)} \\ \Delta w^{(2)} \end{Bmatrix} = [B_z^{(a)}]\{\Delta q\} \quad (11)$$

$$\Delta\gamma_{xz}^{(a)} = \frac{1}{h}[1 -1]\begin{Bmatrix} \Delta u^{(1)} \\ \Delta u^{(2)} \end{Bmatrix} + \frac{1}{2h}[t^{(1)}\ -t^{(2)}]\begin{Bmatrix} \Delta\theta^{(1)} \\ \Delta\theta^{(2)} \end{Bmatrix}$$

$$+ \ \tfrac{z}{h}\ (1-\tfrac{z}{h})]\frac{d}{dx}\begin{Bmatrix} \Delta w^{(1)} \\ \Delta w^{(2)} \end{Bmatrix} = [B_{xz}^{(a)}]\{\Delta q\} \quad (12)$$

$$\{\Delta\varepsilon^{(a)}\} = [B^{(a)}]\{\Delta q\} \quad (13)$$

where $\{\Delta q\}$ represents the nodal displacement increment vector including all displacement elements. It is given as

$$\{\Delta qw\}^T = [\Delta w_1^{(1)},\ \Delta\theta_1^{(1)},\ \Delta w_2^{(1)},\ \Delta\theta_2^{(1)},\ \Delta w_1^{(2)},\ \Delta\theta_1^{(2)},\ \Delta w_2^{(2)},\ \Delta\theta_2^{(2)}]$$

The finite element discretization of the principle of virtual work [6] gives the following incremental equation of equilibrium for each finite element of adhesive bonded structures.

$$([k] + [k_n^{(a)}])\{\Delta q\} = \{f_{n+1}\} + \{r_n\} \quad (14)$$

The stiffness matrix $[k]$ is derived from the stiffness of the adherends and it does not change during step by step calculation while the instantaneous stiffness matrix $[k^{(a)}]$ changes after the yielding in the adhesive layer of the element occurs. Suffix "n" means the step number. $[k]$ consists of two stiffness matrixes,

$[k^{(1)}]$ and $[k^{(2)}]$ as

$$[k] = \begin{vmatrix} [k^{(1)}] & [0] \\ [0] & [k^{(2)}] \end{vmatrix} \tag{15}$$

They consist of the stiffness matrixes of the adherend (1) and (2), and they are as same as the conventional stiffness matrix of the beam element. On the other hand, $[k^{(a)}]$ is given by the following integral.

$$[k^{(a)}] = b \cdot h \cdot \int_0^1 [B^{(a)}]^T [D_{ep}^{(a)}][B^{(a)}]dx \quad \text{(b: width of the joint)} \tag{16}$$

Previously, the Gauss numerical integration in two dimensions having 9 integrating points was employed [7], but in this case, we suppose the adhesive stresses to be constant in the element and one integrating point (at the center of the element) is used in order to save computing time. The vector $\{f_{n+1}\}$ is the external force vector at the (n+1)th stage. The vector $\{r_n\}$ is also the equivalent force vector given by the stresses at the nth stage. Loading and unloading processes can be calculated step by step according to the Yamada's method [4].

E. Failure Criterion for Adhesive Layer

In order to calculate the progressive failure in the adhesive layer, the stress failure criterion must be given, while fracture mechanics or similar ideas [8] are used for the analysis of crack growth in the adhesive layer. In this paper, we use a simple failure criterion well-known as the Mohr-Coulomb failure criterion although other failure theories exist. It is

$$F(\sigma_z^{(a)}, \tau_{xz}^{(a)}) = k_0 + k_1 \cdot \tau_{xz}^{(a)} + k_2 \cdot \tau_{xz}^{(a)2} - \sigma_z^{(a)} = 0 \tag{17}$$

where k_0, k_1 and k_2 are constants determined experimentally.

RESULTS AND DISCUSSION

A. Elasto-Plastic Stress Distribution for Lap Joints Subjected to Tensile Shear Load

In this section, the comparison of elasto-plastic behavior between single and double lap joints under tensile shear load is shown. A typical equivalent stress-strain relation for ductile adhesives is shown in Figure 4. Apparent non-linear behavior is found and it can be represented by the following equation

$$\sigma^{(a)} = c(a + \varepsilon^{p(a)})^n \tag{18}$$

where $\varepsilon^{p(a)}$ is the equivalent plastic strain of adhesive, a, c and n are experimental constants. Dimensions of the joints and boundary conditions for calculation are given in Figure 5. Material data of the joints are given in Table 1. The stress-strain curve is represented bi-linearly in all calculations throughout this paper for saving computing time. The lap part of the joint should be divided into small elements. In this case, 0.1 mm long elements are used nearest

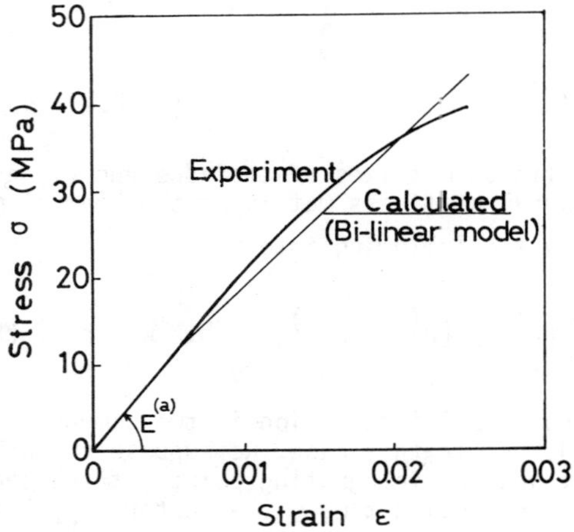

Fig. (4) - Typical stress-strain curve

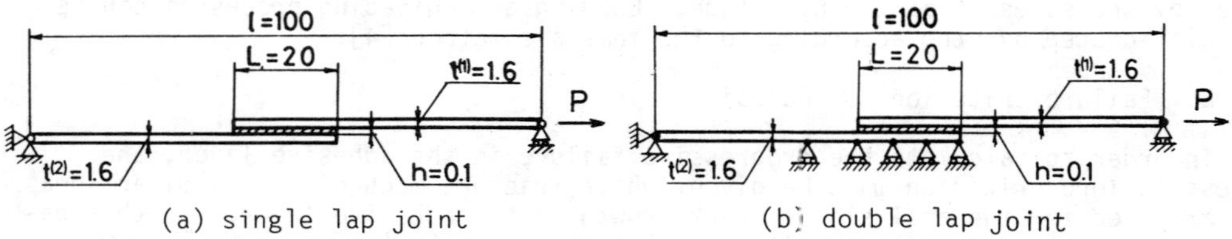

Fig. (5) - Dimensions of the joints

Table 1 - Material data for elasto-plastic analysis

	Adherend	Adhesive
Young's modulus	$E^{(1)} = E^{(2)} = 70$ GPa	$E^{(a)} = 2.14$ GPa
Poisson's ratio	————	$\nu^{(a)} = 0.4$
Thickness	$t^{(1)} = t^{(2)} = 1.6$ mm	$h = 0.1$ mm

a	c	n
0.00905	158	1

$$\sigma^{(a)} = c\,(a + \epsilon^{P(a)})^n$$

the lap ends, but the number of total elements is less than 100.

Figures 6 and 7 show the comparisons of shear stress distribution (divided by the average shear stress $\tau^{(a)}_{mean} = P/A$, A; lap area) and progressive expansion of yield zone in the adhesive layer for single and double lap joints under tensile shear loading. It is found that the shear stress concentration near the lap ends

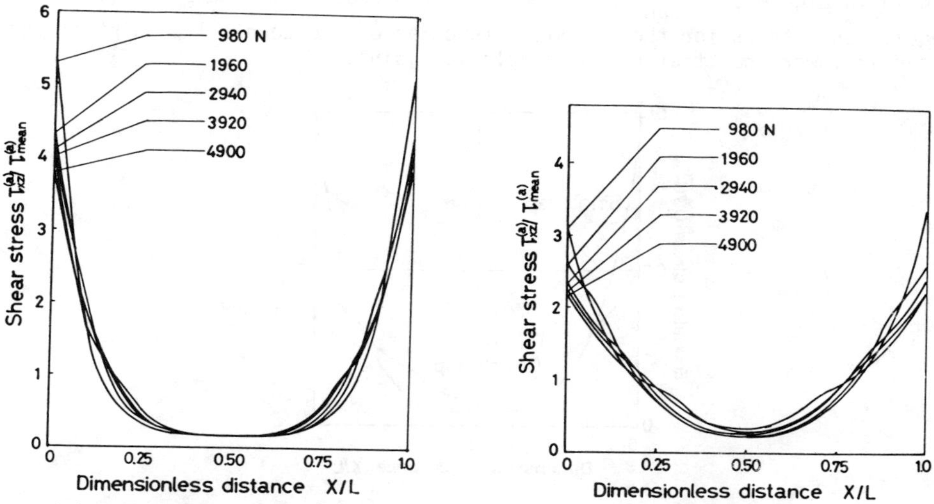

(a) Single lap joint (b) Double lap joint

Fig. (6) - Shear stress distribution in the adhesive layer

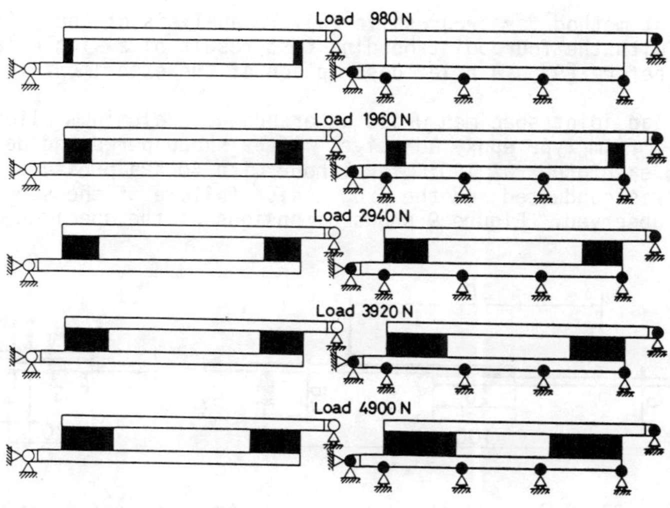

Fig. (7) - Expansion of yield zone

for the single lap joint is higher than that for the double lap joint even at the elasto-plastic stage. The plastic zone for the double lap joint expands wider than for the single lap joint under the same load, but it does not mean that the strength of the double lap joint is lower than that of the single lap joint. The wider plastic zone is due to the uniformity of the equivalent stress distribution as shown in Figure 8 ($\sigma^{(1)}_{mean}$ is the mean fiber stress in the adherend (1)). An equivalent stress for the von Mises function of the double lap joint at the lap end is lower than that of the single lap joint.

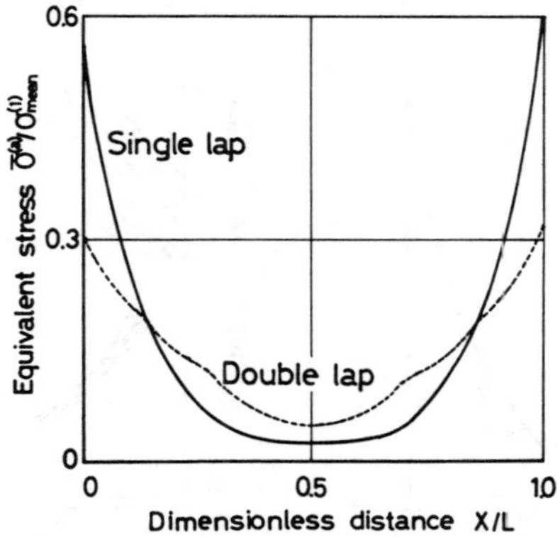

Fig. (8) - Equivalent stress distribution in the adhesive layer

B. Progressive Failure in Adhesive Layer

The present method for progressive failure analysis of the joint is examined by comparing with the four point bending test result of a single lap joint which was reported before [9]. A brief description of the experiment is given below.

A single lap joint specimen whose adherends were aluminum alloy strips was bonded using a film type epoxy adhesive. After sandpapered and degreased, they were glued to each other at 180°C for 1 hour with 40 KPa pressure. A four point bending test was conducted and the progressive failure of the specimen under bending load was observed. Figure 9 shows dimensions of the specimen and locations

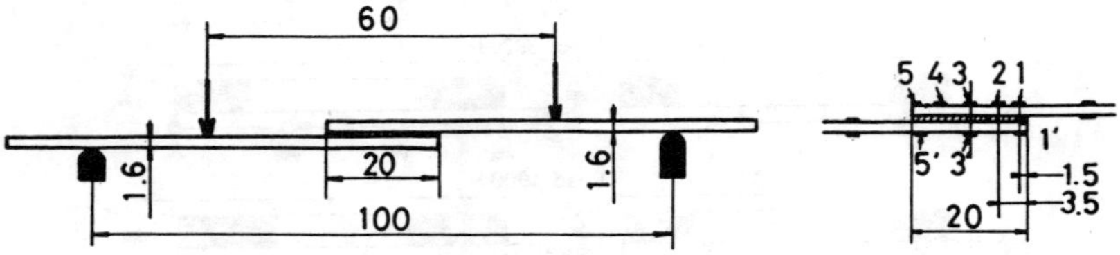

Fig. (9) - Dimensions of the specimen and the four point bending test

of strain gages mounted on both lap surfaces. Deflections of loading points were also measured. Figure 10 shows the relation between the deflection and the applied bending moment M to the joint (it was normalized by unit width of the joint). The deflection at the left side loading point of the specimen was plotted in this figure. The same displacement increments were applied to both loading points for calculation in order to stabilize the step by step calculation. A little difference in reaction forces of two supports were found in the calculated result because the

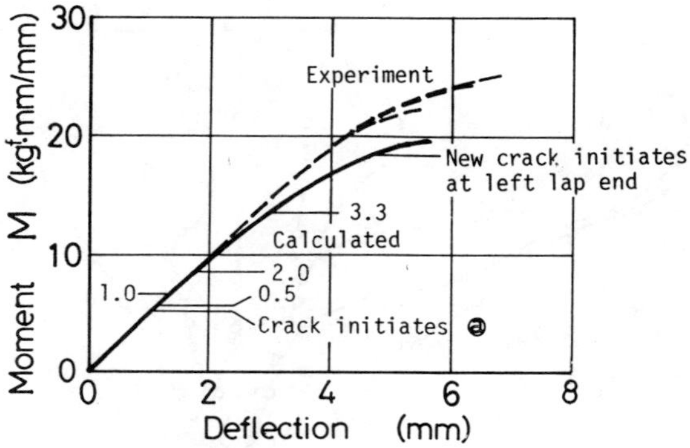

Fig. (10) - Bending moment versus deflection at the left loading point

crack initiated and grew from one lap end (right side) at early stages. Therefore, the mean bending moment for plotting the result was used. Table 2 shows the data for failure criterion used. The calculated result can be corresponded

Table 2 - Failure criterion data of the adhesive

k0	k1	k2
43	0	-0.024

(unit ; MPa)

$$F(\sigma_z^{(a)}, \tau_{xz}^{(a)}) = k0 + k1 \cdot \tau_{xz}^{(a)} + k2 \cdot \tau_{xz}^{(a)2} - \sigma_z^{(a)} = 0$$

with the experimental one although the maximum bending moment at final failure obtained from calculation is lower than that of the experimental one. More satisfactory agreement is found in the failure process of the joint. Calculated result shows that at point a in the figure, the adhesive layer of the first element from the right lap end is failed and the failure zone (crack) in the adhesive layer expands with an increase of the applied bending moment. It becomes 1 mm long at moment = 7.3 kgf.mm/mm. Under such a condition, surface strain of the lower adherend at the strain gage location 1' will decrease. Figure 11 shows the surface strain records of the experiment. The above prediction coincides with the strain trace 1' in the figure.

The calculated result reveals more interesting behaviors in the progressive failure process. Once the crack initiates at the right lap end after the adhesive layer yielded, it expands towards the other side. At the same time, the yielding in the adhesive layer near the left lap end occurs and the plastic zone also expands towards the right lap end without cracking because the normal stress $\sigma_z^{(a)}$ at the left lap end is negative, namely the compressive stress acts. However, when the crack at the right lap end becomes a certain length, a new crack near the left lap end occurs. It is noted that the new crack does not initiate from just the left lap end. It initiates a little inside away from the left lap end since the normal stress becomes tension and the shear stress is still large at that point as can be seen in Figure 12. As soon as the crack opposite side yields the final fracture of the joint occurs. One of the reasons for discrepancy of the final failure load between the experiment and calculated results is due to the ap-

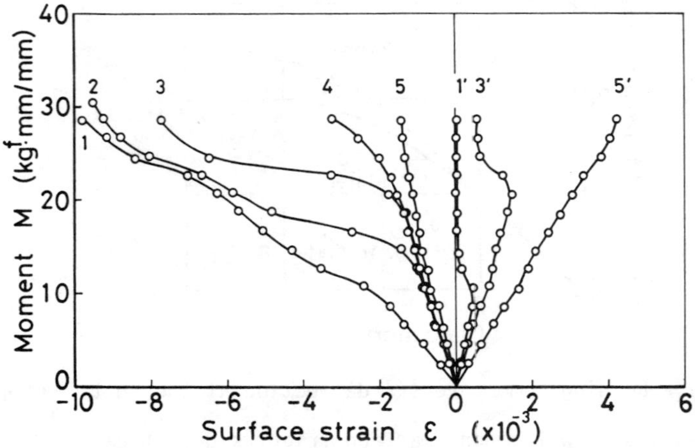

Fig. (11) - Relation between bending moment and surface strain

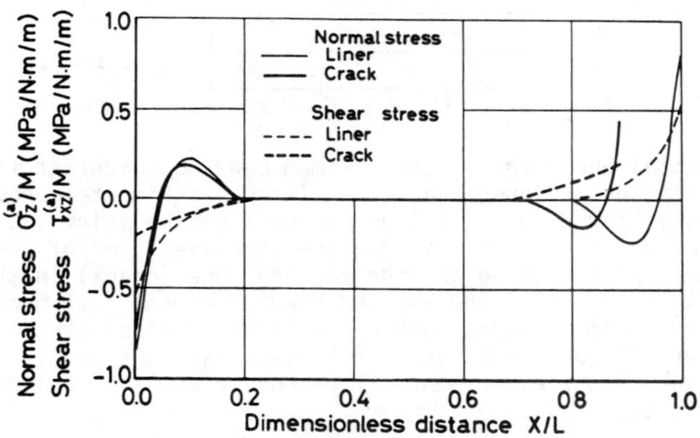

Fig. (12) - Adhesive stress distribution

proximation of the stress-strain curve for the adhesive (bi-linear approximation). The approximated s-s curve may give the higher stress state than in the case of the actual s-s curve because the tangent of final section of the actual s-s curve is lower than that of the approximated model. Therefore, the adhesive stresses near the crack tip would be estimated higher in the above calculation.

Figure 12 shows the calculated stress distribution for two cases. One is a linear calculation with no consideration of yielding and progressive failure of the adhesive layer. Another is the result where the crack becomes 2 mm long. Figure 13 shows the stress trajectory of the adhesive layer at a point 2 mm away from the lap end on the normal stress $\sigma_z^{(a)}$ - shear stress $\tau_{xz}^{(a)}$ map. In this fig-

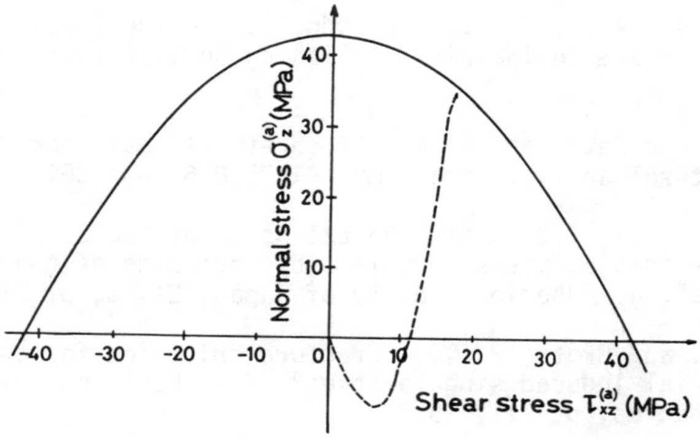

Fig. (13) - Adhesive stress trajectory

ure, a parabolic line shows the failure surface of the adhesive. At this point, the normal stress is negative, namely compression at the first stage. Both stresses ($\sigma_z^{(a)}$ and $\tau_{xz}^{(a)}$) become high with an increase of the bending load. However, the compressive $\sigma_z^{(a)}$ decreases after the failure zone of the adhesive layer grew to be 0.5 mm long. It becomes positive (tension) then the stress state reaches the failure surface.

CONCLUSION

In this paper we show a simple numerical method to analyze elasto-plastic stress distribution and progressive failure in adhesive layer of the bonded joint. The method is based on the finite element procedure and one dimensional approximation for the adherends. An analysis for a single lap joint subjected to four point bending load was conducted and its result was compared with the experimental result. The comparison shows that the present method is reasonable to calculate the elasto-plastic response and progressive failure in the adhesive layer. It is not necessary for this method to divide the joint model into finite elements through the thickness. Total number of elements is less than in case of the two dimensional F.E.M. The stress singularity does not occur.

REFERENCES

[1] Adams, R. D. and Peppiett, N. A., "Stress analysis of lap joints in fiber reinforced composite materials", Fiber Reinforced Materials, Inst. of Civil Eng., London, p. 45, 1977.

[2] Grossfeld, H. and Ishai, O., "The effect of cyclic loading on residual stresses and strains in bonded joints", Int. J. Adhesion and Adhesives, 2, p. 155, 1982.

[3] Harris, J. A. and Adams, R. D., "Strength prediction of bonded single lap joints by non-linear finite element", Int. J. Adhesion and Adhesives, 4, p. 65, 1984.

[4] Yamada, Y., Plasticity Viscoelasticity, (in Japanese), Baifukan Co., p. 75, 1980.

[5] Desai, C. S. and Abel, J. F., "Introduction to the finite element method", Murakoshi Science & Engineering Publishing Co., (translated in Japanese), p. 95, 1974.

[6] Bathe, K. J. and Peterson, F. E., "Theoretical basis for CEL/NONSAP, a non-linear structural analysis program", CR 75.005 (AD/A004989), p. 2-A.1, 1975.

[7] Amijima, S., Fujii, T., et al, "An extension of the simple stress analysis method to the initial stress problems by shrinkage of the adhesive for bonded joints", J. Adhesion Society of Japan, 21, 8, p. 319, 1985.

[8] Gradin, P. A. and Groth, H. L., "Fracture criterion for adhesive joints in term of material induced singularities", Int. Conf. on Numerical Methods in Fracture Mechanics, p. 711, 1984.

[9] Amijima, S., Fujii, T., et al, "A study on adhesive bonded joints and structures under bending load (2nd report: effect of lap length and taper angle on the strength of single lap joints under bending load)", J. Adhesion Society of Japan, 19, 7, p. 265, 1983.

STRESS AND STRENGTH ANALYSIS OF SINGLE-LAP ADHESIVELY BONDED JOINTS

C.-C. Ko

ARL, P.O. Box 8676-1
Taichung, Taiwan 40722, Republic of China

C.-L. Ong

MRDC, CSIST, P.O. Box 8671-1-10
Taichung, Taiwan, Republic of China

and

J. Q. Tarn

National Cheng Kung University
Tainan, Taiwan, Republic of China

ABSTRACT

The adhesively bonded joining of structural components is extremely desirable, because both bolting and riveting will result in the cutting of fibers and the introduction of stress concentrations. In this study, the stress analysis of single lap adhesively bonded joints has been carried out. For this analysis, the adherends are considered as laminate plate whereas the adhesive layer is treated as an isotropic material. In addition, the range of the analysis covered is in the elastic range of the bonded joints.

In order to determine the bonding strength, the Griffith energy balance concept is used to predict the failure of the composite single-lap joints.

INTRODUCTION

Almost every engineering composite structure is constructed by the connection of component members. For the requirement of structural connections, the designer may choose the mechanical fasteners such as rivets and bolts or adhesive joining. A serious disadvantage of mechanical connections is that they do not uniformly distribute the load and almost always result in local stress concentrations. In addition, the drilling of composite laminates causes delamination frequently. These problems can often be significantly reduced by adhesively joining the members.

In this study, the stress analysis and critical shear fracture strength of a single-lap joint of composite adherends under axial, shear and bending loadings will be investigated. Since the anisotropic behavior of composite materials, the theoretical analysis is more complicated than that of isotropic materials on which the past stress analysis research [1-4] are based. Chen and Cheng [2] adopted elasticity concept assuming a stress field which satisfies the equilibrium equations and surface traction continuity. With the help of minimum complementary energy principle, the stress can be calculated for each point in the lap joint. However, the accuracy of this method depends on the assumed stress field. By Yuceoglu and Updike [3,4], as a boundary value problem, a system of simultaneous partial differential equations has been solved numerically for stress analysis [3,4].

For adhesive strength analysis, it is difficult to model mathematically because there are many kinds of adhesives of different properties and bond joint fabrication procedures [5-7].

In this paper, the theory of stress analysis of adhesively bonded structure which consists of two different orthotropic adherends [8] will be extended and applied to laminated adherends. Then using the Griffith energy balance theory, the relationships among critical shear strength, bonding length and the adherend thickness under axial and bending loadings can be obtained. Furthermore, the results of different strength analysis models will be collected and compared for the basis of bond joint designs.

THEORETICAL ANALYSIS

A. Basic Assumptions

1. As shown in Figure 1, the width of y direction is assumed to be infinite

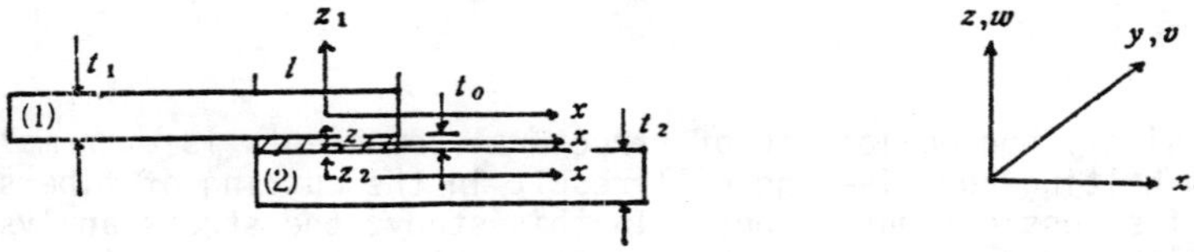

where: t_1: thickness of laminate 1

t_2: thickness of laminate 2

t_0: thickness of adhesive layer

2ℓ: bond length

Fig. (1) - The configuration of a single-lap joint

since the thickness of laminates is far less than that of length and width. Therefore, under axial, shear and bending loads in X-direction, the strain and stress fields are independent of y direction. Consequently, we can regard the problem as a plane strain, i.e., $v = 0$ for the entire structure.

2. The adhesive layer is an isotropic material.

3. The Kirchhoff assumptions are adopted since the bonded composite laminates are thin plates.

4. The variation of stress in thickness direction of the adhesive layer is neglected.

5. The normal stress σ_x of the adhesive layer is neglected.

From the assumptions, the stress state in the adhesive layer may be expressed as:

$$\sigma_x = \sigma(x); \; \tau_{xz} = \tau(x); \; \sigma_x = 0$$

The displacements and strains of adherends may be expressed by local coordinate systems 1 and 2:

$$u_i(x,z_i) = u_{io}(x) - z_i \frac{dw_i(x)}{dx} \tag{1a}$$

$$w_i = w_i(x) \tag{1b}$$

$$\varepsilon_{ix}(x,z_i) = \frac{du_{io}(x)}{dx} - z_i \frac{d^2w_i(x)}{dx^2} \tag{2a}$$

$$\varepsilon_{iy} = \varepsilon_{izi} = \gamma_{ixy} = \gamma_{iyzi} = \gamma_{ixzi} = 0 \tag{2b}$$

where i = 1,2; (x,y,z_i) represent the local coordinate system of laminates.

B. Stress Analysis

Considering a small element of unit width, the equilibrium between the laminate and the adhesive layer of a single-lap joint, Figure 2, gives the following equations.

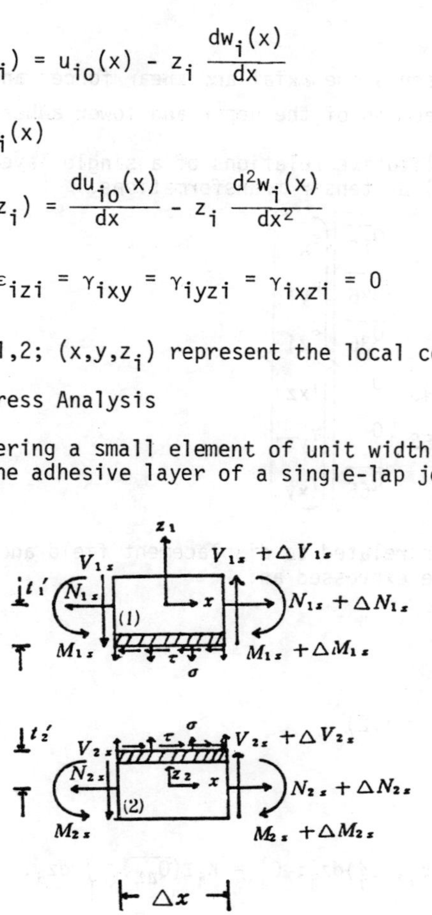

(b)

$$: \; t_1' = \frac{t_1 + t_o}{2}$$

$$t_2' = \frac{t_2 + t_o}{2}$$

$$z_1 = z - t_1'$$

$$z_2 = z + t_2'$$

Fig. (2) - The equilibrium of forces in a single-lap joint

$$\frac{dN_{1x}}{dx} = \tau(x) \tag{3}$$

$$\frac{dN_{2x}}{dx} = -\tau(x) \tag{4}$$

$$\frac{dV_{1x}}{dx} = \sigma(x) \tag{5}$$

$$\frac{dV_{2x}}{dx} = -\sigma(x) \tag{6}$$

$$\frac{dM_{1x}}{dx} = V_{1x}(x) - t_1' \cdot \tau(x) \tag{7}$$

$$\frac{dM_{2x}}{dx} = V_{2x}(x) - t_2' \cdot \tau(x) \tag{8}$$

where N_{ix}, V_{ix}, M_{ix} $(i = 1,2)$ represents the axial and shear forces and bending moments respectively at any cross-section of the upper and lower adherends.

The off-axis stress-strain constitutive relations of a single layer of a laminate can be obtained as equation (9) by tensor transformations.

$$\begin{Bmatrix} \sigma_x \\ \sigma_y \\ \sigma_{zi} \\ \tau_{xzi} \\ \tau_{yzi} \\ \tau_{xy} \end{Bmatrix} = \begin{bmatrix} \overline{Q_{11}} & \overline{Q_{12}} & \overline{Q_{13}} & 0 & 0 & \overline{Q_{16}} \\ \overline{Q_{12}} & \overline{Q_{22}} & \overline{Q_{23}} & 0 & 0 & \overline{Q_{26}} \\ \overline{Q_{13}} & \overline{Q_{23}} & \overline{Q_{33}} & 0 & 0 & \overline{Q_{36}} \\ 0 & 0 & 0 & \overline{Q_{44}} & \overline{Q_{45}} & 0 \\ 0 & 0 & 0 & \overline{Q_{45}} & \overline{Q_{55}} & 0 \\ \overline{Q_{16}} & \overline{Q_{26}} & \overline{Q_{36}} & 0 & 0 & \overline{Q_{66}} \end{bmatrix} \begin{Bmatrix} \varepsilon_x \\ \varepsilon_y \\ \varepsilon_{zi} \\ \gamma_{xzi} \\ \gamma_{yzi} \\ \gamma_{xy} \end{Bmatrix} \tag{9}$$

The stress and moment resultants related to displacement field and cross-sectional rotation angle, β_{ix} may be expressed as:

$$N_{ix} = A_i \frac{du_{io}}{dx} - B_i \frac{d^2 w_i}{dx^2} \tag{10}$$

$$M_{ix} = B_i \frac{du_{io}}{dx} - D_i \frac{d^2 w_i}{dx^2} \qquad (i = 1,2) \tag{11}$$

$$V_{ix} = C_i \cdot (\beta_{ix} + \frac{dw_i}{dx}) \tag{12}$$

where $(A_i, B_i, D_i) = \Sigma(\overline{Q_{11}})_k \int_k (1, z_i, z_i^2) dz_i$; $C_i = K_i \Sigma(\overline{Q_{44}})_k \int_k dz_i$.

A_i, B_i, D_i and C_i are the extensional, coupling, bending and transverse shear stiffness, respectively. K_i is shear coefficient of each lamina, and it equals 2/3 for rectangular cross-sections.

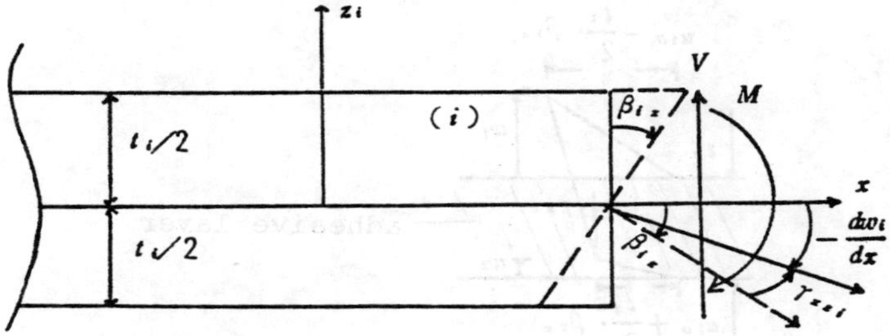

Fig. (3) - The relationships of cross-sectional shear strain rotation angle and slope

If the laminate is a symmetrical lay-up, then B_i equals zero and equations (10) and (11) become

$$N_{ix} = A_i \frac{du_{io}}{dx} \tag{13}$$

$$M_{ix} = -D_i \frac{d^2w_i}{dx^2} \tag{14}$$

Now, neglect the transverse shear strain, γ_{xzi}, then

$$\beta_{ix} = -\frac{dw_i}{dx} \tag{15}$$

Thus, by substituting equations (15) into (14), we get equation (16)

$$\frac{d\beta_{ix}}{dx} = \frac{M_{ix}}{D_i} \tag{16}$$

Now, consider the compatibility relationship between the adhesive layer and the adherend (as shown in Figure 4), the strain field of the adhesive layer can be expressed as follows:

$$\varepsilon_x = (w_1-w_2)/t_o \tag{17a}$$

$$\gamma_{xz} = [u_{10} - \frac{t_1}{2}\cdot\beta_{1x} - (u_{20} + \frac{t_2}{2}\cdot\beta_{2x})]/t_o \tag{17b}$$

$$\varepsilon_x = \frac{1}{2}[\frac{du_{10}}{dx} - \frac{t_1}{2}\cdot\frac{d\beta_{1x}}{dx} + (\frac{du_{20}}{dx} + \frac{t_2}{2}\cdot\frac{d\beta_{2x}}{dx})] \tag{17c}$$

By the assumption of the bonded structure being in a state of plane strain,

$$\varepsilon_y = \gamma_{xy} = \gamma_{yz} = 0 \tag{17d}$$

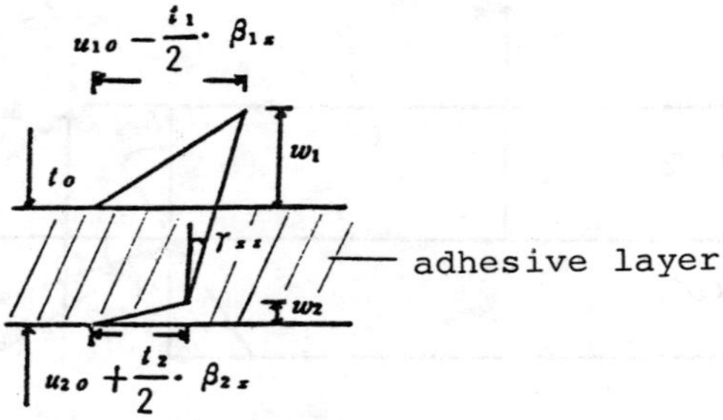

Fig. (4) - The displacements of adhesive layer and adherends

Since $\varepsilon_y = 0$, the stress-strain relation of the adhesive layer is:

$$\sigma = \frac{E\cdot\nu}{(1+\nu)(1-2\nu)}\,\varepsilon_x + \frac{E(1-\nu)}{(1+\nu)(1-2\nu)}\,\varepsilon_x \tag{18}$$

$$\tau = G\cdot\gamma_{xz} \tag{19}$$

where E, G, ν are the Young's modulus, shear modulus and Poisson ratio of the adhesive layer.

Substituting equations (17a), (17c), (13) and (16) into equation (18), equation (20) is derived. Through simple eliminations, equations (3) - (8), (12) - (16) and (20), (21) may be reduced to the following system of differential equations for the adhesive layer stress σ and τ.

$$\sigma = \frac{E\cdot\nu}{2(1+\nu)(1-\nu)}\left(\frac{N_{1x}}{A_1} + \frac{N_{2x}}{A_2} - \frac{t_1}{2D_1}M_{1x} + \frac{t_2}{2D_2}M_{2x}\right) + \frac{E(1-\nu)}{(1+\nu)(1-2\nu)} \times$$
$$\times\,\frac{(w_1-w_2)}{t_o} \tag{20}$$

Combining equations (17b) and (19), we obtain:

$$\tau = \frac{G}{t_o}\left[u_{10} - \frac{t_1}{2}\beta_{1x} - u_{20} - \frac{t_2}{2}\beta_{2x}\right] \tag{21}$$

so that

$$\frac{d^3\tau}{dx^3} + \alpha_1\frac{d\tau}{dx} + \alpha_2\sigma = 0 \tag{22}$$

$$\frac{d^4\sigma}{dx^4} + \beta_1\frac{d^2\sigma}{dx^2} + \beta_2\sigma + \beta_3\frac{d^3\tau}{dx^3} + \beta_4\frac{d\tau}{dx} = 0 \tag{23}$$

where

$$\alpha_1 = \frac{-G}{t_o}\left(\frac{1}{A_1} + \frac{1}{A_2} + \frac{t_1 t_1'}{2D_1} + \frac{t_2 t_2'}{2D_2}\right)$$

$$\alpha_2 = \frac{G}{t_o}\left(\frac{t_1}{2D_1} - \frac{t_2}{2D_2}\right)$$

$$\beta_1 = \frac{E(1-\nu)}{(1+\nu)(1-\nu)} \cdot \left[\frac{\nu}{2(1-\nu)} \cdot \left(\frac{t_1}{2D_1} + \frac{t_2}{2D_2}\right) - \frac{1}{t_o} \cdot \left(\frac{1}{c_1} + \frac{1}{c_2}\right)\right]$$

$$\beta_2 = \frac{E(1-\nu)}{(1+\nu)(1-2\nu)} \cdot \left(\frac{1}{D_1} + \frac{1}{D_2}\right)$$

$$\beta_3 = \frac{-E\nu}{2(1+\nu)(1-2\nu)}\left[\frac{1}{A_1} - \frac{1}{A_2} + \frac{t_1 t_1'}{2D_1} - \frac{t_2 t_2'}{2D_2}\right]$$

$$\beta_4 = \frac{-E(1-\nu)}{t_o(1+\nu)(1-2\nu)} \cdot \left[\frac{t_1'}{D_1} - \frac{t_2'}{D_2}\right]$$

Eliminating σ from equations (22) and (23), we can obtain a seventh degree constant coefficients differential equation of τ.

$$\frac{d^7\tau}{dx^7} + a_5\frac{d^5\tau}{dx^5} + a_3\frac{d^3\tau}{dx^3} + a_1\frac{d\tau}{dx} = 0 \tag{24}$$

where

$$a_5 = \alpha_1 + \beta_1$$

$$a_3 = \alpha_1\beta_1 + \beta_2 - \alpha_2\beta_3$$

$$a_1 = \alpha_1\beta_2 - \alpha_2\beta_4$$

C. Solution of Differential Equation

Let $\tau = e^{mx}$ for equation (24), the characteristic equation of the problem is:

$$m[(m^2)^3 + a_5(m^2)^2 + a_3(m^2) + a_1] = 0$$

The solution of the above equation is:

$$m = 0,\ \pm\phi_1,\ \pm\phi_2 \text{ and } \pm\phi_3$$

Therefore, the solution of the differential equation (24) can be obtained.

$$\tau(x) = k_o + k_1\sinh\phi_1 x + k_2\cosh\phi_1 x + k_3\sinh\phi_2 x + k_4\cosh\phi_2 x + k_5\sinh\phi_3 x + k_6\cosh\phi_3 x \tag{25}$$

Combining equations (22) and (25):

$$\sigma(x) = -\frac{1}{\alpha^2}[(\alpha_1\phi_1 + \phi_1^3)(k_1\cosh\phi_1 x + k_2\sinh\phi_1 x) + (\alpha_1\phi_2 + \phi_2^3) \times$$

$$\times (k_3\cosh\phi_2 x + k_4\sinh\phi_2 x) + (\alpha_1\phi_3 + \phi_3^3)(k_5\cosh\phi_3 x + k_6\sinh\phi_3 x) \quad (26)$$

D. Boundary Conditions

The boundary conditions as shown in Figure 5 are:

$$N_{1x}(\ell) = M_{1x}(\ell) = V_{1x}(\ell) = 0$$
$$N_{2x}(-\ell) = M_{2x}(-\ell) = V_{2x}(-\ell) = 0 \quad (27)$$
$$N_{2x}(\ell) = N;\ M_{2x}(\ell) = M;\ V_{2x}(\ell) = V$$

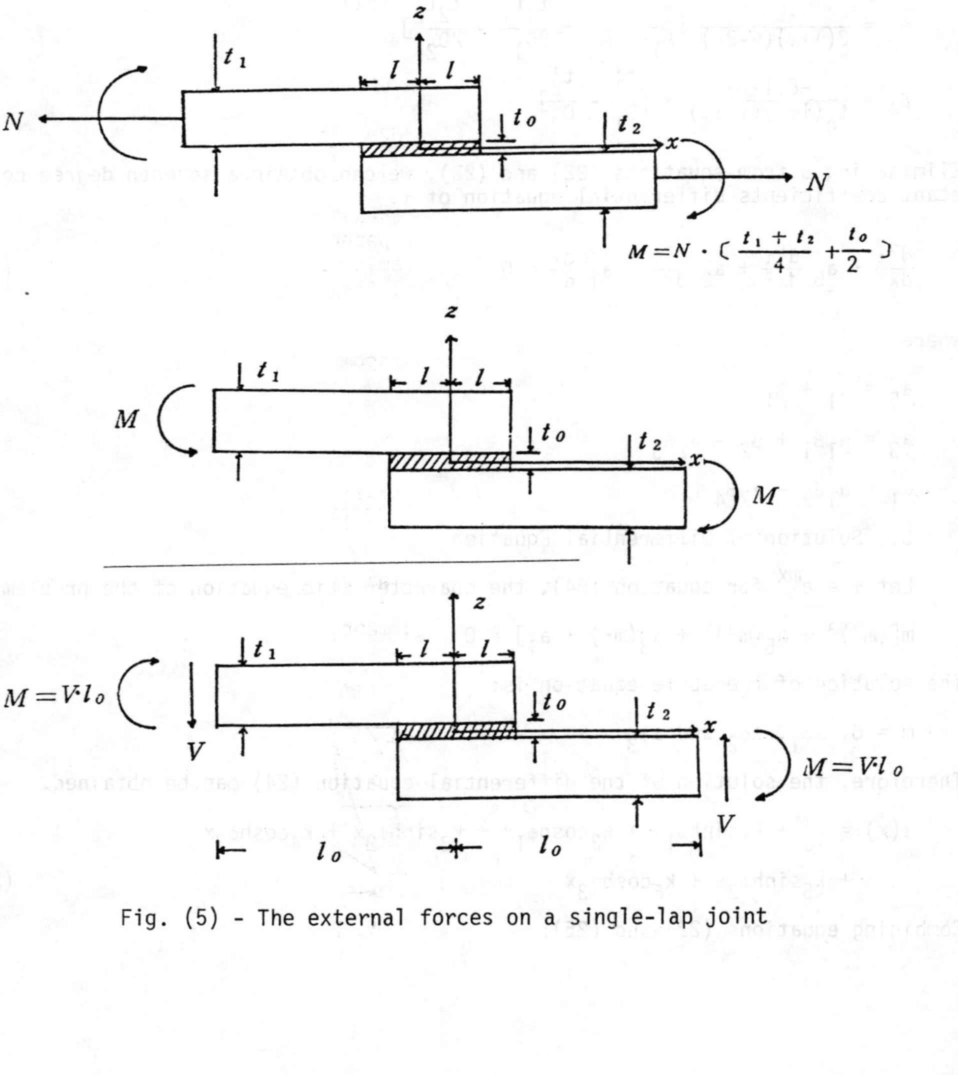

Fig. (5) - The external forces on a single-lap joint

With the application of the above boundary conditions, we can obtain the stress distribution of the adhesive layer. Note that the boundary conditions for adherend 1 at $x = -\ell$ need not be prescribed as they must satisfy the gross equilibrium conditions.

The above boundary conditions can be represented in terms of σ and τ.

$$\int_{-\ell}^{\ell} \tau dx = -N \tag{28}$$

$$\int_{-\ell}^{\ell} \sigma dx = -V \tag{29}$$

$$\int_{-\ell}^{\ell} \sigma x dx = M - \ell V - t_2' N \tag{30}$$

$$\frac{d\tau}{dx}\Big|_{x=\ell} = -\frac{G}{t_o} \cdot \left(\frac{N}{A_2} + \frac{t_2 \cdot M}{2D_2}\right) \tag{31}$$

$$\frac{d^2\tau}{dx^2}\Big|_{x=\ell} = \frac{G}{t_o} \left\{\left(\frac{1}{A_1} + \frac{1}{A_2} + \frac{t_1 t_1'}{2D_1} + \frac{t_2 t_2'}{2D_2}\right)\tau\Big|_{x=\ell} - \frac{t_2 V}{2D_2}\right\} \tag{32}$$

$$\left[\frac{1}{t_o} \cdot \left(\frac{1}{c_1} + \frac{1}{c_2}\right) - \frac{\nu}{4(1-\nu)} \cdot \left(\frac{t_1}{D_1} + \frac{t_2}{D_2}\right)\right]\sigma(\ell) - \frac{(1+\nu)(1-2\nu)}{E(1-\nu)} \frac{d^2\sigma}{dx^2}\Big|_{x=\ell}$$

$$+ \frac{\nu}{2(1-\nu)} \left(\frac{1}{A_1} - \frac{1}{A_2} + \frac{t_1 t_1'}{2D_1} - \frac{t_2 t_2'}{2D_2}\right) \frac{d\tau}{dx}\Big|_{x=\ell} = -\frac{M}{t_o D_2} \tag{33}$$

$$\left[\frac{1}{t_o} \left(\frac{1}{c_1} + \frac{1}{c_2}\right) - \frac{\nu}{4(1-\nu)} \cdot \left(\frac{t_1}{D_1} + \frac{t_2}{D_2}\right)\right] \frac{d\tau}{dx}\Big|_{x=\ell} - \frac{(1+\nu)(1-2\nu)}{E(1-\nu)} \frac{d^3\sigma}{dx^3}\Big|_{x=\ell}$$

$$+ \frac{1}{t_o} \left[\frac{t_1'}{D_1} - \frac{t_2'}{D_2}\right]\tau\Big|_{x=\ell} + \frac{V}{2(1-\nu)} \left(\frac{1}{A_1} - \frac{1}{A_2} + \frac{t_1 t_1'}{2D_1} - \frac{t_2 t_2'}{2D_2}\right) \frac{d^2\tau}{dx^2}\Big|_{x=\ell}$$

$$= -\frac{V}{t_o D_2} \tag{34}$$

The coefficients k_o-k_6 can be determined by substituting equations (25) and (26) into boundary conditions (28) - (34). Furthermore, by equations (3) - (8) and (25) - (27), we can get the stress and moment resultants N_{ix}, V_{ix}, M_{ix} of the adherends.

FAILURE ANALYSIS

The failure modes of a single-lap bond joint of composite laminates can be classified into three categories.

A. Adhesive Layer Failure

The following mechanisms of failure may occur:

1. Adhesive slipping failure caused by shear stress.

2. Adhesive peeling failure caused by normal stress.

B. Cohesive failure.

C. Laminated adherends failure, e.g., delamination.

The frequently happened failure mode considered here is adhesive layer shear failure under axial force and bending moment.

The stress analysis in the last sections shows that there are strong stress concentration at both ends. If τ reaches its maximum value at $x=\ell$ and hence we can assume that a crack initiates at $x=\ell$ of the adhesive layer. The predicted adhesive shear strength can be derived from Griffith energy balance theory which may be expressed as follows:

$$\frac{dW}{dc} + \frac{dU}{dc} - \frac{dL}{dc} = 0 \qquad (35)$$

where $W = 2\gamma c$, the fracture energy and γ is the surface energy per unit fractured area. U is the elastic strain energy between two ends; L is the external work.

According to the model assumed, the total elastic strain energy is composed of four major items, see Figure 6.

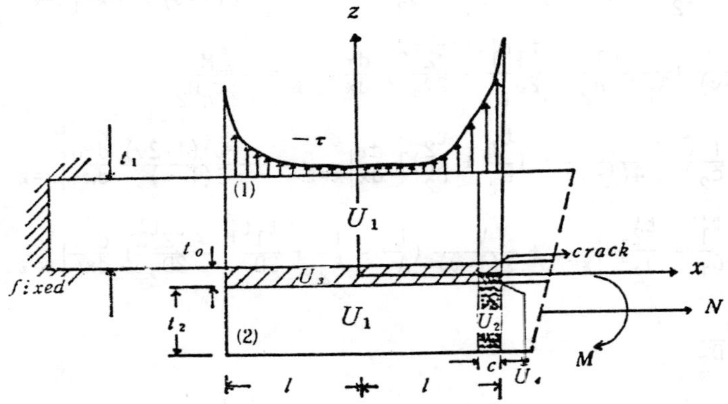

Fig. (6) - The relation of the distribution of shear stress and the elastic strain energy

$$U = U_1 + U_2 + U_3 + U_4$$

$$U_1 = \frac{1}{2}\left[\int_{-\ell}^{\ell-c}\left(\frac{N_{1x}^2}{A_1} + \frac{N_{2x}^2}{A_2}\right)dx\right] + \frac{1}{2}\int_{-\ell}^{\ell-c}\left(\frac{M_{1x}^2}{D_1} + \frac{M_{2x}^2}{D_2}\right)dx$$

$$U_2 = \frac{N^2}{2A_2} \cdot c + \frac{1}{2}\frac{M^2}{D_2} \cdot c$$

$$U_3 = \frac{1}{2}\int_{-\ell}^{\ell-c} \frac{\tau^2}{G} \cdot t_o dx + \frac{1}{2}\int_{-\ell}^{\ell-c} \frac{\sigma^2}{E} \cdot t_o dx \tag{36}$$

where U_1 is the elastic strain energy of the adherends between $x = -\ell$ and $x = \ell - c$; U_2 is the elastic strain energy for the lower adherend between $x = \ell - c$ and $x = \ell$; U_3 is the elastic strain energy for the adhesive layer between $x = -\ell$ and $x = \ell - c$; U_4 is the elastic strain energy for the adhesive layer between $x = \ell - c$ and $x = \ell$. It is negligible since its thickness is far less than those of adherends.

$$L_1 = N[\int_{-\ell}^{\ell-c} \frac{N_{2x}}{A_2} dx + \frac{N}{A_2} \cdot c]$$

$$L_2 = M[\int_{-\ell}^{\ell-c} \frac{M_{2x}}{D_2} dx + \frac{M}{D_2} \cdot c] \tag{37}$$

$$L_3 = N[-\tau \cdot t_o/G]\Big|_{x=-\ell}$$

where L_1 and L_2 are the work done by axial forces and bending moments. They are assumed constant between ℓ-c and ℓ since the crack length c is very small. L_3 is the work done in the adhesive layer by the shear stress.

By defining the critical failure to be the onset of the crack growth and replacing ℓ by ℓ-c in the expressions of N_{ix}, M_{ix}, σ and τ, equation (35) can be used to estimate the shear strength of a single-lap joint.

RESULTS AND DISCUSSION

A. Stress Analysis

The results of this analysis are well agreed with those of Reissner and Goland [1] and Kawata [6], see Figure 7. In addition, Figures 8, 9 and 10 show the variation of stresses (σ,τ) along the bond length of a single-lap joint under axial and shear forces and bending moment. These data show that when the bond length is large, the stresses in the center region of the adhesive layer, i.e., x=0, is very small, and they concentrate at two ends. Furthermore, Figures 11 and 12 display the stress and moment resultants in laminated adherends. The material constants are listed in Table 1.

B. Failure Analysis

1. The critical shear strength, N_{cr} obtained in this study is smaller than that of Kawata's shear model, Figures 13 and 14, since his analysis did not examine the tearing stress resulting from bending of the adherends (eccentricity).

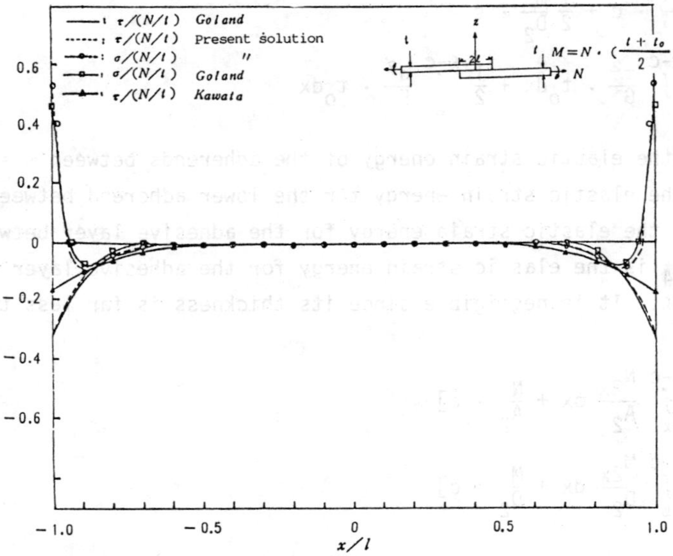

Fig. (7) - Comparisons of the normal and shear stresses in the adhesive layer for a single lap joint under tension obtained from present solutions, Kawata shear model solutions and Goland and Reissner solutions

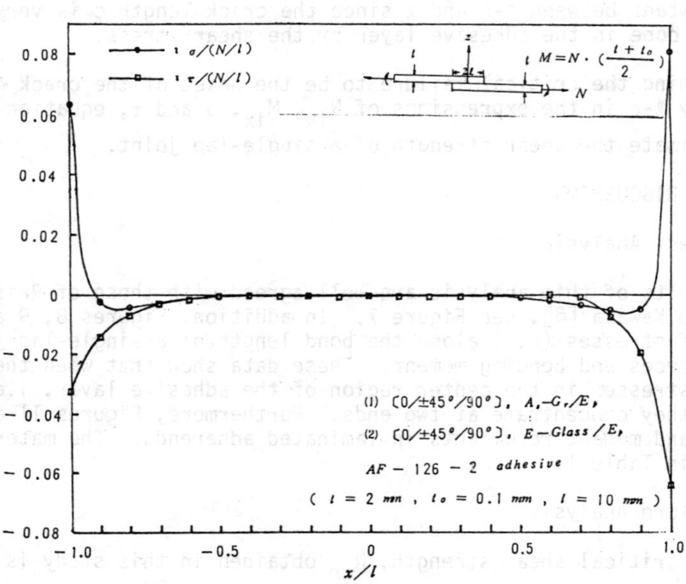

Fig. (8) - The normal and shear stresses in the adhesive layer for a single lap joint under tension

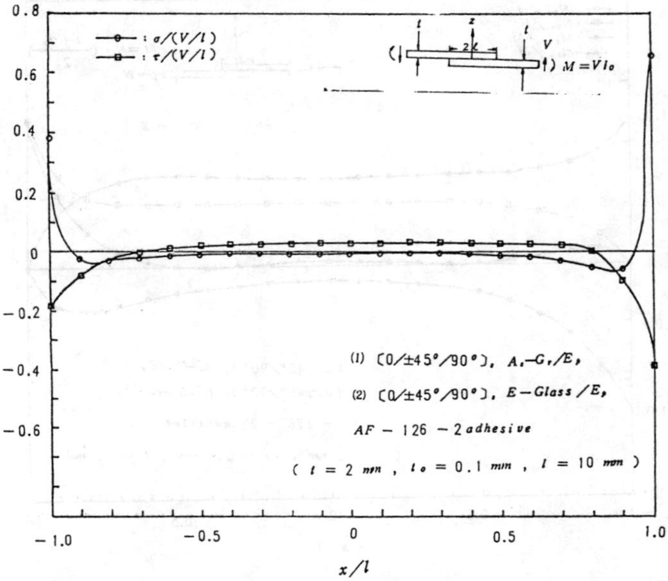

Fig. (9) - The normal and shear stresses in the adhesive layer for a single lap joint under shear

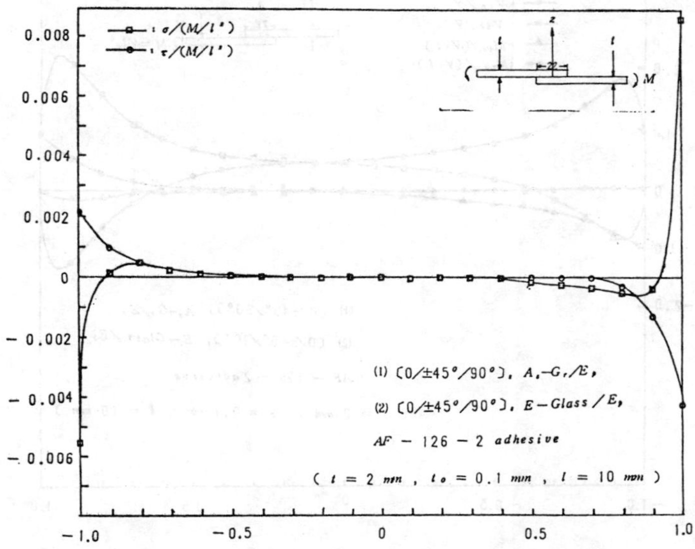

Fig. (10) - The normal and shear stresses in the adhesive layer for a single lap joint under bending

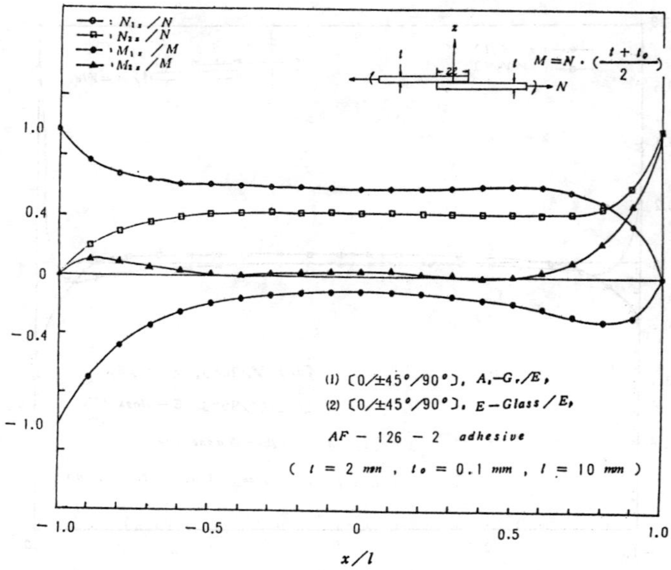

Fig. (11) - The stress and bending resultants in the laminated adherends for a single lap joint under tension

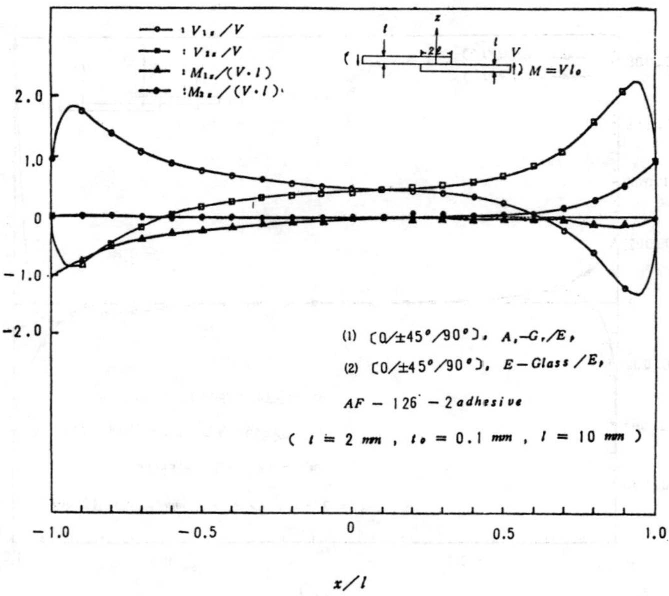

Fig. (12) - The stress and moment resultants in the adhesive layer for a single lap joint under shear

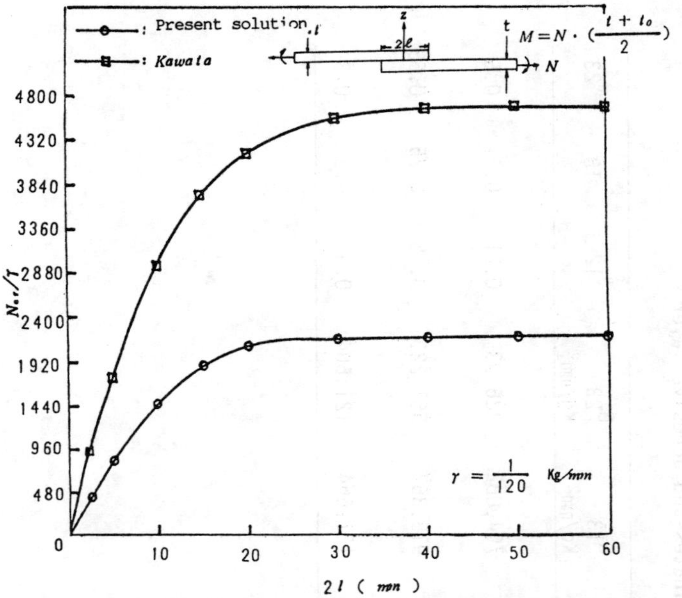

Fig. (13) - Nondimensional shear strength versus overlap length. (E_1 = 21000 Kg/mm^2, E_2 = 7000 Kg/mm^2, $\nu_1 = \nu_2$ = 0.3, E_a = 300 Kg/mm^2, t_o = 0.4 mm, ν_o = 0.3, t = 5 mm)

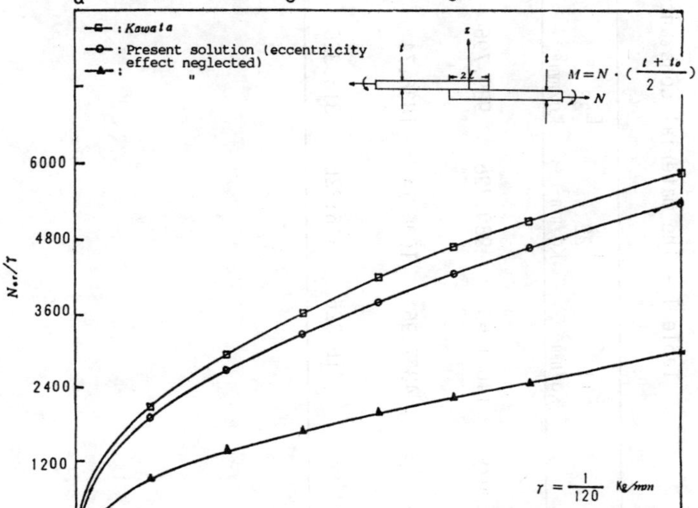

Fig. (14) - Nondimensional shear strength versus adherend thickness. (E_1 = 21000 Kg/mm^2, E_2 = 7000 Kg/mm^2, $\nu_1 = \nu_2$ = 0.3, E_a = 300 Kg/mm^2, t_o = 0.16 mm, ν_a = 0.3, 1 = 15 mm)

Table 1 - The material constants of laminates and adhesive layer

Material	E_{11} Kg/mm^2	E_{22} Kg/mm^2	E_{33} Kg/mm^2	G_{12} Kg/mm^2	G_{13} Kg/mm^2	G_{23} Kg/mm^2	ν_{12}	ν_{13}	ν_{23}
As-Graphite/Epoxy	14081.63	989.796	989.796	704.082	704.082	326.53	0.31	0.31	0.5
E-Glass/Epoxy	4897.96	1836.74	1836.74	918.367	918.367	561.225	0.25	0.25	0.63
AF-126-2 Adhesive	316.326	316.326	316.326	121.664	121.664	121.664	0.3	0.3	0.3

2. The superiority of applying Griffith's theory to predict bonding strength is the clear definition of failure phenomenon. With this phenomenon definition, the critical failure strength is equal to the external forces applied to the bonded structure at the moment of the crack starting to grow.

3. The optimum design of a single-lap bond joint is provided. Generally, the longer bond length provides the better bond strength. However, as found in this study, Figure 13, it is useless to have bond length longer than the "optimum bond length" which is very helpful for the economic bond joint design.

REFERENCES

[1] Goland, M. and Reissner, E., "The stress in cemented joints", J. Appl. Mech., 11, pp. A17-A22, 1944.

[2] Chen, D. and Cheng, S., "An analysis of adhesive-bonded single lap joints", J. Appl. Mech., 50, pp. 105-115, 1983.

[3] Yuceoglu, U. and Updike, D. P., "Bending and shear deformation effects in lap joints", J. of the Engineer Mechanics Division, ASCE, 107, (EM1), pp. 55-76, 1981.

[4] Yuceoglu, U. and Updike, D. P., "Stress analysis of bonded plates and joints", J. of the Engineer Mechanics Division, ASCE, 106, (EM1), pp. 37-56, 1980.

[5] Ikegami, K., Kajiyama, M., Kamiko, S. and Shiratori, E., "Experimental studies of the strength of an adhesively joint in a state of combined stress", J. Adhesion, 10, pp. 25-38, 1979.

[6] Kawata, K. and Takeda, N., "Shearing fracture of single lap joints", Trans. JSCM, 2, (1), pp. 26-29, 1976.

[7] Delale, F., Erdogan, F. and Aydinoglu, M. N., "Stress in adhesively bonded joints: a closed-form solution", J. Composite Materials, 15, pp. 249-271, 1981.

[8] Graves, S. R. and Adams, D. F., "Analysis of a bonded joint in composite tube subject to torsion", J. Composite Material, 15, pp. 211-225, 1981.

[9] Jones, R. M., Mechanics of Composite Material, Scripta Book Co., Washington, D.C., 1975.

[10] Lekhnitskii, S. G., Theory of Elasticity of an Anisotropic Elastic Body, Holden-Day, New York, 1963.

ANALYTICAL MODELLING OF COMPOSITE SANDWICH PANEL

C. H. Lee

Chung Shan Institute of Science and Technology
P.O. Box 8676-1-4, Taichung, Taiwan 400, Republic of China

ABSTRACT

Structural analysis of composite sandwich panel is complex because it entails a host of geometric and material parameters in addition to load transfer behavior. Computations may involve the application of the finite element scheme. Simplifications may also be gained by employing the effective flexural modulus concept [1] and modelling the sandwich panel as an assembly of solid plate elements without considering the effects of the core or rib. The stiffness of the core made of honeycomb or foam will have to be considered should wrinkling of the face skin occur. Similarly, the effect of the spar may also be included. Use can be made of commercialized finite element programs for calculating the effective modulus with sandwich panel skin made of graphite/epoxy, Kevlar/epoxy or metal subjected to prescribed boundary conditions. Quick access to the behavior of horizontal tail, control surface, and other components of airplanes can be made.

INTRODUCTION

Composites are being used more and more as load bearing structural members in military and commercial aircrafts. The Lear 2000 commercial airplane is a case in point where the fuselage and bulkheads are made of composites. Sandwich composite panels are also being used as rudder, horizontal tail, aileron, flap, etc. It is, therefore, pertinent to develop a methodology that can qualify the use of sandwich panel structure for a specified performance during the early stage of preliminary design. The proposed method will be applied to the SAP 6 and DESAP.

EFFECTIVE SOLID PLATE MODEL

The design of composite sandwich panel is involved as it requires a knowledge of several factors. They include the sandwich core material, cell size, facing material, bonding adhesive, setting and curing time, etc. Moreover, the performance may satisfy the service condition in addition to environments. Quality assurance, of course, comes into play.

End Couple. A possible approach is considering the effective flexural modulus E_f of the sandwich panel for each plate element in the finite element modelling and analysis. The behavior must simulate that observed in the actual vehicle in flight. For example, Figure 1(a) shows a sandwich panel in bending while an effective solid plate can be modelled as shown in Figure 1(b). By ig-

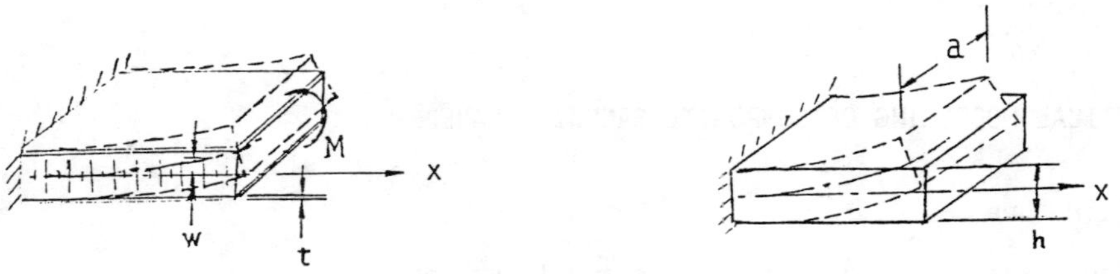

(a) sandwich panel (b) effective solid plate

Fig. (1) - Schematics of sandwich panel modelled by an effective solid plate

noring the core stiffness of the sandwich panel, the face stress σ_f can be approximated as

$$\sigma_f = \frac{E}{E_f}\frac{6M}{ah^2} \tag{1}$$

where t<<h. The corresponding face strain ε_f is

$$\varepsilon_f = \frac{\partial^2 w}{\partial x^2}\left(\frac{h}{2}\right) = \frac{6M}{E_f a h^2} \tag{2}$$

The bending moment is M while E and E_f are, respectively, the facial skin and effective flexural modulus. They are related as

$$\frac{E}{E_f} = \frac{h}{6t} = \frac{I_s}{I_h} \tag{3}$$

In equation (3), $I_s = ah^3/12$ is the moment of inertia for the solid plate and $I_h = ath^2/2$ for the sandwich panel.

Illustrative Example. As an illustrative example, the uniformly loaded cantilever plate in Figure 2 will be considered. The face skin is made of 7075 aluminum with a Young's modulus of 10.5 x 10^6 psi and Poisson's ratio of $\nu = 0.33$. By using Code COP2 [2], E_f for an isotropic or orthotropic sandwich panel can be calculated. For a commercially available SAP 6 panel, E_f refers to E_{fx}, E_{fy} and G_{fxy} which are the effective flexural moduli in x- and y-direction and the effective shear modulus. Shear and extensional effects in the thickness direction can be neglected if the plate is sufficiently thin.

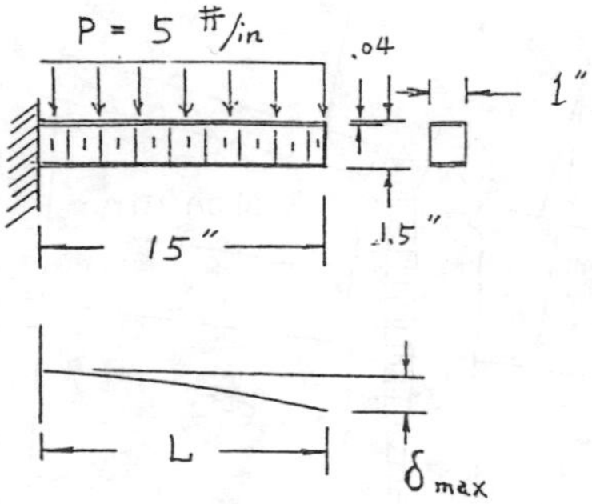

Fig. (2) - Uniformly loaded cantilever plate

Based on the equations in [2], it is found that

$$M = 0.5\ PL^2 = 562.5 \text{ in-lb} \tag{4}$$

This led to

$$\sigma_{max} = \frac{562.5}{t_f h} = 9{,}632 \text{ psi}, \ \delta_{max} = 0.71 \text{ in} \tag{5}$$

For the SAP 6 panel in Figure 2 with 16 composite plate elements*

$$\sigma_{max} = 1{,}486 \text{ psi}, \ \delta_{max} = 0.073 \text{ in} \tag{6}$$

The true skin stress σ_y is

$$\sigma_y = \sigma_{max} \cdot \frac{E}{E_{fx}} = 1{,}486 \left(\frac{10.56 \times 10^6}{1.59 \times 10^6}\right) = 9{,}813 \text{ psi} \tag{7}$$

This is nearly the same as that in equation (5). The errors for σ_{max} and δ_{max} are about 3%.

HORIZONTAL TAIL CONSTRUCTION

A schematic of the horizontal tail structure is shown in Figure 3. The main component consists of a metal honeycomb panel made of aluminum (1/8-5052-.001 x 4.5) that is movable to support the pitching moment. The face skin, spars and

*For the 1-5 in thick solid plate, COP2 [2] gives

$$\sigma_y = E_{fx}\varepsilon_f = E_{fx}\,\frac{h}{2} \cdot \frac{\partial^2 w}{\partial y^2}$$

where $E_{fx} = 1.59 \times 10^6$ psi.

root/tip rib are all made of 7075 aluminum alloy. Reinforced fiberglass is used at the tip fairing and a steel alloy at the pivot. The total weight is approxi-

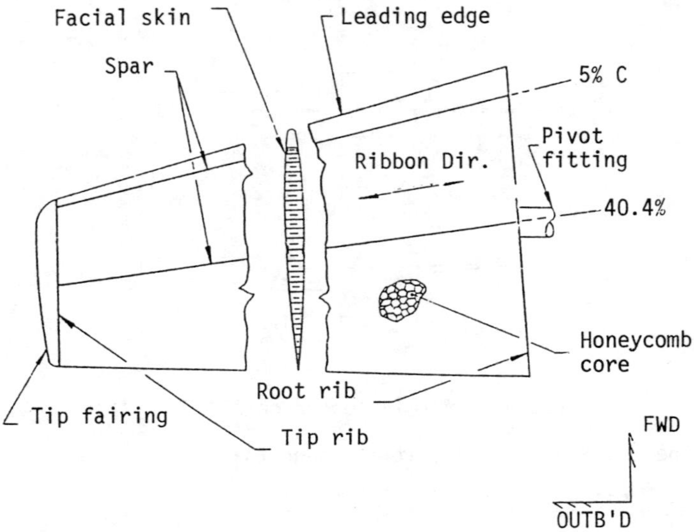

Fig. (3) - Schematic of horizontal tail structure

mately 82 lbs and the deformation at the tip is 4-5 in.

Weight reduction and strength improvement can be made by using Fiberite 1076/976 graphite/epoxy for the cover skin whose properties are listed in Table 1. This composite can operate at a temperature of 160°F and higher and possesses

Table 1 - Fiberite 1076E/976 resin with T300(6K) carbon fiber unit: psi, thickness t = .0055 in

B Value		Room Temperature	70°C Temperature 90% Humidity
Modulus	E_T	16.969×10^6	16.678×10^6
	E_T	1.037×10^6	0.725×10^6
	G_{LT}	0.58×10^6	0.435×10^6
	V_{LT}	0.3	0.3
		Density P = 0.057 16/in³	
Strength	X_t	188,000	144,000
	X_c	-174,000	-140.000
	Y_t	10,400	4.000
	Y_c		- 14.500
	S	11,600	7,250

high fatigue resistance. Reference can be made to [3] (Figures 7.18 and 7.19 on p. 242). Moreover, other hybrids can also be considered. Fiberglass and Nomex honeycomb core may be used. The HFT-1/8-4.0 fiber-glass material made by Hexcel in the USA is recommended. Its properties are given in Table 2. Ser-

Table 2 - The mechanical properties of HFT-1/8-4.0 honeycomb core [2]

Material-Cell-Density	Compressive Stabilized Strength (psi)		Compressive Stabilized Modulus (ksi)	Plate Shear "L" Direction Strength (psi)		Plate Shear "L" Direction Modulus (ksi)	"W" Direction Strength (psi)		"W" Direction Modulus (ksi)
	Typ	Min	Typ	Typ	Min	Typ	Typ	Min	Typ
HFT-1/8-4.0	560	420	46	315	240	37	150	120	12

vice temperature can be as high as 325°F. Refer to the horizontal tail structure in Figure 4 that utilized advanced composite materials.

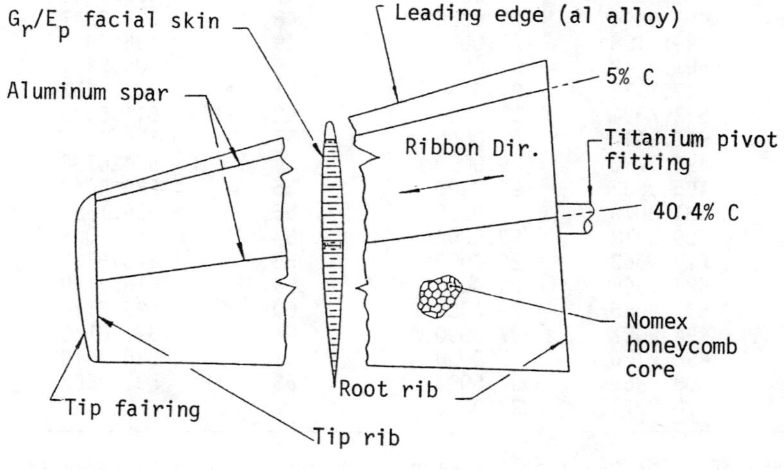

Fig. (4) - Horizontal tail using advanced components

DESIGN AND ANALYSIS

Referring to Figure 4, the horizontal tail section is discretized into 48 elements with 63 nodal points. The x and y coordinates of the nodal points are given in Table 3 and the thicknesses of the elements are listed in Table 4. The static critical pressure in psi on each element is shown in Figure 5. Ignored are the loads transferred from the tail to the main structure.

Preliminary Consideration. In order to ensure that the core will sustain the transverse shear load* which is τ_{aw} = 120 psi for the HFT-1/8-4.0 material with

*The skin of the sandwich panel is assumed not to support any transverse shear.

Table 3 - Nodal point coordinate (in)

Nodal Point	x	y	Nodal Point	x	y
1	486.3075	9.8616	33	520.7330	37.0000
2	490.3229	9.4393	34	527.9713	37.0000
3	497.0225	8.7348	35	537.5731	37.0000
4	505.4666	7.8467	36	497.7611	46.0000
5	517.6979	6.5604	37	500.8744	46.0000
6	527.4425	5.5356	38	505.8221	46.0000
7	538.9432	4.3261	39	512.4562	46.0000
8	487.6191	14.0000	40	521.6303	46.0000
9	491.6392	14.0000	41	528.1225	46.0000
10	498.2658	14.0000	42	537.1957	46.0000
11	506.5939	14.0000	43	501.5643	58.0000
12	518.4397	14.0000	44	504.3377	58.0000
13	527.5848	14.0000	45	508.6556	58.0000
14	538.5375	14.0000	46	514.6546	58.0000
15	489.8377	21.0000	47	522.8268	58.0000
16	493.6594	21.0000	48	528.3241	58.0000
17	499.9188	21.0000	49	536.6925	58.0000
18	507.8762	21.0000	50	505.5260	70.5000
19	519.1376	21.0000	51	507.9452	70.5000
20	527.7024	21.0000	52	511.6073	70.5000
21	538.2440	21.0000	53	516.9446	70.5000
22	492.3732	29.0000	54	524.0732	70.5000
23	495.9682	29.0000	55	528.5342	70.5000
24	501.8078	29.0000	56	536.1683	70.5000
25	509.3418	29.0000	57	511.7063	90.0000
26	519.9353	29.0000	58	513.5730	90.0000
27	527.8368	29.0000	59	516.2119	90.0000
28	537.9085	29.0000	60	520.5170	90.0000
29	494.9087	37.0000	61	526.0175	90.0000
30	498.2770	37.0000	62	528.8619	90.0000
31	503.6969	37.0000	63	535.3507	90.0000
32	510.8074	37.0000			

a density of 4 lb/ft^3, a preliminary check will be made by application of the theory in [2]. Consider the cross-section HS-14 in Figure 6 with thicknesses $h_1, h_2, \ldots, h_6$ which give an average of h_{ave} = 2.5 in. The allowable shear is

$$V_a = 0.7\ \tau_{aw} L h_{ave} = 8{,}820 \text{ lbs} \tag{8}$$

in which L = 24 in. The shear is 7,035 lbs. This gives a safety factor of 1.25. Such a check is made for each of the cross-sections HS14, HS21, etc., as given in Table 5.

Bond strength in resisting transverse shear is τ_{ba} = 450 psi. This gives a shear load allowable

$$V_b = 450(42 \text{ in}) = 18{,}900 \text{ lbs} \tag{9}$$

for each unit depth. A factor of safety of 18,900/7,035 = 2.68 is thus obtained.

Table 4 - Element thickness

Element No.	Thickness (in)	Element No.	Thickness (in)
1	1.9	25	1.5
2	2.9	26	2.2
3	3.8	27	2.9
4	3.5	28	2.8
5	2.1	29	1.7
6	1.2	30	0.9
7	1.8	31	1.3
8	2.8	32	2.0
9	3.7	33	2.6
10	3.4	34	2.5
11	2.0	35	1.6
12	1.1	36	0.8
13	1.7	37	1.1
14	2.6	38	1.8
15	3.6	39	2.3
16	3.3	40	2.1
17	1.9	41	1.5
18	1.0	42	0.2
19	1.6	43	0.9
20	2.4	44	1.4
21	3.2	45	2.0
22	3.0	46	1.75
23	1.8	47	1.2
24	1.0	48	0.7

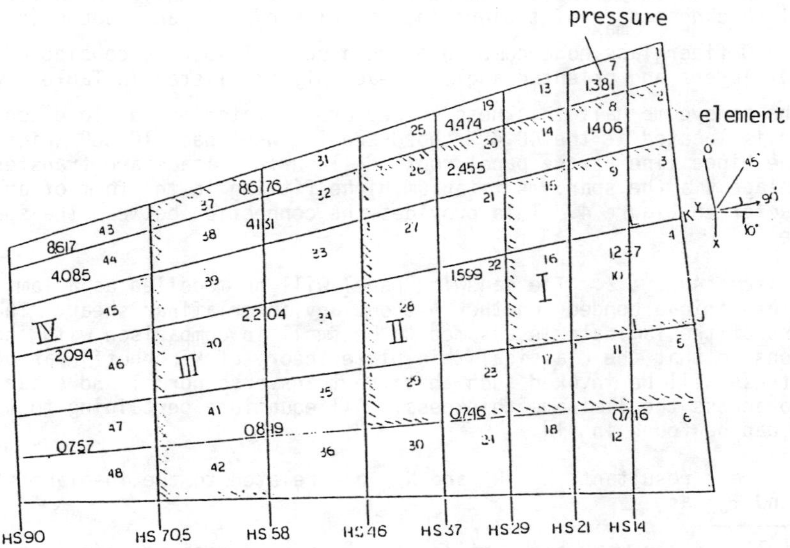

Fig. (5) - Pressure load distribution (psi)

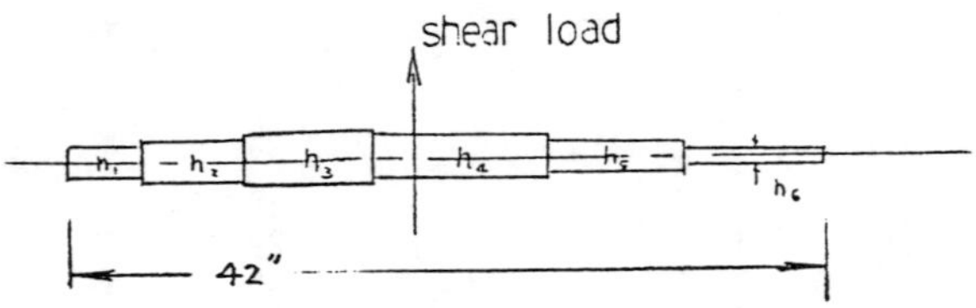

Fig. (6) - Cross section HS-14

Table 5 - Ply orientation angle θ_i

Area	Number of Plies	Orientation Angle θ_i
I	36	$[90_2/45_2/0_2/-45_2/90_2/-45_2/0_2/45_2/90_2/Z_c]_s$
II	28	$[90_2/45/0_2/-45/90_2/-45/0_2/45/90_2/Z_c]_s$
III	24	$[90_2/45/0/-45/90_2/-45/0/45/90_2/Z_c]_s$
IV	12	$[90_2/45/0/-45/90/Z_c]_s$

The above procedure serves as a preliminary check for selecting the core material. Skin thickness must thus be determined accordingly. As a first trial, the full thickness t_{max} of the original metal sandwich panel supported by the HFT-1/8-4.0 fiberglass honeycomb core and a central spar is considered. The number of layers and oriented angle of each ply are listed in Table 5 with Z_c being the honeycomb half thickness. They are dictated by the load conditions. The spar is located at the 40.40% chord and is swept back 10.38° which intersects the hinge line at the panel root. All surface loads are transferred to the fuselage via the spar. A titanium hinge fitting in the form of an "I" is used. Refer to Figure 4. This provides the connection between the spar and fuselage.

Numerical Analysis. The sandwich panel will be modelled as a laminate consisting of laminae bonded together without any interlaminar shear. The total thickness of the laminate is assumed to be small in comparison with the planar dimensions so that the classical thin-plate theory of Kirchhoff* can be used. Plane stress will be invoked such that the transverse normal and shear stresses are zero across the laminate thickness. All equations pertaining to the COP2 program can be found in [1].

The stress resultants N_x, N_y and N_{xy} are related to the in-plane stresses σ_x, σ_y and τ_{xy} as

*The results in the work are confined to the approximations made in the thin plate theory. This includes small deformation and the equivalent shear boundary condition.

$$N_x = t(E_x/E_x^f)\sigma_x$$

$$N_y = t(E_y/E_y^f)\sigma_y \qquad (10)$$

$$N_{xy} = t(G_{xy}/G_{xy}^f)\tau_{xy}$$

in which t is the skin thickness and E_x, E_y and G_{xy} are moduli of the skin. The effective moduli are attached with the superscript f. Refer to Table 6 for the values of N_x, N_y and N_{xy} in equation (10). They are used in the design analysis

Table 6 - The force/unit length on each element

Element No.	N_x	N_y	N_{xy}	Element No.	N_x	N_y	N_{xy}
1	130.472	30.674	79.428	25	344.112	40.632	- 44.381
2	424.403	- 1.909	-214.613	26	769.968	99.262	- 66.189
3	1909.023	- 49.485	-708.056	27	1004.432	169.802	- 72.479
4	1579.816	-298.551	579.102	28	1043.720	135.600	- 52.120
5	361.086	-167.324	177.580	29	575.770	170.792	- 12.241
6	121.852	- 33.151	64.485	30	375.945	101.933	- 7.686
7	382.288	47.179	- 99.224	31	336.486	22.432	- 91.961
8	1039.801	59.201	-173.136	32	528.582	134.606	-125.067
9	1858.282	189.147	-109.935	33	673.885	316.778	- 89.948
10	1643.981	55.847	112.872	34	629.750	207.878	16.883
11	855.299	- 42.294	166.067	35	407.205	108.022	54.456
12	411.444	- 0.760	116.067	36	321.546	3.437	46.455
13	577.987	69.406	-189.777	37	219.034	80.754	- 32.530
14	1260.336	325.245	-256.359	38	364.329	121.935	- 51.926
15	1582.913	677.770	-278.464	39	472.303	161.241	- 62.422
16	1582.395	419.754	446.217	40	430.885	147.889	- 50.888
17	1411.686	357.977	417.694	41	305.486	89.807	- 29.625
18	915.008	177.150	335.507	42	158.606	43.057	- 11.688
19	304.643	83.342	- 19.548	43	46.166	2.424	- 19.139
20	747.961	182.762	- 61.441	44	71.842	23.373	- 26.251
21	1098.446	278.750	- 81.889	45	101.804	23.167	- 33.258
22	1046.754	234.758	- 87.957	46	86.597	17.663	- 22.816
23	595.449	165.609	- 46.215	47	60.286	15.821	- 10.545
24	190.932	55.542	12.605	48	34.953	3.402	- 3.557

as shown in the flow chart given by Figure 7. The allowable stress F_{cw} corresponding to surface wrinkling of the sandwich panel is [4]

$$F_{cw} = 0.43 \left[\frac{E_c G_c (3E_T + E_S)}{4}\right]^{1/3} \tag{11}$$

where the core effective moduli are E_c and G_c while E_S and E_T are the secant and tangent modulus [3] of the face, respectively. Numerical values of the constants

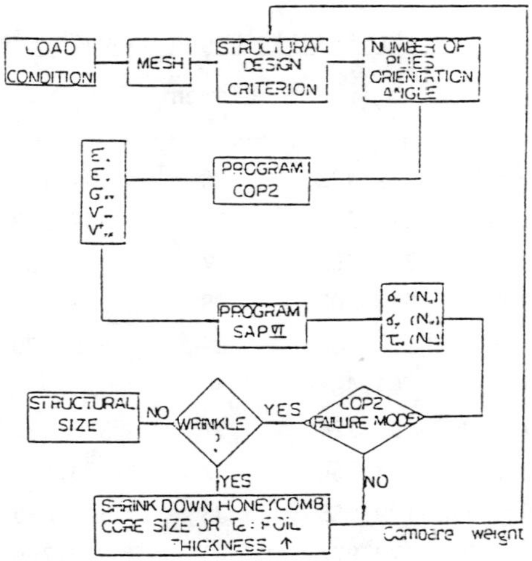

Fig. (7) - Flow chart

in equations (11) can be found in [4]. Face dimpling may also occur but it is not as important as wrinkling if the graphite/epoxy composite is used.

Final results for the deformed shape of the horizontal tail is displayed in Figure 8.

CONCLUSIONS

Analytical modelling of the horizontal tail structure is made by using advanced composites. Analyses are based on the COP2 and SAP 6 program. The titanium hinge fitting is used to guard against galvanic corrosion between the tail and fuselage had the fitting be made of steel. The original weight of 82.5 lbs for the horizontal tail structure is reduced to 68 lbs not counting for the titanium hinge. This is approximately a weight reduction of 20% when aluminum is replaced by composite.

Other materials for the skin and core may also be considered to meet structural performance by using COP2 and SAP 6. Other considerations such as dynamic loading and cost effectiveness must also be made.

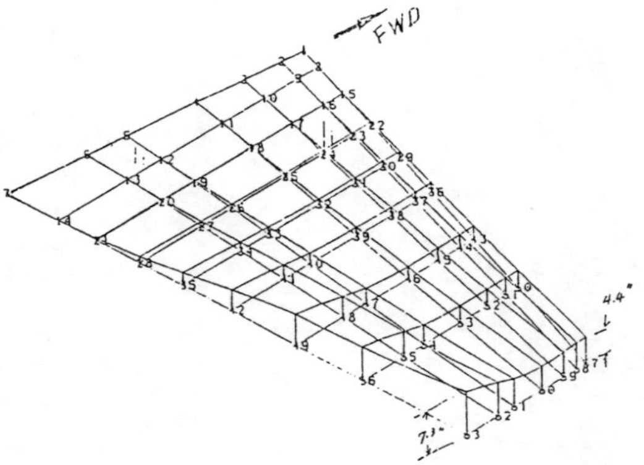

Fig. (8) - The deflection diagram of horizontal tail

REFERENCES

[1] Tsai, S. W. and Hahn, H. T., Introduction to Composite Materials, Technomic Publishing Co., Inc., 1980.

[2] Hexcel Corporation, The Basics on Bonded Sandwich Construction TSB 124", p. 12, 1984.

[3] Agarwal, B. D. and Broutman, L. J., "Analysis and performance of fiber composites", p. 242, 1980.

[4] Bruhn, E. F., Analysis and design of flight vehicle structures", C12.34, 1965.

LIST OF PARTICIPANTS

Professor G. C. Sih
Institute of Fracture and
Solid Mechanics
Lehigh University
Bethlehem, PA. 18015 USA

Mr. D. M. Chou
Institute of Fracture and
Solid Mechanics
Lehigh University
Bethlehem, PA. 18015 USA

Mr. C. K. Chao
Institute of Fracture and
Solid Mechanics
Lehigh University
Bethlehem, PA. 18015 USA

Mr. D. Y. Tzou
Institute of Fracture and
Solid Mechanics
Lehigh University
Bethlehem, PA. 18015 USA

Professor T.-W. Chou
Department of Mechanical
and Aerospace Engineering
University of Delaware
Newark, DE. 19716 USA

Professor H.-W. Liu
Department of Mechanical
and Aerospace Engineering
Syracuse University
Syracuse, NY. 13210 USA

Professor S.-C. Yen
Department of Civil Engineering
Southern Illinois University
Carbondale, IL. 62901 USA

Professor J. N. Yang
George Washington University
Washington, D.C. 20052 USA

Professor J. Pan
The University of Michigan
Ann Arbor, MI. 48109 USA

Professor Y. C. Wu
School of Aeronautics and
Astronautics
Purdue University
West Lafayette, IN. 47907 USA

Professor F. D. Ju
Department of Mechanical
Engineering
The University of New Mexico
Albuquerque, NM. 87106 USA

Professor W. Wang
Northeastern University
6 Jeremiah Circle
Medford, MA. 02155 USA

Professor T. Mura
Department of Civil Engineering
Northwestern University
Evanston, IL. 60201 USA

Professor A. S. Kahn
The University of Oklahoma
Norman, OK. 73019 USA

Dr. L.-C. Hsu
Macromolecular Material, Inc.
P.O. Box 16190
Rocky River, OH. 44116 USA

Dr. S. C. Chou
Army Materials Technology
Laboratory
Watertown, MA. 02172-0001 USA

Dr. T. G. Nieh
Lockheed Palo Alto Research Lab.
Palo Alto, CA. 94304 USA

Dr. P. Peng
Monsanto Company (C57H)
St. Louis, MO. 53167 USA

Professor J. T. Pindera
Institute for Experimental Mechanics
The University of Waterloo
Waterloo, Ontario N2L 3G1
Canada

Dr. R. Jones
Aeronautical Research Laboratory
P.O. Box 4331
Melbourne, Victoria 3001 Australia

Dr. J. J. Wu
US Army European Research Office
223 Old Marylebone Road
London NW1 5TH England

Professor Y. Fujitani
Hiroshima University
Shitami Saijo-cho
Higashihiroshima City 724, Japan

Professor A. Kobayashi
Faculty of Engineering
The University of Tokyo
Tokyo 153, Japan

Professor S. Amijima
Faculty of Engineering
Doshisha University
Kyoto, Japan

Professor T. Fujii
Engineering Center Division
Kobe Steel Ltd.
Kobe, Japan

Minister K. T. Li
Minister Without Portfolio, ROC
c/o 10th Fl., 116 Nanking E. Rd.
Sec. 2, Taipei, Taiwan 10409
ROC

Dr. S.-E. Hsu
The Chinese Society for Materials
Science
P.O. Box 1-26
Lungtan, Taiwan 32500
ROC

Dr. T.-Y. Wung
The Society of Theoretical and
Applied Mechanics, ROC
130 Keelung Rd, Sec. 3
Taipei, Taiwan 10771
ROC

Mr. C. R. Shih
China Shipbuilding Corporation
3 Chung Kang Road
Kaohsiung, Taiwan 81234
ROC

Mr. H.-J. Chen
China Shipbuilding Corporation
9F-1, 5 Chung Cheng 3rd Road
Kaohsiung, Taiwan 80027
ROC

Mr. S.-Y. Hwang
China Shipbuilding Corporation
224 Ho I Road
Keelung, Taiwan 20220
ROC

Mr. H.-C. Chen
Chinese Petroleum Corporation
83 Chung Hua Road
Taipei, Taiwan 10031
ROC

Mr. R.-C. Wang
Kaohsiung Refinery
Chinese Petroleum Corporation
Tsoying, Taiwan 81126
ROC

Mr. L.-C. Chen
Chinese Petroleum Corporation
54 Tung Ning Road
Chutung, Taiwan 31009
ROC

Professor S.-J. Shiao
National Chiao Tung University
Hsinchu, Taiwan 30050
ROC

Professor M.-H. Jen
National Sun Yat-Sen University
Kaohsiung, Taiwan 80424
ROC

Mr. S.-H. Yang
Taiwan Power Company
22nd Fl., 242 Roosevelt Road
Taipei, Taiwan 10763
ROC

Mr. F. Chu
Taiwan Power Company
18th Fl., 242 Roosevelt Road
Taipei, Taiwan 10763
ROC

Ms. K.-M. Wu
Taiwan Power Company
22nd Fl., 242 Roosevelt Road
Taipei, Taiwan 10763
ROC

Ms. B.-F. Lu
Taiwan Power Company
18th Fl., 242 Roosevelt Road
Taipei, Taiwan 10763
ROC

Ms. C. Chang
Taiwan Power Company
18th Fl., 242 Roosevelt Road
Taipei, Taiwan 10763
ROC

Mr. L.-F. Chien
Taiwan Power Company
22nd Fl., 242 Roosevelt Road
Taipei, Taiwan 10763
ROC

Mr. C.-T. Chiu
Taiwan Power Company
16th Fl., 242 Roosevelt Road
Taipei, Taiwan 10763
ROC

Mr. C.-I. Chyan
Taiwan Power Company
16th Fl., 242 Roosevelt Road
Taipei, Taiwan 10763
ROC

Mr. T.-C. Hsu
Taiwan Power Company
16th Fl., 242 Roosevelt Road
Taipei, Taiwan 10763
ROC

Mr. J.-R. Jen
Taiwan Power Company
198 Roosevelt Road
Taipei, Taiwan 10765
ROC

Ms. C.-J. Huang
Taiwan Power Company
22nd Fl., 242 Roosevelt Road
Taipei, Taiwan 10763
ROC

Professor C.-C. Chao
Institute of Aeronautic and Astronautic Engineering
National Cheng-Kung University
Tainan, Taiwan 70101
ROC

Professor T.-L. Shiau
Institute of Aeronautic and Astronautic Engineering
National Cheng-Kung University
Tainan, Taiwan 70101
ROC

Professor L.-J. Lee
Institute of Aeronautic and Astronautic Engineering
National Cheng-Kung University
Tainan, Taiwan 70101
ROC

Professor H.-P. Chen
Institute of Aeronautic and Astronautic Engineering
National Cheng-Kung University
Tainan, Taiwan 70101
ROC

Professor E. Chang
National Cheng-Kung University
Tainan, Taiwan 70101
ROC

Professor S.-C. Lee
National Cheng-Kung University
Tainan, Taiwan 70101
ROC

Professor S.-C. Yu
Department of Earth Science
National Cheng-Kung University
Tainan, Taiwan 70101
ROC

Professor L.-W. Chen
Department of Mechanical Engineering
National Cheng-Kung University
Tainan, Taiwan 70101
ROC

Mr. C. C. Lin
National Cheng-Kung University
Tainan, Taiwan 70101
ROC

Mr. L.-Y. Chen
National Cheng Kung University
Tainan, Taiwan 70101
ROC

Professor T. C. Huang
National Cheng-Kung University
Tainan, Taiwan 70101
ROC

Professor S.-Y. Lee
National Cheng Kung University
Tainan, Taiwan 70101
ROC

Professor H.-T. Lee
National Cheng Kung University
Tainan, Taiwan 70101
ROC

Mr. C.-M. Chang
National Cheng Kung University
Tainan, Taiwan 70101
ROC

Professor H.-C. Shih
National Tsing Hua University
Hsinchu, Taiwan 30043
ROC

Professor T. Hu
National Tsing Hua University
Hsinchu, Taiwan 30043
ROC

Professor S. Lee
National Tsing Hua University
Hsinchu, Taiwan 30043
ROC

Professor C.-C. M. Ma
National Tsing Hua University
Hsinchu, Taiwan 30043
ROC

Professor W.-K. Chin
National Tsing Hua University
Hsinchu, Taiwan 30043
ROC

Mr. C.-C. Wan
National Tsing Hua University
Hsinchu, Taiwan 30043
ROC

Mr. H.-C. Hsia
National Tsing Hua University
Hsinchu, Taiwan 30043
ROC

Mr. B.-Y. Shieh
National Tsing Hua University
Hsinchu, Taiwan 30043
ROC

Mr. J.-S. Hwang
National Tsing Hua University
Hsinchu, Taiwan 30043
ROC

Professor T. P. Pang
National Tsing Hua University
Hsinchu, Taiwan 30043
ROC

Professor C.-C. Chao
National Tsing Hua University
Hsinchu, Taiwan 30043
ROC

Professor C.-C. Lin
National Chung Hsing University
Taichung, Taiwan 40227
ROC

Mr. I.-L. Huang
National Chung Hsing University
Taichung, Taiwan 40227
ROC

Professor D.-W. Ding
Textile Engineering
Feng Chia University
Taichung, Taiwan 40724
ROC

Mr. C.-S. Chiou
Textile Engineering
Feng Chia University
Taichung, Taiwan 40724
ROC

Mr. T.-H. Ko
Textile Engineering
Feng Chia University
Taichung, Taiwan 40724
ROC

Professor C.-H. Koo
Materials Engineering
National Taiwan University
Taipei, Taiwan 10764
ROC

Professor H. J. Keh
Chemical Engineering
National Taiwan University
Taipei, Taiwan 10764
ROC

Professor Y.-H. Lee
Materials Engineering
National Taiwan University
Taipei, Taiwan 10764
ROC

Professor K.-H. Hsieh
Chemical Engineering
National Taiwan University
Taipei, Taiwan 10764
ROC

Professor L.-W. Chen
Chemical Engineering
National Taiwan University
Taipei, Taiwan 10764
ROC

Professor W.-Y. Chiu
Chemical Engineering
National Taiwan University
Taipei, Taiwan 10764
ROC

Mr. P.-H. Lin
Chemical Engineering
National Taiwan University
Taipei, Taiwan 10764
ROC

Professor K.-C. Wu
Applied Mechanics
National Taiwan University
Taipei, Taiwan 10764
ROC

Professor C.-S. Chou
Applied Mechanics
National Taiwan University
Taipei, Taiwan 10764
ROC

Professor M.-K. Kuo
Applied Mechanics
National Taiwan University
Taipei, Taiwan 10764
ROC

Professor Y.-H. Pao
Applied Mechanics
National Taiwan University
Taipei, Taiwan 10764
ROC

Professor J.-S. Chang
Applied Mechanics
National Taiwan University
Taipei, Taiwan 10764
ROC

Professor C.-O. Chang
Applied Mechanics
National Taiwan University
Taipei, Taiwan 10764
ROC

Mr. D.-C. Wu
Applied Mechanics
National Taiwan University
Taipei, Taiwan 10764
ROC

Professor H.-T. Chiu
Polymers and Textiles
National Taiwan Institute of Technology
Taipei, Taiwan 10772
ROC

Professor Y.-E. A. Wu
Polymers and Textiles
National Taiwan Institute of Technology
Taipei, Taiwan 10772
ROC

Professor T.-K. Wu
Union Chemical Laboratories
Industrial Technology Research Institute
1021 Kuan Fu Road
Hsinchu, Taiwan 30042
ROC

Professor S.-N. Tong
Union Chemical Laboratories
Industrial Technology Research Institute
1021 Kuan Fu Road
Hsinchu, Taiwan 30042
ROC

Mr. Y.-H. Wu
Industrial Technology Research
Institute
195 Chung Hsing Road, Sec. 4
Chutung, Taiwan 31015
ROC

Mr. R.-S. Liu
Industrial Technology Research
Institute
195 Chung Hsing Road, Sec. 4
Chutung, Taiwan 31015
ROC

Mr. R.-Y. Wu
Industrial Technology Research
Institute
195 Chung Hsing Road, Sec. 4
Chutung, Taiwan 31015
ROC

Mr. W.-L. Liu
Industrial Technology Research
Institute
195 Chung Hsing Road, Sec. 4
Chutung, Taiwan 31015
ROC

Mr. C.-L. Wu
Industrial Technology Research
Institute
195 Chung Hsing Road, Sec. 4
Chutung, Taiwan 31015
ROC

Mr. C.-H. Chen
Industrial Technology Research
Institute
195 Chung Hsing Road, Sec. 4
Chutung, Taiwan 31015
ROC

Mr. T.-Y. Ye
Industrial Technology Research
Institute
195 Chung Hsing Road, Sec. 4
Chutung, Taiwan 31015
ROC

Mr. K.-J. Chang
Industrial Technology Research
Institute
195 Chung Hsing Road, Sec. 4
Chutung, Taiwan 31015
ROC

Mr. J.-T. Hu
Industrial Technology Research
Institute
195 Chung Hsing Road, Sec. 4
Chutung, Taiwan 31015
ROC

Mr. T.-N. Lung
EMRO
Industrial Technology Research
Institute
195-28, Chung Hsing Road, Sec. 4
Chutung, Taiwan 31015
ROC

Mr. J. Leu
Kunnan Enterprise Ltd.
33 Hsiang Ho Rd., Lee Lin Vil
Taichung, Taiwan 42712
ROC

Mr. T. Chang
Kunnan Enterprise Ltd.
33 Hsiang Ho Rd., Lee Lin Vil
Taichung, Taiwan 42712
ROC

Mr. S.-C. Liao
Kunnan Enterprise Ltd.
33 Hsiang Ho Rd., Lee Lin Vil
Taichung, Taiwan 42712
ROC

Mr. P. Chou
Kunnan Enterprise Ltd.
33 Hsiang Ho Rd., Lee Lin Vil
Taichung, Taiwan 42712
ROC

Mr. Taylor Cheng
Kunnan Enterprise Ltd.
33 Hsiang Ho Rd., Lee Lin Vil
Taichung, Taiwan 42712
ROC

Mr. M.-F. Lin
China Steel Corporation
1 Chung Kang Road
Kaohsiung, Taiwan 81233
ROC

Mr. Y.-F. Chiu
China Steel Corporation
1 Chung Kang Road
Kaohsiung, Taiwan 81233
ROC

Mr. L.-S. Chung
China Steel Corporation
P.O. Box 47-29, Hsiaokang
Kaohsiung, Taiwan 81233
ROC

Professor I.-H. Yang
Mechanical Engineering
Chinese Military Academy
1 Lane 2 Huang Pu 7th Village
Fengshan, Taiwan 83005
ROC

Mr. C.-M. Chen
Fiberglass Reinforced Plastic Society of the Republic of China
12th Fl., 259 Nanking E. Rd., Sec. 3
Taipei, Taiwan 10566
ROC

Dr. Y.-D. Yao
Institute of Physics
Academia Sinica
128 Yen Chiu Yuan Road, Sec. 2
Taipei, Taiwan 11529
ROC

Mr. P. Wu
Science and Technology Advisory Group
The Executive Yuan, ROC
10th Fl., 116 Nanking E. Rd., Sec. 2
Taipei, Taiwan 10409
ROC

Mr. H.-C. Hsieh
MRDC/CSIST
P.O. Box 1-26
Lungtan, Taiwan 32500
ROC

Mr. G.-C. Chang
MRDC/CSIST
P.O. Box 1-26
Lungtan, Taiwan 32500
ROC

Mr. N.-N. Hsu
MRDC/CSIST
P.O. Box 1-26-4
Lungtan, Taiwan 32500
ROC

Dr. S.-J. Yang
MRDC/CSIST
P.O. Box 1-26-7
Lungtan, Taiwan 32500
ROC

Dr. C.-I. Chen
MRDC/CSIST
P.O. Box 1-26-1
Lungtan, Taiwan 32500
ROC

Dr. K.-L. Wang
MRDC/CSIST
P.O. Box 1-26-1
Lungtan, Taiwan 32500
ROC

Mr. K. Yang
MRDC/CSIST
P.O. Box 1-26-1
Lungtan, Taiwan 32500
ROC

Mr. S.-J. Chen
MRDC/CSIST
P.O. Box 1-26-1
Lungtan, Taiwan 32500
ROC

Mr. C.-P. Chang
MRDC/CSIST
P.O. Box 1-26-1
Lungtan, Taiwan 32500
ROC

Mr. T.-I. Hou
MRDC/CSIST
P.O. Box 1-26-1
Lungtan, Taiwan 32500
ROC

Mr. L.-C. Su
MRDC/CSIST
P.O. Box 1-26-1
Lungtan, Taiwan 32500
ROC

Mr. J.-R. Lee
MRDC/CSIST
P.O. Box 1-26-1
Lungtan, Taiwan 32500
ROC

Dr. W.-Y. Chen
MRDC/CSIST
P.O. Box 1-26-2
Lungtan, Taiwan 32500
ROC

Dr. D.-G. Hwang
MRDC/CSIST
P.O. Box 1-26-2
Lungtan, Taiwan 32500
ROC

Mr. G.-S. Ger
MRDC/CSIST
P.O. Box 1-26-2
Lungtan, Taiwan 32500
ROC

Ms. G. Lee
MRDC/CSIST
P.O. Box 1-26-2
Lungtan, Taiwan 32500
ROC

Mr. J.-F. Chang
MRDC/CSIST
P.O. Box 1-26-2
Lungtan, Taiwan 32500
ROC

Mr. S.-W. Leu
MRDC/CSIST
P.O. Box 1-26-2
Lungtan, Taiwan 32500
ROC

Mr. S.-L. Jwo
MRDC/CSIST
P.O. Box 1-26-2
Lungtan, Taiwan 32500
ROC

Mr. C.-H. Liu
MRDC/CSIST
P.O. Box 1-26-2
Lungtan, Taiwan 32500
ROC

Mr. S.-K. Wu
MRDC/CSIST
P.O. Box 1-26-2
Lungtan, Taiwan 32500
ROC

Mr. D. Liao
MRDC/CSIST
P.O. Box 1-26-2
Lungtan, Taiwan 32500
ROC

Mr. R.-H. Wan
MRDC/CSIST
P.O. Box 1-26-2
Lungtan, Taiwan 32500
ROC

Mr. C.-Y. Chu
MRDC/CSIST
P.O. Box 1-26-2
Lungtan, Taiwan 32500
ROC

Mr. W.-J. Wu
MRDC/CSIST
P.O. Box 1-26-2
Lungtan, Taiwan 32500
ROC

Dr. P.-C. Yao
MRDC/CSIST
P.O. Box 1-26-6
Lungtan, Taiwan 32500
ROC

Mr. D.-R. Huang
MRDC/CSIST
P.O. Box 1-26-6
Lungtan, Taiwan 32500
ROC

Mr. C.-W. Hsieh
MRDC/CSIST
P.O. Box 1-26-6
Lungtan, Taiwan 32500
ROC

Mr. C.-S. Hao
MRDC/CSIST
P.O. Box 1-26-6
Lungtan, Taiwan 32500
ROC

Mr. T.-W. Yang
CSIST
P.O. Box 1-2-3
Lungtan, Taiwan 32500
ROC

Mr. C.-S. Liang
CSIST
P.O. Box 1-2-3
Lungtan, Taiwan 32500
ROC

Mr. W.-R. Harn
CSIST
P.O. Box 1-2-3
Lungtan, Taiwan 32500
ROC

Mr. S.-W. Chyuan
CSIST
P.O. Box 1-2-3
Lungtan, Taiwan 32500
ROC

Mr. H.-L. Chong
CSIST
P.O. Box 1-2-3
Lungtan, Taiwan 32500
ROC

Mr. T.-Y. Lee
CSIST
P.O. Box 1-2-3
Lungtan, Taiwan 32500
ROC

Mr. G.-M. Ke
CSIST
P.O. Box 1-2-15
Lungtan, Taiwan 32500
ROC

Dr. H.-Y. Chen
CSIST
P.O. Box 1-22
Lungtan, Taiwan 32500
ROC

Mr. C.-C. Pern
CSIST
P.O. Box 1-22
Lungtan, Taiwan 32500
ROC

Mr. H.-Y. D. Chou
CSIST
P.O. Box 1-22
Lungtan, Taiwan 32500
ROC

Mr. Y.-C. Pan
CSIST
P.O. Box 1-22
Lungtan, Taiwan 32500
ROC

Mr. Y.-K. Chen
CSIST
P.O. Box 1-22
Lungtan, Taiwan 32500
ROC

Mr. J.-D. Tsou
CSIST
P.O. Box 1-4-13
Lungtan, Taiwan 32500
ROC

Mr. P.-H. Sung
CSIST
P.O. Box 1-4-13
Lungtan, Taiwan 32500
ROC

Mr. J.-L. Liang
CSIST
P.O. Box 1-4-13
Lungtan, Taiwan 32500
ROC

Mr. K.-S. Wu
CSIST
P.O. Box 1-6
Lungtan, Taiwan 32500
ROC

Mr. J.-C. Kuo
CSIST
P.O. Box 1-6
Lungtan, Taiwan 32500
ROC

Mr. Y.-S. Lee
INER/CSIST
P.O. Box 3-6
Lungtan, Taiwan 32500
ROC

Mr. C.-S. Su
INER/CSIST
P.O. Box 3-6
Lungtan, Taiwan 32500
ROC

Mr. L.-F. Lin
INER/CSIST
P.O. Box 3-14
Lungtan, Taiwan 32500
ROC

Mr. J. C. Tang
INER/CSIST
P.O. Box 3-14
Lungtan, Taiwan 23700
ROC

Mr. C.-L. Tsao
CSIST
P.O. Box 8763
Sanshia, Taiwan 23700
ROC

Mr. S.-C. Fong
CSIST
P.O. Box 8763
Sanshia, Taiwan 23700
ROC

Mr. T.-Y. Huang
CSIST
P.O. Box 8763
Sanshia, Taiwan 23700
ROC

Mr. Y.-Y. Tsay
CSIST
P.O. Box 8763
Sanshia, Taiwan 23700
ROC

Dr. S.-S. Wang
ARL/CSIST
P.O. Box 8676-1
Taichung, Taiwan 40000
ROC

Dr. R.-C. Chu
ARL/CSIST
P.O. Box 8676-1
Taichung, Taiwan 40000
ROC

Mr. H. Chin
ARL/CSIST
P.O. Box 8676-1
Taichung, Taiwan 40000
ROC

Mr. C.-H. Lee
ARL/CSIST
P.O. Box 8676-1
Taichung, Taiwan 40000
ROC

Mr. L.-J. Hsu
ARL/CSIST
P.O. Box 8676-1
Taichung, Taiwan 40000
ROC

Mr. A.-K. Feng
ARL/CSIST
P.O. Box 8676-1
Taichung, Taiwan 40000
ROC

Mr. C.-C. Ko
ARL/CSIST
P.O. Box 8676-1-3
Taichung, Taiwan 40000
ROC

Mr. Y.-J. Chiou
ARL/CSIST
P.O. Box 8676-1-3
Taichung, Taiwan 40000
ROC

Mr. J. R. Liu
ARL/CSIST
P.O. Box 8676-1-10
Taichung, Taiwan 40000
ROC

Dr. C.-L. Ong
ARL/CSIST
P.O. Box 8676-1-10
Taichung, Taiwan 40000
ROC

Mr. W.-Y. Shu
ARL/CSIST
P.O. Box 8676-1-10
Taichung, Taiwan 40000
ROC

Mr. Y.-P. Bian
ARL/CSIST
P.O. Box 8676-1-10
Taichung, Taiwan 40000
ROC

Mr. M.-F. Sheu
ARL/CSIST
P.O. Box 8676-1-10
Taichung, Taiwan 40000
ROC

Mr. C.-S. Yeh
ARL/CSIST
P.O. Box 8676-1-10
Taichung, Taiwan 40000
ROC

Mr. A.-D. Han
ARL/CSIST
P.O. Box 8676-1-10
Taichung, Taiwan 40000
ROC

Dr. J.-T. Tang
MRDC/CSIST
P.O. Box 1-26-5
Lungtan, Taiwan 32500
ROC

Mr. T.-M. Wu
MRDC/CSIST
P.O. Box 1-26
Lungtan, Taiwan 32500
ROC

Mr. H.-C. Wang
CSIST
P.O. Box 1-2-3
Lungtan, Taiwan 32500
ROC

Mr. C.-K. Chang
CSIST
P.O. Box 1-2-3
Lungtan, Taiwan 32500
ROC

Dr. C.-C. Hsien
CSIST
P.O. Box 1-4
Lungtan, Taiwan 32500 ROC

Mr. C.-L. Yen
Ministry of Education, ROC
5 Chungshan S. Road
Taipei, Taiwan 10040
ROC

Mr. C.-H. Wu
Department of Chemical Engineering
National Chiao Tung University
Hsinchu, Taiwan 30050
ROC

Mr. C.-C. Chu
MRDC/CSIST
P.O. Box 1-26
Lungtan, Taiwan 32500
ROC

Mr. C.-J. Wei
MRDC/CSIST
P.O. Box 1-26
Lungtan, Taiwan 32500
ROC

Mr. H.-F. Kuo
MRDC/CSIST
P.O. Box 1-26
Lungtan, Taiwan 32500
ROC

Mr. W.-H. Chan
MRDC/CSIST
P.O. Box 1-26-5
Lungtan, Taiwan 32500
ROC

Mr. B.-J. Shiau
MRDC/CSIST
P.O. Box 1-26-5
Lungtan, Taiwan 32500
ROC

Ms. H.-H. Chen
MRDC/CSIST
P.O. Box 1-26
Lungtan, Taiwan 32500
ROC

Mr. S.-C. Chen
MRDC/CSIST
P.O. Box 1-26-1
Lungtan, Taiwan 32500
ROC

Mr. C.-H. Liu
MRDC/CSIST
P.O. Box 1-26
Lungtan, Taiwan 32500
ROC

Mr. Y.-J. Wu
MRDC/CSIST
P.O. Box 1-26
Lungtan, Taiwan 32500
ROC

Mr. T.-S. Chou
MRDC/CSIST
P.O. Box 1-26
Lungtan, Taiwan 32500
ROC

Mr. Y.-Y. Yang
MRDC/CSIST
P.O. Box 1-26
Lungtan, Taiwan 32500
ROC

Mr. W.-N. Yeh
MRDC/CSIST
P.O. Box 1-26-2
Lungtan, Taiwan 32500
ROC

Mr. Y.-C. Lo
MRDC/CSIST
P.O. Box 1-26-2
Lungtan, Taiwan 32500
ROC

Mr. J.-C. Liu
MRDC/CSIST
P.O. Box 1-26-2
Lungtan, Taiwan 32500
ROC

Mr. C.-H. Wang
MRDC/CSIST
P.O. Box 1-26-2
Lungtan, Taiwan 32500
ROC

Mr. C.-W. Chang
MRDC/CSIST
P.O. Box 1-26-2
Lungtan, Taiwan 32500
ROC

Mr. C.-J. Lee
MRDC/CSIST
P.O. Box 1-26-2
Lungtan, Taiwan 32500
ROC

Mr. M.-C. Kao
MRDC/CSIST
P.O. Box 1-26-5
Lungtan, Taiwan 32500
ROC

Mr. C.-H. Wang
MRDC/CSIST
P.O. Box 1-26-5
Lungtan, Taiwan 32500
ROC

Mr. T.-C. Yang
MRDC/CSIST
P.O. Box 1-26-5
Lungtan, Taiwan 32500
ROC

Ms. F.-M. Lin
MRDC/CSIST
P.O. Box 1-26-5
Lungtan, Taiwan 32500
ROC

Mr. H.-M. Chen
MRDC/CSIST
P.O. Box 1-26-5
Lungtan, Taiwan 32500
ROC

Dr. B.-C. Fuh
MRDC/CSIST
P.O. Box 1-26-5
Lungtan, Taiwan 32500
ROC

Mr. C.-K. Huang
MRDC/CSIST
P.O. Box 1-26-5
Lungtan, Taiwan 32500
ROC

Mr. F.-T. Pang
MRDC/CSIST
P.O. Box 1-26-5
Lungtan, Taiwan 32500
ROC

Mr. C.-K. Huang
MRDC/CSIST
P.O. Box 1-26-5
Lungtan, Taiwan 32500
ROC

Mr. S.-C. Yang
MRDC/CSIST
P.O. Box 1-26-5
Lungtan, Taiwan 32500
ROC

Mr. J.-Y. Wang
MRDC/CSIST
P.O. Box 1-26-5
Lungtan, Taiwan 32500
ROC

Mr. T.-W. Huang
MRDC/CSIST
P.O. Box 1-26-5
Lungtan, Taiwan 32500
ROC

Ms. C.-M. Ho
MRDC/CSIST
P.O. Box 1-26-5
Lungtan, Taiwan 32500
ROC

Dr. H.-P. Kao
MRDC/CSIST
P.O. Box 1-26-5
Lungtan, Taiwan 32500
ROC

Mr. H.-S. Kao
MRDC/CSIST
P.O. Box 1-26-4
Lungtan, Taiwan 32500
ROC

Ms. S.-H. Peng
MRDC/CSIST
P.O. Box 1-26-4
Lungtan, Taiwan 32500
ROC

Mr. P.-C. Chou
MRDC/CSIST
P.O. Box 1-26-1
Lungtan, Taiwan 32500
ROC

Mr. C.-L. Lin
MRDC/CSIST
P.O. Box 1-26-2
Lungtan, Taiwan 32500
ROC

Mr. H.-J. Fanchiang
MRDC/CSIST
P.O. Box 1-26-5
Lungtan, Taiwan 32500
ROC